How PRODUCTS Are MADE

How PRODUCTS Are MADE

An Illustrated Guide to Product Manufacturing

Volume 2

Kyung-Sun Lim, Editor

Gale Research
An ITP Information/Reference Group Company

ITP
Changing the Way the World Learns

NEW YORK • LONDON • BONN • BOSTON • DETROIT
MADRID • MELBOURNE • MEXICO CITY • PARIS
SINGAPORE • TOKYO • TORONTO • WASHINGTON
ALBANY NY • BELMONT CA • CINCINNATI OH

STAFF

Kyung-Sun Lim, *Editor*

Kristine M. Binkley and Sheila M. Dow, *Associate Editors*

Nicole Beatty and Carley Wellman, *Co-op Assistant Editors*

Maureen Richards, *Research Specialist*

Shanna Heilveil, *Production Assistant*

Cynthia Baldwin, *Art Director*
Bernadette M. Gornie, *Page Designer*
Tracey Rowens, *Cover Designer*

Pamela Hayes, *Photography Coordinator*

Electronic illustrations provided by Hans & Cassady, Inc. of Westerville, Ohio

ISBN 0-8103-8952-5
ISSN 1072-5091

Printed in the United States of America
10 9 8 7 6 5 4

ITP™ Gale Research Inc., an International Thomson Publishing Company.
ITP logo is a trademark under license.

Contents

Introduction

About the Series

Welcome to *How Products Are Made: An Illustrated Guide to Product Manufacturing*. This series provides information on the manufacture of a variety of items, from everyday household products to heavy machinery to sophisticated electronic equipment. You will find step-by-step descriptions of processes, simple explanations of technical terms and concepts, and clear, easy-to-follow illustrations.

Each volume of *How Products Are Made* covers a broad range of manufacturing areas: food, clothing, electronics, transportation, machinery, instruments, sporting goods, and more. Some are intermediate goods sold to manufacturers of other products, while others are retail goods sold directly to consumers. You will find items made from a variety of materials, including products such as precious metals and minerals that are not "made" so much as they are extracted and refined.

Organization

Every volume in this series is comprised of many individual entries, each covering a single product. Although each entry focuses on the product's manufacturing process, it also provides a wealth of other information: who invented the product or how it has developed, how it works, what materials are used, how it is designed, quality control procedures, byproducts generated during its manufacture, future applications, and books and periodical articles containing more information.

To make it easier for you to find what you're looking for, the entries are broken up into standard sections. Among the sections you will find are the following:

- Background
- History
- Raw Materials
- Design
- The Manufacturing Process
- Quality Control
- Byproducts
- The Future
- Where To Learn More

Every entry is accompanied by illustrations. Uncomplicated and easy to understand, these illustrations generally follow the step-by-step description of the manufacturing process found in the text.

Bold faced items in the text refer to other entries in this volume.

A general subject index of important terms, processes, materials, and people is found at the end of the book. Bold faced items in the index refer to main entries.

About this Volume

This volume contains essays on 100 products, arranged alphabetically, and 15 special boxed sections. Written by William S. Pretzer, a manufacturing historian and curator at the Henry Ford Museum & Greenfield Village in Dearborn, Michigan, these boxed sections describe interesting historical developments related to a product. Photographs are also included.

Contributors/Advisor

The entries in this volume were written by a skilled team of technical writers and engineers, often in cooperation with manufacturers and industry associations. The advisor for this volume was William Pretzer.

Suggestions

Your questions, comments, and suggestions for future products are welcome. Please send all such correspondence to:

The Editor
How Products Are Made
Gale Research Inc.
835 Penobscot Building
Detroit, MI 48226

Contributors

Jim Acton
Audra Avizienis
Lawrence H. Berlow
Douglas E. Betts
Carol Brennan
Chris Cavette
Evelyn S. Dorman
David N. Ford
Bridget K. Hall
Loretta Hall
Susan Bard Hall
Gillian S. Holmes
Barry M. Marton
Mary F. McNulty
William S. Pretzer
Peter Psarianos
Renee M. Rottner
Rose Secrest
Laurel M. Sheppard
Andrew M. Sherman
Joel Simon
Douglas Smith
Peter Toeg
Glenn G. Whiteside
Angela Woodward

Acknowledgments

The editor would like to thank the following individuals, companies, and associations for providing assistance with Volume 2 of *How Products Are Made:*

Asphalt: Tom Ritchie, The Ritchie Corporation, Wichita, KS. **Balloon:** Rick Tillotson, Tillotson Corp., Fall River Office, Fall River, MA; John Cameron, Superior Balloon Co., Ferndale, MI. **Barbed Wire:** Scott Gipson, Oklahoma Steel and Wire Company, Madill, OK. **Beeper:** Lisa Coffman and Fred Schmidt, Motorola, Paging Products Group. **Beer:** Miller Visitors Center, Milwaukee, WI. **Bell:** Schulmerich Carillons, Inc., Sellersville, PA. **Bleach:** Sandy Sullivan, Clorox Company. **Business Jet:** Cesar Nolla, The Cessna Aircraft Company, Wichita, KS. **Camera Lens:** Stacie Errera, TAMRON Industries, Inc.; Tom Sobey, SIGMA Corporation of America; Cheryl Van Sise, Leica Camera, Inc., U.S.A. **Carpet:** Carpet and Rug Institute; RBI International Carpet Consultants, both of Dalton, Georgia. **Cat Litter:** Golden Cat Corporation of South Bend, IN, and their public relations firm, Weightman Public Relations of Allentown, PA. **Chalkboard:** Greensteel Division Information Display Technology, Inc. **CNC:** Cincinnati Milicron, Mazak, and Mori Seiki. **Decorative Plastic Laminate:** Formica Corporation. **Dental Floss:** Colgate Oral Pharmaceuticals, a subsidiary of Colgate-Palmolive Co., Canton, MA, and their public relations firm, Mullen, of Wenham, MS; Oral-B Laboratories, a division of The Gillette Company of Redwood City, CA and their public relations firm, Porter/Novelli. **Drywall:** Jack Hodges of Domtar Gypsum. **Fortune Cookie:** Diane Plocher, JFC International Inc.; Gene Yee, Good Fortunes. **Freeze-Dried Food:** Oregon Freeze Dry. **Hay Baler:** John Deere & Company, Moline, IL; Plants and Dealers Worldwide; Antique Balers: Midwest Old Threshers' Association, Mount Pleasant, IA. **Hearing Aid:** Richard C. Brandt and Robert Prifrel, Dalhberg, Inc., Golden Valley, MN. **Industrial Robot:** Staubli Unimation Inc. of Sewickley, PA. **Jelly Bean:** Terri Kaminski and John R. Lux, Brach & Brock Confections, Inc.; Goelitz Confectionery. **Ketchup:** Heinz U.S.A. **Lead:** The Doe Run Company of Herculaneum, MS. **Milking Machine:** Boumatic Division of DEC International Inc., Madison, WI. **Nail:** John Hassall, Inc., Westbury, NY. **Neon Sign:** John Fischer, Signs of the Times. **Pasta:** Meyer Machine Company, San Antonio, TX. **Photographic Film:** Mary K. Bellucco and Alison K. Hofland, Eastman Kodak Company. **Player Piano:** Robert J. Berkman, QRS Music Rolls, Inc.; Jan Kiser and Tom Rogers, PianoDisc, Sacramento, CA. **Road Sign:** 3M Corp. **Rough Terrain Forklift:** Trak International, Port Washington, WI. **Safety Pin:**

Lee Conard, Cooper Hardware & Components, Beacon Falls, CT. **Salt:** Bruce Bertram, Salt Institute. **Ski:** Rossignol Ski Company. **Smoke Detector:** Joe Soltis, Seatt Corporation. **Stained Glass:** The Stained Glass Association of America; The Paul Wismach Glass Co., Paden City, WV. **Sunscreen:** Hawaiian Tropic. **Telescope:** Joe Makemson, Edmund Scientific Company, Barrington, NJ. **Tofu:** Rick Crouse, Clofine Dairy & Food Products Inc., Linwood, NJ. **Violin:** Psarianos Violins Ltd, Troy, MI.

The historical photographs on pages 33, 43, 53, 124, 130, 182, 196, 246, 258, 273, 308, 318, 374, 404, and 463 are from the collections of **Henry Ford Museum & Greenfield Village**, Dearborn, Michigan.

Electronic illustrations in this volume were created by **Hans & Cassady Inc.** of Westerville, Ohio.

How PRODUCTS Are MADE

Acrylic Plastic

Background

Acrylic plastic refers to a family of synthetic, or man-made, plastic materials containing one or more derivatives of acrylic acid. The most common acrylic plastic is polymethyl methacrylate (PMMA), which is sold under the brand names of Plexiglas, Lucite, Perspex, and Crystallite. PMMA is a tough, highly transparent material with excellent resistance to ultraviolet radiation and weathering. It can be colored, molded, cut, drilled, and formed. These properties make it ideal for many applications including airplane windshields, skylights, automobile taillights, and outdoor signs. One notable application is the ceiling of the Houston Astrodome which is composed of hundreds of double-insulating panels of PMMA acrylic plastic.

Like all plastics, acrylic plastics are polymers. The word polymer comes from the Greek words *poly,* meaning many, and *meros,* meaning a part. A polymer, therefore, is a material made up of many molecules, or parts, linked together like a chain. Polymers may have hundreds, or even thousands, of molecules linked together. More importantly, a polymer is a material that has properties entirely different than its component parts. The process of making a polymer, known as polymerization, has been likened to shoveling scrap glass, copper, and other materials into a box, shaking the box, and coming back in an hour to find a working color television set. The glass, copper, and other component parts are still there, but they have been reassembled into something that looks and functions entirely differently.

The first plastic polymer, celluloid, a combination of cellulose nitrate and camphor, was developed in 1869. It was based on the natural polymer cellulose, which is present in plants. Celluloid was used to make many items including **photographic film**, combs, and men's shirt collars.

In 1909, Leo Baekeland developed the first commercially successful synthetic plastic polymer when he patented phenol formaldehyde resin, which he named Bakelite. Bakelite was an immediate success. It could be machined and molded. It was an excellent electrical insulator and was resistant to heat, acids, and weather. It could also be colored and dyed for use in decorative objects. Bakelite plastic was used in radio, telephone, and electrical equipment, as well as counter tops, **buttons**, and knife handles.

Acrylic acid was first prepared in 1843. Methacrylic acid, which is a derivative of acrylic acid, was formulated in 1865. When methacrylic acid is reacted with methyl alcohol, it results in an ester known as methyl methacrylate. The polymerization process to turn methyl methacrylate into polymethyl methacrylate was discovered by the German chemists Fittig and Paul in 1877, but it wasn't until 1936 that the process was used to produce sheets of acrylic safety glass commercially. During World War II, acrylic glass was used for periscope ports on submarines and for windshields, canopies, and gun turrets on airplanes.

Raw Materials

Methyl methacrylate is the basic molecule, or monomer, from which polymethyl methacrylate and many other acrylic plastic polymers are formed. The chemical notation for this material is $CH_2=C(CH_3)$

The most common acrylic plastic is polymethyl methacrylate (PMMA), a tough, highly transparent material with excellent resistance to ultraviolet radiation and weathering. It is sold under the brand names of Plexiglas, Lucite, Perspex, and Crystallite.

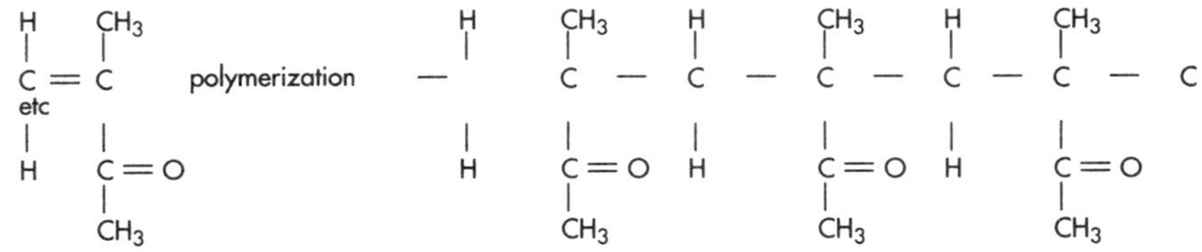

methyl methacrylate (monomer)

polymethyl methacrylate (polymer)

Figure 1

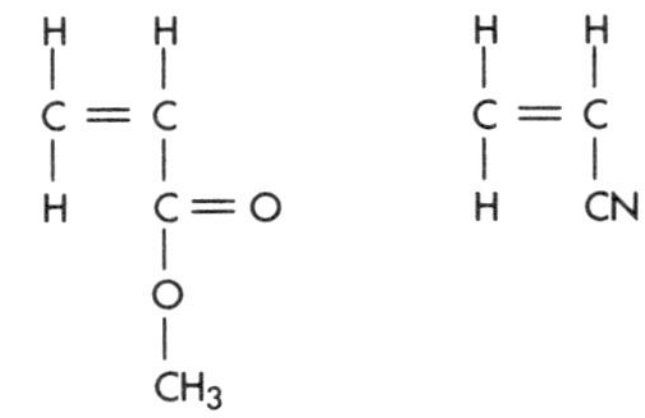

methyl acrylate (monomer)

acrylonitrile (monomer)

Figure 2

Figure 1 shows the polymerization of methyl methacrylate into polymethyl methacrylate (PMMA). Figure 2 shows other acrylic plastic monomers that may be copolymerized with methyl methacrylate.

$COOCH_3$. It is written in this format, rather than the more common chemical notation $C_5H_8O_2$, to show the double bond (=) between the two carbon atoms in the middle. During polymerization, one leg of this double bond breaks and links up with the middle carbon atom of another methyl methacrylate molecule to start a chain. This process repeats itself until the final polymer is formed. (See Figure 1)

Methyl methacrylate may be formed in several ways. One common way is to react acetone [CH_3COCH_3] with sodium cyanide [NaCN] to produce acetone cyanhydrin [$(CH_3)_2C(OH)CN$]. This in turn is reacted with methyl alcohol [CH_3OH] to produce methyl methacrylate.

Other similar monomers such as methyl acrylate [$CH_2=CHCOOCH_3$] and acrylonitrile [$CH_2=CHCN$] can be joined with methyl methacrylate to form different acrylic plastics. (See Figure 2) When two or more monomers are joined together, the result is known as a copolymer. Just as with methyl methacrylate, both of these monomers have a double bond on the middle carbon atoms that splits during polymerization to link with the carbon atoms of other molecules. Controlling the proportion of these other monomers produces changes in elasticity and other properties in the resulting plastic.

The Manufacturing Process

Acrylic plastic polymers are formed by reacting a monomer, such as methyl methacrylate, with a catalyst. A typical catalyst would be an organic peroxide. The catalyst starts the reaction and enters into it to keep it going, but does not become part of the resulting polymer.

Acrylic plastics are available in three forms: flat sheets, elongated shapes (rods and tubes), and molding powder. Molding powders are sometimes made by a process known as suspension polymerization in which the reaction takes place between tiny droplets of the monomer suspended in a solution of water and catalyst. This results in grains of polymer with tightly controlled molecular weight suitable for molding or extrusion.

Acrylic plastic sheets are formed by a process known as bulk polymerization. In this process, the monomer and catalyst are poured into a mold where the reaction takes place. Two methods of bulk polymerization may be used: batch cell or continuous. Batch cell is the most common because it is simple and is easily adapted for making acrylic sheets in thicknesses from 0.06 to 6.0 inches (0.16-15 cm) and widths from 3 feet (0.9 m) up to several hundred feet. The batch cell method may also be used to form rods and tubes. The continuous method is quicker and involves less labor. It is used to make sheets of thinner thicknesses and smaller widths than those produced by the batch cell method.

We will describe both the batch cell and continuous bulk polymerization processes typically used to produce transparent polymethyl methacrylic (PMMA) sheets.

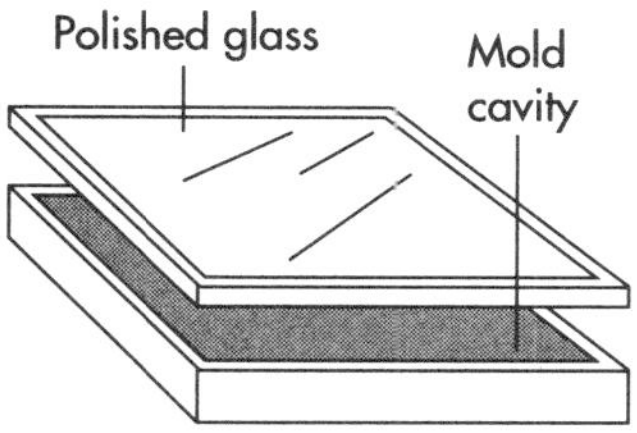

Batch cell bulk polymerization begins with two sheets of polished glass separated by a flexible window-frame spacer. The spacers form a sealed cavity between the plates.

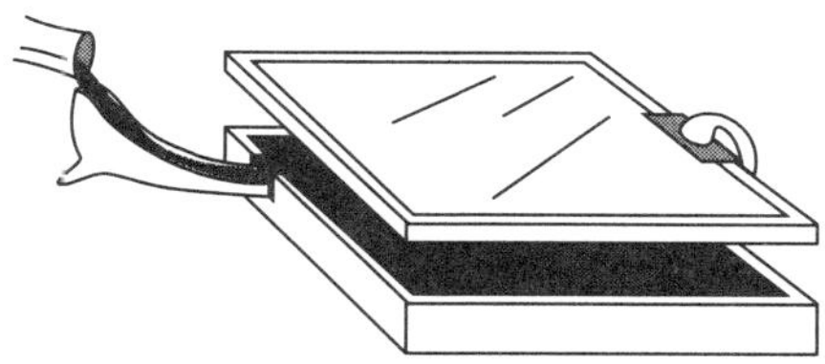

An open corner of the mold cavity is filled with liquid plastic and a catalyst. The mold is sealed and heated.

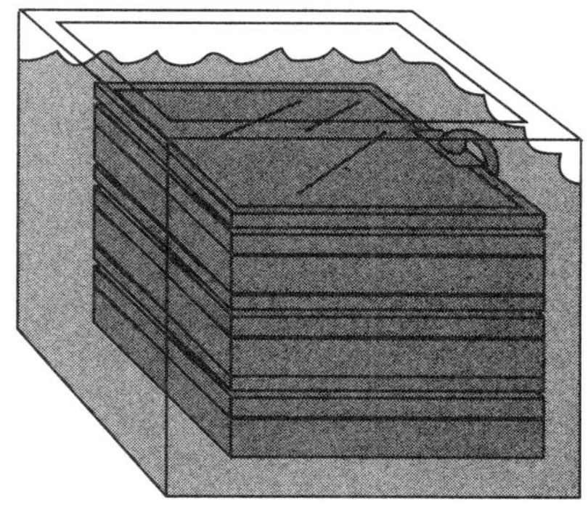

The molds may be placed in water to cool.

Batch cell bulk polymerization

1 The mold for producing sheets is assembled from two plates of polished glass separated by a flexible "window-frame" spacer. The spacer sits along the outer perimeter of the surface of the glass plates and forms a sealed cavity between the plates. The fact that the spacer is flexible allows the mold cavity to shrink during the polymerization process to compensate for the volume contraction of the material as the reaction goes from individual molecules to linked polymers. In some production applications, polished metal plates are used instead of glass. Several plates may be stacked on top of each other with the upper surface of one plate becoming the bottom surface of the next higher mold cavity. The plates and spacers are clamped together with spring clamps.

2 An open corner of each mold cavity is filled with a pre-measured liquid syrup of methyl methacrylate monomer and catalyst. In some cases, a methyl methacrylate prepolymer is also added. A prepolymer is a material with partially formed polymer chains used to further help the polymerization process. The liquid syrup flows throughout the mold cavity to fill it.

3 The mold is then sealed and heat may be applied to help the catalyst start the reaction.

4 As the reaction proceeds, it may generate significant heat by itself. This heat is fanned off in air ovens or by placing the molds in a water bath. A programmed temperature cycle is followed to ensure proper cure time without additional vaporization of the monomer solution. This also prevents bubbles from forming. Thinner sheets may cure in 10 to 12 hours, but thicker sheets may require several days.

5 When the plastic is cured, the molds are cooled and opened. The glass or metal plates are cleaned and reassembled for the next batch.

6 The plastic sheets are either used as is or are annealed by heating them to 284-302°F (140-150°C) for several hours to reduce any residual stresses in the material that might cause warping or other dimensional instabilities.

7 Any excess material, or flash, is trimmed off the edges, and masking paper or plastic film is applied to the surface of the finished sheets for protection during handling and shipping. The paper or film is often marked with the material's brand name, size, and handling instructions. Conformance with applicable safety or building code standards is also noted.

Continuous bulk polymerization

1 The continuous process is similar to the batch cell process, but because the sheets are thinner and smaller, the process times are much shorter. The syrup of monomer and catalyst is introduced at one end of a set of horizontal stainless steel belts running parallel, one above the other. The distance between the belts determines the thickness of the sheet to be formed.

Batch cell bulk polymerization is the most common way to make acrylic plastic sheets because it is simple and easily adapted for making sheets in thicknesses from 0.06 to 6 inches.

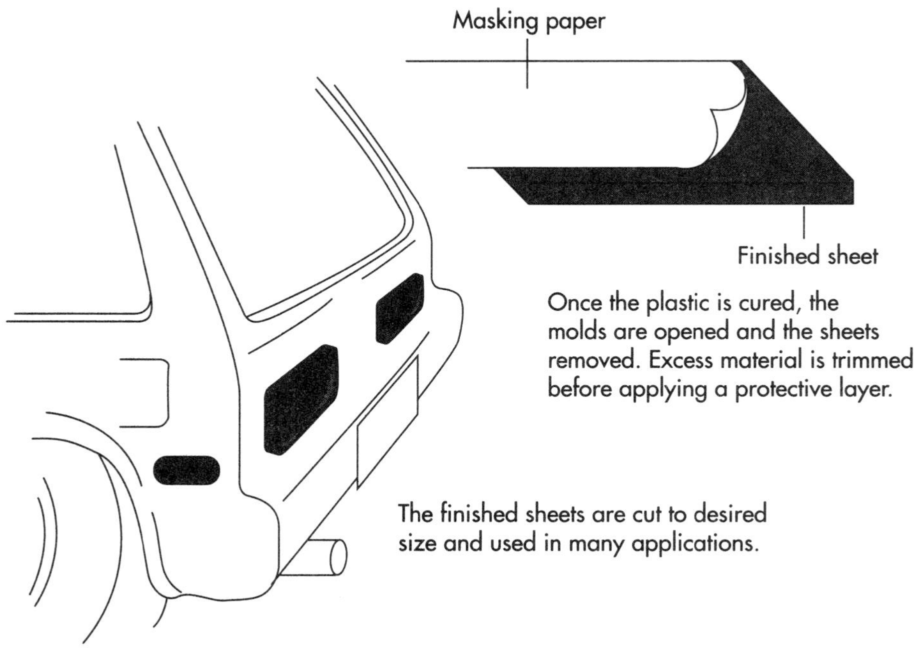

2 The belts hold the reacting monomer and catalyst syrup between them and move it through a series of heating and cooling zones according to a programmed temperature cycle to cure the material.

3 Electric heaters or hot air may then anneal the material as it comes out of the end of the belts.

4 The sheets are cut to size and masking paper or plastic film is applied.

Quality Control

The storage, handling, and processing of the chemicals that make acrylic plastics are done under controlled environmental conditions to prevent contamination of the material or unsafe chemical reactions. The control of temperature is especially critical to the polymerization process. Even the initial temperatures of the monomer and catalyst are controlled before they are introduced into the mold. During the entire process, the temperature of the reacting material is monitored and controlled to ensure the heating and cooling cycles are the proper temperature and duration.

Samples of finished acrylic materials are also given periodic laboratory analysis to confirm physical, optical, and chemical properties.

Toxic Materials, Safety Considerations, and Recycling

Acrylic plastics manufacturing involves highly toxic substances which require careful storage, handling, and disposal. The polymerization process can result in an explosion if not monitored properly. It also produces toxic fumes. Recent legislation requires that the polymerization process be carried out in a closed environment and that the fumes be cleaned, captured, or otherwise neutralized before discharge to the atmosphere.

Acrylic plastic is not easily recycled. It is considered a group 7 plastic among recycled plastics and is not collected for recycling in most communities. Large pieces can be reformed into other useful objects if they have not suffered too much stress, crazing, or cracking, but this accounts for only a very small portion of the acrylic plastic waste. In a landfill, acrylic plastics, like many other plastics, are not readily biodegradable. Some acrylic plastics are highly flammable and must be protected from sources of combustion.

The Future

The average annual increase in the rate of consumption of acrylic plastics has been about 10%. A future annual growth rate of about 5% is predicted. Despite the fact that acrylic plastics are one of the oldest plastic materials in use today, they still hold the same advantages of optical clarity and resistance to the outdoor environment that make them the material of choice for many applications.

Where To Learn More

Books

Baird, Ronald J. and David T. Baird. *Industrial Plastics.* The Goodheart-Willcox Company, 1986.

Cowie, J.M.G. *Polymers: Chemistry and Physics of Modern Materials.* Chapman and Hall, 1991.

Encyclopedia of Polymer Science and Engineering, Volume 1. John Wiley & Sons, 1985.

Rosato, Dominick V. *Plastic Encyclopedia and Dictionary.* Hanser Publications, 1993.

Seymour, Raymond B. *Engineering Polymer Sourcebook.* McGraw-Hill, 1990.

Smith, Edward H., ed. *Mechanical Engineer's Reference Book.* Society of Automotive Engineers, 1994.

—*Evelyn S. Dorman/Chris Cavette*

Aluminum Beverage Can

American can makers produce about 100 billion aluminum beverage cans a year, equivalent to one can per American per day.

Background

Ninety-five percent of all **beer** and **soft drink** cans in the United States are made of aluminum. American can makers produce about 100 billion aluminum beverage cans a year, equivalent to one can per American per day. While almost all food cans are made of steel, aluminum's unique properties make it ideal for holding carbonated beverages. The typical aluminum can weighs less than half an ounce, yet its thin walls withstand more than 90 pounds of pressure per square inch exerted by the carbon dioxide in beer and soft drinks. Aluminum's shiny finish also makes it an attractive background for decorative printing, important for a product that must grab the attention of consumers in a competitive market.

Aluminum was first identified as an element in 1782, and the metal enjoyed great prestige in France, where in the 1850s it was more fashionable than even gold and silver for jewelry and eating utensils. Napoleon III was fascinated with the possible military uses of the lightweight metal, and he financed early experiments in the extraction of aluminum. Although the metal is found abundantly in nature, an efficient extraction process remained elusive for many years. Aluminum remained exceedingly high-priced and therefore of little commercial use throughout the 19th century. Technological breakthroughs at the end of the 19th century finally allowed aluminum to be smelted cheaply, and the price of the metal fell drastically. This paved the way for the development of industrial uses of the metal.

Aluminum was not used for beverage cans until after World War II. During the war, the U.S. government shipped large quantities of beer in steel cans to its servicemen overseas. After the war most beer was again sold in bottles, but the returning soldiers retained a nostalgic liking for cans. Manufacturers continued to sell some beer in steel cans, even though bottles were cheaper to produce. The Adolph Coors Company manufactured the first aluminum beer can in 1958. Its two-piece can could only hold 7 ounces (198 g), instead of the usual 12 (340 g), and there were problems with the production process. Nevertheless, the aluminum can proved popular enough to incite Coors, along with other metal and aluminum companies, to develop better cans.

The next model was a steel can with an aluminum top. This hybrid can had several distinct advantages. The aluminum end altered the galvanic reaction between the beer and the steel, resulting in beer with twice the shelf life of that stored in all-steel cans. Perhaps the more significant advantage of the aluminum top was that the soft metal could be opened with a simple pull tab. The old style cans required the use of a special opener popularly called a "church key," and when Schlitz Brewing Company introduced its beer in an aluminum "pop top" can in 1963, other major beer makers quickly jumped on the band wagon. By the end of that year, 40% of all U.S. beer cans had aluminum tops, and by 1968, that figure had doubled to 80%.

While aluminum top cans were sweeping the market, several manufacturers were aiming for the more ambitious all-aluminum beverage can. The technology Coors had used to make its 7-ounce aluminum can relied on the "impact-extrusion" process,

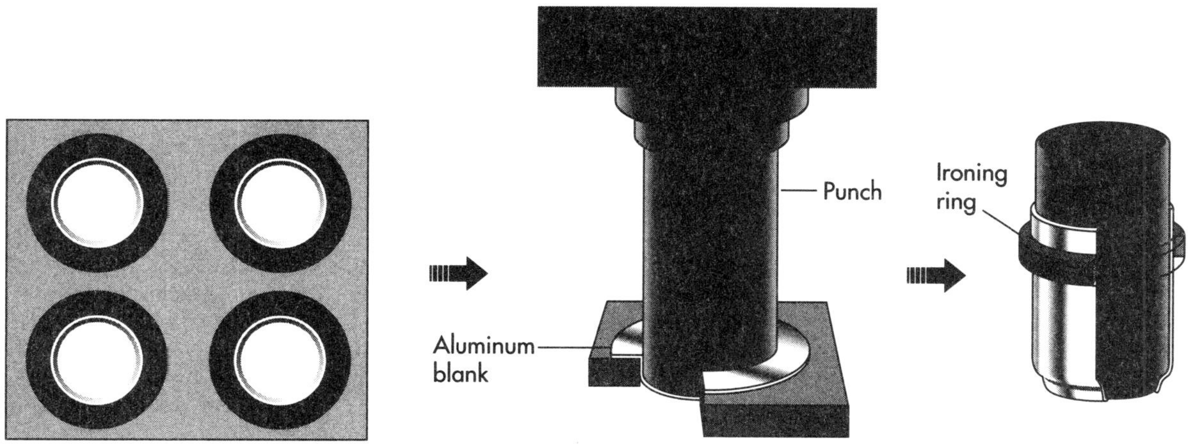

First a blank 5.5" in diameter is cut from a thin sheet of metal. This is then drawn up on the sides to make the diameter 3.5".

The small cup is transferred to a punch machine, and the cup is drawn again to a diameter of 2.6".

A sleeve holds the cup in place as three ironing rings stretch and thin the cup wall. The cup is now about 5" high. Another punch presses against the base of the cup, creating an inward bulge.

where a punch driven into a circular slug formed the bottom and sides of the can in one piece. The Reynolds Metals company introduced an all-aluminum can made by a different process called "drawing and ironing" in 1963, and this technology became the standard for the industry. Coors and Hamms Brewery were among the first companies to adopt this new can, and PepsiCo and Coca-Cola began using all-aluminum cans in 1967. The number of aluminum cans shipped in the U.S. rose from half a billion in 1965 to 8.5 billion in 1972, and the number continued to increase as aluminum became the nearly universal choice for carbonated beverages. The modern aluminum beverage can is not only lighter than the old steel or steel-and-aluminum can, it also does not rust, it chills quickly, its glossy surface is easily imprintable and eye-catching, it prolongs shelf life, and it is easy to recycle.

Raw Materials

The raw material of the aluminum beverage can is, of course, aluminum. Aluminum is derived from an ore called bauxite. U.S. aluminum producers import bauxite, primarily from Jamaica and Guinea. The bauxite is refined and then smelted, and the resulting molten aluminum is cast into ingots. The aluminum base for beverage cans consists mostly of aluminum, but it contains small amounts of other metals as well. These are typically 1% magnesium, 1% manganese, 0.4% iron, 0.2% silicon, and 0.15% copper. A large portion of the aluminum used in the beverage can industry is derived from recycled material. Twenty-five percent of the total American aluminum supply comes from recycled scrap, and the beverage can industry is the primary user of recycled material. The energy savings are significant when used cans are remelted, and the aluminum can industry now reclaims more than 63% of used cans.

The Manufacturing Process

Cutting the blank

1 The modern method for making aluminum beverage cans is called two-piece drawing and wall ironing. The process begins with an aluminum ingot which was cast to be about 30 inches (76 cm) thick, then rolled into a thin sheet. The first step in the actual manufacture of the can is to cut the sheet into a circle, called a blank, that will form the bottom and sides of the can. Each blank is 5.5 inches (14 cm) in diameter. Some material is neces-

The modern method for making aluminum beverage cans is called two-piece drawing and wall ironing, first introduced by Reynolds Metals company in 1963.

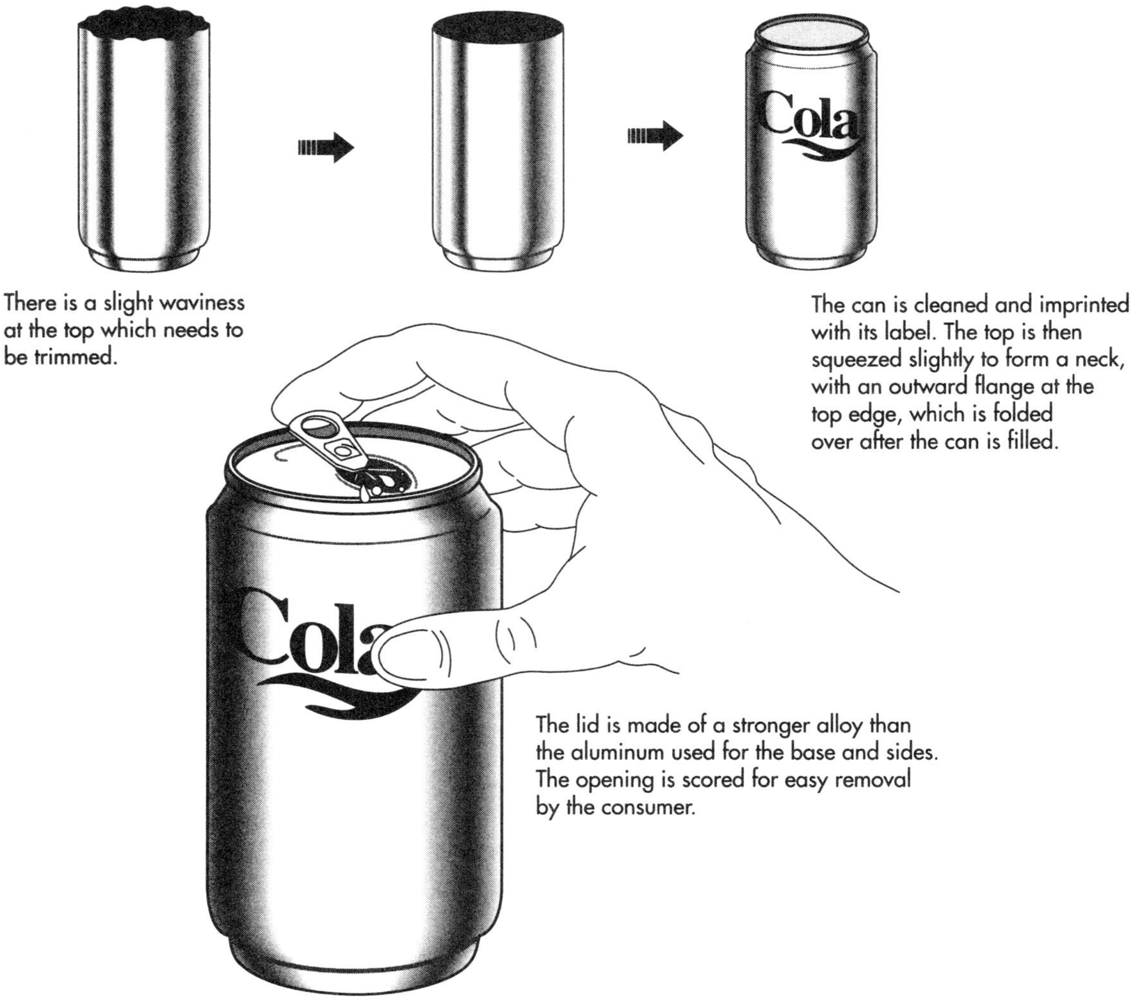

The small ripples at the top of the metal are called "ears". "Earing" is an unavoidable effect of the crystalline structure of the aluminum sheet.

sarily lost between each circle, but manufacturers have found that minimum aluminum is lost when the sheets are wide enough to hold two staggered rows of seven blanks each. About 12-14% of the sheet is wasted, but can be reused as scrap. After the circular blank is cut, it is "drawn" or pulled up to form a cup 3.5 inches (8.9 cm) in diameter.

Redrawing the cup

2 The small cup resulting from the initial draw is then transferred to a second machine. A sleeve holds the cup precisely in place, and a punch lowered swiftly into the cup redraws it to a diameter of about 2.6 inches (6.6 cm). The height of the cup increases simultaneously from the initial 1.3 to 2.25 inches (3.3 to 5.7 cm). The punch then pushes the cup against three rings called ironing rings, which stretch and thin the cup walls. This entire operation—the drawing and ironing—is done in one continuous punch stroke, which takes only one fifth of a second to complete. The cup is now about 5 inches (13 cm) high. Then another punch presses up against the base of the cup, causing the bottom to bulge inward. This shape counteracts the pressure of the carbonated liquid the can will contain. The bottom and lower walls of the can are also a little thicker than the upper walls, for added strength.

Trimming the ears

3 The drawing and ironing process leaves the can slightly wavy at the top. These

small ripples in the metal are called "ears." "Earing" is an unavoidable effect of the crystalline structure of the aluminum sheet. Aluminum companies have studied this phenomenon extensively, and they have been able to influence the placement and height of the ears by controlling the rolling of the aluminum sheet. Nevertheless, some material is lost at this stage. About a quarter inch is trimmed from the top of the can, leaving the upper walls straight and level.

Cleaning and decorating

4 The drawing and ironing process leaves the outer wall of the can with a smooth, shiny surface, so it does not require any further finishing such as polishing. After the ears are trimmed, the can is cleaned and then imprinted with its label. After the can is decorated, it is squeezed in slightly at the top to a make a neck, and the neck is given an outward flange at the very top edge, which will be folded over once the lid is added.

The lid

5 The lid is made of a slightly different alloy than the aluminum for the base and sides of the can. The inward bulge of the bottom of the can helps it withstand the pressure exerted by the liquid inside it, but the flat lid must be stiffer and stronger than the base, so it is made of aluminum with more magnesium and less manganese than the rest of the can. This results in stronger metal, and the lid is considerably thicker than the walls. The lid is cut to a diameter of 2.1 inches (5.3 cm), smaller than the 2.6-inch (6.6 cm) diameter of the walls. The center of the lid is stretched upward slightly and drawn by a machine to form a rivet. The pull tab, a separate piece of metal, is inserted under the rivet and secured by it. Then the lid is scored so that when the tab is pulled by the consumer, the metal will detach easily and leave the proper opening.

To ensure that the cans are made properly, they are automatically checked for cracks and pinholes. One in 50,000 cans is usually found to be defective.

Filling and seaming

6 After the neck is formed, the can is ready to be filled. The can is held tightly against the seat of a filling machine and a beverage is poured in. The lid is added. The upper flange formed when the can was given its neck is then bent around the lid and seamed shut. At this point, the can is ready for sale.

Byproducts/Waste

Some aluminum is lost at several points in the manufacturing process—when the blanks are cut and the ears are trimmed—but this scrap can be reused. Cans which have been used and discarded by consumers can also be reused, and as mentioned above, recycled material makes up a significant percentage of the aluminum used for beverage cans. The savings from recycling are quite significant to the industry. The major expense of the beverage can is in the energy needed to produce the aluminum, but recycling can save up to 95% of the energy cost. Can producers also try to control waste by developing stronger can sheet so that less aluminum goes into each can, and by carefully controlling the manufacturing process to cut down on loss through earing. The lid of the typical can is smaller in diameter than the walls in order to conserve the amount of aluminum that goes into it, and as worldwide demand for beverage cans continues to grow, the trend is to make the lid even smaller. A new can introduced in 1993 with a lid a quarter-inch smaller in diameter than most cans can save manufacturers $3 per thousand. This figure seems small until it is multiplied by the hundreds of millions of cans produced each day in the U.S. It becomes clear that any small savings in raw materials or energy can be a major step in conserving both money and resources.

The Future

Worldwide production of aluminum beverage cans is steadily increasing, growing by several billion cans a year. In the face of this rising demand, the future of the beverage can seems to lie in designs that save money and materials. The trend towards smaller lids is already apparent, as well as smaller neck diameters, but other changes may not be so obvious to the consumer. Manufacturers employ rigorous diagnostic techniques to study can sheet, for example, examining the crystalline structure of the metal with X-ray diffraction, hoping to discover better ways of casting the ingots or rolling the sheets. Changes in the composition of the aluminum alloy, or in the way the alloy is cooled after casting, or the thickness to which the can sheet is rolled may not result in cans that strike the consumer as innovative. Nevertheless, it is probably advances in these areas that will lead to more economical can manufacture in the future.

Where To Learn More

Book

Smith, George David. *From Monopoly to Competition: The Transformations of Alcoa, 1888-1986*. Cambridge University Press, 1988.

Periodicals

Hosford, William F. and John L. Duncan. "The Aluminum Beverage Can." *Scientific American*, September 1994, pp. 48-53.

Larson, Melissa. "New Ideas Come In Cans." *Packaging*, April 1993, pp. 30-31.

Singh, S. Paul. "Internal Gas Pressure on the Compression Strength of Beverage Cans and Plastic Bottles." *Journal of Testing and Evaluation*, March 1993, pp. 129-31.

—*Angela Woodward*

Ammunition

Background

Small arms ammunition, or cartridges, are used in a variety of firearms ranging from pistols to rifles and shotguns to heavier automatic weapons sometimes called machine guns. The term "bullet" is commonly used to describe the cartridge, when in fact, it actually only refers to the projectile. The correct terminology for the cartridge components are bullet, case, primer, and propellant or gunpowder. Each component is manufactured separately and then assembled into the cartridge. Specifications for the size, shape, ignition type, and ballistic performance have been standardized for the majority of military and civilian ammunition, but there are many obsolete and one-of-a-kind "wildcat" cartridges still found. Small arms ammunition includes cartridges with a bullet diameter, or caliber, of up to 0.75 inch (.750 caliber). The bulk of the production is for cartridges with bullets of .45 caliber or smaller.

Until the 19th century, the only way to load a weapon was to first pour the powder into the barrel, then place a greased cloth patch around a lead bullet and ram the bullet down the barrel to the powder with the ramrod. A flintlock produced a small spark, or a percussion cap produced a small explosive flash to ignite the powder which fired the patched bullet. This was a very slow process and often produced an inaccurate shot. After repeated firing, the barrel became fouled with powder residue to the point that loading became impossible.

In the early 1800s, gun manufacturers realized that increased accuracy and rate of fire could only be achieved by redesigning the way the bullet, powder, and igniter were loaded into the weapon. The first successful new design was made in 1848 by Christian Sharps. His design utilized an opening, or breech, at the base of the barrel closest to the person firing the weapon. The breech could be manually closed to seal off the end. With Sharp's design, the bullet was loaded into the open breech, followed by a powder charge held in a paper bag. When the breech was closed, the bag was cut open. This exposed the powder which could then be ignited by the percussion cap.

In 1852, a cartridge with a metal case was developed by Charles Lancaster of England. It held the powder inside the case with the bullet on one end. About the same time, another Englishman, Colonel Boxer, and an American, Hiram Berdan, also developed a metal case cartridge that incorporated an igniter, or primer, inserted into the center of the base of the case. The primer contained a small amount of impact-sensitive explosive that could be set off when struck by a pin—known as the firing pin—that was part of the weapon. The concept of the center-fire metal case cartridge developed by Boxer and Berdan has survived to the present day and is the basis for modern small arms ammunition design.

Raw Materials

Bullets are made of a lead alloy, often containing tin and antimony. Some bullets have a thick jacket of copper over the outside for improved performance.

Cases are made of brass, steel, or aluminum. Brass is the most common. Shotgun shells are often made with a polypropy-

The term "bullet" is commonly used to describe the cartridge, when in fact, it actually only refers to the projectile. The correct terminology for the cartridge components are bullet, case, primer, and propellant or gunpowder.

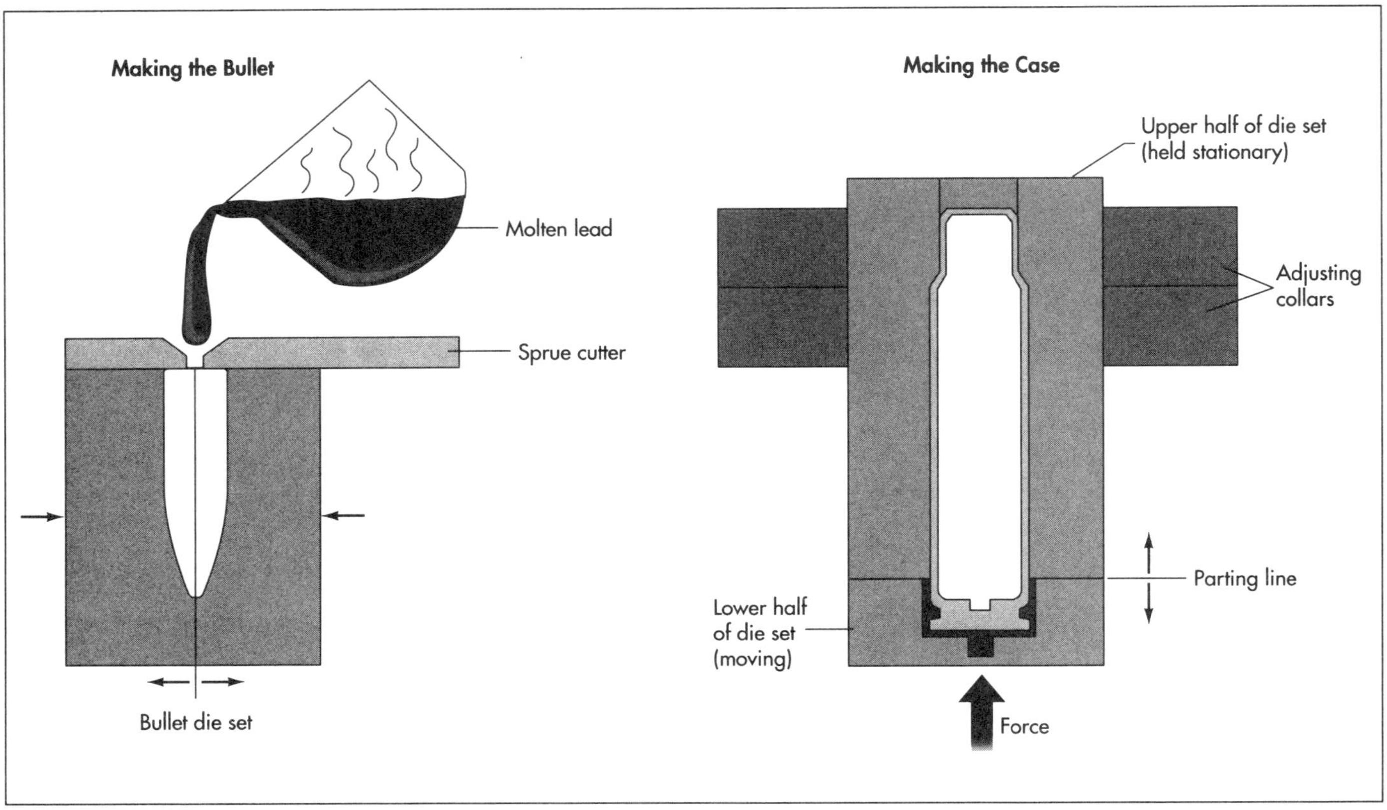

Many handgun and rifle bullets used for competition shooting are cast using conventional casting methods. The molten lead is poured into the bullet mold cavity, cooled quickly, and then extracted from the mold. The typical brass case is formed from annealed sheet by drawing with a multiple punch and die set.

lene plastic case attached to a metal base. A few handgun cartridge cases have been made of plastic, but have not received wide acceptance.

Primers are made of a copper or brass alloy cup with a brass anvil and are filled with an impact-sensitive lead styphnate igniter. The metal parts of the primer are usually nickel-plated to resist corrosion.

Propellants can vary from black gunpowder to a more modern smokeless powder which contains nitrocellulose. Propellants are carefully formulated to ignite and create an expanding gas that accelerates the bullet down the barrel. The expansion rate, physical size and shape of the powder particles, and the stability of the propellant are all important factors in the chemical formula used to produce it.

Bullet Design & Manufacture

Bullets can be made by several different processes. Smaller .22 caliber bullets are usually **lead** and are pressed, or cold formed, into shape. A small piece of thick lead wire is cut to the correct length and formed into the bullet shape by a die set in an automatic press. High production rates can be achieved by this type of automated process. Many handgun and rifle bullets used for competition shooting are cast using conventional casting methods. The molten lead is poured into the bullet mold cavity, cooled quickly, and then extracted from the mold. The point at which the lead enters the cavity (or "sprue") is trimmed away as the bullet is extracted. Both cold-formed and cast bullets may be further improved by copper plating. The plating process electrically deposits a thin layer of copper on the outside of the bullet, protecting the lead from oxidation and providing a harder surface to engage the grooves, or rifling, in the barrel which give the bullet a spin to improve accuracy. Copper also reduces the lead fouling of the rifling after firing, allowing the firearm to maintain accuracy after firing many rounds.

To improve bullet performance and accuracy, the "jacketed" bullet was developed. This is a family of bullets that use a substantial brass or copper outer shell, usually filled with lead by casting or cold forming, and having several different configurations for specific performance criteria. Some examples are FMJ (full metal jacket), JHP (jacketed hollow point), and JSP (jacketed soft point), each with options such as boat-tail design, controlled expansion, tracer, incendiary, and armor-piercing. The brass outer shell of these bullets engage the rifling tightly upon firing, providing a close fit for improved accuracy. Designed to further improve accuracy, the boat-tail bullet has the base reduced in diameter to improve air flow and stability in flight. The soft nose and hollow point bullets are designed to expand upon striking the target to intensify their impact.

Specialized bullets are sometimes found in military applications. Armor-piercing bullets can be solid brass or copper jacketed steel core. These can penetrate engine blocks and aircraft frames, damaging and incapacitating mechanisms inside. Tracers have a small amount of a phosphorus compound in their base. Upon firing, the phosphorous ignites and burns with a bright light. At night they can be seen streaking away from the firing position towards the target, allowing the shooter to track the bullet in flight and make aiming adjustments. Incendiary bullets contain small amounts of magnesium, which, like phosphorous, burns when ignited, but stays burning for a longer time and causes ignition of fuels or ammunition upon impact at the target.

Case Design & Manufacture

Nearly all small arms ammunition cases are of brass alloy. Some use aluminum, steel, or plastic, but the brass case is most popular and easiest to manufacture.

The design of the case is determined by the firearm in which the ammunition is used. The typical brass case is formed from annealed sheet by drawing with a multiple punch and die set. The first stage of the multiple die set forms the metal, the second stretches the metal deeper, the third forms the rim, and so on. Each step stretches the metal slightly farther until the final stage produces an accurately formed case. The cases are trimmed to length and the primer hole is punched. Heat treating and stress relieving are performed to selected types of cases to improve durability. This is accomplished in large batch ovens, where baskets of cases are heated with enough temperature to gently soften the metal without distorting it. When cooled, the metal is "relaxed" and better able to take the punishment of firing. Some handgun caliber cases are nickel plated for durability in reloading, corrosion resistance, and for appearance. Each case is stamped with information such as caliber, manufacturer, munitions codes, and year of manufacture.

Primer Design & Manufacture

The primer consists of two metal parts and a small amount of explosive compound. Primers come in different sizes depending on the firearm. Using a small pistol primer as an example, the cup is usually about 0.125 inch (0.32 cm) in diameter and 0.125 inch (0.32 cm) tall, and made of soft copper or brass. Inside is placed a small amount of the impact-sensitive explosive lead styphnate, and pressed into the opening is a triangle-shaped piece called the anvil. When struck by the firing pin, the center of the cup collapses, squeezing the explosive between its inner surface and the anvil. The explosive ignites and shoots a flame through the flash hole, igniting the propellant to fire the cartridge.

The Manufacturing Process: Cartridge Assembly

The assembly process for the cartridge components begins with a thorough cleaning and polishing of the case by a vibratory finisher. The finisher works by vibrating a corn byproduct (dried and ground corncobs) with a polishing compound around the cases, creating a high luster. Thus prepared, they are ready for final assembly. This is how a typical center-fire metal cartridge is assembled:

Sizing the case

1 The cases are fed into a loading press which first sizes the case. This sizing

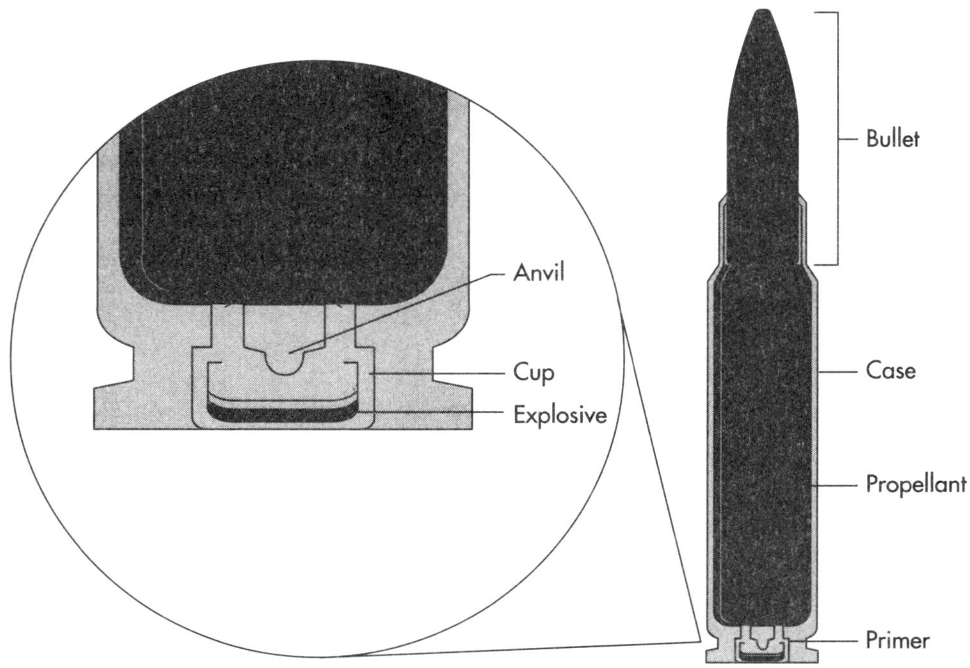

The cartridge is assembled by sizing the case, inserting the primer, adding propellant, and seating the bullet.

forms the metal case to standard dimensions. The case must be within 0.001 inch for it to function correctly.

Inserting the primer

2 The primer is then pressed into the case primer hole flush with the base. The primer must be flush or the cartridge will not feed properly in the weapon magazine, causing a "jam." At the same time, the mouth of the case is slightly expanded, in preparation for receiving the bullet.

Charging the case

3 The case is "charged," or filled with the correct amount of propellant. This step is of utmost importance, for miscalculation or double charging could be disastrous.

Assembling the bullet

4 The bullet is firmly seated into the open end of the case. The bullet has a coating of lubricant to prevent corrosion and assist in the assembly process. The bullet is then crimped into the case to give the correct overall length of the cartridge. The crimp reduces the diameter of the open end of the case and captures the bullet tightly, sealing the assembly together so moisture cannot invade the powder.

The press used to assemble cartridges must feed each component accurately and in the correct sequence. Otherwise, cases could be unprimed, powder left out, or bullets seated incorrectly. Any of these could result in a misfire or loss of accuracy at the minimum and, at worst, cause the firearm to blow apart upon firing. In each stage of the process, special dies perform the important assembly function. The dies are made of tooling carbide for long life, and have close adjustments to produce quality ammunition.

After assembly, the finished cartridges are packaged, usually 50 to a box, and prepared for shipment to the shooter.

Quality Control

Most manufacturers shoot thousands of their own cartridges as part of their quality control programs and processes. The accuracy, pressure, reliability, velocity, and consistency are all recorded. The weapons used for this are specially made, highly accurate, and equipped with data-gathering electronics. Each production run of a particular cartridge is given a "lot code." This number, printed on the ammunition box, allows ammunition to be inventoried and traced. Should a particular lot show problems in the field, that group can be recalled and replaced using the lot code system.

The Future

Small arms ammunition will be available in its present form for the foreseeable future. Its function will continue to be to propel a projectile over a distance to strike a target. Variations in the material and design of this ammunition will be in response to the specific needs of the many groups of small arms users.

The military will continue to develop ammunition which can penetrate and incapacitate a wide variety of targets ranging from humans to sophisticated electronic equipment. Currently, they are investigating "non-lethal" weapons and ammunition which will incapacitate a target without destroying it. Small arms weapons in this category include hand-held chemical lasers to knock out electronic sensors, and foam guns which shoot a sticky foam that envelops the target. These non-lethal devices would supplement, not replace, the conventional small arms weapons and ammunition.

Police are also interested in non-lethal weapons and ammunition. Rubber bullets that impact without penetration are already in use for riot control. Another device is a shotgun which fires a small bean bag. When fired at a close range, the bean bag hits with the impact of a punch to momentarily incapacitate the target.

Hunters will want ammunition which hits accurately and kills with a single shot. Much of the development of commercial small arms ammunition has been in this area, and has included many variations in powder loads and bullet configuration.

Target shooters will continue to develop ammunition which offers excellent accuracy and repeatability for competition shooting.

Where To Learn More

Books

Barnes, Frank C. *Cartridges of the World*, 6th ed. DBI Books, Inc., 1989.

Hackley, F.W. *History of Modern U.S. Military Small Arms Ammunition.* Macmillan, 1967.

Periodicals

Gresham, Grits. "Nosler and His Partition." *Sports Afield*, December 1992, p. 40.

Langreth, Robert. "Softkill." *Popular Science*, October 1994, pp. 66-69.

—Douglas E. Betts

Antilock Brake System

Generations of drivers were taught to handle skidding on slippery roads by quickly pumping the brakes during a skid. With the introduction of antilock brake systems, electronic controllers automatically adjust for the loss of friction and enable the driver to resume control of the vehicle.

Background

Stopping safely is one of the most important functions a motor vehicle can perform. Failure of the brake system will almost invariably result in property damage, personal injury, or even death. Consequently, a great deal of consideration has been given to improving the brake system in trucks and passenger cars over the last nine decades. One of the latest improvements is an antilock brake system which, as the name suggests, prevents a vehicle's brakes from locking up and skidding during hard stops on wet or icy roads.

The problem of skidding reveals the one overwhelming weakness of all motor vehicle braking systems: they depend strongly on the coefficient of static friction between the tire and the road. If for any reason the tire momentarily loses its adhesion to the road while the brakes are applied, the friction of the brakes against the drums or rotors locks the wheel solidly and the tire begins skidding across the road. In this condition, the braking force of that wheel is dependent on the sliding friction between the tire and the road, which is much less than the static friction. Under wet or icy conditions, the sliding friction is reduced even further, resulting in significantly longer stopping distances. In addition, when the front wheels are in this condition, they cannot be used to steer the vehicle; regardless of the angle of the front wheels, the vehicle continues to skid in whatever direction its momentum sends it until either the driver releases the brakes or the vehicle collides with something solid enough to bring it to a halt.

Generations of drivers were taught to handle this condition by quickly applying and releasing, or pumping, the brakes during a skid. However, this training was often lost during panic situations. Additionally, even the calmest and best-trained driver could not pump the brakes faster than two or three times per second, which limited the effectiveness of the technique.

A better way to handle skids while braking had been used on aircraft for decades before it was introduced in ground vehicles. Aircraft were subject to the same low-traction conditions as were cars and trucks, and a skidding aircraft—already only marginally steerable—was truly a danger to its occupants and those around it. To combat this problem, many aircraft were equipped with antilock brake systems, known as ABS, which prevented the braking wheels from locking up and skidding on slippery runways.

At first, this was accomplished through elaborate and expensive hydraulic controls which cycled the brakes on and off rapidly, permitting the airplane to be steered under slippery conditions while still allowing a large measure of stopping ability. Later, electronic controls permitted anti-lock action that was more responsive to actual ground conditions.

As the electronic and hydraulic portions of aircraft ABS became smaller and less expensive, truck and automobile manufacturers began to take interest. At first, antilock brake systems were developed only for heavy-duty trucks. Large semi trucks—truck tractor-trailer combinations weighing up to 80,000 pounds (36,364 kg)—were especially hazardous to traffic around them when they skidded since they not only moved out of the driver's control, but also articulated, or jack-knifed, and frequently rolled over. Today, antilock brake systems are standard on many cars and trucks.

Regardless of manufacturer or the type of vehicle, all antilock brake systems operate in a similar manner. Wheel speed sensors are placed on each wheel that is to be controlled. Each speed sensor usually has a toothed wheel, like a gear, that rotates at the same speed as the vehicle wheel or axle. Mounted close to, but not touching this toothed wheel, is a permanent **magnet** wrapped with a coil of wire, called the pick-up coil (see illustration). As each tooth rotates past the permanent magnet, it causes the magnetic field to concentrate and increase slightly. This, in turn, induces a small pulse of current in the coil of wire. The number of pulses per second are directly proportional to the speed of the wheel. The faster the wheel turns, the faster the teeth pass the magnet and the higher the pulse rate.

The pulsed output from the wheel speed sensors goes to an electronic controller, which monitors each wheel's speed relative to the speed of the other wheels. As long as the brakes are not being applied and all of the monitored wheels are rotating at roughly the same speed, the system takes no action. If, however, the brakes are being applied and one or more of the monitored wheels suddenly begins to reduce speed at a higher rate than the others—indicating a loss of traction with the road and an imminent wheel lockup and skid—the controller then activates the antilock system.

The antilock brake system on any vehicle is simply an additional monitoring and controlling function superimposed on the existing vehicle brake system. ABS is not a second brake system, nor does it replace the vehicle brake system. When all four wheels on an automobile are monitored and controlled, the system is called a four-channel ABS. If the front two wheels plus the rear axle (but not each rear wheel individually) are to be controlled, the system is called a three-channel ABS. On heavy trucks with two rear drive axles, the ABS is commonly a four-channel system which controls the front wheels and two of the four rear wheels. Trailers pulled by heavy truck tractors may also have their own separate ABS which must interconnect with the ABS on the tractor.

In an automobile, the brakes are actuated by hydraulic pressure. The ABS controller operates solenoid valves built into the high pressure side of the master brake cylinder. These valves are normally open and do not interfere with braking. When the controller senses that a wheel is locking up while braking, it first activates a solenoid to close a valve in the affected wheel's brake line which prevents the pressure from increasing any further. If the locked wheel continues to lose speed, the controller activates a second solenoid which bleeds pressure off the affected brake line, in effect releasing the brake for that wheel regardless of whether the driver is still pushing on the brake pedal. As soon as the wheel regains traction and its speed increases, the solenoids are de-activated, and normal braking resumes. Of course, if the conditions are such that the wheel starts to skid again, the brake will promptly begin to lock up and the ABS will take over. This cycle is repeated 12 to 15 times per second until either the road condition changes or the driver releases the brakes. The driver will be able to detect this rapid cycling as a vibration felt through the brake pedal, but will not have to take any action. The ABS will minimize the skid and will allow the driver to maintain directional control of the vehicle.

The brakes on a heavy truck are actuated by air pressure, rather than hydraulic pressure. The antilock brake system on a truck works in a manner similar to the ABS on an automobile, except the antilock air pressure control valves are located on the vehicle frame rail, near each wheel.

Design

An antilock brake system is designed for a specific vehicle application. A truck which does not pull a trailer, like a cement mixer, would have a slightly different ABS than a truck tractor which pulls one or more trailers. Likewise, an antilock brake system for a trailer would have a different design.

ABS for automobiles may be even more specific and may be designed for a particu-

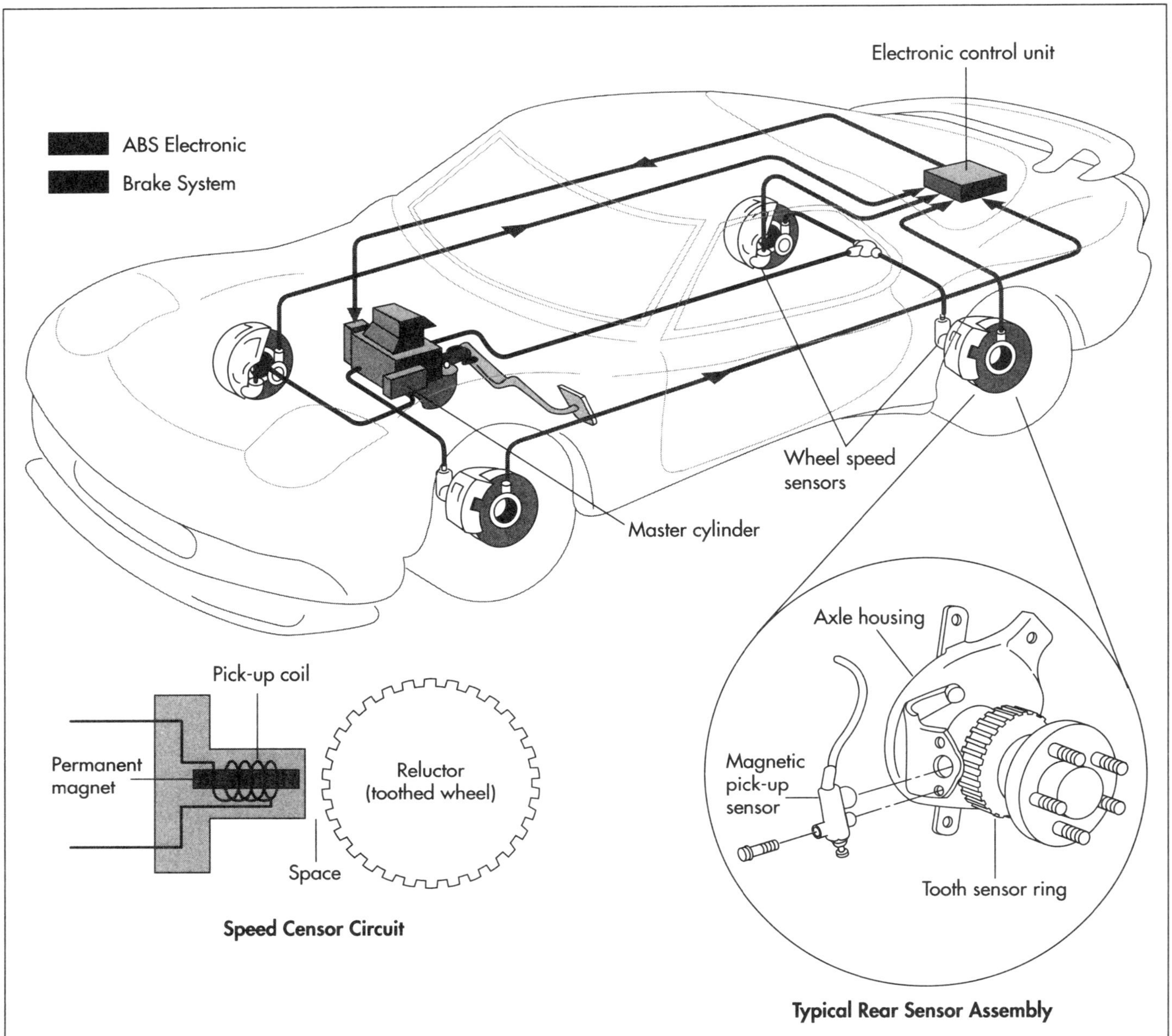

Speed Censor Circuit

Typical Rear Sensor Assembly

Regardless of manufacturer or the type of vehicle, all antilock brake systems operate in a similar manner. Wheel speed sensors are placed on each wheel that is to be controlled. Each speed sensor usually has a toothed wheel that rotates at the same speed as the vehicle wheel or axle. If the brakes are applied and one or more of the monitored wheels suddenly begins to reduce speed at a higher rate than the others, the controller activates the antilock system.

lar brand name and model of car. Since ABS components must fit and function along with existing vehicle components on each model, the design and manufacturing process of a new antilock brake system is carried out in partnership between the automobile manufacturer and the ABS supplier.

Raw Materials

The toothed wheel or gear in the speed sensor is made of soft **iron**, usually cast. Iron is chosen because of its high magnetic permeability and low magnetic reluctance. Magnetic reluctance is roughly equivalent to electrical resistance, and sometimes the toothed wheel is called the reluctor. The function of the toothed wheel is to allow the permanent magnet's field to easily pass through each tooth to cause a momentary concentration of field strength which induces a current in the pick-up coil. The pick-up coil has a permanent magnet in the core, wrapped with a coil of copper wire.

The controller usually employs transistors known as hot-side drivers which control the power side of the circuit rather than the ground side. These transistors produce more heat than is usual in an electronic circuit. Rather than being placed in a plastic or stamped steel housing, they are attached to a cast aluminum housing with a finned heat sink to dissipate the heat.

The hydraulic brake pressure solenoids used in automobiles have a standard construction of copper coil elements with steel valves and bodies. They are housed in the same casing as the brake system master cylinder which is usually cast from aluminum.

The electrical wiring is copper, often with cross-linked polyethylene insulation. To prevent radio frequency interference (RFI), in which high-power radio signals might be received through the wiring and cause the system to activate, all wiring is either shielded or the wires are run as twisted pairs to cancel out the effects of radio waves. Connectors are plastic with internal copper contacts.

The Manufacturing Process

The manufacturing process for antilock brake systems consists of manufacturing the component parts and then installing those parts on the vehicle. The parts are built in one plant, then packaged and shipped to a vehicle assembly plant for installation. This is a typical process for an automobile antilock brake system.

Making the master brake cylinder

1 The master cylinder, including the base for the solenoid body, is cast as a single unit. The seating and sealing surfaces are machined smooth and the connection ports are threaded.

2 The individual primary and secondary pistons, solenoid coils, reservoir caps and seals, pressure accumulator, and any metering and proportioning valves are installed. The solenoid body has a cover which attaches to the master cylinder with four or more screws and is sealed with a gasket.

Making the wheel speed sensors

3 The toothed wheel is cast from iron. Minor machining may be required at the mounting points.

4 The pick-up coils are wound around the permanent magnet core in a machine called a coil winder. The entire assembly is encased, or potted, in plastic resin with an electrical connector attached.

Making the controller

5 The electronic controller components are soldered to a **printed circuit board**.

6 The board is connected inside a protective housing and mounted to the cast aluminum heat sink base. External electrical connections are provided for the input wiring from each speed sensor and the output wiring to the solenoids in the master brake cylinder.

Installing the ABS

7 In the automobile assembly plant, the steel tubing brake lines are installed in the framework of the body. They run from the partition between the engine compartment and the occupant compartment, called the firewall, to the vicinity of each wheel. The electrical wires for the ABS are also run from the vicinity of each wheel to the controller location and from the controller to the firewall.

8 The brake master cylinder is bolted to the firewall in the engine compartment near the brake pedal. The brake lines are attached to the appropriate ports on the solenoid body, and the electrical wires are connected.

9 The toothed sensor wheels are pressed onto the outer constant velocity joints or the ends of the axle spindles so that they ride just inside the wheels. Once the axles are attached to the frame, the brake lines are attached and the pick-up coils are installed so that the end of the coils are close to the toothed wheels. The pick-up coils are then electrically connected to the wires to the controller.

10 The controller is installed either under the instrument panel or in the vehicle's trunk. The electrical connections are made, including the power connection from the vehicle battery through the fusebox.

Quality Control

The idea of an electronic system being able to take over the operation of a vehicle's brakes is disturbing to some people. For this reason, the operation of the system is thoroughly tested beforehand, and the quality of the installation is constantly reviewed.

In addition, all antilock brake systems are designed to be fail-safe—that is, any failure of any component will cause the system to fail in such a manner as to still allow the overall safe operation of the brakes.

The Future

There is a strong possibility that the federal government will mandate the use of antilock brakes on certain vehicles in the near future. ABS has been in use for several years, and evidence mounts regarding its benefits—specifically its ability to improve vehicle stopping distances and to maintain vehicle directional control under extremely slick road conditions.

These findings are not without controversy, however. Initial claims of the benefits of ABS were significantly overstated, and many drivers have found that ABS offers them little or no advantage in their particular situation. In this respect, the controversy is a little like the one that surrounded seat belts.

Additional systems have been developed that enhance the benefits of the basic ABS. One of these systems is automatic traction control, called ATC. ATC uses the same components as ABS, but works at the other end of the speed spectrum—getting a vehicle started under slippery conditions. In operation, it senses each wheel's speed to detect when one or more wheels "break loose" and start to spin. When that happens, it applies the brake on that wheel 12 to 15 times per second to let it slow down and regain traction. In demonstrations, vehicles have been held by blocks on an ice-covered grade. When the vehicles start and the blocks are pulled away, the vehicle without ATC spins its wheels and slowly slides backwards down the grade, while the ATC-equipped vehicle pulls its way up the ice.

It is expected that ABS, along with other new vehicle products, will continue to increase in popularity as the price goes down and the benefits become more apparent.

Where To Learn More

Periodicals

Chaikin, Don. "How It Works: Antilock Brakes." *Popular Mechanics,* July 1993, p. 88.

Kirk, John. "Antilock Brake Systems." *Home Mechanix,* March 1993, pp. 16-18.

Kott, D. "Anti-Lock Braking Systems." *Road and Track*, February 1994, p. 110.

Murray, C. "Making a Comeback in the Big Rigs (ABS)." *Popular Science,* July 1993, p. 38.

Sherman, Don. "Controlling Traction." *Popular Science*, April 1994, pp. 87-90, 96-97.

—*Joel Simon*

Asphalt Cement

Background

Asphalt is a heavy, dark brown to black mineral substance, one of several mixtures of hydrocarbons called bitumens. Asphalt is a strong, versatile weather and chemical-resistant binding material which adapts itself to a variety of uses. Asphalt binds crushed stone and gravel (commonly known as aggregate) into firm, tough surfaces for roads, streets, and airport runways. Asphalt, also known as mineral pitch, is obtained from either natural deposits such as native asphalt or brea or as a byproduct of the petroleum industry (petroleum asphalt). Prehistoric animal skeletons have been preserved completely intact in natural asphalt deposits, one of the most famous being the La Brea Tar Pits in Los Angeles, California.

Asphalt is one of the world's oldest engineering materials, having been used since the beginning of civilization. Around 6000 B.C. the Sumerians had a thriving shipbuilding industry that produced and used asphalt for caulking and waterproofing. As early as 2600 B.C. the Egyptians were using asphalt as a waterproofing material and also to impregnate the wrappings of mummies as a preservative. Ancient civilizations widely used asphalt as a mortar for building and paving blocks used in temples, irrigation systems, reservoirs, and highways. The asphalts used by early civilizations occurred naturally and were found in geologic strata as either soft, workable mortars or as hard, brittle black veins of rock formations (also known as asphaltic coal). Natural asphalts formed when crude petroleum oils worked their way up through cracks and fissures to the earth's surface. The action of the sun and wind drove off the lighter oils and gases, leaving a black residue. Natural asphalts were extensively used until the early 1900s. The discovery of refining asphalt from crude petroleum and the increasing popularity of the automobile served to greatly expand the asphalt industry. Modern petroleum asphalt has the same durable qualities as naturally occurring asphalt, with the added advantage of being refined to a uniform condition free from organic and mineral impurities.

Most of the petroleum asphalt produced today is used for highway surfacing. Asphalt paving material is a dull black mixture of asphalt cement, sand, and crushed rock. After being heated, it is dumped out steaming hot onto the roadbed, raked level, and then compacted by a heavy steamroller. Asphalt is also used for expansion joints and patches on concrete roads. Airport runways, tennis courts, playgrounds, and floors in buildings all use asphalt as well. Light forms of petroleum asphalt called road oils are sprayed on roadways to settle dust and bind gravel. Another major use of asphalt is in asphalt shingles and roll roofing, which usually consists of felt saturated with asphalt. The asphalt helps to preserve and waterproof the roofing material. Other applications for asphalt include the following: waterproofing tunnels, bridges, dams and reservoirs; rustproofing and soundproofing metal pipes and automotive underbodies; and soundproofing walls and ceilings.

Obtained from either natural deposits such as native asphalt or brea or as a byproduct of the petroleum industry, asphalt is a strong, versatile weather and chemical-resistant binding material which adapts itself to a variety of uses.

Raw Materials

The raw material used in modern asphalt manufacturing is petroleum, which is a natu-

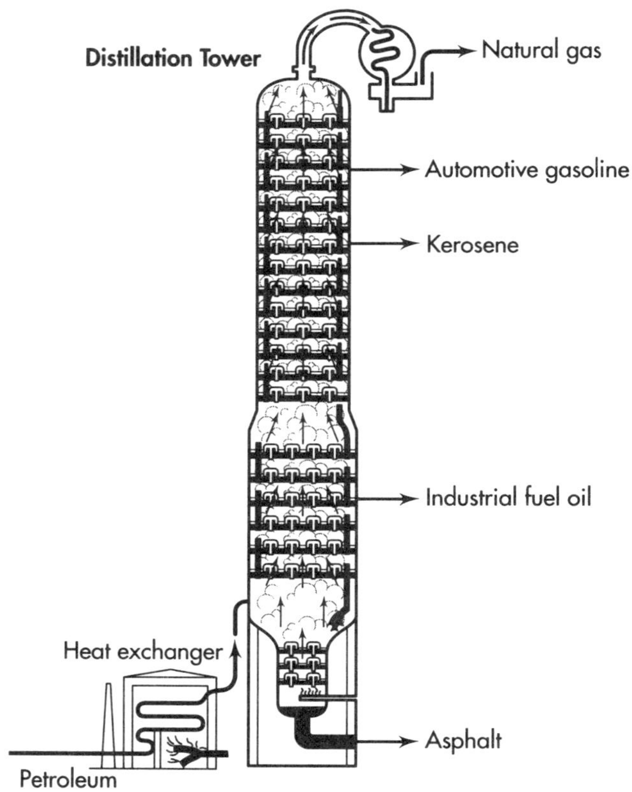

A fractional distillation tower is used to separate different compounds in petroleum. Asphalt is essentially the heavy residue of the oil refining process.

rally occurring liquid bitumen. Asphalt is a natural constituent of petroleum, and there are crude oils which are almost entirely asphalt. Oil wells supply the crude petroleum to the oil refineries, where it is separated into its various components or fractions.

The Manufacturing Process

Crude petroleum is separated into its various fractions through a distillation process at the oil refinery. After separation, these fractions are further refined into other products which include asphalt, paraffin, **gasoline**, naphtha, lubricating oil, kerosene, and diesel oil. Since asphalt is the base or heavy constituent of crude petroleum, it does not evaporate or boil off during the distillation process. Asphalt is essentially the heavy residue of the oil refining process.

Distilling the crude

1 The refining process starts by piping the crude petroleum from a storage tank into a heat exchanger or tube heater where its temperature is rapidly raised for initial distillation. It then enters an atmospheric distillation tower where the lighter and more volatile components, or fractions, vaporize and are drawn off through a series of condensers and coolers. It is then separated for further refining into gasoline (considered a "light" distillate), kerosene (considered a "medium" distillate), diesel oil (considered a "heavy" distillate), and many other useful petroleum products.

The heavy residue from this atmospheric distillation process is commonly called topped crude. This topped crude may be used for fuel oil or further processed into other products such as asphalt. Vacuum distillation may remove enough high boiling fractions to yield what is called a "straight run" asphalt. However, if the topped crude contains enough low volatile components which cannot be economically removed through distillation, solvent extraction—also known as solvent deasphalting—may be required to produce asphalt cement of the desired consistency.

Cutting back

2 Asphalt may next be blended or "cut back" with a volatile substance, resulting in a product that is soft and workable at a lower temperature than pure asphalt cement. When the cut-back asphalt is used for paving or construction, the volatile element evaporates when exposed to air or heat, leaving the hard asphalt cement. The relative speed of evaporation or volatility of the cutting agent determines whether a cut-back asphalt is classified as slow, medium, or rapid-curing. Heated asphalt cement is mixed with residual asphaltic oil from the earlier distillation process for a slow-curing asphalt, with kerosene for medium-curing, and with gasoline or naphtha for the rapid-curing asphalt.

Emulsifying

3 The asphalt cement may also be emulsified to produce a liquid that can be easily pumped through pipes, mixed with aggregate, or sprayed through nozzles. To emulsify, the asphalt cement is ground into globules 5 to 10 microns and smaller (one micron is equal to one millionth of a meter). This is mixed with water. An emulsifying agent is added, which reduces the tendency of the asphalt and water to separate. The emulsifying agent may be colloidal clay, soluble or insoluble silicates, soap, or sulphonated vegetable oils.

Pulverizing

4 Asphalt may also be pulverized to produce a powdered asphalt. The asphalt is crushed and passed through a series of fine mesh sieves to ensure uniform size of the granules. Powered asphalt can be mixed with road oil and aggregate for pavement construction. The heat and pressure in the road slowly amalgamates the powder with the aggregate and binding oil, and the

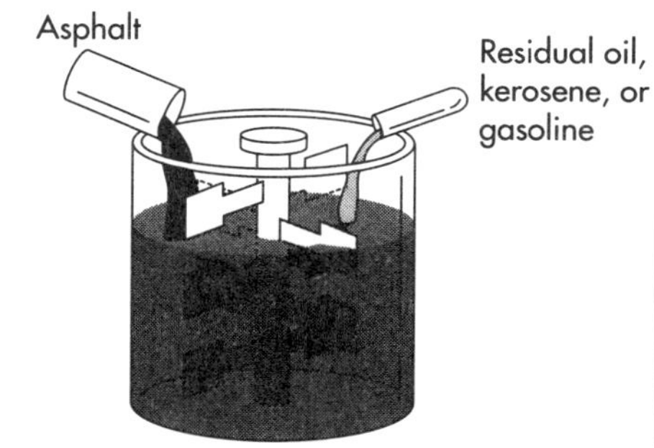

The asphalt is "cut back" by blending it with a volatile substance to make it softer and more workable. The volatile element evaporates when exposed to air.

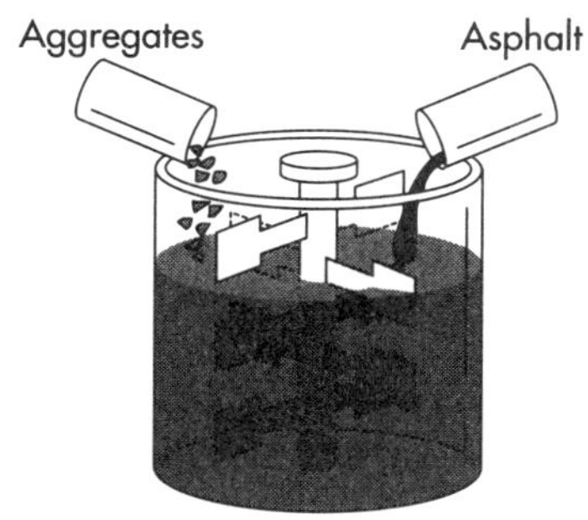

To make paving mixtures, aggregates are mixed with asphalt.

substance hardens to a consistency similar to regular asphalt cement.

Air Blowing

5 If the asphalt is to be used for a purpose other than paving, such as roofing, pipe coating, or as an undersealant or waterproofing material, the asphalt may be oxidized, or air blown. This process produces a material that softens at a higher temperature than paving asphalts. It may be air blown at the refinery, at an asphalt processing plant, or at a roofing material plant. The asphalt is heated to 500°F (260°C). Then air is bubbled through it for one to 4.5 hours. When cooled, the asphalt remains liquid.

Asphalt Paving Mixtures

Since asphalt cement is a major constituent used in road paving, the following is a brief

While still hot, asphalt is spread on the desired surface.

There are two types of asphalt mixes: hot-mix and cold-mix. Hot-mix asphalt (HMA) is commonly used for heavier traffic areas while cold-mix asphalt is used for secondary roads.

description of how asphalt paving mixtures are produced. Asphalt paving mixes made with asphalt cement are usually prepared at an asphalt mixing facility. There are two types of asphalt mixes: hot-mix and cold-mix. Hot-mix asphalt (HMA) is more commonly used while cold-mix asphalt (generally mixes made with emulsified or cut-back asphalts) is usually used for light to medium traffic secondary roads, or for remote locations or maintenance use. Hot-mix asphalts are a mixture of suitable aggregate coated with asphalt cement. The term "hot-mix" comes from the process of heating the aggregate and asphalt before mixing to remove moisture from the aggregate and to obtain sufficient fluidity of the asphalt cement for proper mixing and workability.

6 Asphalt cement and aggregate are combined in a mixing facility where they are heated, proportioned, and mixed to produce the desired paving mixture. Hot-mix facilities may be permanently located (also called "stationary" facilities), or it may be portable and moved from job to job. Hot-mix facilities may be classified as either a batch facility or a drum-mix facility, both can be either stationary or portable. Batch-type hot-mixing facilities use different size fractions of hot aggregate which are drawn in proportional amounts from storage bins to make up one batch for mixing. The combination of aggregates is dumped into a mixing chamber called a pugmill. The asphalt, which has also been weighed, is then thoroughly mixed with the aggregate in the pugmill. After mixing, the material is then emptied from the pugmill into trucks, storage silos, or surge bins. The drum-mixing process heats and blends the aggregate with asphalt all at the same time in the drum mixer.

7 When the mixing is complete, the hot-mix is then transported to the paving site and spread in a partially compacted layer to a uniform, even surface with a paving machine. While still hot, the paving mixture is further compacted by heavy rolling machines to produce a smooth pavement surface.

Quality Control

The quality of asphalt cement is affected by the inherent properties of the petroleum crude oil from which it was produced. Different oil fields and areas produce crude oils with very different characteristics. The refining method also affects the quality of the asphalt cement. For engineering and construction purposes, there are three important factors to consider: consistency, also called the viscosity or the degree of fluidity of asphalt at a particular temperature, purity, and safety.

The consistency or viscosity of asphalt cement varies with temperature, and asphalt is graded based on ranges of consistency at a standard temperature. Careless temperature and mixing control can cause more hardening damage to asphalt cement than many years of service on a roadway. A standardized viscosity or penetration test is commonly specified to measure paving asphalt consistency. Air-blown asphalts typically use a softening point test.

Purity of asphalt cement can be easily tested since it is composed almost entirely of bitumen, which is soluble in carbon disulfide. Refined asphalts are usually more than 99.5% soluble in carbon disulfide and any impurities that remain are inert. Because of the hazardous flammable nature of carbon disulfide, trichloroethylene (TCE), which is also an excellent solvent for asphalt cement, is used in the solubility purity tests.

Asphalt cement must be free of water or moisture as it leaves the refinery. However, transports loading the asphalt may have moisture present in their tanks. This can cause the asphalt to foam when it is heated above 212°F (100°C), which is a safety hazard. Specifications usually require that asphalts not foam at temperatures up to 347°F (175°C). Asphalt cement, if heated to a high enough temperature, will release fumes which will flash in the presence of a spark or open flame. The temperature at which this occurs is called the flashpoint, and is well above temperatures normally used in paving operations. Because of the possibility of asphalt foaming and to ensure an adequate margin of safety, the flashpoint of the asphalt is measured and controlled.

Another important engineering property of asphalt cement is its ductility, which is a measure of a material's ability to be pulled, drawn, or deformed. In asphalt cements, the presence or absence of ductility is usually more important than the actual degree of ductility because some asphalt cements with a high degree of ductility are also more temperature sensitive. Ductility is measured by an "extension" test, whereby a standard asphalt cement briquette molded under standard conditions and dimensions is pulled at a standard temperature (normally 77°F [25°C]) until it breaks under tension. The elongation at which the asphalt cement sample breaks is a measure of the ductility of the sample.

Byproducts/Waste

Environmental protection laws have developed stringent codes limiting water flows and particulate and smoke emissions from oil refineries and asphalt processing plants. Not only dust but sulfur dioxides, smoke, and many other emissions must be rigorously controlled. Electrostatic precipitators, primary dust collectors using single or multiple cone cyclones, and secondary collection units consisting of fabric filter collectors commonly called "baghouses" are all required equipment to control emissions. Hydrocarbons formed in asphalt production, if unchecked, create odoriferous fumes and pollutants which will stain and darken the air. Pollutants emitted from asphalt production are controlled by enclosures which capture the exhaust and then recirculate it through the heating process. This not only eliminates the pollution but also increases the heating efficiency of the process.

Higher costs of asphalt cement, stone, and sand have forced the industry to increase efficiencies and recycle old asphalt pavements. In asphalt pavement recycling, materials reclaimed from old pavements are reprocessed along with new materials. The three major categories of asphalt recycling are 1) hot-mix recycling, where reclaimed materials are combined with new materials in a central plant to produce hot-mix paving mixtures, 2) cold-mix recycling, where reclaimed materials are combined with new materials either onsite or at a central plant to produce cold-mix base materials, and 3) surface recycling, a process in which the old asphalt surface pavement is heated in place, scraped down or "scarified," remixed, relaid, and rolled. Organic asphalt recycling agents may also be added to help restore the aged asphalt to desired specifications.

Because of solvent evaporation and volatility, use of cutback asphalts, especially rapid-cure cutback asphalts which use gasoline or naphtha, is becoming more restricted or prohibited while emulsified asphalts (in which only the water evaporates) are becoming more popular because of cost and environmental regulations.

The Future

Increasing economic and environmental needs will bring many new technical refinements to recycling old asphalt pavements, such as using microwaves to completely break down the pavement. Microwaves heat the crushed rock in asphalt pavement faster than the surrounding cement, which is then warmed by the radiant heat from the rock. This method prevents the asphalt cement from burning.

Alternative sources of raw material are being researched, such as the production of synthetic asphalt from the liquefaction of sewage sludge. To ensure consistent product quality, new methods are being developed for manufacturing modified asphalts and emulsions. Many new tests are being developed to help characterize asphalts, such as high-performance gel-permeation chromatography (HP-GPC), which allows many properties to be studied and the results compiled in only a few minutes. New processes, more efficient mixing and milling units, in-line liquid mass flow meters, on-line monitoring systems, and new safety equipment are some other areas being investigated for improvement.

Polymer-modified asphalt crack sealers are gaining in popularity, and many other asphalt modifiers are being developed. Modifiers are added to control pavement rutting, cracking, asphalt oxidation, and water damage. Some commercially available asphalt modifiers are polymers, including elastomers, metal complexes, elemental sulfur, fibers, hydrated lime, Portland cement, silicones, various fillers, and organic antistrip agents. Many of these modifiers have not been extensively used and are being researched for further development. It might even be possible one day to have "smart asphalt cements" by mixing in certain asphalt friction modifiers which would allow it to change characteristics depending on whether moisture was present. In conjunction with **antilock brakes**, automatic traction controls, and airbags, this could serve to save many lives on our nation's highways.

Where To Learn More

Books

Asphalt Construction Handbook. 6th ed. Barber-Greene, A Division of Caterpillar Paving Products Inc., 1992.

Bitumens, Asphalts, and Tar Sands. Elsevier North-Holland, 1978.

Periodicals

"The Long and Recyclable Road." *High Technology Business*, September-October 1989, p. 4.

Planche, Jean-Pascal, Helen W. King, and Gayle N. King. "A Refiner's Approach to Asphalt Characterization." *Nature*, April 18, 1991, pp. 24-25.

"Physical and Chemical Characteristics of Synthetic Asphalt Produced from the Liquefaction of Sewage Sludge." U.S. Environmental Protection Agency, Municipal Environment Research Laboratory: Center for Environmental Research Information, 1982.

Other

Asphalt Institute, Executive Office and Research Center. Research Park Drive, P.O. Box 14052, Lexington, KY 40512-4052. (606) 288-4960.

—*Glenn G. Whiteside*

Balloon

Background

A balloon is an air-tight bag made out of a light material that can be inflated with air or gas. Toy balloons are available in all kinds of shapes, sizes, and colors to delight children and adults at birthday parties and other festive occasions.

Balloons were first invented in France in the late 18th century. Two papermakers, Jacques and Joseph Montgolfier, discovered that when **paper** bags are filled with hot air, the bags rise. Quick to realize the potential of this, they began experimenting with balloons of various materials such as paper, cloth, and **silk**. They made the first public demonstration of a lighter-than-air balloon in June 1783, with a 35-foot (11 m) diameter balloon made of cloth lined with paper. Later that year, Jacques Charles flew a balloon made of silk coated with a rubber varnish and filled with hydrogen, a gas that is lighter than air. These early demonstrations attracted a great deal of excitement, and balloons were soon put to many uses in science, sport, and war.

The rubber toy balloon as we know it today is different from the early balloons in that it is made entirely of rubber. A practical way of making such formed rubber products required several discoveries and inventions. These developments took place gradually over the years since the first rubber factory in the world was established near Paris in 1803.

Natural latex is a mixture of small globules of rubber substance suspended in water (much like milk). When it is exposed to air, heat, or certain chemicals, it coagulates or clots together. The globules of rubber lump together and separate from the watery portion of the latex, eventually forming an elastic, solid material. To improve its strength, resilience, and resistance to hot and cold temperatures, rubber is vulcanized or cured by various methods, such as mixing with certain chemicals or treating with heat.

The idea of making a product out of rubber is an old one. The natives of South America created bottles and other articles by coating molds made of earth long before Europeans began experimenting with rubber in the mid-1700s. In 1830, the Englishman Thomas Hancock patented a process for creating products by pouring latex over molds or dipping molds into a latex mixture—the forerunner of the modern technique of producing dipped products such as rubber gloves and **condom**s.

In 1921, a method of retarding the coagulation of liquid latex was developed. This method enabled rubber makers to transport raw latex in a liquid form more easily to manufacturing centers around the world. This in turn led to new processes for making rubber goods. In the early 1920s, a number of patents were granted in England for processes that allowed molds to be dipped in liquid latex. In 1931, the first modern latex balloon was created by Neil Tillotson in his attic. He sold 15 of his "Tilly Cat" balloons (shaped like a cat's head, complete with whiskers printed on with dye) for the Patriot's Day parade in Massachusetts in April 1931, and formed a company that still makes balloons today.

Balloons were first invented in France in the late 18th century. Two papermakers, Jacques and Joseph Montgolfier, experimented with various materials such as paper, cloth, and silk to create a lighter-than-air balloon. In June 1783 they flew a 35-foot diameter balloon made of cloth lined with paper.

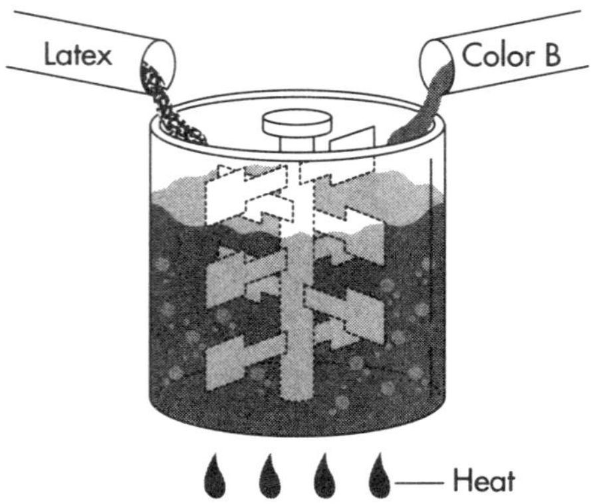

To make colored balloons, a pigment is mixed into latex.

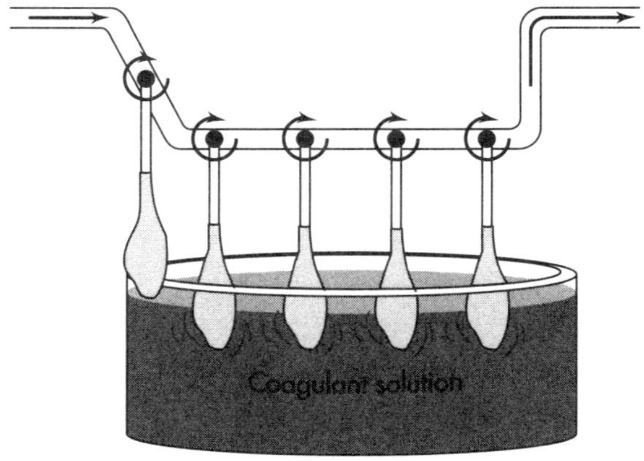

The forms in the shape of a deflated balloon are heated and immersed in a tank of coagulant solution.

Although rubber can be made synthetically, natural latex—a white or yellowish opaque liquid similar in appearance to milk—is preferred for its great elasticity.

Raw Materials

Although rubber can be made synthetically, natural latex is preferred for its great elasticity. It can be stretched to seven or eight times its original length and still return to its former shape. Synthetic rubber has not proven to be as elastic and resilient as natural latex.

Raw, natural latex is a white or yellowish opaque liquid, similar in appearance to milk. Latex is the secretion of certain plants, in particular the Hevea tree originally found in Brazil. The most important sources of natural rubber today are plantations in Malaysia and Africa.

Producers of rubber must harvest the raw material from these trees, which involves scoring the trees with shallow cuts and letting the sap ooze from the cuts into buckets. The latex is collected in large containers, filtered to remove foreign particles, and mixed with alkali to prevent coagulation. It is then shipped in liquid form to processing centers in different parts of the world.

Latex must be mixed with additives before it can be used in industrial processes. Certain chemicals are mixed in to achieve a desired thickness, rate of drying, and other properties. Other chemicals (collectively known as antidegradants) are added to slow the oxidation and decomposition of the rubber. To give it color, pigments are mixed into the latex. The pigments may be fine metal oxide powders or organic dyes.

The Manufacturing Process

In essence, the process of making a toy balloon involves dipping a mold into liquid latex. The mold, or form, is shaped like a deflated balloon.

The earliest balloon forms were disposable, made from cardboard attached to dowels. Modern forms are reusable and usually made from stainless steel, aluminum, or porcelain. The forms must be smooth and polished. A number of such forms are attached upside down to a board or rack. The boards are moved mechanically from one station to another in the factory.

To be efficient in terms of cost and number of balloons produced, balloon manufacture has become a highly automated, continuous loop process. Balloons are made in batches, all of the same color and size, since changing the color and form is time-consuming and requires manual intervention. Manual intervention is usually only needed for setting up a run and then later for packaging the finished product, and for dealing with occasional mechanical problems that may arise.

Preparing the latex

1 Prior to its use, the latex may need to be colored. This involves mixing a pigment into the latex. It may be done at the balloon factory, or the balloon maker may purchase already-pigmented latex from a supplier.

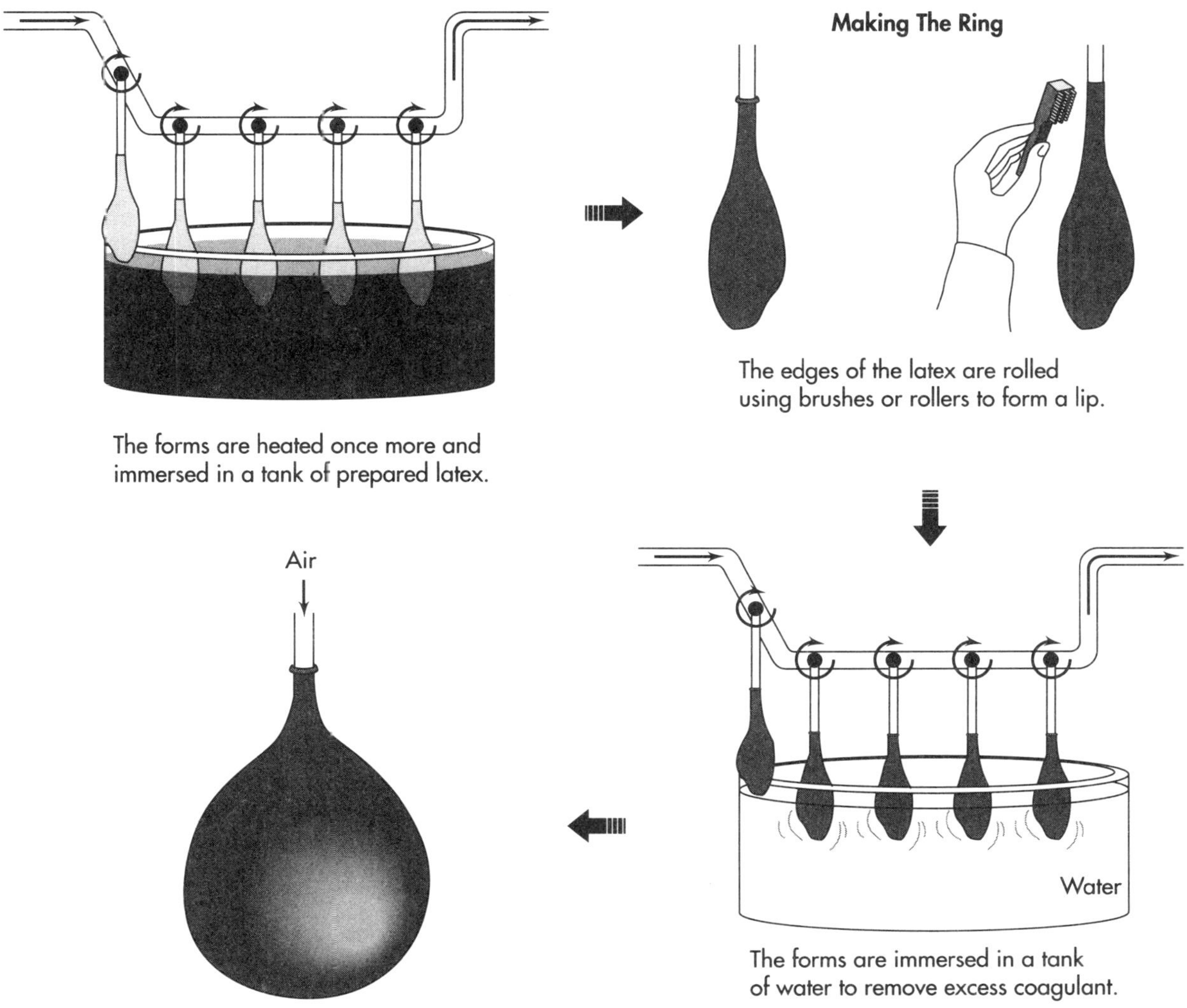

The forms are heated once more and immersed in a tank of prepared latex.

The edges of the latex are rolled using brushes or rollers to form a lip.

The forms are immersed in a tank of water to remove excess coagulant.

To remove the finished balloon from the form, a spray of air or water is injected.

2 The latex must be poured into tanks into which the forms will be dipped. The tanks are kept at a certain temperature and may include stirring mechanisms to keep the latex circulating to avoid settling.

Dipping the forms

3 The balloon forms are first heated, then immersed in a tank of coagulant solution for a few seconds. When the forms are immersed in the liquid latex, the coagulant will cause the rubber to gel in a thin sheet around the forms. A commonly used coagulant solution is a mixture of water, a calcium-based salt, **soap**, and talc powder. The **salt** is the actual coagulant; the soap helps the latex spread in an even film, and the talc helps ease the removal of the rubber from the forms in a later step.

4 The forms are heated to a temperature between 100°F (38°C) and 200°F (93°C), and then immersed in a tank of colored latex. The coagulant causes the latex to coat the forms. The longer the forms are left in the tank, the thicker the coating that sticks to them. For balloons, a very thin layer of latex is desired, so the forms are immersed only for a few seconds. The forms must be inserted and removed at carefully controlled speeds to avoid trapping air bubbles and to achieve an even, thin coating.

Making the ring

5 A lip is formed on the neck of the balloon by rolling the edges of the rubber using brushes or rollers. This creates the ring seen around the opening of the balloon.

Removing excess coagulant

6 Next, the forms are immersed in a tank of leaching solution (often plain water) to dissolve and leach away excess coagulant from the rubber.

Curing the rubber

7 The rubber on the forms must be dried and cured. The method used varies among manufacturers. Some balloon makers use a latex that already contains a vulcanizing agent, in which case the rubber is dried at a moderate temperature. Other makers induce vulcanization by putting the rubber-coated forms into an oven and curing for as long as an hour.

Removing the balloons

8 The balloons are then mechanically removed from their forms. One approach is to blow them off using a spray of water or air and collecting the balloons in a basket or net.

9 If the balloons are removed using a spray of water, they are next placed in a centrifuge, where excess water is removed by spinning the balloons around at high speed.

10 The balloons are then dried in large tumble dryers.

Printing and packaging

11 Next, the balloons may either be packaged, or first printed and then packaged. If they are packaged directly, they are moved on a conveyor belt past a counting device and placed into bags. When an appropriate number of balloons has been placed in each bag, the bags are sealed.

12 Printing designs on balloons, such as logos or faces, actually involves several steps. First, the balloons must be inflated in order to allow even printing. This requires a worker to manually place each balloon on the inflating device. Next, a pattern is carefully printed on each balloon. Finally, the balloons are removed and passed on to the packaging stage.

Quality Control

The balloon manufacturing environment must be strictly controlled in order to achieve high quality and consistency. Throughout the manufacturing process, computer-based instrumentation records and controls air humidity, air temperature, latex tank temperature, the temperature in the ovens, dryers, and other parameters.

The latex and other chemicals used in the process must be carefully formulated for specific properties, and carefully maintained. For example, the latex must have certain viscosity and speed of drying. The tanks in which it is held must have devices to keep the latex circulating to avoid forming a "skin," and to prevent ingredients from settling.

Byproducts/Waste

It is in the manufacturers' best interests to waste as little rubber as possible because the cost of latex is high compared to the selling price of individual balloons. Balloon makers also reclaim much of the coagulant that ends up in the leaching solution. Unfortunately, what is not reclaimed ends up as liquid waste in the environment. The amount of chemical waste that can be released by a factory is regulated by government laws. Balloons also result in some waste after they are manufactured because they are invariably thrown away after they deflate or pop. However, because latex is natural, it eventually breaks down into other substances.

Safety Concerns

Toy balloons can be a source of joy, but they can also be unexpectedly hazardous. Young children have been known to die from accidentally choking on balloons. Latex balloons may also end up in water, where they eventually lose their color and can resemble jellyfish. Sea animals such as whales and turtles have attempted to eat them and have died because the latex clogs their digestive systems.

The Future

The toy balloon industry is very competitive. Manufacturers are constantly looking for ways to make the process more automatic and efficient, especially by reducing manual intervention. Currently the most labor-intensive portions are the printing and packaging steps. Increasing automation in these steps is an area for potential future improvement.

In recent years balloons made of metal films have become popular. The manufacturing process of these balloons is very different. They are made from a sandwich of two swatches of mylar—a **polyester** film—often circular in shape, which are sealed together around the edges. A small opening is left through which the balloon may be inflated. Because the material is initially flat, these balloons can be printed more easily than balloons made of rubber. The foil can be made very shiny and reflective, allowing for very bright designs. They are stronger and more durable than rubber balloons, but for some uses, this is also a disadvantage. For example, they cannot be twisted into various shapes nor can they be filled with water. The foil also takes much longer to degrade in the environment than rubber.

Where To Learn More

Books

Barlow, Fred W. *Rubber Compounding: Principles, Materials, and Techniques.* Marcel Dekker, Inc. 1988.

Coates, Austin. *The Commerce in Rubber: The First 250 Years.* Oxford University Press, 1987.

Hofmann, Werner. *Rubber Technology Handbook.* Oxford University Press, 1989.

—Renee M. Rottner

Barbed Wire

Steel wire was first used for fencing during the settling of the American West in areas where wood was scarce. Early wire fences consisted of single strands which were easily broken in cold weather or by wandering cattle.

Background

Barbed wire is a fencing material consisting of a metal cable with regularly spaced sharp projections. The cable usually consists of two wires twisted around each other to add strength and to allow the cable to expand and contract with temperature changes without breaking. The sharp points, called barbs, usually consist of short pieces of wire twisted around one or both of the cable wires.

Fences of various kinds have been used since the earliest days of agriculture 10,000 years ago. Fences have been built from wood, earth, stone, and living plants (hedges in Europe and cactus in Latin America). Metal was not used for fencing until steel wire became available in the 19th century.

Short lengths of wire were first made at least 5,000 years ago by hammering pliable metals such as gold. By the year 1000, longer lengths of wire were made by pulling rods of soft metal, such as alloys of **lead** and tin, through a die of harder metal, such as **iron**. In modern times, until the middle of the 19th century, most wire was made from wrought iron. By 1870 improvements in steelmaking made it possible to produce large amounts of steel wire for the first time.

Steel wire was first used for fencing during the settling of the American West in areas where wood was scarce. Early wire fences consisted of single strands which were easily broken in cold weather or by wandering cattle. In 1860, Frenchman Leonce Eugene Grassin-Baledans patented the use of twisted strands of sheet metal with projecting points as a "fence protector." A similar method was patented in the United States in 1867 by Alphonso Dabb. That same year Lucien Smith and William Hunt received patents for single-stranded wire with barbs. In 1868 Michael Kelly invented the first double-stranded barbed wire, but the first commercially successful barbed wire was patented by Joseph Farwell Glidden of DeKalb, Illinois, in 1874. Similar patents were filed that same year by Jacob Haish and Leonard Ellwood, both also of DeKalb. After twenty years of legal battles, the United States Supreme Court decided in Glidden's favor, and he is often thought of as the "inventor" of barbed wire.

The use of barbed wire increased tremendously in the 1870s and 1880s, with some unfortunate side effects. In the severe winters of 1885-1886 and 1886-1887 thousands of cattle froze to death when they were unable to break through barbed wire "drift fences" intended to keep them from straying too far south. Conflicts between ranchers who wanted unfenced pastures and farmers who wanted fenced croplands escalated into fence-cutting, land-grabbing, and violent range wars. Eventually the conflict subsided when it became clear that barbed wire was becoming necessary as humans and cattle increased in number.

Barbed wire was adapted for military use during the Boer War and used in enormous quantities during World War I. Although barbed wire is often used for security, agriculture still accounts for 90% of its use. Many people collect antique barbed wire, with some rare specimens selling for hundreds of dollars. Hundreds of collectors attend the annual Barbed Wire Festival in La Crosse, Kansas, home of the Barbed Wire Museum.

Raw Materials

Barbed wire is usually made of steel, which is an alloy of iron and a small amount of carbon. The raw materials required to manufacture steel are iron ore, coke (a carbon-rich substance produced by heating coal to a high temperature in the absence of air), and limestone. To prevent rusting, the steel wire is usually coated with **zinc**. Sometimes the steel is coated with aluminum, and occasionally the barbed wire itself is made of aluminum.

The Manufacturing Process

Making steel ingots

1 Iron ore, coke, and limestone are heated in a blast furnace by hot pressurized air. The coke produces heat (to melt the iron ore) and carbon monoxide (which reacts with iron oxides in the ore to release iron). The limestone reacts with impurities in the iron ore such as sulfur to form slag, which is removed. The final product of the blast furnace is pig iron, which contains at least 90% iron, 3-5% carbon, and various impurities.

2 To convert pig iron into steel, the impurities and most of the carbon must be removed. (Iron without carbon is much weaker than steel, but iron with too much carbon is brittle.) Various methods exist to purify pig iron, the most common of which is the basic oxygen steel (BOS) process. In this method oxygen is blasted into molten pig iron under high pressure. Carbon is released as carbon monoxide, and the impurities are released as slag. The remaining molten steel is poured into molds and allowed to cool into ingots weighing thousands of pounds each.

Making billets

3 A billet is a bar of steel with a square cross-section whose dimensions are usually less than about 6 inches (15 cm) by 6 inches (15 cm). (If the dimensions are

An advertising broadside from about 1800 notes that 100,000 miles of barbed wire fencing was installed the year before. (From the collections of Henry Ford Museum & Greenfield Village.)

A seemingly simple invention, the barbed wire has had an interesting history. Before its introduction, early American farmers relied on European traditions to create an adequate enclosure for different types of farming. They had tried common fields without enclosures, "dead" fences of stone or timber, "live" fences of hedge plants, or simple wire fences. Eventually they were replaced by barbed wire, which was affordable, relatively simple to install and maintain, did not monopolize scarce local resources like timber, and proved highly effective in controlling livestock.

Barbed wire was an immediate cause and central weapon in the infamous "range wars" between cattlemen and farmers. Barbed wire was also the center of controversy as various inventors and manufacturers battled over patent rights and licenses and ultimately formed the Barbed Wire Manufacturers Union to establish prices. Farmers charged manufacturers with price fixing and monopolistic practices, and they banded together against the manufacturers.

Barbed wire factories also brought industry to rural America. DeKalb, Illinois, was a sleepy farm hamlet sixty miles west of Chicago until local inventor-entrepreneurs like Joseph Glidden and Jacob Haish became heavily involved in barbed wire production in the 1870s. Factories with new kinds of industrial jobs changed the town's economic base and demographics, while expanded railroad facilities brought it that much closer to Chicago. The addition of a normal school and teacher's college in the 1890s confirmed the transformation and DeKalb, also known as "Barb City," began the 20th century a virtual outpost of the great metropolis.

William S. Pretzer

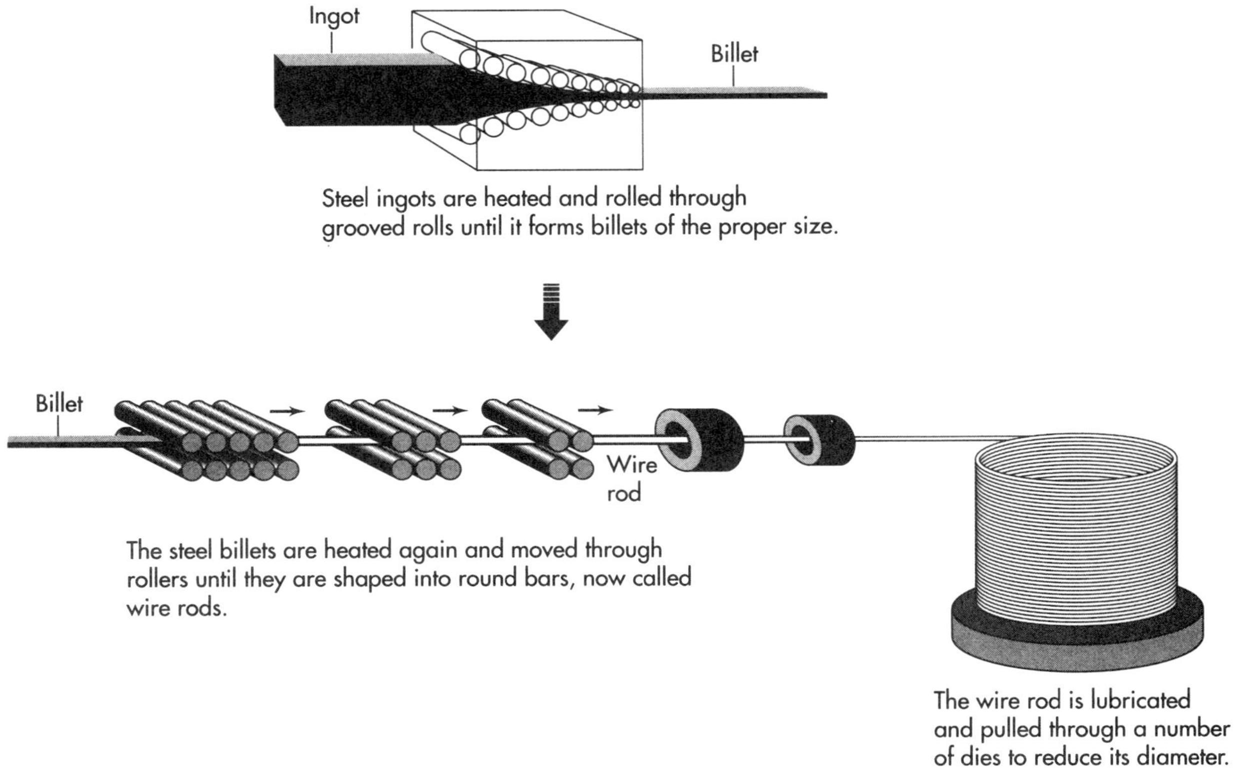

To make barbed wire, iron ore, coke, and limestone are heated in a blast furnace to produce pig iron. The pig iron is purified and converted to steel.

larger, the bar is known as a bloom; if the cross-section is rectangular rather than square, the bar is known as a slab.) A steel ingot is heated until it is about 2192°F (1200°C), then rolled back and forth between grooved rolls until it has reached the proper size. Giant shears cut the billet to the desired length; then it is allowed to cool. It is also possible to form billets directly from molten steel by pouring it through a water-cooled copper mold to shape it, then spraying it with water to solidify it.

Making wire

4 The steel billet is again heated and rolled until it has been shaped into a round bar 0.2 inch (5.6 mm) in diameter, known as a wire rod. The wire rod is rolled into a coil weighing as much as 3,969 pounds (1,800 kg), which is shipped to the wire manufacturer.

5 Scale, a surface coating of iron oxide which forms on steel when it is heated, must be removed from the wire rod. This can be done by soaking it in hot, dilute acid (sulfuric acid at 180°F [82°C] or hydrochloric acid at 140°F [60°C]) and rinsing it with water, a process known as pickling. Scale can also be removed by bending the wire rod back and forth between a series of wheels to break it off, or by blasting it off with fast-moving particles of sand, aluminum oxide, or cast-iron shot.

6 The clean wire rod is coated with lime, borax, or phosphate. This coating prevents rusting, neutralizes any remaining traces of acid, and helps lubricants adhere to the wire rod when it is made into wire.

7 One end of the coated wire rod is shaped to a point. This end is inserted like a thread going through the eye of a needle into a die consisting of a very hard central nib made of tungsten carbide surrounded by a steel holder. The wire rod is lubricated with oil or soap and is pulled through the die to reduce its diameter. This process is known as drawing. A series of dies are used to reduce the wire rod from its

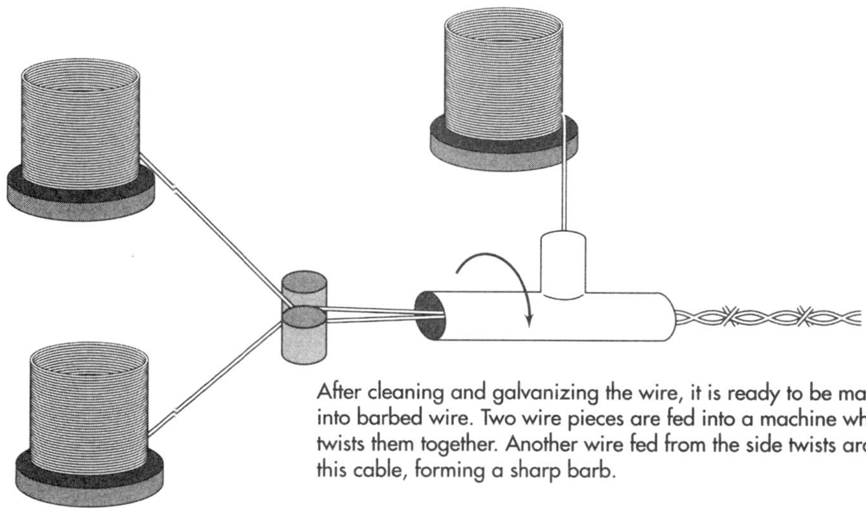

After cleaning and galvanizing the wire, it is ready to be made into barbed wire. Two wire pieces are fed into a machine which twists them together. Another wire fed from the side twists around this cable, forming a sharp barb.

original size to the desired size of the wire. For barbed wire, the diameter is typically 0.097 inch (2.5 mm). Usually about six or seven dies are needed to reach this size.

8 Drawing the wire causes it to become hard and stiff. To make it pliable, it is heated, a process known as annealing. Wire can be annealed by heating it in a bath of molten salt, a bath of molten lead, or in a furnace containing nitrogen. All these methods prevent scale from forming by protecting the steel from oxygen.

Making barbed wire

9 Wire to be made into barbed wire is usually galvanized (coated with zinc) to protect it from corrosion. The wire must be perfectly clean and dry to be properly galvanized. First it is cleaned in a bath of hot, dilute hydrochloric acid, then rinsed with hot water. It then passes through a solution of hot zinc chloride or ammonium chloride to prevent rust from forming as it is dried. After drying, the wire passes through a bath of molten zinc. Excess zinc is wiped off and the coated wire is allowed to cool. (Sometimes the wire is coated with aluminum instead in a similar way.) Wire can also be coated with zinc by a process known as electrogalvanizing. The wire is given a negative electric charge and passed through a solution of zinc sulfate or some other zinc salt. The positive zinc ions are attracted to the negative wire and form a coating.

10 A single automated machine performs all the steps needed to transform galvanized wire into barbed wire. Two wires are fed into the machine and twisted together to form the cable. Another wire is fed into the machine from the side and

twisted around one or both of the cable wires. This wire is cut at an angle on both sides to form a two-point barb. Two wires are twisted and cut together if four-point barbs are needed. The barbed wire is pulled along a set distance (usually 4 or 5 inches [10 or 13 cm]), and the process is repeated to space the barbs evenly. The barbed wire is wound onto spools and cut into 1,319-foot (402 m) lengths. These spools are then loaded onto trucks and shipped to the customer.

Quality Control

Standards for barbed wire have been established by the American Society for Testing and Materials. Manufacturers of barbed wire use the tests described in these standards to ensure their customers that they are purchasing a quality product.

One spool of barbed wire out of every 50 is selected for testing and inspection. First the dimensions are measured for accuracy. The diameter of the cable wires and the barbs must not vary more than 0.5 inch (0.13 mm). The barbs must extend at least 0.37 inch (9.5 mm) from the center of the cable. At least 93.5% of the spaces between the barbs must be within 0.74 inch (19 mm) of the desired length. (100% accuracy in barb spacing is impossible due to small movements of the barbs during handling.) A 25-foot (7.6 m) sample of the barbed wire must contain at least 69 barbs if they are spaced 4 inches (10 cm) apart and at least 55 barbs if they are spaced 5 inches (13 cm) apart. The wire on the spool must be at least 1,319 feet (402 m) long.

A strength test is performed on a 4-foot (1.2 m) sample of the barbed wire. The sample is pulled by a measured force until it breaks. It must be able to withstand a force of at least 4,230 newtons.

For galvanized barbed wire, another 4-foot (1.2 m) sample is tested for its zinc coating. The sample is weighed, then the zinc is removed with hydrochloric acid. By weighing the sample again and noting the difference in the two weights, the amount of zinc coating can be determined. A similar procedure is used to measure the zinc coating on the barbs. The minimum weight required varies with the diameter of the cable wires. For the most common diameter (0.097 inch or 2.5 mm), each line wire and each barb must be coated with at least 3.2 ounces (90 g) of zinc per square meter (11 sq ft) for a Class 1 coating or at least 8.6 ounces (245 g) per square meter (11 sq ft) for a Class 3 coating. Standard Grade barbed wire has a Class 1 or a Class 3 coating on the line wires and a Class 1 coating on the barbs. Chain Link Fence Grade barbed wire must have a Class 3 coating on the line wires and the barbs.

The Future

Although the classic barbed wire fence is still commonly used on farms, it is slowly being replaced by more advanced products such as woven wire fences (similar to chicken wire, with crossing horizontal and vertical wires) and electric fences. For military and security use, barbed wire may become obsolete with the recent development of barbed tape, a flat, thin strip of metal which has been cut to produce clusters of sharp points. Perhaps some day barbed wire will exist only in museums and private collections.

Where To Learn More

Books

Hooper, Meredith. *Everyday Inventions.* Taplinger, 1976.

McGannon, Harold, ed. *The Making, Shaping, and Treating of Steel.* Herbick and Held, 1971.

Periodical

Evans, Charlotte. "Time Was When the Wild West Danced to the 'Devil's Rope'." *Smithsonian*, July 1991, pp. 72-83.

—Rose Secrest

Baseball Bat

Background

When the game of baseball was first played, sticks were used to hit the ball. By the time the game had been officially organized as a team sport, the players either whittled their own bats or bought them from a wood turner. League specifications set in 1863 were broad: any type of wood was permissible and the bats had to be round, not more than 2.5 inches (6.5 cm) in the thickest part. There were no length restrictions. Early bats ranged in weight from 48-50 ounces (1361-1417 g) with an average handle circumference of 4.5 inches (11.4 cm). The hefty weight meant home runs were rare. By the 1960s, however, players such as Hank Aaron were using shorter, lighter bats to smash balls into the centerfield seats. Aaron's bat measured 35 inches (89 cm) long and 33 ounces (979 g) in weight.

Modern baseball rules limit bat lengths to 42 inches (107 cm) and the diameter to 2.75 inches (7 cm). There are no weight restrictions. The bats must be made of wood with no metal, cork, or other type of reinforcement inserted into the bat's center. Over the years, several major league players have tried to use a reinforced bat. A particularly colorful controversy surrounded a bat used by Albert Belle of the Cleveland Indians. Belle's bat was confiscated during a game between the Indians and the Chicago White Sox in July of 1994. The bat was stored in the umpires' locker room at Comiskey Park until it could be tested the next day. However, it disappeared overnight. It was returned anonymously the next day and found to have a corked center. In spite of protests, Belle received a temporary suspension. The mystery of the bat's disappearance and reappearance has yet to be revealed.

Hillerich & Sons, a Kentucky wood-turning shop, was the first company to devote a full-time operation to the manufacturing of baseball bats. According to company lore, in 1884, John "Bud" Hillerich, the son of the company's founder, was attending a Louisville Eclipse baseball game when a player named Pete "Old Gladiator" Browning broke his bat. Bud invited Browning back to the shop where Bud custom-made a new bat from a piece of white ash. During the next day's game, Browning pounded three hits in three at-bats using the new bat. And the rest, as they say, is history. The ensuing requests for custom-made bats from other players helped Bud convince his father to add bat manufacturing to the family business. The company named its new product the "Louisville Slugger." (The company became Hillerich & Bradsby in 1911 when Frank Bradsby, a sporting goods magnate, joined the firm.)

Baseball players are notoriously particular about their bats, and have been so throughout the sport's history. Frank Frisch, who played in 50 World Series games for the New York Giants and the St. Louis Cardinals, cured his bats during the off-season by hanging them like sausages in a barn. Boston Red Sox slugger Ted Williams bathed his bats in alcohol to keep them cool during his many hitting streaks. Williams was also known to visit lumberyards looking for pieces of wood with narrow growth rings. The legendary Babe Ruth preferred his bats to have pin knots in the barrels.

Modern baseball rules limit bat lengths to 42 inches and the diameter to 2.75 inches. There are no weight restrictions. The bats must be made of wood with no metal, cork, or other type of reinforcement inserted into the bat's center.

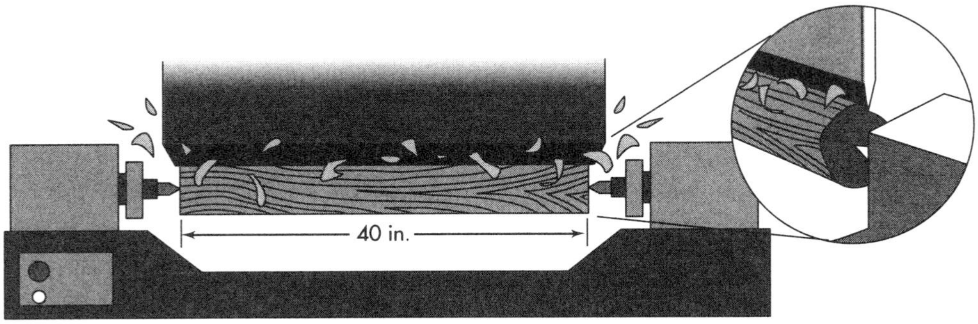

40-inch splits of wood are placed on a lathe which shaves the rough edges. These pieces are called billets.

The billets are bundled and sent to the lumberyard, where they are "seasoned" for 6-24 months. This seasoning removes the sap and gum and strengthens the wood.

Traditionally, forty- to fifty-year-old ash trees are used to make baseball bats because of their strength, flexibility, and light weight.

Raw Materials

Traditionally, ash trees from Pennsylvania and upstate New York are used to make baseball bats. The ash is valued for its strength, flexibility, and light weight. The best trees are those that grow in dense clusters where they are protected from the wind and forced to grow straight up towards the sunlight. Forty to fifty years of growth is required to bring an ash tree to the preferred trunk diameter of 14-16 inches (36-41 cm). Each tree yields approximately 60 bats.

When a tree has reached the proper height and width, a forester marks it with spray paint. A log cutter then uses a chain saw to bring down the tree. The top branches are removed and left in the forest. The tree trunks are sawed into 10-16-foot (3-5 m) lengths, loaded on a truck, and taken to the mill. At the mill, the logs are inspected for knots and uneven grains. Only half of what is cut in the forest is ultimately used to manufacture baseball bats. The logs that make the grade are rolled to a hydraulic wedge that cuts them into 40-inch (101 cm) splits.

The Manufacturing Process

Turning the splits into billet

1 A mill worker places each split onto an automatic lathe that shaves the rough edges off as it turns the wood. The billets, as they are now called, are inspected again for straightness of grain. The billets are stacked and strapped together into six-sided bundles. Workers paint the ends with a protective preservative to keep the wood from fraying or rotting. The bundled billets are then trucked to the lumberyard of the bat manufacturer.

Seasoning the billets

2 The billets that arrive at the lumberyard are considered "green" wood because they still contain sap and gum. In order to strengthen the wood, the sap and gum must be removed by an air-drying process called "seasoning." To achieve the proper seasoning, the billets are simply stacked in the yard for a period of six months to two years.

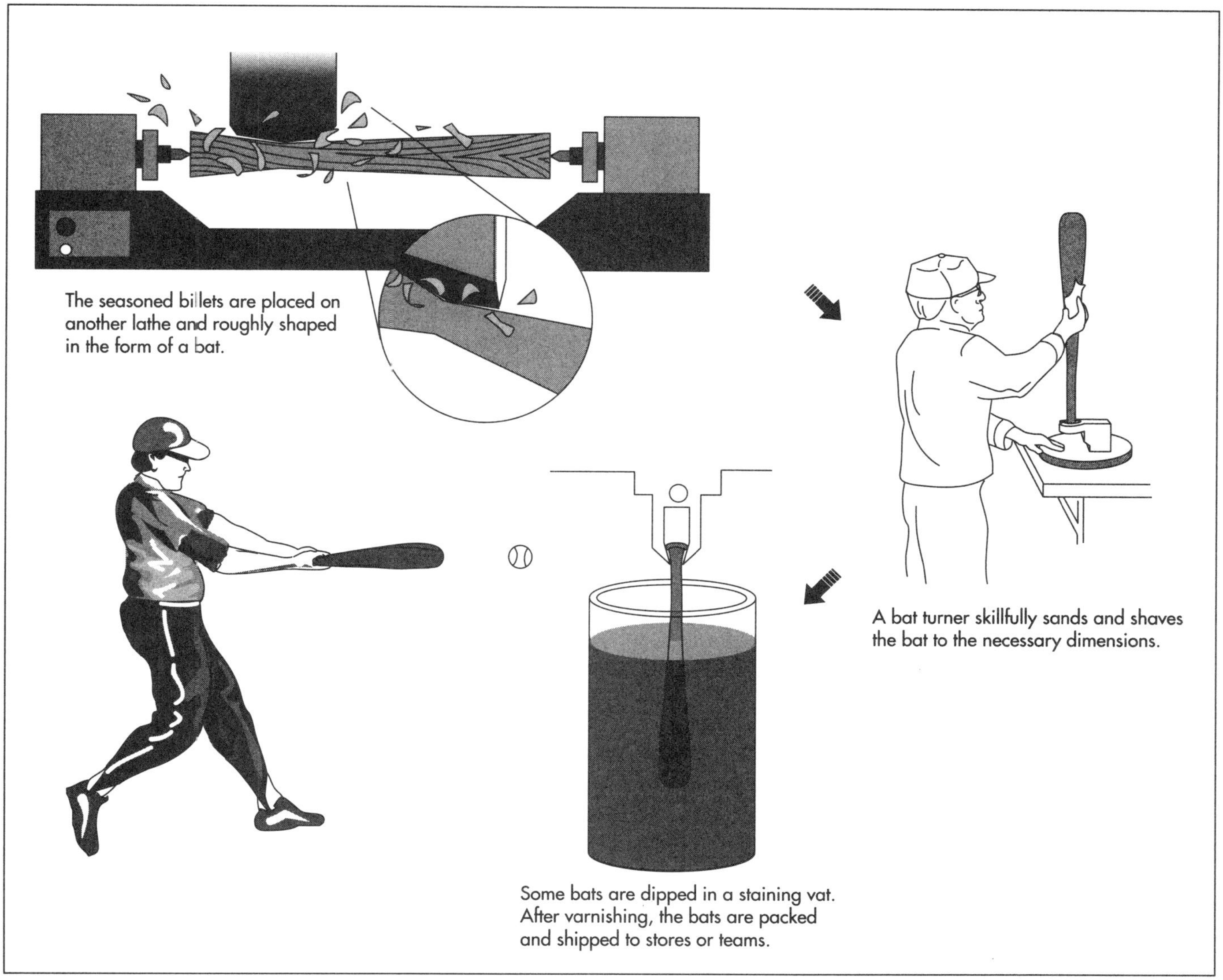

The seasoned billets are placed on another lathe and roughly shaped in the form of a bat.

A bat turner skillfully sands and shaves the bat to the necessary dimensions.

Some bats are dipped in a staining vat. After varnishing, the bats are packed and shipped to stores or teams.

Shaping and sanding the billets

3 When the billets have dried completely, they are weighed and inspected for quality. A worker places each billet on an automatic lathe and shapes it into a rough baseball bat shape with a narrowed neck. The bat forms are sanded, inspected once more, and then sorted according to weight.

Matching the bat to the model

4 The bat manufacturer keeps a model of each bat made, typically identified by the baseball player who initially ordered it. When a player or team places an order, the order may look like this: six Johnny Bench models, ten Hank Aarons, four Mickey Mantles.

The plant workers who create the final product are called bat turners. They are highly skilled artisans who have been specially trained for the intricate work. When an order is placed, the bat turner selects a billet from the storage bin that fits the called-for weight and length. The billet is placed on a lathe. The model bat is placed on a rack above and behind the lathe.

The bat turner revolves the billet slowly on the lathe, sanding and shaving it to an exact replica of the model. Using calipers, the bat turner measures the billet every 1-2 inches (2.54-5 cm) and weighs it repeatedly until it is perfect.

The plant workers who create the final product are called bat turners. They are highly skilled artisans who have been specially trained for the intricate work. When an order is placed, the bat turner selects a billet from the storage bin and creates a replica of the desired model.

Branding, staining, and varnishing the bats

5 The bat is branded with the company trademark and the signature of the player associated with the model. The trademark is placed one-quarter of a turn from the sweet spot (the ideal spot where the ball should strike the bat). If the order calls for staining, the bat is dipped into a staining vat. All of the bats are then varnished, packed into cartons, and shipped to the player or team.

Quality Control

The structural integrity of the baseball bats are monitored through repetitive impact testing. Some factories have compressed-air cannons that shoot baseballs at precise points on the bat. High-speed cameras record the impact while accelerometers measure the velocity. In other plants, robotic arms whack the balls off over-sized golf tees. Inspectors collect data on the frequency of bending and how the balls travel off the bat.

The Future

In spite of manufacturers' assurances that the supply of ash trees is not decreasing, the development of composite and aluminum bats continues. The wood composite bat typically consists of a plastic foam core surrounded by woven layers of resin-impregnated synthetic fibers. One of the newest innovations is a bat made of "lanxide," a ceramic-enforced material. Proponents of non-wood bats point to their resistance to breakage. These bats also greatly alter hitting power: a player's batting average increases markedly with an aluminum bat.

Although the composite and aluminum bats are popular with amateur and college baseball players, they are required to use all-wood bats if they advance to the major leagues. It is doubtful that Major League Baseball will ever allow anything but pure wood for bats. The sport is steeped in tradition, and the use of aluminum or composite materials would alter the record books dramatically.

Where To Learn More

Book

Arnow, Jan. *Louisville Slugger: The Making of a Baseball Bat*. Pantheon Books, 1984.

Periodicals

Ashley, Steven. "High Tech Up at Bat." *Popular Science*, May 1992, pp. 108-11, 122-24.

Curreri, Joe. "Romance of the Bat." *Antiques and Collecting Hobbies*, May 1992, p. 26.

Pesky, Greg. "Expanded Coverage." *Sporting News*, June 1993, p.32.

"Lanxide Shows Mettle with New Material." *Sporting Goods Business*, July 1993, p. 36.

—Mary F. McNulty

Bathtub

Background

Though humans have bathed since prehistoric times, baths served a primarily religious, social, or pleasurable function far more often than a hygienic one. The Greeks had bathrooms, complete with water supply and drainage, as early as 1700 B.C. The Romans developed bathing as a central social habit in the third century B.C., constructing elaborate public buildings of enormous size with several rooms for disrobing, exercise, and bathing.

After the fall of the Roman Empire, bathing declined in popularity in Europe, though it did survive as a part of monastic routine, and in Muslim countries. The Muslim public bathhouse included a dressing room, cold bath, and warm bath clustered around a domed, central steam chamber. Public baths regained popularity in Europe in the 11th and 12th centuries. In private homes, bathing was done in wooden tubs set up in bedrooms, but some castles and palaces had permanent bathrooms. In fact, Henry III of England had hot and cold running water installed in the bathhouse at his Westminster palace.

In the 18th century, it became fashionable to spend a season at a watering place (such as Bath in England) but only 19th-century research into hygiene made a virtue of bathing. Bathing took place in primitive and usually portable cold baths at schools and institutions. Though permanent tubs were installed in bedrooms during the mid-19th century, plumbing was nonexistent and tubs had to be emptied by hand. Only after World War I did plumbing and bathtub production allow the bath with running water to become a permanent installation in the home.

Bathtubs are now part of the plumbing fixtures and fittings industry, which totaled around $5.7 billion in the U.S. in 1994. This market is shared about equally between fixtures and fittings. Over three million bathtubs, including whirlpool baths and hot tubs, are shipped per year. Plumbing fixtures are classified into three industries according to the materials from which they are made: vitreous, metal, and plastic. In recent years changing consumer tastes have displaced other materials in favor of plastics for bathtubs, whirlpool baths, and lavatory sinks. Sixty-two percent of bathtubs, 92% of whirlpool baths, and 28% of lavatories are made out of plastic. Besides plastic, the standard material for bathtubs is enameled cast **iron** or steel. Bathtubs must be manufactured according to standards established by the American National Standards Institute.

The Manufacturing Process: Enameled Bathtubs

Raw Materials

The metal base for bathtubs is made of gray cast iron (containing carbon, silicon, manganese, phosphorus, and sulfur), titanium steel, zero carbon steel, or partially decarburized steel. These compositions have been specially designed for enameling. The enamel is made from a frit or glass that consists of a variety of raw materials, both manufactured chemicals and natural minerals. These include clay, feldspar, barium

Though humans have bathed since prehistoric times, baths served a primarily religious, social, or pleasurable function far more often than a hygienic one. The Romans developed bathing as a central social habit in the third century B.C., constructing elaborate public buildings of enormous size with several rooms for disrobing, exercise, and bathing.

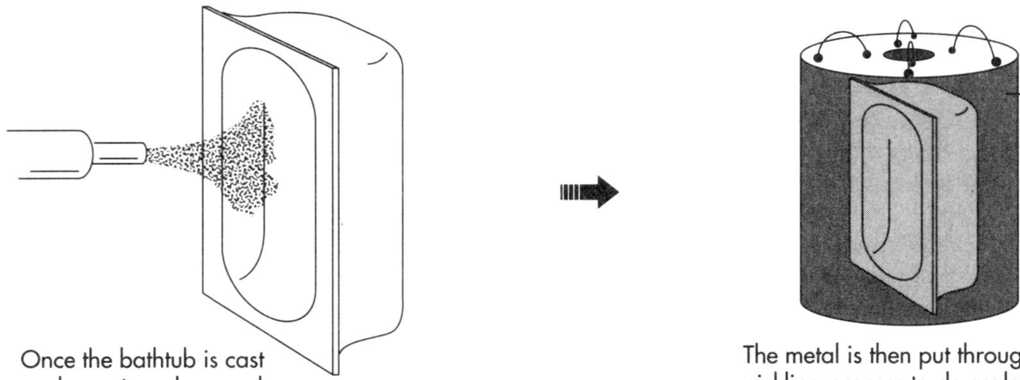

Once the bathtub is cast or drawn into sheet and formed, it is cleaned by shot-blasting. In this process, iron grit is projected onto the tub to remove dirt.

The metal is then put through an acid pickling process to de-scale the surface. The metal is rinsed and sometimes given a nickel dip.

The metal base for bathtubs is made of gray cast iron, titanium steel, zero carbon steel, or partially decarburized steel. These compositions have been specially designed for enameling.

carbonate, boric acid, limestone, fluorspar, sand, and other oxides.

Preparing the metal

1 The metal is either cast into molds (gray cast iron) or drawn into sheet and formed (steel). Before enameling it must then be cleaned. Cleaning of castings is carried out by blasting the surface with chilled iron grit, sometimes after preliminary annealing (heating at 1350-1500°F [730-820°C]). The grit, or "shot," is projected through nozzles using compressed air, or flung against the surface by centrifugal force using a special machine. A combination of both methods may be used. This blasting removes any dirt or sand particles from the mold from the metal surface.

Sheet metal must undergo a more complex cleaning process. First the surface is cleaned by shot-blasting after annealing to release stresses and to remove any grease deposits. Degreasing is done with chemical cleaners, first with an organic solvent, followed by a hot alkaline solution. The organic solvent removes most of the grease and oil from the metal surface. The alkaline solution removes the remaining film of oil and leaves a surface ready for acid pickling.

Acid pickling uses hydrochloric or sulfuric acid or a combination of these acids. This process de-scales the surface, which helps to form a strong bond between the iron and ground-coat enamel. Pickling is followed by rinsing in running water. The next step is sometimes a nickel dip, which uses a solution of nickel sulfate and boric acid to coat the metal with a layer of nickel. This layer also helps to form a good bond with the enamel. The nickel dip is followed by a thorough rinsing of the ware and another dip in a neutralizer solution. This solution consists of soda ash and borax in water and removes any traces of acid, as well as prevents rust. After neutralizing, the metal is dried as quickly as possible to prevent rusting.

Preparing the enamel

2 After the raw materials are carefully weighed and mixed together in precise amounts, the enamel frit is prepared by melting the batch in furnaces of rotary or continuous type, fueled by oil or gas. When melting is completed, the molten enamel is run out of the furnace in a thin stream into a tank of cold water, which produces small fragments. For continuous furnaces, the molten frit is run between water-cooled rolls, which chills the frit. The frit is then dried and stored in bags. Before the enamel is applied, it must be ground into a powder using a ball mill, with blocks of enamel as the grinding media. If applied wet, the enamel is milled with additions of clay and water to make a slip or slurry.

The enameling process

3 Dry enamel is dusted on the metal surface, which has been previously heated to a temperature above the melting point of the enamel. The powder melts on contact with the hot article, forming a continuous coating. Firing in a furnace produces a smooth, porcelain-like surface.

Wet enamel is applied by a dipping process using a large open tank. After dipping, the coated part is allowed to drain, producing a thin uniform coating. The dipping tank uses a recalculating system to recycle the enamel. The wet enamel may also be applied using automatic spraying equipment.

After enameling, the coated parts are dried using cabinet or conveyor dryers heated by gas burners, steam or waste-heat from the enameling furnaces, or infrared lamps. Firing takes place in furnaces heated by metallic heating elements. Two coatings of enamel are normally applied, a ground coat followed by a cover coat. The cover coat takes longer to fire.

The Manufacturing Process: Plastic Bathtubs

Raw Materials

Plastic bathtubs are made out of a number of different polymer materials, including ABS (acrylonitrile-butadiene-styrene), acrylic resins, or glass-fiber reinforced **polyester**. The glass-polyester type dominates the tub-shower market. Special additives may be incorporated into the polymer material to improve fire-resistance. To give a cultured-marble appearance, marble chips or dust can also be added.

The starting materials for most plastics are petrochemicals—chemicals made from petroleum and natural gas. These chemicals are mixed with other special chemicals (stabilizers, fillers, dyes or pigments, and plasticizers) in steel tanks and then heated to cause a chemical reaction that converts the mixture into the desired polymer composition. The resulting polymer material is cooled and dried to form a powder, beads, or pellets, depending on the specific process and material.

This 1920s bathtub folded up to save space and came with its own hot water heater. (From the collections of Henry Ford Museum & Greenfield Village.)

The bathroom is not usually the place one thinks of to illustrate the principles of mechanization. Yet the transformation of bathing facilities aptly illustrates Western society's obsession with efficiency and mechanization. Baths in ancient Greece and especially in the Roman Empire were much more elaborate and technologically sophisticated—and less private—than the simple "outhouses" of 20th-century rural America. The obvious attention given to the design, construction, and maintenance of bathing facilities indicates just how integral the activities of the bath have been to cultural identity for centuries.

In 19th-century America, bathing was not commonplace. In the 1880s, probably five out of six city dwellers had no proper bathing facilities, just the use of a pail and sponge. During the last half of the century, numerous efforts were made to encourage communal bathing facilities. Reformers also advocated the use of showers in private homes as well as public facilities with only modest success. Essentially, Americans did not consider bathing all that necessary for general health nor did they associate it with the more complete mental and physical therapy sessions (hot-air or steam baths, massage, gymnastics) common in other cultures. For upper-class Americans, a trip to a spa might occur once a year, but certainly not once a week.

The bathtub was considered a luxury well into the 20th century. The real increase in the appearance of bathtubs did not come until the 1920s, with the extension of central water systems. Following the form of hotel buildings developed in the 1880s, houses, tenements, and apartments were increasingly built with separate rooms devoted to the bath. The production of sanitary enameled bath fixtures (toilets, sinks, and tubs) doubled between 1921 and 1923. The relatively standardized, mass-produced tub meant for private, familial use soon came to dominance.

William S. Pretzer

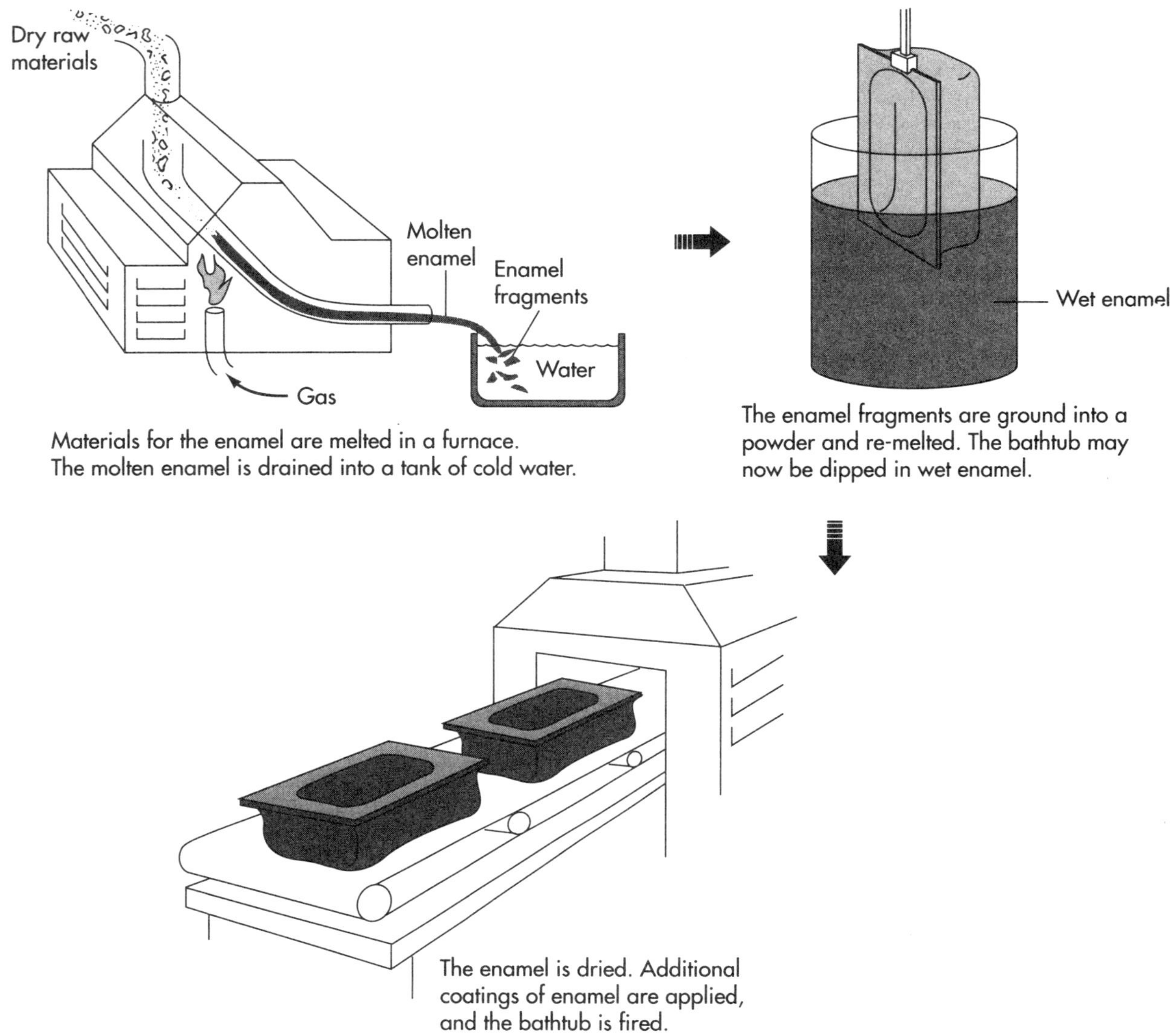

Plastics fall into two groups: thermoplastic and thermosetting. Thermoplastics soften or melt on heating and include vinyl and styrene polymers. Thermosetting plastics, although moldable when produced as simple polymers, are converted by heat and pressure, sometimes using a hardener, to a cross-linked infusible form. Plastics that fall into this category include epoxy resins and polyesters.

Forming Processes

A number of processes are used to form plastics into shapes, including calendaring (squeezing between rollers to form films), casting, compression molding, and injection molding (melting and forcing into a cooled mold).

In injection molding, a reciprocating screw machine is generally used, which melts the plastic, pressurizes the melt to inject it into a closed mold, closes the mold so the plastic can solidify, and opens the mold to allow removal of the part. Injection occurs as the screw is hydraulically forced forward in the barrel. During the cooling and solidification period, the screw begins to rotate and melt new material for the next part.

Plastic Bathtub

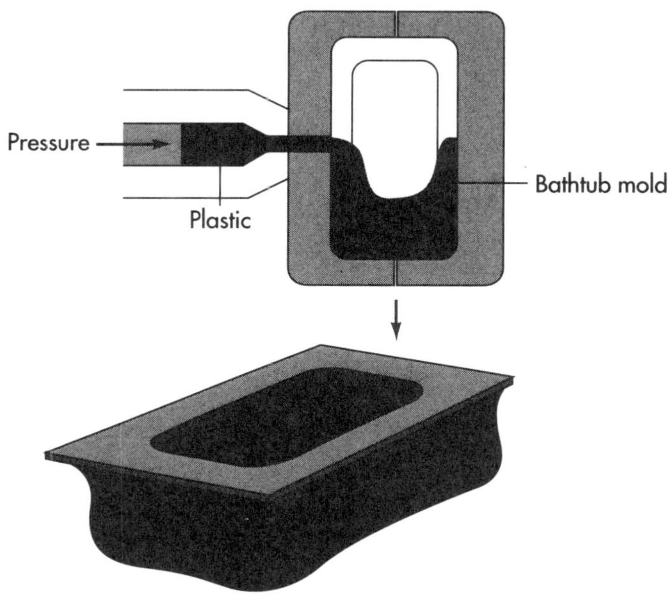

Plastic bathtubs can be made by injection molding, which injects molten plastic into a mold cavity.

Compression molding is used to make both plastic and fiber reinforced bathtubs. For reinforced bathtubs, a mixture of fibers, polyester resin, and pigments, fillers, and other additives are mixed together and formed into a sheet. In this process, a specified amount of resin filler paste is placed onto a plastic carrier film using a special machine. The carrier film is passed under a chopper, which cuts glass roving into short lengths. After the glass falls to the resin bed, another carrier film with another layer of paste is added on top, sandwiching the glass between the two layers. This sandwich structure is passed through a series of compaction rollers to thoroughly mix the fibers and resin.

After the material—called a sheet molding compound—thickens, the carrier film is removed and the sheet material is cut into charges, which are placed in matched metal die molds made of machined steel. High pressure is applied, which heats the material so that it flows to all areas of the mold. Heat from the mold activates the catalyst, which achieves curing. Once cured, the part is then removed from the mold.

Two processes called hand lay-up and spray lay-up are used to make **fiberglass** bathtubs. Hand lay-up is a method in which successive plies of reinforcing material or resin-impregnated reinforcement are positioned in a mold by hand. Cure occurs at room temperature with no applied pressure. Special tools are used to work out air bubbles and ensure complete wetting of the fiber, if the polymer is added separately. The spray lay-up process is faster than hand lay-up and involves feeding a stream of chopped fibers into a spray of liquid plastic in a mold cavity. The direction of the fibers is random and the process is usually automated. After lay-up is completed, the plastic must solidify or cure in a reasonable time at room temperature, which occurs via chemical reactions.

The Future

Since new housing construction is the principal source of demand for plumbing products, the timing and magnitude of the revival of construction activity and the overall economy are pivotal factors in determining the direction of bathtub shipments. Over the past decade, the proportion of new single-family houses with 2.5 or more baths has doubled to about 44%. This has obviously increased the demand for bathtubs, which should continue as long as the economy in general and construction activity improve.

Bathtub design is also undergoing an evolution. With the passage of the American Disabilities Acts, bathtubs that accommodate the handicapped are being patented, and this trend should continue. Taking advantage of the molding capabilities of plastic, manufacturers are also designing one-unit bathtub and shower. Safety is another important design factor, and prefabricated slip-resistant surface coatings have been developed.

Other design improvements will continue in the area of leak prevention. Such designs include providing a moisture barrier unit between the bathtub and the subfloor, or designing the bathtub so that the wall and base are an integral part of the bathtub. The latter approach relocates all lines of contact between the bathtub and surrounding adjacent surfaces so that they are covered and protected from water penetration.

Where To Learn More

Books

Mazzurco, Philip. *Bath Design: Concepts, Ideas, and Projects.* Watson-Guptill Publications, 1986.

Wright, Lawrence. *Clean and Decent: The Fascinating History of the Bathroom and the W.C.* Routledge & Paul, 1960, reprinted 1980.

Yegul, Fikret. *Baths and Bathing in Classical Antiquity.* MIT Press, 1992.

Other

Plumbing Manufacturers Institute, 800 Roosevelt Rd., Building C, Glen Ellyn, IL 60137, (708) 858-9172.

—Laurel M. Sheppard

Beeper

Background

A pager, or beeper, is a small, battery-powered device that alerts the person carrying it when someone is trying to reach them by telephone. The beeper utilizes electronic components sensitive to an FM radio signal and will beep or otherwise sound, flash, or vibrate to alert its carrier. Originally only used by doctors and certain businesspersons to notify them of an urgent call, beepers have become more common in the last two decades. By 1992, an estimated 2.9 million people in the U.S. alone carried beepers, and the number of users worldwide was much larger.

The first radio paging device was used in the New York City area in 1950. Two decades after its introduction, there were approximately 32,600 pagers in use. The pagers manufactured during this early era were often large and somewhat unreliable. Technology led to improvements in electronic circuitry, and small, more reliable pagers came on the market around 1972. During the next decade, the devices became more popular as the price came down. By 1981, there were an estimated one million users in the U.S. Originally, pagers could receive a radio signal only in a specific geographic area. A caller in New York City, for example, might have a difficult time paging someone in Boston. Today, several pager companies offer the ability to page people across country, and worldwide paging is on the horizon.

The typical beeper contains an FM receiver, a tone-decoding device, and an audio amplifier. When alerted, the carrier can access the telephone number of the person trying to reach him or her. More complex beepers have alphanumeric capability and are able to display names of and messages from callers. The industry is also developing small, hand-held devices that are much more sophisticated versions of the simple beeper. These future communication devices will be able to both send and receive alphanumeric messages via satellite transmission.

Beepers can be either rented or purchased outright. In both cases, a contract for access to a local paging network is mandatory; this is typically a one- or two-year commitment with a set monthly fee. If the beeper is rented, the user pays one fee that covers the cost of the device as well as access to the network—this can range from $14 to $20 per month. A consumer who buys to own pays for the cost of the beeper—beginning at about $90—and then pays a separate monthly fee to access the paging network. Such access is roughly $7-10 per month. A beeper must be activated when the purchase or rental agreement is complete and the contract with the paging network agreed upon. A special phone number for the beeper is designated and programmed into its circuitry by the retail salesperson. The beeper number may then be given out to friends, family, and colleagues. When someone dials the number, a radio signal reaches the device and activates the alert signal. This is accomplished by one of two methods: the call to the beeper is automatically transmitted to the paging network control center, where automatic processing equipment alerts the pager. In the second method, the user calls the beeper number, waits for a tone, then punches in the num-

Originally only used by doctors and certain businesspersons to notify them of an urgent call, beepers have become more common in the last two decades. By 1992, an estimated 2.9 million people in the United States carried beepers.

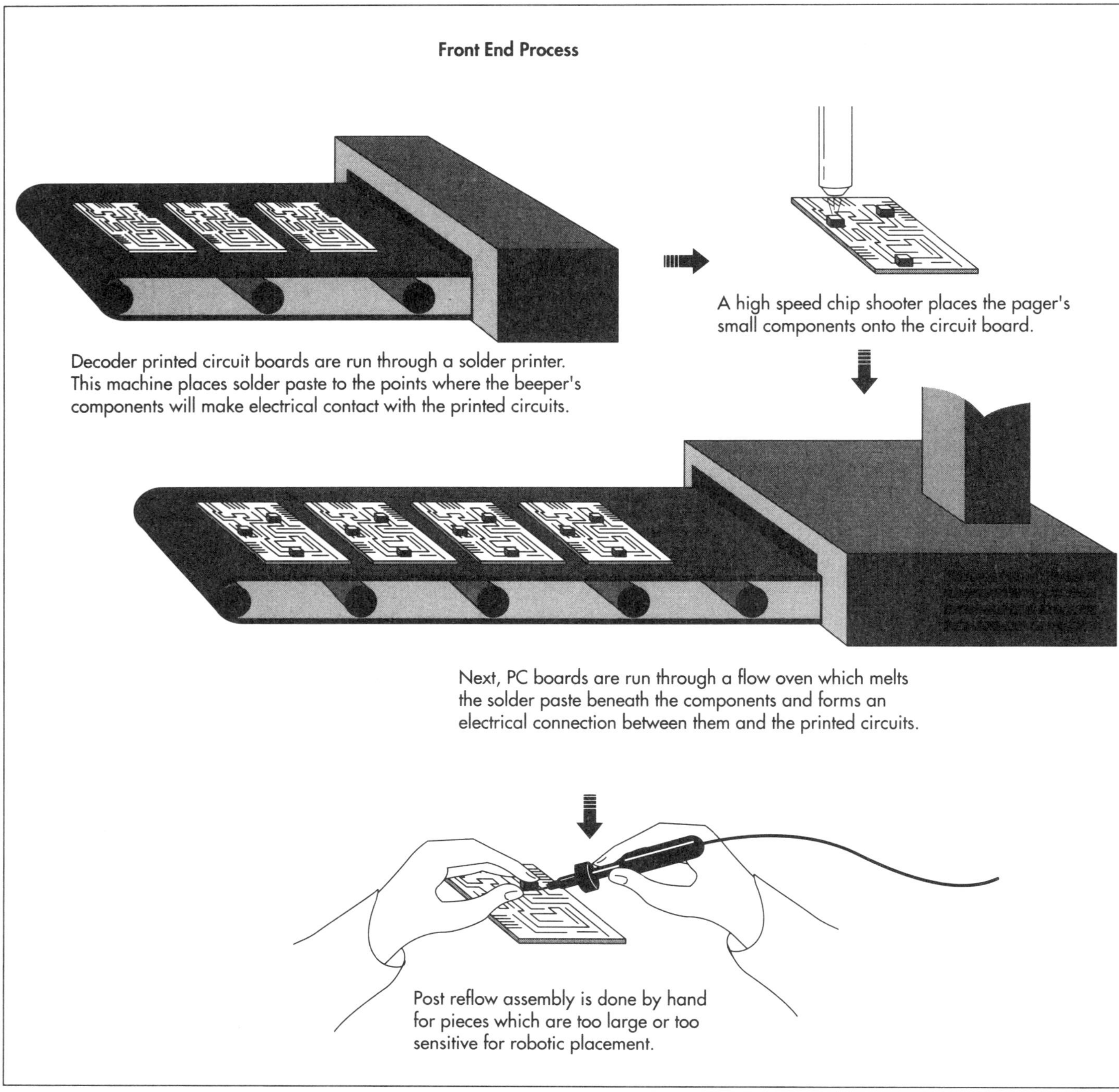

The basic design of pagers has changed very little since they were introduced, but advances in electronic components and circuitry have significantly reduced the size and weight of the overall package.

ber of the phone from which he or she is calling.

How It Works

In operation, a typical pager works as follows. Each pager is assigned an individual telephone number and a specific radio frequency on which it can receive signals. When someone dials the beeper telephone number, the telephone call is routed through the telephone lines to the paging service office. At the paging service, a controller device checks its memory for the radio frequency assigned to the beeper being called, then activates a radio transmitter to broadcast a short signal on that frequency. These frequencies are within the VHF (very high frequency) or UHF (ultra high frequency) FM radio bands. Within the beeper, there is a receiver circuit and a decoder circuit. The receiver circuit receives the signal, separates it from other signals, amplifies it, and passes it on to the decoder. The decoder verifies that the number dialed matches its own unique, programmed telephone number, then activates the beeper or other alert circuit. The decoder also processes the second part of the signal, which contains the telephone number or message of the person calling. A basic pager simply displays this telephone number on a small display screen. More sophisticated pagers can display the number and a short message. Some pagers have voice capability and can give a ten-second voice message that the caller has made.

Raw Materials

Materials used in the manufacture of pagers range from metals to ceramics, **paper**, rubber, and plastics. The outer housing of the pager is usually made of a resilient, high-impact plastic. It is made at the plant by injection molding machines. The shells are usually black, but marketing efforts to attract younger users have produced brightly colored ones as well. The entire beeper unit, including the battery, generally weighs only few ounces.

Design

The basic design of pagers has changed very little since they were introduced, but advances in electronic components and circuitry have significantly reduced the size and weight of the overall package. The addition of numerous options to the basic pager function has increased the complexity of the circuits. Some of these options include voice message capability and the ability to transmit short messages from the pager back to the calling party.

The Manufacturing Process

The typical manufacturing process for pagers involves two distinct steps. These steps are known as the "front end process" and the "back end process." Both processes utilize computer integrated manufacturing, or CIM. CIM is a network of physical hardware linked with computers using special software, or programs. CIM is designed to assist the production operators in tracking each stage of the manufacturing process and ensuring that the pager is built with the proper components and options specified by the customer.

The Front End Process

1 A group, or array, of one or more receiver or decoder **printed circuit boards** is run through a machine called a solder printer. Printed circuit boards (PCB) are thin, stiff pieces of electrically non-conducting material onto which a pattern of electrically-conducting material has been bonded to form numerous circuit paths, much like flat wires. Electronic components are attached to both sides of the board and interconnected by the printed circuitry. The solder printer is programmed to apply a small amount of solder paste to the points where the components will make electrical contact with the printed circuits.

2 The PCBs are then processed through another machine known as a high-speed chip shooter. The shooter quickly and accurately places, or shoots, the majority of the pager's small components onto each board.

3 After a visual inspection, the PCBs are run through a series of robotic placement machines which place the larger and odd-shaped parts onto the boards. These include the **integrated circuit** (IC) chips, oscillators, and crystal filters.

4 Once all these parts have been placed, the boards are run through a reflow oven. In this oven, the solder paste melts, or reflows, to form an electrical connection between the components and the printed circuits.

5 The board arrays are then given a final visual inspection for defects and are sent through the singulation process in which the individual boards are cut from the array and labeled.

6 Some pagers require parts which must be placed by hand in a process called "post reflow assembly." These parts are generally too large for robotic placement or too sensitive to the extreme heat of the reflow oven. The liquid crystal display (LCD) is attached to the decoder board, and the decoder board with the LCD is placed

Back End Process

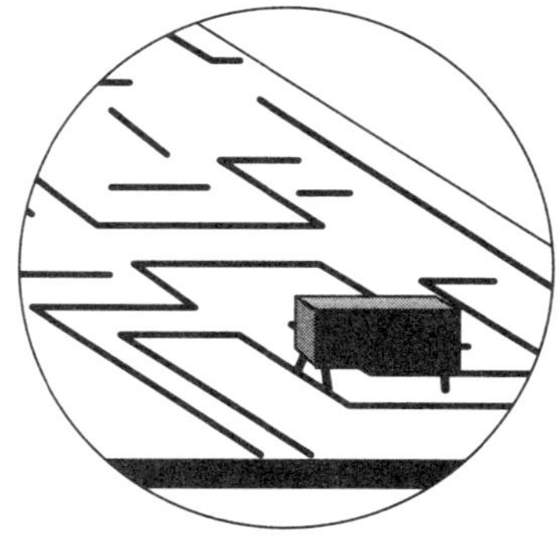

A crystal is soldered to the receiver board.
This device is activated by specific radio frequency.

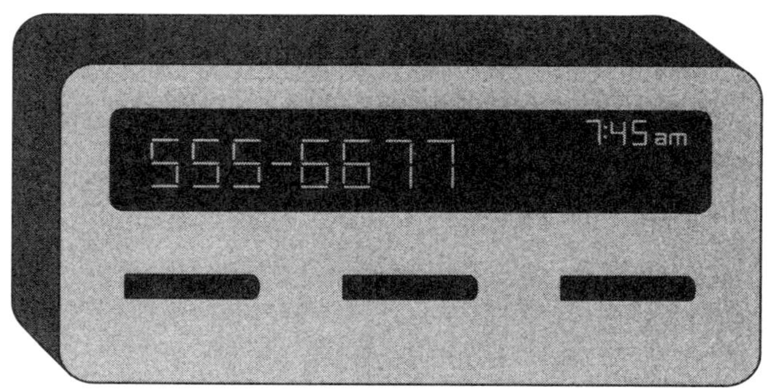

The receiver board is connected to the decoder board. The pager is then sent to a programming station for identification number and special options.

into the plastic outer housing. The housing also contains the on-off switch, belt fastener clip, and battery compartment with electrical contacts.

At this point, the pagers have none of their specific features. They have not been assigned a radio frequency, nor have they been programmed with a unique telephone number or any customer options. All of that will happen in the back end process.

The Back End Process

7 First, a crystal is soldered to the receiver board. A crystal is an electronic component which can be activated only by a specific radio frequency. The crystal is selected to match the radio frequency assigned to the pager.

8 The receiver board is then connected to the decoder board inside the pager housing, and the housing is sent through a laser which is programmed to etch the pager identification number and other data onto the back cover of the housing.

9 The pager is sent to the programming station where it is automatically programmed with a unique identification number and a large spectrum of options selected by the customer.

10 After programming, the pager crystal and filter frequencies are tuned at a manual tuning station. This ensures the pager will receive the exact radio frequency assigned to it and prevents reception of messages intended for other pagers on other

frequencies. Sometimes this process is done automatically.

11 In the final assembly station, the back cover is secured onto the decoder board, a battery is inserted, and the pager is turned on. After passing a radio frequency test and a final visual inspection, the pager is packed and sent to the customer.

Quality Control

In addition to the visual inspections and electronic tests, the entire pager manufacturing process is monitored by the CIM system. This system can alert production workers if any component has been inadvertently omitted or if any function has not been programmed into the device.

Manufacturers also perform rigorous testing of new designs under extremes of temperature, vibration, and impact. This simulates conditions such as leaving the pager in the sun in a locked car or accidentally dropping the pager on a hard surface.

The Future

Recent developments in pager technology include the ability to transmit a limited number of pre-programmed responses back from the pager to the caller via radio signals. For example, this would allow the pager user to send a message of "buy" or "sell" in response to an urgent inquiry from his or her stockbroker. Two-way communication via pager is expected to expand in the near future.

Another recent development is a pager card which can receive and store a large amount of information. Looking something like a thick credit card, this device signals the user when a message is received. The user then plugs the card into a computer, and the pager displays the message on the screen. Some models have up to 512 kilobytes of memory. Future developments in pagers will result in even greater range, including worldwide paging, and greater information retrieval capabilities.

Although basic pagers will probably continue to be popular because of their low price, competing technologies may limit further growth. Cellular telephones and wireless personal digital assistant (PDA) devices, for example, have the capability to transmit voice, fax, and e-mail via radio networks. As the price of these devices and services becomes more competitive, many current pager users may decide to upgrade to obtain the benefits of increased information communication.

Where To Learn More

Book

Lewis, Geoffrey E. *Newnes Communications Technology Handbook.* Newnes, 1994.

Periodicals

Chien, Phillip. "Letter to a Beeper." *Popular Mechanics,* April 1995, pp. 50-53, 124.

Cooke, Jim. "Build the Phone Pager." *Popular Electronics,* November 1993, pp. 37-40, 94.

Henkoff, Ronald. "Keeping Motorola on a Roll." *Fortune,* April 18, 1994, pp. 67-78.

Kantra, Suzanne and Marcelle M. Soviero, eds. "What's New: Omni-Pager." *Popular Science,* February 1994, p. 12.

Kirschner, Suzanne Kantra and Bob Sillery, eds. "What's New: Talking Back." *Popular Science,* April 1995, p. 12.

"What's New: Pen-Size Pager." *Popular Science,* April 1995, p. 20.

Therrien, L. "Pagers Start to Deliver More than Phone Numbers." *Business Week,* November 15, 1993, p. 180.

—Carol Brennan/Chris Cavette

The Manufacturing Process section of this entry was written with the help of Fred Schmidt at Motorola.

Beer

On a per capita basis, Germans consume the most beer at about 40 gallons per person per year. Beer drinkers in the U.S. rank fourteenth in the world, with American breweries producing approximately 156,900 million barrels of beer a year.

Background

The family of beverages generally referred to as "beer" has been brewed for centuries. Beers are obtained by the yeast fermentation of malted cereal grains, to which hops and water have been added. Brewing has evolved from a cottage craft into a modern industry where large breweries export their beers worldwide. On a per capita basis, Germans consume the most beer at about 40 gallons (151 l) per person per year. Beer drinkers in the U.S. rank fourteenth in the world, with American breweries producing approximately 156,900 million barrels of beer a year. Each barrel is the equivalent of 117 liters or approximately 31 gallons.

The true origin of beer can only be conjectured. Early attempts at brewing occurred around 7000 B.C. in Mesopotamia. The Egyptians and Greeks also brewed alcoholic beverages by various methods, but the term "beer" did not appear in these early languages. The Babylonians offered brewing recipes, and there are various references to beer in the Bible. The English word "beer" seems to stem from the Celtic word "beor," which referred to a malt brew made by monks at a North Gaul monastery. In the Middle Ages, monasteries were the leading producers of beer, and monks are credited with many early brewing techniques, such as the addition of hops to improve the aroma and help preserve the beer. The distinction between ales, lagers, and darker bock beers began to appear in French and Irish writings in the 13th century. It is generally accepted that the modern beers as we know them today date to the 1600s.

Beer brewing was already a thriving industry in Europe when the United States declared its independence in 1776. European immigrants brought their brewing skills to America and founded a thriving beer industry. Some technological advancements—the yeast separator, for example—made mass production of beer possible. Bottled beer was introduced in 1875 by the Joseph Schlitz Brewing Company in Milwaukee, Wisconsin, a city famed for its breweries. Canned beer first came on the market in the 1930s. The American beer market today is dominated by several large companies such as Miller and Anheuser Busch, though microbreweries and brew pubs that produce their own brands are becoming increasingly popular.

Raw Materials

Beer requires these ingredients for brewing: properly prepared cereal grain (usually barley and corn or rice), hops (scientific name *Humulus lupulus*), pure water, and brewer's yeast. Each ingredient can affect flavor, color, carbonation, alcohol content, and other subtle changes in the beer. Grains are carefully stored and handled to promote highest quality. Hops are a form of cultivated perennial hemp, and the useful portions of the vine, the sticky cones, are developed from the bloom. About 35 pounds (16 kg) of barley malt and 15 pounds (7 kg) of grain are used to make each 31-gallon barrel of beer. Large quantities of pure water are extremely important not only as an ingredient, but for maintaining the cleanliness of the brewing equipment. In beer, water high in lime or iron can interfere with the fermentation process and discolor the final product. Yeasts are fungi, which are microorganisms that reduce sugars to alcohol by fermentation. Some types of brewer's yeast are closely guarded trade secrets.

Outside of the beer itself, the process also requires various acids and cleaning chemicals to maintain and sterilize the brewing equipment. The finished product also requires packaging, which includes cardboard products for boxes, aluminum for cans, glass for bottles, and stainless steel for kegs and other commercial dispensing equipment. The majority of the brewing equipment is stainless steel, with the exception of the brew kettles, which are copper.

The Brewing Process

Malting

1 Fully ripened barley grains are "steeped," or soaked in cold water until they are fully saturated. The water is changed once a day, and after 45-72 hours the grains are placed in shallow tanks. The grain is aerated and stirred, which causes it to germinate, releasing enzymes such as malt diastase. Malt diastase converts the starches contained in the grain to sugar for fermentation. As soon as the germination is adequately complete, usually six days, the grain is roasted to stop the germination process. The exact point at which the roasting starts and ends affects the flavor and color of the beer. The product at this point is referred to as malt.

Preparing the mash

2 The malt is crushed using iron rollers and transferred to the mash tank (or "tun"). This tank is a large copper or stainless steel vessel that mixes the malt with warm water until it is of porridge-like consistency. This mixture is called mash. After mixing with similarly prepared cereal grains, the temperature of the mash is raised incrementally from 100-170°F (38-77°C) so that the enzymes react. The enzymes break down the starch in the grain and convert it to simple sugars. Later, the yeast will convert the sugars into alcohol. Once complete, the mash is allowed to sit undisturbed so the solids can descend to the bottom of the tank.

The automatic canning machinery dwarfs the workers in this 1970s brewery canning room. (From the collections of Henry Ford Museum & Greenfield Village.)

While amateur brewers swap recipes at will, the commercial recipes for beer are held tightly as any state secret. Until recent decades, the production of beer, like wine, was a wonderful combination of art, science, and luck. At the heart of the process has been the brewmaster, a traditional handcraftsman wrapped in the lab coat of a scientist and carrying the clipboard of a production engineer. In the 20th century, corporate breweries have evolved into an intriguing combination of flow production in the brewing process and automated canning, bottling, and warehousing.

In the 19th century, the brewing industry flourished as numerous brewmasters drew on their European heritages and functioned as chemists, biologists, engineers, inventors, and salesmen. The combination of local ingredients, water quality, and the brewmaster's traditions and skill meant that many regions, even locales, could have their own brands. Before mechanical refrigeration, pasteurization, and rapid transportation facilities, national distribution was, of course, impossible. One result of this was that the United States has always enjoyed a wide variety of regional beers. In 1867 there were breweries in every state and territory, an astonishing total of 3,700; in 1934 there were still over 800 in operation; in 1994 there were about 500. After Prohibition and with the development of steel cans for beer in 1935, breweries shifted their focus away from primary interest in bars and toward home consumption.

Despite the seeming pervasiveness of national brands from the mega-breweries supported by their huge advertising budgets, this tradition of hundreds of local brands continues. In recent years it has even been augmented by the proliferation of so-called "micro-breweries" which often display the brewing equipment as part of the decor of a drinking establishment and distribute their products primarily on-site.

William S. Pretzer

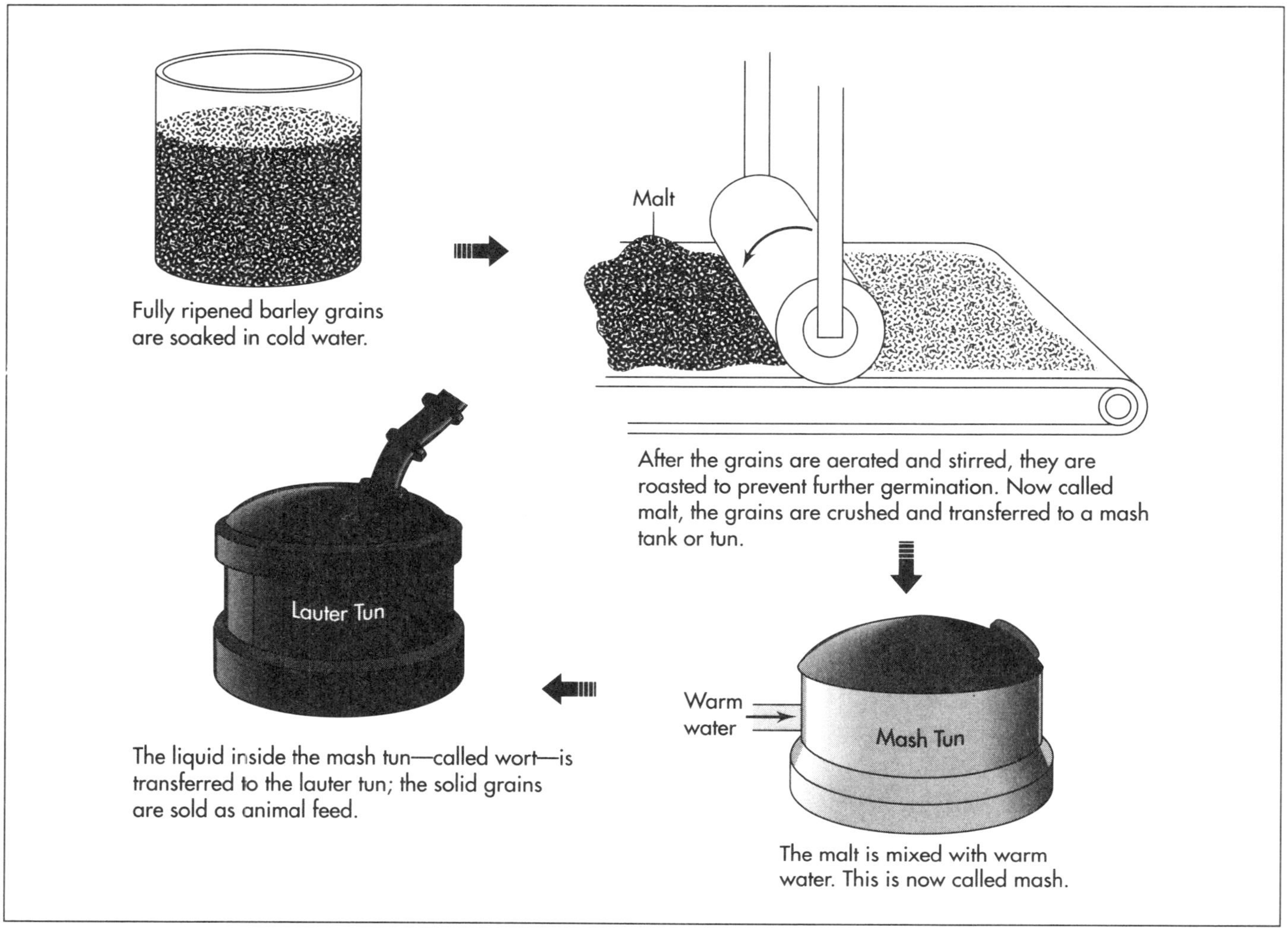

Beer requires these ingredients for proper brewing: prepared cereal grain (usually barley and corn or rice), hops, pure water, and brewer's yeast. Each ingredient can affect flavor, color, carbonation, alcohol content, and other subtle changes in the beer.

Brewing the wort

3 The liquid contained in the mash is transferred into another tank called a lauter tun. This is accomplished by drawing the liquid out through the bottom layer of mash solids, which acts as a filter. Hot water is added to the top of the mash tank to rinse the remaining liquid, now called wort, from the mash. The solid remains of the grain are dried and sold by the brewery as animal feed. The wort travels on to the brew kettles, where it is boiled to sterilize it, and where the carefully prepared hops are added. The addition of the hops is important because they contribute to the bitterness of the beer. The brew kettles are the most impressive equipment in the process. Gleaming copper, they can be 7-12 feet (2-3.6 m) in diameter and two stories high. Steam usually provides the heating energy to the brew kettles. After brewing is complete, the finished wort is filtered again and pumped to the fermentation tanks.

Fermenting

4 In the fermentation tanks, the atmosphere must be carefully controlled to prevent any "rouge" bacteria from interfering with the yeast. Carefully maintained yeast (approximately one pound per barrel of wort) is added to the wort, and the temperature of the mixture is slowly reduced over a period of days to between 50°F and 60°F (10-15°C). In this temperature range, the yeast grows, consuming the sugar in the wort, and bubbles of carbon dioxide form. The wort has now become beer. The new beer is filtered and transferred once more into the aging casks, where the temperature is controlled at 33°F (1°C) for 2-24 weeks. The shorter storage time produces a pale lager beer while the European lagers (called

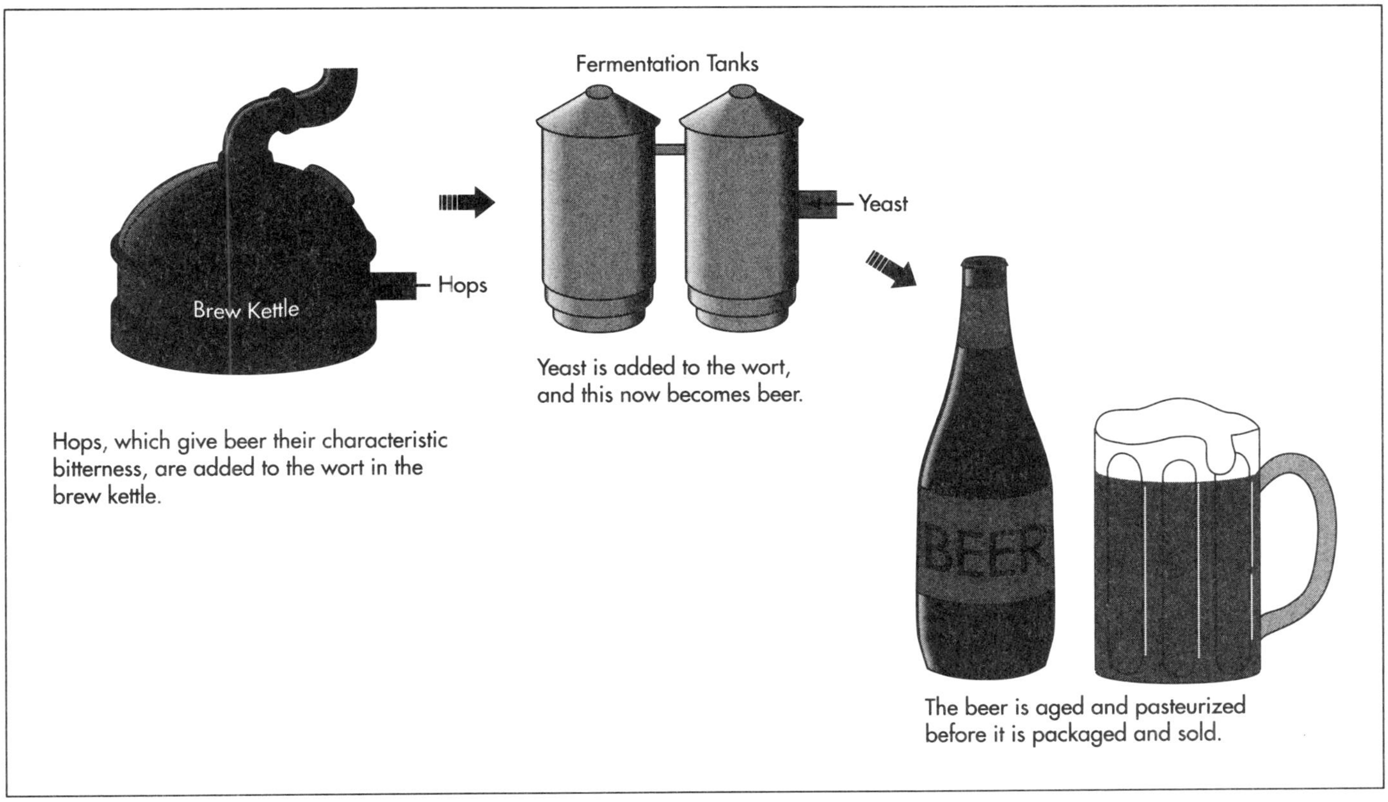

Pilsner) are aged longer to increase the alcohol content.

Pasteurizing

5 After aging, the beer can be pasteurized to kill the remaining yeast and prevent further alcohol production. This is accomplished by heating the beer above 135°F (57°C). This process, named after Louis Pasteur, is widely known for preserving milk. Interestingly, Pasteur originally developed this process to preserve beer in the 1860s. Pasteurization, however, is not used in the production of genuine draft beers. These beers are also known as "ice" beers, since they must be kept refrigerated to preserve their flavor and slow the remaining yeast activity. Many consider the draft beers best in aroma as well as taste.

Packaging

6 Whether packaged into cans, bottles, or kegs, the beer is always moved gently through the maze of piping in the bottling area. This is to preserve the natural carbonation. During bottling, additional carbon dioxide gas from the fermentation kettles is used to improve the aroma of the beer. High-speed packaging lines can process thousands of cases of beer per day, and with modern computerized control, the inventory can be tracked throughout the distribution network. Most beer is delivered from local distributors who have purchasing contracts with the major breweries.

Most beer is available in the following package sizes: "pony" cans and bottles of about 8 fluid ounces, standard 12-ounce cans and bottles, 16- and 32-ounce jumbo cans, 40-ounce "picnic" bottles, 8-gallon "pony" kegs, and the standard 16-gallon beer keg. Other novelty and party packages are also available. Cans and bottles are packed in 6, 8, 12, or 24 each to a box or case. Most states require a deposit at point of sale to encourage the return of the bottles and cans.

When beer is dispensed from the keg, a pressure apparatus called a "tapper" is used to apply a light pressure of carbon dioxide (usually 2-6 PSI) to the tapper head for dispensing.

Byproducts/Waste

Beer brewing produces several byproducts that can be used by other industries. During the malting of the barley, rootlets form on the grain and drip off. These can be collected and used for animal feed. The hops that is filtered out from the finished wort can also be collected and used again as fertilizer. The residual yeast from the brewing process is a rich source of B vitamins. It can be put to use by pharmaceutical companies to make vitamins or drugs, or used as a food additive. Used beer cans and beer bottles are routinely recycled.

The Future

Recently, concern among citizens' groups over the excessive consumption of alcoholic beverages by some individuals has initiated additional government regulation of beer. New warnings have been added to labels, warning of impaired driving, hazards to pregnant women, and other health ailments associated with alcohol consumption. Reduced tolerance for drunk driving, for example, encouraged many brewing companies to advocate responsible consumption. As a result, certain states have established laws to control the alcoholic content of beer for sale within their jurisdiction. The beer industry will continue to contend with these large social issues.

Much research is currently conducted in the area of plant engineering. Brewery researchers are manipulating the genes of barley and other common grains to increase their resistance to disease and to encourage helpful mutations. This genetic research also extends to improving the yeast. Current research is aimed at producing yeast strains that resist contamination and to making new varieties of yeast that can ferment carbohydrates, which common yeasts cannot process.

The brewing industry is also making advances in the area of rapid testing for contaminants. New technology such as DNA probes and protein and chromosome fingerprinting is being developed by brewers to detect microorganisms that can adversely affect the brewing process. Some of this technology is already in use in medical science for drug screening, AIDS testing, and pregnancy testing. Brewers are eager to adapt this cutting edge research to the beer industry.

Where To Learn More

Books

Briggs, D.E. and J.S. Hough. *Malting and Brewing Science*, vols. 1 & 2. Chapman and Hall, 1981.

Hough, J.S. *Biotechnology of Brewing*. Cambridge University Press, 1985.

Periodicals

Altany, David. "Seeds to Suds." *Industry Week*, May 4, 1992, p. 40+.

Booth, Stephen A. "The Local Beer Returns." *Popular Mechanics*, November 1992, p. 44+.

Hummel, George. "The Quest for Homebrew." *Mother Earth News*, December-January 1994, p. 54+.

Polaneczky, Ronnie. "Beer Here! How Home Brewers Are Foamenting a Revolution." *Philadelphia Magazine*, December 1994, p. 57+.

—Douglas E. Betts

Bell

Background

Since prehistoric times bells have been used to herald significant events. Bells call the faithful to worship and toll the time. The sound of a bell can express great joy, sound a warning, or signal mourning. Bells have also been rung to bring on or stop the rain, keep evil spirits at bay, invoke curses, and lift spells.

Bells hold an honored place in religious ceremonies. In both Buddhism and Christianity, bells are blessed before each ceremony. In Roman Catholicism, bells are symbols of paradise and the voice of God. The Russian Orthodox and the Chinese employ bells to speak to spirits or God.

Bells hold an honored place in religious ceremonies. Bells have been rung to bring on or stop the rain, keep evil spirits at bay, invoke curses, and lift spells.

Bells are also revered as patriotic symbols, and it was not unusual for invading conquerors to capture and silence the town bell. In the U.S., the great symbol of the American republic is the Liberty Bell.

The Chou Dynasty, which reigned in China from 1122 to 221 B.C., was particularly known for its superior bell founding. European bell founding occurred much later and originated in medieval monasteries. The first European bells resembled cow bells: **iron** plates that had been hammered square and then riveted together. By the 15th century, founders began to experiment with bell shape and tone. Secular bellmakers gained prestige in the Renaissance with the flourishing of Gothic architecture which featured grand bell towers.

In the 17th century, Belgium and the Netherlands emerged as the leaders in bell founding. Dutch brothers Francois and Pierre Hemony are generally credited with developing the bell into a sophisticated musical instrument. The Hemonys worked with a blind musician named Jacob Van Eyck on a tuning system for the five separate and distinct tones contained in each bell's ring. After the deaths of Francois and Pierre and that of their star pupil, Caes Noorder, in the 18th century, the art suffered a decline. It was not until the 20th century that tuning techniques once again gained excellence.

Bell shapes vary by country and culture. The sides can be straight, convex, concave, or hemispherical. East Asian bells tend to be barrel-shaped while Western bells are tulip-shaped with a bulge near the rim. Chinese bells often have lotus-shaped rims. Bells of Western cultures are generally struck by an interior metal striker as the bell swings back and forth. Asian bells are non-swinging and are usually struck manually on the outside with a wooden mallet.

Raw Materials

While decorative bells can be made of such materials as horn, wood, glass, and clay, bells that are designed to ring or to play music are cast in a bronze alloy of approximately 77% copper and 23% tin. This combination produces a tough, long-lasting material that resists rusting. Bell founders must be careful not to mix in more than 25% tin or the bell will be brittle and susceptible to cracking. It is not unusual for old bells to be melted down and the metal re-used to cast new bells.

The Manufacturing Process

The craft of casting bells has remained essentially the same since the 12th century.

The bellmaker designs the shape of the bell and creates wooden templates called strickle boards. The boards are used to create the inner and outer molds of the final bell.

A model of the outer shell, called the false bell, is made of stone and covered with sand and molten wax. A steel mantle is lowered onto the false bell.

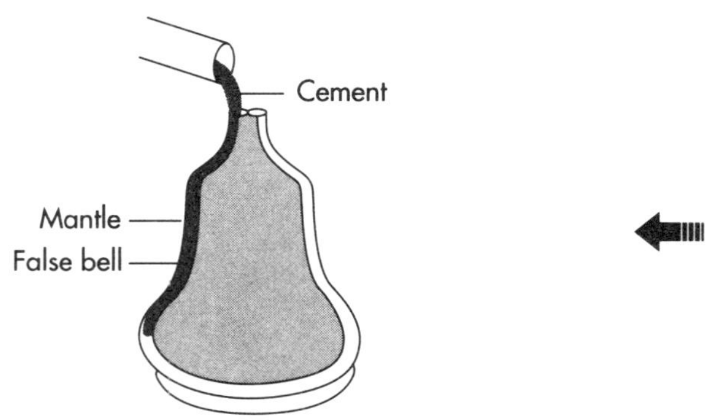

The space between the mantle and false bell is filled with cement. After the cement has hardened, the mantle is lifted and the false bell chipped away.

The one singular innovation was the invention of the tuning machine in the 19th century. Prior to that time, the proper tone was achieved by chipping the sides of the bell with a hammer and chisel. This procedure carried a high risk of damaging the bell. The tuning machine, which is essentially a vertical lathe, has reduced that risk. Electronic tuning machines have increased the bell founder's ability to test the accuracy of the bell's tone. All in all, however, creating a bell is still very much a hands-on process.

Calculating the bell design

1 Using the specifications submitted by the purchaser, the bellmaker determines the shape that the bell will need to take in order to resonate with the proper number of vibrations. After estimating the required weight, the bellmaker orders the metal. These are painstaking measurements that can take several weeks of calculations to accomplish.

Making the bell pattern or template

2 The bellmaker cuts out two wooden templates called "strickle boards." One of the boards matches the dimensions of the outer bell (called the case or cope); the other matches that of the inner bell (called the core). These templates are used to construct the mold.

Constructing the mold

3 An exact stone model of the outer bell, sometimes called a false bell, is covered first with sand or loam, then with molten wax. Figures and inscriptions, also made of wax, are applied by hand. The false bell is

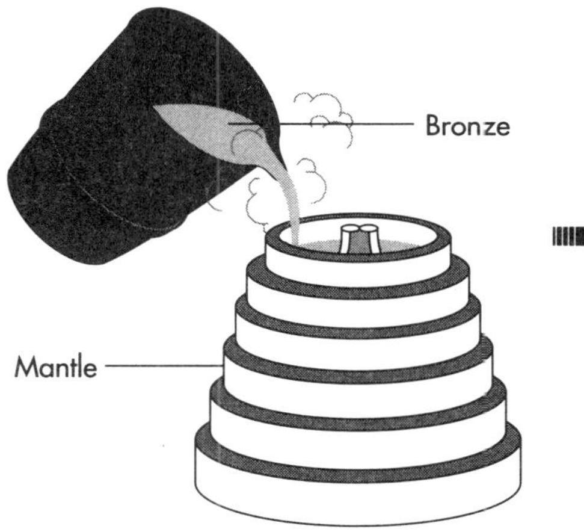

The mantle and the outer mold are lowered over a prepared inner mold. Molten bronze is poured into the mantle, which fills the space between the outer and inner molds. The bell is cooled.

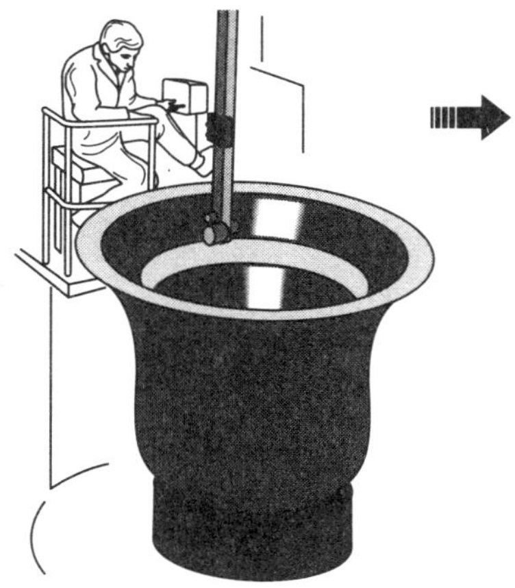

The bell is tuned upside down by a lathe operator who grinds metal off the lower edges.

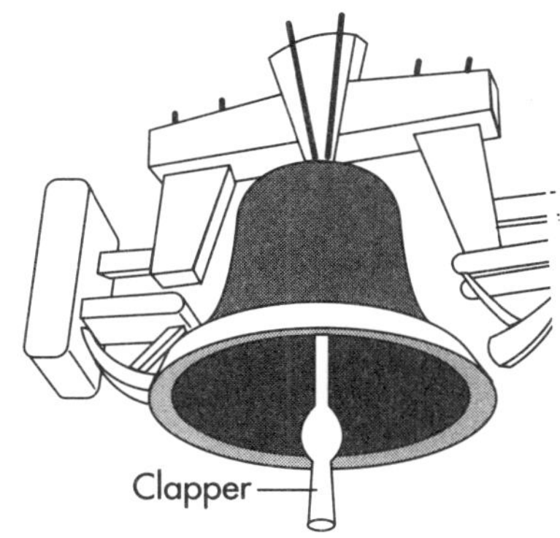

Using mounting bolts and supports, the prepared clapper is fastened to the bell.

painted over with three coats of very fine, fireproof clay. It is then enclosed in a steel mantle that has been lowered by rope pulleys.

The space between the false bell and the mantle is filled with cement. After the cement has hardened, the mantle is lifted off the cement mold. The false bell, under the mold, is chipped away. Any remaining scraps of the false bell are removed with a blow torch. The mold is then set over a coke fire to melt the remaining wax and to evaporate any water that has accumulated.

A model of the inner bell is constructed of stone and coated with fireproof cement. It is then smoothed to remove any irregularities.

Casting the bell

4 After the mantle has been cleaned. it is again lowered over the outer bell model. The mantle and the outer bell mold are then lowered over the inner mold. The outer and inner sections are clamped together, leaving a space between them, and set into a pit.

Ingots of bronze are melted in oil burners and heated to a temperature of approximately 1150°F (1100°C). The molten metal is skimmed to remove impurities and then poured into drums. The drums are carried to the pit and carefully tipped so that the hot metal flows into the space between the two molds. Holes in the top of the mantle allow gases to escape. If the gases remained in the metal, the bell would be too porous and easily cracked.

The bell is allowed to cool for several days. Large bells can take as much as a week to cool completely. Small bells, usually classified as those under 500 pounds (227 kg), can be removed from the molding pit the next day.

Tuning the bell

5 The bell is cast with slightly thicker sides so that the bell can be ground as it twirls slowly upside down on a circular lathe to acquire the precise tone . The bell tuner is highly skilled; it takes years of experience to know just how much metal to remove. The bell tone is tested frequently during the tuning process using an electronic device that registers the vibrations as the bell is struck. If the tone is too low, the lathe operator grinds more metal off the lower edge of the bell. If the tone is too high, the bell is thinned with a file.

Fitting the clapper into the bell

6 The clapper is manufactured in much the same manner as the bell itself. Special care is given to cast the clapper at the proper weight. A clapper that is too lightweight will not bring out the true tones of the bell. A heavy clapper might cause the bell to crack.

Holes are drilled into the top of the bell. Using mounting bolts and supports, the clapper is fastened to the bell.

Quality Control

Great care is taken to calculate the precise weight and size of the bell before it is cast. If the finished bell does not meet specifications, it is completely melted down and recast. Should a bell crack at a future date, it might be welded and patched, but that is rare. The bell is more likely to be retired, as in the case of the Liberty Bell, or it is melted down and recast.

Where To Learn More

Book

Yolen, Jane. *Ring Out! A Book of Bells.* The Seabury Press, 1974.

Brochure

Royal Eijsbouts. Schulmerch Carillons, Inc., Carillon Hill, Sellersville, PA 18960, (215) 257-2771.

—Mary F. McNulty

Bicycle

Background

Bicycles are one of the world's most popular modes of transportation, with some 800 million bicycles outnumbering cars by two to one. Bicycles are also the most energy-efficient vehicle—a cyclist burns about 35 calories per mile (22 calories per km), while an automobile burns 1,860 calories per mile (1,156 calories per km). Bicycles are used not only for transportation, but for fitness, competition, and touring as well. They come in myriad shapes and styles, including racing bikes, all-terrain bikes, and stationary bicycles, as well as unicycles, tricycles, and tandems.

Available in myriad shapes and styles, including racing bikes, mountain bikes, and stationary bicycles, as well as unicycles, tricycles, and tandems, bicycles are one of the world's most popular modes of transportation, outnumbering cars by two to one.

History

As far back as 1490, Leonardo da Vinci had envisioned a machine remarkably similar to the modern bicycle. Unfortunately, da Vinci did not attempt to build the vehicle, nor were his sketches discovered until the 1960s. In the late 1700s a Frenchman named Comte de Sivrac invented the Celerifere, a crude wooden hobby horse made of two wheels and joined by a beam. The rider would sit atop the beam and propel the contraption by pushing his or her feet against the ground.

In 1816 the German Baron Karl von Drais devised a steerable hobby horse, and within a few years, hobby-horse riding was a fashionable pastime in Europe. Riders also discovered that they could ride the device with their feet off the ground without losing their balance. And so, in 1840, a Scottish blacksmith named Kirkpatrick Macmillan made a two-wheel device that was operated by a treadle. Two years later he traveled as many as 40 miles (64 km) at a stretch during a record 140-mile (225 km) round trip to Glasgow. A couple decades later, a Frenchman, Ernest Michaux, designed a hobby horse that utilized cranks and rotating pedals connected to the front axle. The Velocipede, made with wooden wheels and an **iron** frame and tires, won the nickname of the "boneshaker."

The 1860s proved to be an important decade for bicycle improvements with the inventions of ball-bearing hubs, metal-spoked wheels, solid rubber tires, and a lever-operated, four-speed gearshift. Around 1866 an unusual version of the Velocipede was created in England by James Stanley. It was called the Ordinary, or Penny Farthing, and it had a large front wheel and a small rear wheel. The Ordinaries were soon exported to the U.S. where a company began to manufacture them as well. These bicycles weighed a hefty 70 pounds (32 kg) and cost $300—a substantial sum at the time.

By 1885, another Englishman, John Kemp Starley, created the Rover Safety, so called since it was safer than the Ordinary which tended to cartwheel the rider over the large front wheel at abrupt stops. The Safety had equally sized wheels made of solid rubber, a chain-driven rear wheel, and diamond-shaped frame. Other important developments in the 1800s included the use of John Boyd Dunlop's pneumatic tires, which had air-filled inner tubes that provided shock absorption. Coaster brakes were developed in 1898, and shortly thereafter freewheeling made biking easier by allowing the wheels to continue to spin without pedaling.

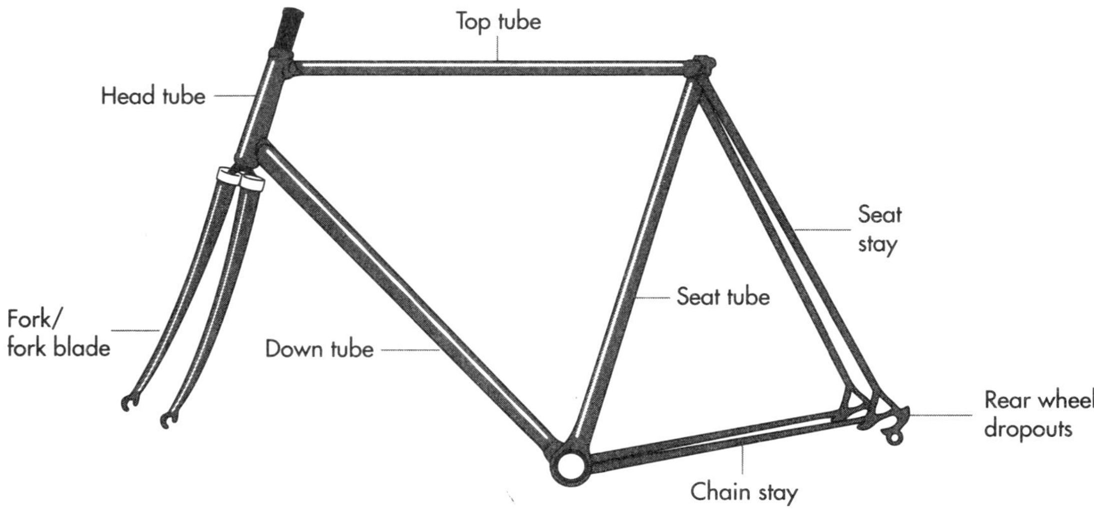

The most important part of the bicycle is the diamond-shaped frame, which links the components such as wheels, derailleurs, brakes, and chains together.

The frame consists of the front and rear triangles, the front really forming more of a quadrilateral of four tubes: the top, seat, down, and head tubes. The rear triangle consists of the chainstays, seatstays, and rear wheel dropouts. Attached to the head tube at the front of the frame are the fork and steering tube.

During the 1890s bicycles became very popular, and the basic elements of the modern bicycle were already in place. In the first half of the 20th century, stronger steel alloys allowed thinner frame tubing which made the bicycles lighter and faster. Derailleur gears were also developed, allowing smoother riding. After the Second World War, bicycle popularity slipped as automobiles flourished, but rebounded in the 1970s during the oil crisis. About that time, mountain bikes were invented by two Californians, Charlie Kelly and Gary Fisher, who combined the wide tires of the older balloon-tire bikes with the lightweight technology of racing bikes. Within 20 years, mountain bikes became more popular than racing bikes. Soon hybrids of the two styles combined the virtues of each.

The Raw Materials

The most important part of the bicycle is the diamond-shaped frame, which links the components together in the proper geometric configuration. The frame provides strength and rigidity to the bicycle and largely determines the handling of the bicycle. The frame consists of the front and rear triangles, the front really forming more of a quadrilateral of four tubes: the top, seat, down, and head tubes. The rear triangle consists of the chainstays, seatstays, and rear wheel dropouts. Attached to the head tube at the front of the frame are the fork and steering tube.

For much of the bicycle's history the frame was constructed of heavy, but strong, steel and alloy steel. Frame material was continually improved to increase strength, rigidity, lightness, and durability. The 1970s ushered in a new generation of more versatile alloy steels which could be welded mechanically, thereby increasing the availability of light and inexpensive frames. In the following decade lightweight aluminum frames became the popular choice. The strongest metals, however, are steel and titanium with life-expectancy spanning decades, while aluminum may fatigue within three to five years.

Advances in technology by the 1990s led to the use of even lighter and stronger frames made of composites of structural fibers such as carbon. Composite materials, unlike metals, are anisotropic; that is, they are strongest along the axis of the fibers. Thus, composites can be shaped into single-piece frames, providing strength where needed.

The components, such as wheels, derailleurs, brakes, and chains, are usually made of stainless steel. These components are generally made elsewhere and purchased by the bicycle manufacturer.

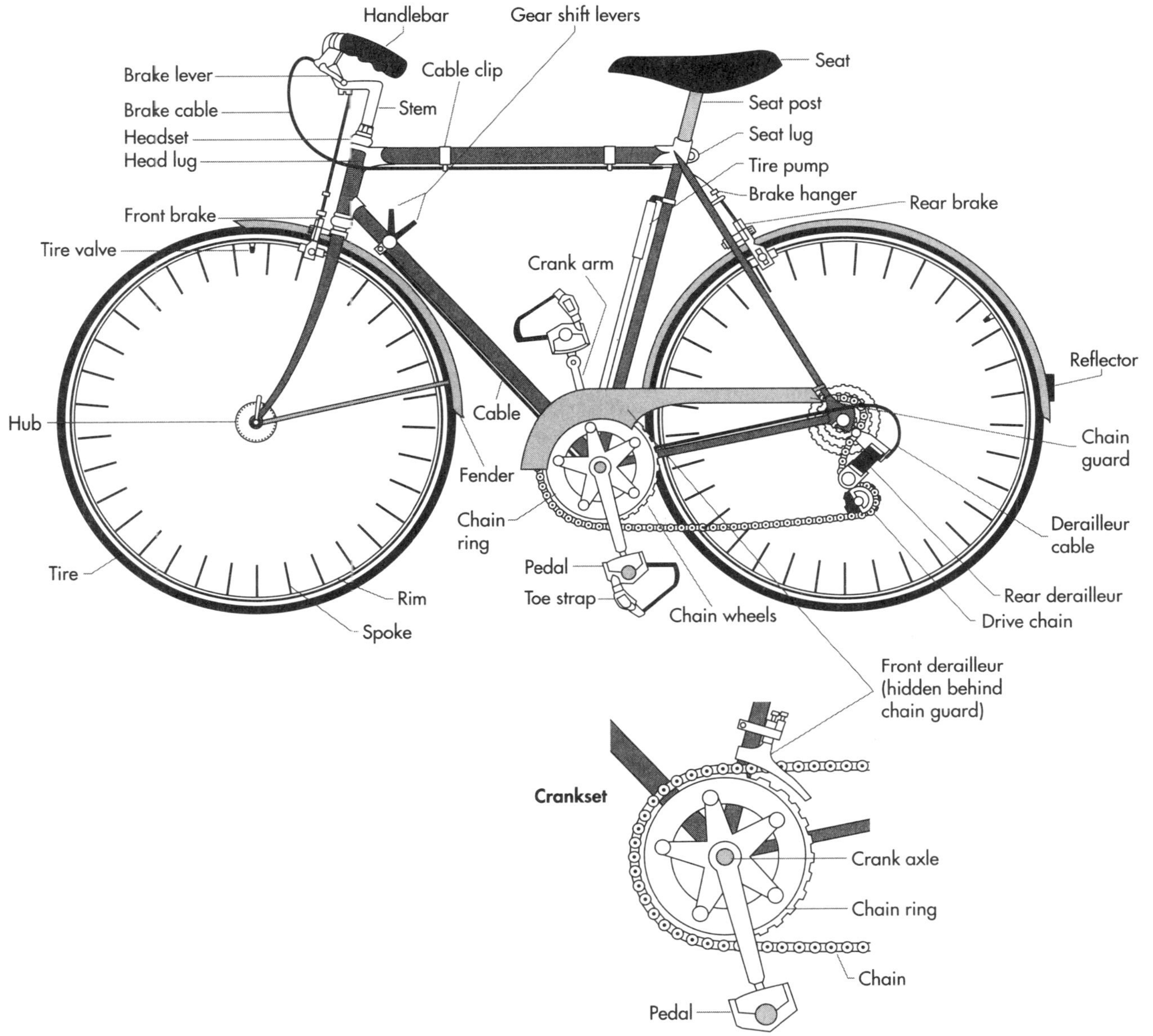

The Manufacturing Process

Seamless frame tubes are constructed from solid blocks of steel that are pierced and "drawn" into tubes through several stages. These are usually superior to seamed tubes, which are made by drawing flat steel strip stock, wrapping it into a tube, and welding it together along the length of the tube. Seamless tubes may then be further manipulated to increase their strength and decrease their weight by butting, or altering the thickness of the tube walls. Butting involves increasing the thickness of the walls at the joints, or ends of the tube, where the most stress is delivered, and thinning the walls at the center of the tube, where there is relatively little stress. Butted tubing also improves the resiliency of the frame. Butted tubes may be single-butted, with one end thicker; double-butted, with both ends thicker than the center; triple-butted, with different thicknesses at either end; and quad-butted, similar to a triple, but with the center thinning towards the middle. Constant thickness tubes, however, are also appropriate for certain bikes.

The tubes are assembled into a frame by hand-brazing or welding by machine, the former being a more labor-intensive process and therefore more expensive. Composites may be joined with strong glue or plastic binders. The components are generally manufactured by machine and may be attached to the frame by hand or machine. Final adjustments are made by skilled bicycle builders.

Assembling the Frame

Tailoring the tubes

1 The metal is annealed, or softened by heating, and hollowed out to form "hollows," or "blooms." These are heated again, pickled in acid to remove scale, and lubricated.

2 The hollows are measured, cut, and precision mitered to the appropriate dimensions. Frame sizes for adult bicycles generally run from 19-25 inches (48-63 cm) from the top of the seat post tube to the middle of the crank hanger.

3 Next, the hollows are fitted over a mandrel, or rod, attached to a draw bench. To achieve the right gauge, the hollows pass through dies which stretch them into thinner and longer tubes, a process called cold drawing.

4 The tubes may be shaped and tapered into a variety of designs and lengths. The taper-gauge fork blades may have to pass through more than a dozen operations to achieve the correct strength, weight, and resilience.

Brazing, welding, and gluing

5 Tubes can be joined into a frame either by hand or machine. Frames may be brazed, welded, or glued, with or without lugs, which are the metal sleeves joining two or more tubes at a joint. Brazing is essentially welding at a temperature of about 1600°F (871°C) or lower. Gas burners are arranged evenly around the lugs which are heated, forming a white flux that melts and cleans the surface, preparing it for brazing. The brazing filler is generally brass (copper-**zinc** alloy) or silver, which melt at lower temperatures than the tubes being joined. The filler is applied and as it melts, it flows around the joint, sealing it.

Aligning and cleaning

6 The assembled frames are placed into jigs and checked for proper alignment. Adjustments are made while the frame is still hot and malleable.

7 The excess flux and brazing metals are cleaned off by pickling in acid solutions and by washing and grinding the brazing until it is smooth.

8 After the metals have cooled, further precision alignments are made.

Finishing

9 The frames are painted, not only to create a more finished appearance, but also to protect the frame. The frame is first primed with an undercoat and then painted with a colored enamel. Paint may be applied by hand-spraying or by passing the frames through automatic electrostatic spraying rooms. The negatively charged frames attract the positively charged paint spray as the frames rotate for full coverage. Finally, transfers and lacquer are applied to the frame. Chrome plating may also be used instead of paint on components such as the fork blades.

Assembling the Components

Derailleurs and gear shift levers

10 Depending on the style of bicycle, the gear shift levers are mounted either on the down tube—popular on racing bikes—on the stem, or on the handlebar ends. A cable is attached, which extends to the front and rear derailleurs. Front derailleurs, which move the chain from one drive sprocket to another, may be clamped or brazed onto the seat tube. Rear derailleurs may be mounted with bolt-on hangers or integral hangers.

Handlebars, stems, and headsets

11 Handlebars may be raised, flat, or dropped. They are bolted to the bicy-

cle stem which is then fitted into the head tube. The headset components, including bearings, cups, and locknuts, are attached to the head tube. The headset allows the fork to turn inside the head tube and thus makes steering easier.

Brakes

12 The brake levers are mounted to the handlebars. Cables extend to the brakes and are fastened to the calipers. Tape, made of plastic or cloth, can then be attached to the handlebars and the ends are plugged.

Saddles and seat posts

13 Seat posts are generally steel or aluminum alloy and are bolted or clamped into position. The saddle is generally made of molded padding and covered with nylon or plastic materials. Although leather was the norm for saddles for a long time, it is less commonly used today.

Cranksets

14 The crankset supports the pedals and transfers power from the pedals to the chain and rear wheel. Cranksets consist of steel or aluminum alloy crank arms, chain rings, and the bottom bracket assembly of axle, cups, and bearings. They are attached with bolts and caps into the bottom bracket of the bicycle frame. The pedals are then screwed to the ends of the crank arms.

Wheels, tires, and hubs

15 Wheel manufacturers conform to the International Standards Organization (ISO) system for wheel diameter and tire sizes. Wheels may be constructed by machines, which roll steel strips into hoops that are welded into rims. The rims are drilled to accept spokes, which are laced one round at a time between the rim and hub flange.

16 A wheel must be trued, or straightened, in radial and lateral directions to achieve uniform tension. Next, the rim liner, tire, and inner tube are attached. The chain may also be fitted onto the bicycle.

17 Rear wheels are fitted with a freewheel, consisting of several cogs and spacers, which frees the rear wheel from the crank mechanism when the rider stops pedaling.

18 Wheels are attached to the bicycle frame by means of an axle which runs through the hub of the wheel. The axle may be tightened with bolts at the ends or with quick-release skewers.

The Future

The future for bicycles looks promising as we approach the 20th century. Developments in bicycle technology in the 1990s have led to advances in human-powered vehicles (HPVs) design. Most HPVs are low-slung recumbents, which are more aerodynamic than conventional bicycles and therefore reduce drag and increase speed. Recumbents are also safer, and many provide cargo room and weather protection. A hybrid of the bicycle and automobile called the Ecocar began to surface on European streets by the 1990s. Designed by a Dutch surgeon, Wim Van Wijnen, it provided weather protection, safety, luggage room, easy maintenance, comfort, and speed.

The use of computer technology greatly enhanced the design capabilities of manufacturers and designers. Designers are able to simulate various forces working on the bicycle, such as pedaling and road shock. Computer-generated programs make testing simpler, and variations of designs are modified more easily and quickly.

Where To Learn More

Books

Ballantine, Richard and Richard Grant. *Richards' Ultimate Bicycle Book.* Dorling Kindersley, 1992.

The Bicycle Builder's Bible. TAB Books Inc., 1980.

Bicycle Magazine's Complete Guide to Bicycle Maintenance and Repair, revised ed. Rodale Press, 1990.

Watson, Roderick and Martin Gray. *The Penguin Book of the Bicycle,* Allen Lane Pub., 1978.

Periodicals

Brown, Stuart F. "The Anybody Bike." *Popular Science*, August 1991, pp. 58-59, 89.

Schwartz, David M. "Over Hill, Over Dale, On a Bicycle Built for...Goo." *Smithsonian*, June 1994, pp. 74-86.

Soviero, Marcelle M. "Easy Riders." *Popular Science*, May 1993, pp. 84-87.

—*Audra Avizienis*

Bleach

Background

Bleach is a chemical compound derived from natural sources used to whiten fabrics. Bleach works by the process of oxidation, or the alteration of a compound by the introduction of oxygen molecules. A stain is essentially a chemical compound, and the addition of bleach breaks down the molecules into smaller elements so that it separates from the fabric. Detergent and the agitation of the washing machine speed up the cleaning process. The disinfecting properties of bleach work in the same manner—germs are broken down and rendered harmless by the introduction of oxygen. In industry, different forms of bleach are used to whiten materials such as **paper** and wood, though most bleach is used to launder textiles.

History

Humans have been whitening fabrics for centuries; ancient Egyptians, Greeks, and Romans bleached materials. As early as 300 B.C., soda ash, prepared from burned seaweed, was used to clean and whiten cloth. During the Middle Ages, the Dutch perfected the bleaching of fabrics in a process called crofting, whereby fabrics were spread out in large fields for maximum sunlight exposure. Textile mills as far away as Scotland shipped their material to the Netherlands for this bleaching. The practice quickly spread throughout Europe, and bleaching fields were documented in Great Britain as early as 1322. In 1728 a bleaching company using Dutch methods went into business in Galloway, Scotland. In this process, the fabrics were soaked in a lye solution for several days, then "bucked," or washed clean. The fabrics were then spread out on the grass for weeks at a time. This process was repeated five or six times until the desired whiteness was achieved. Next, the fabric was treated with sour milk or buttermilk, and again bucked and crofted. This method was lengthy and tedious, and it monopolized large tracts of land that could have been used for farming.

Late in the 18th century, scientists discovered a chemical that had the same effect as crofting, but yielded much quicker results. In 1774, Swedish chemist Karl Wilhelm Scheele discovered the chemical element chlorine, a highly irritating, green-yellowish gaseous halogen. In 1785, the French scientist Claude Berthollet found that chlorine was an excellent whitening agent in fabrics. Some mill operators attempted to expose their fabrics to chlorine gas, but the process was so cumbersome and the fumes so strong that these attempts were soon abandoned.

Near Paris, in the town of Javel, Berthollet began a small facility for the manufacture of a new product called "Eau de Javelle." The bleaching powder consisted of potash (soda ash) which had absorbed chlorine gas. In 1799, another bleaching powder was invented by Scottish chemist Charles Tennant. In the early years of the Industrial Revolution, his patented lime powder was widely used to whiten a variety of fabrics and paper products. To make the bleaching powder, slaked lime (lime treated with water) was spread thinly over the concrete or lead floor of a large room. Chlorine gas was pumped into the room to be absorbed by the lime. Though an effective whitener, the powder was chemically unstable. It was

A stain is essentially a chemical compound, and the addition of bleach breaks down the molecules into smaller elements so that it separates from the fabric.

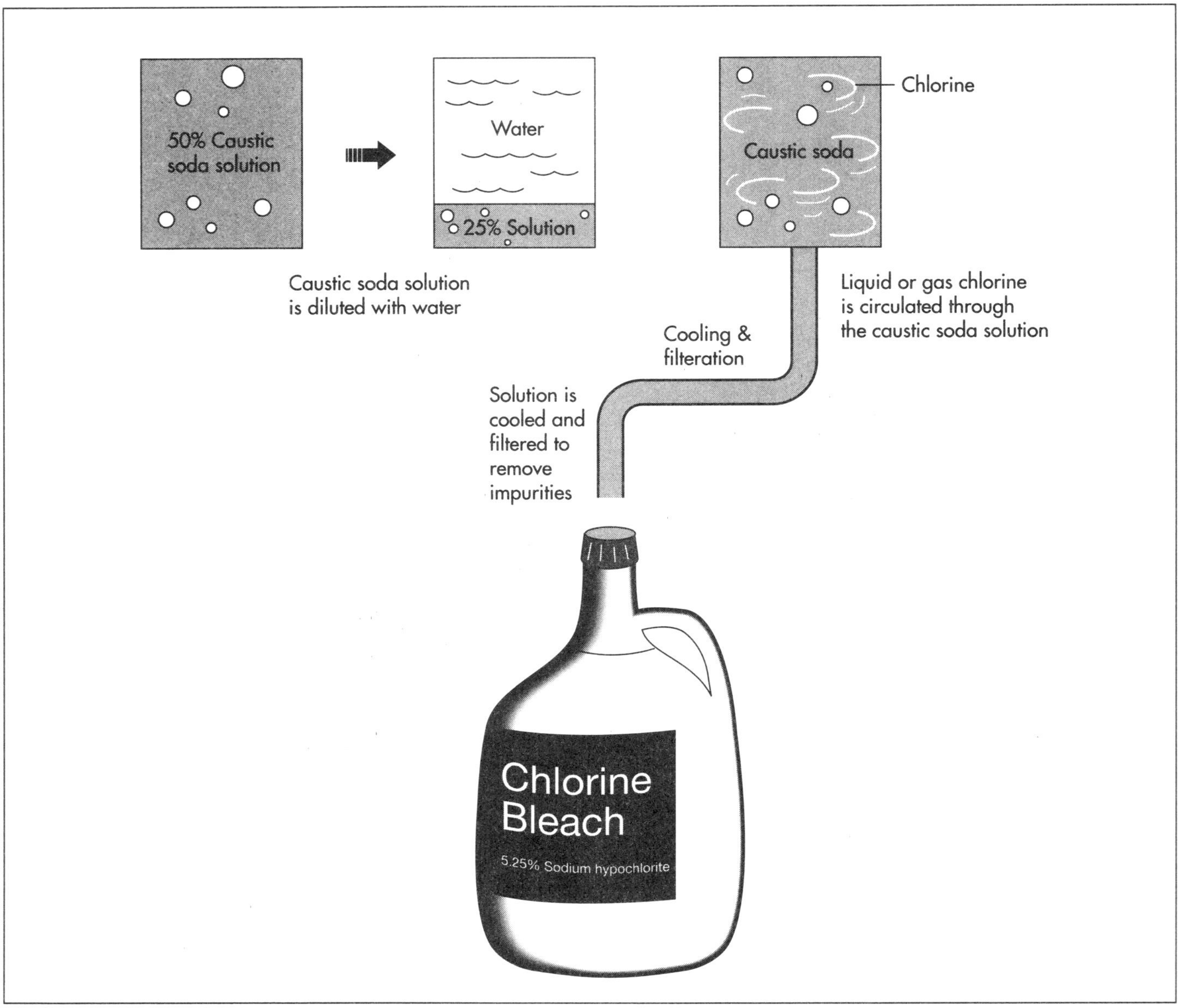

The raw materials for making household bleach are chlorine, caustic soda, and water. The chlorine and caustic soda are produced by putting direct current electricity through a sodium chloride salt solution in a process called electrolysis.

commonly used until around World War I, when liquid chlorine and sodium hypochlorite solutions—the forerunners of modern household bleach—were introduced. About this time, researchers found that injecting **salt** water with electrical current broke down the salt (sodium chloride) molecules and produced a compound called sodium hypochlorite. This discovery enabled the mass production of sodium hypochlorite, or chlorine, bleach.

Types of Bleach

Today, bleach is found in nearly every household. It whitens fabrics and removes stains by a chemical reaction that breaks down the undesired color into smaller particles that can be easily removed by washing. The two types of household bleach are chlorine bleach and peroxide bleach. Peroxide bleach was introduced in the 1950s. Though it helps to remove stains, especially in higher wash temperatures, it will not bleach most colored materials and does not weaken fabrics, as does sodium hypochlorite bleach. Peroxide bleach does not disinfect and is commonly added to laundry detergents which are advertised as color-safe. It also has a longer shelf life than chlorine bleach. Peroxide bleach is more commonly used in Europe, where washing machines are manufactured with inner heating coils that can raise the water temperature to the boiling point.

The more common form of household bleach in the U.S. is chlorine bleach. It is most effective in removing stains and disinfecting fabrics. Chlorine bleach is cheap to manufacture and effective in both warm and hot wash temperatures. However, it has strong chemical properties which can weaken textile fibers.

The disinfecting properties of chlorine bleach can also be useful outside the laundry. Chlorine bleach disinfects drinking water where groundwater contamination has occurred, as it is a powerful germicide. It was first used to sanitize drinking water in New York City's Croton Reservoir in 1895, and is approved by the government for sanitizing equipment in the food industry. In recent years, bleach has been promoted by community health activists as a low-cost method of disinfecting the needles of intravenous drug users.

Raw Materials

The raw materials for making household bleach are chlorine, caustic soda, and water. The chlorine and caustic soda are produced by putting direct current electricity through a sodium chloride salt solution in a process called electrolysis. Sodium chloride, common table salt, comes from either mines or underground wells. The salt is dissolved in hot water to form a salt solution, which is then treated for impurities before it is reacted in the electrolytic cell.

The Manufacturing Process

The manufacture of sodium hypochlorite bleach requires several steps. All the steps can be carried out at one large manufacturing facility, or the chlorine and caustic soda can be shipped from different plants to the reactor site. Both chlorine and caustic soda are hazardous chemicals and are transported according to strict regulations.

Preparing the components

1 Caustic soda is usually produced and shipped as a concentrated 50% solution. At its destination, this concentrated solution is diluted with water to form a new 25% solution.

2 Heat is created when the water dilutes the strong caustic soda solution. The diluted caustic soda is cooled before it is reacted.

The chemical reaction

3 Chlorine and the caustic soda solution are reacted to form sodium hypochlorite bleach. This reaction can take place in a batch of about 14,000 gallons or in a continuous reactor. To create sodium hypochlorite, liquid or gaseous chlorine is circulated through the caustic soda solution. The reaction of chlorine and caustic soda is essentially instantaneous.

Cooling and purifying

4 The bleach solution is then cooled to help prevent decomposition.

5 Often this cooled bleach is settled or filtered to remove impurities that can discolor the bleach or catalyze its decomposition.

Shipping

6 The finished sodium hypochlorite bleach is shipped to a bottling plant or bottled on-site. Household-strength bleach is typically 5.25% sodium hypochlorite in an aqueous solution.

Quality Control

In the bleach manufacturing facility, the final sodium hypochlorite solution is put through a series of filters to extract any leftover impurities. It is also tested to make certain that it contains exactly 5.25% sodium hypochlorite. Safety is a primary concern at manufacturing plants because of the presence of volatile chlorine gas. When the chlorine is manufactured outside the reactor facility, it travels in liquid form in specially designed railroad tank cars with double walls that will not rupture in the event of a derailment. On arrival at the plant, the liquid chlorine is pumped from the tank cars into holding vats. As a safety measure, the tank cars have shutoff valves that work in conjunction with a chlorine detection system. In the event of a chlorine leak, the detection system triggers a device on the tank that automatically stops the transmission of the liquid in 30 seconds.

Inside the facility, chlorine vats are housed in an enclosed area called a car barn. This enclosed room is equipped with air "scrubbers" to eliminate any escaped chlorine gas, which is harmful to humans and the environment. The vacuum-like scrubber inhales any chlorine gas from the enclosed area and injects it with caustic soda. This turns it into bleach, which is incorporated into the manufacturing process. Despite these precautions, safety and fire drills are scheduled regularly for plant personnel.

Special Considerations in Packaging

Household sodium hypochlorite bleach was introduced to Americans in 1909 and sold in steel containers, then in glass bottles. In the early 1960s, the introduction of the plastic jug brought a cheaper, lighter, and nonbreakable packaging alternative. It reduced transportation costs and protected the safety of workers involved in its shipping and handling. Additionally, the thick plastic did not permit ultraviolet light to reach the bleach, which improved its chemical stability and effectiveness. In recent years, however, plastic containers have become an environmental concern because of the time it takes the material to decompose in a landfill. Many companies that depend on plastic packaging, including bleach manufacturers, have begun to reduce the amount of plastic in their packaging or to use recycled plastics. In the early 1990s, Clorox introduced post-consumer resins (PCR) in its packaging. The newer bottles are a blend of virgin high-density polyethylene (HDPE) and 25% recycled plastic, primarily from clear milk jug-type bottles.

Consumer Safety

The bleach manufacturing industry came under fire during the 1970s when the public became concerned about the effects of household chemicals on personal health. Dioxin, a carcinogenic byproduct of chemical manufacturing, is often found in industrial products used to bleach paper and wood. In its final bottled form, common sodium hypochlorite bleach does not contain dioxins because chlorine must be in a gaseous state for dioxins to exist. However, chlorine gas can form when bleach comes into contact with acid, an ingredient in some toilet-bowl cleaners, and the labels on household bleach contain specific warnings against such combination.

In addition to the danger of dioxins, consumers have also been concerned about the toxicity of chlorine in sodium hypochlorite bleach. However, the laundry process deactivates the potentially toxic chlorine and causes the formation of salt water. After the rinse water enters the water system through the household drain, municipal water filtration plants remove the remaining traces of chlorine.

Where To Learn More

Periodicals

Ainsworth, Susan. "Resurgence in Demand Reviving Market for Sodium Chlorite." *Chemical & Engineering News*, March 22, 1993, pp. 11-12.

Grime, Keith and Allen Clauss. "Laundry Bleaches and Activators." *Chemistry and Industry*, October 15, 1990, pp. 647-49.

—Carol Brennan

The manufacturing section of this entry was written with the help of Clorox Company.

Bread

History

Wheat and barley were two of the earliest plants to be cultivated, and primitive people living as early as 5000 B.C. are known to have eaten these grains. Eventually it was discovered that adding water to the grain made it more palatable, and people experimented with cooking the grain and water mixture on stones that had been heated in a fire. In this manner, porridge and flat breads were developed.

The ancient Egyptians were known to grow barley and wheat. Excavations of their cities revealed that they enjoyed flat breads with nearly every meal. It is likely that leavened, or raised, bread was discovered accidentally when a wheat and water mixture was left in a warm place, causing the naturally occurring yeast to produce a puffed-up dough. It is also possible that a piece of leftover dough was mixed into a new batch, producing the same results.

Cooking the dough in an oven over an open fire produced an even better grade of bread. The first ovens were clay structures in which a wood fire was burned. When the wood had completely burned, the ashes were scooped out from an opening on the side of the oven. The wheat dough was placed inside the oven and then the opening was sealed. By the time the oven had cooled, the bread was baked.

The Romans are credited with inventing grinding methods by rubbing grain between two stones. Eventually, the manual grinding process was replaced by a mechanical one in which one stone revolved on top of a lower, perpendicular and stationary stone. In the beginning, the wheel stones were driven by cattle or slaves. Later, water mills or windmills provided the power.

Grinding was a time-consuming process and for centuries, leavened bread remained a pleasure reserved for the wealthy. White bread was an even rarer commodity. In fact, a family's social and economic status could be determined by the type of bread they ate. The poorest families ate the dark whole-grain bread. Ironically, nutritionists today favor whole-grain breads over those made with white flour.

Bread making remained primarily a home-based function well into the Middle Ages. About that time, some families, particularly those without ovens of their own, began to take their dough to small local bakeries to have the dough shaped and baked. As towns and villages sprang up throughout the countryside, bakeries flourished and home baking decreased significantly. These local bakeries had large brick ovens heated by wood or coal. The dough was moved in and out of the ovens with a long-handled wooden shovel called a "peel." Many small, independent bakeries still employ peel ovens although they have since been converted to use gas or oil fuel.

In the late 18th century, a Swiss miller invented a steel roller mechanism that simplified the grinding process and led to the mass production of white flour. Charles Fleischmann's development of an easy-to-use, dependable packaged yeast later further simplified the baking process. During the 20th century, scientific and technical innovations have made it possible for large bread factories to control the complex physical, chemical, and biological changes inherent in

It is likely that leavened, or raised, bread was discovered accidentally when a wheat and water mixture was left in a warm place, causing the naturally occurring yeast to produce a puffed-up dough.

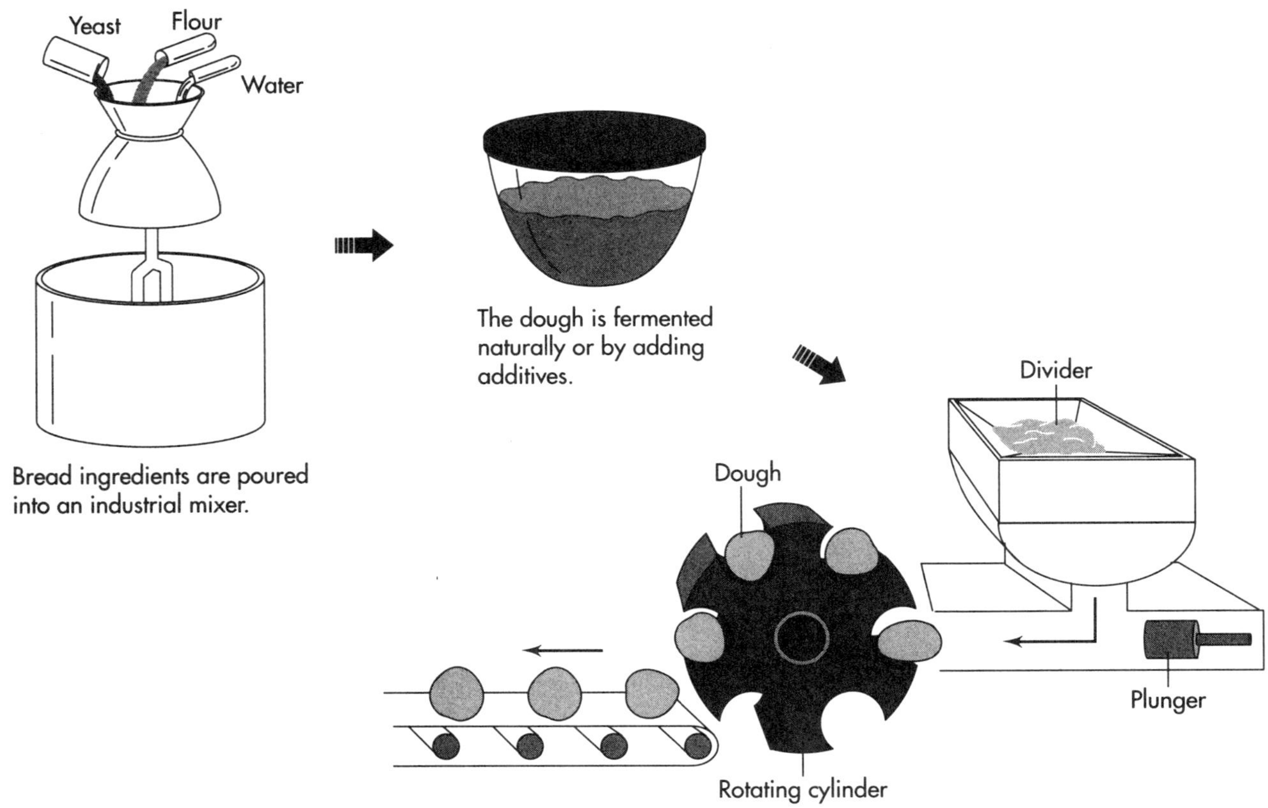

The dough is divided into pre-determined weights by a machine.

bread making. High-speed machinery can now accomplish the kneading and ripening processes in a matter of seconds.

For some time, bread was thought to be fattening, and many people avoided it in their daily diet. Studies showed, however, that it was toppings such as **butter** that accounted for most of the fat-induced calories. In fact, bread is an excellent source of low-fat, complex carbohydrates. The renewed interest in bread has led to consumers' taste for a variety of bread types. No longer is sliced white bread the norm. Grocery store shelves now offer myriad wheat breads and multigrain breads.

Raw Materials

Bread is made with three basic ingredients: grain, water, and bakers' yeast. The harvested grain is ground according to the type of bread being made. All grains are composed of three parts: bran (the hard outer layer), germ (the reproductive component), and endosperm (the soft inner core). All three parts are ground together to make whole wheat and rye breads. To make white flour, the bran and the germ must be removed. Since bran and germ contain much of the nutrients in grain, the white flour is often "enriched" with vitamins and minerals. Some white flour has also been fortified with fiber and calcium.

The grinding takes place at grain mills, which sell the grain to bakeries in bulk. The bakeries keep the grains in storage sacks until they are ready to be used. In the baking factory, water and yeast are mixed with the flour to make dough. Additional ingredients such as salt, fat, sugar, honey, raisins and nuts are also added in the factory.

The Manufacturing Process

Mixing and kneading the dough

1 The sifted flour is poured into an industrial mixer. Temperature-controlled water

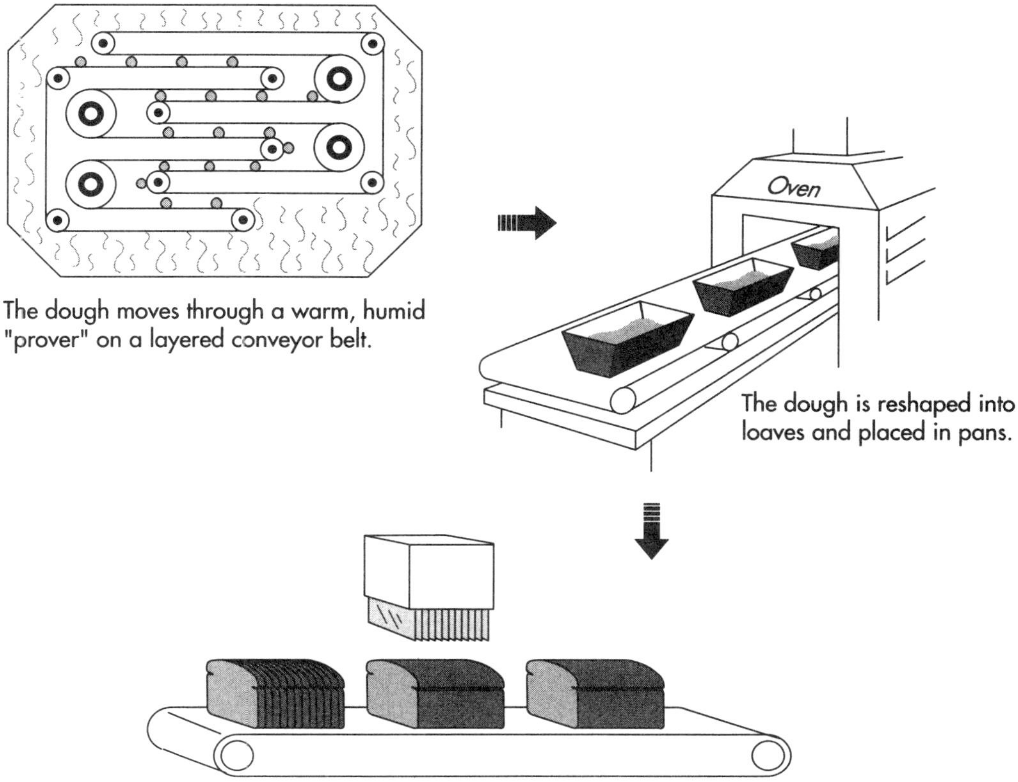

The dough moves through a warm, humid "prover" on a layered conveyor belt.

The dough is reshaped into loaves and placed in pans.

The bread cools as it moves along the conveyor belt. Vertical serrated blades slice the loaf.

is piped into the mixer. This mixture is called "gluten" and gives bread its elasticity. A pre-measured amount of yeast is added. Yeast is actually a tiny organism which feeds off the sugars in the grain, and emits carbon dioxide. The growth of the yeast produces gas bubbles, which leaven the bread. Depending on the type of bread to be made, other ingredients are also poured into the mixer. Modern mixers can process up to 2,000 pounds (908 kg) of dough per minute.

2 The mixer is essentially an enclosed drum that rotates at speeds between 35 to 75 revolutions per minute. Inside the drum, mechanical arms knead the dough to the desired consistency in a matter of seconds. Although modern bread production is highly computerized, the ability of the mixing staff to judge the elasticity and appearance of the dough is critical. Experienced personnel will be able to determine the consistency by the sound of the dough as it rolls around the mixer. The mixing process takes about 12 minutes.

Fermentation

3 Three methods are used to ferment the dough. In some plants, the high-speed machinery is designed to manipulate the dough at extreme speeds and with great force, which forces the yeast cells to rapidly multiply. Fermentation can also be induced by the addition of chemical additives such as l-cysteine (a naturally occurring amino acid) and vitamin C. Some breads are allowed to ferment naturally. In this instance, the dough is placed in covered metal bowls and stored in a temperature-controlled room until it rises.

Division and gas reproduction

4 After the dough has fermented, it is loaded into a divider with rotating blades that cut the dough into pre-deter-

mined weights. A conveyer belt then moves the pieces of dough to a molding machine. The molding machine shapes the dough into balls and drops them onto a layered conveyer belt that is enclosed in a warm, humid cabinet called a "prover." The dough moves slowly through the prover so that it may "rest," and so that the gas reproduction may progress.

Molding and baking

5 When the dough emerges from the prover, it is conveyed to a second molding machine which re-shapes the dough into loaves and drops them into pans. The pans travel to another prover that is set at a high temperature and with a high level of humidity. Here the dough regains the elasticity lost during fermentation and the resting period.

6 From the prover, the pans enter a tunnel oven. The temperature and speed are carefully calculated so that when the loaves emerge from the tunnel, they are completely baked and partially cooled. While inside the tunnel, the loaves are mechanically dumped from the pans onto shelves. The baking and cooling process lasts approximately 30 minutes.

Slicing and packaging

7 The bread continues to cool as it moves from the oven to the slicing machine. Here vertical serrated blades move up and down at great speeds, slicing the bread into consistently sized pieces.

8 Metal plates hold the slices together while picking up each loaf and passing it to the wrapping machine. Pre-printed plastic bags are mechanically slipped over each loaf. At some bakeries, workers close the bags with wire twists. Other plants seal the bags with heat.

Quality Control

Commercial bread making is held to strict government guidelines regarding food production. Further, consumer preferences compel bread producers to maintain a high quality standard of appearance, texture, and flavor. Therefore, quality checks are performed at each step of the production process. Producers employ a variety of taste tests, chemical analyses, and visual observation to ensure quality.

Moisture content is particularly critical. A ratio of 12 to 14% is ideal for the prevention of bacteria growth. However, freshly baked breads have a moisture content as high as 40%. Therefore it is imperative that the bakery plants be kept scrupulously clean. The use of fungicides and ultraviolet light are two popular practices.

Where To Learn More

Periodicals

"'Breadworks' Program Takes the Work Out of Breadmaking." *Bakery Production and Marketing*, December 1993, p. 27+.

Denny, Sharon. "Enough Dough for Everyone." *Current Health*, December 1994, p. 25+.

Harper, Roseanne. "Bred to Rise." *Supermarket News*, January 30, 1995, p. 21+.

Ryan, Nancy Ross. "Flour Power." *Restaurants and Institutions*, October 15, 1994, p. 137+.

—Mary F. McNulty

Bungee Cord

Background

Bungee cord is made of one or more strands of an elastic material, usually rubber, bound together by a fabric covering. It is commonly used as a tie-down for luggage or equipment carried on the outside of a vehicle. Bungee cord is also used by the military to absorb the opening shock of the large cargo parachutes when dropping heavy loads such as tanks. The development of long, heavy-duty bungee cord for the military has led to the recreational sport of bungee jumping. In this sport, the participant jumps from an elevated structure while wearing a harness attached to one end of a long bungee cord with the other end attached to the structure.

The term "bungee" or "bungie" is thought to be British slang for india-rubber. Some references to the india-rubber originally used for erasing pencil marks on paper call it "india-bungie." Another source claims the term was derived from the Anglo-Indian word "bangy" referring to the colloquial term for a yoke carried on the shoulder with two equal loads suspended by cords front and rear. In either case, the concepts of an elastic material and load-bearing cords both apply to the modern bungee cord.

The history of bungee jumping as a sport or test of courage is believed to date back 1500 years to Pentecost Island in what is now the Republic of Vanuatu in the South Pacific. According to local legend, a wife felt she was being mistreated by her husband and fled, taking refuge in a tall tree. As her husband was climbing the tree in pursuit, she secretly tied vines around her ankles. When he tried to grab her, she jumped. He jumped after her and fell to his death. The springy vines broke her fall and she lived. After that, the men of the island decided that no woman should ever trick them again, and they began to practice jumping with vines. In time, it became a test of courage, with the bravest men jumping from a height of 80 feet (24 m) to have their heads brush the ground before the vines completely stopped their fall.

Modern bungee jumping using elastic bungee cords started in the late 1970s. On April Fools' Day in 1979, the Oxford Dangerous Sports Club of Britain caught the public's attention when members bungeed off the 245-foot (75 m) Clifton Bridge in Bristol, England. Since that time, bungee jumping has become a commercialized sport with thousands of participants.

Although modern bungee jumping started in the late 1970s, the history of bungee jumping as a sport or test of courage is believed to date back 1500 years to Pentecost Island in what is now the Republic of Vanuatu in the South Pacific.

Raw Materials

The elastic material of a bungee cord is usually made of natural or synthetic rubber. Natural rubber, sometimes called latex rubber, has excellent extensibility (the ability to be extended), resilience (the ability to regain its original shape after being extended), and tensile strength (the ability to be extended under load without breaking). For these properties, it makes an excellent material for bungee cords. Natural rubber has the disadvantages of having only fair resistance to air and the ultraviolet radiation in sunlight. Synthetic rubbers, such as neoprene, have better resistance to air and sunlight, but less resilience and tensile strength than natural rubber. The military specification (mil-spec) for bungee cords allows either natural rubber or syn-

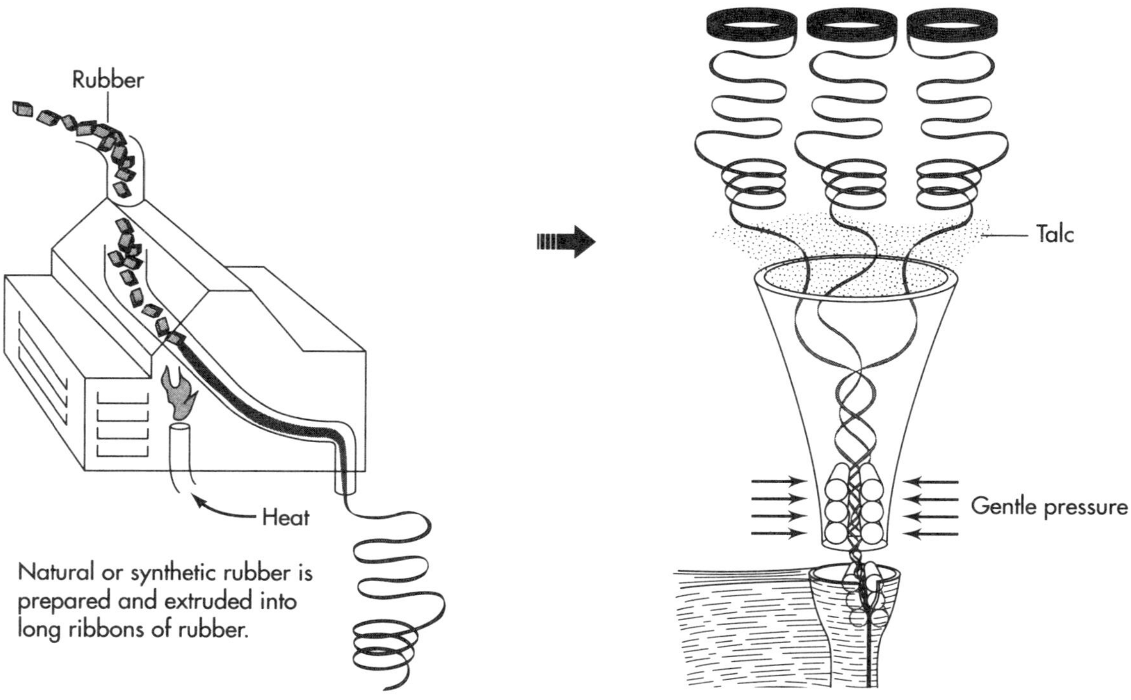

Natural or synthetic rubber is prepared and extruded into long ribbons of rubber.

Rubber strands are partially unwound from their coils and fed through a braider, which squeezes and stretches the strands together.

thetic rubber, or a mixture of both. Reclaimed rubber may not be used for mil-spec cords. Natural rubber is widely used for cords used in bungee jumping.

The fabric covering for the bungee cord may be braided from cotton or nylon yarn. Commercial bungee cord usually has a single layer of nylon covering which is more resistant to abrasion and has a higher tensile strength. Mil-spec bungee cord is required to have two layers of cotton covering. Some cords used for bungee jumping have cotton covering, the same as the mil-spec cord. Other specially designed bungee jumping cords have a braided covering of natural rubber.

Design

Standard commercial bungee cord is available in diameters from about 0.25-inch to 0.62-inch (0.64-1.6 cm). Mil-spec cord is available from 0.25-inch to 0.87-inch (0.64-2.2 cm) in diameter. These cords require no new design work other than an occasional change in the colors of the outer covering. (See Quality Control). Bungee jumping cord is usually about 0.62-inch (1.6 cm) in diameter and is usually individually designed by each bungee jumping company to its specifications.

Custom-made bungee cord may be required for special applications. When one light truck manufacturer decided to demonstrate the strength of their product's frame, they sent the vehicle bungee jumping off a bridge. This stunt was featured in a commercial on national television and required a set of nine specially designed bungee cords, each 5 inches (13 cm) in diameter and 100 feet (30 m) long.

The Manufacturing Process

Some bungee jumping companies design and manufacture their own cord. In general, this is a proprietary process which they keep secret. The following describes a typ-

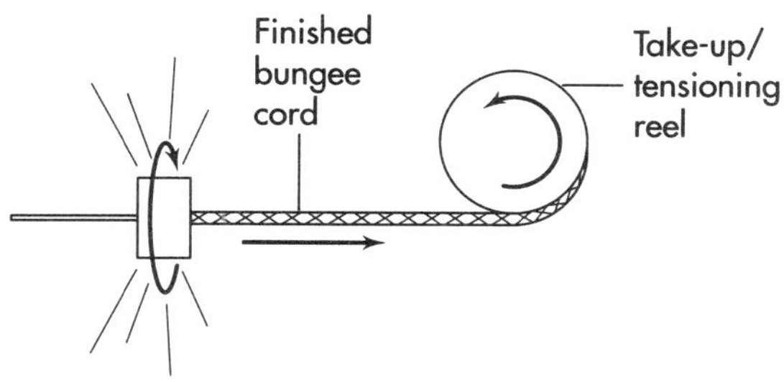

The braider also weaves yarn around the stretched rubber. The end fittings differ depending on how the cord will be used.

J-hook end fitting

ical process by which commercial or mil-spec bungee cords are manufactured.

Extruding the rubber ribbons

1 Natural or synthetic rubber is prepared and extruded into long ribbons of rubber. These ribbons are approximately 0.09 inch to 0.12 inch (0.24-0.32 cm) thick, 0.25 inch (0.64 cm) wide and up to 100 feet (30 m) long. The extruder consists of a heated cylinder into which the rubber is placed. One end of the cylinder moves under pressure to force the rubber out through a small hole in the other end. The hole, or die, is in the shape of the desired ribbon cross-section. When the rubber ribbons have cooled, they are coiled and shipped to the bungee cord manufacturer.

Preparing the rubber ribbons

2 The number of ribbons, or strands, in a bungee cord determines the diameter of the cord and the overall tensile strength rating. The appropriate number of rubber strands are partially unwound from their coils. To prevent the individual strands from sticking to each other in hot weather, they are coated with finely powdered talc or soapstone. This is done continuously as the strands are being unwound from the coils during the braiding process (steps 3 and 4 below).

Braiding the cover

3 The free ends of the rubber strands are brought together and manually fed through a machine called the braider. At the input end of the braider, they pass through a roller or other device which gently squeezes them into a bundle. At the other end, another roller or a take-up reel pulls the strands through the machine. By adjusting the pressures and speeds of the input and output devices, the rubber strands are placed under tension and slightly stretched while they pass through the machine. This reduces the diameter of the bundle of rubber strands to allow the fabric covering to be wound tightly.

4 The braider weaves the covering yarn around the stretched rubber bundle as it passes through the machine. The yarn must be woven in a tight enough pattern to prevent dirt from entering gaps between the yarn threads when the cord is fully extended. If two or more layers of covering are required, they are woven one after the other. The outer covering yarn may be colored and woven in a pattern for age dating or decorative purposes. (See Quality Control).

Shipping

5 The finished bungee cord is cut to length, coiled, and placed in cardboard boxes for shipping. Some cords are first placed in dark plastic bags as further protection against ultraviolet radiation during handling and storage.

Attaching the end fittings

6 The end fittings may be attached by someone other than the bungee cord manufacturer. In the case of commercial bungee cord used as luggage and equipment tie downs, another manufacturer or distributor cuts the bungee cord into the required lengths, doubles each end over and tightly coils a stiff wire around the ends with a wire coiler machine. The other end of this wire is usually bent into a j-shaped hook to allow the ends to be secured. Bungee cord used for bungee jumping may have the ends bent around a non-metallic eye and wrapped with a strong, waxed string, called whipping, that is wound tightly while the cord is stretched. Other end fittings may involve sewing the cord to fabric webbing.

Quality Control

Bungee cord is subjected to different levels of quality control depending on the final application. These usually consist of visual inspection, testing, and labeling or color coding.

Visual Inspection

All bungee cords are given a visual inspection for defects in the rubber strands and covering during manufacture. Defects include broken strands, improperly woven covering and noticeable stains on the covering.

Testing

Commercial bungee cord rarely requires any testing. Mil-spec cord, on the other hand, must undergo a rigorous series of tests, including size and weight measurements, tensile strength, percent elongation under various loads, and a number of extension-contraction flex cycles. For example, a 0.62-inch (1.6 cm) diameter mil-spec cord must weigh 14 pounds per 100 feet (or 6 kg per 30 m), support 250-350 pounds (113-159 kg) at 100% elongation, have a minimum breaking strength of 500 pounds (227 kg), and be able to endure a minimum of 50,000 flex cycles. Manufacturers of bungee cord for bungee jumping have their own standards which may include subjecting the cord to a number of full flex cycles and measuring the change in force versus extension to ensure the cord does not overextend.

Labeling and Color Coding

Bungee cord performance can be affected by age. Cord built to military specifications must be shipped to the end user within six months of the date it was manufactured. To ensure compliance with this requirement, mil-spec cord uses a color-coded outer covering to indicate the date of manufacture. The main color indicates the year as follows: red (1992), blue (1993), yellow (1994), black (1995), and green (1996). For succeeding years, the cycle of colors is repeated starting with red again for 1997. To further define the date of manufacture, a second, minor color is incorporated into the outer covering as follows: red (January-March), blue (April-June), green (July-September), and yellow (October-December).

Some bungee jumping companies which make their own cord use a different color-coding system to identify the load capacity of the cord rather than the age. This color is often sewn into the webbing attached to the harness end of the cord to ensure that the proper capacity cord is matched to the weight of the jumper. The age of the cord is controlled by periodic testing and regular replacement cycles.

The colors on the outer covering of commercial bungee cords have no significance and are for decorative purposes only.

The Future

Commercial bungee cord is a simple, low-cost product with numerous uses. It will continue to be used for the foreseeable future. Likewise, the military is expected to continue to use the current bungee cord design in numerous applications.

Bungee cord used for bungee jumping has evolved into a specialty product. Because of concerns with liability, many commercial manufacturers no longer manufacture or sell cord for bungee jumping. Companies that promote bungee jumping as a sport now manufacture their own cord to their own specifications. The specifications and manufacturing processes vary from one company to another and are considered highly proprietary. The primary concern, of course, is safety for the jumper. To that end, these companies will continue to have very tight controls over the manufacturing, handling, testing, and replacement process for their cords.

Where To Learn More

Book

Frase, Nancy. *Bungee Jumping for Fun and Profit.* ICS Books, 1992.

Periodicals

Gromer, Cliff. "Jumpin' Jimmy: To Prove the Strength of Its Frame, GMC Truck Goes Bungee Jumping." *Popular Science*, March 1993, pp. 56-57.

Sterba, James P. "Vanuatu Visitors See Origins of Bungee Are Sort of Darwinian." *Wall Street Journal*, June 1, 1993, p. A1.

—Chris Cavette

Business Jet

Accommodating 5 to 15 passengers, business jet aircraft—also known as bizjet—are chiefly used by business executives and government officials.

Background

Business jet aircraft (also known as "bizjets") typically carry 5-15 passengers and are primarily used as transportation by business executives and government officials.

In the early days of flight, before World War I, aircraft were constructed entirely of wood and canvas. They were shaped and joined by skilled craftsmen, many of whom were drawn from other woodworking trades. Every aircraft was unique, reflecting many different thoughts and constant design changes. The beginning of World War I brought a sudden demand for thousands of aircraft. This meant that factories had to accommodate large-scale manufacture and assembly of aircraft components by unskilled workers. Small companies grew into major manufacturers capable of producing many different types of aircraft in large numbers.

The techniques for building aircraft evolved gradually during the years between the wars. Wood and canvas changed to aluminum as the principal structural material while designs improved and records were set and broken. Monoplanes (single wing aircraft) were becoming more popular than biplanes (two wing aircraft). More powerful and reliable aircraft engines were continually being developed to increase payloads and ranges. Because of the increased reliability and improved comfort, aircraft became a more acceptable form of freight and passenger transport.

The aircraft industry had to gear up once again for the mass production of aircraft with the approach of World War II. A great many more aircraft were produced than during World War I by the leading powers, the U.S., Britain, Italy, Germany, and Japan. The aircraft industry had spread worldwide, and changed dramatically during the five years of conflict. Piston aircraft engines became larger and more complex and were produced in large quantities, while the jet engine was also being developed and tested. The development of radar and other sophisticated electronics had also taken place, eventually forming the large avionics (aeronautical electronic equipment) industries of today.

The corporate jet aircraft industry had its beginnings in the mid-1950s with the introduction of Rockwell's Sabreliner and Lockheed's JetStar models. LearJet entered the business jet market in the early 1960s with its Model 23, and was followed by Cessna in the late 1960s with its Citation 500 model. Today there are at least eight different U.S. and international aircraft companies marketing business jets.

Raw Materials

The principal material used in modern aircraft manufacturing is aluminum sheet, billet, and castings, but the use of composite materials is rapidly increasing. Composite materials are structural materials made up of two or more contrasting components, normally fine fibers or whiskers in a bonding resin. Composites such as carbon epoxies, graphite, **fiberglass**, carbon fiber reinforced plastics (CFRP), boron fiber reinforced plastics (BFRP), and glass reinforced plastics (GRP) enable manufacturers to build aircraft that are lighter and stronger than aluminum models. Steel alloys, titanium, stainless steel, and magnesium castings are also used, but in much smaller quantities.

The Manufacturing Process

There are six major subassemblies which make up an aircraft: 1) the fuselage or body, 2) the empennage or tail assembly, 3) the wings, 4) the landing gear assemblies, 5) the powerplant or jet engine, and 6) the flight control systems and instruments.

Just as in automotive manufacturing, the aircraft industry uses assembly lines for manufacturing. The production volume is much lower in aircraft, but the idea is the same. In aircraft manufacturing, a series of "positions" and "setbacks" are used to indicate the stage of the aircraft assembly. For example, if 16 positions are used to manufacture an aircraft, the 16th position would be the beginning of assembly, starting with either the nose section or wing spar buildups, and the 1st position would entail the installation of the engines and nacelle assemblies (the "nacelle" is the streamlined body which houses the engine). Position 0 indicates that the plane is "out the door" (OTD) and ready for pre-flight inspection and flight test. "Setbacks" indicate the stage a subassembly or "buildup" is within a position. For example, a wing assembly may only encompass one position, but within this position there may be three setbacks. Regardless of position or setback, assembly work is constantly ongoing. Even though one position may have more priority than others, other positions are simultaneously assembled so that both assemblies will be ready for mating at the proper time. The painting and work on the interior of the aircraft—adding seats and cabinets, for example—are done last as they can vary from aircraft to aircraft.

The production of an aircraft relies on the precise and accurate alignment and mating of each one of the major subassemblies. For subassembly production and assembly mating, a series of floor assembly jigs (FAJs) are used. These jigs hold, support, and locate the individual workpieces or subassemblies until they can be riveted, bonded, or bolted in place. Rigidity of the assembly jigs is critical to prevent misalignment, so most of these tools are large and heavy. Some of the jigs are permanently installed, while others are on rollers so they can be moved to the assembly line when needed.

Fuselage Assembly

The fuselage group is the first main assembly to be produced. The fuselage group consists of the nose structure assembly, forward cabin structure assembly, aft cabin structure assembly, and the tailcone assembly. The aircraft is essentially assembled from the back forward.

1 The first part of the fuselage to be assembled is the aft cabin barrel assembly (see illustration marked "Position 4, Setback 0"). The cabin barrel is assembled in the vertical direction in a floor assembly jig. The cabin barrel jig incorporates all the frame assemblies, skins, and supporting structures necessary to complete the aft cabin barrel assembly. Details and subassemblies are provided with tooling holes and locators set to contour templates which define the aircraft's *loft* or contour. Next, an aft cabin intermediate jig is used to assemble three primary components: the aft cabin barrel, the aft pressure bulkhead (which serves as the boundary of the pressurized section of the fuselage), and the wing attach fittings.

2 The nose and forward cabin structures are assembled next (see "Position 3, Setback 1"). The nose section jig assembles the forward frame wheelwell assembly, skin assemblies, and supporting structures. The forward cabin buildup jig assembles the windshield frame, cabin door frame, forward pressure bulkhead, supporting structure, and skins.

3 The forward and aft cabin sections are now mated using a cabin mate jig. Both cabin sections are located in the jig through the use of tooling holes which coordinate both the forward and aft pressure bulkheads (see "Position 3, Setback 0").

4 While the cabin sections are being built, the upper and lower tailcone sec-

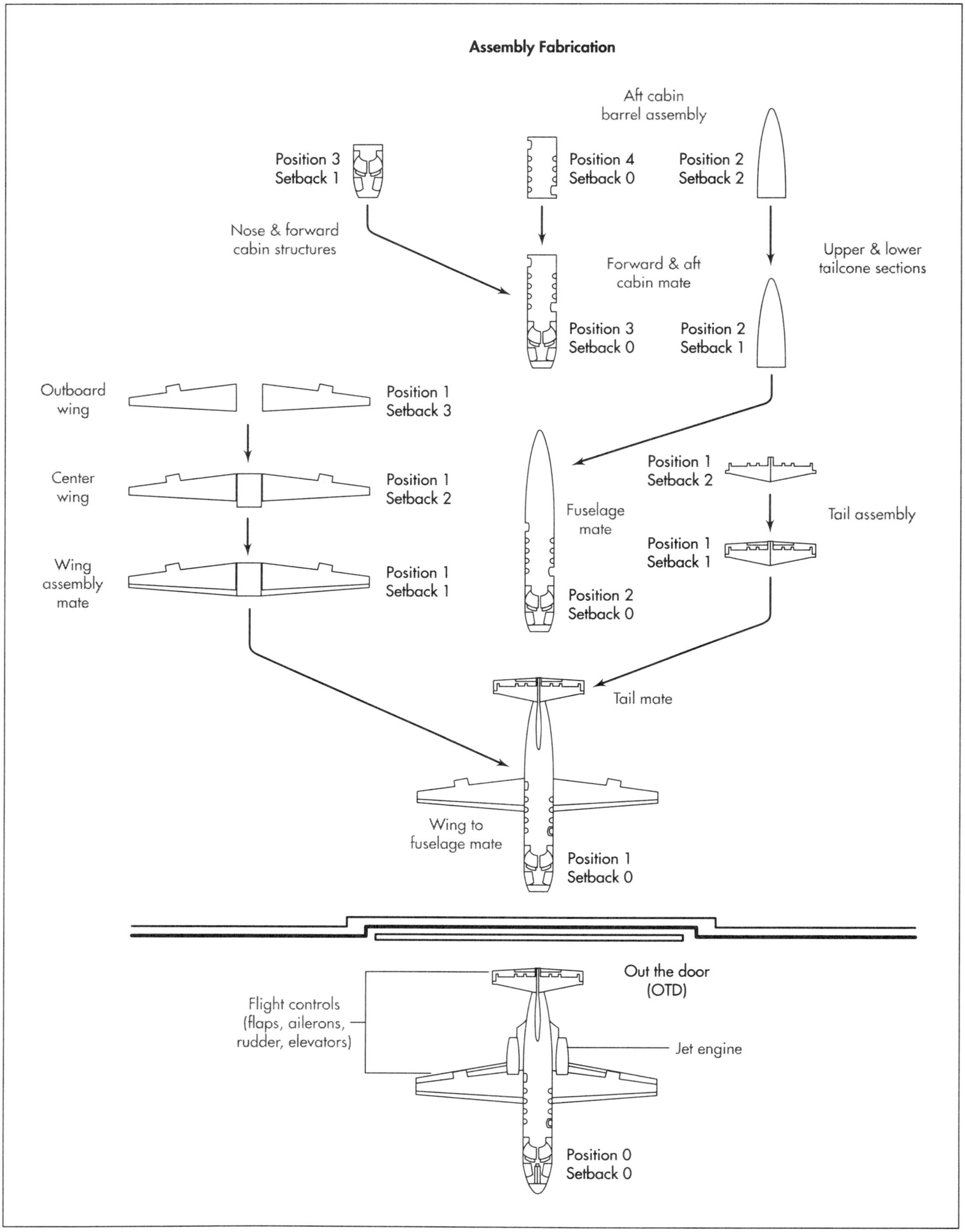
Assembly Fabrication
Aft cabin
barrel assembly
Position 3
Setback 1
Position 4
Setback 0
Position 2
Setback 2
Nose & forward
cabin structures
Forward & aft
cabin mate
Upper & lower
tailcone sections
Position 3
Setback 0
Position 2
Setback 1
Outboard
wing
Position 1
Setback 3
Center
wing
Position 1
Setback 2
Position 1
Setback 2
Fuselage
mate
Tail assembly
Position 1
Setback 1
Wing
assembly
mate
Position 1
Setback 1
Position 2
Setback 0
Tail mate
Wing to
fuselage mate
Position 1
Setback 0
Out the door
(OTD)
Flight controls
(flaps, ailerons,
rudder, elevators)
Jet engine
Position 0
Setback 0

tions are also being assembled. The tailcone mate jig is used to connect and align the upper and lower tailcone subassemblies (see "Position 2, Setback 2" and "Position 2, Setback 1").

5 The three primary fuselage sections, nose, forward and aft cabin assembly, and tailcone are located and assembled using a fuselage mate jig. The forward and aft cabin sections are loaded into the jig first, followed by the nose and tailcone sections. Engine mount brackets, forward and aft, are now installed onto the structural engine beams which extend out from the fuselage. Mounting holes are also aligned. These will be used to attach the vertical stabilizer to the tailcone and the aft canted bulkhead (the aft canted bulkhead "caps" off the end of the tailcone section). (See "Position 2, Setback 0").

Empennage or Tail Assembly

The empennage or tail assembly is the next section to be assembled. It consists of the vertical fin, rudder, horizontal stabilizer, and elevators. The rudder is the primary control surface for *yaw* or side to side movement usually used to turn the aircraft. Two elevators are mounted on the trailing edge of the horizontal stabilizer and are used to control the *pitch* or up and down motion of the aircraft.

1 The horizontal stabilizer frame buildup jig is used to assemble the leading edge and spar assemblies, along with the vertical attach fittings, stringers (aluminum extrusions which are used to provide structural support for sheet metal skins), skins, and supporting structures (see "Position 1, Setback 2").

2 Elevator frame buildup, trim tab assembly, and skinning jigs are used to assemble the right and left hand elevators. The trim tabs are movable control surfaces attached on the trailing edge of the elevators, used to hold the aircraft in level flight during cruise conditions (somewhat analogous to cruise control in a car). After the elevator frame and trim tabs are constructed, the skinning jig is then used to assemble the frame and trim tab assemblies along with the tip, leading, and trailing edge skins.

3 The vertical fin buildup jig is used to assemble the leading edge, spar, and bonded skin assemblies, along with the horizontal attach side plates and the supporting structure required to complete the vertical fin section. The fastener locations in the tailcone are established by the airframe alignment jig to ensure the vertical fin's relationship to the wing and engine attach points.

4 Rudder frame buildup, trim tab assembly, and skinning jigs are utilized in assembling the rudder assembly. After the rudder frame and trim tab is completed, the skinning jig is then used to assemble the frame and trim tab assemblies, along with the leading and trailing edge skins.

5 The empennage section of the aircraft is completed after the elevators, horizontal stabilizer, vertical stabilizer, and rudder are assembled (the rudder is usually installed last along with the flight control systems). (See "Position 1, Setback 1"). The empennage section is then mated to the aircraft tailcone section (see "Position 1, Setback 0").

Wing Assembly

The wing assembly is next and typically consists of the center wing section, outboard wing sections, and aileron and flap assemblies. The ailerons are movable control surfaces, usually hinged to the outer wing, which help provide control in *roll* about the longitudinal axis of the plane. The flaps are movable control surfaces, mounted inboard on the wing, which are able to hinge downward. These increase low-speed lift and add drag, allowing the aircraft to make steep approach landings without gaining excess airspeed.

1 Aileron frame buildup and skin and rivet jigs are used to assemble the left and right hand aileron assemblies. After the aileron frame is completed, the skin and rivet jig is used to load the aileron frame, skin and doublers (used for extra strength), then rivet the assembly complete. The aileron frame is located by pinning the hinge bearings and the inboard and outboard rib webs (the ribs are primary structural members running across the aileron).

The ailerons are usually installed last, along with the flight control instruments and flaps.

2 Flap frame buildup and skin jigs are used in constructing the left and right hand flap assemblies. The flap frame is completed first. Then the flap skin jig assembles the bonded upper skin and trailing edge skin, flap spar section, leading edge assembly and end ribs and interconnect clevises.

3 The building of the outboard wing section involves the use of many different jigs for drilling, riveting, and buildup. The main tool used is the outboard wing buildup jig, which assembles the forward outboard wing assembly, rear spar assembly, trailing edge bonded skin assemblies, and the supporting structure (see "Position 1, Setback 3").

4 The construction of the center wing section also requires the use of many different buildup jigs. The primary tool used here is the center wing buildup jig, which assembles the center section subassembly, wheelwell structure, rib and skin assemblies, and the supporting structure (see "Position 1, Setback 2").

5 The wing assembly mate jig assembles both the left and right outboard wings with the center wing. The wing sections and center section are located in the jig by locators and contour boards. The center section is loaded first, followed by the left and right outboard wings (see "Position 1, Setback 1"). The completed wing assembly is then mated to the fuselage section (see "Position 1, Setback 0").

Landing Gear Assembly

There are two different landing gear assemblies: the nose and main landing gears. Both use retraction systems which are electrically controlled and hydraulically actuated. The main landing gear (MLG) is usually a trailing link type, and retracts inboard into the wing. The nose landing gear (NLG) retracts forward into the fuselage nose section, and is enclosed by doors. The landing gears are assembled away from the main assembly line and are brought to the line when needed, usually when the fuselage and wings are being mated (see "Position 1, Setback 0").

Powerplant-Jet Engine

A business jet is typically powered by two turbofan jet engines located on each side of the rear fuselage in nacelle assemblies. The nacelle assemblies consist of an inlet section, a cowl or outer housing, an exhaust nozzle section, and a bleed air system, which diverts hot air to the wing and nacelle leading edges for deicing. Bleed air is also used for cabin heating and pressurization. The large sheet metal panels which form the cowl are typically roll formed. Some of the other sheet metal parts, such as the nose cap on the nacelle inlet section, are formed using a female die in a draw press. Nacelle assemblies are built separately away from the line and then brought back for installation (see "Position 1, Setback 0").

Flight Control Systems

The flight control systems are usually installed last, along with the ailerons, flaps, and rudder. There are many different flight control systems which go into a modern aircraft. The following is a partial list of the major systems: aileron control system; aileron trim system; speedbrake system; flap interconnect system; rudder control system; rudder trim control system; elevator control system; elevator trim control system; pressurization system; windshield anti-ice system; wing anti-ice system; oxygen system; pitot static system. (See "Position 1, Setback 0").

Out the Door

Before the aircraft leaves the factory, all electrical and mechanical systems undergo a functional test. Examples of items checked are fuel calibration, hydraulic systems, gear blow down and lock, warning lights and horns, and avionics. After the engines and flight control systems are installed, the aircraft is ready to go out the door for engine testing and flight test. The aircraft is put through numerous performance and systems tests before it is approved for delivery to the customer. Before delivery, the aircraft is sent to be painted, after which the interior is finished. (See "Position 0, Setback 0").

Quality Control

The quality of aircraft depends on good design, documentation, and electronic record keeping to meet Federal Aviation Administration (FAA) regulations and certification requirements. The windshields, wing leading edges, engines and other critical components must meet the FAR 25 (Federal Aviation Regulation) bird strike requirements before the aircraft is certified for commercial use. Many different forms and checklists are used throughout the manufacturing process to detail the history of each part made. Various laboratory tests and standardized aerospace material specifications have been developed specially for aircraft. To check how well bonded panels have adhered, they are placed in a water tank for ultrasonic testing. Stress testing is used extensively. A section of the aircraft is assembled and then placed in a test fixture which simulates actual use under varying conditions. Some of the tests are run until the parts fail, to see if the design safety factor is acceptable.

Byproducts/Waste

Environmental protection laws have developed stringent codes limiting water flows and emissions from aircraft manufacturing facilities. In compliance with federal laws, aircraft companies have been using fewer solvents and looking for better ways to clean parts, such as steam vapor degreasing systems. Aluminum chips and scrap material are the major byproducts of the aircraft industry, and are recycled.

The Future

Technological change is a major driving force in the evolution of aircraft manufacturing. Many developments underway involve computerized controls and automation designed to improve economy and quality and lower energy consumption and pollution. More assembly operations, such as riveting, may become completely automated. "Smart" sensors—sensors with predictive abilities involving fuzzy logic and artificial intelligence—are becoming more prevalent. Artificial intelligence or "fuzzy controls" enable the sensors to predict changes needed in the settings due to changes in load or production volume. In addition to these developments, increasing economic and environmental needs will bring further technical refinements to aircraft manufacturing.

Where To Learn More

Books

Bright, Charles D. *The Jet Makers.* The Regents Press of Kansas, 1978.

Gunston, Bill. *Jane's Aerospace Dictionary*, 3rd edition. Jane's Information Group Ltd, 1988.

Phillips, Almarin, A. Paul Phillips, Thomas R. Phillips. *Biz Jets: Technology and Market Structure in the Corporate Jet Aircraft Industry.* Kluwer Academic Publishers, 1994.

Porter, Donald J. *The Cessna Citations.* TAB Books, 1993.

Rowe, Frank Joseph and Craig Miner. *Borne on the South Wind: A Century of Aviation in Kansas.* The Wichita Eagle and Beacon Publishing Co., 1994.

Todd, Daniel. *The World Aircraft Industry.* Auburn House Publishing Company, 1986.

Winant, John H. *Keep Business Flying: A History of the National Business Aircraft Association, Inc. 1946-1986.* The National Business Aircraft Association, Inc., 1989.

—Glenn G. Whiteside

Butter and Margarine

Butter is made of butterfat, water, air, and sometimes salt. Margarine is an inexpensive alternate to butter, made from oil or a combination of oils through the process of hydrogenation.

Background

Butter is a soft, yellow-hued, edible emulsion of butterfat, water, air, and sometimes salt. It is made from the churning of cream and is used as a spread as well as an important ingredient in cooking and baking. Margarine is an inexpensive alternate to butter, made from oil or a combination of oils through the process of hydrogenation. Many people prefer margarine over butter because it is lower in fat and cholesterol than butter.

Butter

References to butter date back to as early as the ninth century B.C. in India, but its "invention" is credited to the nomadic tribes of Asia around 3500 B.C., although the first batch probably came about by accident. It is assumed that when the people of these tribes strapped bags containing milk onto their persons or saddles as nourishment for a journey, the resulting motion of the ride churned the milk. If the weather was cold enough, a bit of fat rose to the top of the bag and the result came to be butter. On the other hand, if the weather was too warm, the result was what came to be cheese. The use of butter eventually spread westward when these Asian peoples invaded the lands of the Near East and Europe.

Eventually, butter began to be manufactured in vertical churns by other developed civilizations. A picture of a primitive churn is evident in a Sumerian bas-relief dating from around 3500 B.C. One of the first written references to the substance comes from India in the form of a ninth-century etiquette manual. It suggested that Hindu brides be given milk, honey, and butter on the day of their wedding. The manuscript also mentioned greasing the wheel of the bridal carriage with butter to insure a trouble-free marriage. Because the cow is regarded as a sacred animal in Hindu religion, butter has long played an important role in Indian cuisine, and is mentioned specifically in religious tracts. In neighboring Tibet, butter made from the milk of yak was sometimes smeared on religious statues.

Soon butter became common to cultures that relied on the domesticated cow for sustenance, but it could also be made from the milk of sheep and goats. Although Greeks and Romans were not fond of rich, dairy-based foods, the word "butter" is derived from the Greek term *buturon*, meaning "cow's cheese." The term later found its way into Latin as *butyrum*. Celts and later the Vikings eventually became devotees of the substance, and by the late Middle Ages it was a staple in the diets of many Europeans and a valued trading commodity. The dairy product has also been considered a mystical salve during certain periods of history. For centuries, the people of Brittany placed butter near a person suffering from cancer to absorb the disease.

The first printed instructions for making butter can be traced back to a 14th-century Venetian cookbook. By the 17th century, butter was traded on routes that included England, Brittany, Flanders, and Iceland. The butter produced in Vanves, France, was thought to be the most exquisite during this era. In the southern regions of Europe, where olive oil remained the predominant cooking oil, some people believed butter

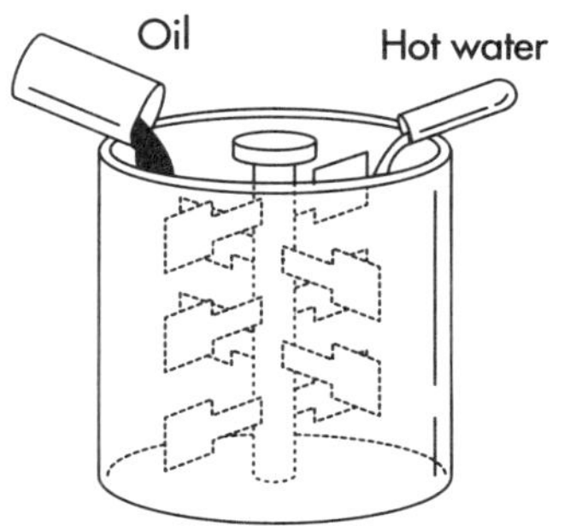

The oil is first "washed" by mixing with hot water, separating it, and drying it under a vacuum.

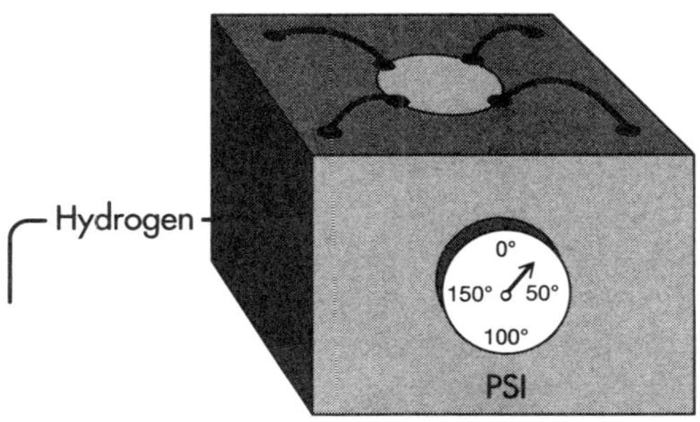

The oil is hydrogenated to ensure correct consistency. Hydrogen gas is added to the oil under pressurized conditions.

These illustrations show the commercial manufacture of margarine. Guidelines for margarine production dictate that margarine contain at least 80% fat. The oils used in the production can be derived from a variety of animal and vegetable sources. Its aqueous content may be milk, water, or a soy-based protein fluid.

caused leprosy. The dairy product eventually became a prohibited item for fast days as decreed by the Roman Catholic Church, although a dispensation could be purchased for those who simply could not go without it. In Rouen, France, legend has it that a "butter tower" was financed solely by such dispensations granted for eating butter on fast days.

Butter manufacturing in the U.S. dates from the time of the first colonists who brought cows with them to North America. Families who owned their own cows usually made butter themselves. Milk from the cow was left alone until its cream rose to the top. The cream was then skimmed off and left to cool, and the temperature was raised to about 70°F (21°C) a day before the butter-making process was to begin. Heating the cream ripened it, and it was then cooled several degrees. Next, the cream was placed in a wooden device, sometimes barrel-shaped or otherwise cylindrical, and mixed with the help of a paddle. This process generally took at least 30 minutes. The leftover cream in the churn was buttermilk. If the cows were eating grass, the butter possessed a yellowish cast, but during winter, when they were getting other types of feed, it was white. The butter was then rinsed with cool water, "worked over" a bit more, then salted for taste.

Margarine

Margarine is similar in taste and appearance to butter but possesses several distinct differences. Margarine was developed as a substitute for butter. By the 19th century, butter had become a common staple in the diet of people who lived off the land, but was expensive for those who did not. Louis Napoleon III, a socialist-minded emperor of mid-century France, offered a reward to anyone who could produce an acceptable,

The continuous-flow process is the most commonly used method in the manufacture of margarine. If milk is used as the liquid base, it is joined with salt and an emulsifying agent in a chamber. An emulsifier works by decreasing the surface tension between the oil globules and the liquid mixture, thereby helping them form chemical bonds more easily. The result is a substance that is neither wholly liquid nor wholly solid.

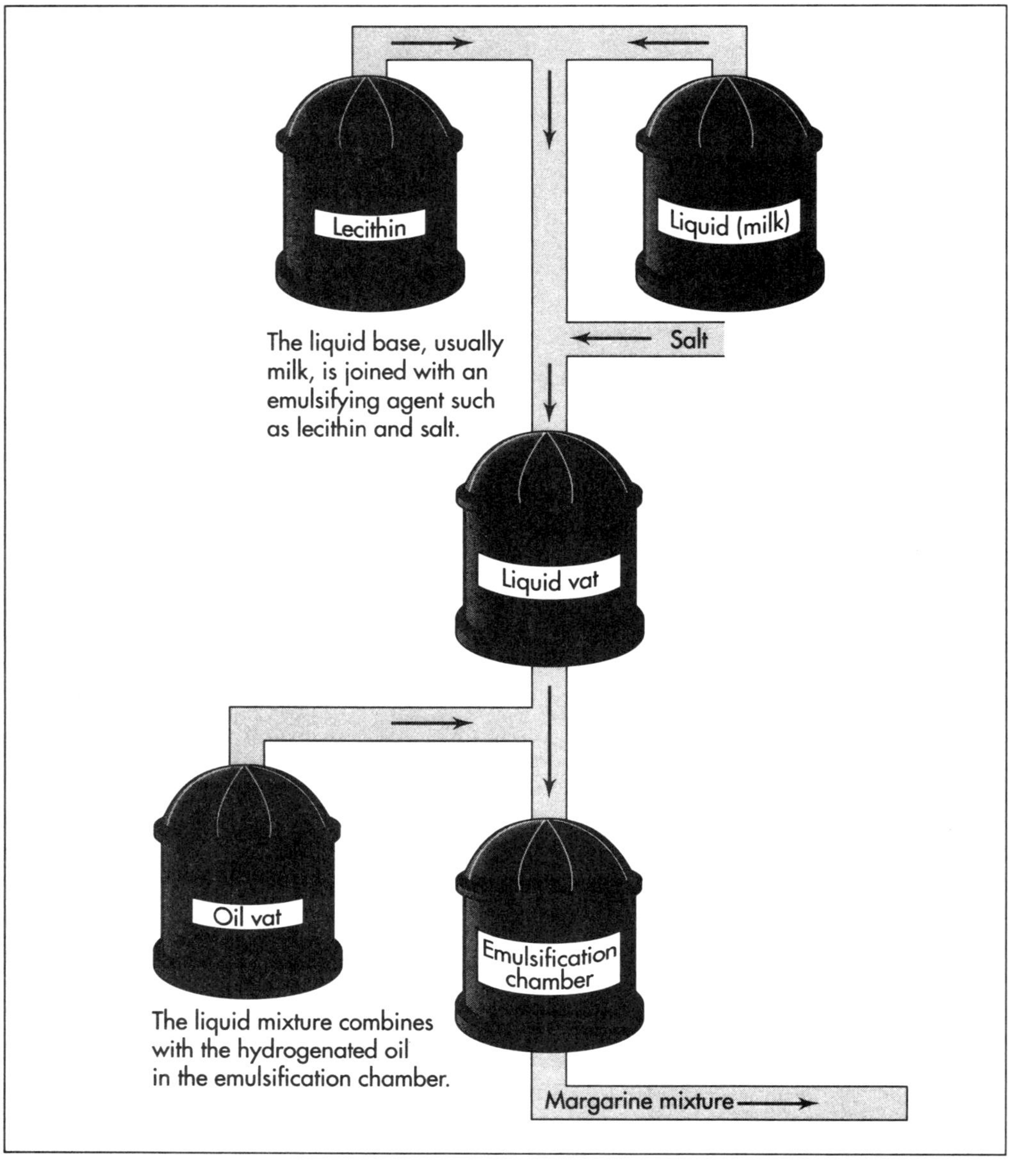

affordable alternative. Hippolyte Mege-Mouriez won the 1869 competition for the item he named margarine after its primary ingredient, margaric acid. The margaric acid had only recently been discovered in 1813 by Michael Eugene Chevreul and derived its name from the Greek term for pearls, *margarite*, because of the milky drops that Chevreul noticed in his invention. In modern times it is manufactured from an oil or combination of oils through the process of hydrogenation, a method perfected around 1910. This process helps animal or vegetable oils emulsify, or turn from a liquid substance into a fatty one of a semi-solid state.

In the U.S., butter was the preferred taste for many years, and until relatively recent times, margarine suffered from a poor brand image. A well-organized dairy cartel campaigned against margarine, fearing competition from the margarine industry. At about 1950, Congress repealed taxes on butter substitutes which had been in effect for several decades. The so-called "Margarine Act" was also heralded for at last defining margarine: "all substances, mixtures and compounds which have a consistency similar to that of butter and which contain any edible fats and oils other than milk fat if made in imitation or semblance of butter." Part of margarine's acceptance into the diets of Europeans and Americans came from rationing during times of war. Butter was scarce, and margarine, or oleo, was the best substitute. Today, margarine

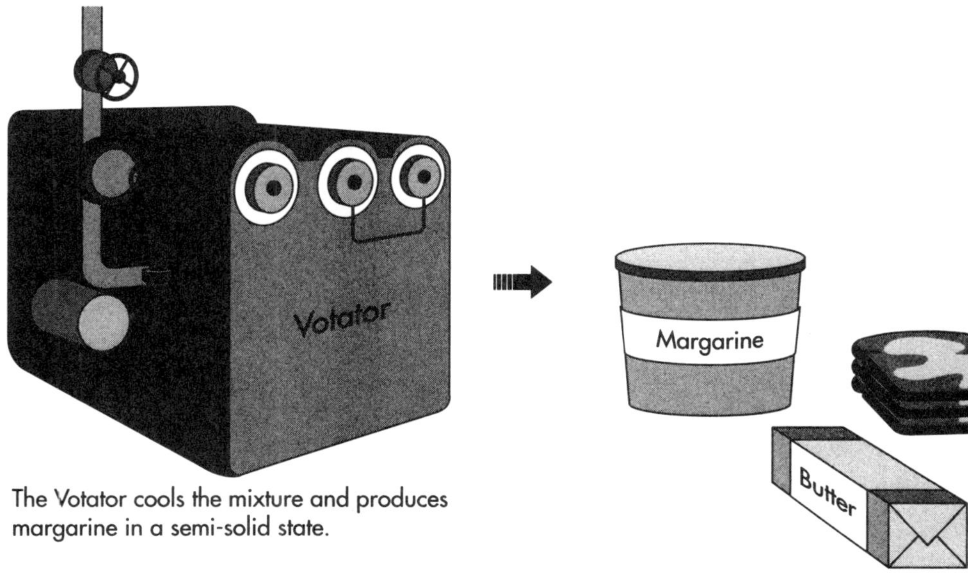

The Votator cools the mixture and produces margarine in a semi-solid state.

has become a nearly interchangeable substitute for butter and provides less fat and cholesterol than butter at a lower cost.

Raw Materials

Butter is made from dairy milk and salt. Margarine, on the other hand, can be made from a variety of substances. The first of these is any edible animal or vegetable oil such as corn oil or sunflower oil. Its liquid component can be made from milk, water, or sometimes a liquid protein mixture derived from soybean.

Butter Manufacture

Preparation

1 For many years the major creameries for butter manufacturing were located in the states of the Eastern seaboard, but the flourishing of a more industrialized agriculture in the Midwest led to the predominance of butter-making facilities there. The modern butter-making process begins when fresh cow's milk from dairy farms is brought into the facility. The product is inspected, classified into different groups according to its adjudged quality, and then filtered to remove impurities. Then the milk is separated by means of centrifugal force. It is pumped into a large, cylindrical, vertical rotator device. When turned on, this rotator spins the liquid until the cream rises to the top. The cream is then fed into large stainless steel vats and heated to 180°F (82°C) for about 30 minutes in the pasteurization process to remove any lingering bacteria. The pasteurized cream is then left to cool.

Churning

2 The cream is placed in a large, mechanical churn usually made of aluminum. Some of these industrial-sized churns can make 1,500-5,000 pounds (681-2270 kg) of butter at a time. When the churn is activated, it tumbles the cream, much like the motion of a clothes dryer, while a worker watches the process through a small glass window on the churn. After about 45 minutes, small granules of butter begin to form, and the butter and buttermilk are separated. Salt is added, and the mixture is churned further. When this process is completed, a stainless steel mobile device sometimes called a "boat" is placed adjacent to the opening of the mechanical churn. The door of the churn is opened, and the butter begins to spill out into the boat; activating the churn removes the rest. It is then wrapped into 64-pound (29 kg) cartons and sent to the distributor. There, the butter is repackaged for consumer and food-service industry use.

Since the 1930s, the Votator has been the most commonly used apparatus in U.S. margarine manufacturing. In the Votator, the margarine emulsion is cooled and occasionally agitated to form semi-solid margarine.

Margarine Manufacture

Margarine can be made from a variety of animal fats and was once predominantly manufactured from beef fat and called oleomargarine. Unlike butter, it can be packaged into a variety of consistencies, including liquid. No matter what the form, however, margarine must meet strict government content standards because it is a food item which government analysts and nutritionists consider to be easily confused with butter. These guidelines dictate that margarine be at least 80% fat, derived from animal or vegetable oils, or sometimes a blend of the two. Around 17-18.5% of the margarine is liquid, derived from either pasteurized skim milk, water, or soybean protein fluid. A slight percentage (1-3%) is salt added for flavor, but in the interest of dietary health some margarine is made and labeled salt-free. It must contain at least 15,000 units (from the U.S. Pharmacopeia standards) of vitamin A per pound. Other ingredients may be added to preserve shelf life.

Preparation

1 When the ingredients arrive at the margarine manufacturing facility, they must first undergo a series of preparatory measures. The oil—safflower, corn, or soybean, among other types—is treated with a caustic soda solution to remove unnecessary components known as free fatty acids. The oil is then washed by mixing it with hot water, separating it, and leaving it to dry under a vacuum. Next, the oil is sometimes bleached with a mixture of bleaching earth and charcoal in another vacuum chamber. The bleaching earth and charcoal absorb any unwanted colorants, and are then filtered out from the oil. Whatever liquid is used in the manufacturing process—milk, water, or a soy-based substance—it too must undergo preparatory measures. It also undergoes pasteurization to remove impurities, and if dry milk powder is used, it must be checked for bacteria and other contaminants.

Hydrogenation

2 The oil is then hydrogenated to ensure the correct consistency for margarine production, a state referred to as "plastic" or semi-solid. In this process, hydrogen gas is added to the oil under pressurized conditions. The hydrogen particles stay with the oil, helping to increase the temperature point at which it will melt and to make the oil less susceptible to contamination through oxidation.

Combining the ingredients

The continuous-flow process is the most commonly used method in the manufacture of margarine. If milk is used as the liquid base, it is joined with salt and an emulsifying agent in a chamber. The emulsifying agent ensures that the emulsification process—chemically defined as a suspension of small globules of one liquid in a second liquid—takes place. An emulsifier works by decreasing the surface tension between the oil globules and the liquid mixture, thereby helping them form chemical bonds more easily. The result is a substance that is neither wholly liquid nor wholly solid but rather a combination of the two called semi-solid. Lecithin, a natural fat derived from egg yolk, soybean, or corn, is one typical emulsification agent used in margarine manufacturing.

3 In the initial step, the liquid, salt, and lecithin are mixed together into one tank opposite another vat holding the oils and oil-soluble ingredients. In the continuous-flow process, the contents of the two vats are fed on a timed basis into a third tank, typically called the emulsification chamber. While the blending process is taking place, the equipment's sensors and regulating devices keep the mixture's temperature near 100°F (38°C).

Agitation

4 Next, the margarine mixture is sent to a device called a Votator, the brand name for the most commonly used apparatus in U.S. margarine manufacturing. It has been standard equipment to the industry since the 1930s. In the Votator, the margarine emulsion is cooled in what is referred to as Chamber A. Chamber A is divided into a trio of tubes that successively decrease its temperature. Within two minutes the mixture has reached 45-50°F (7-10°C). It is then pumped into a second vat called Chamber B. There it is occasionally agitated but generally left to sit still and form its semi-solid

state. If it needs to be whipped or otherwise prepared for special consistency, the agitation is done in Chamber B.

Quality Control

Quality control is an obvious concern at modern food-processing facilities. Unclean equipment and shoddy methodology can lead to a mass bacterial contamination that could disrupt the stomachs and even lives of thousands of consumers within a matter of days. The U.S. government, under the auspices of the Department of Agriculture, maintain specific industrial hygiene codes for modern creameries and margarine manufacturing plants. Inspections and fines for poorly maintained equipment or unclean conditions help keep companies in compliance.

Butter is graded by USDA inspectors at the creamery. They inspect each batch, test it, taste it, and assign a score to it. They give a maximum of 45 points for flavor, 25 for body and texture, 15 points for color, 10 for salt content, and 5 for packaging. Thus, a perfect batch of butter can receive a score of 100 points, but usually the highest number assigned to a package is 93. At 93, butter is classified and labeled Grade AA; a batch that receives a score below 90 is considered inferior.

Guidelines for margarine production dictate that margarine contain at least 80% fat. The oils used in the production can be derived from a variety of animal and vegetable sources but all must be fit for human consumption. Its aqueous content may be milk, water, or a soy-based protein fluid. It must be pasteurized and contain at least 15,000 units of vitamin A. It may also contain a salt substitute, sweeteners, fatty emulsifiers, preservatives, vitamin D, and coloring agents.

Where To Learn More

Book

Riepma, S. F. *The Story of Margarine*. Public Affairs Press, 1970.

Periodicals

"New Generation of Foods with Reduced Fat." *Food Engineering*, January 1990, p. 23.

"Breakthrough in Fats and Oils Technology Boosts Product Quality and Process Flexibility." *Food Engineering*, May 1994, pp. 63-64.

Kleyn, Dick H. "Textural Aspects of Butter." *Food Technology*, January 1992, pp. 118-21.

—Carol Brennan

Button

The earliest buttons date to prehistoric times, and in spite of millennia of change in fashion and manufacturing techniques, the button has endured as the most common fabric fastener.

Background

The earliest buttons date to prehistoric times, and in spite of millennia of change in fashion and manufacturing techniques, the button has endured as the most common fabric fastener. Though buttons were used for thousands of years, the buttonhole was not invented until sometime in the 13th century. The buttonhole is thought to have been brought to Europe from the Middle East by knights returning from the Crusades, and its advent led to a surge in button use. Buttons became a staple of men's fashion in the Renaissance, when jackets often featured rows of buttons from chin to waist, sleeves were tightly buttoned from elbow to wrist, and trousers too sported buttons at the waist, knee, or thigh. Guilds of buttonmakers were in existence in Paris in the 13th century, where buttons were made out of a variety of materials including wood, bone, brass, pewter, gold, and silver.

By the 18th century, the button industry flourished all across Europe, and artisans developed many different techniques for making them. The court of Louis XIV of France set the fashion for intricate buttons of precious metals and jewels and fabric buttons of embroidered cloth. English manufacturers invented steel buttons, and glass or glass and metal buttons were popular in France. Many artists famous in other trades also lent their skills to the button industry. The French painter Antoine Watteau made buttons, and some of the leading names in fine china such as Wedgwood, Limoges, and Staffordshire are also associated with fine buttons.

By the late 18th century, buttons began to be made in factories. Metal buttons were punched out by dies, and die-makers were prohibited from emigrating from England, so that they would not take their trade secrets abroad. Nevertheless, the technology spread, and buttons began to be mass-produced in metal, glass, and other materials. Extravagant buttons were still popular elements of 19th-century fashion. Die-makers turned out complex designs using scenes from plays, novels, and nursery rhymes, and Wagner operas and the operettas of Gilbert and Sullivan were routinely commemorated with buttons depicting scenes and characters.

By the early 20th century, the prevailing style was much simpler, reflecting the more sedate look of the growing white-collar class. Inexpensive matched shirt buttons for men and women were available in five-and-dime stores around 1910. Plastic buttons became widely available in the 1930s, though most typical shirt buttons were still made of sea shells or other natural materials. World War II brought many advances in plastic technology. Acrylic buttons were actually made from material left over from the manufacture of bomber gun turrets. The button industry converted almost entirely to plastic after the war. Plastic buttons could be made by a variety of methods. They could be mold cast, where plastic slugs cut from a long rod are placed in a two-part mold. The mold is closed, and heat and pressure applied to finish the button. Another process is injection molding. In this method, melted plastic is forced into a mold with a button-shaped cavity. Outlined below is the most common process for mak-

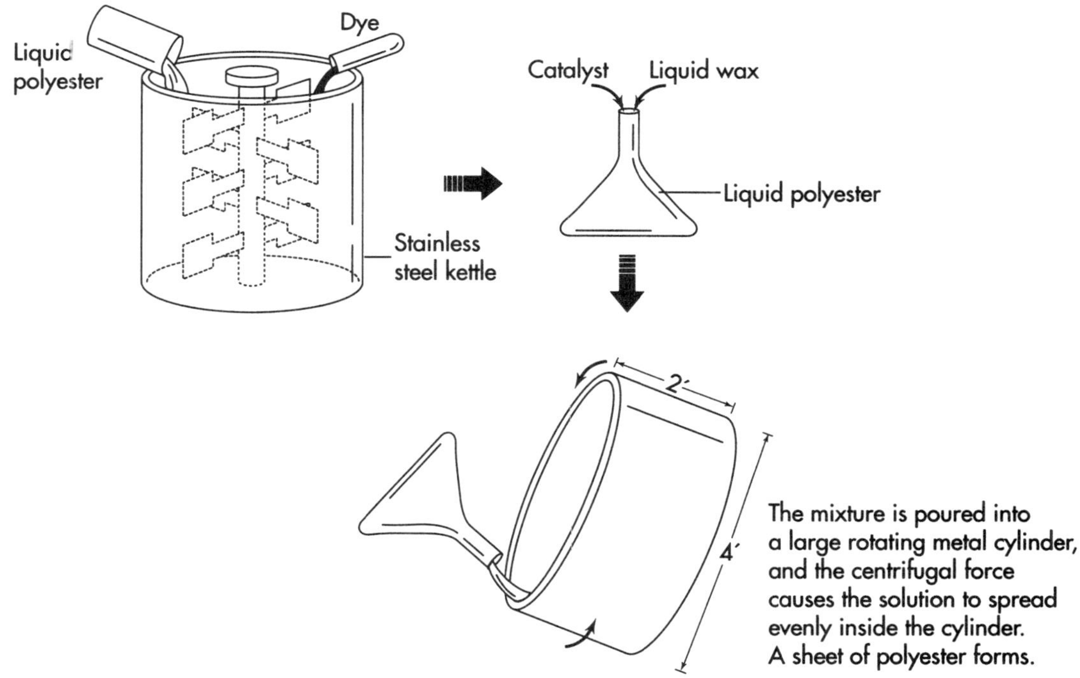

ing plastic buttons: die cutting from cylinder-cast **polyester**.

Raw Materials

Buttons are still made from natural products, but these require more work by hand than do plastic buttons, and some formerly common button materials are no longer widely available. For instance whale ivory, elephant ivory, or tortoiseshell buttons cannot be made in the U.S. because of laws enacted to protect endangered animals. Horn buttons are made from cow and buffalo hooves and horns, but button aficionados claim that modern horn is of poor quality and color because the animals graze on low-quality grass. Antique horn buttons are often streaked and come in a variety of colors, whereas modern horn is a duller light or dark brown. Horn buttons are still an element of the best quality men's fashion, but they cost as much as a dollar a piece, compared to the half a cent price of a standard button. Mother-of-pearl buttons, derived from sea shells, are still prized for their luster. But after World War II, the divers in the South Pacific islands who provided most of the mother-of-pearls began to charge much more for their dangerous labor, and the price of the material rose drastically. Glass buttons, which were widely imported from Germany in the middle of this century, are now much less common as well. The glass buttons were factory made, but they required a lot of hand work under unpleasantly hot conditions, and this industry too dwindled after World War II.

The common material for buttons is polyester, which is a special kind of plastic with properties that make it suitable for buttons. A variety of chemical dyes are added to the polyester to make different colors. To make buttons with the pearlescent sheen of shell buttons, red carbonate is added to the polyester. Black buttons are made with the addition of carbon black, and white buttons are made with titanium. The button making process also requires a chemical catalyst that hardens the polyester, and wax.

The Manufacturing Process

Mixing the polyester

1 Polyester arrives at a button factory in liquid form. At the start of the manufacturing process, polyester is drained from its

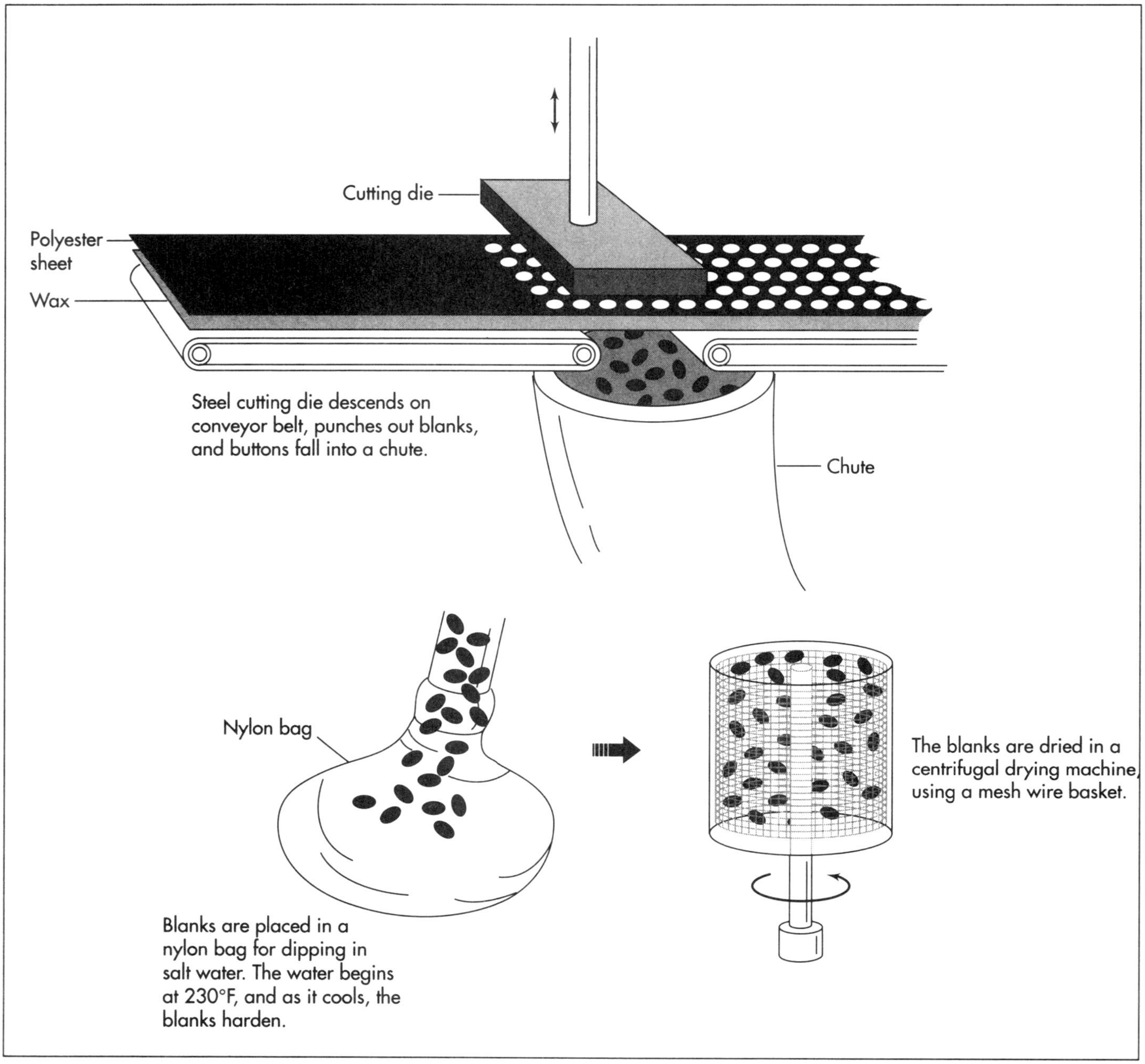

storage tank and measured into a stainless steel kettle. Then dye is added, if the buttons are to be any color other than the natural translucence of the polyester. After the dye is mixed in, the liquid polyester is poured into a 3-gallon (11 l) metal beaker. The catalyst and liquid wax are added.

Pouring into the cylinder

2 The mixture of polyester, catalyst, and wax is then poured into a large rotating metal cylinder. The cylinders are made of steel and lined with chrome, and are typically 2 feet (61 cm) long and 4 feet (122 cm) in diameter. The cylinders lie on their sides on rollers which rotate the drums at 250 rpm. The polyester solution is slowly poured into the rotating interior of the cylinder, and the centrifugal force of the rotation causes the solution to spread, lining the drum with an even sheet. A greater amount of polyester is used for thicker buttons, and less for thinner ones. A 2-inch (5 cm) lip around the ends of the cylinder prevents the polyester from leaking out.

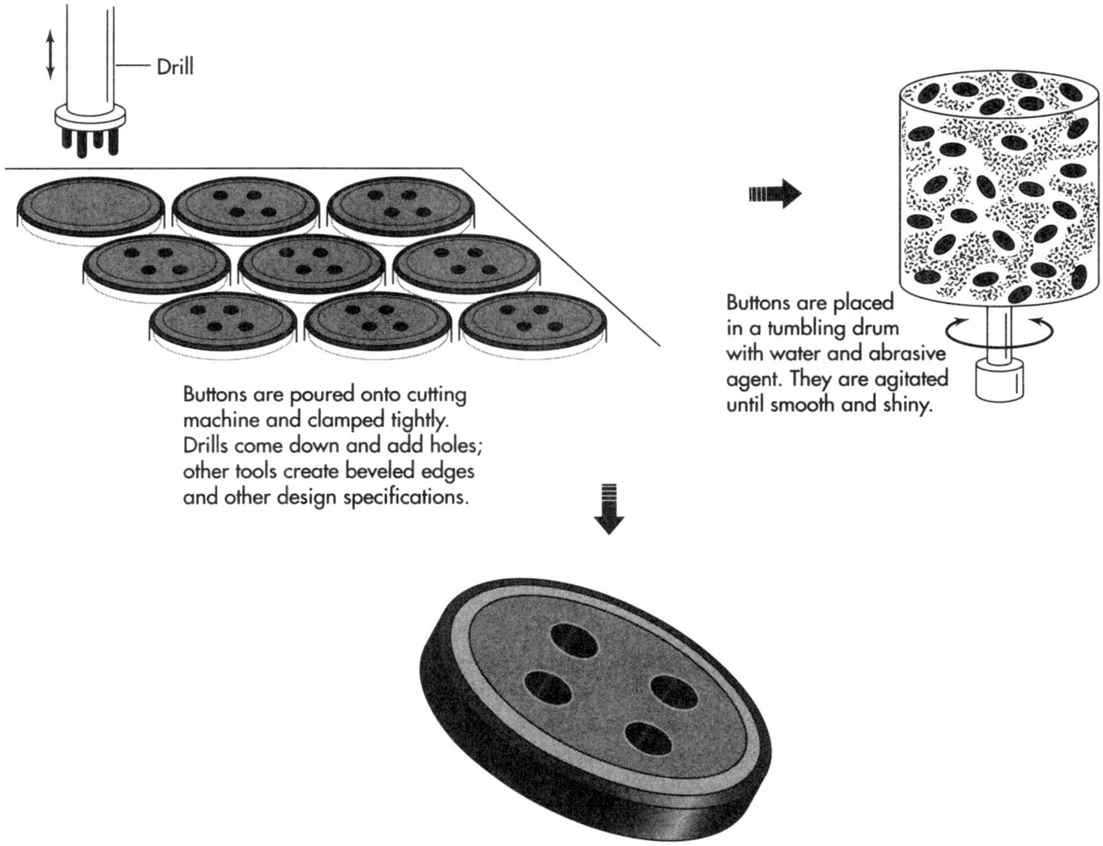

Hardening the sheet

3 As the polyester rotates in the cylinder, it begins to interact with the chemical catalyst and harden. The wax rises to the top of the sheet, and also sinks to the bottom, so that the hardening polyester is eventually held between two layers of wax. This process is completed after 20 minutes of rotation. The resulting polyester sheet has changed from its liquid state to a crumbly solid likened to the consistency of stale cheese.

Cutting the sheet

4 When the sheet has reached the proper hardness, the drum is stopped and the sheet is cut. Then it is rolled out of the cylinder onto a wooden tube. The wax makes it easy to remove from the drum, but the material is still very delicate. The top layer of wax is then peeled off, and the sheet is transferred to a blanking machine.

Cutting the blanks

5 The blanking machine moves the polyester sheet along on a conveyor belt. As the sheet passes along the belt, circular steel cutting dies descend and punch out button-sized circles, called blanks. Buttons come in standard sizes, and different diameter dies can be loaded into the blanking machine, depending on the size needed. After the blanks are cut, they fall into a chute, and the punched out sheet of polyester rolls beneath the chute. Cutting the blanks from the sheet takes from two to four minutes, depending on the size of the buttons being made.

Cooling the blanks

6 The blanks at this stage are hot, because the polyester is still reacting with the catalyst, releasing heat. So at this point the blanks are removed from the chute and poured into a nylon bag. The bag is then lowered into a tank of salt water, which is

heated to 230°F (110°C). The blanks float in the salt water for 15 minutes. The water slowly cools, and the polyester blanks harden. Next, the nylon bag is transferred to a cold water tank, and the blanks reach their final state of hardness. After the hot and cold baths, the blanks are dried in a centrifugal drying machine, which spins them in a wire mesh basket.

Styling the blanks

7 The blanks are now ready to be cut into their finished button shape. The exact design of the button can be specified by a clothing manufacturer, and the button maker must make a steel cutting tool according to the design he is given. A different cutting tool is needed, for example, to make a beveled edge or a flat one, or to make a slightly concave button. When the appropriate cutting tool is in place, the buttons are poured into a hopper at the top of the cutting machine. The blanks fall into a holder where they are clamped tightly and moved toward the cutting tool. The spinning blade advances and cuts the button, then retracts. Next, the button moves beneath a set of drills, which create the holes. Like the cutting tool, the drills must be designed to conform to the clothing manufacturer's specifications. The design specifies not only two holes or four holes, but the diameter of the holes and the distance between them as well. After the buttons pass beneath the drill, they are sucked by vacuum out of the holder and into a box beneath the machine. Hundreds of buttons a minute can be made this way, though the number varies according to the size of the button and the complexity of the design.

Finishing the buttons

8 After the buttons are cut and drilled, they have rough or sharp edges, scratches, and tool marks. They are placed into hexagonal tumbling drums, which contain water, an abrasive material, and a foaming agent. The drums spin for up to 24 hours. The buttons bounce around in the drum until they are smooth and shiny. After tumbling, the buttons are washed and dried.

Quality Control

After the buttons are completely finished, they are placed on a conveyor belt and visually inspected for defects. The inspector must check each button for flaws and remove any cracked or mis-cut ones. The buttons are now ready for packaging and sale.

The Future

The 20th century has seen entirely new clothing fasteners such as the zipper and velcro, and we can now manufacture stretchy fabrics that require no fasteners at all. Nevertheless, the button does not seem in danger of fading away. It is both utilitarian and fashionable, and will likely long be with us. However, button technology is not entirely staid. One recent development is a button of superior strength, a ceramic button made of zirconium oxide. Beer magnate Joseph Coors Jr. decided in 1989 that there was a need for an indestructible button, and he used a ceramics research unit at the Adolph Coors Company to develop this new product. The resulting Diamond Z button debuted in 1993. It is said to be harder than steel, with 2.5 times steel's flexing strength. These men's shirt buttons are fired at 3200°F (1760°C), then polished and coated with an ivory-like finish. The proof of the Diamond Z's indestructibility is a "drop test" where a heavy pointed rod falls down a long tube onto the button. The button can withstand this rigorous ordeal as well as the everyday wear and tear of repeated washing and ironing. The Diamond Z button is, however, quite expensive to make compared to the ordinary polyester button, and for that reason it is not likely to displace the existing technology.

Where To Learn More

Books

Epstein, Diana and Millicent Safro. *Buttons*. Harry N. Abrams, Inc., 1991.

Fink, Nancy and Maryalice Ditzler. *Buttons: The Collector's Guide to Selecting, Restoring and Enjoying New and Vintage Buttons*. Running Press, 1993.

Periodicals

Berendt, John. "The Button." *Esquire*, September 1989, p. 72.

Coy, Peter. "Coors: From Beer to Superstrong Buttons." *Business Week*, July 12, 1993, p. 149.

—*Angela Woodward*

Camera Lens

Photography is America's favorite hobby, and today a whole range of cameras and camera lenses are available to suit the needs and budgets of almost everyone from the occasional picture taker to the dedicated amateur to the working professional.

Background

The camera lens is an invention that attempts to duplicate the operation of the human eye. Just like the eye, the lens sees an image, focuses it, and transmits its colors, sharpness, and brightness through the camera to the **photographic film**, which, like our memory, records the image for processing and future use. Lenses are made of optical glass or plastic. They focus light rays by refracting or bending them so that they meet or converge at a common point.

A simple lens "sees" well through its center, but its vision around the edges tends to blur. Blurring, color changes, distortion of lines, and color halos around objects are caused by defects in the lens called aberrations. Some aberrations can be corrected in the simple lens by shaping one or both surfaces so they are aspheric; aspheric curves vary like the curves of a parabola, rather than staying constant like the curvature of a sphere. A camera lens reduces the effects of aberrations by replacing a simple lens with a group of lenses called lens elements, which are lenses of different shapes and distances of separation. The lens becomes more complex as greater correction of vision is achieved. The lens will also be more complex depending on the size of the aperture—the opening that allows light to pass through—and the range of angles it "sees." Lens design used to rely on the optician's art and considerable experimentation. Today, computer programs can adjust the shaping and spacing of lens elements, determine their effects on each other, and evaluate costs of lens production.

Lens elements are usually described by their shape. The convex lens curves outward; a biconvex lens curves outward on both sides, and a plano-convex lens is flat on one side and outwardly curved on the other. There are also concave lenes, biconcave, and plano-concave lenses. The elements are not necessarily symmetrical and can curve more on one side than the other. Thickening the middle of the lens relative to its edges causes light rays to converge or focus. Lenses with thick edges and thin middles make light rays disperse. A complex camera lens contains a number of elements specially grouped. The combination of the composition, shape, and grouping of the elements maximizes the light-bending properties of the individual elements to produce the desired image. The lens is focused by moving it nearer or farther from the film or focal plane. The lens can be twisted, causing the lens elements to move in and out along a spiral screw thread machined into the casing of the lens. Twisting the lens also moves a scale on the casing that shows the distance of the best focus.

The stop or diaphragm is a specialized part of the lens. In simple cameras, the stop is a fixed stop or a ring of black sheet metal that is permanently set in front of the lens. Box cameras, studio cameras, and some cameras of European manufacture use a sliding stop, which is a strip of metal that slides across the front of the lens between grooves. It has two or more holes of different sizes that are the apertures. Lenses with a variable stop have a machined ring on the outside of the lens mount, printed with f-stop numbers. By turning this ring, the diaphragm can be opened or closed. This iris diaphragm works much like the iris of the eye in allowing adjustments for varied light conditions.

The lens in a compact camera is usually a general-purpose lens with a normal focal length that takes pictures of an image the way our eyes see it. Lenses designed for special purposes are used with more advanced cameras. Telephoto lenses work much like binoculars or **telescope**s, and make a distant image appear closer. Wide-angle lenses make the image appear farther away; a panoramic lens is a special kind of wide-angle lens that is useful for taking pictures of broad expanses of scenery. Some disposable cameras are equipped with panoramic lenses. A fish-eye lens is also a special kind of wide-angle lens that deliberately distorts the image so the central part is enlarged and the outer image details are compressed. Fish-eye lenses cover very wide angles like horizon-to-horizon views. Another special purpose lens is the variable-focus lens, also called a "zoom" lens. It uses moveable lens elements to adjust the focal length to zoom closer to or farther away from the subject. These lenses are complex and may contain 12 to 20 lens elements; however, one variable-focus lens may replace several other lenses. Some compact cameras also have limited zoom, telephoto, or wide-angle features. The single-lens reflex (SLR) camera is made so that the photographer sees the same view as the lens through the viewfinder. This enables the photographer to plan the image that will appear on film with the flexibility of a variety of interchangeable lenses.

History

The camera lens evolved from optical lenses developed for other purposes, and matured with the camera and photographic film. In 1568, a Venetian nobleman, Daniel Barbaro, placed a lens over the hole in a camera box and studied sharpness of image and focus. His first lens was from an old man's convex spectacles. The astronomer Johann Kepler elaborated on Barbaro's experiments in 1611 by describing single and compound lenses, explaining image reversal, and enlarging images by grouping convex and concave lenses.

In the 1800s, the first box cameras had a lens mounted in the opening in the box. The lens inverted the image on a light-sensitive plate at the back of the box. There was no shutter to open the lens; instead, a lens cap was removed for several seconds or longer to expose the plate. Improvements in the sensitivity of the plate necessitated ways of controlling the exposure. Masks with different sized openings were made for insertion near the lens. The iris diaphragm was also developed to control the aperture. Its metal leaves open and close together to form a circular opening that can be varied in diameter.

In 1841, Joseph Petzval of Vienna designed a portrait lens with a fast aperture. Previously, lenses made for daguerreotype cameras were best suited for landscape photography. Petzval's lens allowed portraits to be taken ten times faster, and the photograph was less likely to be blurred. In 1902, Paul Rudolph developed the Zeiss Tessar lens, considered the most popular ever created. In 1918, he produced the Plasmat lens, which may be the finest camera lens ever made. Rudolph was followed shortly by Max Berek, who designed sharp, fast lenses that were ideal for miniature cameras.

Other essential developments in lens history include lens coating technology, use of rare-earth glass, and calculation methods made possible by the computer. Katharine B. Blodgett developed techniques for thin-coating lenses with soap film to remove reflection and improve light transmission in 1939. C. Hawley Cartwright continued Blodgett's work by using coatings of metallic fluorides, including evaporated magnesium and calcium that were four-one-millionths of an inch thick.

Design

Design of a camera lens begins by identifying the photographer who will use it. When the market is identified, the lens designer selects the optical and mechanical materials, the optical design, the appropriate method for making the mechanical parts, and, for auto focus lenses, the type of interface between the lens and camera. There are conventions or patterns for the different categories of lenses, including macro, wide-angle, and telephoto lenses, so some design aspects are standardized. Advancements in materials give designers many challenging

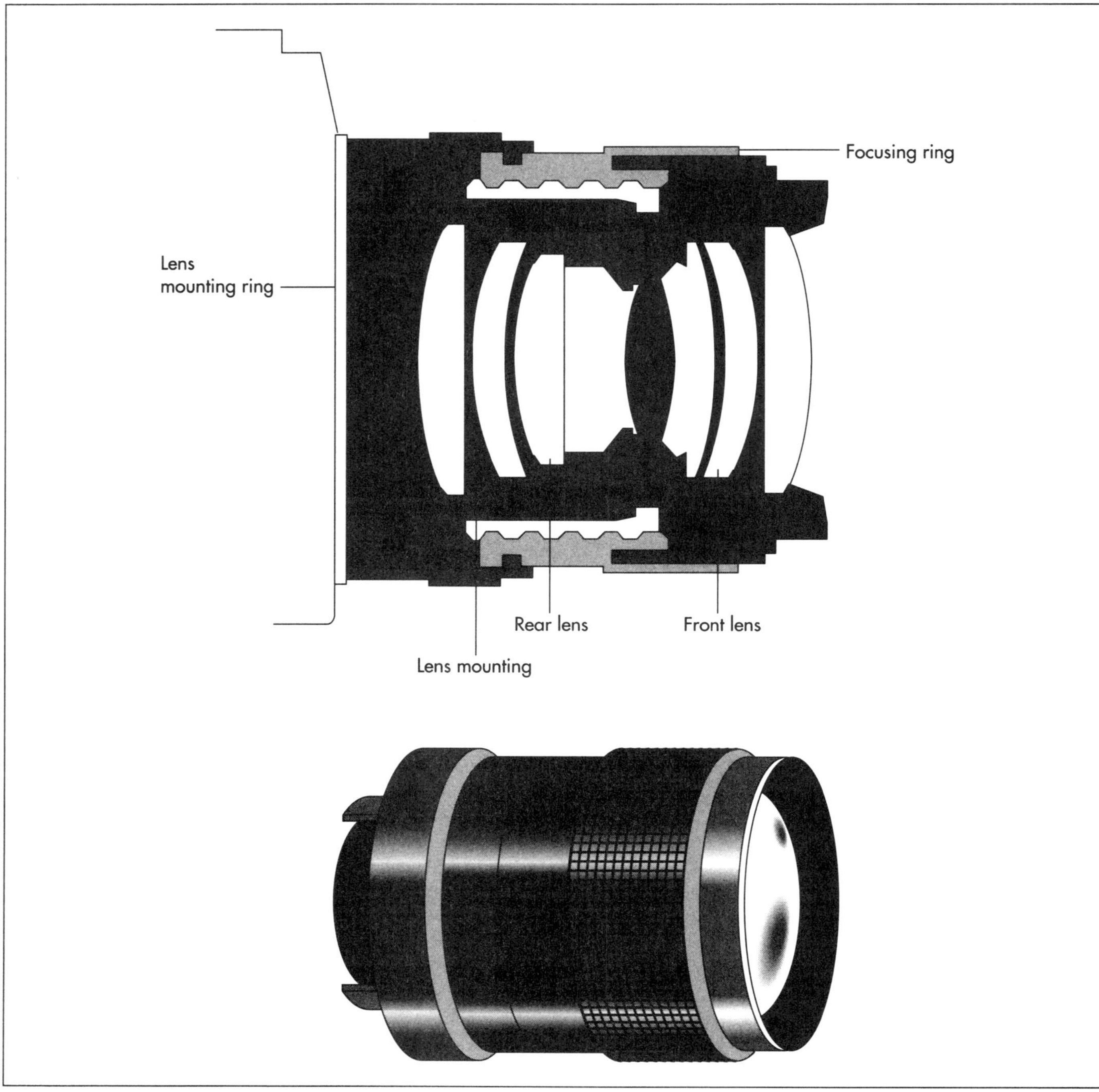

A group of lenses called lens elements, which are of different shapes and distances of separation, make up the camera lens. Lens design used to rely on the optician's art and considerable experimentation. Today, computer programs can adjust the shaping and spacing of lens elements, determine their effects on each other, and evaluate costs of lens production.

options, however. In selecting materials, the engineer must consider a range of metals for the components and various types of glasses and plastics for the lenses, all the while mindful of the final cost to the photographer.

When the designer has completed the design, its performance is tested by computer simulation. Computer programs that are specific to lens manufacturers tell the designer what kind of image or picture the lens will produce at the center of the image and at its edges for the range of lens operation. Assuming the lens passes the computer simulation test, the criteria for performance that were chosen initially are reviewed again to confirm that the lens meets the needs identified. A prototype is manufactured to test actual performance. The lens is tested under varying temperature and environmental conditions, at every aperture position, and at every focal length for zoom lenses. Target charts in a laboratory are photographed, as are field conditions of varying light and shadow. Some lenses are aged rapidly in laboratory tests to check their durability.

Additional design work is needed if the lens focuses automatically, because the auto focus (AF) module must work with a range of camera bodies. The AF module requires both software and mechanical design. Extensive prototype testing is performed on these lenses because of their complex functions and because the software is fine-tuned to each lens.

Raw Materials

The raw materials for the lenses themselves, the coating, the barrel, or housing for the camera lens, and lens mounts are described below in the manufacturing section.

The Manufacturing Process

Grinding and polishing lens elements

1 Optical glass is supplied to lens manufacturers by specialized vendors. Usually, it is provided as a "pressed plate" or sliced glass plate from which the elements are cut. The glass elements are shaped to concave or convex forms by a curve generator machine that is a first-step grinder. To reach the specifications for its shape, a lens goes through a sequence of processes in which it is ground by polishing particles in water. The polishing particles become smaller in each step as the lens is refined. Curve generation and subsequent grinding vary in speed depending on the frailty, softness, and oxidation properties of the optical materials.

After grinding and polishing, the elements are centered so that the outer edge of the lens is perfect in circumference relative to the centerline or optical axis of the lens. Lenses made of plastic or bonded glass and resin are produced by the same processes. Bonded materials are used to make lenses with non-spherical surfaces, and these lenses are called "hybrid aspherics." The aspherical surfaces of these lenses are completed during centering.

Coating lenses

2 Formed lenses are coated to protect the material from oxidation, to prevent reflections, and to meet requirements for "designed spectrum transmission" or color balance and rendition. The lens surfaces are carefully cleaned before coating. Techniques for applying coatings and the coatings themselves are major selling points for a manufacturer's lenses and are carefully guarded secrets. Some types of coatings include metal oxides, light-alloy fluorides, and layers of quartz that are applied to lenses and mirrors by a vacuum process. Several layers of coating may be applied for the best color and light transmission, but excessive coating can reduce the light that passes through the lens and limit its usefulness.

Producing the barrel

3 The barrel includes the chassis that supports the various lens elements and the cosmetic exterior. Metal mounts, grooves, and moving portions of the lens are critical to the performance of the lens, and are machined to very specific tolerances. Lens mounts may be made of brass, aluminum, or plastic. Most metal barrel components are die-cast and machined. Metal mounts last longer, maintain their dimensions, can be machined more precisely, and can be dismantled to replace elements, if necessary. Plastic mounts are less expensive and of lighter weight. If the barrel is made of engineering plastic, it is produced by a highly efficient and precise method of injection molding. The interior surfaces of the barrel are also coated to protect them and to prevent internal reflection and flare.

Assembling the lens

4 Other parts of the lens, such as the diaphragm and auto focus module, are produced as subassemblies. The iris diaphragm is constructed of curved leaves cut out of thin sheets of metal. The metal leaves are held in place by two plates. One plate is fixed, the other moves, and has slots for sliding pins. These slide the leaves back

toward the barrel to open the diaphragm or into the center to close the opening as the f-stop ring is turned. The diaphragm assembly is fastened into place when the lens mount is attached to the end of the barrel. The auto focus is also added, the optical elements are positioned, and the lens is sealed. After final assembly, the lens is adjusted and inspected rigorously. It must meet the design standards for optical resolution, mechanical function, and auto focus response. Lenses may also be tested by subjecting them to shocks, dropping, and vibration.

Quality Control

Approaches to lens manufacture vary greatly among companies. Some use full automation including **industrial robot**s to make their products, others use large assembly lines, and still others pride themselves on hand-crafting. Quality and precision are essential to lens production, however, regardless of manufacturing approach. Incoming materials and components are rigorously inspected for quality and compliance with engineering specifications. Automated processes are also inspected constantly and subjected to tolerance checks. Hand-craftsmanship is performed only by skilled artisans with long years of training. Quality control and stress tests are incorporated in each manufacturing step, and elements and components are measured with precise instruments. Some measuring devices are laser-controlled and can detect deviations of less than 0.0001-millimeter in a lens surface or in lens centering.

The Future

Camera lenses are enjoying new developments in many areas. The consumer's interest in the best photos for the lowest cost has led to disposable cameras with simple but effective lenses. Lenses for professional photographers and for specialized uses such as high-performance binoculars or telescopes are made with exotic and "non-preferred" glasses that are more sensitive, expensive, and harder to obtain than traditional materials. These are called "abnormal dispersion" materials because they merge all the colors in the light passing through the lens to produce the best images, rather than allowing colors to disperse like a simple lens. Water and other liquids also bend light, and scientists have identified liquids that are abnormally dispersive and can be trapped between layers of ordinary glass to produce the same image quality as exotic optical glass. The ordinary or "preferred" glass (preferred because of low cost and workability) is bonded around the liquid with flexible silicone adhesive. The resulting "liquid lens" may replace several elements in a professional-quality lens. It also reduces the coating required and the amount of lens polishing needed because the liquid fills imperfections in the glass. The cost of the lens is reduced, and the light transmission properties are improved. Lens makers in the U.S., Japan, and Europe are preparing to produce liquid lenses in the near future.

Where To Learn More

Books

Bailey, Adrian and Adrian Holloway. *The Book Of Color Photography.* Alfred A. Knopf, 1979.

Collins, Douglas. *The Story of Kodak.* Harry N. Abrams, Inc., Publishers, 1990.

Sussman, Aaron. *The Amateur Photographer's Handbook.* Thomas Y. Crowell Company, 1973.

Periodicals

Coy, Peter, ed. "A Clear-Eyed View from Liquid Camera Lenses." *Business Week,* January 17, 1994, p. 81.

From Glass Plates to Digital Images. Eastman Kodak Company, 1994.

"Photographic Lenses." *Photographic,* April 1991, pp. 56-57.

"Liquid Lens." *Popular Science,* May 1994, p. 36.

—Gillian S. Holmes

Carpet

Background

Carpet is a textile floor covering that is distinguished from the more general term "rug" by being fixed to the floor surface and extending wall to wall. The earliest peoples covered the floors of their dwellings with animal skins, grass, or, later, woven reed mats. When people learned how to spin cotton and wool, woven mats of these materials largely replaced earlier coverings. Around 3000 B.C. Egyptians sewed brightly colored pieces of woolen cloth onto linen and placed it on their floors.

The first carpets of note were woven by nomads. The thick carpets were easy to transport and were placed over the sand floor of tent dwellings. Early looms were similarly easy to transport. Two forked branches were joined by a crosspiece holding the suspended warp, and a wooden bar was used to flatten binding weft threads, while the loose warp ends formed the carpet's pile. The Pazyryk carpet has been documented as the earliest hand loom carpet, dating back from 500 B.C. and discovered in a tomb located in the Alti Mountains in Central Asia.

From these early beginnings, carpet weaving rose to its highest art form in Turkey, Iran, India, and China. Using cotton, linen, or hemp as the foundation, and wool or silk as the luxurious pile, weavers would make a knot out of the pile thread, then form a row of knots that was tightly beaten down. The process was time-consuming: some of the finest handmade carpets have as many as 2,400 knots per square inch (372 knots per sq cm). The brilliant colors of these ancient carpets came from natural dyes such as madder, indigo, genista, woad, and ocher. Some weavers added alum to these dyes to fix the color, and a few wove gold and precious jewels into their carpets.

While Europeans for centuries eagerly received carpets ready-made from the Middle East, carpet making itself did not find a firm foothold on the continent until France imported Moorish weavers around A.D. 1300. By 1600, carpet guilds were flourishing around Aubusson and Savonnerie. England also imported Persian weavers, as well as French ones, and by 1700 both Wilton and Axminster, known for their wool, were chartered carpet-making towns. Carpet making in Europe started with the "Brussels weave" in France and Flanders. This weave is formed by putting yarn over rods to create uncut loops. Wilton carpets are cut by a blade that replaces the rod in the Brussels weave. In 1801, Joseph M. Jacquard invented a device for handlooms that used punch cards to place up to six varieties of yarn colors in textiles, thus increasing production. This technique was adopted for carpet looms in 1825.

The first carpet factory in the U.S. was built by William Sprague in Philadelphia in 1791. His looms, based on English inventions, could make 27-inch (69-cm) runners that could be sewn together to make larger carpets. By 1800, 6-8 yards (7-9 m) of carpet could be made in a day. Erastus Bigelow built a mill in 1825 in Clinton, Massachusetts, and invented the power loom in 1839, which doubled carpet production. He also invented the first broadloom in 1877. Power looms improved over the years; soon one loom could make 75 yards (82 m) of high quality carpet a day.

Carpet weaving rose to its highest art form in Turkey, Iran, India, and China. Using cotton, linen, or hemp as the foundation, and wool or silk as the luxurious pile, early weavers would make a knot out of the pile thread, then form a row of knots that was tightly beaten down. Some of the finest handmade carpets have as many as 2,400 knots per square inch.

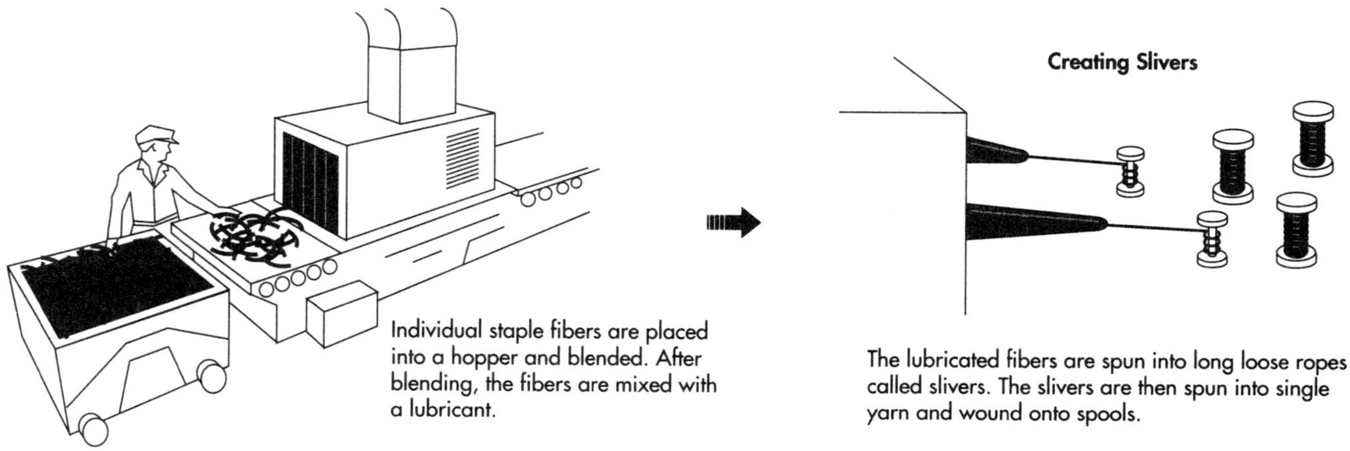

Individual staple fibers are placed into a hopper and blended. After blending, the fibers are mixed with a lubricant.

The lubricated fibers are spun into long loose ropes called slivers. The slivers are then spun into single yarn and wound onto spools.

Synthetic yarns arrive at the carpet manufacturer either in staple fiber form or bulk continuous filament form. The staple fibers, which average 7 inches long, are loose, individual strands that arrive in bales.

Carpet production changed dramatically at the beginning of the 20th century, beginning inauspiciously with a burst of tufted bedspread production in Dalton, Georgia, led by young entrepreneur Catherine Evans Whitener. Tufting is the process of punching yarn into a ground fabric to create many uncut loops at a very fast pace. Tufted bedspread factories dominated the Dalton area by World War II, and they soon began producing tufted rugs as well. Demand for these roughly made rugs was as great as that for the bedspreads. At first using cheap, readily available cotton before switching to synthetic yarns, the number of Dalton carpet makers grew as they produced great amounts of relatively easy-to-make broadloom tufted rugs and, eventually, carpets. Carpet, once a luxury, became affordable for most Americans. Today, carpet makes up 72% of all flooring, with tufted carpet being 91.5% of production, and the city of Dalton is responsible for over 70% of the world's production of carpet.

Raw Materials

Carpet consists of dyed pile yarns; a primary backing in which the yarns are sewn; a secondary backing that adds strength to the carpet; adhesive that binds the primary and secondary backings; and, in most cases, a cushion laid underneath the carpet to give it a softer, more luxurious feel.

Ninety-seven percent of pile yarns today are made up of synthetic polymers; the rest of the yarns are wool and comprise the more expensive, woven carpet. Synthetics are plastics such as nylon (which is in 66% of all carpet), acrylics (15%), **polyester** (less than 15%), and polypropylene (less than 5%). These pile yarns are dyed using a variety of organic chemical compounds, or occasionally, organometallic complexes.

Both the primary and secondary backing are largely made of woven or nonwoven polypropylene, though some secondary backing may still be made of jute, a natural fiber that, when woven, looks like burlap. The adhesive used to bind the backings together is almost universally synthetic rubber latex. The most common padding is rebond (bonded urethane), though various forms of synthetic latex, polyurethane, or vinyl might be used instead. Rebond is recycled scrap urethane that is chopped into uniformly sized pieces and pressed into layers. Although rare, some carpet cushioning is made up of horse hair or jute. A plastic top sheet is usually added to the top to insure a smooth surface against the carpet.

The Manufacturing Process

Since most carpet in the U.S. is tufted; earlier methods of weaving carpet, such as Wilton and Axminster, are ignored in the following account.

Preparing the yarn

1 Synthetic yarns arrive at the carpet manufacturer either in staple fiber form or bulk

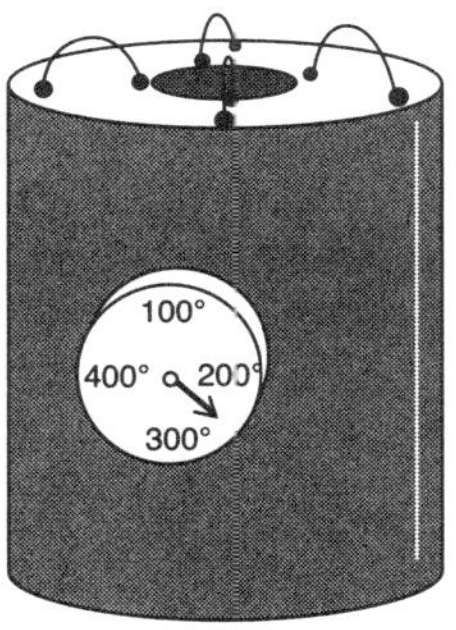

The yarn is steamed and heated. It is cooled before it is transported to the tufting machine.

Tufting the Carpet

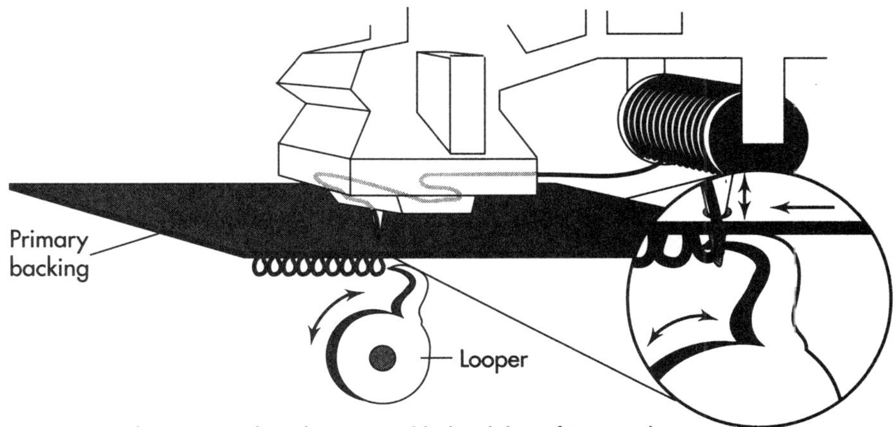

The yarn is placed on a creel behind the tufting machine which feeds into a needle. As the needle lowers and pierces the primary backing, a looper grabs the yarn and keeps it below the backing. The process is repeated.

Dyeing the Carpet

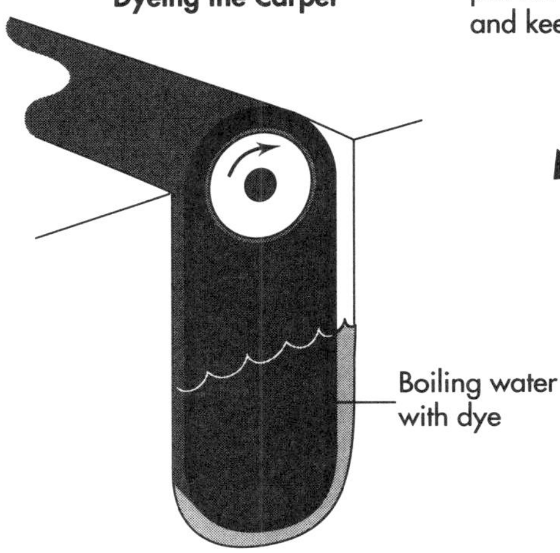

The ends of the carpet are stitched together and continuously rolled through a vat containing the dye. Next the carpet will be steamed to fix the color and rinsed and dried.

continuous filament form. The staple fibers, which average 7 inches (18 cm) long, are loose, individual strands that arrive in bales. Several bales are blended together into one batch in a hopper. Then, after lubrication, they are spun into long, loose ropes called slivers by a carding machine. The slivers are then pulled, straightened, and spun into single yarn that is wound onto spools. Both the single-ply staple fibers (now spun into filament) and the bulk continuous filament must now be twisted together to form thicker two-ply yarn suitable for tufting. The yarns are then steamed to bulk them, and then heated to 270-280°F (132-138°C). This heat setting causes the yarn to maintain its shape by fixing its twist. After cooling, these yarns are wound onto tubes and transported to the tufting machines.

Dyeing the yarn

2 Most carpets are dyed after tufting, yet sometimes the yarns are dyed first. The methods include putting 500-1,000 pounds (227-455 kg) of fiber into pressurized vats

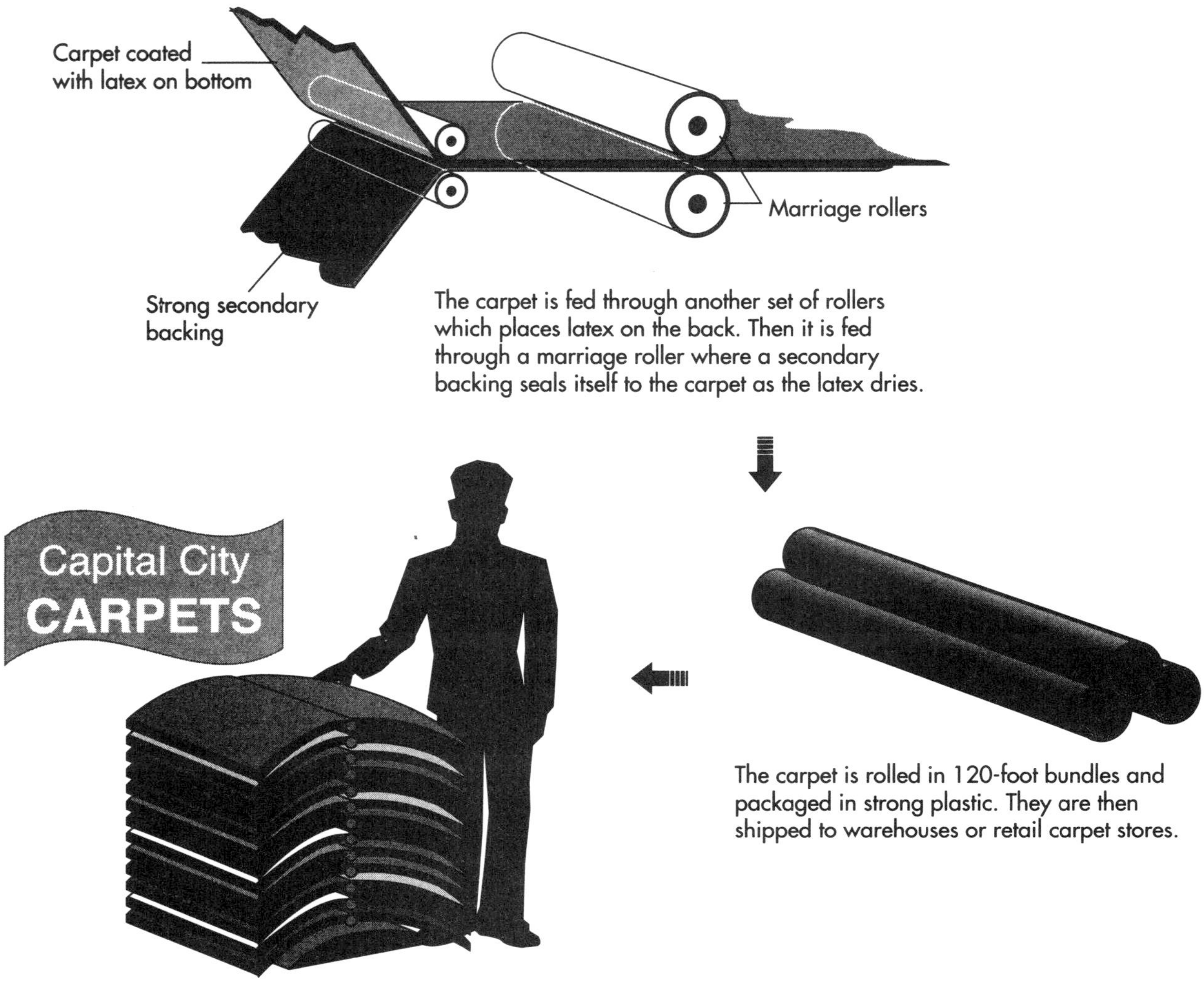

through which treated dyes are circulated, or passing the fiber continuously through the bath, or passing skeins of yarn through the vat of dye. The yarn can also be put on forms, and the heated dyes can then be forced under pressure from inside the forms to color the yarn. Another method passes the yarn through printing rollers, while yet another involves knitting the yarn onto a form that is then printed with dyes before the yarn is unraveled. All yarn that has been dyed is then steamed, washed, and dried.

Tufting the carpet

3 The yarn is put on a creel (a bar with skewers) behind the tufting machine, then fed into a nylon tube that leads to the tufting needle. The needle pierces the primary backing and pushes the yarn down into a loop. Photoelectric sensors control how deeply the needles plunge into the backing, so the height of the loops can be controlled. A looper, or flat hook, seizes and releases the loop of yarn while the needle pulls back up; the backing is shifted forward and the needle once more pierces the backing further on. To make cut pile, a looper facing the opposite direction is fitted with a knife that acts like a pair of scissors, snipping the loop. This process is carried out by several hundred needles (up to 1,200 across the 12 foot [3.7 m] width), and several hundred rows of stitches are carried out per minute. One tufting machine can thus produce several hundred square yards of carpet a day.

Dyeing the tufted carpet

4 For solid color carpeting, carpet of several standard roll lengths is sewn together to make a continuous roll, which is then fed into a vat. The vat is filled with water, which is first heated before dyes and chemicals are mixed in. The mixture is then slowly brought to a boil and cooked for four hours. Another method of making solid color carpet is to sew several rows together to make one continuous roll, which is then fed under rods that bleed the color into the pile. After dyeing, the carpet is then steamed to fix the color, excess color is washed off, and the carpet is dried and put on a roll.

5 To make printed carpet of various designs, white carpet passes under screens in which holes in the desired pattern have been cut. The desired color is squeegeed through the holes in the screen, and the carpet is advanced 36 inches (91 cm) to a different screen that applies a new color in a different design through the screen. Up to eight colors can be applied with this method.

6 Another method of dyeing printed carpet is to pass it under embossed cylinders that have raised portions in a design, which press color into the carpet. Each cylinder provides a different design for a different color. After dyeing, the printed carpet is steamed, excess dyes are washed off, and the carpet is then dried and put onto rolls to go to the finishing department.

Finishing the carpet

7 The ends of the dyed carpet are first sewn together to form a continuous belt. This belt is then rolled under a dispenser that spreads a coating of latex onto the bottom of the carpet.

At the same time, a strong secondary backing is also coated with latex. Both of these are then rolled onto a marriage roller, which forms them into a sandwich and seals them together.

The carpet is then placed in an oven to cure the latex.

8 The completed carpet is then steamed, brushed, vacuumed, and run through a machine that clips off any tufts that rise above its uniform surface. The carpet is then rolled into 120 foot (37 m) lengths that are then packaged in strong plastic and shipped to either the carpet manufacturer's inventory warehouse or to a retail carpet store.

Quality Control

Every piece of carpet that is tufted is inspected to see if any tufts are missing. One person with a single needle tufting gun shoots pile yarn wherever holes are found. Each piece of carpet is then inspected. The manufacturer checks that the piece is of the proper dimensions and that the tuft height is of the desired length. The static shock potential is also tested.

Most states require a flammability test. A prepared 9 x 9 inch (23 x 23 cm) specimen is placed on a steel plate that has a hole 8 inches (20 cm) in diameter in its middle. A methenamine tablet is ignited in the center. If the charred portion in seven out of eight trials does not reach to the circumference of the hole, the carpet passes. Another important test determines the carpet's resistance to wear. A specimen of carpet is placed in a drum and beaten with a steel ball that has rubber studs on it for 20,000 to 50,000 revolutions. The carpet should look fairly new after this test. To test how the carpet's color stands up to sunlight, a standard light source that simulates sunlight is directed at a specimen, which is then rated according to the number of units of exposure required to produce visible loss of color.

The mass per unit area of pile yarn is a significant test because pile density determines the feel of the carpet. First, the synthetic yarn is removed from the carpet, either by physical means (it is ripped off the primary backing) or chemical means (it is dissolved off). The yarn is then dissolved in a solvent, then dried in an oven to remove the solvent. The dry residue is then weighed and checked to see if the mass is as specified for that type of carpet. Each type of synthetic fiber has its own recipe. Nylon, the most commonly used synthetic yarn, is dissolved in hydrochloric acid and dried 15 minutes at 77°F (25°C).

Backing fabrics and carpet padding are tested for strength by being pulled in a vise until they break. The primary backing's strength is checked both before and after tufting. The delamination strength of the secondary backing is also tested by determining at what force the secondary backing can be pulled away from the primary backing.

Part of the quality control process is up to the customer, who must select carpet of the proper strength and durability for the amount of traffic expected in the room, vacuum regularly, and have the carpet professionally deep cleaned at least once a year.

Where To Learn More

Books

Deaton, Thomas M. *Bedspreads to Broadloom: The Story of the Tufted Carpet Industry*. Tapestry Press, 1993.

Ellis, Robert Y. *The Complete Book of Floor Coverings*. Charles Scribner's Sons, 1980.

Garstein, A.S. *The How-To Handbook of Carpets*. Van Nostrand Reinhold, 1979.

Revere, Glenn. *All About Carpets: A Consumer Guide*. TAB Books, 1988.

Shoshkes, Lila. *Contract Carpeting*. Billboard, 1974.

Von Rosensteil, Helene. *American Rugs & Carpets*. William Morrow & Co., 1978.

Periodicals

"New Technology in Carpeting." *American Dyestuff Reporter*, February 1992, p. 10.

"Magic Carpets." *The Economist*, November 7, 1992, p. 108.

"Carpet Machine Bonds Yarn to Backing Without Sewing." *Machine Design*, January 21, 1988, p. 56.

"Floor Covering." *Textile World*, June 1993, p. 60+.

—*Rose Secrest*

Cat Litter

Background

H. Edward Lowe is considered the father of the cat litter industry. In January 1947, Lowe had a thriving building supply business in Cassopolis, Michigan. One of Lowe's neighbors asked for his assistance in "finding something absorbent" to put in her cat's box other than the sand ashes that she had been using. His answer was dried ground clay. Nearly 50 years later, the cat litter industry has grown into a multi-million dollar business. Sales for 1994 were estimated at $767 million, and are expected to increase significantly throughout the decade.

Until Lowe's invention of "kitty litter," a product that enabled cat owners to have an indoor cat box, most owners had little choice but to let their house cats out. The advent of cat litter brought with it the possibility of a more domesticated cat that no longer needed to be let out. In 1995, approximately 34.1 million Americans, or nearly one-third of all households, are cat owners.

Cat box fillers can be categorized into two types: conventional and clumping. Although the size of the granules varies, conventional cat box filler can be described as gravel-like in texture. Most brands claim to be dust-free or dust-reduced and to have a deodorizing agent. With conventional litter, the contents of the cat box must be discarded every time the box is cleaned in order to ensure that the cat has a thoroughly fresh supply of litter.

Clumping litter was introduced in 1989 and accounts for approximately 30% of today's cat box filler market. The smaller granule litter sticks together when it comes in contact with liquid, hence the term "clumping." When clumping litter is used, the cat owner can remove the clumps, then add more litter to the box, eliminating the need to refill it with every cleaning. The major drawback of clumping litter appears to be its tendency to "track," that is for the finer granules to stick to the cat's paws when it leaves the box.

Raw Materials

Cat box fillers made from absorbent clay account for approximately 95% of all cat litter. Clay, a naturally-occurring, non-metallic substance, is composed of a combination of aluminum silicates and minerals. Light-colored clays are more popular than the darker clays because the latter tends to become muddy when wet and cannot absorb additional moisture.

Although most commercial cat box fillers use an absorbent clay as their base, anything that can absorb moisture theoretically can be used as cat litter.

Wood

Recycled waste products from the lumber industry are used to make alternative cat litter. A blend of cedar chips and hardwood, for example, is lighter than clay-based litters, and its cedar scent absorbs litter box odors. Another type is made from the waste products of aspen lumber. In this manufacturing process, the sawdust and bark are finely ground, then heated to 1200°F (649°C) in a kiln-type vat. This causes the resin or tree sap to bind the wood together as the material is extruded through a screen to form quarter-inch pellets. The pellets are then passed through a cooling tank on a covered conveyor before they are packaged.

The development of today's modern, clay-based cat litter is credited to H. Edward Lowe, a building supplier who invented clay cat litter for his neighbor's cat.

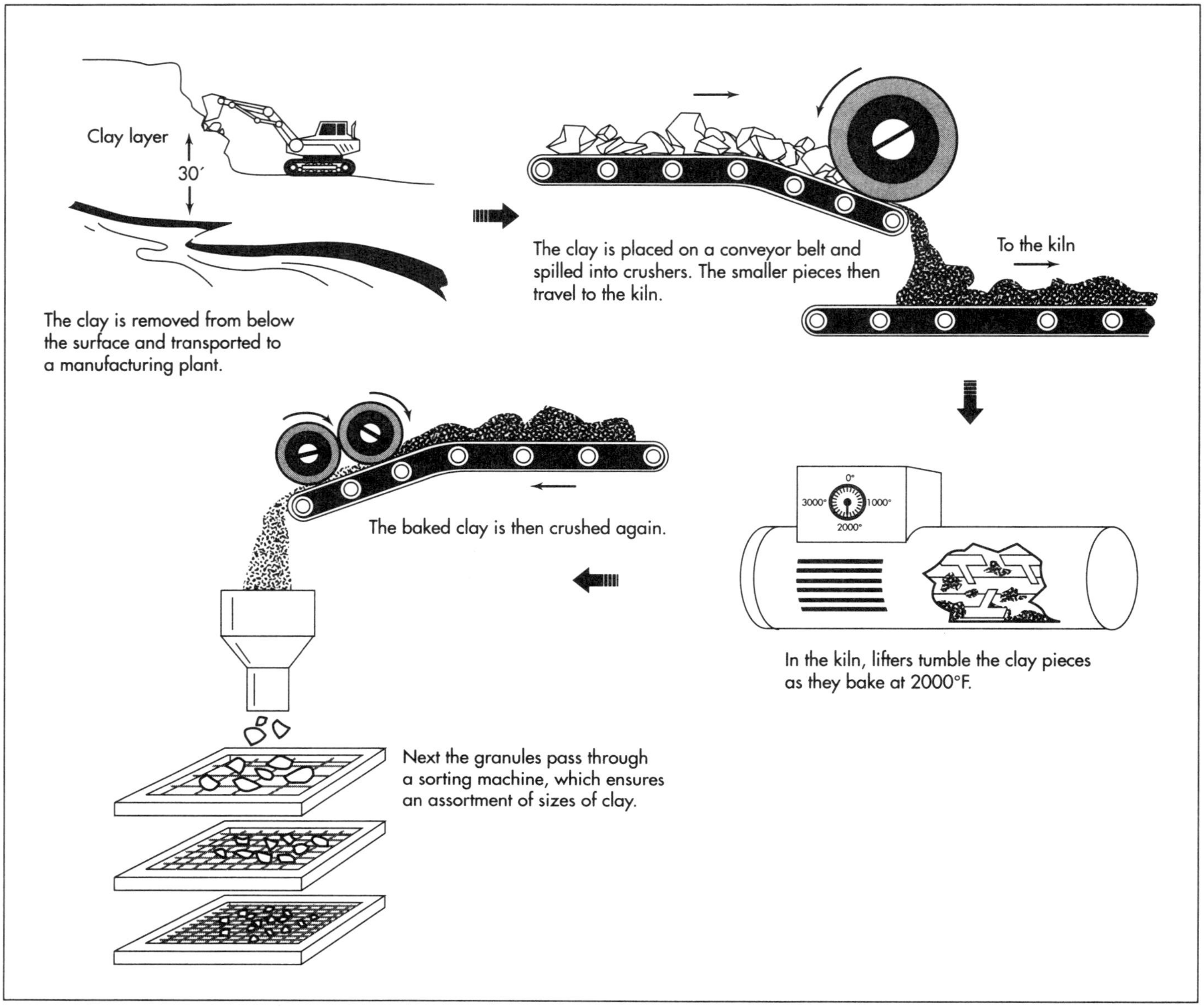

While cat litter can be made from wood, paper, grain, corncobs, citrus, and grass, 95% of all cat litter is made from clay.

Paper

The natural absorbency of **paper** makes it a viable alternative to traditional cat box fillers. To produce "paper" litter, post-consumer **newspaper** is re-pulped, de-inked, and screened. During the re-pulping process, long fibers are processed back into paper, but short fibers are considered waste. The short fibers are de-watered, pelletized, dried, combined with an odor-control agent, and packaged as cat litter. Like those made from wood, paper litter is also lightweight, about 40% less than clay-based litters. This paper-based litter is also flushable in small quantities.

Grain

The byproducts of hard grains are also used as alternative cat litter. The manufacturer mills hard grains, similar to those used to make **pasta**, to a consistency required for mixing and pelletizing. Fragrance, small pieces of corncob, and sometimes catnip are added. An advantage to grain products is that the enzymes in the mixture naturally neutralize the cat's waste odors.

Corncobs

Recycled corncobs can be made into pellets for use as cat litter. During the manufacturing process, the light outer portion and the

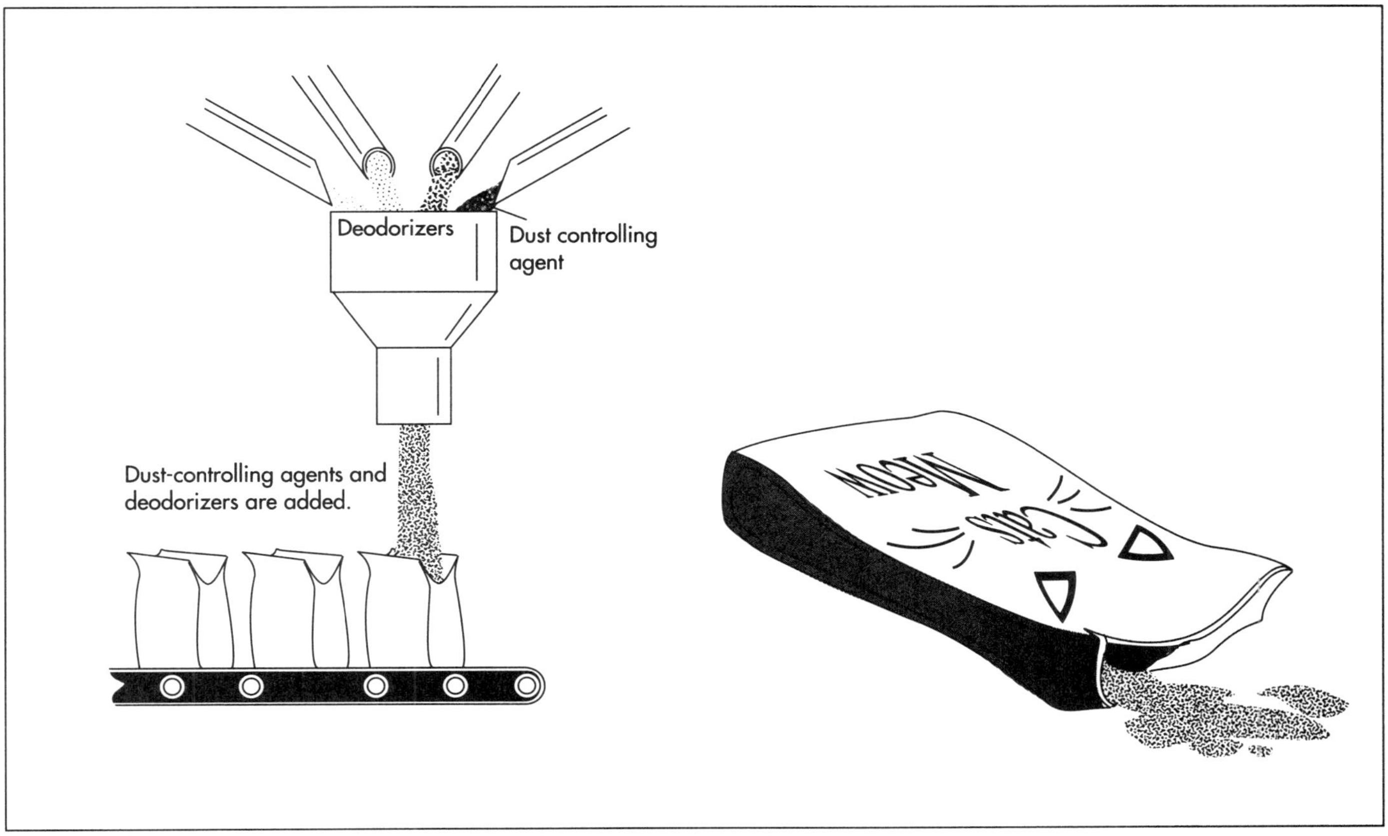

Odor-control, tracking, and dust are three areas of concern for cat litter manufacturers. Many companies experiment with different ingredients and test out the litter on their own cats before selling it to pet owners.

spongy center are ground, heated, and processed into pellets, which are reground and screened. Compared to the same volume of clay product, the material weighs about half as much and is five times as absorbent. Other advantages are that the litter clumps and is flushable.

Citrus

Recycled waste from the citrus fruit industry makes fresh-smelling cat box fillers. The manufacturer takes peels of de-juiced citrus fruits, presses them, and dries them in natural gas-fired kilns. A screening process removes the dust and fine pieces from the final product. Citrus cat litter is highly absorbent, flushable, and biodegradable. In addition, the residual citric acid in the dried peel naturally neutralizes waste odors.

Grass

Finally, cat litter has been made from northern red wheat straw grasses. Bales of straw are ground and chopped into tiny pieces before they are pelletized. The pellets are vacuumed twice to remove dust and then packaged. The resulting product is biodegradable.

The Manufacturing Process

Cat litter made from wood, paper, grain, corncobs, citrus, and grass account for only 5% of today's market. The rest—about 95% of all cat litter—is clay based. Cat box fillers contain few ingredients, and the entire production process is performed by the manufacturers who mine, dry, and size their own clay.

Gathering the clay

1 Various kinds of clay are found throughout the United States, especially in Alabama, southern California, southern Illinois, eastern Kansas, Nevada, Oregon, southeastern Texas, and Wyoming. Clay is found approximately 30-40 feet (9-12 m) below the surface, and earth movers scoop it from open pits. The raw clay is then transported to the plant.

Drying

2 The clay is loaded onto conveyor belts and spilled into giant crushers. The crushers break the clay into smaller pieces, then deposit them onto another conveyor belt, which feeds the clay into a kiln.

3 The kiln resembles a horizontal chimney or sewer pipe, 100 feet (30 m) or more in length. Inside the kiln, where temperatures can reach 2000°F (1093°C), lifters tumble the clay as it bakes. The clay continues its baking process until it has traveled the entire length of the kiln.

4 As the clay emerges from the kiln, it is fed into a second crusher. Here, rollers crumble the clay to its final consistency.

Sorting

5 Next, the granules pass through an enclosed machine that contains a series of screens stacked one on top of another. Here, the various sizes of granules are sorted. Because litter is less absorbent when granules are the same size, the sorting process ensures that an assortment of sizes are represented.

6 Clumping litter, which are ground to a smaller size than conventional litter, are frequently blended with sodium bentonite, a naturally swelling clay known for its absorbent qualities. A dust-controlling agent is then added to the ground clay to prevent the dust, created when particles rub against one another, from becoming airborne. Most cat litter manufacturers also add deodorizers to prevent odors.

Quality Control

Odor-control, tracking, and dust are three areas of concern in the manufacture of cat box fillers. Many companies keep their own colonies of cats to test their products' effectiveness in these three areas. Feedback from cat owners is also important to manufacturers, who usually provide toll-free telephone numbers on the products' packaging.

The Future

Inventors like Theodore Kiebke are trying new materials to develop a better cat box filler. As an alternative to the traditional clay cat litter, he invented "wheat litter," which clumps, does not contain silica dust, and does not track. He experimented with different flours and corn starches before he hit upon durum wheat, which is used to make pasta. He mixed it with regular clay litter, and, after a few refinements, wheat cat litter was created. The product is billed as non-toxic, 100% biodegradable, almost dust free (and totally free of silica dust), scoopable, odorless, and has minimal tracking problems.

Another new product in the cat box filler industry is called "indicating cat litter." This product changes color to indicate the pH level (a measure of acidity or alkalinity) of a cat's urine. Knowing the cat's pH level is important in managing feline lower-urinary-tract disease (FLUTD), a condition that, in its most serious form, can kill domestic cats.

Despite obvious benefits, indicating cat litter has its disadvantages as well. For instance, the litter works best immediately after the cat has used its box; if the litter box is not checked for several hours, the urine clump dries out and the color disappears. Sprinkling the dried clump with distilled water, however, will bring back the color.

Where To Learn More

Periodicals

"Cracking the Cat Litter Code." *Catnip: A Newsletter for Caring Cat Owners.* Tufts University School of Veterinary Medicine. May 1993, pp. 4-6.

"In or Out?" *Catnip: A Newsletter for Caring Cat Owners.* Tufts University School of Veterinary Medicine. January 1995, p.1.

"Wheat Cat Litter." *Catnip: A Newsletter for Caring Cat Owners.* Tufts University School of Veterinary Medicine. February 1995, p. 7.

"The Cat Box Filler Industry." *Golden Cat Corporation News Release*. February 23, 1995, pp. 1-5.

Shojaim, Amy D. "How Cat Litter Is Made." *Cat Fancy Magazine*, October 1994, pp. 12-14, 16, 18-19.

—*Susan Bard Hall*

Cathode-Ray Tube

The first cathode-ray tube was used as an oscilloscope to view and measure electrical signals. Today, the best-known application of a cathode-ray tube is as the picture tube in a television.

Background

A cathode-ray tube, often called a CRT, is an electronic display device in which a beam of electrons can be focused on a phosphorescent viewing screen and rapidly varied in position and intensity to produce an image. Probably the best-known application of a cathode-ray tube is as the picture tube in a television. Other applications include use in oscilloscopes, radar screens, computer monitors, and flight simulators.

The cathode-ray tube was developed in 1897 by Ferdinand Braun of Strasbourg in what was then the French-German region of Alsace-Lorraine. It was first used as an oscilloscope to view and measure electrical signals. In 1908, A.A. Campbell-Swinton of England proposed using a CRT to send and receive images electronically. It wasn't until the 1920s, however, that the first practical television system was developed. The concept for a color cathode-ray tube was proposed in 1938 and successfully developed in 1949.

Although General Electric introduced their first television set for home use in 1928, commercial television broadcasting remained an experimental technology with only limited range and audience. It took until the late-1940s before television networks had established themselves sufficiently to start a boom in consumer sales. Black-and-white television sets gave way to the first color sets in the 1960s. In the following decades cathode-ray tubes for televisions got both larger and smaller as manufacturers sought to satisfy consumer wants. Recent developments have included tubes with flatter faces, sharper corners, and higher resolution for better viewing.

A CRT consists of three basic parts: the electron gun assembly, the phosphor viewing surface, and the glass envelope. The electron gun assembly consists of a heated metal cathode surrounded by a metal anode. The cathode is given a negative electrical voltage and the anode a positive voltage. Electrons from the cathode flow through a small hole in the anode to produce a beam of electrons. The electron gun also contains electrical coils or plates which accelerate, focus, and deflect the electron beam to strike the phosphor viewing surface in a rapid side-to-side scanning motion starting at the top of the surface and working down. The phosphor viewing surface is a thin layer of material which emits visible light when struck by the electron beam. The chemical composition of the phosphor can be altered to produce the colors white, blue, yellow, green, or red. The glass envelope consists of a relatively flat face plate, a funnel section, and a neck section. The phosphor viewing surface is deposited on the inside of the glass face plate, and the electron gun assembly is sealed into the glass neck at the opposite end. The purpose of the funnel is to space the electron gun at the proper distance from the face plate and to hold the glass envelope together so that a vacuum can be achieved inside the finished tube.

The CRT used in a color television or color computer monitor has a few additional parts. Instead of one electron gun there are three—one for the red color signal, one for blue, and one for green. There are also three different phosphor materials used on the viewing surface—again, one for each color. These phosphors are deposited in the form of very small dots in a repeated pattern across the screen—red, blue, green, red, blue, green, and so on. The key to a color CRT is a piece of perforated metal, known as the shadow mask, which is placed between the electron guns and the viewing screen. The perforations in the shadow mask are aligned so that the red gun can fire electrons at only the phosphor dots which produce the red color, the blue gun at the blue dots, and the green gun at the green dots. By controlling the intensity of the beam for each color as it scans across the screen, different colors can be produced on different areas of the screen, thus producing a color image. To give an idea of how small the perforations and dots have to be, a 25-inch (63 cm) color television picture tube may have a shadow mask with 500,000 perforations and 1.5 million individual phosphor dots.

Design

The electron gun must be designed for each new application. New screen sizes, new overall glass envelope dimensions, and new image resolution requirements all require a new gun design. Brighter images may require higher power accelerating coils. Finer image resolution may require improved beam focusing coils or plates. While the basic design remains the same, the details are constantly refined.

Likewise the basic design of the phosphor viewing surface is fairly well defined, but the details may change. New image resolution requirements may require a new method of depositing the phosphor dots on the face plate, which in turn may require new material processing techniques. The search for truer colors may result in new material formulations. The amount of time the phosphors emit light, or glow, after being struck by the electron beam is also important and is controlled by the chemical composition of the phosphor. This property is called persistence. In a color television, the electron beam scans the screen 25 times per second. If the persistence is longer than one twenty-fifth of a second (0.04 second), the image would show two scans at the same time and would appear blurred. If the persistence is shorter than this time, the image from the first scan would have disappeared before the second scan came along, and the image would appear to flicker.

Even the glass envelope requires extensive design. Strength, radiation absorption characteristics, temperature tolerance, impact resistance, dielectric properties, and optical clarity are a few of the design criteria used when designing the glass components. Computers may be used to perform finite element analysis to evaluate the stresses in complex envelope shapes. This technique divides the part into a finite number of smaller, more easily definable pieces, or elements, and then performs the calculations for each element to spot unacceptably high stress concentrations. Using the computer, dimensions for contours and wall thickness can easily be adjusted until a satisfactory design is achieved.

Raw Materials

Cathode-ray tubes use an interesting and varied assemblage of raw materials. In many cases, it is the raw materials, not the design or manufacturing process, that determine the performance characteristics of the finished product.

The electron gun is made from a variety of metal pieces. The cathode, or electron emitter, is made from a cesium alloy. Cesium is used as a cathode in many electronic vacuum tube devices because it readily gives off electrons when heated or struck by light. In a CRT, the cathode is heated with a high resistance electrical wire. The accelerating, focusing, and deflection coils may be made from small diameter copper wire. A glass tube protrudes from the rear of the electron gun assembly and is used to evacuate the air from the finished CRT.

The phosphor viewing surface is formed from a continuous layer of a single material in monochromatic CRTs, or is composed of individual dots of three different materials in color CRTs. Zinc sulfide is a common phosphor material. The color is determined by adding a very small amount of material called an activator. Zinc sulfide with 0.01% silver activator emits a blue light. When a 0.001% copper activator is used, it produces a green light. A 50/50 mixture of zinc sulfide and cadmium sulfide with a 0.005% silver activator produces a yellow light. Red light can be produced by adding silver or copper to zinc sulfide mixed with a

A CRT consists of three basic parts: the electron gun assembly, the phosphor viewing surface, and the glass envelope. The electron gun assembly consists of a heated metal cathode surrounded by a metal anode. The phosphor viewing surface is a thin layer of material which emits visible light when struck by an electron beam. The glass envelope consists of a relatively flat face plate, a funnel section, and a neck section.

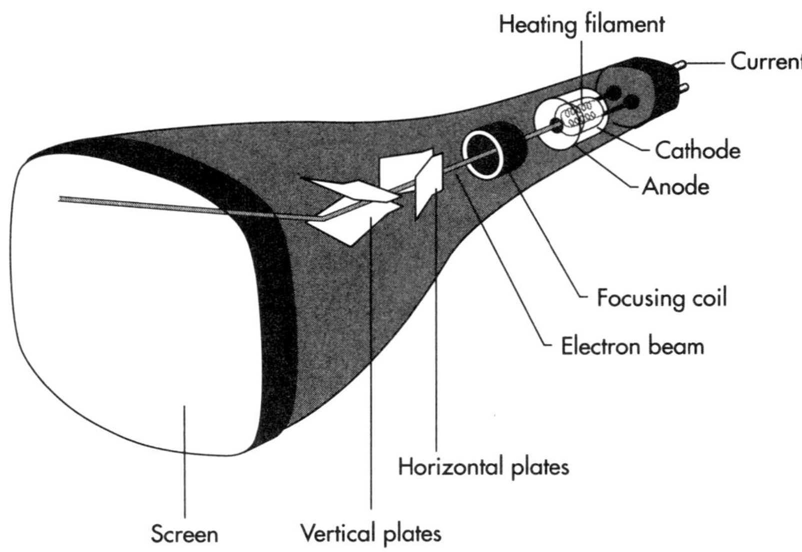

high percentage of cadmium sulfide. The phosphors are usually ground into a fine powder before they are applied to the inside of the face plate.

The glass envelope uses slightly different raw materials for each of its three component parts. The basic raw material for all of the glass components is silica. Alumina may be added to adjust the flow properties of the molten glass when forming it. Various oxides are used to lower the melting temperature. Barium oxide, strontium oxide, and lead oxide are used to provide radiation protection in the neck and funnel. The face plate, on the other hand, must have a minimum of lead oxide to prevent a discoloration phenomenon known as electron or x-ray browning. Neodymium oxide may be used on the face plate to enhance the contrast of the viewed picture.

In color CRTs, the shadow mask is usually made from a thin sheet of a nickel alloy.

The Manufacturing Process

The glass envelope or its components are usually formed at a glass manufacturing facility and shipped to the cathode-ray tube manufacturer who forms the phosphor viewing screen, fabricates and assembles the electron gun, and assembles the finished CRT.

Forming the glass envelope

1 The glass ingredients are weighed and mixed prior to melting. The glass is melted in gas-fired furnaces about 500-3,000 square feet (46-279 sq m) in size. If this is a continuous process, new ingredients are added to maintain a constant level as the molten glass flows out of the furnace to the forming areas. Before forming, the molten glass must be cooled somewhat and made uniform in temperature throughout.

2 The face plate is normally pressed into the desired shape by dropping a gob of molten glass into a mold and pressing on the gob with a plunger. The funnel can be formed either by pressing or by centrifugal casting. In the casting method a gob of molten glass drops into a mold, which then spins rapidly to spread the glass uniformly over the inside surface of the mold. A grooving disk near the top of the mold cuts the soft glass at the desired height so that the excess glass can be removed easily. The neck is made from glass tubing, and one end is flared to facilitate insertion of the electron gun.

3 In a monochromatic CRT the three glass components are joined together

before they are shipped to the CRT manufacturer. In a color CRT only the neck and funnel are joined, and the face plate is shipped separately for further processing. The glass components are usually joined by heating the mating surfaces to a high temperature with gas jets or electric heaters.

Applying the phosphors

4 In monochromatic CRTs the phosphor viewing surface is coated on the inside of the glass face plate. This is done by preparing a liquid suspension of the phosphor and pouring a measured amount into the neck of the glass envelope along with a gelling agent. After about 20 minutes, the coating has set and the excess liquid is poured off. The process for color CRTs is more complicated. First the shadow mask is made by applying a light-sensitive coating to the thin mask material, exposing it to light through a perforated template, and then etching away the exposed coating with an acid to form the millions of holes. The mask is then pressed into a slightly curved shape and attached just behind the face plate. The face plate is placed in a centrifuge and the inside surface is coated with the green phosphor material. The centrifuge spins the face plate to ensure an even coating of phosphor. A strong ultraviolet light is shown through the mask to harden the green phosphor material into hundreds of thousands of dots. The remaining material is then washed off. This process is repeated to form the red and blue phosphor dots, with the ultraviolet light being shifted a small amount each time. When this process is finished, the glass face plate is joined to the funnel. On color tubes, the phosphor dots are sensitive to high temperatures, so instead of using high-temperature gas jets, a mixture of chemical solvent and powdered glass, called a frit, is applied to the joint. This acts like a glass "solder," and the joint can be sealed at a much lower temperature.

Assembling the electron gun

5 The metal components of the electron gun are precision formed. If coils are used they are wound from fine copper wire. Some electron guns use metal plates instead of coils, and these plates are stamped and formed. The components are assembled either by hand or with automated machines in a clean environment. The glass tube is sealed into the base, and the base is welded into the gun assembly.

Final assembly and packing

6 The inside of the glass envelope neck is lubricated with graphite, and the electron gun is inserted and aligned. The neck is then sealed around the gun. A vacuum pump is attached to the glass tube extending from the rear of the gun, and the inside of the CRT is evacuated of air. When the proper vacuum has been achieved, the glass tube is heated and quickly pinched closed to form a seal.

7 The finished CRT is tested for performance and carefully packed to prevent damage. Because the CRT is under a high vacuum, any fracture in the glass envelope could result in an inward explosion known as an implosion.

Quality Control

Although the operating principle of a cathode-ray tube is simple, the manufacturing process requires strict controls and precise alignments. The phosphor materials must be extremely pure to achieve the desired colors. Even a tiny variance in the amount of activator used can result in a significant change in color. Likewise, when you consider that a color television CRT requires the placement of over a million tiny dots side by side on the viewing surface, even a small error in alignment could be disastrous.

Byproducts and Recycling

The principal byproduct of CRT manufacturing is scrap glass. Much of this glass is recycled. Recycled glass with a high content of lead oxide is used to provide radiation protection in CRT funnels and has completely replaced previous sources of lead oxide for this application.

The Future

The worldwide market for cathode-ray tubes was estimated at nearly 400 million units in 1994 and is expected to grow at a 6% annual rate through 2000. The color television market is expected to grow at a 5% annual rate, while the color computer monitor market is expected to grow at a 20% rate. In the television market, the demand for larger television picture tubes with higher image resolution is expected to continue.

One important trend is the development of high definition television (HDTV), which has scanning rates more than twice that of conventional systems. This will require new electron gun designs as well as new glass materials and technologies to handle the doubled radiation rate.

Where To Learn More

Books

Braithwaite, Nicholas and Graham Weaver, eds. *Electronic Materials.* Butterworths, 1990.

Connelly, J.H. and D.J. Lopata. *Engineered Materials Handbook, Volume 4.* ASM International, 1991.

Haider, Z. *Television Glass Bulb Design and Manufacturing Developments, Glass Production and Technology International.* Sterling Publications, Ltd., 1992.

Periodicals

Fleischmann, Mark. "The Big Picture." *Popular Science*, November 1994, pp. 82-85, 92-95.

Meeks, T. "Inside the CRT: Monitor Technology Explained." *PC Novice,* July 1993, pp. 40-43.

—*Laurel M. Sheppard/Chris Cavette*

Chalkboard

Background

The chalkboard is a flat, vertical writing surface on which anything can be inscribed by means of a piece of chalk. The device is generally used for educational purposes, but it can also be found in the workplace, the home, and restaurants. While chalkboards can be manufactured from a variety of materials, porcelain enamel is the most common material used in today's chalkboard.

The origins of the chalkboard date back to the early decades of the 19th century. The forerunner of the chalkboard was the small, paddle-shaped hornbook. This item had been in use in schools of medieval England, and by the time of the Revolutionary War era in colonial America, it was carried by legions of students. The hornbook was a strip of wood with a piece of **paper** fastened onto it. On the paper were a variety of learning aids in small print. A typical hornbook would carry both the Lord's Prayer and the alphabet, and a translucent sheet of animal horn covered the paper. The hornbooks were small objects, sometimes with a hole at the bottom so they could be tied on a string and worn about the neck.

Eventually the hornbook evolved into the reading board. This was a strip of about 15 inches (38 cm) in length, also containing the alphabet and other learning aids, that was hung at the front of the late 18th-century classroom. From the reading board came the concept of one general chalkboard for all students in the classroom to both view and use.

The chalkboard of modern times was patented in 1823. It was developed by a leading educator of the day, Samuel Reed Hall. A minister, Hall founded Vermont's Concord Academy, one of the first formal training schools for American teachers. The early chalkboards were simple pine boards painted black. In other cases, a combination of lime, plaster of Paris (a white powdery substance), and lampblack (fine black soot) was spread on the classroom wall.

The origins of the chalkboard date back to the early decades of the 19th century. The forerunner of the chalkboard was the small, paddle-shaped hornbook, on which a piece of paper was fastened.

Raw Materials

Most modern chalkboards are made of porcelain enamel. In this particular manufacturing process, a tough and durable material such as steel is used as the base. There are several thicknesses of steel used in the manufacture of chalkboards, but the most common is 22-gauge. Another crucial element is silica, a crystalline compound derived from quartz or similar minerals. Found in the crust of the earth, silicon is a tough compound and is called silica when combined with oxygen. Silica is found in most rocks and is a common ingredient in many glass and ceramic products. The surface of a chalkboard is usually a blend of inorganic compounds such as a powdered glass opacifier and oxides, an organic element that provides color to the coating material.

Design

Chalkboards can be manufactured in a variety of sizes, styles, and colors. The most common hues are green and black, although shades of brown, blue, and gray are also available. They can be customized during the manufacturing process to include special graphic elements. A music department of a college or university, for example,

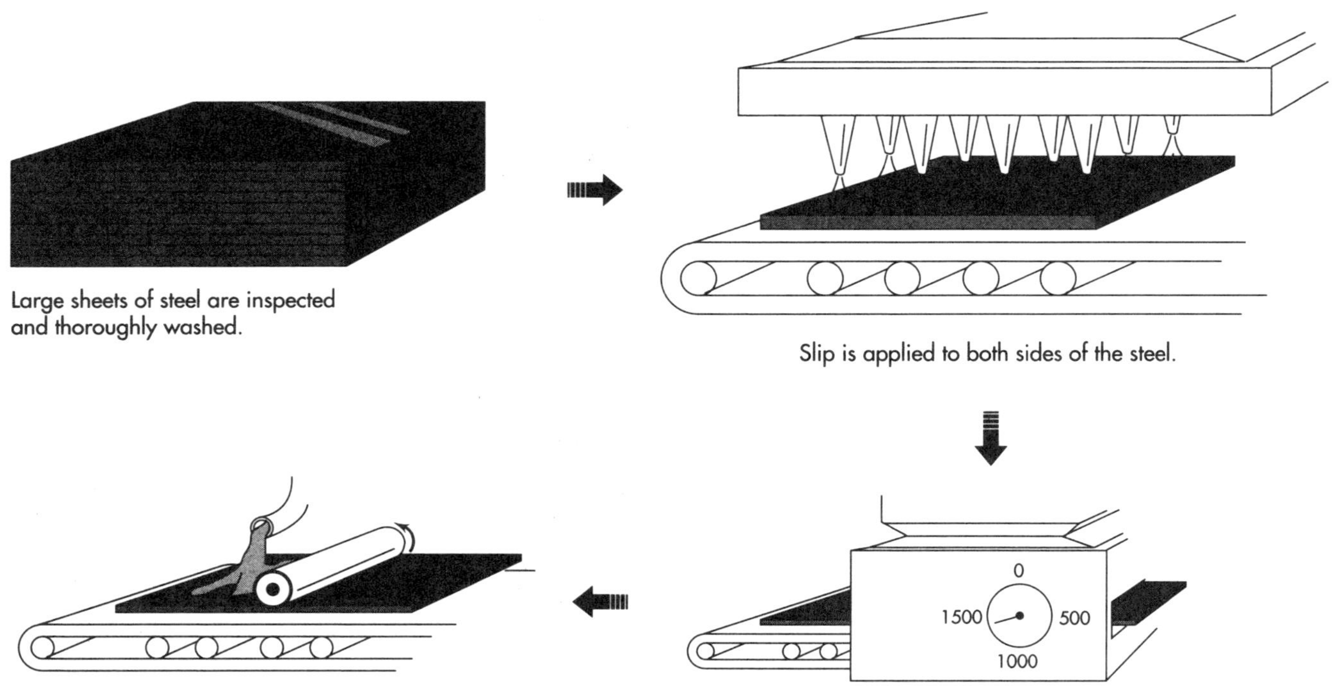

Large sheets of steel are inspected and thoroughly washed.

Slip is applied to both sides of the steel.

The dried, coated sheets of steel are fed into a furnace chamber where the slip fuses with the steel.

A surface-coating compound is then applied for smoother texture. Oxides for color may be added.

might request classroom chalkboards with musical staffs imprinted on the surface. A basketball team might use a chalkboard with a court layout to go over game strategies. Such lines are typically painted on the surface, but may also be fused onto the enamel during the manufacturing process. The size of the board may be as large as 120" x 48" for classroom use; 42" x 25" for basketball court layout; or 72" x 48" for stand-alone, moveable boards.

The Manufacturing Process

Preparation of steel

1 The manufacture of chalkboards begins when large sheets of steel in desired sizes enter the manufacturing facility from an outside supplier. This steel is cold-rolled and inspected for irregularities upon arrival. Next, the large sheets are sent into a chemical washer. This chamber washes, rinses, and dries the steel. When this step is completed, the steel is again inspected for flaws and imperfections.

Applying slip

2 Next, a slip is applied to the sheets of steel. A slip is a mixture of clay or another organic compound applied to a surface during the manufacture of porcelain or other ceramics. In this case, the slip is usually made from silica, and applied to both sides of the steel sheet by passing it through a coating chamber. The coating must be at least 0.0025 inches (.062 mm) thick. The slip is set aside to dry. The sheets once again pass through an inspection process before they are transferred to the ground coat furnace area.

Firing

3 This ground coat area of the manufacturing facility typically houses a large furnace chamber. The sheets of steel are fed into the chamber and subjected to high temperatures. This softens the steel and allows fusion of the slip with the steel. This is a crucial step in all porcelain manufacturing and fabrication of industrial ceramics.

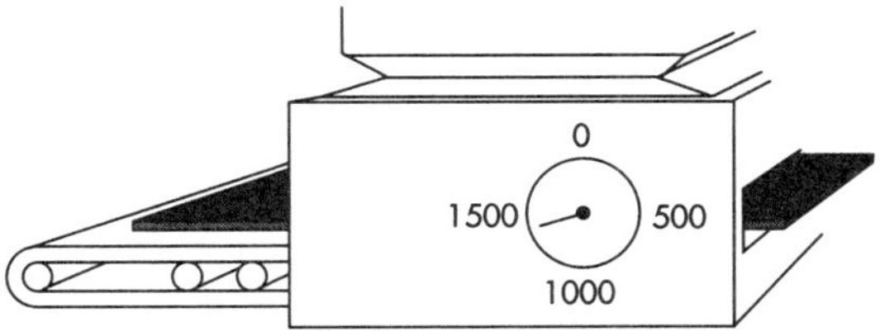

The sheet is placed through another furnace to fuse the ground coat with the surface coat.

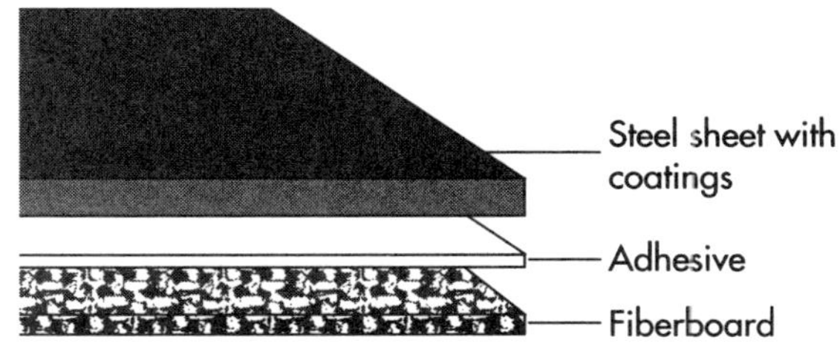

The chalkboard is laminated to fiberboard with special adhesives.

Applying surface compounds

4 Once the newly porcelained material leaves the furnace, it is treated with a surface-coating compound. Typically, this compound is derived from glass opacifiers and imparts a smoother texture to the board. Oxides for color may also be added. Again, this coating must be at least 0.0025 inches (.062 mm) thick. The boards are once again sent to a drying area. After they are completely cooled and dried, they are once again inspected for surface blemishes and uniformity of color.

Fusing the coats

5 Next, the boards are placed in a cover coat furnace. The purpose of this heating process is to fuse the first ground coat with the surface coat. A temperature of at least 1200°F (649°C) is needed to successfully complete this process. Next, the chalkboards are passed through a cooling chamber, which gradually reduces the temperature of the steel so that the flat sheets do not buckle or weaken, which might occur if left to cool by themselves.

Final surface preparation

6 Next, the surface of the chalkboard is laminated onto a fiberboard. This backing material must be at least 0.44 (11 mm) inch thick. A special adhesive is used for this application. In the next few steps, the finishing touches are put on the board. A wood or aluminum trim is added to the edges to make a border, and accessories such as chalktrays, map rails and hooks, and flag holders are attached.

Quality Control

The manufacturing of porcelain enamel chalkboards falls under the category of industrial ceramics, and manufacturers of the product adhere to standards set by the Porcelain Enamel Institute. One important guideline of this organization is its gloss standard. This is measured by a 45-degree gloss meter. According to the specifications, the gloss of a chalkboard cannot exceed three units as measured by the meter. This assures uniformity of writing surface. Further quality specifications as to durability are also detailed in Porcelain Enamel Institute guidelines.

The Future

The future of chalkboards is limited. Manufacturers of the product are diversifying into the making of dry-erase boards, which are smooth polypropylene surfaces. Special markers are used to write on them, and they can be erased by a piece of cloth. They are replacing standard chalkboards, particularly in business settings, because chalk dust is seen as a health hazard to humans and harmful to sensitive electronic and computer equipment.

Where To Learn More

Book

Good, Harry G. *A History of American Education.* Macmillan, 1973.

Periodicals

Dennis, Lee. "Old Schoolhouse: Collectibles Get Top Grades." *Antiques & Collecting Hobbies*, October 1990, p. 64+.

Gilbert, Steve. "Schoolhouse Blackboard." *Workbench*, December-January 1992, p. 62+.

—Carol Brennan

Cigarette

Background

Tobacco smoking is a practice which has changed little since American natives first stuffed the tobacco they cultivated in the hills of what is today modern Mexico into hollow reeds. As the practice spread through the Americas, different cultures wrapped their tobacco in vegetable leaves or corn husks, or put it in pipes for smoking. Spanish explorers enjoyed smoking and returned to the Old World with cigars (tobacco wrapped in tobacco leaves). In the beginning of the 16th century, beggars in Seville, Spain developed the first paper-rolled cigarettes when they collected discarded cigar butts, shredded them, and rolled them in scraps of **paper**. Although the Spanish elite first dismissed them as recycled garbage, these *cigarillos*, or little cigars, eventually gained popularity during the 18th century. Cigarette smoking spread to Italy and Portugal, and eventually to the rest of Europe and into Asia.

As cigarette use spread, the cultivation of tobacco gained in popularity. The Spanish, who had begun to cultivate the plant in the West Indies around 1530, soon transplanted it to their own native soil. Jean Nicot, the Portuguese ambassador to France, introduced tobacco to that country in the 1560s. The ambassador's surname later formed the basis for tobacco's botanical name, *nicotiana*, and the French coined the term "cigarette." In 1612, John Rolfe of Virginia began the commercial cultivation of tobacco, which became the first and most important export of the English colonies. In fact, French and English smokers soon came to prefer the mild taste of Maryland and Virginia tobacco to their homegrown varieties.

At first, all cigarettes were rolled manually, whether by the individual smoker or by shop workers, who rolled and glued cigarettes before they were packaged. Baron Josef Huppmann was an integral figure in modernizing early cigarette production. He established the Ferme cigarette factory in St. Petersburg, Russia in 1850 and opened a branch in Dresden, Germany in 1872. Ten years later he also established the Monopal cigarette works in New York City. In the 1850s, Englishman Robert Peacock Gloag manufactured cigarettes with Turkish tobacco and yellow tissue paper. Gloag's method used a thin metal tube to feed crushed tobacco into a paper cylinder, forming a cigarette.

In the U.S., cigarettes continued to be produced manually until the late 1800s. To make a cigarette, the worker sat in front of a table containing a small trench the length of a cigarette. The rolling paper was placed in the trench so its edges were slightly above the tabletop, and a pinch of shredded tobacco was placed in the paper. The worker, wearing a piece of felt over the palm of the hand, rubbed the felt over the trench until it caught an edge of the paper. Continuing the motion, the worker rolled the cigarette into shape and sealed it with paste. A good roller could make almost 40 cigarettes per minute using this method.

In 1880, James A. Bonsack was granted a U.S. patent for a cigarette machine that uniformly fed tobacco onto a continuous strip of paper. It mechanically formed, pasted, closed, and cut cigarettes with a rotary

In the beginning of the 16th century, beggars in Seville, Spain developed the first paper-rolled cigarettes when they collected discarded cigar butts, shredded them, and rolled them in scraps of paper.

This 1914 pamphlet collected testimony opposing smoking by boys. (From the collections of the Henry Ford Museum & Greenfield Village.)

In 1914 Henry Ford outraged the tobacco industry by publishing a widely publicized booklet condemning smoking. The pamphlet, entitled *The Case Against the Little White Slaver*, contained testimonials from doctors, lawyers, ministers, and employers, among others, on the delitrious effects of smoking.

Ford prefaced his attack on cigarettes by soliciting a letter from Thomas Edison that read: *"The injurious agent in cigarettes comes principally from the burning paper wrapper. The substance thereby formed is called 'acrolein.' It has a violent action on the nerve centers, producing degeneration of the cells of the brain, which is quite rapid among boys. Unlike most narcotics, this degeneration is permanent and uncontrollable. I employ no person who smokes cigarettes."*

For Ford, as for many of the others, cigarettes posed more of a moral threat than a physical one. Ford aimed his attack at young boys, hoping to dissuade them from taking up the habit. He wrote of cigarette smokers congregating in saloons and pool halls, and linked smoking to criminal activity. He produced testimony from university presidents that smokers seldom excelled in academics.

The Ford Motor Co. was one of the few businesses that forbid smoking on its premises. Not only were factory workers prohibited from lighting up, but Ford dealerships, all 7,000 around the world, banned smoking by employees, customers, or visitors. It was considered a victory for the workers when smoking was finally permitted by Henry Ford II after his grandfather's death in 1947. In 1949 the right to smoke was made part of the contract developed during formal bargaining between the company and the United Auto Workers union.

William S. Pretzer

blade. Six years later this machine was refined by William O'Brien and James B. Duke, and it produced 4,000,000 cigarettes per day and reduced costs by 50 cents per every 1,000 made. During the World War I era, the longstanding popular bias against female smokers began to diminish, providing a new market for the tobacco companies. Packaging machines were developed during the early 1900s, and, in 1931, moisture-proof cellophane that preserved the freshness of the cigarettes was introduced. Also in the 1930s, seed flax, an herb commonly cultivated in the U.S., was discovered to be a viable new source of cigarette paper. This discovery and the erection of a cigarette paper plant in North Carolina enabled the U.S. cigarette industry to flourish after the end of World War II.

Cigarettes and Health

As the popularity of cigarette smoking increased in U.S., the federal government and private agencies began to investigate the hazards of smoking. Tar, a residue present in tobacco smoke, was found to contain at least a dozen carcinogens. When cigarette smoke is drawn into the mouth, throat, and lungs, the tar condenses to form brownish deposits on the walls of the airways. Nicotine is a toxic alkaloid that is both narcotic and addictive. It occurs naturally in tobacco, although the percentage varies depending on the growing conditions and curing methods. The nicotine contained in tobacco first stimulates and later depresses the central nervous system. It also increases heart rate, blood pressure, and the heart's need for oxygen.

In 1964, the federal government published its first report on smoking and human health. The latest studies released by the Surgeon General and the Environmental Protection Agency (EPA) state that cigarette smoking increases the incidence of heart disease, cancer of the larynx, esophagus, and mouth, and birth defects in pregnant women. They also detail the newly studied effects on female smokers and the carcinogenic properties of secondhand smoke.

In response to these concerns, several cigarette manufacturers introduced a number of

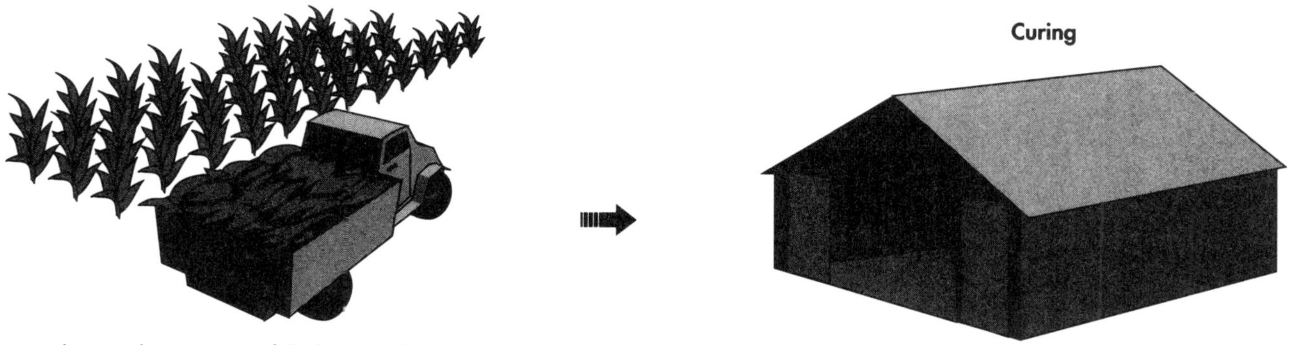

Tobacco plants are carefully harvested in fields and transported to curing areas.

In air curing, stalks are hung in a barn and dried naturally.

alternative cigarettes, including menthol, filter-tipped, and low-tar cigarettes. Menthol cigarettes smell and taste "cooler" because they are flavored with a substance found in mint oil, although they pose the same health risks. Filters help block some materials from entering the body, but their effectiveness varies from brand to brand, and even low-tar cigarettes expose the body to potentially harmful levels of tar. Recently, manufacturers have sought to reduce the amount of nicotine in cigarettes as well.

Raw Materials

The most important component of cigarettes is tobacco, which grows in two varieties: *Nicotiana tabacum*, or cultivated tobacco, and *Nicotiana rustica*, or wild tobacco. Native to the western hemisphere, the plant is now widely grown in countries such as China, India, Brazil, the former Soviet Union, Turkey, and the U.S. About one third of the tobacco cultivated in the U.S. is exported. North Carolina is the leading domestic grower, followed by Kentucky, South Carolina, Tennessee, Virginia, and Georgia, all of which have favorable soil and climates for tobacco growing. The plant does best in light and sandy loam soils that drain well and permit good aeration. The tobacco plant requires a frost-free growing season of 100-130 days; thus, it tends to be cultivated within 50 degrees latitude of the equator.

Cigarette rolling papers use seed flax mixed with paper pulp to produce a thin, flammable paper. The filters are made of synthetic, cotton-like fibers that catch particles as they are drawn through the length of the cigarette. The finished cigarettes are packaged in hard or soft cardboard boxes and wrapped in protective cellophane.

The Manufacturing Process

Growing the tobacco

1 Tobacco is initially grown in outdoor frames called seedbeds. In warm regions, the frames are covered with mulch or a cotton top sheet; in cooler regions, glass or plastic shields are installed to protect the plants. After 8-10 weeks, when the seedlings are almost 10 inches (25 cm) tall, they are transplanted to the fields. Although transplanting machines are available, the vast majority of the world's tobacco plants are still planted manually. As the plants grow, the heads are broken off by hand so the leaves will grow fuller, a process called *topping*. The plants stay in the field 90-120 days before they are harvested.

Harvesting the tobacco

2 Tobacco plants are harvested by one of two methods, *priming* or *stalk-cutting*. In the priming method, the leaves are gathered and brought to a curing barn as they ripen. In the stalk-cutting method, the entire plant is cut and the plants are allowed to

The most important component of cigarettes is tobacco, which grows in two varieties: *Nicotiana tabacum*, or cultivated tobacco, and *Nicotiana rustica*, or wild tobacco. Although transplanting machines are available, the vast majority of the world's tobacco plants are still planted by hand.

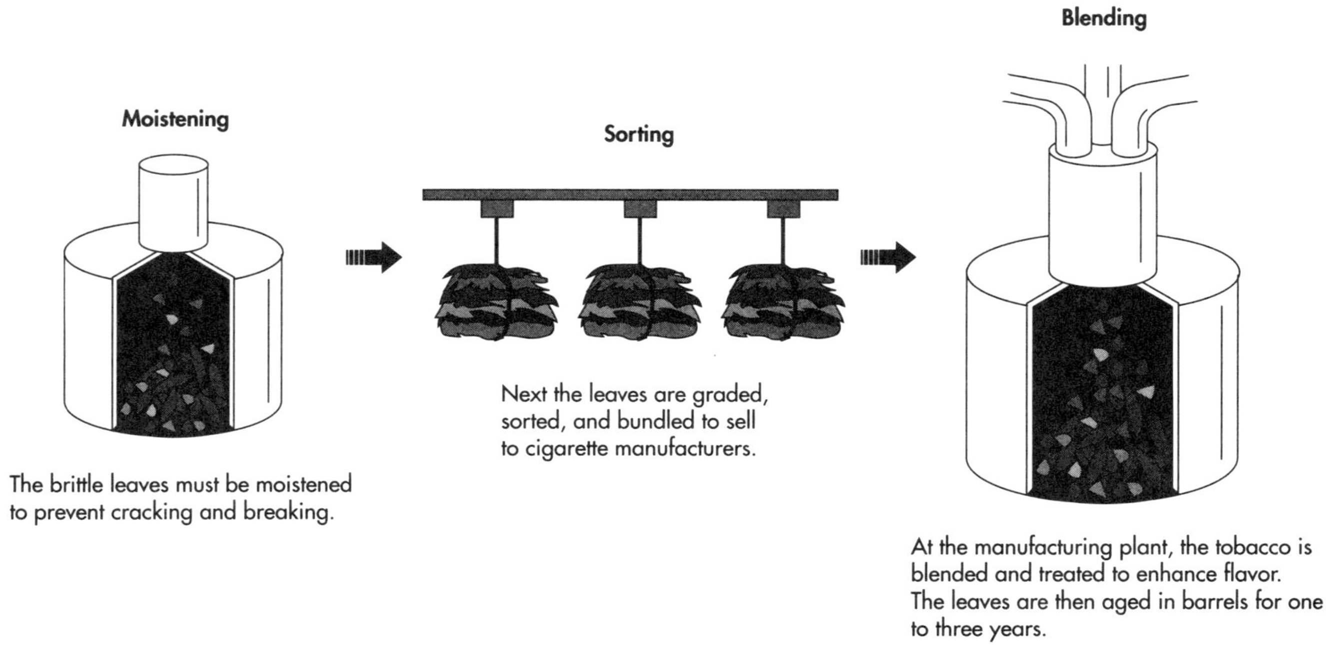

wilt in the field before being taken to the curing barn.

Curing the leaf

3 Next, the leaves are carefully, gradually dried in a specially constructed barn by air curing, flue curing, or fire curing. *Air curing* uses natural weather conditions to dry tobacco. Stalks are hung in a barn with ventilators that can be opened and closed to control temperature and humidity. Artificial heat is used only during cold or excessively humid weather. The stalks are hung for four to eight weeks.

4 *Flue curing* is done in small, tightly constructed barns that are artificially heated. The heat comes from flues (metal pipes) that are attached to furnaces. Open oil and gas burners are sometimes used, but this method is problematic because smoke cannot come in direct contact with the tobacco. Flue curing takes about four to six days.

5 *Fire curing* dries tobacco with low-burning wood fires whose smoke comes in direct contact with the leaves, thus producing a smoky flavor and aroma. The tobacco is allowed to dry naturally in the barn for three to five days before it is fire-dried for 3-40 days.

Moistening and stripping

6 Unless humid weather conditions eliminate the need, the brittle, cured tobacco leaves must be conditioned in moistening chambers so they do not break when they are handled. After moistening, the tobacco is stripped. During this process, the leaves are sprayed with additional moisture as a precaution against cracking or breaking.

Sorting and auctioning

7 After the leaves are moistened and stripped, they are sorted into grades based on size, color, and quality, and tied in bundles for shipment. The farmers then bring the tobacco to warehouses, where it is placed in baskets, weighed, graded once again by a government inspector and, finally, auctioned to cigarette manufacturers.

Conditioning, aging, and blending

8 After they have purchased and transported the material to their factories, manufacturers treat and age the tobacco to enhance its flavor. First, the manufacturer

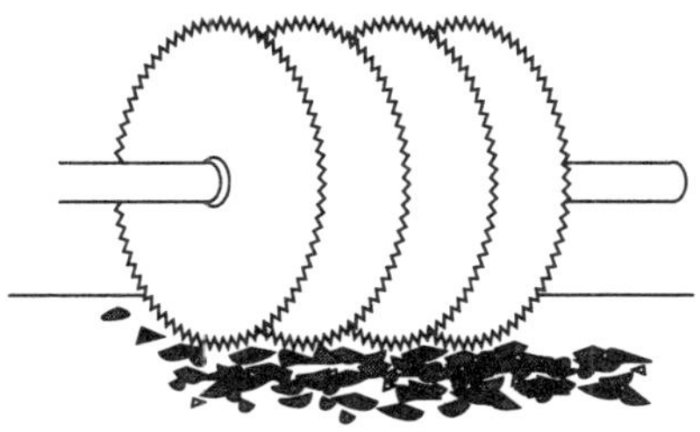

The leaves are shredded, and fruit juices or menthol may be added.

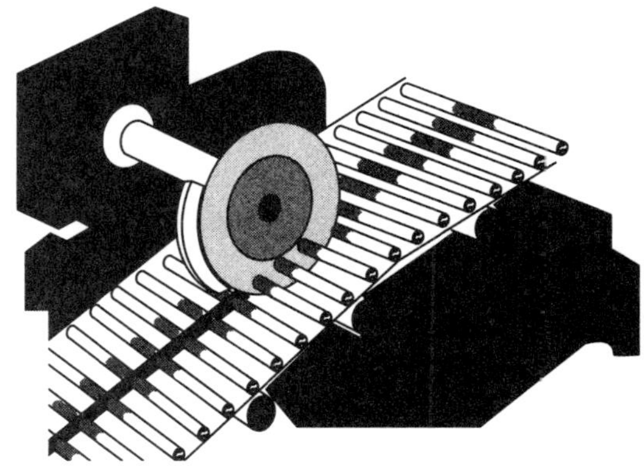

The tobacco is dispersed over a continuous roll of cigarette paper. A machine cuts the paper, seals the edge, and adds filters.

redries the tobacco. This involves completely drying the leaves by air and then adding a uniform amount of moisture. Packed into barrels called hogsheads, the tobacco is then aged for one to three years, during which period it develops its flavor and aroma. After it is aged, the tobacco leaves are again moistened and the stalks and other wastes removed. Leaves from different types of tobacco are mixed to create a particular flavor.

Making the cigarettes

9 After blending, the tobacco leaves are pressed into cakes and mechanically shredded. Materials such as fruit juices or menthol are added to give additional flavor. The final shredded tobacco is then dispersed over a continuous roll of cigarette paper. A machine rolls the shredded tobacco into the paper and cuts it to the desired length. A device then grabs each cigarette and fastens a filter in one end. Modern cigarette machines can produce 25-30 cigarettes a second.

Packaging

10 The final stage of cigarette manufacture is packaging. The completed cigarettes are packed 20 to a package. The hard or soft packs are mechanically sealed in cellophane and hand-placed in cartons.

The Future

As more and more evidence suggests that cigarette smoking harms both smokers and those around them, efforts to limit smoking in the U.S. have intensified. The federal government continues to take an active role in smoking and public health. Recent Surgeons General have all publicly opposed the practice as an unnecessary health risk. Numerous studies into the effects of smoking have been funded, and government-sponsored public service announcements are being aired and printed. Many such announcements are tailored towards children, as are many anti-smoking school and public health programs. To raise revenues and discourage younger smokers, several states have approved large cigarette taxes. Employers have also banned smoking within their buildings to curb secondhand smoke.

Though the market for tobacco products has eroded significantly since the U.S. government began publishing reports on the dangers of cigarette smoking, smoking is still common among Americans. Although a smaller percentage of American men smoke today than thirty years ago, a greater percentage of smokers are women and teenagers. Large cigarette manufacturers have also been cultivating markets in developing nations, particularly the Far East.

Where To Learn More

Book

Grise, Vernier. *The U.S. Tobacco Industry.* U.S. Department of Agriculture Economics Research Service, 1988.

Periodicals

Colford, Steven W. "What High Court Ruling Means for Cigarette Ads." *Advertising Age*, June 29, 1992.

Mallory, Maria. "For Reynolds, Where There's Smokeless...." *Business Week*, March 27, 1995, p. 39.

Manning, Anita. "Smoke Report May Change Public Habits." *USA Today*, January 7, 1993, p. A-1.

Marwick, Charles. "Advocates Say Smoke-Free Eventually May Result from More Curbs, Taxes on Tobacco Use." *JAMA*, February 10, 1993, p. 724.

"Tobacco Industry Fights Anti-Smoking Tax Plan." *New York Times*, October 25, 1992, pp. 1-21.

—Jim Acton/Kristine M. Krapp

CNC Machine Tool

Background

CNC or "computer numerical controlled" machines are sophisticated metalworking tools that can create complicated parts required by modern technology. Growing rapidly with the advances in computers, CNCs can be found performing work as lathes, milling machines, laser cutters, abrasive jet cutters, punch presses, press brakes, and other industrial tools. The CNC term refers to a large group of these machines that utilize computer logic to control movements and perform the metalworking. This article will discuss the most common types: lathes and milling machines.

History

Although wood-working lathes have been in use since Biblical times, the first practical metalworking lathe was invented in 1800 by Henry Maudslay. It was simply a machine tool that held the piece of material being worked, or workpiece, in a clamp, or spindle, and rotated it so a cutting tool could machine the surface to the desired contour. The cutting tool was manipulated by the operator through the use of cranks and handwheels. Dimensional accuracy was controlled by the operator who observed the graduated dials on the handwheels and moved the cutting tool the appropriate amount. Each part that was produced required the operator to repeat the movements in the same sequence and to the same dimensions.

The first milling machine was operated in much the same manner, except the cutting tool was placed in the rotating spindle. The workpiece was mounted to the machine bed or worktable and was moved about under the cutting tool, again through the use of handwheels, to machine the workpiece contour. This early milling machine was invented by Eli Whitney in 1818.

The motions that are used in machine tools are called "axis," and are referred to as "X" (usually left to right), "Y" (usually front to back), and "Z" (up and down). The worktable may also be rotated in the horizontal or vertical plane, creating a fourth axis of motion. Some machines have a fifth axis, which allows the spindle to pivot at an angle.

One of the problems with these early machines was that they required the operator to manipulate the handwheels to make each part. Besides being monotonous and physically exhausting work, the ability of the operator to make identical parts was limited. Slight differences in operation resulted in variation of the axis dimensions, which, in turn, created poorly fitting or unusable parts. Scrap levels for the operations were high, wasting raw materials and labor time. As production quantities increased, the number of usable parts produced per operator per day were no longer economical. What was needed was a means to operate the motions of the machine automatically. Early attempts to "automate" these operations used a series of cams that moved the tools or worktable through linkages. As the cam rotated, a link followed the surface of the cam face, moving the cutting tool or the workpiece through a series of motions. The cam face was shaped to control the amount of linkage movement, and the rate at which the cam turned controlled the feedrate of the tool. These early machines were difficult to set correctly, but once set, they offered excellent repeatability for their day.

Growing rapidly with the advances in computers, CNCs can be found performing work as lathes, milling machines, laser cutters, abrasive jet cutters, punch presses, press brakes, and other industrial tools.

This 1980s CNC chucking machine shows the technology's combination of machine tool and computer. (From the collections of Henry Ford Museum & Greenfield Village.)

The story of how computers were first linked to production machinery is full of intrigue and controversy. It illuminates how intertwined industry, universities, and the military have been in the 20th century. The story also exemplifies how difficult it is to attribute many innovations to a single individual or institution. Sorting out who did what when and with what impact is a complex enterprise.

In 1947, John Parsons headed up an aeronautical manufacturing firm in Traverse City, Michigan. Faced with the increasing complexity of parts' shapes and the mathematical and engineering problems they entailed, Parsons searched for ways to reduce his firm's engineering costs. He asked the International Business Machine Corp. to allow him to use one of their main frame office computers to make a series of calculations for a new helicopter blade. Eventually, Parsons made an arrangement with Thomas J. Watson, the legendary president of IBM, whereby IBM would work with the Parsons Corporation to create a machine controlled by punched cards. Soon Parsons also had a contract with the Air Force to produce a machine controlled by cards or tape (like a player piano) that would cut contour shapes like those in propellers and wings. Parsons then went to engineers at the Massachusetts Institute of Technology Servomechanism Laboratory for help with the project. MIT researchers had been experimenting with various types of control processes and had experience with Air Force projects dating back to World War II. In turn, the MIT laboratory saw this as an opportunity to expand their own research into control and feedback mechanisms. The successful development of computer-numerical-control machine tools was then undertaken by university researchers seeking to meet the demands of military sponsors.

William S. Pretzer

Some have survived to this day and are called "Swiss" machines, a name synonymous with precision machining.

Early Design to Present Day Operation

The modern CNC machine design grew out of the work of John T. Parsons during the late 1940s and early 1950s. After World War II, Parsons was involved in the manufacture of helicopter rotor blades, which required precise machining of complex shapes. Parsons soon found that by using an early IBM computer, he was able to make much more accurate contour guides than were possible using manual calculations and layouts. Based on this experience, he won an Air Force contract to develop an "automatic contour cutting machine" to produce large wing section pieces for aircraft. Utilizing a computer card reader and precise servomotor controls, the resulting machine was huge, complicated, and expensive. It worked automatically, though, and produced pieces with the high degree of accuracy required by the aircraft industry.

By the 1960s, the price and complexity of automated machines had been reduced to the point where they found applications in other industries. These machines used direct current electric drive motors to manipulate the handwheels and operate the tools. The motors took electrical instructions from a tape reader, which read a paper tape approximately 1 in (2.5 cm) in width that was punched with a select series of holes. The position and sequence of the holes allowed the reader to produce the necessary electrical impulses to turn the motors at just the precise time and rate, which in effect operated the machine just like the human operator. The impulses were managed by a simple computer that had no "memory" capability at the time. These were often called "NC," or Numerical Controlled machines. A programmer produced the tape on a typewriter-like machine, much like the old "punch cards" used in early computers, which served as the "program." The size of the program was determined by the feet of tape needed to be read to produce a specific part.

With advances in integrated electronics, the tape was eliminated, or used only to load the program into magnetic memory. In fact, the capacity of the memory of modern CNC machines is still sometimes referred to as "feet of memory."

The modern CNC machine works by reading the thousands of bits of information stored in the program computer memory. To place this information in the memory, the programmer creates a series of instructions that the machine can understand. The program may consist of "code" commands, such as "M03" which instructs the controller to move the spindle to a new position, or "G99," which instructs the controller to read an auxiliary input from some process inside the machine. Code commands are the most common way to program a CNC machine tool. However, the advancement in computers has allowed the machine tool manufacturer to offer "conversational programming," where the instructions are more like plain words. In conversational programming, the "M03" command is entered simply as "MOVE," and the "G99" command is simply "READ." This type of programming allows faster training and less memorizing of the code meanings by the programmers. It is important to note, however, that most conversational machines still read code programs, since the industry relies on that form of programming quite heavily.

The controller also offers help to the programmer to speed up the machine use. In some machines, for example, the programmer can simply type in the location, diameter, and depth of a feature and the computer will select the best machining method for producing the feature in the workpiece. The latest equipment can take a computer-generated engineering model; calculate the correct tool speeds, feeds, and paths; and produce the part without a drawing or program ever being created.

Modern Design and Raw Materials

The mechanical components of the machine must be rigid and strong to support the quickly moving parts. The spindle is usually the strongest part and is supported by large bearings. Whether the spindle holds the work or the tool, an automatic clamping feature allows the spindle to rapidly clamp and unclamp during the program run.

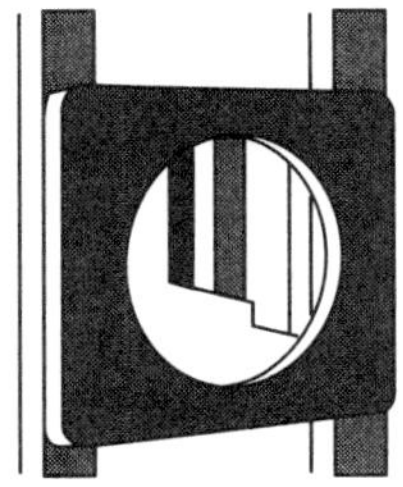

Base of machine is welded together.

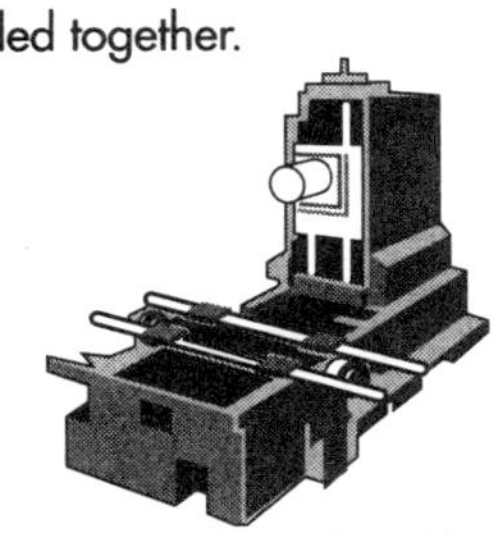

The ways are ground and hardened for the machine components and worktable to move on.

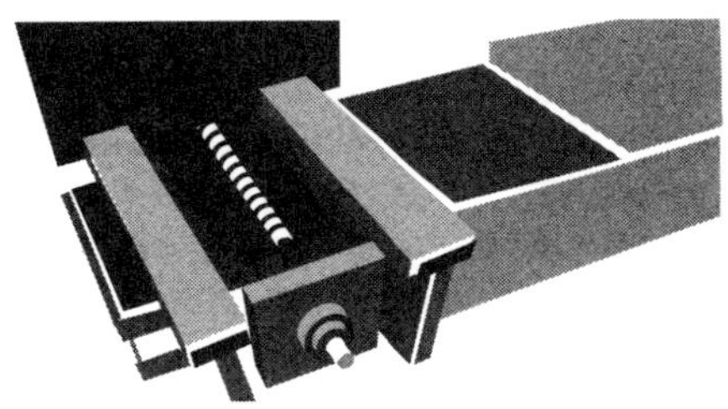

Ballscrews are bolted to the base with the bearing-mount bolted to the worktable or spindle-carrier.

Attached to the side of the machine is a magazine of different tools. A transfer arm, sometimes called the tool bar, removes a tool from the machine, places it into the magazine, selects a different tool from the magazine, and returns it to the machine through instructions in the program. Typical cycle time required for this procedure is two to eight seconds. Some machines may contain up to 400 tools in large "hives," each automatically loaded in sequence as the program runs.

The bed or worktable of the machine is supported on hardened steel "ways" which are usually protected by flexible guards.

Cast **iron** or Meehanite used to be the material of choice for metalworking machines. Today, most machines make liberal use of weldments of hot-rolled steel and wrought

Cast iron or Meehanite used to be the material of choice for metalworking machines. Today, most machines make liberal use of weldments of hot-rolled steel and wrought products such as stainless steel to reduce cost and allow fabrication of more intricate frame designs.

The mechanical components of the machine must be rigid and strong to support the quickly moving parts. The spindle is usually the strongest part and is supported by large bearings. Whether the spindle holds the work or the tool, an automatic clamping feature allows the spindle to rapidly clamp and unclamp during the program run.

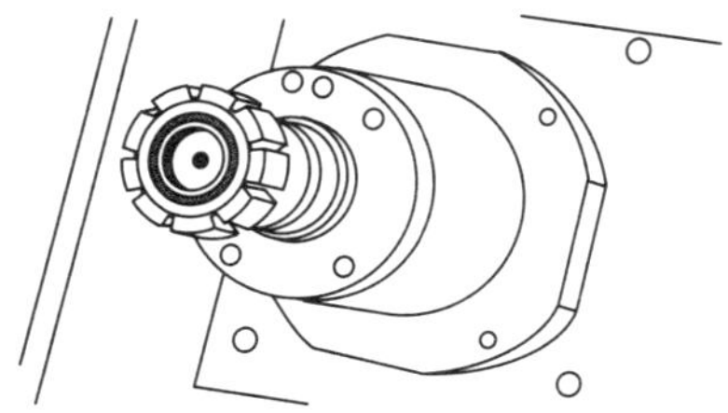

Once the spindle is machined and ground, it is mounted to the drive motor and bolted to a moveable spindle carrier.

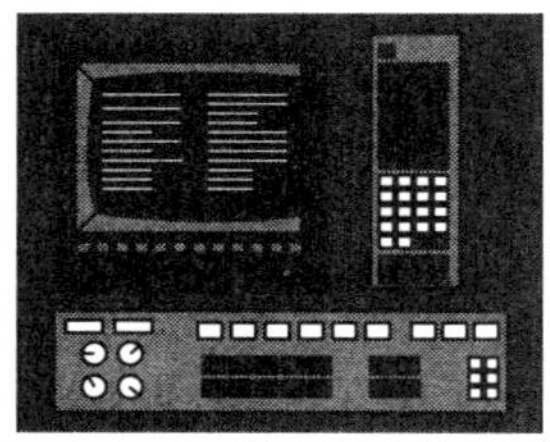

The computer is housed in a separate enclosure. This is to protect the sensitive electronics.

products such as stainless steel to reduce cost and allow fabrication of more intricate frame designs.

Some machines are designed as cells, which means they have a specific group of parts they are designed to manufacture. Cell machines have large tool magazines to carry enough tools to do all of the various operations on each of the different parts, large worktables or the ability to change worktables, and special provisions in the controller for data inputs from other CNC machines. This allows the CNC machine to be assembled with other similarly equipped machines into a Flexible Machining Cell, which can produce more than one part simultaneously. A group of cells, some containing 20 or 30 machines, is called a Flexible Machining System. These systems can produce literally hundreds of different parts at the same time with little human intervention. Some are designed to run day and night without supervision in what is referred to as "lights out" manufacturing.

The Manufacturing Process

Until recently, most machining centers were built to customer specifications by the machine tool builder. Now, standardized tooling design has allowed machines to be built for stock or later sale, since the new designs can perform all the needed operations of most users. The cost of a new CNC machine runs from about $50,000 for a vertical center to $5 million for a Flexible Machining System for engine blocks. The actual manufacturing process proceeds as follows.

Welding the base

1 The base of the machine is either cast or welded together. It is then heat treated to remove casting or welding stresses and to "normalize" the metal for machining. The base is fixtured into a large machining center, and the mounting areas for the ways are machined to specification.

2 The ways are ground flat, bolted, and pinned to the base.

Bolting the ballscrews

3 The mechanisms that move the bed or spindle are called ballscrews. These change rotary motion of the drive motors into linear motion and consist of a screw shaft and support bearings. As the shaft turns, a bearing mount follows the spiral grooves in the shaft and produces a very accurate linear movement that moves either the worktable under the spindle, or the spindle carrier itself. These ballscrews are bolted to the base with the bearing mount bolted to the worktable or spindle carrier.

Mounting the spindle

4 The spindle is machined and ground, mounted to its drive motor, and then bolted to the movable spindle carrier. Each axis of motion has a separate ballscrew and set of ways in most machining centers.

The controller

5 The computer, or controller, is an electronic assembly separate from the rest of the machine. It has a climate-controlled enclosure mounted on the side of the frame or in an operator's console. It contains all of the operating memory, computer boards, power supplies, and other electronic circuitry to operate the machine. Assorted wiring connects the controller to the machine motors and positional slides. The

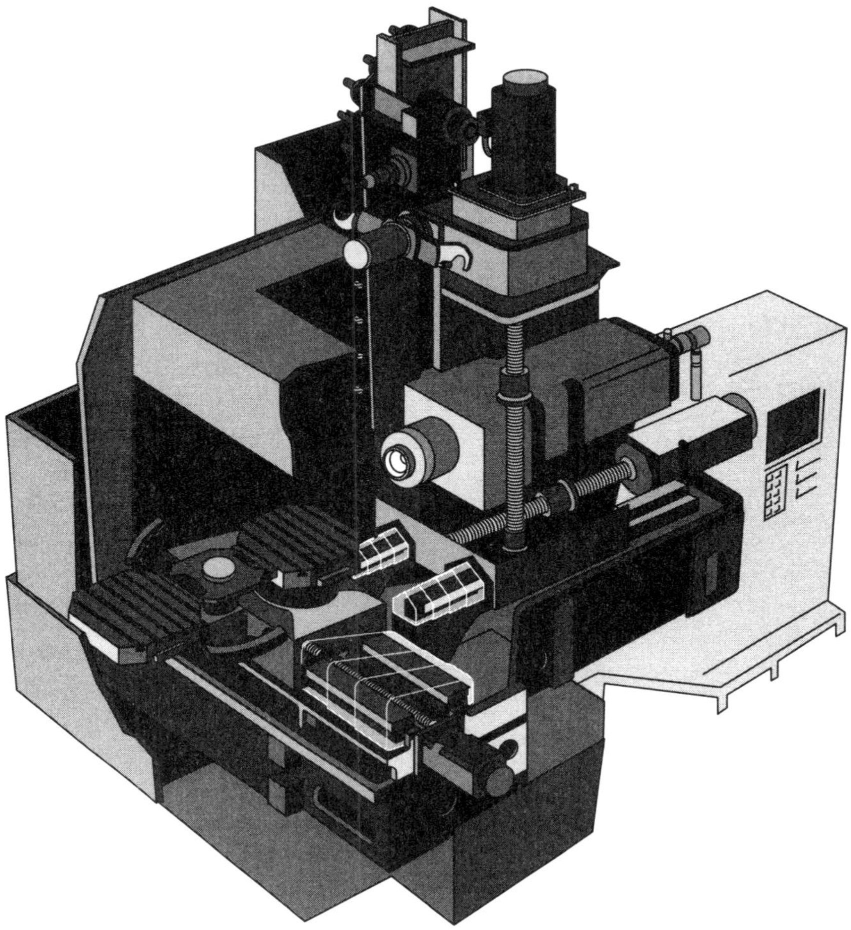

The assembled machine is test run for accuracy. After testing, the finished machine is painted and prepared for shipment.

slides continuously send the axis location information to the controller, so the exact position of the worktable in relationship to the spindle is always known. The front of the controller has a video screen that displays the program information, position, speeds and feeds, and other data required for the operator to monitor the machine's performance. Also on the front panel are the data entry keys, data connection ports, and start-stop switches.

6 The assembled machine is test run for accuracy. Each machine has slight physical differences that are mathematically corrected in the computer operating system. These correction values are stored in a separate memory, and the machine checks these continuously. As the machining center wears from use, these parameters can be recalibrated to assure accuracy. After testing, the finished machine is painted and prepared for shipment.

Quality Control

Quality in a machining center must be built in from the design through delivery and setup. Careful instruction to the operators is also important to prevent a crash, the unintentional collision of the work with the tool. Crashes can result in tool damage or machine failure. Many controllers have subprograms to sense an impending crash and place the machine into emergency stop. All CNCs are shipped with special handling to avoid shocks, and are set up carefully by factory-trained technicians. The original correction factors are recorded for later reference. Complete programming, operation, and maintenance manuals are provided.

The Future

The future of CNC machines is exploding. One idea under development is a spider-like machine whose spindle is suspended by six telescoping ballscrew struts. The struts are like the ways in a conventional machine, but they are round with the ballscrew assembly in the center. The motions of the spindle are controlled by a sophisticated computer performing millions of calculations to assure proper part contour. Costing several million dollars to develop and using high level, proprietary mathematics, this machine promises to perform previously unheard of operations in metal machining. Advancement in computers and artificial intelligence will make CNC machines of the future faster and easier to operate. This will not come cheaply, and the price of sophisticated CNC machines will be beyond the reach of many companies. It will, however, reduce the prices of the basic CNC machines performing the original three-axis movements.

Where To Learn More

Books

Noble, David F. *Forces of Production.* Alfred A. Knopf, 1984.

Olivo, C. Thomas. *Machine Tool Technology and Manufacturing Processes.* C. Thomas Olivo Associates, 1987.

Periodicals

Adams, Larry. "CNC Machining Centers: Do It All." *Wood & Wood Products*, June 1994, pp. 101-102.

Herrin, Golden E. "The Next 40 Years of NC." *Modern Machine Shop*, April 1994, p. 154.

Koelsch, James R. "Milling, Drilling, and Boring." *Manufacturing Engineering*, August 1994, pp. 65-81.

Morningstar, D. "Anatomy of a Machine Tool: Controls, Servos and Electromechanical Mysteries." *Tooling and Production*, June 1993, pp. 68-70.

Proctor, P. "New Flexible Bed Replaces Fixed Tools." *Aviation Week and Space Technology*, November 14, 1994, p. 52.

Pye, Andy. "Turning Towards CNC." *Engineering*, January 1994, pp. 26-27.

—Douglas E. Betts

Coin

History

Human civilizations have long used metals as a medium for exchange. In addition to their long-lasting properties, metals lend themselves easily to melting and casting. As early as 1000 B.C., the Chinese were using a type of metal token to represent payment. These artifacts have been labeled "spade" and "key" money because of their resemblance to a digging tool and to the modern-day Yale key. Both types bore denominations and were cast from molds. Although the ancient Egyptians did not mint coins, gold weights and rings were used to trade for products and services.

The first record of Western coins did not occur until 700 B.C., in western Asia Minor. Evidence of coins made from a naturally occurring alloy of gold and silver called electrum were found in the foundation of the temple to Artemis at Ephesus on the banks of the Aegean Sea. King Croesus of Lydia, who ruled from 560 to 546 B.C., has been credited with creating a bi-metallic system of pure gold and pure silver coins. These early coins typically carried imprints of animals, such as bulls, birds, insects, or mythical creatures. Engravings of vegetables were also popular. Imprints were stamped on one side of the coins with a tool bearing that particular design. Coin design was elevated to an art form during this period, and elaborately imprinted coins were afforded a high status. Many Greek cities vied for the distinction of having the most beautifully designed coins.

Alexander the Great built mints throughout his kingdom, from Macedonia to Babylon. He instituted uniform weights and types. It was during Alexander's reign that the coin portrait rose to popularity. Rulers, gods, and goddesses were the portraits of choice. By the fourth and fifth centuries A.D., engravers in Italy, and particularly in Sicily, were generally recognized as the experts in coin design. So revered was their skill that the engravers began signing their work.

Before the advent of the Industrial Age, the striking of coins was accomplished manually. A round blank of metal was placed over an anvil that had been fitted with an imprinted die. Another die was affixed to a pestle, which was then placed on top of the blank. The coin maker held the pestle in place with one hand and then brought a two-pound hammer down on top of the pestle. Remarkably, this resulted in seven tons of pressure, which forced impressions into both sides of the blank. The high relief typical of early Greek coins sometimes required two or three blows to achieve the desired effect. Heating the blank before striking often reduced the number of required strikes. This method allowed one coin to be struck every two seconds.

Raw Materials

Each country institutes strict guidelines for the composition of its currency. The outside vendors who provide the metal or "stock" to the mint must follow these guidelines to the letter. Originally, the U.S. penny (or cent) was composed of 95% copper and 5% **zinc**. In 1982, this composition was changed to a copper-plated zinc. A zinc alloy with traces of copper constitute the core of the coin, while the outer surface is electroplated with copper. Five-cent coins are composed of cupronickel, an alloy of

Early coins typically carried imprints of animals such as bulls, birds, insects, or mythical creatures. It was not until the reign of Alexander the Great that coin portraits rose to popularity.

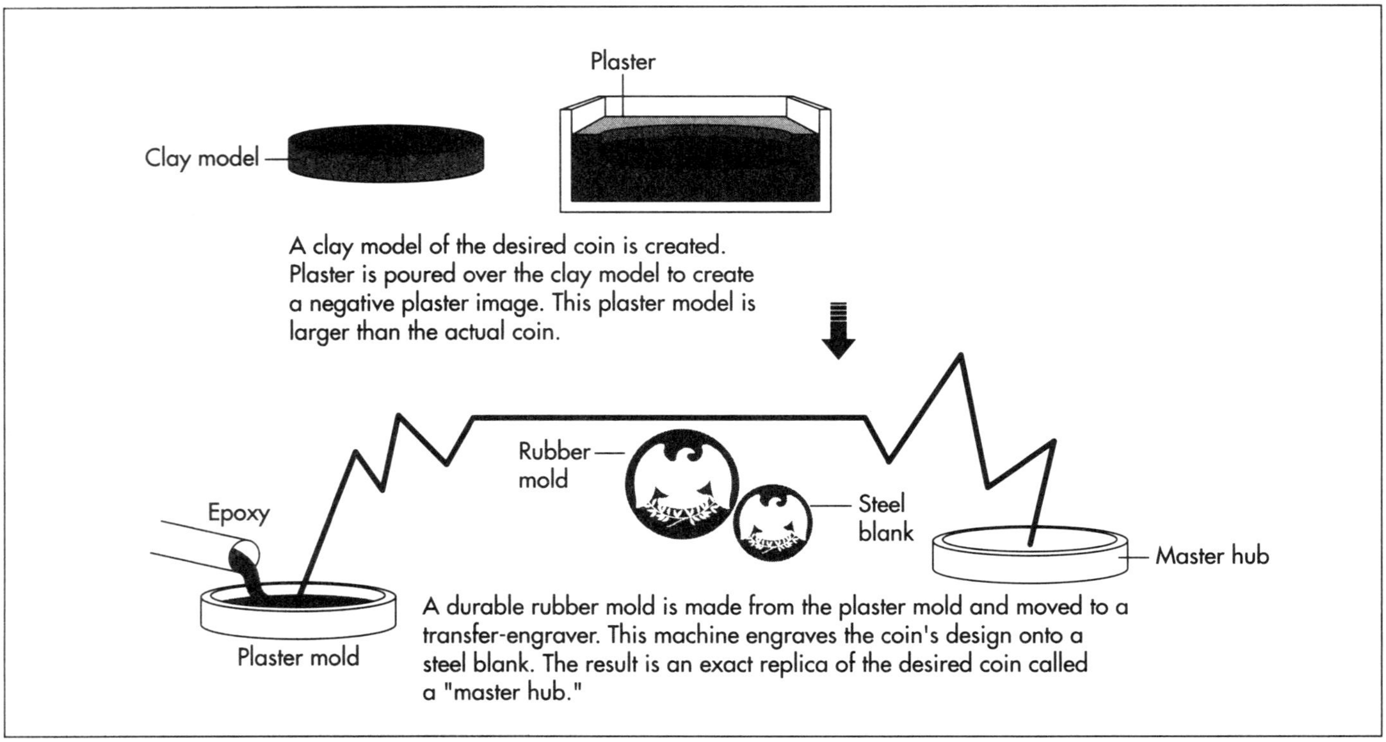

75% copper and 25% nickel. Dimes, quarters, half dollars, and dollar coins are made from three layers of metal that have been bonded or "cladded" together. The outer layer is 75% copper and 25% nickel, while the core is pure copper.

In the factories of the outside vendors, the metal alloys are melted in furnaces and poured into rectangular molds. When the stock cools, it is rolled under pressure to the appropriate thicknesses. The rolling process causes the stock to harden excessively, requiring the application of a process called annealing. In this process, a series of heatings and coolings softens the stock and brings it to the consistency needed for shaping and stamping. The rectangular sheets of metal are cut into strips approximately 13 inches (33 cm) wide and 1,500 feet (457 m) long, and then rolled into coils. The mints purchase the coils according to their needs.

The Manufacturing Process

Molding and engraving the master hub

1 When a new coin has been commissioned, sculptors employed by the mint develop a set of sketches. When one particular sketch has been approved and refined, the sculptor creates a clay model. The model can be anywhere from three to twelve times larger than the actual coin.

2 Plaster is poured over the clay model to create a negative, or reverse, plaster model. The words of the inscriptions are carved into the plaster in reverse. The sculptor repeats this process several times until the plaster model is perfect.

3 Next, a durable rubber mold is made by pouring epoxy into the plaster mold. The epoxy mold is mounted onto a transfer-engraver. At one end of the transfer-engraver, a stylus traces the epoxy mold. As the stylus moves, a ratio bar in the middle of the engraver reduces the design to the actual coin size. This reduced size is communicated to a carbide tool at the opposite end, which then cuts the design into a steel blank. The result is a positive replica called a "master hub." The sculptors examine the master hub and remove any imperfections.

Creating the working dies

4 Heat-treated metal is placed under a computerized lathe, where it is smoothed

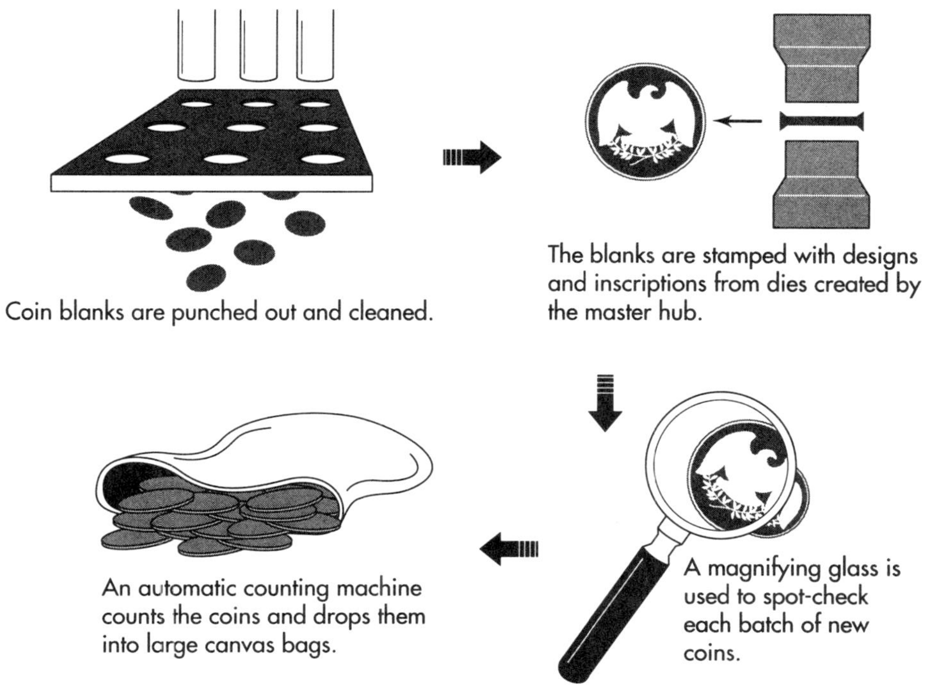

and polished into a precisely measured blank die. The master hub is pressed into the die. The result is called the "master die." The master die is used to create working hubs and working dies. The master hubs and dies are then placed in storage.

Punching out the blanks

5 The appropriate coil of metal is fed through a blanking press, which punches out round disks corresponding in size to the coin to be minted. The blanks are cut at a speed of 400 strokes per minute. The leftover scraps of metal are shredded and recycled for future use.

Annealing and pickling the blanks

6 The blanks are subjected to another annealing process and then placed in industrial washing machines and dryers. The lubricants used in these various processes cause the blanks to become stained and oxidized.

7 The blanks are next placed into revolving tubs or barrels filled with an acidic pickling agent. As the blanks are tossed together in the tubs, they become burnished.

Sorting and weeding the blanks

8 The blanks are sifted through a "riddler," a metal sheet fitted with holes that match the exact size of the particular coin to be minted. In this manner, misshapen and odd-sized blanks are weeded out.

Striking the coins

9 The perfect blanks are carried by conveyer belt to the coining press, where they are stamped with designs and inscriptions. A steel collar is inserted into the press around one of the dies. The die for the reverse side is loaded into the upper arm of the press. Hundreds of tons of air pressure push the blank into the collar. At the same time, the overhead die is forced down into the collar and onto the blank. The impact causes the impressions to form on both sides of the blank. The press releases the newly minted coin, and it moves along a conveyer belt to the inspection line.

In some instances, the collar has grooves to make the ridged edges on the coin. Otherwise, the grooves are made after the striking process, on a tool called an upsetting mill. The size of the press varies from single capacity to ones that stamp four coins simultaneously. Single-striking presses generally stamp 400 coins per minute, with pressure loads up to 180 tons. Multiple presses can crank out 120 coins per minute under 250 tons of pressure.

Inspecting and sorting

10 The press operator spot-checks each batch of new coins with a magnifying glass. The coins move through another riddler that sorts out blanks that have become misshapen or dented during the striking process.

Counting and bagging

11 An automatic counting machine spits out a predetermined amount of coins and drops them into large canvas bags. The bags are sewn shut, loaded onto pallets, and then moved by forklift trucks to storage vaults.

Quality Control

Inspections are carried out at many points throughout the engraving and manufacturing process. Alloys are analyzed using x-ray fluorescent spectrometers or chemical processes. The surface condition of the blanks is checked frequently for maximum center line average. The diameters of the blanks are measured with gauges such as micrometers. Weights are controlled by weighing a specific number of coins against a standard weight plus a pre-determined allowance.

The Future

In the mid-1990s, the U.S. made preparations to join other industrialized countries in the use of a dollar coin instead of a paper bill. Although backers point to the savings that the switch would bring, and environmentalists extol the virtues of phasing out the dollar bill, traditionalists see the dollar bill as a well-entrenched symbol of the United States. Unions and trade associations representing the paper industry also voiced opposition to the new coin.

Elimination of the penny has also gained support in recent years. Ironically, the American public's view of the penny as worthless has caused millions of people to stockpile them in jars and boxes at home, to be traded in for larger denominations at a later date. This has led to a shortage of pennies in the commercial arena. Decisions about eliminating coins are intensely political, attesting to the continuing symbolic power of the metallic coin.

Where To Learn More

Periodicals

"The Art of Money." *The Economist*, January 15, 1994, p. 91+.

Georges, Christopher. "House Republicans, Believing Change is Due, Consider Plan to Insert Coin in Place of $1 Bill." *The Wall Street Journal*, April 18, 1995, p. A22.

"Noncents." *The New Republic*, August 8, 1994, p. 7+.

Other

How Coins Are Designed and Produced. The Department of the Treasury, United States Mint, Philadelphia, Pennsylvania, June 1992.

U.S. Mint: Its History and Coinage. The Department of the Treasury, United States Mint, Philadelphia, Pennsylvania.

—Mary F. McNulty

Condom

Background

Condoms are thin sheaths worn by men during sexual intercourse to prevent pregnancy and venereal infections. According to the 1995 National Survey of Family Growth, conducted by the National Center for Health Statistics in Hyattsville, Maryland, male condoms or prophylactics are the third most popular form of birth control—preceded only by female sterilization (29.5%) and birth control pills (28.5%)—with usage at 17.7%. They are also one of the most effective: research indicates that with correct use, failure rates are 2-3%. Most condoms are made of latex rubber, but they can also be made from lamb cecum or polyurethane.

In addition to their contraceptive value, condom use has been found effective in preventing the spread of sexually transmitted diseases. In 1986, the U.S. Surgeon General endorsed the use of condoms as the only currently available effective barrier against the transmission of Acquired Immunodeficiency Syndrome (AIDS). The spread of many other sexually transmitted diseases, such as chlamydia and gonorrhea, can also be virtually eliminated with the use of a latex condom. With the government touting the health benefits of condom use, manufacturers openly advertise their products, and retailers stock condoms in visible, accessible locations. Condoms, previously kept behind the prescription counter, are now found on most store shelves. Today in the U.S., 450 million condoms are sold each year.

Despite the wide variety of styles, there are few differences among the many latex condoms available on the market today. They can be straight-sided, contoured, ribbed, sensitive, or smooth. They may be treated with lubricants or spermicides. They can be blunt-ended or have a reservoir tip. Because the condoms undergo stringent testing before they are sold, quality is generally not a marketable issue. Hence, manufacturers attempt to build brand loyalty and market their products to specific target consumers.

Condoms made from lamb cecums—the blind pouch in which the intestines begin and into which the ileum opens from one side—are also available. However, they are more expensive than latex condoms, and while they prevent pregnancy, "skin" condoms are ineffective in preventing the transmission of sexually transmitted diseases. In 1994, the Food and Drug Administration (FDA) approved a polyurethane condom for sale in the U.S. The new condom has not been extensively tested for effectiveness in preventing pregnancy and sexually transmitted diseases.

History

The first recorded use of condoms was in Egypt in 1350 b.c. In 1564, the Italian anatomist Fallopius described a linen condom used to prevent venereal disease. The term *condom* is actually a corruption of the name of an 18th-century British physician, Dr. John Conton, who provided condoms to France's King Charles II. The legendary lover Giovanni Casanova (1725-1798) used pieces of sheep intestine to protect himself against venereal disease. The first condom manufacturer in the U.S. was Schmid Laboratories. In 1883, Julius Schmid, a former

Condoms, previously kept behind the prescription counter, are now found on most store shelves. Today in the U.S., 450 million condoms are sold each year.

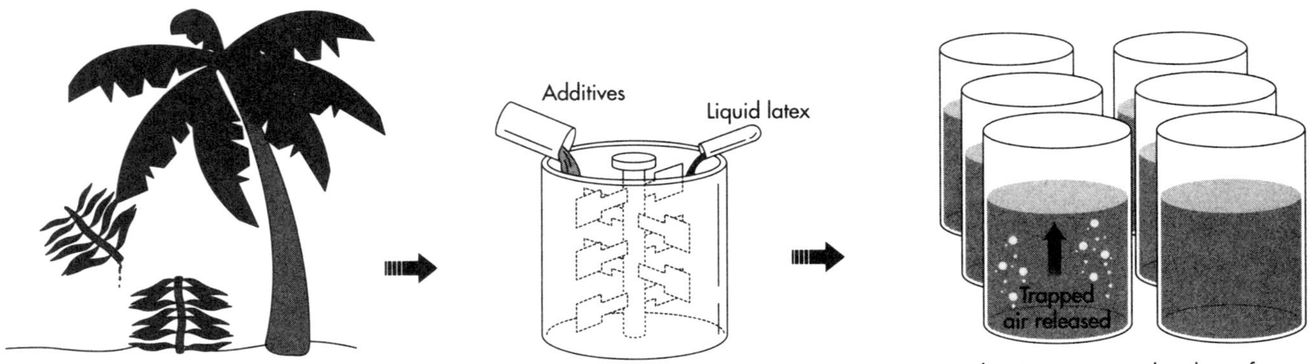

Rubber latex is extracted from the milky fluid of various tropical plants.

The latex and chemical additives are combined to form a paste.

The mixture is stored in drums for vulcanization. This also allows air to escape.

While condoms made from lamb cecums are available, most condoms are made from rubber latex.

sausage skin-maker, acquired a business that manufactured bottle seals from animal membranes. Five years later, Schmid used his experience with sausage casings and capping skins to manufacture prophylactic sheaths from lamb cecum.

Even as Schmid was marketing his skin condoms, technology was progressing to allow thinner, more pliable, and less expensive condoms to became available. Vulcanization, the chemical linking of rubber particles that was originally developed in 1839 for use in automobile tires, made condoms strong, durable, and fit for consumer use. A form of rubber called latex was developed in the 1930s; this new material, combined with a mechanized dipping process, facilitated the mass production of condoms and lowered manufacturing costs.

Raw Materials

The first condoms manufactured by Julius Schmid were formed from the cecum of lambs. As of 1990, condoms made from lamb cecum accounted for 5.5% of the market, and because of their higher price, for 20% of retail sales. This manufacturing process remains relatively unchanged since Schmid first manufactured condoms: the cecums are washed, defatted, and salted. The raw skins are then shipped to the finishing plants. New Zealand, which raises large numbers of sheep, is the primary source and initial processing center for most "skin" condoms.

Latex condoms account for most of today's market. Because rubber latex is a natural material, it can vary greatly in strength and elasticity. Manufacturers add chemicals to the latex to stabilize and standardize the composition of the latex. Many brands also add talc, lubricants, or spermicides to the condoms before they are packaged.

The Manufacturing Process

Collecting the raw materials

1 Rubber latex is obtained from the milky fluid produced by various tropical plants. Latex is actually an emulsion or dispersion of tiny rubber particles in water, and ingredients added to the latex must be able to attach to the rubber particles during compounding.

Compounding

2 Next, chemical additives are mixed to form a paste. This paste is then blended with the liquid latex in a process called compounding.

Storage

3 The latex and chemical compound is then unloaded into drums for storage, where it remains for approximately seven days. During this period, vulcanization chemically strengthens the bonds of the rubber. The storage time also allows any

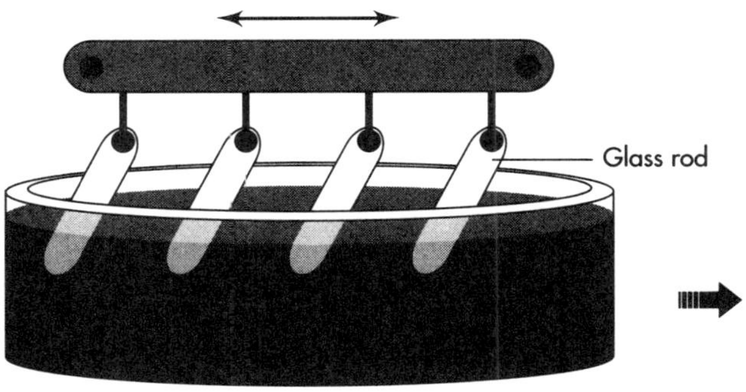

The latex compound is transferred to a dipping machine. Glass rods move back and forth through the compound, and as the mandrels rotate, latex coats the surface.

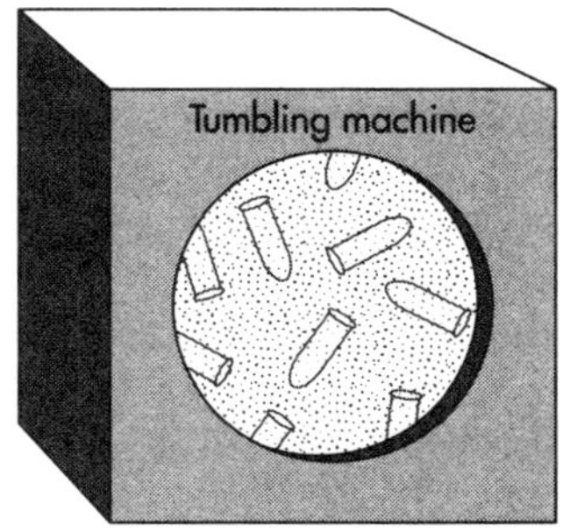

The condoms are dried and then placed in tumbling machines, where they are coated with powder to prevent the rubber from sticking to itself.

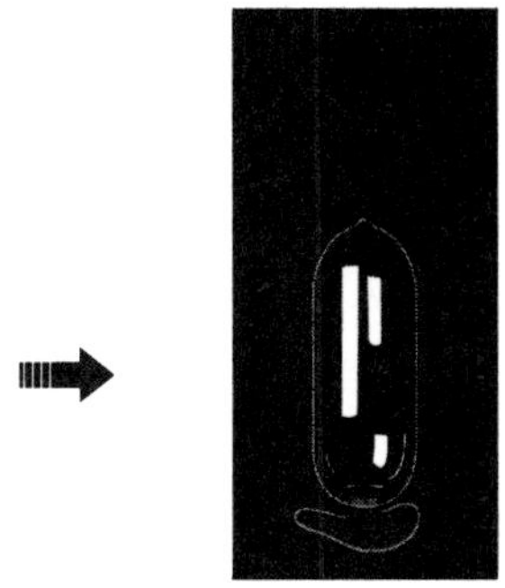

The condoms are put through rigorous tests before they are rolled by a machine and packaged.

air, which might have been trapped in the mixture during compounding, to escape.

Dipping

4 The compound is then added to the dipping or condom-forming machine. The dipping machine is a long, hooded machine approximately 100 feet (30.5 m) in length. Thick tempered glass rods move along a closed belt between two circular gears. The belt drags the rods, which are called mandrels, through a series of dips into the latex compound. The mandrels rotate to spread the latex evenly. Several coats are required to build the condom to its required thickness. Between each dip, the latex is hot air dried.

5 After the final dipping and drying, the condoms automatically roll off the mandrels. A machine shapes and trims the ring of latex at the base of each condom.

Tumbling

6 Next, the condoms are put in a tumbling machine, where they are coated with talc or another similar powder to prevent the rubber from sticking to itself.

Testing

7 After a curing period of several days, the condoms are sampled by batch and tested for leaks and strength. The first such test is the inflation test, in which the condom is filled with air until it bursts. Condoms are required to stretch beyond 1.5 cubic feet, about the size of a watermelon, before bursting. This test is considered most important because the elasticity of the condom keeps it from tearing during intercourse.

8 In the water-leakage test, the condom is filled with 10 ounces (300 ml) of water and inspected for pin-sized holes by rolling it along blotter paper.

9 Condoms are also tested electronically. This involves mounting each condom on a charged stainless steel mandrel. The mandrel is passed over by a soft, conductive brush. If pin holes are present, a circuit will be established with the mandrel, and the machine will automatically reject the condom.

Packaging

10 Condoms that have successfully passed these tests are rolled by a machine. Rolling the condom makes it easier to package and use. Lubricant and spermicide may be applied by a metering pump just before the top wrap is added in the foiling process.

Quality Control

Condoms are classified as Class II Medical Devices. According to the Medical Device Amendments of 1976 of the FDA, the FDA is required to inspect each condom manufacturing plant at least once every two years. All electrical and mechanical equipment must be impeccably maintained. Condom-dipping machines are designed to operate continuously; if they remain idle, their mechanisms can get clogged and rust. During any downtime, partially cured compound cannot be left in the dip tank because it could contaminate future production.

All condoms sold in the U.S. must comply to specifications that were voluntarily developed by condom manufacturers and adopted by the FDA. Condom measurements can range from 5.8-7.8 inches (150-200 mm) in length, 1.8-2.1 inches (47-54 mm) in width, 0.001-0.003 inches (0.03-0.09 mm) in thickness (although most condoms range between 0.002 and 0.0024 inches), and the weight cannot exceed 0.07 ounces (2 grams). Additionally, physical characteristics must include a minimum tensile strength of 15,000 pounds psa and elongation before breakage of 625%.

The FDA reviews U.S. company records and spot checks batches for cracking, molding, drying, or sticking latex. The organization also tests every lot of imported condoms. Upon sampling, lots will not pass inspection if they reveal greater than 4% failure with respect to the above dimensions, 2.5% failure with respect to tensile strength and elongation, and 0.4% failure due to leakage.

The Future

Manufactured by Chicago-based Female Health Co., the Reality condom for women has been on the market and available through family-planning clinics in the U.S. since August 1994. It has been sold in 12 European countries since 1993. The female condom is a long polyurethane sheath with one open ring and one closed ring that is anchored between the women's cervix and vagina. According to Female Health Co., these condoms are 40 times stronger than latex; each costs approximately $3, compared to about $.64 for male latex condoms.

Research, started in 1988, lead to the development of the new polyurethane male condom, which also went on the market in 1994. The new condom is said to be just as strong but only one-tenth as thick as the latex condom. It is recommended for people who are sensitive to latex condoms.

Where To Learn More

Book

Murphy, James S. *The Condom Industry in the United States.* McFarland and Company, Inc., 1990.

Periodicals

"How Reliable Are Condoms?" *Consumer Reports,* May 1995, pp. 320-25.

Goldberg, Stephanie B. "Birth Control Update: Specialists Recommend Reviewing Choices As Life Changes." *Chicago Tribune,* March 5, 1995, pp. 1, 6.

—Susan Bard Hall

Contact Lens

Background

The contact lens is a device worn in the eye to correct vision, although some people wear colored contact lens to enhance or change their eye color. The thin plastic lens floats on a film of tears directly over the cornea. For some forms of eye disease, contact lenses correct vision better than conventional spectacles. Many people prefer contact lenses over glasses for cosmetic reasons, and active sports enthusiasts prefer contact lens because of the freedom it provides them. There are basically three types of lenses: soft, hard, and gas-permeable. Soft contact lenses are usually more comfortable to wear, but they also tear more easily than hard contact lenses. Hard lenses also tend to "pop" out more frequently. Gas-permeable lenses are a compromise between the hard and soft, allowing greater comfort than hard lenses but less chance of tearing than soft lenses. Contacts are usually worn during the day and taken out every night for cleaning. Extended-wear lenses allow users to leave in their contacts for longer periods of time, even when they're sleeping. More recently, one-a-day contact lenses are gaining popularity among lens wearers. These contacts are worn for only one day and thrown away, eliminating the hassle of cleaning them every night.

History

The first contact lens was made by German physiologist Adolf Fick in 1887. Fick's lens was made of glass and was a so-called scleral lens because it covered the sclera, the white part of the eye. By 1912, another optician, Carl Zeiss, had developed a glass corneal lens, which fit over the cornea. Two scientists, Obrig and Muller, introduced a plastic scleral lens in 1938. It was made of the material commonly known as Plexiglas. Because it was lighter than glass, the Plexiglas lens was easier to wear. The first plastic corneal lens was made by Kevin Touhy in 1948.

To fit these early lenses, an impression was made of the patient's eyeball, and the lens was formed in the resulting mold. This procedure was doubtlessly uncomfortable, and the lenses themselves were often problematic to wear. Scleral lenses deprived the eye of oxygen, and many of these earlier lenses slipped out of place or popped out of the eye, and were often, oddly enough, difficult to remove. Touhy's first corneal lens had a diameter of 10.5 millimeters, and in 1954 Touhy reduced the diameter further to 9.5 millimeters, resulting in better wearability. Around this time the Bausch & Lomb company developed the keratometer, which measures the cornea, and eliminated the need for eyeball impressions.

The first successful soft contact lenses were developed by chemists in Czechoslovakia. In 1952, professors in the Department of Plastics at the Technical University in Prague set themselves a task of designing a new material that was optimally compatible with living tissue. They did not set out to create contact lenses, but by 1954 the team of Czech scientists had invented what is called a "hydrophilic" (for its affinity to water) gel, a polymer plastic that was suitable for eye implants. The scientists immediately recognized the new

The first successful soft contact lenses were developed by chemists in Czechoslovakia in 1952. When other scientists doubted the feasibility of using a polymer plastic to make contact lenses, one of the chemists, Otto Wichterle, produced 5,500 pairs of contact lenses in his home.

Contacts can be formed by filling small molds with liquid plastic and allowing them to harden. These "buttons" are then cut to correct prescriptions.

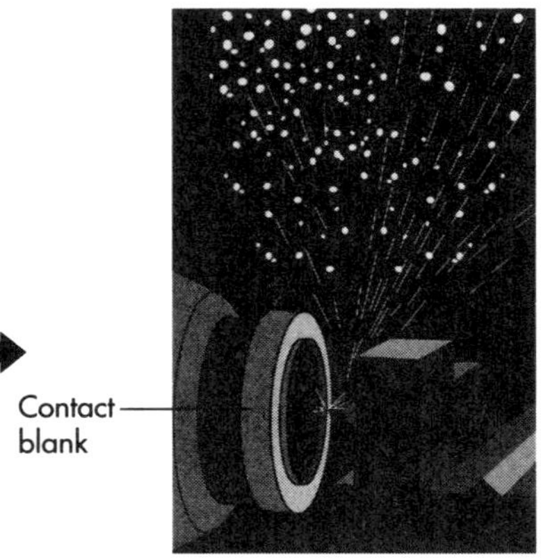

Contacts can also be made by cutting a blank from a solid plastic rod. The blank is placed on a lathe and sometimes cut with a diamond.

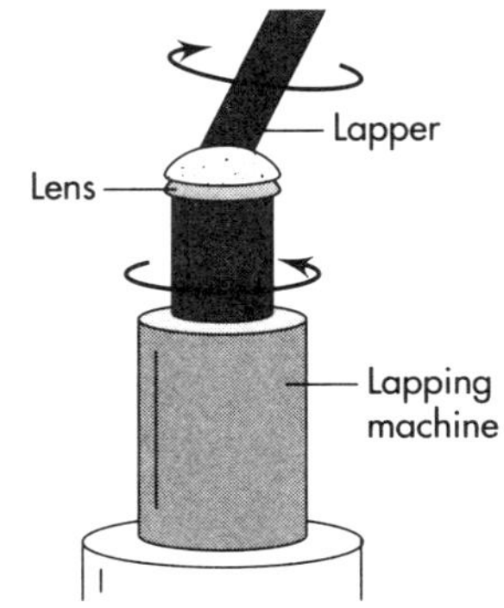

A lapping machine rotates the lens against a lapper spinning in the opposite direction. An abrasive compound on the lapper polishes the lens' surface.

plastic's potential as a corrective lens, and they began experimenting on animals. These efforts were met with scorn by their colleagues in the optics field, but one of the scientists, Otto Wichterle, was undaunted and began perfecting soft contact lenses in his kitchen. Wichterle and his wife produced 5,500 pairs of contact lenses from their home for testing in 1961, and their success eventually got the attention of the wider scientific community. The American firm Bausch & Lomb licensed the technology and launched their Softlens in 1971. That first year alone, the firm sold about 100,000 pairs, and soft contact lenses have had great appeal with the public ever since.

Raw Materials

The raw material for contact lenses is a plastic polymer. (A polymer is a blend of materials created by linking the molecules of different chemical substances.) Hard contact lenses are made of some variant of polymethyl methacrylate (PMMA). Soft contact lenses are made of a polymer such as poly hydroxyethyl methacrylate (pHEMA) that has hydrophilic qualities, that is, it can soak up water and still retain its shape and optic functions. The science of lens material is always being updated by lens manufacturers, and the specific material of any contact lens may differ depending on the maker.

The Manufacturing Process

Contact lenses may be produced by cutting a blank on a lathe, or by a molding process. The forming of the lens involves shaping the plastic into specified curvatures. The major curves of the lens are named the *central anterior curve* (CAC) and the *central posterior curve* (CPC). The CAC refers to the overall curve of the side of the lens that faces out. This outer contour produces the correct refractive change to fit the patient's visual needs. The CPC is the concave inner side of the lens. This conforms to the measurements of the patient's eye. Usually these two curves are formed first, and the lens is then called semi-finished. The lens is deemed finished when peripheral and intermediate curves are formed, and the edge is shaped.

Molding method

1 Molding the lens can be carried out in several different ways. The lenses first developed in Prague were spin-cast. Three different fluids were poured into open rotating molds. The outside curvature of the lens was shaped by the mold, and the inside curvature was formed according to the speed of the rotation of the mold. The centrifugal force of the spinning mold led to the polymerization of the fluids so that the molecular chains linked to form the required hydrophilic plastic. A more reliable mass-

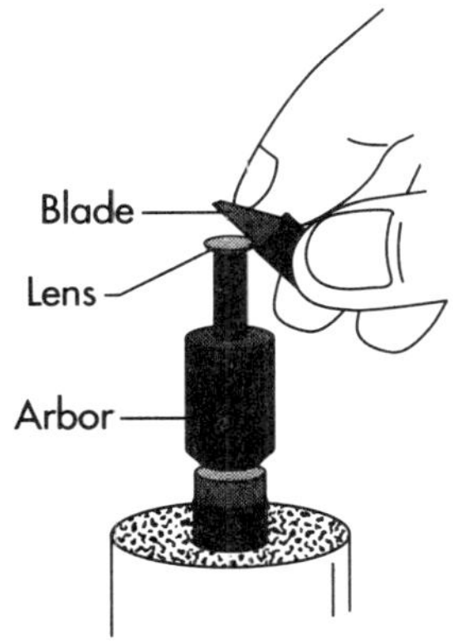

The lens is mounted on an arbor, and further cuts and curves are made with a razor blade.

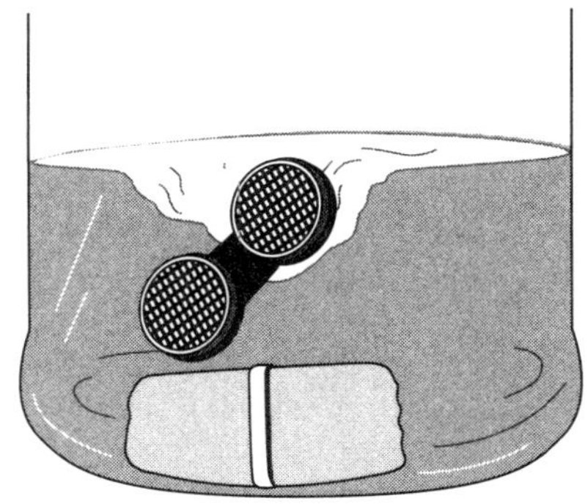
The lens is placed in a mixture of boiling water and salt for several hours to soften the lens.

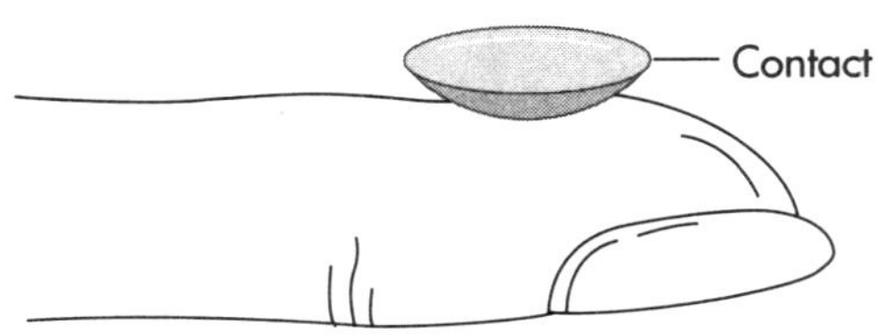

The finished contacts undergo several quality control procedures before they are sent to the user.

production method is injection molding. In injection molding, the molten plastic is injected into the mold under pressure. Then the lens is removed from the mold and cooled. The lens is then finished on a lathe. It is also possible to produce lenses entirely through molding, that is, they need no lathe cutting. This is a recent development, made possible through highly automated, computer controlled mold production.

Lathe process

2 The initial forming of the lens can also be done by cutting on a lathe. First a blank is made. The blank is a circle only slightly larger than the size of the finished lens. This can be cut from a plastic rod, or stamped from a plastic sheet. Next the blank is fastened to a steel button with a drop of molten wax. The button is then centered on a lathe, which begins to spin at high speed. A cutting tool, which may be a **diamond** or a laser, makes concave cuts in the blank to form the CPC. Indicators on the lathe measure the depth of the cuts to guide the lens operator.

The button holding the blank is next moved to a lapping machine. The lapping machine holds the blank against a lapper, which is a revolving disk coated with an abrasive compound. The shape of the lapper matches the CPC of the lens. The lapping machine spins the blank in one direction, and the lapper in the other. It also moves the blank in a small figure eight motion. The abrasion polishes the lens surface.

The polished lens is then mounted on a steel shaft called an arbor. The end of the arbor has been ground to match the CPC so the lens will fit on the shaft. The arbor is installed in a lathe, and the operator makes convex cuts in the lens to form the other major curve, the CAC. Now this side of the lens is polished, and the lapper is modified to fit the convex CAC. When this second side of the lens is polished, the lens is considered semi-finished.

Finishing

3 The contact lens requires several more curves to be ground before the lens will fit exactly on the patient's eye. The final curves are the peripheral anterior and posterior curves and the intermediate anterior and posterior curves, which govern the shape of the lens nearest and next-nearest the edge. The lens is mounted on an arbor again by suction or with double-sided tape. The arbor is installed in the lathe or grinding machine. These shallower cuts may be ground with emery paper or cut with a razor blade. The diameter of the lens may also be trimmed at this time.

Quality control

4 Quality control is very important for contact lenses, since they are medical devices and they must be custom fit. The lenses are inspected after each stage of the manufacturing process. The lenses are examined under magnification for anomalies. They are also measured by means of a shadow graph. A magnified shadow of the lens is cast on a screen imprinted with a graph for measuring diameter and curvature. Any errors in the lens shape show up in the shadow. This process may be automatically performed by computer.

Packaging

5 After the lens has passed inspection, it is sterilized. Lens are boiled in a mixture of water and salt for several hours to soften the lens. Next, the lens are packaged. Standard packaging for lenses is a glass vial, filled with a saline solution and stoppered with rubber or metal. The hydrophilic material of soft contact lenses soaks up the saline solution, which is similar to human tears, and becomes soft and pliable. The lenses in this state are ready to wear.

The Future

The material for contact lenses is the subject of much research. Scientists are investigating different chemical recipes that may give plastic more desirable characteristics. One polymer currently being researched is a silicon-oxygen compound called siloxane. Siloxane forms a thin, flexible film and admits oxygen through to the eye 25 times better than current standard soft lenses. There are disadvantages to this compound, however: siloxane does not wet easily and it attracts lipids (fats) to its surface, causing it to cloud. Researchers have found a way to add flourine molecules to the siloxane compound, causing the material to resist lipids. Then they chemically attach a wetting agent, which changes its molecular shape when boiled in a saline solution, so that the material can soak up water like traditional soft lens. This material may ultimately lead to extended-wear contacts that can be worn for weeks at a time.

Researchers are also investigating new polymers that can be used for scleral lenses. For most people, corneal lenses are the norm, but the large scleral lenses are useful for patients with severely damaged corneas. Depending on the eye problem, some patients cannot regain their sight without a corneal transplant, but scleral lenses may help patients avoid eye surgery. Scleral lenses rest on the white part of the eye and form a vault over the cornea itself. This space over the cornea is filled with artificial tears, which serve to smooth out the cornea's damaged surface. In the past, scleral lenses have been uncomfortable because they do not allow enough oxygen to the eye, but investigations into new materials are focusing on more oxygen-permeable lenses.

Material for oxygen-permeable lenses has also been experimented on the space shuttle *Endeavour*. The designers of the experiment believe that microgravity conditions would promote a lens material that repels debris better and processes oxygen more effectively than polymers made in traditional labs. If commercially feasible, a new generation of contact lenses may be manufactured in space.

Where To Learn More

Book

Ruben, Montague, ed. *Soft Contact Lenses: Clinical and Applied Technology*. John Wiley & Sons, 1978.

Periodicals

"Making Eye Contact." *Ad Astra*, September-October 1993, p. 5.

"This Contact Lens Is A Sight For Sore Corneas." *Business Week*, April 20, 1992, p. 94.

"Extending Extended-Wear Contacts." *Science News*, September 5, 1992, p. 153.

—Angela Woodward

Cowboy Boots

Designed for men who spent virtually their entire day in the saddle, cowboy boots are notoriously uncomfortable to walk in, and though adjustments have been made over the years, the boots remain unsuited for almost any work a cowboy or a rancher has to do on foot. Today, they are largely a fashion item.

Background

Traditional cowboy boots have narrow toes, high heels that slope under the foot, and leather tops that reach halfway up the shins. Designed for men who spent virtually their entire day in the saddle, cowboy boots are notoriously uncomfortable to walk in, and though adjustments have been made over the years, the boots remain unsuited for almost any work a cowboy or a rancher has to do on foot. Cowboy boots have also led a long double life as fashion accessories, beginning in the early 20th century, when Western life and work done on the open range were first mythologized in movies. Most cowboy boots that are manufactured now are not sold to people who will ever wear them on a horse, and the boots are valued more for the image they have acquired than the work they were originally intended to do.

History

In the 19th century, Anglo-Americans moving into the Southwest found people of Spanish descent already working the cattle that roamed the West Texas plains. Branding cattle and the practice of roping them from horseback were originally developed in Mexico, as was the basic design of what we now call the Western saddle. Though American settlers adopted many of their tools and methods, the Mexican *vaqueros* wore short, flat boots that were not well suited to the demands of their profession. Cowboy boots are direct descendants of the Northern European riding boot, and they may be the only contribution these settlers made to the essentially Spanish tradition of working cattle from horseback.

The Northern European riding boot was adapted for use on the range by German bootmakers who settled in Texas during the second half of the 19th century. The original German bootmakers designed a boot to meet the requirements of working in stirrups. Narrow toes made it easier for horsemen to put their feet in and out of stirrups while mounting and dismounting, and the high heel prevented the foot from slipping all the way through the stirrup and getting caught there. A foot caught in a stirrup could be especially dangerous if a cowboy were thrown out on the range, where he could be dragged for miles by a running horse. The length of the leather tops reduced chafing from stirrup leathers, and the boots also had high, reinforced arches, designed to make standing in the stirrups less strenuous. All of these features make the boots difficult to wear while working on the ground; they are particularly hard to run in, and when not on horseback many cowboys and ranchers today wear sneakers or a boot called a *roper*, with a round toe, a low heel, and a softer, more flexible sole.

Modern bootmakers divide into two categories: custom shops, where boots are made individually and much of the work is done by hand, and fully automated factories. The large industrialized companies, such as Tony Lama and Justin Industries, were originally family businesses that developed from smaller shops. The custom shops that remain in operation are often staffed by family members, and there the craftsmen are traditionally divided into "top men" and "bottom men." The former group cut, decorate, and assemble the

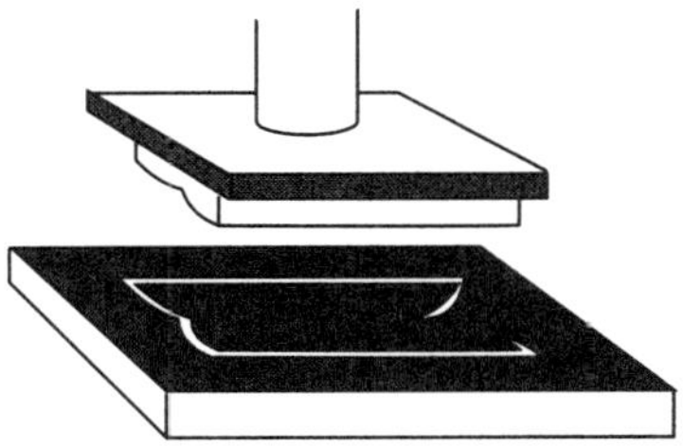

In most factories, the individual pieces of the boot are cut out using metal dies. Custom bootmakers cut them by hand.

To create patterns on the boots, stencils of the design are laid on the leather and sprinkled with white powder. An operator then follows the powder marks with a sewing machine.

upper parts of the boot, and the latter group shape the heels and soles. Top men are largely responsible for how a boot looks and bottom men for how comfortable it is to wear. Texas still remains the center for the manufacture of cowboy boots. Though a factory such as Tony Lama's in El Paso may produce thousands of pairs a week and a custom shop such as Charlie Dunn's in Austin may produce only few, the basic steps are the same.

Raw Materials

The most widely used material for cowboy boots is calfskin, which is both easier to work with and more durable than cowhide. Most calfskins used in bootmaking actually come from Europe rather than the U.S., as few Americans eat veal and the skins of European calves are less likely to be scarred by brands or **barbed wire**. In the 1990s, however, changing eating habits have brought on a worldwide shortage of leather: fewer people are eating beef and so fewer cattle are being raised to any age. Though calfskin is the most common material, cowboy boots are also made from pigskin, horsehide, and kangaroo skin. For dress boots, bootmakers use a variety of exotic leathers including the skins of armadillos, ostriches, sharks, alligators, eels, lizards, and large snakes such as pythons.

The Manufacturing Process

Cutting out the pieces

1 Once the leather has been selected, the process of making the top part of the boot begins by cutting out the individual pieces. This might be done by hand in custom shops, but in factories it is done by metal dies, which work like cookie cutters. The top of a boot consists of three parts: the part that covers the top of the foot, the part that encloses the back of the heel, and the part that fits around the bottom of the shins. These are called, respectively, the *vamp*, the *counter*, and the *uppers*. The vamp is like the top and sides of an ordinary man's shoe, but it is one piece instead of several, without lace holes and a separate tongue. The counter covers what the vamp does not. The uppers are cut in two pieces, one for the front and one for the back, designed to join each other at the sides. At this stage, the lining for the inside of each of these pieces is cut out and then glued into place. The lining is particularly important for boots made of fragile skins such as snake or eel, for the leather backing will provide most of the strength.

Decorating the pieces

2 If the boot is to have any kind of stitched decoration—whether a simple

The most widely used material for cowboy boots is calfskin, which is both easier to work with and more durable than cowhide. They are also made from pigskin, horsehide, and kangaroo skin.

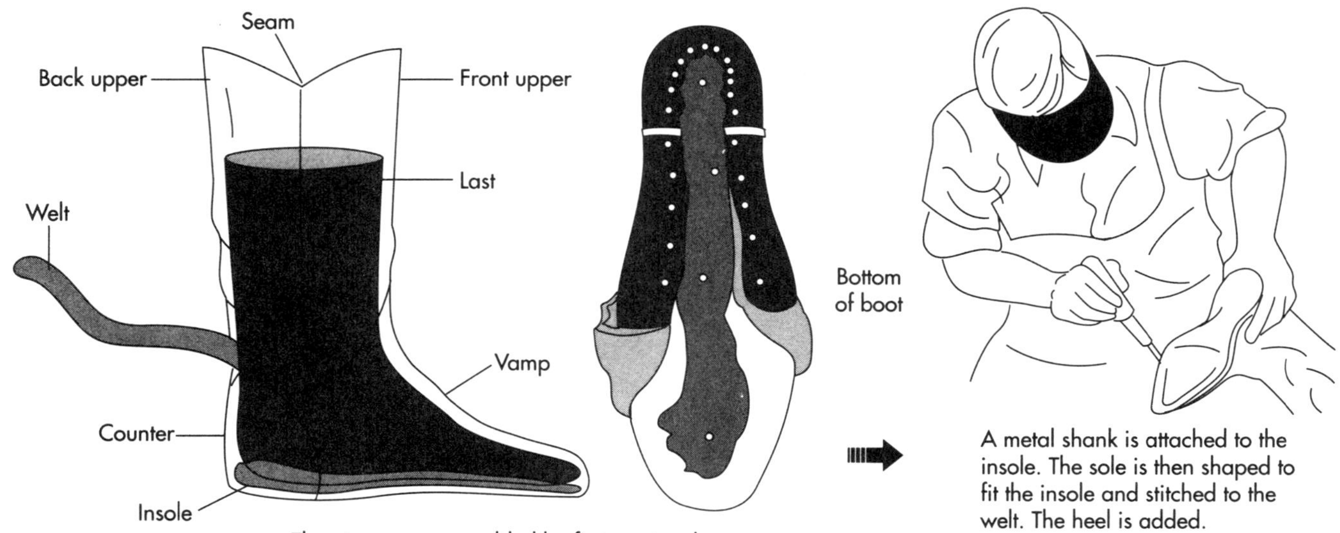

The pieces are assembled by first sewing the vamp to the front upper and sewing the counter to the back upper. The front and back are then sewn together. The insole is nailed to the last, vamp, and counter.

A metal shank is attached to the insole. The sole is then shaped to fit the insole and stitched to the welt. The heel is added.

pattern, or an elaborate picture such as a yellow rose, an oil derrick, or the state of Texas—this is done before the pieces are assembled. In custom shops, the design is sketched on a paper pattern or stencil and then outlined with a series of small holes. This stencil is laid over each piece and then sprinkled with a marking agent such as white powder, so the design can be followed by someone operating a sewing machine. Factories tend to use computerized sewing machines for this task, with preprogrammed designs, so marking the leather isn't necessary. Any additional colors the design requires are dyed into the leather at this stage.

Assembling the top of the boot

3 The boot is initially assembled in two halves, front and back. The vamp, the part that covers the top and sides of the foot, is sewn to the front upper; the counter, the part that covers the heel, is sewn to the back upper. Excess leather around the stitching is then trimmed off. A long strip of leather called the *welt* is then attached to the back of the counter and left hanging there. The welt will be used at a later stage to attach the top of the boot to the sole.

4 The front and back halves of the boot are then glued and sewn together. These seams are made first on the inside, so the boot initially takes shape inside out, like a shirt. Leather is, however, harder to work than cotton, and once the seams are made the top of the boot must be soaked in water until the leather is flexible enough to be turned right outside out again. At the end of this stage, the top part of the boot is complete; in a custom shop, the work of a top man would now be done.

Attaching the insole

5 The first step in building the bottom of the boot is attaching the insole to the vamp and the counter. The *insole*, in any shoe or boot, is the part you see when looking down inside it; it often bears the imprint of the maker's name. A key component in this part of the process is the last. The *last* is basically a model of a foot—an anatomically accurate version of a shoetree—which is left inside the boot during the rest of the manufacturing process. In a factory, lasts are standard sizes and generally made of molded plastic. In custom shops, they are made of hardwood and adjusted to the precise shape of an individual's foot. The workshop of these bootmakers may contain

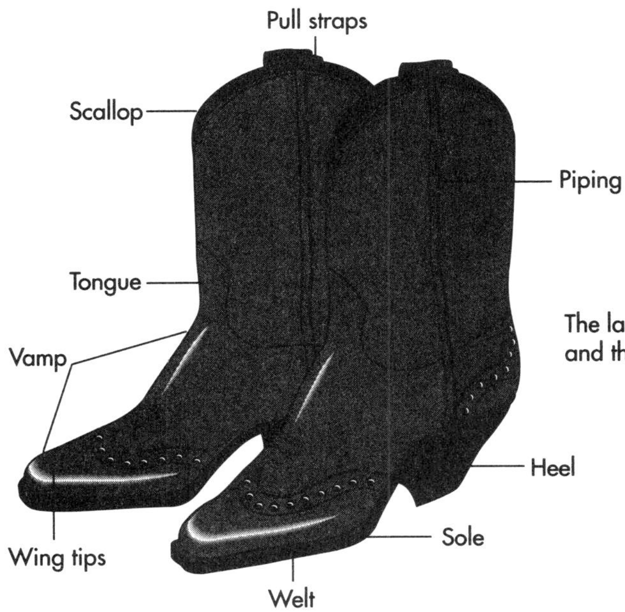

The last inside the boot is removed, and the boot is waxed and polished.

thousands of lasts, hanging from the walls and ceilings, available for the customers they expect to reorder.

6 The insole is first tacked to the last. Then the vamp and the counter are nailed over the insole into the last—first in front, at the toe, and then working around on both sides towards the back. At this point, a stiff piece of leather is inserted at the front of the vamp to reinforce the boot at the toe. The welt, which has been hanging on the back of the counter since the top was first made, is then sewn onto the vamp and the insole. The boot is now almost complete, lacking only a heel and a sole.

Assembling the sole

7 Because the welt now binds together the insole, the vamp, and the counter, the nails that tacked the leather to the last are no longer necessary. The nails are removed, but the last remains inside the boot until it is finished. A metal shank is then attached to the insole, to reinforce the high arch; it is held in place by a piece of leather. The sole is shaped to fit the insole and then stitched to the welt. The heel is then nailed on, and then both the heel and the sole are shaped by sanding.

Finishing process

8 The finishing process gives the boot its final appearance. The last is removed and a boot tree is used to make fine adjustments in the shape of the boot. Seams are trimmed and stray threads cut short. Final dyes are applied, if necessary, and then the boot is waxed and polished. The boots are checked for quality at this stage, though the standards of quality control vary between factories and custom shops. In a custom shop, how the boot fits an individual customer will distinguish an approved product from one that may need to be reworked or rejected. But with both kinds of manufacturing, the number of stitches per inch is important, as is the quality of the leather, and the strength of the welt.

The Future

The manufacture and marketing of cowboy boots, like western wear in general, experienced dramatic cyclical changes in the 1980s and 1990s. Most bootmakers still remember the consequences of one boom-and-bust period in western wear, the so-called "urban cowboy" fad of the early 1980s, when many companies expanded their production capacity only to see demand plummet. One manufacturer, Justin

Industries in Fort Worth, Texas, was only saved from bankruptcy by its investments in other sectors of the economy. Another growth cycle in fashion sales of cowboy boots began at the end of that decade, with actors, rock stars, and fashion models wearing them, as well as politicians and businessmen. Though slower, growth in this cycle lasted longer. In the mid-1990s, some retailers and manufacturers speculated that demand for cowboy boots had peaked, but others saw prospects for growth in the increasing popularity of country music.

Some industry observers feel that new marketing strategies can guarantee continued growth in the sales of cowboy boots, but these cycles may also be what characterizes the new life of this particular commodity. In a world where few people work on horseback anymore, where even most of a typical rancher's day is no longer spent in the saddle, cowboy boots are now largely a fashion item.

Where To Learn More

Books

DeLano, Sharon and David Rieff. *Texas Boots.* Viking Press, 1981.

Morris, Michele. *The Cowboy Life.* Simon and Schuster, 1993.

Periodicals

Holly, Susan. "A Country Twist: Business Profiting from Country Fads." *Nation's Business*, March 1993, pp. 8-13.

McKay, Deidre. "Western Bootmakers Vigorously Scratch Their Niches." *Footwear News*, July 10, 1994, p. 10.

Sheehy, Sandy Granville. "Boots to Live and Die In." *Town and Country Monthly*, March 1994, p. 10.

Thomas, Les. "These Boots Are Made for Texas." *Southern Living*, December 1994, p. 40.

—*Douglas Smith*

Crayon

Background

The earliest form of recorded communication is graphic art, beginning with drawings scratched on the walls of caves by prehistoric peoples. Cave dwellers had limited drawing tools, perhaps only pieces of carbon left over from a cooking fire. Today's graphic communicator can, of course, purchase a variety of far more efficient tools, including the common crayon.

Crayons are made from paraffin, a waxy substance derived from wood, coal, or petroleum. Paraffin was produced commercially by 1867, and crayons appeared around the turn of the century. The early crayons were black and sold mainly to factories and plants, where they were used as waterproof markers. Colored crayons for artistic purposes were introduced in Europe around the same time, but like the black crayons, they contained materials that were toxic (usually charcoal and wax) and thus were not appropriate for children. The Binney & Smith Company, who still make crayons, had a canny grasp of the American educational market, having previously marketed dustless chalk for **chalkboard**s. This company sold its first package of eight colored crayons, suitable for use in schools by children, in 1903.

Raw Materials

Crayons are made of paraffin mixed with various chemical pigments. Paraffin is delivered to the crayon factory in liquid form; delivery trucks must maintain a warm enough temperature to keep the paraffin from hardening (paraffin becomes liquid at about 135°F [57°C]).

Because paraffin will not mix with water or water mixtures, the pigments are in powdered form, although they may have been made from a water mix and then dried. Pigments are made by suppliers following formulas dictated by the crayon manufacturer. Individual pigments are made of chemicals mixed together in wooden tanks and forced through filters to remove excess water, leaving chunks of the individual pigments. The pigments are then kiln-dried for several days. After drying, the chunks of pigment are mixed according to the formula for the desired color, pulverized into a powder, and blended for color consistency. The mixes are sent to the crayon factory.

Over the past 10 years, additional ingredients have been added to crayons. One of the most popular is glitter, small pieces of reflective material that make the crayon-produced work shine as it catches and reflects random beams of light. Perfumes and other scents can also be added to the mix.

Binney & Smith Company of Pennsylvania, makers of Crayola Crayons, manufactures more than two billion crayons a year.

The Manufacturing Process

Crayon manufacturing is a simple process, but one which is still relatively labor intensive. In the American market, the predominant manufacturer is the Binney & Smith Company of Pennsylvania, which manufactures more than two billion crayons a year.

Mixing the batch

1 The paraffin is pumped into supply tanks outside the crayon factory. Each tank holds about 17,000 gallons (65,875 l) of liquid. When the process begins, the paraffin is piped into small, heated tubs with a volume of about 6 gallons (23 l) (about the size

Paraffin is pumped into supply tanks outside the crayon factory. Each tank holds about 17,000 gallons of liquid. When the process begins, the paraffin is piped into small, heated tubs with a volume of about 6 gallons.

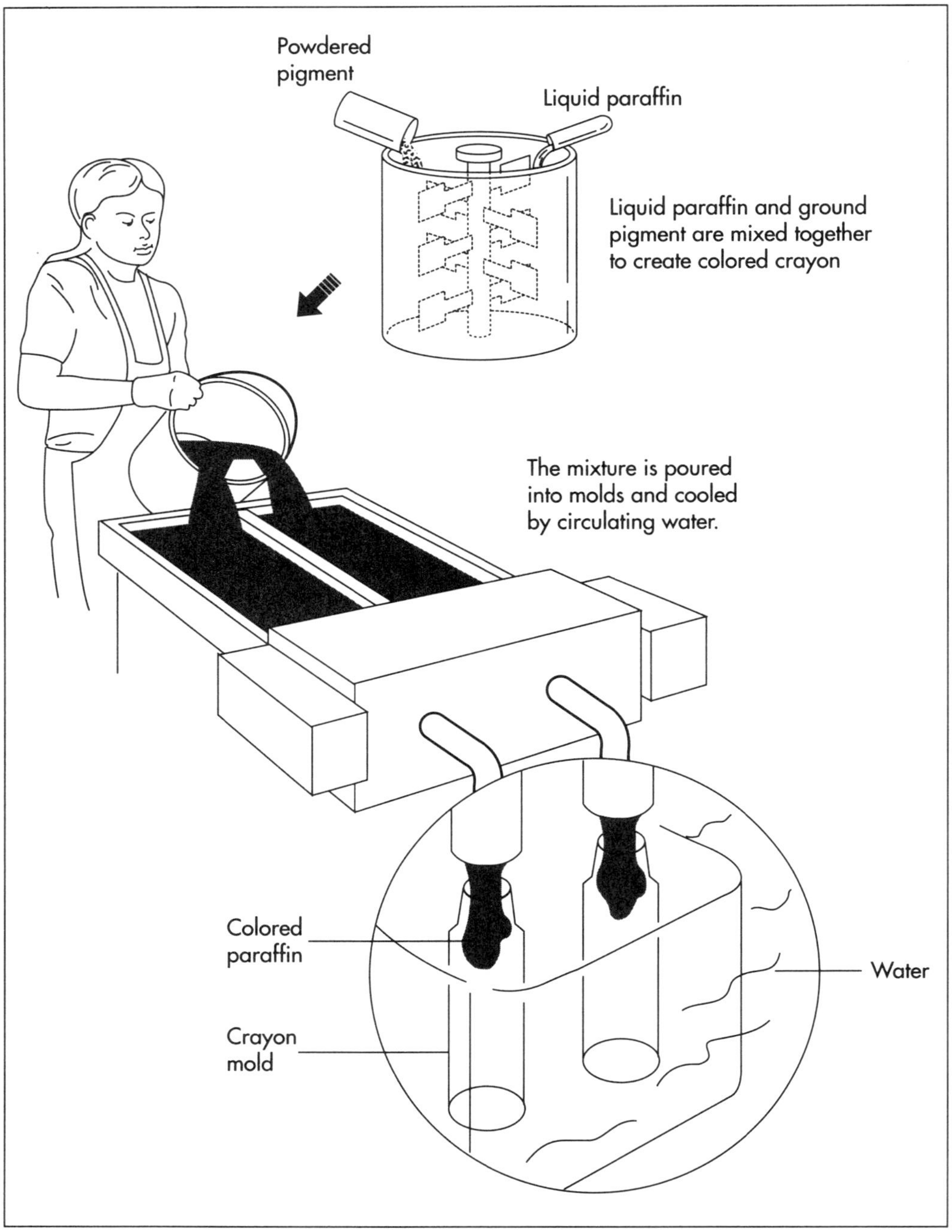

of a home washing machine). At the same time, or soon after, a pigment or mixture of pigments is added to give a batch of crayons a particular color.

Mixing and molding

2 The paraffin "wets" the pigments as they are mixed, and the mixing disperses the pigment uniformly. When the combination of paraffin and color is fully blended, the batch is automatically pumped out of the tubs into molds of the desired shape. Rotating blades inside the mold prevent the formation of bubbles and lumps as the mixture cools. The molds are cooled as water circulates around them. The color of the crayon, and thus the type and quantity of pigment used, can affect the time needed for cooling. Some colors will be ready for the next step within four minutes, other colors can take as much as seven minutes. The molds used for the crayons might hold as many as 2,400 forms.

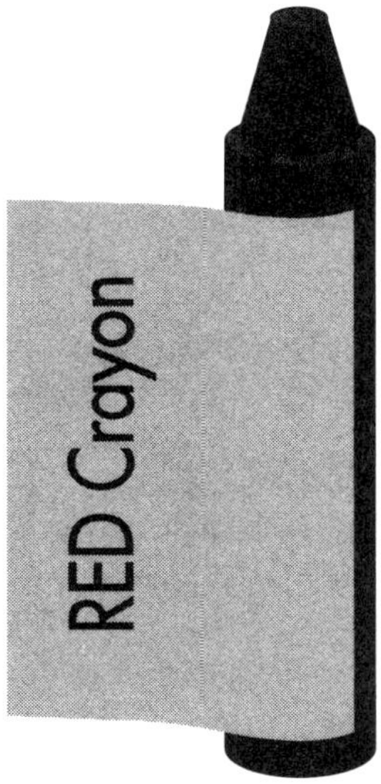

After cooling, crayons are pushed out of the molds and inspected. Crayons that pass inspection are wrapped with paper and packaged.

Until recently, the paraffin-pigment mixtures were poured by hand from the tubs into a bucket and then into molds. Newer machinery now automates the process and pumps the mixture directly into the molds. In some large, older factories, both processes might be used.

Inspection and quality control

3 After cooling, the crayons are automatically pushed out of the molds and inspected. Inspectors examine each crayon for breaks and chips, as well as signs of bubbles in the cooled crayon, a condition that occurs if mixing has not been complete. Rejected crayons are returned to the tubs for remelting and recasting.

Wrapping and boxing

4 Crayons that have passed the quality control inspection are automatically placed into racks where they are wrapped with paper labels; most manufacturers use a double wrapping of paper to give the crayons added strength. Crayons are automatically filled into boxes and sent to wholesalers.

In the early days of crayon manufacture, an entire factory floor might be devoted to the production of a single color for a day or more. Following that color's molding, machines would be cleaned and a new color would be made. In today's factory, the demand for crayons is so huge, and the number of different colors so great (easily more than 100), that individual vats and molds are dedicated to only one or only a few colors, and the production lines run day and night.

Environmental Concerns

In the early 1990s, the Binney & Smith Company found great success marketing crayons with food scents that matched the colors and names of its crayons. Responding to concerns that young children would try to eat the crayons, the company determined to produce only crayons that smelled like non-edible objects such as flowers. All crayons sold in the U.S. are nontoxic. Although the ingestion of a large quantity of paraffin might result in a stomach ache, long-term effects are not likely to occur. U.S. law requires that all art materials sold in the U.S. be nontoxic. Most toxicology evaluations in the U.S. are performed under the auspices of the Arts and Crafts Materials Institute in Boston, Massachusetts. Each formula for any art product is submitted to the Institute and evaluated by a toxicologist. In addition to the out-and-out toxicity of an ingredient, a broad range of possible effects are also investigated. Testing is sometimes required, for example, to evaluate the interactions of individual ingredients within a single product or to determine whether a product will cause skin irritation. All materials must be evaluated at least every five years, and any change in the formula triggers a new evaluation.

Where To Learn More

Book

Pluckrose, Henry. *Crayons.* Franklin Watts, 1987.

Periodical

Baggerman, Lisa. "Creating Crayons." *Art Materials Today*, July-August 1994, pp. 54-62.

—Lawrence H. Berlow

Decorative Plastic Laminate

Background

Decorative plastic laminate is a durable flat sheeting material used in home and industrial furnishings. It is most familiar under the Formica brand name. The Formica Corporation is the world's largest manufacturer of plastic laminate. Other well known manufacturers include the Premark Corporation and DuPont.

Decorative laminate is commonly used to surface kitchen counters, table tops, and cabinetry because of its resistance to stains, scratches, and heat. The laminate sheets are made up of three layers: the bottom layer of brown paper coated with phenolic resin, a second layer of paper decorated with the desired pattern, and a third layer of clear sheet. Both the second and third layers are coated with melamine resin.

History

Plastic laminate was first used as an insulating material for industrial products. Its invention is credited to two engineers, Herbert A. Faber and Daniel J. O'Conor. In the early 1900s, these two joined the Westinghouse company in Pittsburgh as part of a dream team of scientists working on insulating material and phenolic laminate resins. Phenolic laminate resins had been developed around this time by the Belgian chemist Leo Baekeland. Baekeland condensed phenol and formaldehyde to produce the first totally synthetic plastic, Bakelite. The material was resistant to heat, water, chemicals, and electric current, thus having the properties to replace hard rubber and shellac for electrical insulation. Baekeland's many experiments included impregnating paper with Bakelite resin and then compressing it under molds at high pressure and temperature in a process known as thermosetting. The two Westinghouse engineers worked in this same vein. They began by impregnating heavy canvas with Bakelite resin, and by 1913 they had applied for a patent for a flat laminate sheet made from Bakelite and paper. Faber called the new plastic laminate formica: "for" (in place of) "mica" (mineral used as electrical insulation material).

Faber and O'Conor left Westinghouse to found their own company in 1913, the Formica Insulating Company in Cincinnati, Ohio. The new company produced rings and tubes of plastic laminate for electrical insulating purposes, but by 1914, Faber and O'Conor were using a press to churn out flat laminate sheets. The laminate was widely used in radio sets in commercial shipping and naval vessels to insulate coils, tuners, and other parts. But plastic laminate was soon used for its decorative properties as well because its flawless, uniform character was the perfect radio exterior. By 1921, the laminate manufactured by the Formica Insulating Company had been integrated into the manufacture of home radios as well as ship radios.

In 1927, Faber and O'Conor discovered that by adding decorative paper through a lithographic printing process, their laminates could be made with patterns that simulated wood grains and marble. As the laminate became more colorful and decorative, its market expanded. Faux marble laminate was popular for soda fountains in the 1930s, and a woodgrain laminate was used in place of aluminum inside airplanes in the

Commonly used to surface kitchen counters, table tops, and cabinetry because of its resistance to stains, scratches, and heat, decorative plastic laminate is most familiar under the Formica brand name.

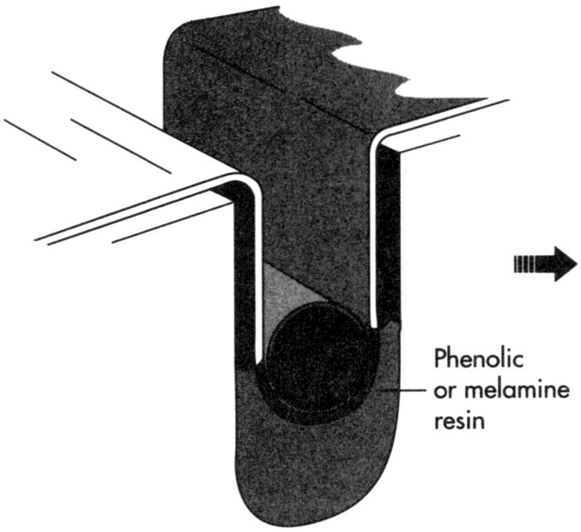

Ribbons of paper are sent through vats containing phenolic or melamine resin.

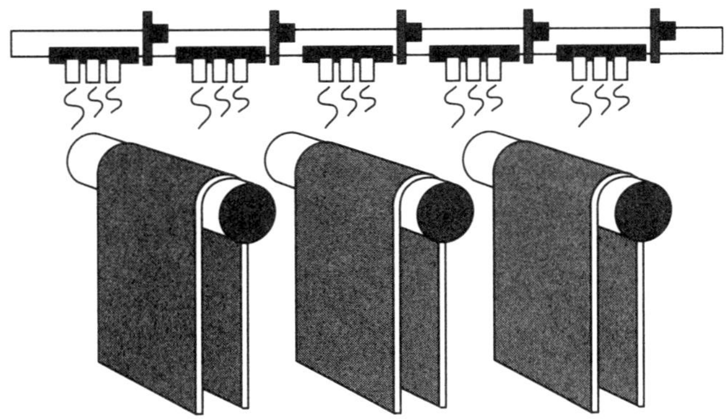

The resin-impregnated sheets are then put into a drying chamber.

Decorative plastic laminate sheeting is made of resins that react with aldehydes during the thermosetting process.

1940s. Manufacturing improvements soon enabled plastic laminate to resist **cigarette** burns, and the material became more attractive, colorful, and durable, spurring its use by manufacturers of kitchen and dining furniture.

Raw Materials

Decorative plastic laminate sheeting is made of resins that react with aldehydes during the thermosetting process. The resins are laminated onto layers of kraft paper topped with a decorative sheet. Kraft paper is the same brown paper used in grocery bags. The first plastic laminates were made with phenolic resin, a polymer of formaldehyde and phenol. Phenolic resins produce only dark colors. In the 1930s, a urea-based resin called melamine was developed that produced a clear surface. In the modern manufacturing process, the top two layers of paper are impregnated with melamine resin, and the lower layers use phenolic.

The Manufacturing Process

Impregnating the paper

1 The process begins by soaking strips of paper in resin. Decorative plastic laminates can be made in different grades or thicknesses, depending on its intended use. There may be from 7-18 layers of paper combined into the final sheet. The bottom layers are kraft paper. The paper comes in ribbons of different widths, commonly of three, four, or five feet. The kraft paper is run through a "bath tub" or vat containing phenolic resins. The paper for the top layer of the sheet is translucent. This is run through a vat of melamine resin. The layer just beneath the top is the decorative layer. This is a sheet of paper printed with the color or design that will show through the clear top layer for the desired surface pattern. This sheet is also run through a melamine vat.

Drying

2 The resin-impregnated sheets are then put into a drying chamber. Next, they are cut and stacked in layers. The clear layer and the decorative layer are on top of the kraft paper.

Thermosetting

3 The layers of paper are then loaded onto a flat-bed hydraulic press for final curing. The press compresses the sandwich of resin-soaked paper at 1,400 psi, while heating it to a high temperature. The heat catalyzes a reaction in the resins. The phenol (or melamine) and formaldehyde mole-

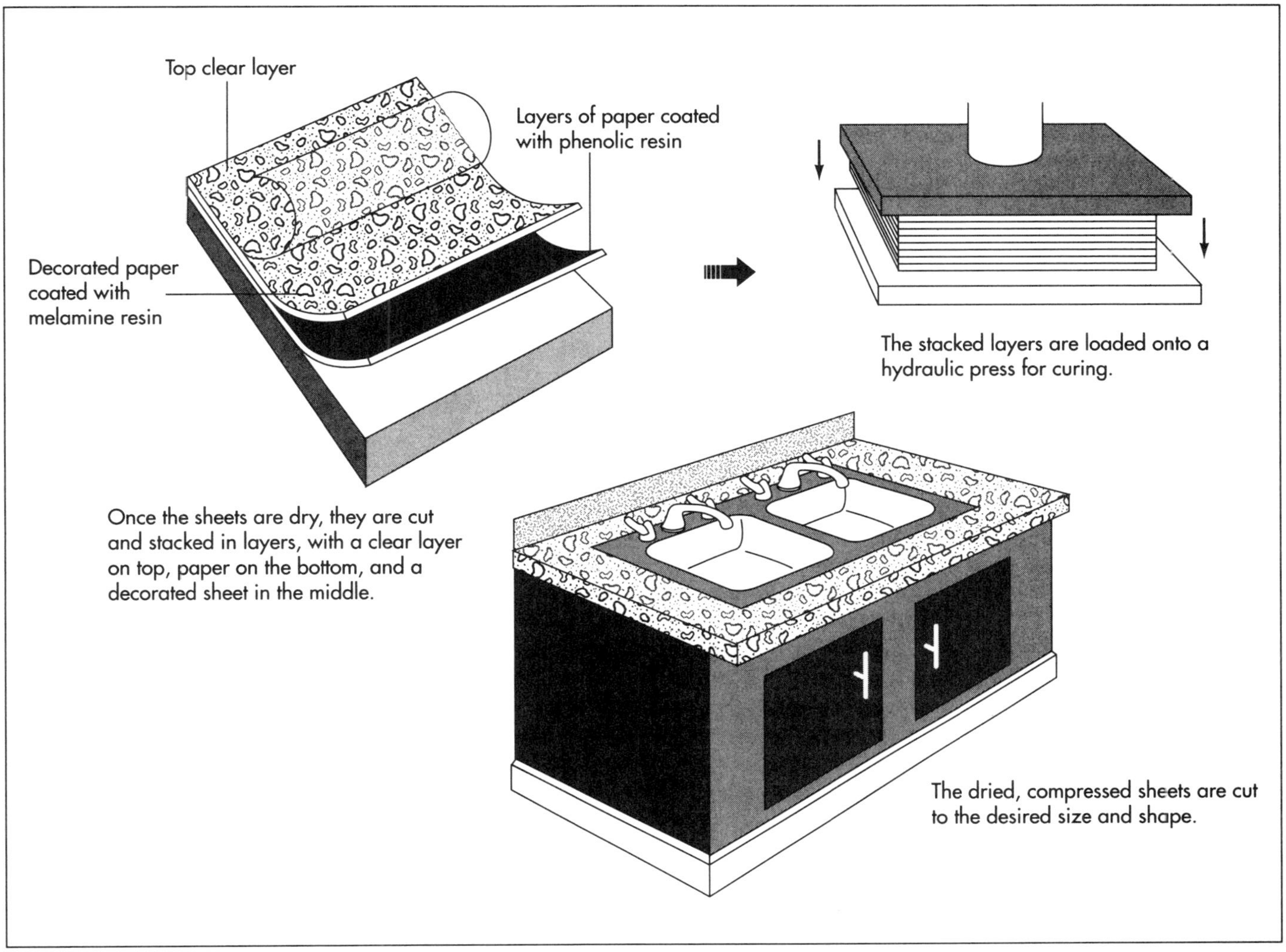

cules attach to each other in an alternating-chain fashion, releasing water molecules in the process. The resins flow together and then set. Thermosetting converts the paper sheets into one single, rigid laminated sheet. This sheet is dry and insoluble, and it cannot be shaped or molded, even at high temperatures.

Finishing

4 The dry sheet is cut into the desired size and shape. It may also be bonded to a building material such as plywood, flakeboard, fiberboard, or metal.

Byproducts/Waste

The plastic laminate manufacturing process produces several byproducts, some of which are considered hazardous. Toxic emissions emanate from phenolic resins during the laminating process, and acrylic resins and hardeners used in applying plastic laminates to surfaces are also considered hazardous. Decorative plastic laminate itself is not considered a "recyclable" plastic. However, at least one major manufacturer has taken steps to reduce harmful waste and emissions. By switching from solvent-based to water-based phenolic resins, the amount of toxins released during lamination can be reduced. Recent changes in the composition of melamine have also virtually eliminated alcohol emissions from this type of resin as well. Control devices such as so-called packed column scrubbers also reduce particulate emissions into the air.

Paper and laminate residue generated during the manufacturing process are burned in power boilers. This reduces the amount of waste sent to landfills. Metal-based pigments used in decorative papers also create a waste problem, as these can be hazardous. The leading manufacturer of plastic laminate has for this reason reduced the use of such pigments, and plans to totally eliminate the use of metal-based pigments in the future. Although used laminate is not recyclable, some companies have collated old laminates into new sample sets suitable for distribution.

Where To Learn More

Books

Cowie, J.M.G. *Polymers: Chemistry and Physics of Modern Materials,* 2nd edition. Chapman and Hall, 1991.

Lewin, Susan Grant, ed. *Formica and Design: From the Counter Top to High Art.* Rizzoli International Publications, 1991.

Periodicals

Jones, Thomas H. "Repairing Plastic Laminate." *Home Mechanix*, April 1990, p. 18+.

Wicks, Harry. "Working with Plastic Laminates." *Home Mechanix*, August 1988, p. 25+.

—*Evelyn S. Dorman*

Dental Floss

Background

Dental floss is a thin filament used to remove debris caught between teeth and between teeth and gums. In 1994, Americans used more than 2.5 million miles of dental floss, the equivalent of circling the earth more than 100 times.

The use of dental floss helps to remove plaque, a sticky, gel-like substance made of bacteria that forms on teeth and between teeth, as well as on the tooth surface below the gum line. If the plaque is not removed, it hardens and is then called tartar. If tartar is allowed to accumulate, gingivitis, or an inflammation of the gums, usually accompanied by redness, swelling, and bleeding, can result. Eventually, gums begin to separate from the teeth, forming "pockets" that frequently become infected. If this goes unchecked, the bone that supports the teeth is destroyed, resulting in tooth loss. To avoid this, adults and children over age 10 are advised to floss at least once a day. Flossing disturbs bacteria, stopping it before it can create plaque and ultimately cause gum and bone disease.

Floss is available in string or ribbon form, and can be lightly waxed, waxed, or unwaxed. It is also available in several flavors such as cinnamon, mint, bubble-gum, and plain. Ribbon floss is the most effective choice when there are ample spaces between the teeth; since baby and children's teeth are widely-spaced, ribbon floss is the most common selection for children. On the other hand, when teeth have contact points, that is, when they come in contact with one another, the preferred choice is the narrower or string floss. Waxed or lightly waxed is recommended for use between crowded or crooked teeth.

Raw Materials

Dental floss is commonly made out of one of two polymers (synthetic compounds), either nylon or Teflon. Nylon is defined as a fiber-forming substance of a long-chain synthetic polyamide. A polyamide is a compound characterized by more than one amide group; an amide is a chemical related to ammonia. Teflon is the trade name of the polymer polytetrafluoroethylene, or PTFE. Other raw materials are the coatings, which may be wax, flavors, and various proprietary ingredients which vary with the manufacturer.

The Manufacturing Process

Filament extrusion and twisting (for nylon)

1 To make floss from nylon, nylon salt is polymerized and the resulting polymer is poured out as a ribbon. Next, it is cut into small chips, flakes, or pellets. These are blended, remelted, and pumped through spinnerets to form filaments. When the nylon cools, the filaments solidify and regather to form a yarn. The ends of the filaments are combined to create one strand of floss. Correctly twisting the nylon is a vital part of this process; the turns average between 2.5 and 3.5 per inch. Twisting adds strength to the floss as well as greatly reducing fraying or breaking.

Since floss consists of many filaments, it can be produced in different "decitexes."

In 1994, Americans used more than 2.5 million miles of dental floss, the equivalent of circling the earth more than 100 times.

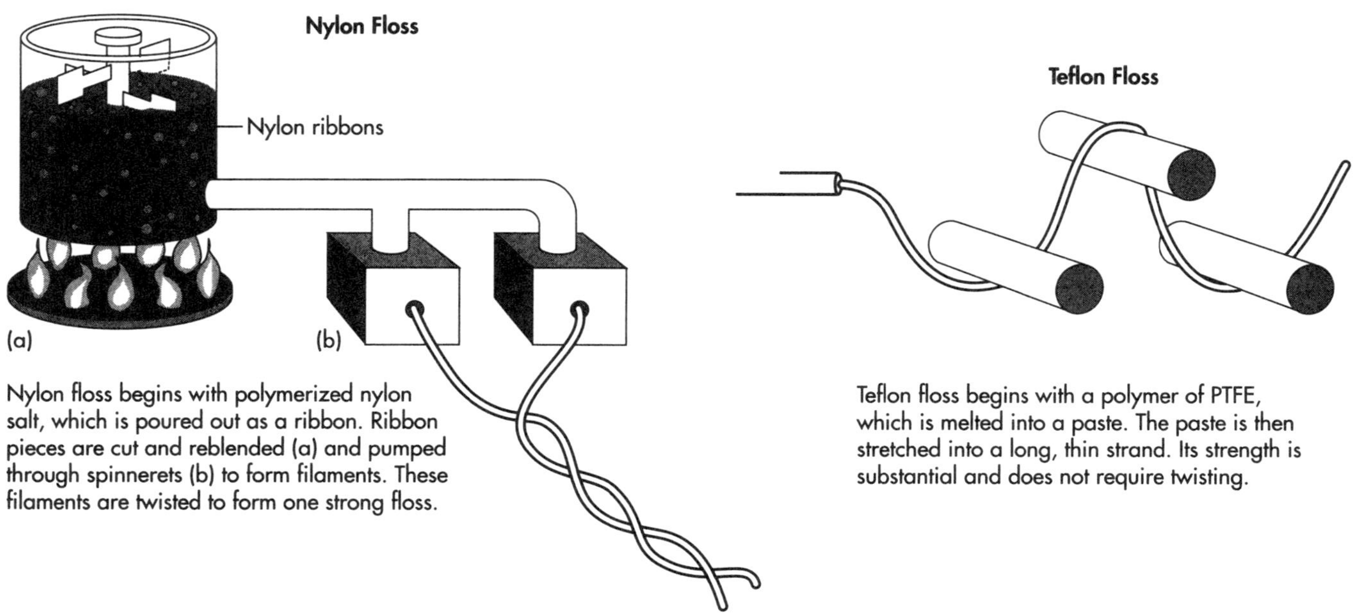

Nylon floss begins with polymerized nylon salt, which is poured out as a ribbon. Ribbon pieces are cut and reblended (a) and pumped through spinnerets (b) to form filaments. These filaments are twisted to form one strong floss.

Teflon floss begins with a polymer of PTFE, which is melted into a paste. The paste is then stretched into a long, thin strand. Its strength is substantial and does not require twisting.

Dental floss is commonly made out of one of two synthetic compounds: nylon or Teflon.

Decitex is defined as the weight of 10,000 meters of unwaxed nylon in grams. Floss is also measured by "Denier." Denier is defined as the weight of 9,000 meters of uncoated floss in grams. There is a direct correlation between the numerical value of the decitex or denier and the thickness of the strand of floss: if this value increases, the thickness of the strand also increases.

Making floss with Teflon

2 To produce a Teflon dental floss, a polymer of polytetrafluoroethylene (PTFE) is formulated. Next, the polymer is melted into a paste and stretched into a long, thin strand. The polymer is then expanded into one or more directions. The PTFE is cut, forming different deniers. After the PTFE is processed, its tensile strength is substantial. Unlike multi-filament nylon flosses, PTFE is a monofilament, which does not shred or break easily. Therefore, twisting is not required for this process. The rates of stretching during the manufacturing process give this floss its added strength.

Filament coating

3 This process allows manufacturers to differentiate their products by permeating the floss with distinctive and proprietary coatings. The coating takes place in emulsion baths, where the floss is pulled through waxes, flavors, and other desired coatings at a consistent velocity.

Bobbin winding

4 Floss bobbins are next wound in one of two shapes: cylindrical or roll type. Winding the floss bobbins requires the use of equipment that transfers the yarn from one type of spool to another type. A cylindrical bobbin is dispensed through a tube or rectangular-shaped package. It can be wound where the floss is pulled through the center or pulled from the outer layer first. The advantage to this type of bobbin is that it can accommodate more yardage. The roll bobbin is the more conventional type of floss bobbin. It is dispensed through flat containers by pulling from the outer layer only.

Case molding

5 The standard case for nylon flosses is an angled, two-piece construction, usually polypropylene, with an insert that holds the floss spool. Recently, many manufacturers have opted to upgrade their floss dispensers by using a one-piece "clam shell" construction that incorporates a window to gauge product use as well as grooves to facilitate gripping.

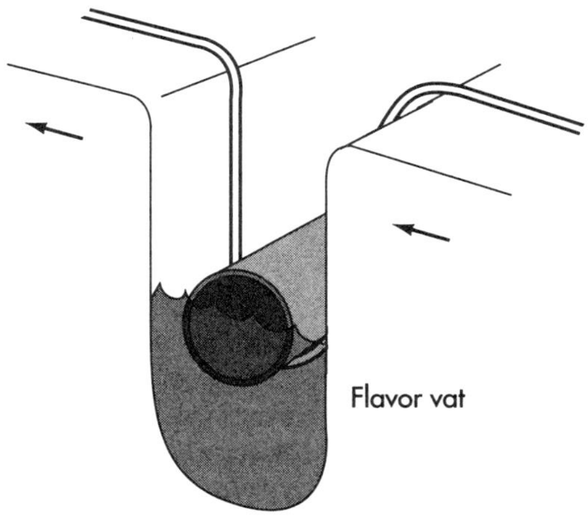

The floss is sent through an emulsion bath and coated with wax and flavorings.

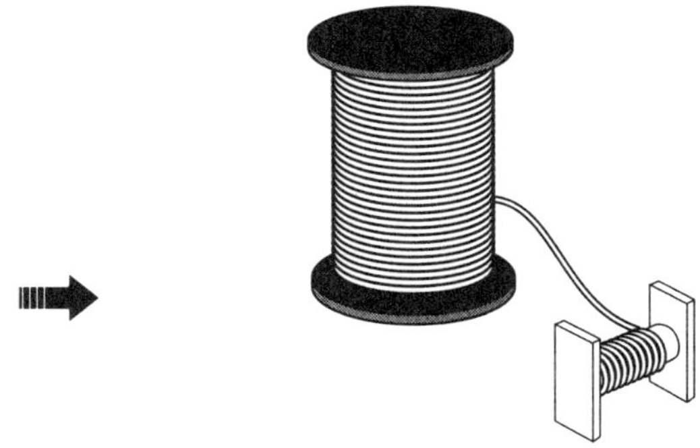

The bobbins are wound by transferring the floss from a large spool to individual spools.

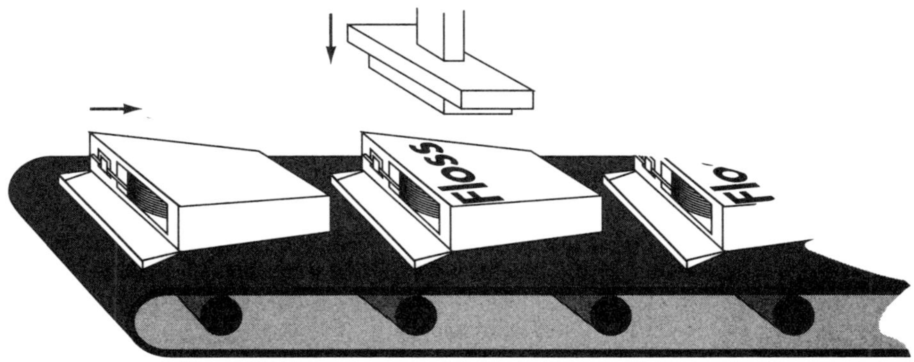

Floss bobbins are manually assembled into the container and labeled by a machine.

Assembly

6 Floss bobbins are usually assembled manually into the floss container. The bobbin cores are cut to separate them. Next, they are placed into the container, the floss is threaded around a metal cutting clip, and the cover is closed. The package is now ready for labeling, or if the container is pre-decorated, it is ready for packing and shipping.

Decoration

7 There are three basic methods of decoration: pad printing, labeling, and thermal transfer. Pad printing is the most prevalent form of decorating used in the U.S. Labeling is the preferred choice for most product exported to Europe. Thermal transfer generates an image quality equal to or better than the pad printing image with the advantage that it allows for a greater range of colors and designs.

Packaging

8 Marketers prefer blister packaging (a plastic mold affixed to a cardboard backing, which can hang on a display peg) because it prominently displays the package and eliminates the need for paper packaging, which is deemed better for the environment. Another packaging trend is to co-pack floss along with a **toothbrush**, toothpaste, or mouthwash.

The Future

Two leading manufacturers have recently developed dental flosses with new types of filament. Oral-B Laboratories introduced Oral-B ULTRA FLOSS. Unlike conventional or ordinary dental floss, which has a series of straight nylon strands, ULTRA FLOSS features an ultra strong filament, containing a patented network of interlocking fibers that resists shredding and fraying. ULTRA FLOSS' woven, spongy texture also works differently than conventional floss; it stretches thin to fit easily between tight teeth spaces, then springs back to its original thickness to trap plaque in its filament. ULTRA FLOSS is soft for sensitive gums, gentler on the fingers, and pre-measured into 18-inch (46 cm) segments, the length recommended by the American Dental Association.

John O. Butler Company introduced Butler-Weave, a dental floss that acts like dental tape. This smooth, shred-resistant floss spreads out when pulled between teeth, providing more surface contact with the tooth for effective plaque removal. In addition, its thin, flat profile glides easily between tight contacts.

Where To Learn More

Books

Foster, Malcolm S. *Protecting Our Children's Teeth: A Guide to Quality Dental Care from Infancy through Age Twelve.* Insight Books (a division of Plenum Publishing Corporation), 1992, pp. 186-90.

Holt, Robert Lawrence. *Straight Teeth: Everything You Need to Know About Orthodontics, Including How to Avoid Costly Treatment for You and Your Children Through Preventive Care.* William Morrow and Company, 1980, pp. 44, 121-35.

Siegel, Dorothy. *The Encyclopedia of Health: Dental Health.* Chelsea House Publishers, 1994, pp. 70, 75, 105-106.

Periodicals

Bedell, Thomas. "Looks: Floss, Anyone?" *Men's Health,* December 1991, pp. 20-21.

Sangiorgio, Maureen P. "Flossing Right: The Real Truth about the Most Underrated Tooth Saver." *Prevention,* June 1992, pp. 108, 135-36.

—Susan Bard Hall

Diamond

Background

The diamond is the hardest natural substance known. It is found in a type of igneous rock known as kimberlite. The diamond itself is essentially a chain of carbon atoms that have crystallized. The stone's unique hardness is a result of the densely concentrated nature of the carbon chains. Like other igneous rocks, kimberlite was formed over the course of thousands of years by volcanic action that occurred during the formation of the earth's crust. Kimberlite is located inside these former spheres of volcanic activity—often near mountain ranges—in vertical shafts that extend deep inside the earth. Inside the kimberlite are intermittent deposits of diamonds, one of several minerals present. However, not all kimberlite contains diamond. Other stones often found with diamonds are mica, garnet, and zircon. Kimberlite may be blue-grey in hue—thus termed *blue ground*—or if exposed to air it may have a yellowish cast and is called *yellow ground.*

History

It is thought that diamonds were first discovered in India about 6,000 years ago in the riverbeds of the region. Traders were responsible for bringing the gems as far east as China and as far west as Rome during the classical and early medieval eras. The Chinese were the first to harness the unusually tough nature of the gem and used it as a tool to cut other stones. Pliny the Elder, a Roman scholar, wrote about the diamond in the first century. The word itself stems from the Greek term *adamas* which means "invincible" or "unconquerable."

From the earliest days, the diamond has been imbued with mystery and superstition. Because they were so rare—at first found only in India—it became a commonly held superstition that the diamond lent its wearer special powers. They were worn in battle to insure victory and sometimes invoked as an antidote to poison. Other superstitions associated with the stone included the caveat that placing it in the mouth would bring on a loss of teeth. In other cases, finely ground diamond, made into a powder, was thought to be an effective poison. Indeed, experts agree that even in a pulverized form, the unique sharpness of the mineral would tear minuscule holes in the digestive tract. Because it is both the hardest and one of the rarest natural substances, diamonds have always fetched exceedingly high prices. The extreme value of the stone also made it a portable form of wealth in times of warfare and upheaval.

The actual mining of diamonds as an industry can be traced back to India to around 800 to 600 B.C. India was the only known source of the rocks for over a thousand years, until they were unearthed in Borneo around A.D. 600. During the Middle Ages, the diamond was overshadowed by some of the more colorful gems like the ruby and emerald. These other stones found their way into the jewelry of the rich and powerful of Europe more easily than the diamond. Additionally, gem-cutting techniques had not yet been developed to unleash the brilliance of the stone. Diamonds were usually left in their natural state or shaped by a rudimentary cut. In the 17th century, however, a Venetian lapidary named Vincenzo Peruzzi developed the so-called brilliant cut. This cut revealed the intricacies and the natural perfection of the stone.

From the earliest days, the diamond has been imbued with mystery and superstition. Because they were so rare—at first found only in India—it became a commonly held superstition that the diamond lent its wearer special powers.

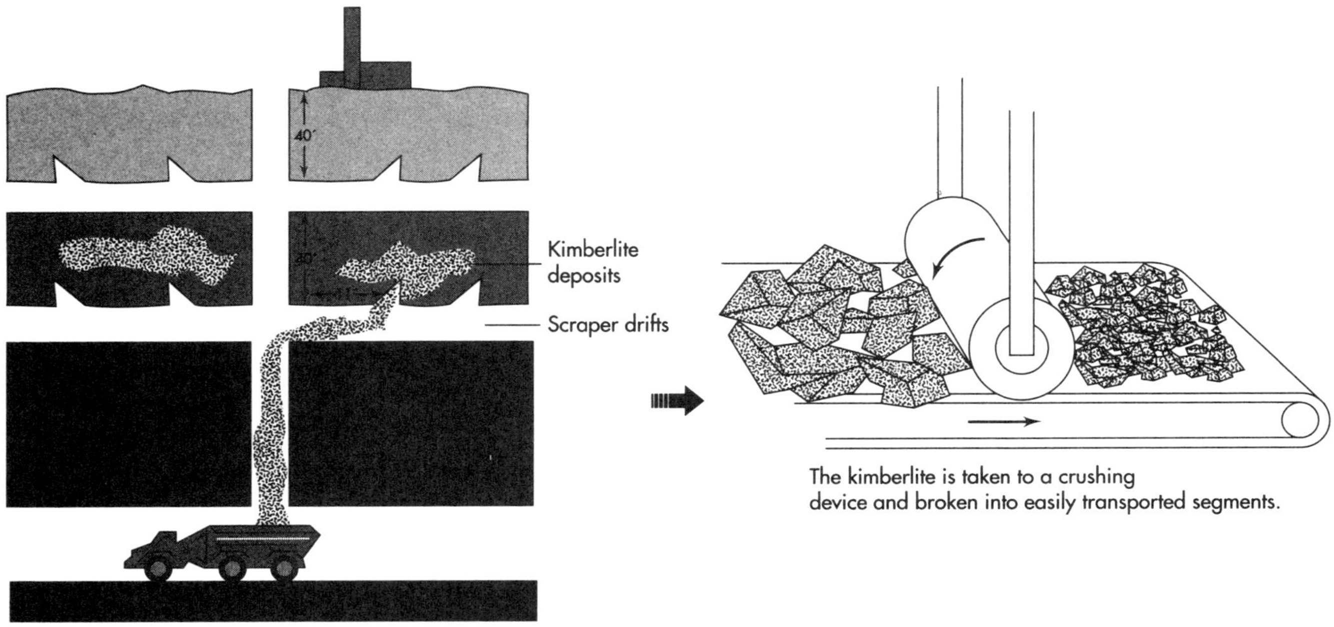

The kimberlite is taken to a crushing device and broken into easily transported segments.

Block caving is used to excavate the kimberlite deposits.

Diamond is found in a type of igneous rock known as kimberlite. Like other igneous rocks, kimberlite was formed over the course of thousands of years by volcanic action that occurred during the formation of the earth's crust. Inside the kimberlite are intermittent deposits of diamonds, one of several minerals present.

In the 18th century, diamond deposits were discovered in Brazil in small quantities, and later in Australia, Russia, and the United States. Brazilian gems were first taken to India and shipped to Europe as Indian diamonds, since people considered non-Indian gems less valuable. In the 20th century, an American mine near Murfreesboro, Arkansas, was open for novelty public mining for a small fee. High-quality diamonds have been found in Siberia, but the extremely cold temperature has made large-scale mining unfeasible.

In 1866 the world's largest cache of diamonds was discovered in South Africa. Some children had found a rock and brought it home, and a curious neighbor passed it on to a trader, who gave it to a geologist. It was discovered to be a diamond of enormous size and worth a small fortune. South Africa soon experienced a diamond rush, and shanty towns sprang up with the influx of prospectors. Eventually, the various mines and mine companies of the region were consolidated under the control of the DeBeers organization. With the DeBeers Consolidated Mines, Ltd., a Central Selling Organization, and a Diamond Trading Company, this conglomerate controls about 80% of the world's diamond output. Contemporary diamond mining is centered at Kimberley, South Africa, and carried out by DeBeers. Every six weeks or so, representatives of the DeBeers Diamond Trading Company invite a special list of diamond wholesalers—less than a hundred worldwide—to London to view preselected lots of the gem. This is the only method by which South African DeBeers diamonds come onto the market.

Industrial Applications

In modern times diamonds have become indispensable to industry. Automobile magnate Henry Ford was the first to uncover the contemporary industrial uses of the stone. He sponsored research into its applications for the manufacturing sector, especially as a low-cost abrasive, and the Detroit area became a hub for dealers of diamond tools. The aircraft industry followed the lead of the automotive sector, becoming an avid user of diamond-based products. Diamonds used for industrial applications are usually of a lower grade than those found in the gemstone market, but they retain the same properties of hardness and durability. Dia-

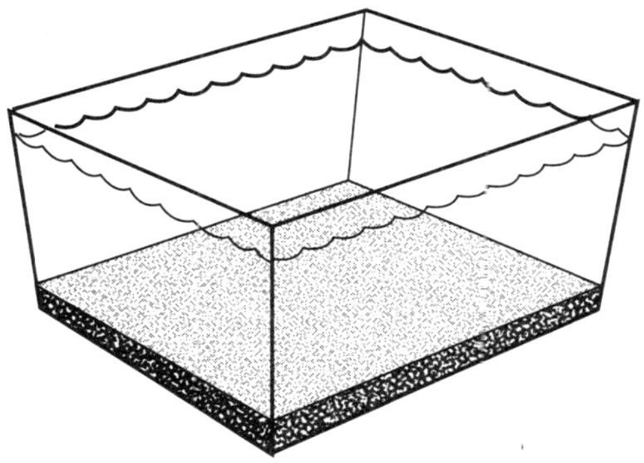

Roughly crushed and milled kimberlite is mixed with water to separate the diamond-containing rocks. The light tailings rise to the top. and the concentrate falls to the bottom.

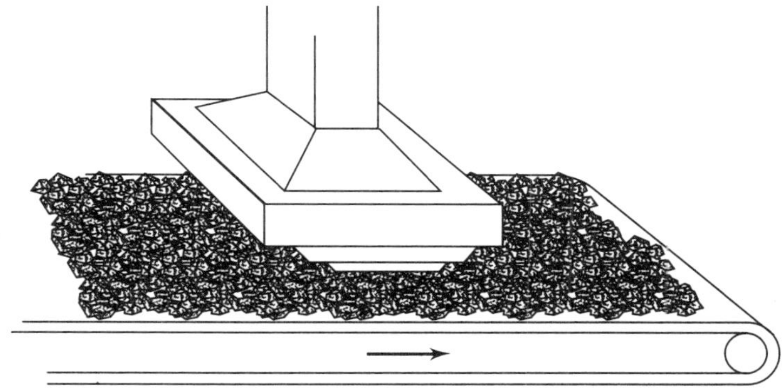

One of several methods, an x-ray machine further separates the diamond from effluvial material.

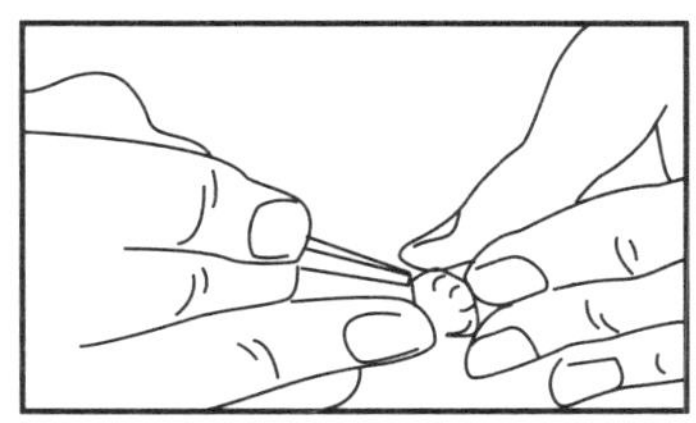

The diamond is cut and crafted for use in industry or jewelry.

mond tools last much longer than those made from other sources and offer a nearly unmatched precision in cutting other substances. Additionally, such tools work faster and much more quietly than other alternatives.

Tools made from industrial diamonds are used in the mirror and optical manufacturing fields as well as in gas and oil drilling endeavors. In the textile industry, devices made from diamonds are used to cut patterns. In medicine, cutting instruments made from diamonds are used to cleanly slice bone and tissue. The construction industry uses diamond tools in the grinding and cutting of concrete and pavement. Diamonds are also used to make needles for stereo record players.

Physical Characteristics

Diamonds are chains of carbon. Carbon is one of the most common substances on the planet. In one form it is simple graphite, used in pencils, but in its crystallized form, it takes an altogether different appearance as diamond. On the scale used by mineralogists to measure the hardness of minerals, diamonds rate ten on a scale of one to ten. Diamonds are measured in carats, the standard unit of measurement for gemstones. One carat is roughly equal to one-fifth of a gram. The carat can be further divided into points based on a scale of 100. One of the reasons diamonds are so prized is because the light they absorb is reflected directly back outward, if the stone has been properly cut. The unusual crystal structure of the gem allows this high degree of refractability. Because of their structure, diamonds are also excellent conductors of electrical current.

Structurally, the diamond can be described as an octahedron. This means that there are double four-sided pyramids of carbon chains inside that meet one another at the bases. Cubes or dodacahedrons—a twelve-sided shape—are also found within the stone. Sometimes small triangular pockets called trigons can be observed.

Diamonds are found in nature in a variety of hues. Colorless or white diamonds are the most common, while some tinted stones are rare and valuable. The shades may be yellow, blue, pink, green, or amber. In South Africa it is common to see orange diamonds as jewelry, but this is a custom that has not made its way into the rest of the world. Some of the world's most famous diamonds are the colored ones—the heavy Dresden Green, for instance, and the infamous Hope Diamond. The latter, blue in color, is thought to hold certain negative energy, and many unexplained deaths have been associated with its owners. It is now in the collection of the Smithsonian Institution in Washington, DC.

Extraction and Refining

Diamonds are mined either from the kimberlite pipes below the earth's surface, or from alluvial deposits. Alluvial (riverbed) deposits occurred when volcanic action carried kimberlite and other minerals from the center of activity to naturally forming irrigation systems. Such diamonds are found quite near the earth's surface. In alluvial mining, considerable amounts of sand must first be removed from the area. The sand and other such components are called overburden, and large mechanical scrapers are used to move it out of the way. Underneath the overburden lies a gravel bed, and bulldozers scoop the gravel up and set it aside in piles.

The piles are then taken to a screening plant, where the diamonds are extracted. In alluvial mining, it is sometimes necessary to reach the bedrock underneath the gravel bed—or sometimes even below the bedrock itself—in order to unearth the diamond deposits. The bedrock must be thoroughly searched. Sometimes an enormous vacuum device called a Vacuveyer is used for this purpose. As the mining process moves along in a horizontal fashion, the removed overburden is again deposited to fill over the excavated sites.

Below-ground mining of kimberlite for diamond also requires moving enormous quantities of rock and other material in order to unearth gems, but on a much larger scale than alluvial mining. For one part diamond uncovered, it is estimated that 15 to 30 million parts waste must be moved out of the way. Unlike mining endeavors for gold or other substances, engineers cannot determine beforehand whether an area has a large abundance of diamond.

Mining

1 Block caving is the most commonly used method in excavating diamonds from kimberlite deposits. This method offers the highest yield and thus is the most cost effective. First, a large vertical hole is excavated, typically 1,750 feet (533 m) in diameter. Levels are placed approximately every 40 feet (12 m). Along these levels are horizontal tunnels known as scraper drifts. In the drifts, there are small inclined cone-shaped openings at intervals of every 11 feet (3 m) or so. These openings are roughly four feet by four feet. When a horizontal slice is cut above the cones—usually about six feet (1.8 m) in height—the kimberlite begins to break off and fall into the cone and into the scraper drift. The material is then pushed onto trucks. The trucks travel underground through the mining area and take the collected kimberlite to a crushing device.

Crushing

2 In the crushing operation, which occurs in the below-ground mining facilities, large chunks of kimberlite are broken up into more easily transportable segments. After an initial crushing, the kimberlite passes through a grizzly, or a set of iron bars. If the crushed chunks do not pass through the grizzly, they are still too large, and they are sent back for further crushing. The crushed kimberlite is then taken above the surface for further processing. When no more kimberlite is found entering the cones, the area is depleted and work moves on to a lower level.

Separating

3 The actual diamonds must be separated from the rock that surrounds them.

Crushing or milling the excavated material is the first step, but this is done in a rudimentary form so as not to damage the potential gems inside. Next, a gravity-based device is used to sort the diamond-containing portions—called the concentrate—from the tailings, or the filler rock. One of the most commonly used methods to separate the two is a type of washing pan developed in South Africa in the 1870s. Decomposed kimberlite and water—in a mixture known as a puddle—is put into the pan. The mixture's viscosity is a crucial element, because the lighter particles will rise to the top, but the diamonds and other heavy minerals will descend to the bottom of the pan.

Another method of uncovering diamonds uses media separators. A stew called a slurry is made up—typically consisting of water added to the crushed concentrate and tailings. Ferro-silicon powder, which has a heavy density, is also added.

The slurry may be put into one of three types of media separators. The first is a cone-shaped tank, with a cone-shaped agitating element inside. The agitator moves around the sides of the tank, but leaves enough room so that the lighter tailings can rise to the top and the heavier elements sink to the bottom. In a lifting-wheel type of media separator, a wheel is filled halfway with slurry. Paddles inside it agitate the mixture, and lift the heavy particles from the bottom and separate them from the rest of the mixture. The third type of media separator is known as a hydrocyclone. It is a large vat that spins around, and through centrifugal force, the heavier, diamond-rich particles are separated.

Greasing

4 After this rudimentary separation, the concentrate moves to a greasing area, another innovation in diamond manufacturing developed in South Africa in the late 19th century. Mixed with water, the kimberlite-and-diamond mixture is placed on a greased belt or table. This device is usually slanted and vibrated. The method operates on the premise that diamonds newly excavated will not become wet when brought into contact with water. Instead they will stick to the grease. Petroleum jelly is usually the preferred substance on the grease belt or table. The water then carries away the remaining non-diamond particles. The diamond-laden concentrate is then swept off the table and boiled to remove the traces of grease. In a newer method, X-ray technology is used to determine which of the concentrate is diamond and which is effluvial material.

Cutting

5 Chunks of diamond eventually become small, perfectly shaped gemstones commonly used in engagement rings and other jewelry. Since diamond is the hardest known substance, diamond dust must be used to cut the stone. In cutting, a minuscule groove is incised into the surface of the diamond, and a cleaving iron is inserted into the groove. With a quick, forceful blow, the diamond should split perfectly along its naturally occurring planes. The lapidary determines further cuts by marking them off on the surface with ink. Next, a diamond saw, oiled with the unusual combination of diamond dust and olive oil, is rotated vertically on the surface of the raw gem. This device divides the diamond into new segments. These parts are then fed into a lathe-like device for grinding.

The Future

Diamonds are a finite resource. The fate of Indian diamonds is a good example of what the future might hold for the South African diamond-mining industry. From the first discovery of the gems in India until relatively recently, it is thought that over 12 million carats originated from India. By the mid-20th century, the resources were nearly depleted, and India was producing only about 100 carats annually. Diamonds will continue to be used in industry and high-technology enterprises, but synthetically produced facsimiles—first manufactured in 1953—may accomplish some of the tasks originally the exclusive province of the real stone. These "manufactured" gems have the same properties of hardness and durability, and while they will never be as popular as the real diamond for adornment purposes, they are well suited for industrial applications.

Where To Learn More

Book

Arem, Joel A. *Gems and Jewelry*, 2nd ed. Geoscience Press, 1992.

Periodicals

Austin, Gordon T. "Diamond." *American Ceramic Society Bulletin*, May 1990, p. 854-55.

"More Australian Diamonds?" *Engineering and Mining Journal*, November 1992, p. 62.

"Diamond Exploration—The Trace Element Revolution." *Engineering and Mining Journal*, July 1994, p. 7.

Galli, Giulia, Richard M. Martin, and Roberto Car. "Melting of Diamond at High Pressure." *Science*, December 14, 1990, p. 1547-49.

—*Carol Brennan*

Drywall

Background

Drywall is a construction material consisting of thin panels of gypsum board. The board is composed of a layer of gypsum rock sandwiched between two layers of special paper. Drywall makes for a much more efficient method of construction than the common earlier technique of applying wet plaster to a gypsum lath.

In addition to being easy to install, drywall provides a measure of fire protection to buildings. Gypsum contains large amounts of water bound in crystalline form; 10 square feet (1.0 sq m) of gypsum board contains over 2 quarts (2 l) of water. When exposed to fire, the water in the gypsum board vaporizes; the temperature of the panel remains at 212°F (100°C) until all of the water is released, protecting the underlying wood framework. Even after all of the water evaporates, the gypsum itself will not burn and continues to provide substantial fire protection.

Plaster made from gypsum has been used as a construction material for thousands of years. In fact, plaster applied at least 4,000 years ago to walls inside the Great Pyramids of Egypt is still in good condition. Today drywall panels are widely utilized in modern construction around the world.

Raw Materials

The primary component of drywall is the mineral gypsum. It is a light-density rock found in plentiful deposits worldwide. Each molecule of gypsum (or dihydrous calcium sulfate) is composed of two molecules of water (H_2O) and one of calcium sulfate ($CaSO_4$). By weight the compound is 21% water, but by volume it is nearly 50% water.

Because the water present in gypsum is in crystalline form, the material is dry. Although ice, another form of crystalline water, becomes a liquid at room temperature, the water bound in the gypsum molecules remains solid unless it is heated to 212°F (100°C), at which point it changes to a gaseous state and evaporates.

Gypsum, called *gypsos* by the ancient Greeks, is one of the most useful minerals known to man. In its pure form it is white, but impurities often give it colors like gray, brown, pink, or black. Ancient Assyrians called it alabaster and made sculptures from it. Today, pulverized gypsum is used for a wide variety of applications. It is an ingredient in some brands of toothpaste and is used as a filler in products such as paint, cosmetics, and drugs. Automotive window glass is secured in a bed of gypsum while it is being polished. Gypsum is applied to farmland as a fertilizer and soil conditioner. An excellent source of calcium, it is used to fortify foods such as **bread**s. It is even used to create simulated snowstorms in motion pictures.

Gypsum that has been crushed and heated to remove 75% of its water content is known as plaster of Paris. When water is added to this fine white powder, the resulting material is easily molded into any desired shape. Upon drying, the reconstituted gypsum regains its rock-like qualities while retaining the desired shape. Besides its use in making gypsum board, this material is used to make sculptures, pottery, dishes, bathroom fixtures, and casts for broken bones.

The primary component of drywall is the mineral gypsum, a light-density rock found in plentiful deposits worldwide. By weight the compound is 21% water, but by volume it is nearly 50% water, an aspect of gypsum that provides a measure of fire protection to buildings lined with gypsum drywall.

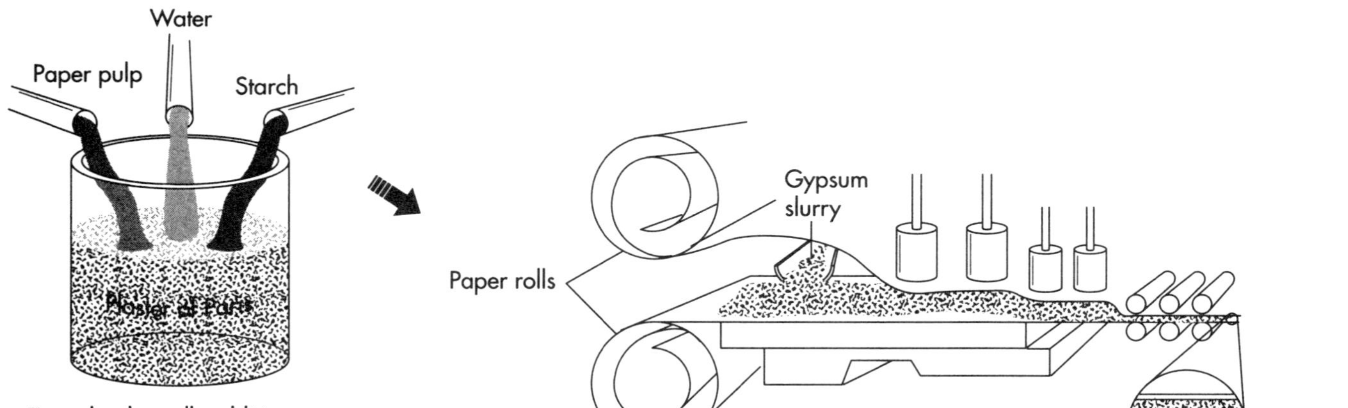

To make drywall, additives are added to gypsum powder (plaster of Paris). Water is then mixed in to form a slurry.

The gypsum slurry is poured onto two layers of paper and passed through a system of rollers.

In its pure form, gypsum is white, but impurities often give it colors like gray, brown, pink, or black. Gypsum that has been crushed and heated to remove 75% of its water content is known as plaster of Paris. When water is added to this fine white powder, the resulting material is easily molded into any desired shape.

Millions of tons of gypsum are mined each year in North America, and gypsum board is the principal product in which it is used. Besides the newly mined material, up to 20% of the gypsum used to manufacture drywall can be recycled from waste generated at the manufacturing plant or at construction sites. Gypsum produced as a byproduct of the flue-gas desulfurization process at electric power plants provides an economical, environmentally sound raw material for making high-quality gypsum board.

Two types of **paper** are used in the production of most drywall, and both types are made from recycled **newspaper**. The ivory manila face paper, when properly primed, readily accepts most paints and other types of wall finishing products. The gray back paper can be laminated with aluminum foil to produce a special type of drywall that resists the flow of water vapor in environments like bathrooms. Specialized varieties of gypsum board might be made with different types of paper; for instance, some papers are made to be moisture resistant to various degrees, while another type of highly absorbent paper is designed to accept a thin coat of plaster veneer after installation.

The Manufacturing Process

Fabrication of drywall consists of placing the gypsum core material between two layers of paper, drying the product, and finishing it into panels of standard size.

Blending of additives

1 Depending on the variety of wallboard being produced, certain additives are blended with the plaster of Paris that will form the core of the drywall. Each additional ingredient amounts to less than one-half of one percent of the amount of gypsum powder. Starch is added to help the paper facings adhere to the core, and paper pulp is added to increase the core's tensile strength (resistance to lengthwise pressure). Unexpanded vermiculite is added when producing fire-resistant grades of gypsum board; in some cases clay is also added.

2 Water is added to the plaster of Paris mixture to form a slurry of the proper consistency. An **asphalt** emulsion and/or a wax emulsion is added to achieve the desired level of moisture resistance in the final product. A foaming agent such as a detergent is included, and during the mixing process air is entrained into the material. The finished gypsum panel will be over 50% air; this minimizes the board's weight and makes it easier to cut, fit, and nail or screw to the framing. Glass fibers are added to the wet core material when making fire-rated gypsum board.

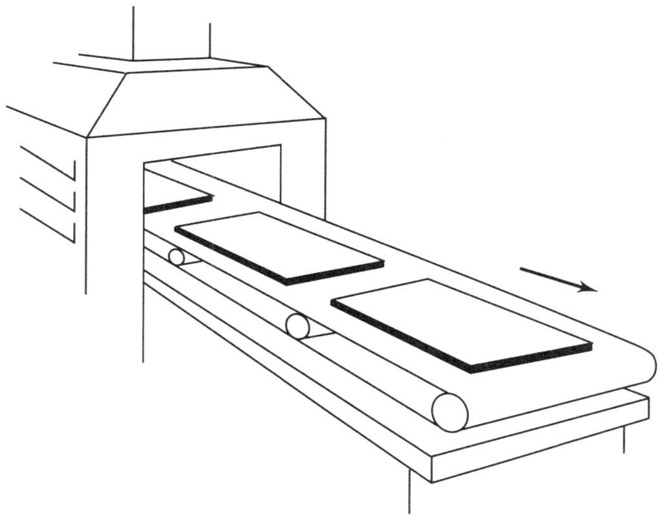

After the panels are cut and the edges shaped, they are fed through a drying oven.

The panels are inspected and bundled.

Making the sandwich

3 The gypsum slurry is poured onto a layer of paper that is unrolling onto a long board machine. Another layer of paper unrolls on top of the slurry. The sandwich then passes through a system of rollers that compact the gypsum core to the proper thickness. The most common thicknesses are 0.37 inch (9.5 mm), 0.5 inch (12.7 mm), and 0.62 inch (15.7 mm).

Finishing the edges

4 Automated assembly lines in gypsum board plants range from 300-800 feet (93-247 m) long. As the drywall continues along the conveyor belt, the edges are formed. Various shapes of edges are possible, depending on the final use of the panel. Options include the traditional square edge, a tongue and groove type, tapered and/or beveled edges, and even rounded edges.

5 The face paper is wrapped snugly around each edge and sealed to the back paper.

Cutting the panels

6 By the time the edges have been shaped, the plaster core has set sufficiently for a knife to slice the continuous strip into standard panel sizes. The board, generally 48 inches (1219 mm) or 54 inches (1572 mm) wide, is usually cut into panels that are 8 feet (2400 mm) or 12 feet (3600 mm) long.

The drying process

7 The panels are transferred to a conveyor line that feeds them through a long, drying oven. At one plant, for exam-

ple, the gas-fired oven is 470 feet (143 m) long. Panels enter the oven at 500°F (260°C) and are exposed to gradually decreasing levels of heat during the 35-40 minutes they travel through the system. Humidity and temperature are carefully controlled in the dryer.

The finished product

8 After emerging from the drying oven, the drywall panels are visually inspected before being bundled into "lifts" of 30 or 40 boards and transferred to the warehouse to await shipment. Each board is labeled with a UPC bar code that is used for warehouse inventory, billing, and price scanning at the retail level.

Product Evolution

Since the invention of gypsum board at the turn of the century, there has been gradual progress in making it lighter in weight while improving its performance characteristics. In the late 1950s, standard gypsum board (not fire-rated) weighed 2 pounds per square foot (9.8 kg per sq m); the various kinds of standard gypsum board now average about 1.6 pounds per square foot (8 kg per sq m). This not only makes handling and installation easier, but decreases shipping costs as well.

Fire-resistant and moisture-resistant gypsum boards were developed in the late 1950s and early 1960s. Another innovation came in 1988 with the development of controlled density (CD) ceiling board. In this product, the core is compressed in such a way as to create thin, dense layers of gypsum on both sides of a standard density core. Although CD board is 0.5 inch (12.7 mm) thick, it is more resistant to sagging than conventional gypsum board that is 0.62 inch (15.7 mm) thick.

Another area of investigation involves better ways of disposing of wallboard waste. During building construction drywall scrap is generated, both as trimmings from panels cut to fit required shapes and as damaged panels that cannot be used. An estimated 1.7 million tons (1.5 billion kg) of gypsum board waste material was deposited in landfills in the United States in 1990. Research has begun in the area of pulverizing this material and using it as a soil treatment rather than simply discarding it. It appears that the effects are very similar to those achieved with gypsum products manufactured specifically for agricultural use.

Where To Learn More

Periodicals

"TU Electric Pioneers FGD-Gypsum Production for Use in Wallboard." *Power*, April 1988, pp. 33-34.

White, Edwin H. and Mark E. Burger. "Construction Drywall as a Soil Amendment." *BioCycle*, July 1993, pp. 70-71.

Other

Standing, H.A. "The Story of Gypsum." Domtar Gypsum Company, 1-800-DOMTAR1.

—Loretta Hall

Dynamite

Background and Raw Materials

Dynamite is a commercial explosive used mainly for demolition and mining. Invented in 1866 by Alfred Bernhard Nobel (1833-1896), it is more accurately described as the packaging of nitroglycerin, a highly poisonous explosive liquid, or other volatile compounds such as sensitized ammonium nitrate. Dynamites can be packed in measured charges, transported easily, and, with the proper detonator, exploded safely. Because a dynamite explosion creates a "cool flame," which is less likely to ignite methane and coal dust mixtures present in mines, dynamites are frequently used in coal mining operations.

History

Alfred Nobel, his father Immanuel, and younger brother Emil began experimenting with nitroglycerin near Stockholm in 1862. Discovered by Italian chemist Ascario Sobrero in 1846, nitroglycerin was highly unstable and difficult to handle, and accidental explosions were not uncommon. One such accident killed Emil, among others, at a plant in 1864. Despite the personal tragedy, Alfred continued his work with this dangerous liquid, working on a boat in the middle of a lake before conducting his experiments in a factory. In 1866 he discovered that mixing nitroglycerin with *kieselguhr* (diatomaceous earth) stabilized and reduced the volatility of the explosive. Diatomaceous earth is formed by the fossil remains of a single-celled plankton called diatoms, and the result is an absorbent material which "soaks up" the nitroglycerin. Alfred named the product "dynamite"—derived from the Greek "dynamis" meaning "power"—and received a patent for the process in 1867. Nobel went on to develop several other explosives and propellants, including smokeless powder ballistite. He held over 355 patents and his considerable fortune provided the financial basis for the Nobel Prize, which is awarded "to those who, during the preceding year, shall have conferred the greatest benefit on mankind."

Dynamite is classified as a Secondary High Explosive, which means a detonator of Primary or Initiating High Explosive (mercury fulminate, for example) is utilized to set off the main charge. Dynamite is considered a commercial explosive, as opposed to TNT (trinitrotoluene) explosives, which are considered military munitions explosives. The first large scale use of dynamite for construction purposes was in the creation of the Hoosac Tunnel, completed in 1876.

Process Design and Facilities

Dynamite manufacture is highly regulated and the process strictly controlled to prevent accidental detonations. The equipment used is specially designed to reduce the exposure of the mixture to heat, compaction forces, or ignition sources. Bearings in the product mixers, for example, are mounted outside of the apparatus frame to prevent contact with the explosive mixture. Buildings and storage areas (called magazines) are constructed at great distances from other structures and with specialized heating, ventilation, and electrical systems. These buildings are "hardened" with bullet-

A commercial explosive used mainly for demolition and mining, dynamite was invented by Alfred Bernhard Nobel in 1866 after a series of accidents and experimentations.

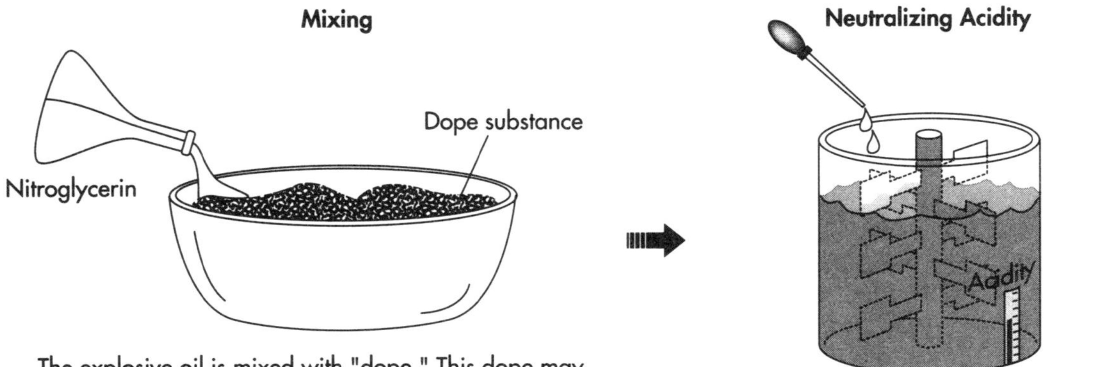

The explosive oil is mixed with "dope." This dope may be made of wood pulp, sawdust, flour, or starch, or a combination of substances.

Very small amounts of antacid are added to neutralize any acidity present in the dope. The mixture is monitored closely until the correct ingredient level is attained.

Dynamite manufacturing can be described as the safe packaging of nitroglycerin, a highly poisonous explosive liquid.

resistant roofs and walls and extensive security systems. Other important precautions include thorough inspection systems which insure correct mixing, grading, packaging, and inventory control. Employees are also highly trained to work with the explosives, and special health precautions are required. Exposure to nitroglycerin commonly produces throbbing headaches, although an immunity to the toxic effects can develop. Interestingly, nitroglycerin is also used in medicine to treat some forms of angina and other ailments. In the body, it acts as a vasodilator and relaxes muscle tissue.

The Manufacturing Process

The process begins with the compound liquid such as nitroglycerin (explosive oil), a "dope" substance, and an antacid. Ethylene glycol dinitrate, composing approximately 25-30% of the explosive oil, is used to depress the freezing point of the nitroglycerin. This allows the dynamite to be safely used at low temperatures. In fact, nitroglycerin in a semi-frozen state with both liquid and solid present is actually more sensitive and unstable than either frozen or liquid state alone. In that semi-solid state, nitroglycerin is extremely dangerous to handle.

Mixing the oil

1 The explosive oil is carefully added to a mechanical mixer, where it is absorbed by the "dope," which can be either diatomaceous earth (now no longer used), wood pulp, sawdust, flour, starch, and/or other carbonaceous substances and combinations of substances.

Neutralizing acidity

2 Approximately 1% antacid such as calcium carbonate or zinc oxide is added to neutralize any acidity present in the dope. The mixture is monitored carefully and when the correct ingredient level is attained, the mixture is ready for packaging into the various forms. This process produces what is termed "straight dynamite," in which the dope does not contribute to the explosive strength of the dynamite. For example, 40% straight dynamite contains 40% nitroglycerin and 60% dope; 35% straight dynamite contains 35% nitroglycerin and 65% dope. In some cases, sodium nitrate is mixed with the dope, which acts as an oxidizer and gives additional strength to the explosive.

Packaging dynamite

3 The appearance of dynamite typically resembles a round cartridge approxi-

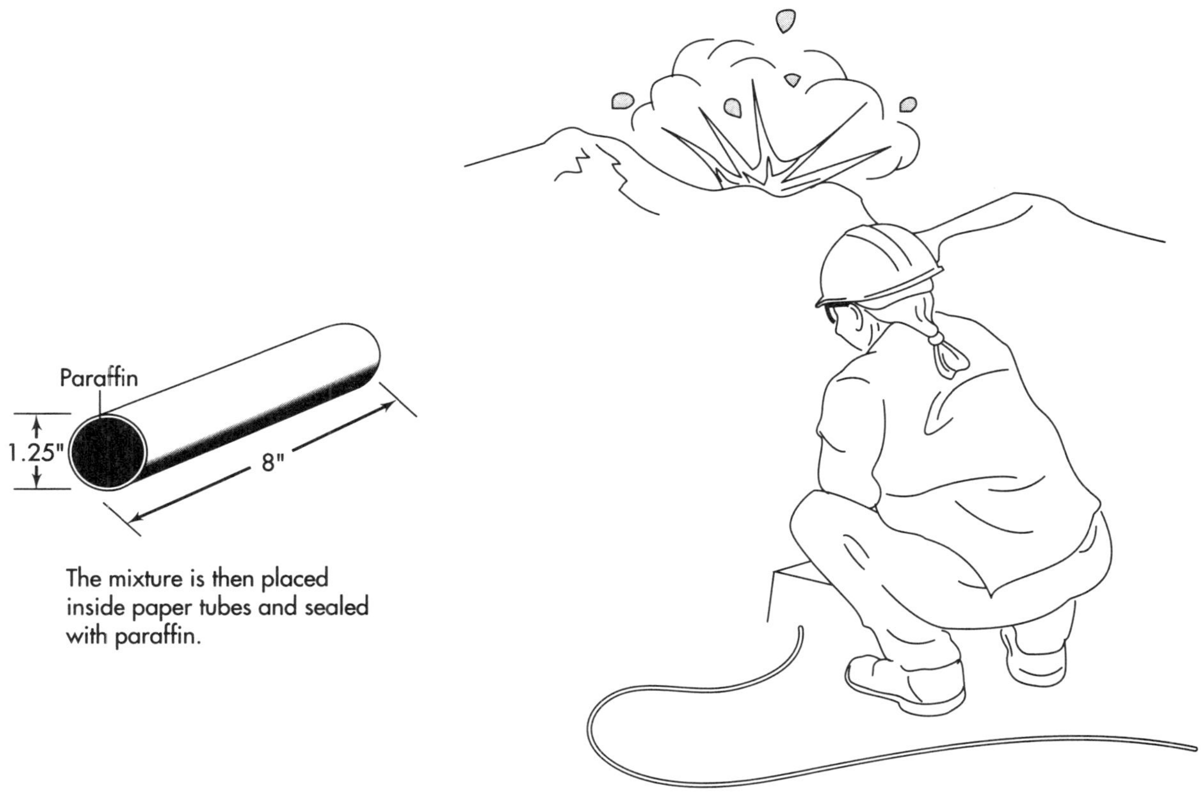

The mixture is then placed inside paper tubes and sealed with paraffin.

Dynamite is transported under strict guidelines and its application carefully monitored at the blasting site.

mately 1.25 inches (3.2 cm) in diameter and 8 inches (20 cm) long. This type is produced by pressing the dynamite mixture into a paper tube sealed with paraffin. The paraffin enclosure protects the dynamite from moisture and, being a combustible hydrocarbon, contributes to the explosive reaction. Dynamite can also exist in many other forms, from smaller sizes of cartridges for specialized demolition work to large 10-inch (25 cm) diameter charges that are used for large strip mining operations. Regulations limit the length of these big charges to 30 inches (76 cm) and the weight to 50 pounds (23 kg). Dynamite is also available as a bag powder and in a gelatinized form for underwater use.

Dynamites are also made using other substances besides nitroglycerin. For instance, replacing a larger portion of the explosive oil with ammonium nitrate can increase the explosive strength of the dynamite. This form of dynamite is referred to as ammonia dynamite.

Quality Control

Accurate dynamite strength measurement and testing by detonation assure safe performance of the explosive. The relative strength of dynamite is graded by comparison to straight dynamite and by the percentage of weight of the explosive oil. For example, ammonia dynamite is compared to straight dynamite and is graded accordingly. Fifty percent ammonia dynamite is equal in explosive strength to 50% straight dynamite. In this instance, the "50%" reflects the strength comparison rather than the explosive content.

After manufacture and batch testing of the dynamite, it is dispensed to the job site

The appearance of dynamite typically resembles a round cartridge approximately 1.25 inches in diameter and 8 inches long. The paraffin enclosure protects the dynamite from moisture and, being a combustible hydrocarbon, contributes to the explosive reaction.

under strict transportation and storage regulations.

Application

The following brief example is one of many scenarios for the proper application of dynamite. It must be noted that no one but a certified blasting expert with the correct procedures and equipment should ever attempt to detonate dynamite.

In this example, a rock formation must be blasted to make way for a construction project. The first step in the blasting procedure is to determine the size of the charge by various means, including charts, calculations, and the blaster's experience. Close examination of the affected area and surrounding terrain is made to determine the safe zone. Signs are placed a minimum of 1000 feet (305 m) outside the safe zone to warn the public of the blasting. Radio transmitters are turned off and locked to prevent accidental firing of the electric detonators. The charge is then withdrawn from the magazine and transported to the blast site using closed and secure trucks. The detonators are brought to the job site in a separate vehicle.

The charges are unloaded and placed into the blast holes drilled in the rock formation. They slide into the blast hole by air pressure or by tamping with wooden or plastic rods. The blaster takes great care that the leadwires to the detonators are shorted together until all charges have been placed. This provides a short circuit path for the wiring which prevents accidental ignition. Only the blaster is allowed to make the final electrical connections to the main firing switch.

During this time, a 5-foot (1.5 m) gap in the wiring immediately ahead of the main switch is used as a "lightning gap," another safety practice to eliminate the possibility of static electricity setting off the charges. Once all of the preparation for the blast is complete, a warning horn sounds a one-minute series of blasts prior to the detonation signal. At this time, the final connections to the firing switch are made. At one minute to detonation, a series of short horn blasts are sounded. The blaster then unlocks the main switch and detonates the charges. After the explosion, all electrical circuits to the blasting equipment are once again locked into the safe positions, and the area is inspected for misfired charges and general safety. A prolonged horn blast signals the all clear.

Byproducts/Waste

Explosives manufacture and use contribute some measure of hazardous waste to the environment. Nitroglycerin produces several toxic byproducts such as acids, caustics, and oils contaminated with heavy metals. These must be disposed of properly by neutralization or stabilization and transported to a hazardous waste landfill. The use of explosives creates large amounts of dust and particulate from the explosion, and, in some cases, releases asbestos, **lead**, and other hazardous materials into the atmosphere. Also, uncontrolled or improperly calculated explosions may rupture nearby tanks and pipelines, releasing their contents into the environment as well.

The Future

Since their development in the 1950s, advanced forms of plastic explosives and shaped charges have replaced dynamite. These explosives are now referred to as blasting agents, since their stability is improved and require a more powerful primer to detonate. One of the most common blasting agent is ANFO, or ammonium nitrate and fuel oil. ANFO is readily available, considerably cheaper than dynamite, and can be mixed on site. However, concrete demolition crews requiring relatively small charges still use dynamite as the blasting agent.

Where To Learn More

Books

Grady, Sean M. *Devices of Controlled Destruction.* Lucent Books, 1994.

Keller, J.J. *OSHA Safety and Compliance Manual*, 1992.

Lewis, Richard, Sr. *Sax's Dangerous Properties of Industrial Materials*, 8th ed. Van Norstrad Rienhold, 1993.

Meyer, Rudolf. *Explosives*. VCH Publishers, 1992.

Porter, Samuel J. *Explosives and Blasting Agents, Fire Protection Handbook*. 15th ed., 1981.

Sickler, Robert A. *Explosive Principles: An Essential Guide to Understanding Explosives and Detonations*. Paladin Press, 1992.

—Douglas E. Betts

Elevator

The modern elevator is a direct descendant of a design first shown by Elisha G. Otis at the New York World's Fair in 1853. A notable feature of the Otis elevator, and the principal reason for its popular acceptance, was a safety device that immediately engaged and held the elevator in the event the hoisting cables broke.

Background

An elevator is a platform, either open or enclosed, used for lifting people or freight to upper floors within a building. Elevators are a standard part of any tall commercial or residential building. In recent years, the introduction of the Federal Americans with Disabilities Act has required that many two-story and three-story buildings be retrofitted with elevators.

Manually operated elevators were used for lifting freight in warehouses and manufacturing plants as early as the 1600s. The modern elevator is a direct descendant of a design first shown by Elisha G. Otis at the New York World's Fair in 1853. A notable feature of the Otis elevator, and the principal reason for its popular acceptance, was a safety device that immediately engaged and held the elevator in the event the hoisting cables broke. The first elevators were operated by steam power to turn the cable drums. In 1871, the first hydraulic elevators were introduced using water pressure as the source of power. At first the hydraulic rams were one-piece, which meant a hole had to be dug under the elevator shaft as deep as the elevator was to be high. Later multiple-section, telescoping hydraulic rams allowed shallower holes. In many cities hydraulic power for these early elevators was supplied by power companies which installed and maintained networks of hydraulic piping throughout the city. The first commercially successful electric elevator was installed in 1889, and electricity quickly became the accepted source of power.

Electric-powered elevators offered two significant advantages. First, electric power was clearly becoming universally available, and any building likely to be equipped with an elevator would also have electric power. Second, hydraulic elevators were severely limited in the height to which they could rise, while electric elevators, using a simple cable and pulley system, had virtually no height limit. For many years, electric elevators used either direct current (DC) motors or alternating current (AC) motors. Today, almost all elevators use one of two types of AC motors: the most common are geared motors for elevators moving at speeds up to 500 feet per minute (153 m per minute), while direct-drive motors are used for elevators moving at higher speeds. Some modern high-speed elevators move at up to 2,000 feet per minute (610 m per minute).

Control systems on early elevators required human operators to regulate the speed of the lift and descent, to stop the elevator at each floor, and to open and close the doors. In the 1950s automatic pushbutton control systems replaced manual controls. In the 1970s electromechanical controls were gradually replaced with solid state electronic controls.

Riding in a small box hundreds of feet in the air would be a disconcerting experience if one were not convinced of its safety. Electric elevators are equipped with two primary safety mechanisms: a governor which controls the elevator's speed by controlling the speed of the cable pulleys, and the emergency brake which consists of jaws that grip the elevator guide rails in the event the cables break. Elevators also include electromechanical door interlocks to prevent the elevator from operating if the door is not completely closed and to protect passengers from being trapped by the closing door. The same door interlocks also prevent the outer doors on each floor from opening if the elevator is not present. Most elevators are equipped with a telephone, and sometimes a trap door in the ceiling, so that passengers can call for help or escape if an elevator becomes stuck between floors.

Design

Elevators themselves are simple devices, and the basic lifting systems have not changed much in over 50 years. The control systems, however, have changed substantially to improve safety and speed of operation. Elevators are designed for a specific building, taking into account such factors as the height of the building, the number of people traveling to each floor, and the expected periods of high usage.

Most elevators use counterweights which equal the weight of the elevator plus 40% of its maximum rated load. This counterweight reduces the weight the motor must lift and ensures that the elevator cannot fall out of control while the cable is intact. In a lifting drum installation, a hoist cable runs down from a drive drum attached to the hoist motor, around a large pulley on the top of the elevator, up to a second pulley hanging from the roof of the elevator shaft, and down again to the counterweight. In a traction drum installation, the cable runs from the elevator, up and once around a drive drum attached to the hoist motor, then back to the counterweight. The elevator, called the car, and the counterweight each run in their own sets of guide rails. A second governor cable runs from the car up to a governor pulley, then down to a tension pulley at the bottom of the elevator shaft, and up to the car again. This cable rotates the governor pulley at a speed directly proportional to the speed of the car. In the event of excessive car speed, the governor uses another cable to activate the emergency brake jaws which grip the guide rails and slow the car to a stop.

A ramped bar on the side of the elevator shaft activates a series of switches on the outside of the car to slow and stop the car at the proper floor. As the car approaches the desired floor, the ramp activates the slow-down switch, which signals the hoist motor to reduce speed. When the car is aligned with the outer door opening, the ramp activates a limit switch to stop the car. If the door interlock switches also sense that the car is in the proper location, the electric door opening motor is activated to open both the inner car door and the outer floor door.

Modern commercial buildings commonly have multiple elevators with a unified control system. The object of the control system is to minimize the average time any passenger spends from the time the elevator call button is pushed to the arrival of the first available elevator. Different systems use different levels of sophistication. The simplest systems use a single up and down button on each floor regardless of the number of elevators. When a passenger calls for an elevator, the controller sends the nearest elevator that is traveling in the desired direction. The approach of an elevator car is signaled by an illuminated arrow above the elevators doors pointing up or down.

In more sophisticated systems, the controller monitors the elevator call system for a set, or bank, of elevators operating side by side. The operation zone of these elevators is divided into sectors, with each sector being made up of adjacent floors. When a car has answered a call and completed the designated run, it becomes available to answer another call. At this point, depending on the controller's programming, the car may be returned to a designated "home" floor, or may be sent to the sector furthest from other operating or available cars to cover that sector. When a call is received, the controller automatically compares the location of all the cars in the bank and sends the nearest one.

Controllers can also be programmed to respond differently at different times of the day. For example, the elevator controller in a busy office building will receive a preponderance of calls from the ground floor in the morning, when workers are arriving and need to go to their workplaces on the upper floors. In that case, the controller will be programmed to send all unassigned cars to the ground floor, rather than have them return to a home floor in their sector.

The elevator in the new Lord and Taylor department store in New York City looked like this on opening day in 1873. (From the collections of Henry Ford Museum & Greenfield Village.)

The elevator is one of those inventions whose "ripple effect" is often overlooked. Just think of the practicality of any building over eight or ten stories without an elevator. Then imagine a modern city without buildings over ten stories! Along with structural steel and reinforced concrete, the elevator was essential to the development of the modern skyscraper and thus to the common form of the modern urban center.

The elevator's practical impact was almost matched by its symbolic impact. The 1880s were years of immense urban growth, and the influx of newcomers to the cities included middle-class career people as well as factory workers. With property values skyrocketing in the cities, the middle-class families could not afford single family homes. Apartment building owners promoted apartment living with advertisements of "high-tech" amenities: hot and cold running water, telephone systems, central gas for cooking and lighting, fully equipped bathrooms, and elevators.

Moreover, with all these modern conveniences, apartment living captured the middle-class imagination as the embodiment of a new organization of domestic duties. Buildings came with centralized heating, ventilating, and plumbing systems; some had kitchens in the basement which would prepare food for individual apartment dwellers; some even had a centralized vacuum system with nozzles in each room connected to a pump in the basement.

The elevator was even extolled as a contributor to democracy. In an elevator-equipped building, it made little difference which floor one lived on; every floor was equally accessible. By contrast, in Europe, wealthy families were generally found on the middle floors where they did not have to climb many flights. Poorer families were usually confined to the basement or the upper floors.

William S. Pretzer

Later in the day, a different set of instructions can be used to send unassigned elevators to different sectors, since passengers leaving the building will be much more evenly distributed among the floors than in the morning.

All modern elevators also have special override controls that firefighters can activate with a key to make elevators go directly to a specific floor without intermediate stops.

Raw Materials

The elevator car itself is constructed with a steel framework for durability and strength. A set of steel beams above the car, called the crosshead, span the elevator shaft from side to side and hold the pulley for the hoist cable. A steel structure, called the sling, extends down the sides of the car from the crosshead and cradles the floor, or platform. The sides of a passenger elevator car are usually made from steel sheet and are trimmed on the inside with decorative paneling. The floor of the car may be tiled or carpeted. Handrails and other interior trim may be made from stainless steel for appearance and wearability. A suspended ceiling is usually hung below the actual top of the car and may contain fluorescent lighting above plastic diffuser panels. The elevator controls, alarm buttons, and emergency telephone are contained behind panels in the front of the car, next to the doors.

Steel guide rollers or guide shoes are attached to the top and bottom of the sling structure on each side to run along the guide rails. The guide rails are also steel and are attached to the interior walls of the elevator shaft which runs from the top of the building to the bottom. The emergency brake mechanism consists of two clamping faces which can be driven together by a wedge to squeeze on the guide rail. The wedge is activated by a screw turned by a drum attached to the emergency cable.

Elevator hoisting cable usually consists of six or more strands, each of which consist of a number of separate steel wires. The strands may be twisted around a hemp center which serves as a cushion and also contains a lubricant.

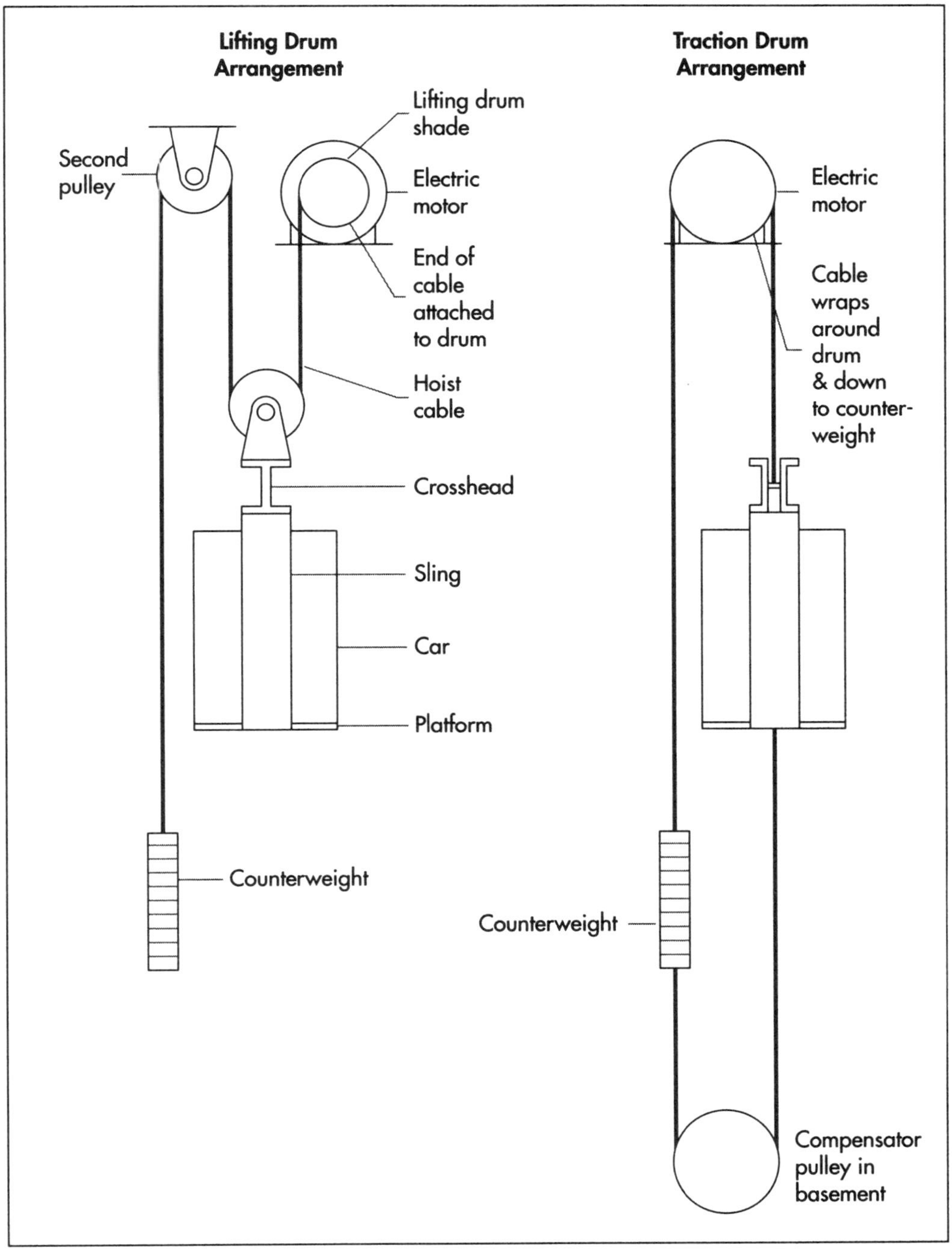

In a lifting drum installation, a hoist cable runs down from a drive drum attached to the hoist motor, around a large pulley on the top of the elevator, up to a second pulley hanging from the roof of the elevator shaft, and down again to the counterweight. In a traction drum installation, the cable runs from the elevator, up and once around a drive drum attached to the hoist motor, then back to the counterweight.

The electric hoisting motors are specifically designed for elevator service and may drive the hoisting drum through a gearbox, both of which are purchased parts.

The Manufacturing Process

1 The elevator cars are built at the elevator manufacturer's plant using standard metal cutting, welding, and forming techniques. If the cars will be exposed to the weather during building construction, the interior trim may be installed after the building is finished.

2 The rest of the elevator is assembled on the building site. The building design integrates the elevator shaft from the beginning, and the shaft grows as the building is erected. The walls of the shaft are poured concrete, and the shaft straightness and

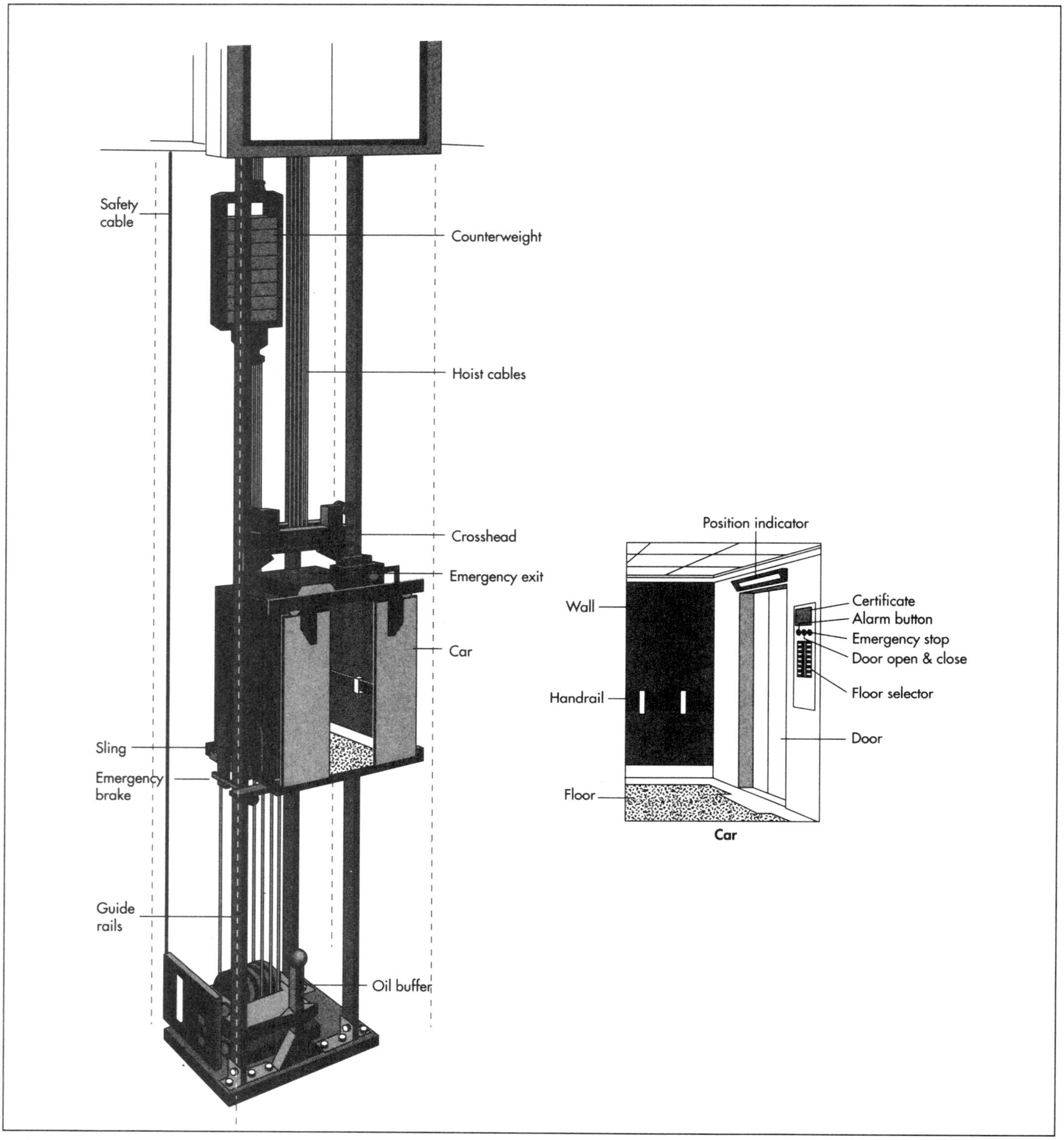

Most elevators use counterweights which equal the weight of the elevator plus 40% of its maximum rated load. This counterweight reduces the weight the motor must lift and ensures that the elevator cannot fall out of control while the cable is intact.

other dimensions are carefully monitored as each floor goes up.

3 Guide rails, switch ramps, service ladders, and similar support equipment are bolted into the shaft after the shaft walls are complete, but before the shaft is roofed.

4 While the shaft is still open at the top, a crane raises the counterweight to the top of the building and lowers it into the shaft along its rails.

5 The crane then lifts the elevator car and inserts it partly into the shaft. The

guide wheels connect the car to the guide rails, and the car is carefully lowered to the bottom of the shaft.

6 The shaft is then roofed over, leaving a machine room above the shaft. The hoist motor, governor, controller, and other equipment are mounted in this room, with the motor located directly over the elevator car pulley.

7 The elevator and governor cables are strung and attached, the electrical connections completed, and the controller programmed.

Quality Control

Each elevator installation in the United States must meet the safety standards of the American National Standards Institute and the American Society of Mechanical Engineers. These standards may be incorporated into local building codes, or the local codes may have their own safety standards. The state must inspect, rate, and certify each passenger elevator installation before it goes into operation and must reinspect on a regular basis thereafter.

The Future

Elevators have not changed substantially in many years and are unlikely to do so in the near future. Electronic controls will continue to improve in ways that are evolutionary and not very dramatic. Control systems are being developed that will learn from past traffic patterns and use this information to predict future needs in order to reduce waiting times. Laser controls are coming into use, both to gauge car speed and distance, as well as to scan building floors for potential passengers.

Where To Learn More

Books

Elevator Technology. Published for the International Association of Elevator Engineers by Ellis Horwood. Halsted Press, 1986.

Ford, Barbara. *The Elevator*. Walker and Company, 1982.

Periodicals

Evans, Barrie. "Fuzzy Logic for Smarter Lifts." *Architect's Journal*, May 18, 1994, pp. 24-25.

Richards, Kristen. "Design Files." *Interiors*, February 1991, pp. 22-23.

—Joel Simon

Fat Substitute

Low calorie, low fat, fat free, or no fat foods are being mass marketed to legions of Americans who are fighting the proverbial battle of the "bulge." One development in this battle has been the introduction of fat substitutes.

Background

Fat. No one likes it; no one wants it. Diets high in fat have been linked to serious health problems and weight gain. The battle to limit fat consumption is now a billion dollar industry. Low calorie, low fat, fat free, or no fat foods are being mass marketed to legions of Americans who are fighting the proverbial battle of the "bulge." One development in this battle has been the introduction of fat substitutes. Simplesse®, a protein-based fat substitute, is one such product being promoted to help diet-conscious consumers fight weight gain.

History

Simplesse is a brand-name fat substitute made by NutraSweet Company of Deerfield, Illinois, but it was not the first "fake fat" to be manufactured. In the early 1970s, Proctor & Gamble developed Olestra® while conducting research to develop premature infant formula. Olestra was made by chemically binding a sugar to the fatty acids in vegetable oil to create a substance known as sucrose polyester that looks, tastes, and feels like real fat. Sucrose polyester cannot be broken down by digestive enzymes so it passes through the body without giving off calories.

Olestra was used in ice cream, cooking oil, salad dressings, baked goods, deep-fried foods, and snack chips. Recently, Olestra's safety as a food additive has been questioned by the Center for Science in the Public Interest. The Center asserted that Olestra is a chemically altered substance that produces "a new molecule that is totally foreign to the body." Olestra's patent expired in 1994, and Proctor & Gamble still awaits approval by the Food and Drug Administration (FDA). Meanwhile, the company has developed Caprenin®, a reduced-calorie fat that is similar to cocoa butter. Caprenin uses two unsaturated fats and acids from coconut and palm kernel oils and is only partially absorbed by the body.

In 1984, inventors Norman S. Singer, Shoji Yamamoto, and Joseph Latella of London, Canada, filed for a U.S. patent for Simplesse. According to the patent (patent #4,734,287), Simplesse is made of egg white and whey, a milk byproduct obtained through a microparticulation process. This process turns the milk mixture into small particles that resemble the texture of fat. Compared to real fat, Simplesse has less than one-third of the calories. NutraSweet compared the fat and calories of a super-premium vanilla ice cream containing 16% butterfat with a similar dessert using Simplesse. A four-ounce ice cream serving contained 19 grams of fat, 97 milligrams of cholesterol, and 274 calories. The same-size serving of Simple Pleasures®, containing Simplesse, had less than 1 gram of fat, 14 milligrams of cholesterol, and 120 calories. However, unlike other fat substitutes, Simplesse cannot withstand heat and therefore cannot be used in cooked items.

In 1990, the FDA approved Simplesse for use as a thickener or texturizer in frozen desserts. The FDA approval confirmed that this product met its "generally recognized as safe" (GRAS) standards. Simplesse is marketed as a food ingredient and is currently used in 18 products worldwide.

Raw Materials

Simplesse is made out of egg whites, milk and whey proteins, pectin, and citric acids.

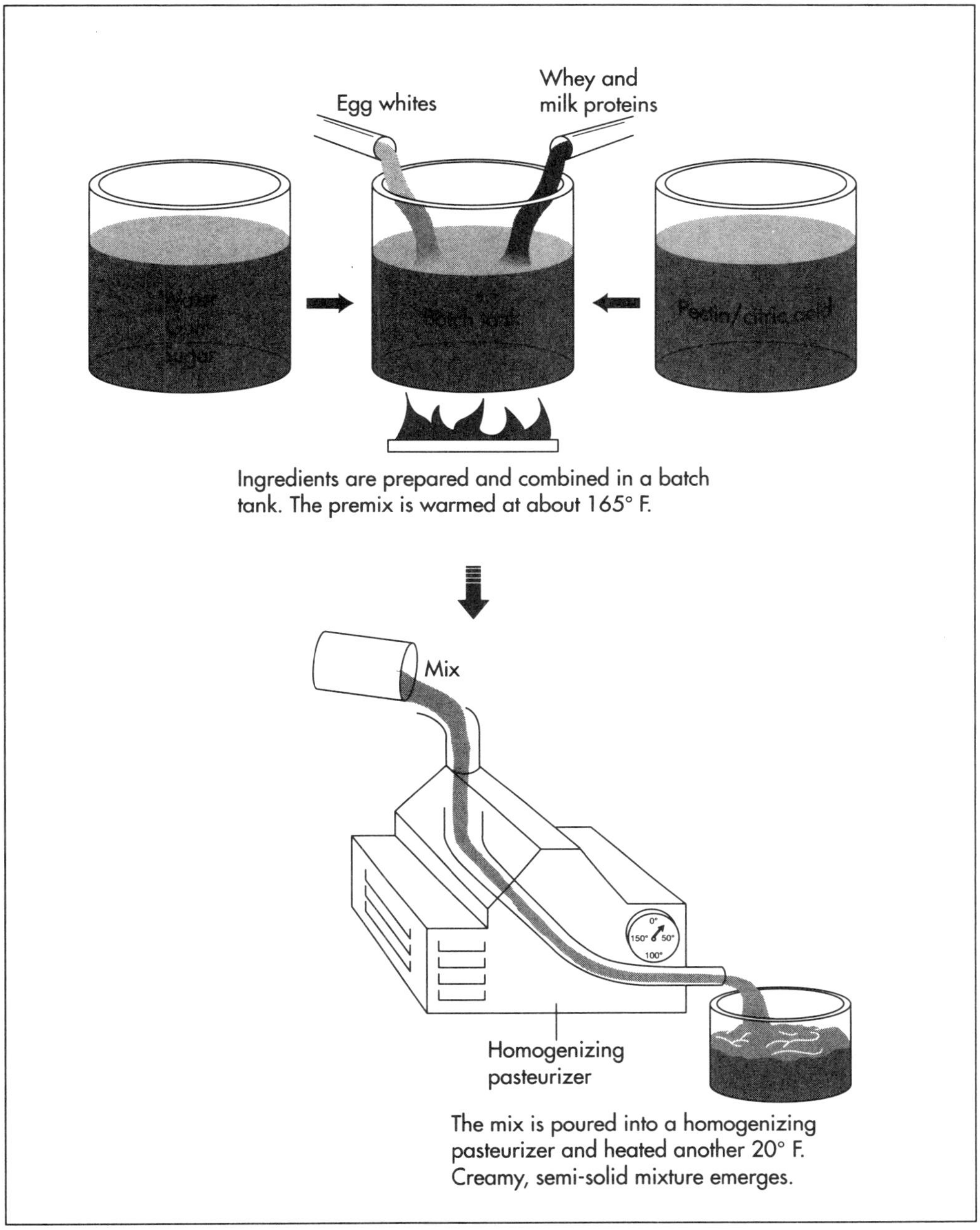

Ingredients are prepared and combined in a batch tank. The premix is warmed at about 165° F.

The mix is poured into a homogenizing pasteurizer and heated another 20° F. Creamy, semi-solid mixture emerges.

Simplesse, a brand name fat substitute made by NutraSweet, is made out of egg whites, milk and whey proteins, pectin, and citric acids.

Whey is a byproduct of cheesemaking. These materials are combined, homogenated, and heated in a microparticulation process whereby the proteins are shaped into round particles that have the texture of fat and that roll smoothly over one another.

The Manufacturing Process

The microparticulation process is one that dates back to about 1000 b.c. The ancient Apicius cookbook detailed a mixture of eggs, milk, and honey that formed a custard-like gel (tyropatinam) that was eaten along with cheese. Microparticulation does the same for Simplesse, which is said to create a "creamy" sensation when eaten. After microparticulation, there are about 50 billion balls or particles of the egg white and whey and milk protein substance per teaspoon. To the tongue, Simplesse feels smooth. Norman Singer, Simplesse's chief inventor, says the process rearranges the molecules similar to "winding them up like spaghetti on a fork." Simplesse is considered to be virtually free of cholesterol.

Mixing the ingredients

1 Water, gum, and sugar are mixed in a dry blender and then moved to a batch tank. Egg whites, whey protein concentrate, and skim milk proteins are added. The pH, or acidity, levels are adjusted to the mix as it sits in the batch tank.

2 Lethecin, pectin, and citric acid are completely dissolved and dispersed so that no particles larger than 1 micron (one millionth of a meter) can be found in the premix before homogenizing pasteurization. The pectin/citric acid mix is then added to the egg white, whey, and milk proteins.

Heating

3 The premix, a thin, pourable liquid, is warmed by a heat exchanger to a temperature just below the coagulation region of the protein. This is about 165°F (74°C) for egg whites.

4 The warmed mix is poured into the homogenizing pasteurizer and heated another 20°F (-6.7°C) in less than 10 seconds. During this time, the mixture is continuously exposed to a uniformly turbulent field of homogenizing force. Half of the protein denatures as a gel and forms beadlets of about 1 to 3 microns in diameter. Sometimes the mixture is poured through a holding tube before it is cooled.

5 Upon exiting the pasteurizer, the mixture is cooled to become creamy, smooth, opaque, and semisolid. Simplesse is the end product.

Quality Control

Whey is a key ingredient in Simplesse, but it is difficult to convert industrially. Whey is obtained from cheesemaking and contains about 90% water. Removing this excess water is costly, and the water is not easily disposed of without causing environmental concerns. One alternative is to heat the whey to denature and coagulate it. Once done, whey could then be separated into other byproducts to defer processing and disposal costs.

Because Simplesse uses natural and not synthetic products, it is presumed safe to use. However, care must be taken to manufacture the skim milk, whey proteins, and egg proteins in sterile conditions to prevent bacteria from contaminating the end product.

The Future

Fat replacers had a market value of about $100 million in 1991, but this is expected to triple by 1996. The market for reduced-fat food products is estimated at nearly $30 billion, making the future of fat substitutes highly favorable. At issue, however, is making fat substitutes more palatable and accessible to consumers. NutraSweet has asserted that full use of Simplesse could decrease American fat consumption by 14% and cholesterol intake by 5%. As of yet, Simplesse has not been approved for use in cooked, baked, or fried food because it breaks down under heat.

Where To Learn More

Books

"No-Fat Foods: Less than Meets the Eye?" *Consumer Reports*, May 1993, p. 279-81.

DeMoss, Virginia. "Ersatz Fat." *Bicycling*, June 1990, pp. 112-14.

Kantrowitz, Barbara. "A Heavyweight Fuss Over the New 'Fake Fat'."

Newsweek, March 5, 1990, p. 41.

Monmaney, Terence and Karen Springen. "The Cholesterol Connection." *Newsweek*, February 8, 1988, p. 56-58.

Segal, Marian. "Fat Substitutes: A Taste of the Future?" *FDA Consumer*, December 1990, p. 25-27.

Singer, Norman S. and John Michael Dunn. "Protein Microparticulation: The Principle and the Process." *Journal of the American College of Nutrition*, 1990, pp. 388-97.

Springen, Karen. "How Low Can They Go?" *Newsweek*, August 20, 1990, pp. 66-67.

Steacy, Anne. "Fat That Will Not Fatten." *Maclean's*, February 15, 1988, p. 50.

Street, Robin. "Finally 'Fake Fat' Foods." *Better Homes and Gardens*, June 1990, p. 32.

Thayer, Ann. M. "Food Additives." *C&EN*, June 15, 1992, pp. 26-44.

—Evelyn S. Dorman

Fiberglass

Fiberglass can be formed into yarn much like wool or cotton, and woven into fabric. Fiberglass textiles are commonly used as a reinforcement material for molded and laminated plastics. Fiberglass wool, a thick, fluffy material, is used for thermal insulation and sound absorption.

Background

Fiberglass refers to a group of products made from individual glass fibers combined into a variety of forms. Glass fibers can be divided into two major groups according to their geometry: continuous fibers used in yarns and textiles, and the discontinuous (short) fibers used as batts, blankets, or boards for insulation and filtration. Fiberglass can be formed into yarn much like wool or cotton, and woven into fabric which is sometimes used for draperies. Fiberglass textiles are commonly used as a reinforcement material for molded and laminated plastics. Fiberglass wool, a thick, fluffy material made from discontinuous fibers, is used for thermal insulation and sound absorption. It is commonly found in ship and submarine bulkheads and hulls; automobile engine compartments and body panel liners; in furnaces and air conditioning units; acoustical wall and ceiling panels; and architectural partitions. Fiberglass can be tailored for specific applications such as Type E (electrical), used as electrical insulation tape, textiles and reinforcement; Type C (chemical), which has superior acid resistance, and Type T, for thermal insulation.

Though commercial use of glass fiber is relatively recent, artisans created glass strands for decorating goblets and vases during the Renaissance. A French physicist, Rene-Antoine Ferchault de Reaumur, produced textiles decorated with fine glass strands in 1713, and British inventors duplicated the feat in 1822. A British silk weaver made a glass fabric in 1842, and another inventor, Edward Libbey, exhibited a dress woven of glass at the 1893 Columbian Exposition in Chicago.

Glass wool, a fluffy mass of discontinuous fiber in random lengths, was first produced in Europe at the turn of the century, using a process that involved drawing fibers from rods horizontally to a revolving drum. Several decades later, a spinning process was developed and patented. Glass fiber insulating material was manufactured in Germany during World War I. Research and development aimed at the industrial production of glass fibers progressed in the United States in the 1930s, under the direction of two major companies, the Owens-Illinois Glass Company and Corning Glass Works. These companies developed a fine, pliable, low-cost glass fiber by drawing molten glass through very fine orifices. In 1938, these two companies merged to form Owens-Corning Fiberglas Corp. Now simply known as Owens-Corning, it has become a $3-billion-a-year company, and is a leader in the fiberglass market.

Raw Materials

The basic raw materials for fiberglass products are a variety of natural minerals and manufactured chemicals. The major ingredients are silica sand, limestone, and soda ash. Other ingredients may include calcined alumina, borax, feldspar, nepheline syenite, magnesite, and kaolin clay, among others. Silica sand is used as the glass former, and soda ash and limestone help primarily to lower the melting temperature. Other ingredients are used to improve certain properties, such as borax for chemical resistance. Waste glass, also called cullet, is also used as a raw material. The raw materials must be carefully weighed in exact quantities and thoroughly mixed together (called batching) before being melted into glass.

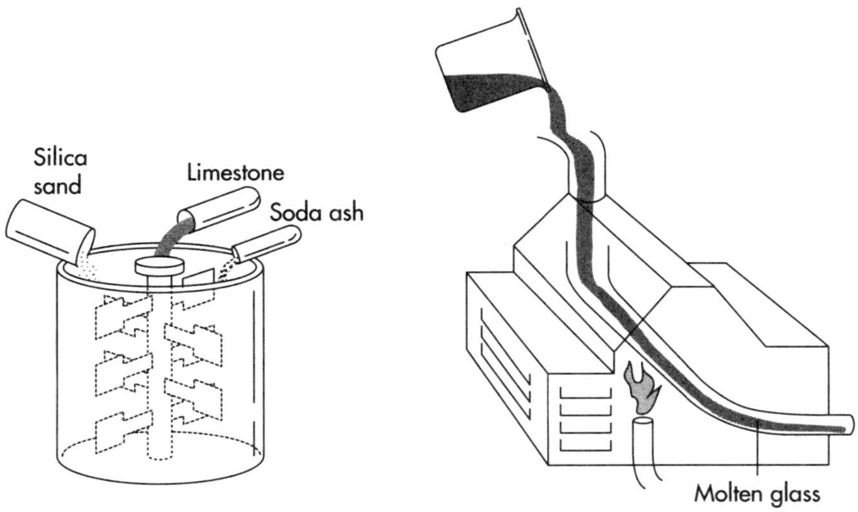

Fiberglass raw materials are mixed together and melted.

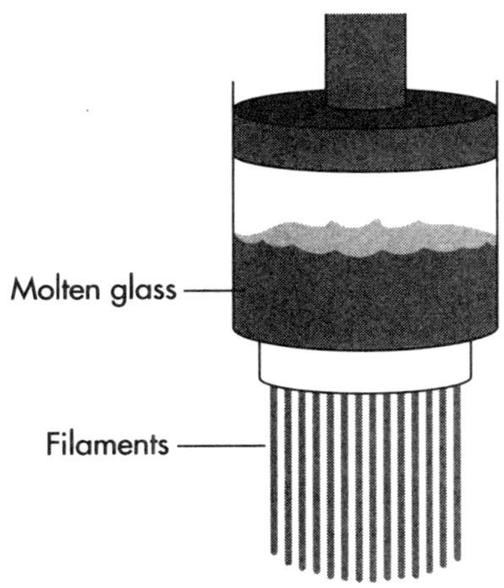

The molten glass is passed through platinum bushings that contain hundreds of holes. Fine filaments are extruded.

The Manufacturing Process

Melting

1 Once the batch is prepared, it is fed into a furnace for melting. The furnace may be heated by electricity, fossil fuel, or a combination of the two. Temperature must be precisely controlled to maintain a smooth, steady flow of glass. The molten glass must be kept at a higher temperature (about 2500°F [1371°C]) than other types of glass in order to be formed into fiber. Once the glass becomes molten, it is transferred to the forming equipment via a channel (forehearth) located at the end of the furnace.

Forming into fibers

2 Several different processes are used to form fibers, depending on the type of fiber. Textile fibers may be formed from molten glass directly from the furnace, or the molten glass may be fed first to a ma-

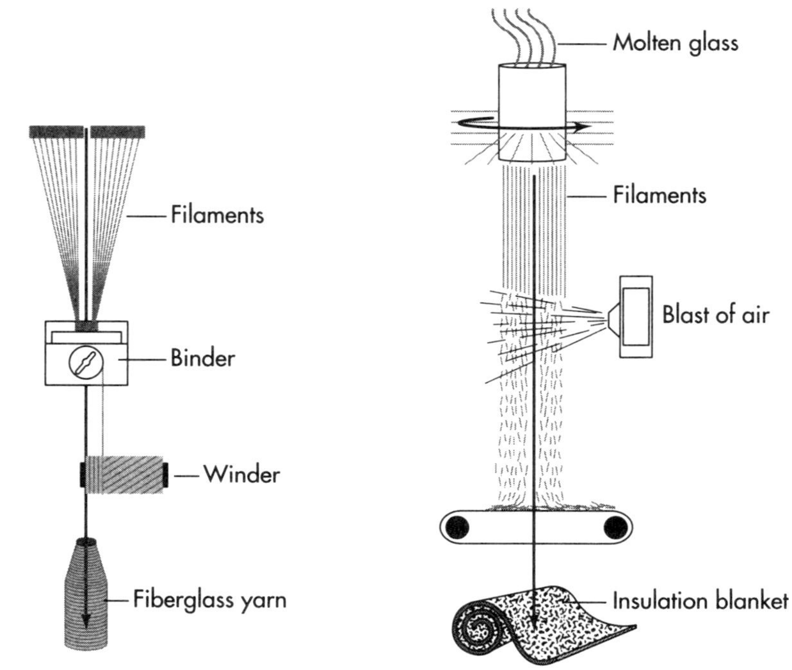

Depending on the usage, the filaments can be wound like yarn or converted to a fleecy mass for insulation.

chine that forms glass marbles of about 0.62 inch (1.6 cm) in diameter. These marbles allow the glass to be inspected visually for impurities. In both the direct melt and marble melt process, the glass or glass marbles are fed through electrically heated bushings (also called spinnerets). The bushing is made of platinum or metal alloy, with anywhere from 200 to 3,000 very fine orifices. The molten glass passes through the orifices and comes out as fine filaments.

Continuous-filament process

3 A long, continuous fiber can be produced through the continuous-filament process. After the glass flows through the holes in the bushing, multiple strands are caught up on a high-speed winder. The winder revolves at about 2 miles (3 km) a minute, much faster than the rate of flow from the bushings. The tension pulls out the filaments while still molten, forming strands a fraction of the diameter of the openings in the bushing. A chemical binder is applied, which helps keep the fiber from breaking during later processing. The filament is then wound onto tubes. It can now be twisted and plied into yarn.

Staple-fiber process

4 An alternative method is the staple-fiber process. As the molten glass flows through the bushings, jets of air rapidly cool the filaments. The turbulent bursts of air also break the filaments into lengths of 8-15 inches (20-38 cm). These filaments fall through a spray of lubricant onto a revolving drum, where they form a thin web. The web is drawn from the drum and pulled into a continuous strand of loosely assembled fibers. This strand can be processed into yarn by the same processes used for wool and cotton.

Chopped fiber

5 Instead of being formed into yarn, the continuous or long-staple strand may be chopped into short lengths. The strand is mounted on a set of bobbins, called a creel, and pulled through a machine which chops it into short pieces. The chopped fiber is

formed into mats to which a binder is added. After curing in an oven, the mat is rolled up. Various weights and thicknesses give products for shingles, built-up roofing, or decorative mats.

Glass wool

6 The rotary or spinner process is used to make glass wool. In this process, molten glass from the furnace flows into a cylindrical container having small holes. As the container spins rapidly, horizontal streams of glass flow out of the holes. The molten glass streams are converted into fibers by a downward blast of air, hot gas, or both. The fibers fall onto a conveyor belt, where they interlace with each other in a fleecy mass. This can be used for insulation, or the wool can be sprayed with a binder, compressed into the desired thickness, and cured in an oven. The heat sets the binder, and the resulting product may be a rigid or semi-rigid board, or a flexible batt.

Protective coatings

7 In addition to binders, other coatings are required for fiberglass products. Lubricants are used to reduce fiber abrasion and are either directly sprayed on the fiber or added into the binder. An anti-static composition is also sometimes sprayed onto the surface of fiberglass insulation mats during the cooling step. Cooling air drawn through the mat causes the anti-static agent to penetrate the entire thickness of the mat. The anti-static agent consists of two ingredients—a material that minimizes the generation of static electricity, and a material that serves as a corrosion inhibitor and stabilizer.

Sizing is any coating applied to textile fibers in the forming operation, and may contain one or more components (lubricants, binders, or coupling agents). Coupling agents are used on strands that will be used for reinforcing plastics, to strengthen the bond to the reinforced material.

Sometimes a finishing operation is required to remove these coatings, or to add another coating. For plastic reinforcements, sizings may be removed with heat or chemicals and a coupling agent applied. For decorative applications, fabrics must be heat treated to remove sizings and to set the weave. Dye base coatings are then applied before dying or printing.

Forming into shapes

8 Fiberglass products come in a wide variety of shapes, made using several processes. For example, fiberglass pipe insulation is wound onto rod-like forms called mandrels directly from the forming units, prior to curing. The mold forms, in lengths of 3 feet (91 cm) or less, are then cured in an oven. The cured lengths are then de-molded lengthwise, and sawn into specified dimensions. Facings are applied if required, and the product is packaged for shipment.

Quality Control

During the production of fiberglass insulation, material is sampled at a number of locations in the process to maintain quality. These locations include: the mixed batch being fed to the electric melter; molten glass from the bushing which feeds the fiberizer; glass fiber coming out of the fiberizer machine; and final cured product emerging from the end of the production line. The bulk glass and fiber samples are analyzed for chemical composition and the presence of flaws using sophisticated chemical analyzers and microscopes. Particle size distribution of the batch material is obtained by passing the material through a number of different sized sieves. The final product is measured for thickness after packaging according to specifications. A change in thickness indicates that glass quality is below the standard.

Fiberglass insulation manufacturers also use a variety of standardized test procedures to measure, adjust, and optimize product acoustical resistance, sound absorption, and sound barrier performance. The acoustical properties can be controlled by adjusting such production variables as fiber diameter, bulk density, thickness, and binder content. A similar approach is used to control thermal properties.

The Future

The fiberglass industry faces some major challenges over the rest of the 1990s and beyond. The number of producers of fiberglass insulation has increased due to American subsidiaries of foreign companies and improvements in productivity by U.S. manufacturers. This has resulted in excess capacity, which the current and perhaps future market cannot accommodate.

In addition to excess capacity, other insulation materials will compete. Rock wool has become widely used because of recent process and product improvements. Foam insulation is another alternative to fiberglass in residential walls and commercial roofs. Another competing material is cellulose, which is used in attic insulation.

Because of the low demand for insulation due to a soft housing market, consumers are demanding lower prices. This demand is also a result of the continued trend in consolidation of retailers and contractors. In response, the fiberglass insulation industry will have to continue to cut costs in two major areas: energy and environment. More efficient furnaces will have to be used that do not rely on only one source of energy.

With landfills reaching maximum capacity, fiberglass manufacturers will have to achieve nearly zero output on solid waste without increasing costs. This will require improving manufacturing processes to reduce waste (for liquid and gas waste as well) and reusing waste wherever possible. Such waste may require reprocessing and remelting before reusing as a raw material. Several manufacturers are already addressing these issues.

Where To Learn More

Books

Aubourg, P.F., C. Crall, J. Hadley, R.D. Kaverman, and D.M. Miller. "Glass Fibers, Ceramics and Glasses," in *Engineered Materials Handbook*, Vol. 4. ASM International, 1991, pp. 1027-31.

McLellan, G.W. and E.B. Shand. *Glass Engineering Handbook*. McGraw-Hill, 1984.

Pfaender, H.G. *Schott Guide To Glass*. Van Nostrand Reinhold Company, 1983.

Tooley, F.V. "Fiberglass, Ceramics and Glasses," in *Engineered Materials Handbook*, Vol. 4. ASM International, 1991, pp. 402-08.

Periodicals

Hnat, J.G. "Recycling of Insulation Fiberglass Waste." Glass Production Technology International, Sterling Publications Ltd., pp. 81-84.

Webb, R.O. "Major Forces Impacting the Fiberglass Insulation Industry in the 1990s." *Ceramic Engineering and Science Proceedings*, 1991, pp. 426-31.

—Laurel M. Sheppard

Fire Engine

Background

The term "fire truck" is commonly used as a generic expression to describe a fire-fighting vehicle. Technically, a "fire truck" is a vehicle equipped with ladders and is used mainly to gain access to elevated portions of a structure or to provide a means of applying an elevated stream of water. "Fire engine" is a vehicle with a pump and is used primarily to pump water. A "fire wagon" carries large amounts of hose and is primarily used to lay hose as a complement to a pumper. "Fire apparatus" is the proper generic term for all of these vehicles. This entry will focus on the manufacture of a fire engine. From our first toys and books as toddlers to the everyday newscasts showing fire fighters in action, the fire apparatus remains one of the most familiar and impressive examples of technology in our lives. Uncontrolled fire was one of the greatest fears until quite recently. Early attempts to quell fires were merely bucket brigades, that is, lines of citizens handing water buckets to the fire, which was often ineffective against a fully involved building. Some attempts at increased water application were hand-operated piston pumps whose hoses pumped water from a holding tank or a pond. (These early hoses were made of leather with copper rivets; cotton hoses came into use in the 1800s.) Eventually wheels were added to the apparatus, but it was still pulled and operated by the firemen. Volunteer fire departments were established to man the equipment and fight the fire.

With the appearance of property insurance, insurance companies created fire departments and spent considerable time improving the fire apparatus. By the 1860s, the steam engine was used to operate the piston pump, and it was pulled to the fire by horses. Other attempts to pressurize the hoseline were chemical tanks, which used acid combined with soda dissolved in water to start a chemical reaction that would produce carbon dioxide. In this process, the carbon dioxide expanded, pressurizing the tank and propelling the entire mixture out of the hoseline and onto the fire. All of these designs were practically obsolete after the introduction of the centrifugal pump in early 1900. After the advance of the automobile, the internal combustion engine became the primary power source for the fire engine. Adaptation of the truck frame to accommodate the pump and tank completed the transition to the present-day fire truck apparatus.

Design

The basic design of the fire apparatus begins with a thorough review of the fire load and geographical terrain of the area the fire department will be responding to. The vehicle's ease of operation, adequate response speed, and equipment storage and deployment are all important factors to consider. The National Fire Protection Association (NFPA) has compiled guidelines for apparatus design based upon these and other variables. In addition, the fire fighters responsible for the apparatus also contribute to the design of the vehicle.

Most fire apparatuses are purchased by tax-supported governments, cities, towns, townships, and counties. In small departments, the design and approval process can take up to a year, mainly because of funding issues. When medium and large departments choose a new apparatus, funding is usually already established, and designs can

From our first toys and books as toddlers to the everyday newscasts showing fire fighters in action, fire apparatus remains one of the most familiar and impressive examples of technology in our lives.

This 1835 volunteer fire company's hose cart from Cleveland was decorated with sterling silver fittings to reflect their status. (From the collections of Henry Ford Museum & Greenfield Village.)

Fire was a plague in many early American cities. Americans responded with a variety of weapons: volunteer fire departments; inventions that prevented, retarded, or fought fire; and insurance cooperatives and plans that protected against losses caused by fire. Through the early and mid decades of the 19th century, a major fire was a calamity but also a social affair.

A city's volunteer fire companies often illustrated the city's social, ethnic, and demographic composition. Fire companies represented the loyalty and comraderies of special groups. Neighborhood pride, political rivalries, ethnic hostility, and class animosity were all evident in the volunteer fire companies. Companies received public acclaim for their efforts and competed with one another for those public accolades. At the first signal of a fire, companies would race to the scene pulling their hose carts and hand-pumped engines—not necessarily to catch the fire in its early stages. The commander of the first company to arrive would be in charge of the entire operation with absolute control over rival companies. Members fought with one another on the way and at the site, sometimes forgetting even to fight the fire! Once at the scene, more fighting could occur and volunteers were threatened as much by other firefighters as they were by the fire itself.

Members took great pride in their uniforms and equipment because these represented the group's social status as much as they did its fire-fighting competence. Equipment and uniforms were displayed during public ceremonies and parades where competition for social status was particularly evident in pre-Civil War America. Hose carts and engines were display pieces as well as functional equipment.

William S. Pretzer

be approved and purchased in less time. A fire engine can cost from $50,000 to over $750,000.

The basic fire apparatus has evolved into specialized units for particular types of fires and response. In the 1950s, pumpers, ladder trucks, and other miscellaneous designs such as small tankers and hose trucks were common, though airport crash trucks and large rural tanker trucks were also in use. Today, fire apparatuses are tailored to meet many kinds of specific hazards. Due to this wide variety, the rest of this article will discuss the manufacture of the basic combination pumper apparatus for small- and medium-size communities.

The Manufacturing Process

The manufacture of a fire engine remains a custom, almost one-of-a-kind operation. Typically, a group of workers is assigned to the fabrication of the body and assembly of the truck frame. A separate group performs body finishing, then the apparatus is wired, equipped, and tested by a third, or "equipment," group. Aside from the body, manufacture of the fire apparatus is typically an assembly process.

Chassis

1 The selection and purchase of the truck chassis is based upon the tactical application of the apparatus. The truck frame may be "bare," having only the engine, axles, springs, frame members, steering, and brake systems installed. The apparatus builder may choose, however, to use a "cab and chassis," where the frame has the engine, cab, seats, front sheet metal, and wiring already completed. In either case, the truck frame is usually ordered from a well-established truck builder such as General Motors, Ford, International, Freightliner, or Peterbilt. The apparatus builder advises the truck manufacturer of the intended design so special options concerning the performance of the frame can be made. It is critical that the proper selection of suspension be made to support the average 35,000-pound (15,890 kg) fire engine. In our example, we have chosen the cab and chassis frame design.

First a cab and chassis is chosen, with the engine, cab, seats, and wiring already completed.

A water pump is selected from a manufacturing company and bolted to the chassis. The tank is attached immediately behind the pump.

Body

2 After the truck frame is received and inspected, the fabrication of the body (or "coach," as it is sometimes called) begins. Primarily sheet steel is used for body panels and supports, although aluminum and some stainless steel are also incorporated. Sheet steel is approximately 0.06 inch (1.5 mm) thick and comes in sheets 48 inches (1.2 m) and 96 inches (2.4 m) long. It can also be supplied in rolls of the same width and weighing 2,000 pounds (908 kg) or more. Each door panel, support, body panel, and equipment tray is cut and bent using shearing machines and press brakes. Holes for electrical equipment and piping are also punched at this time. The body is typically metal inert gas (Mig) welded together, with the doors and access panels bolted for ease of replacement. Many manufacturers protect the body from corrosion by dipping each panel into a rust inhibiting sealer. The interior and other hidden areas of the doors and cabinets are pre-painted before assembly. While the body construction is underway, the pump and tank are mounted to the frame.

Pump

3 The pump is selected to perform specifically for the practices of the fire department. Essentially, the pump is the only reason for the fire engine. All of the other features are practically useless if the pump does not perform adequately. Most firepumps today are centrifugal pumps, and are manufactured by separate companies and purchased by the apparatus builder. A typical firepump can pump 1,000 gallons (3,785 l) of water per minute. This flow can safely provide enough water to fight a residential or small commercial building fire. Pump size can be restricted by the available water supply, particularly in rural areas where there are no fire hydrants. Pumps are usually cast **iron**, with bronze for the rotating impeller and steel gears in the drive unit. Our truck shall have the 1,000 gallon per minute pump bolted to the frame near the center, and driven by an auxiliary output shaft (or "power take off") from the truck transmission. This type of arrangement is referred to as a "midship" pump apparatus. The pump sits across the frame of the truck, and is bolted through the support castings to the upper frame flange. Alignment with the transmission is important to reduce vibration.

Tank

4 The water tank for this example apparatus would typically be a polypropylene tank holding 500-1,000 gallons (1,892-3,785 l) of water. This is bolted to special

The basic design of the fire apparatus begins with a thorough review of the fire load and geographical terrain of the area the fire department will be responding to.

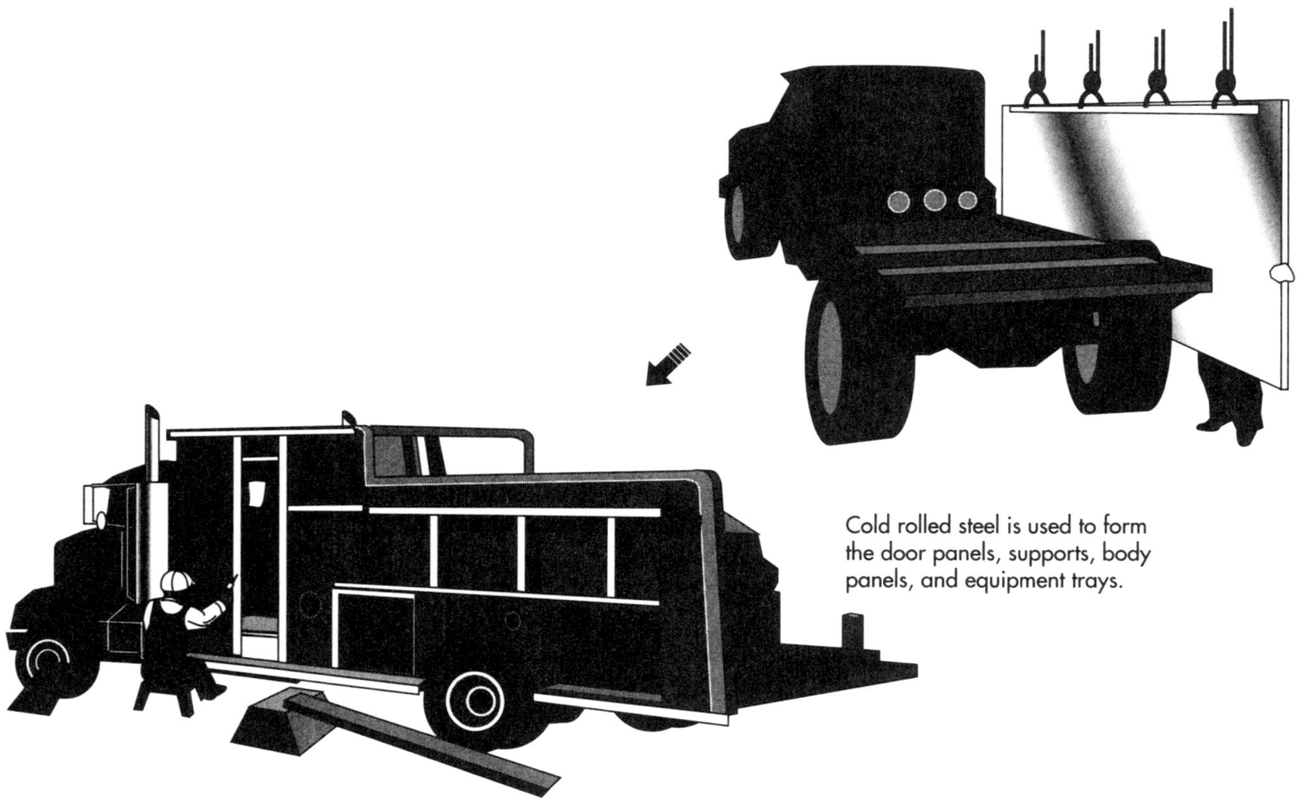

supports before being attached to the frame immediately behind the pump. Connective piping to the pump and filling openings is also installed, typically of galvanized steel construction. Tanks may be square, round, or oval, and many are "hot-welded" together out of sheet material. The hot-welding process uses a high temperature stream of air to melt the plastic pieces at their joint, where the material mixes and fuses together. Most tanks contain perforated plates or "baffles" to reduce the sloshing of the water while driving.

Occasionally, a "quick dump" is installed in the tank of an apparatus designed for use in rural locations. This large valve "dumps" the water from the main tank into a folding portable tank, carried to the fire on a separate tanker truck. This allows the tanker trucks to shuttle water from long distances and empty into the portable tank, where large suction hoses feeding directly into the pump allow the flow to be increased to maximum capacity. These quick dump valves empty the 1,500 gallons (5,677 l) of water in 45 seconds.

Assembly

5 When the body is completed, it is mounted to the frame over the pump and tank, and bolted using neoprene or rubber vibration strips. Along each side and above the tank are the equipment and hose compartments. They are formed into the body with doors and covers for protection. Related equipment, such as hose nozzles, as well as tactical equipment, such as forcible entry tools and self-contained breathing units, are stored in the side equipment compartments. They are sized and equipped to allow the fastest and easiest deployment of the equipment.

The hose beds carry the fire hose. The top bed usually contains 50 foot (15 m) long 2.5 inch (76 cm) diameter sections of fire

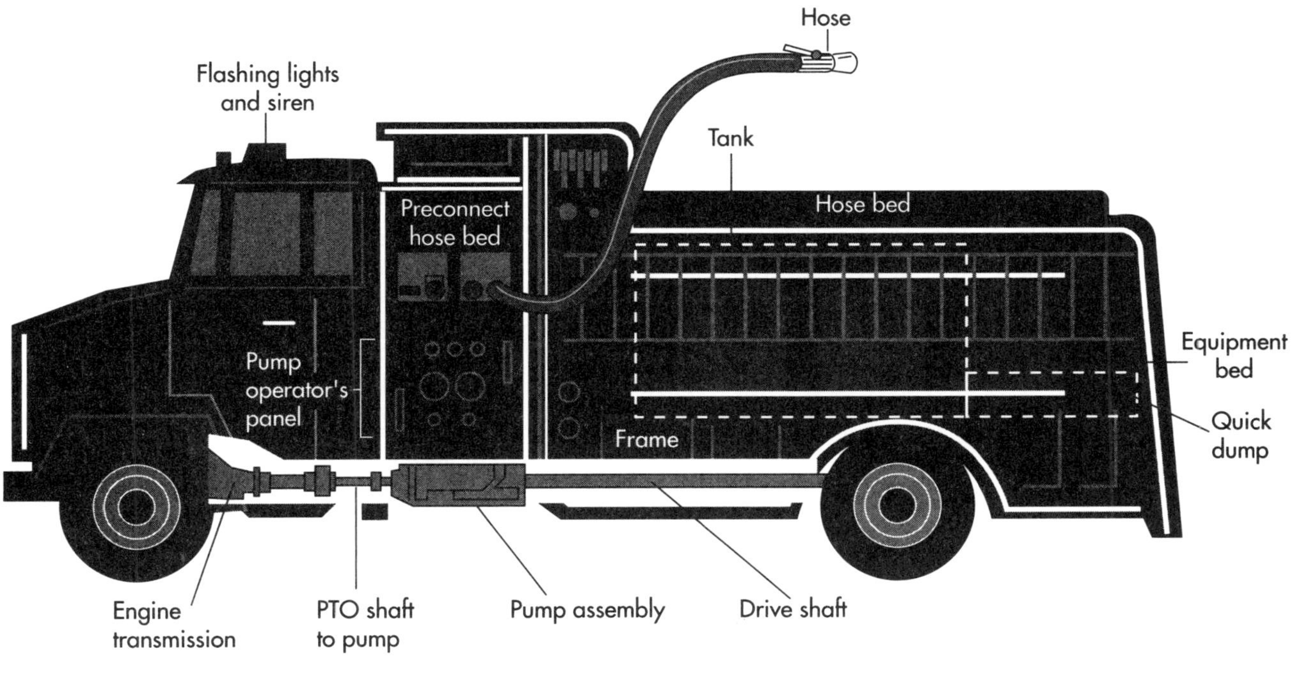

hose, connected together to form hoselines reaching up to 1,000 feet (305 m). This hose is folded in such a manner as to allow it to be pulled from its bed with ease and speed. Shorter sections of suction hoses are also carried on the apparatus. These suction hoses are made of a firm material so they do not collapse from the reduced pressure when coupled to the suction of the pump. They cannot be folded, so they are usually placed into hose troughs specially designed for suction hoses. All other types of fire hoses are collapsible and for use on the pressure side of the pump only.

In front of the tank, and above the pump, are the attack hoseline beds. These beds contain 1.5 inch (3.8 cm) diameter firehoses preconnected to the output of the pump with the nozzles ready for action. Two or four beds of "preconnects" are common. These will be the first hoses used in the incident, therefore they are the most important. Behind the tank is the reel booster. Here, a one-inch (2.5 cm) diameter hose on a power roll-up reel is stored. This booster hose is used for cleaning the apparatus and equipment, and for cooling the outside of the apparatus when it is near a large fire. It is also occasionally used to put out small trash or grass fires. This reel is handy because the hose can be quickly rewound by simply pressing a button. All other hoses on the fire truck must be washed and packed back into their beds by hand after the fire incident.

Ladders for the fire service are extra heavy duty. Though made out of aluminum, they are physically heavy, so they are typically carried in racks above the curb-side (right side) equipment compartments. Power racks have been developed to lower the ladder to waist height, but most commonly the fire fighters simply lift the ladder off of the rack and proceed to the fire. Ladders, like pumps, are also purchased from other suppliers and installed by the apparatus builder.

Painting

6 Once the body has been assembled to the truck frame, the equipment mounting holes are located and drilled, and any additional holes or passages are cut into the panels. This allows the painting operation to seal the exposed edges of holes and other openings. The exterior of the apparatus is washed and sanded in preparation for painting. The interior of the doors and compartments have already been painted. The out-

After the last accessories and wirings are installed and individually tested, the apparatus builder submits the finished truck to an independent inspection agency. Upon successful testing, the apparatus and the builder are awarded a certificate of performance.

side can be painted in matching colors. The painting process includes a primer surfacer to fill small sanding marks and surface defects and a sealer to improve paint adhesion. Fire trucks used to be all red, but some experiments with yellow, blue, and white have been done to increase visual identification. Today, the NFPA recommends yellow or the standard Fire Engine Red. The type of paint is usually a tough enamel or synthetic to resist burning embers and wear from fire service. Hardening agents are added to the paint to improve the shine and durability. After painting, the ladders and accessories can be installed using stainless steel fasteners.

Installing wires

7 The modern fire apparatus can be complicated. Radio systems, cellular phones, computers, and cellular fax machines are all finding their way into the fire service. Even our example apparatus will require several hundred feet of wiring to operate the warning lights, siren, radios, bay lights, generator, flood lights, pump electricals, and other systems. Most apparatuses have two independent battery systems, which must be wired into the apparatus after painting. The pump operator's control panel, containing the pressure gauges and valves used to control the pump, will also be installed.

Quality Control

After the last accessories are installed and individually tested, the apparatus builder submits the finished truck to an independent inspection agency. The agency takes delivery of the truck equipped just as it would be placed into service. It operates and inspects all of the apparatus systems. The pump is operated at maximum capacity for two hours to assure that this vital component performs properly. Upon successful completion, the apparatus and the builder are awarded a certificate of performance. In many areas, the apparatus cannot be legally delivered without this certificate.

The Future

Many new technologies are being applied to the modern fire apparatus. New tank materials are increasing in strength while reducing weight, allowing for more water capacity. Some large departments are using computers and cellular communications to handle information about hazardous materials, structure design, and geographics of the fire district. Perhaps the most exciting of the new technologies in the fire apparatus is the increased use of Class A foam solutions to attack structure and vegetation fires. This involves a separate on-board foam concentrate tank and a mixing proportioner to combine the foam concentrate with the water in the proper amount. Class A foam coats burning surfaces and absorbs heat significantly better than water. When used with compressed air from a separate air compressor on the vehicle, this technology is known as a Compressed Air Foam System, or CAFS. The compressed air pushes the water/foam stream to a much greater distance, and the resulting foam clings like shaving cream. In addition to these advances, firefighters are becoming better trained and are customizing their equipment to fit the types of fires they encounter.

Where To Learn More

Books

Buff, Sheila. *Fire Engines: Motorized Apparatus Since 1900*. Longmeadow Press, 1994.

Fire Protection Handbook, 17th Ed., 1991. National Fire Protection Association, Quincy, Mass.

Periodicals

Cottet, Jack L. "Enclosed Cabs for Fire Apparatus." *Fire Engineering*, October 1990, p. 50-52.

"Agile Fire Engine Borrows from Racing Car Design." *Engineering*, September 1990.

—*Douglas E. Betts*

Fireworks

Background

A firework is a device that uses combustion or explosion to produce a visual or auditory effect. Modern pyrotechnics also includes devices similar to fireworks, such as flares, matches, and even solid-fuel rocket boosters used in spaceflight.

The earliest ancestors of fireworks were paper or bamboo tubes filled with finely ground charcoal and sulfur used in China two thousand years ago. These tubes produced a flash of fire and smoke when ignited, but no explosion. True fireworks did not exist until saltpeter was added to the mixture to create black powder, the first chemical explosive, one thousand years later. Black powder was probably first made in China, but some scholars suggest that it may have been invented by the Arabs.

The earliest ancestors of fireworks were paper or bamboo tubes filled with finely ground charcoal and sulfur used in China two thousand years ago. True fireworks did not exist until saltpeter was added to the mixture to create black powder, the first chemical explosive, one thousand years later.

The Chinese used black powder for fireworks, signals, and weapons such as bombs and rockets. Black powder was introduced to Europe in the 14th century as an explosive for both fireworks and guns. It was applied to mining and roadbuilding projects by the late 17th century. Black powder was used for gunpowder until it was replaced by nitrocellulose in the late 19th century, and (for industrial purposes) by **dynamite** in the early 20th century, but it is still used in fireworks today.

Fireworks in China evolved from simple firecrackers to the extravagant displays witnessed by European explorers in the 16th century. In Europe, fireworks began with military explosives adapted for use in celebrating victories and progressed to the elaborate productions designed by Italian pyrotechnists in the 16th, 17th, and 18th centuries. (Even today, most of the large firework companies in the United States are run by Italian-American families.) These Italian fireworks were usually shown on lavishly decorated wooden sets, often floating on bodies of water, both for safety and to reflect the beautiful displays. On the other hand, German fireworks of the time were usually shot into the air, much like today's fireworks.

Although the firework displays of the Italian masters were extremely complex and impressive works of art, the technology of the time limited their color and brightness. During the 19th century, the introduction of aluminum and magnesium greatly increased the brightness, while the development of potassium chlorate by the French chemist Claude-Louis Berthollet (who was trying to improve the gunpowder used by Napoleon's troops) made it possible to produce more intense colors.

Fireworks came to the New World with the earliest settlers, and have been used to celebrate Independence Day, July 4, since the earliest days of the United States. During the early 20th century these fireworks became bigger, more powerful, and dangerous. Between 1900 and 1930 more than 4,000 people were killed by fireworks. Federal and state governments began regulating the use of fireworks in the 1930s. Explosives are classified as Class A (dangerous substances such as dynamite and TNT), Class B (fireworks used for professional displays) and Class C (smaller fireworks intended for private use.) Class C fireworks must not contain more than 50 milligrams of explosive. Some states allow all Class C

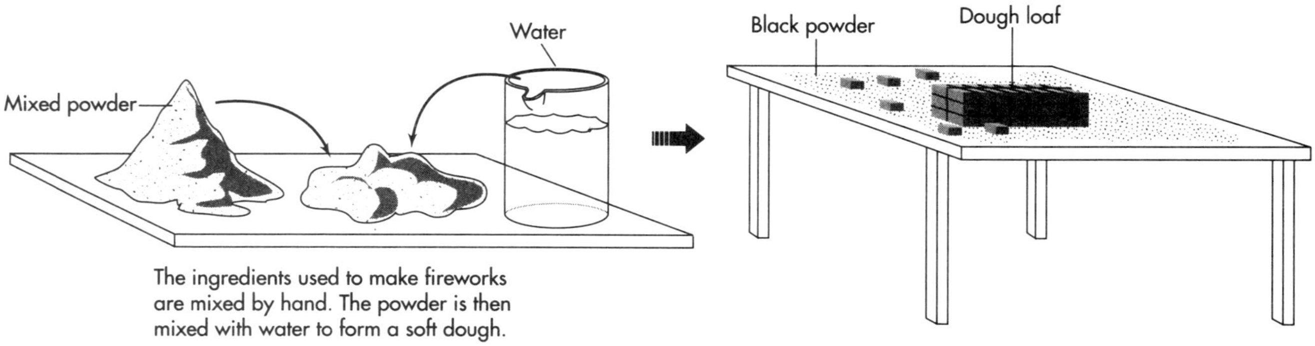

The ingredients used to make fireworks are mixed by hand. The powder is then mixed with water to form a soft dough.

The dough is shaped into loaves and cut into dice. They are sprinkled with black powder and allowed to dry.

fireworks, some allow only "Safe and Sane" fireworks (Class C fireworks that do not move or leave the ground), and some states or counties and cities ban the private use of all fireworks. Some very dangerous fireworks, such as cherry bombs, M-80s, and silver salutes, are banned in all states, but continue to be made and sold illegally. Most firework deaths and injuries in the United States today are caused by these illegal devices.

While the private use of fireworks is heavily restricted, public displays have become more and more elaborate. Computers are used to time fireworks precisely, so they can be choreographed in time to music. Lasers are sometimes used to produce unique visual effects. Today fireworks are made and displayed around the world, particularly in Europe, Latin America, the United States, and Japan.

Raw Materials

A modern firework consists of a shell of plastic, papier-mache, or heavy paper surrounding compartments separated by cardboard. A small compartment at the base of the shell contains black powder to propel the firework into the sky from a mortar made of **iron**, aluminum, plastic, or heavy cardboard. A larger compartment contains chunks of a mixture of chemicals that produce light and color when heated. These chunks are known as stars. In American and European fireworks the stars are mixed with black powder inside a cylindrical compartment. The black powder explodes to ignite the stars and scatter them across the sky. In Asian fireworks the stars surround the black powder in a spherical compartment to produce a more symmetrical display. Instead of black powder and stars a compartment may contain flash powder, which produces a sudden bright light and loud bang. The various compartments in a firework are attached to fuses made of threads mixed with grains of gunpowder.

Black powder consists of a mixture of saltpeter (potassium nitrate), charcoal, and sulfur in a 75 to 15 to 10 ratio by weight. Flash powder consists of a mixture of potassium chlorate or potassium perchlorate, sulfur, and aluminum. Stars consist of a fuel that burns to provide heat, a coloring agent that provides color when heated, and an oxidizer to burn the fuel. Fuels may be slow-burning such as charcoal, dextrin (derived from corn starch), or red gum (a tree secretion) to produce a dim, long-lasting display, or fast-burning, such as aluminum, magnesium, or titanium, to produce a bright, short-lasting display. Sugar may be used as a fuel to produce smoke. Coloring agents include aluminum, magnesium, or titanium (white), carbon or iron (orange), sodium compounds (yellow), copper compounds (blue), strontium carbonate (red), and barium nitrate or barium chlorate (green). Oxidizers are highly reactive oxygen-containing compounds such as potassium perchlorate or ammonium perchlorate. They also contain chlorine, which reacts with the copper, strontium, and barium compounds in the

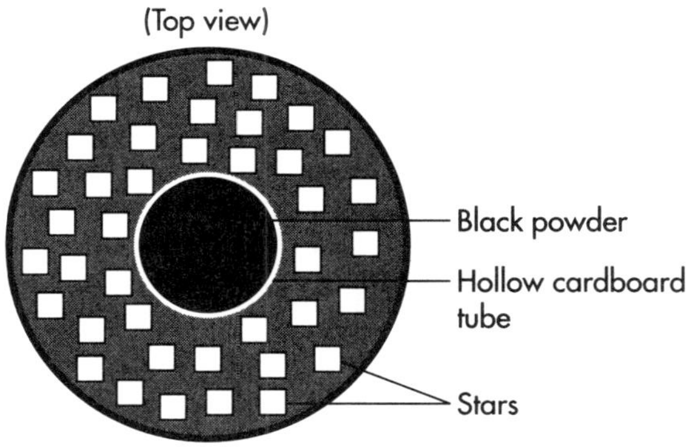

The dried dice are called stars. They are placed in a cylindrical container around a cardboard tube, which is filled with black powder.

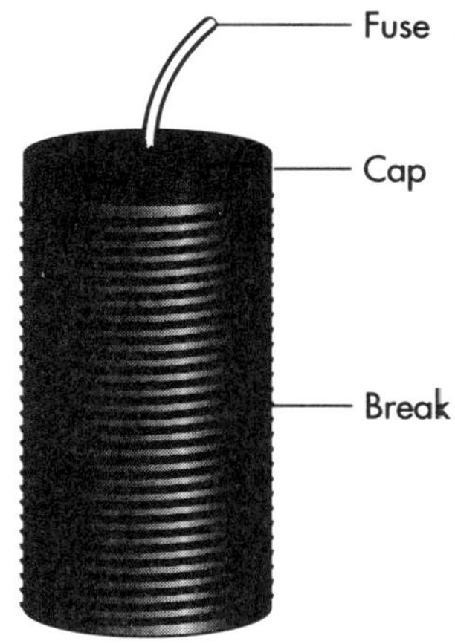

Now called a break, the filled container is capped and wrapped in string. A time fuse is inserted on top.

coloring agents to produce the unstable chlorides of these elements which actually provide the color.

The Manufacturing Process

Making the stars

1 The ingredients used to prepare stars are obtained from chemical supply companies and stored in barrels. At the time of mixing the chemicals are scooped out of the barrels, weighed, and sifted twice through brass screens to remove lumps. (Brass is used because it does not produce sparks.) The sifted powders are placed on a large sheet of paper and gently mixed by hand. The powders may also be mixed inside a rotating drum or a stationary container with rotating paddles. These devices must be used with great care to avoid generating heat through friction or trapping bits of powder between moving parts.

2 The mixed powder is placed in barrels and taken from the mixing room to the cutting room. Water is mixed with it to form a soft dough. Lumps of the dough are scooped into large paper-lined wooden molds shaped like loaves of bread. The dough is packed firmly into the mold with a wooden mallet. (The wet dough is much safer to manipulate than the dry powder.)

3 The loaves of dough, weighing about 35 pounds (16 kg) each, are unmolded onto a workbench covered with heavy cardboard sprinkled with black powder. The loaves are cut in one direction to form slices then cut in the other direction to form dice. The dimensions of the dice may be anywhere from 0.06-2 inches (0.16-5 cm). The black powder adheres to the wet dice and will help them burn when the firework is ignited. The dice are allowed to dry on paper-covered screens.

Making the breaks

4 The dried dice are now stars. They are moved to the packing room to be placed into cardboard containers. A hollow cardboard tube is placed in the center of the cylindrical container and stars are gently poured around it. A large container may hold as many as 900 stars (about 4.4 pounds [2 kg]). When the container is full, black powder is poured inside the hollow tube and the tube is removed. The powder fills the spaces between the stars, and will serve to ignite and scatter them. A paper cap is placed on the filled container, now called a "break."

5 The break is wrapped with heavy string, a process known as spiking. Spiking is done by tying one end of a large spool of string to the break and winding the string around it. When the break is completely cov-

ered, the string is cut and tied. Some breaks are not spiked, but are made instead of plastic or heavier cardboard to withstand the stress of being launched. A time fuse (a short, slow-burning fuse that causes the break to explode a certain amount of time after it is launched) is inserted into the break, and it is wrapped in heavy paper. The wrapped breaks are moved to the pasting room to be wrapped in heavy, paste-soaked **paper**, then allowed to dry for about two days. The paper hardens as the paste dries to form a strong, tight seal.

6 Some breaks, known as salutes, are filled with flash powder rather than stars and black powder. Flash powder is mixed in much the same way as the chemicals used to make stars. It is then poured into cardboard containers that are thicker and stronger than other breaks. This allows more pressure to build up before the salute bursts, resulting in a louder bang. These salutes are then spiked and pasted like other breaks.

Making the shells

7 The dry breaks are moved to the finishing room to be assembled into shells. The simplest shells consist of a small compartment of black powder combined with a single break. Due to their spherical structure, Asian shells always contain only one break. Because American and European shells are cylindrical, more than one break can be stacked together, so that the shell will display multiple bursts of different colors when it explodes. Multi-break shells usually consist of a small compartment of black powder, three or four colored breaks, and a salute. Some large shells contain as many as 10 breaks, and at least one gigantic shell has been made holding 22 breaks. The shell is assembled by stacking the components together, attaching a starting fuse (a long, fast-burning fuse used to ignite the black powder that launches the firework), wrapping them in heavy paper, and tying the package together with string. The completed firework is then labeled and stored until needed.

Making small fireworks

8 Small fireworks, intended for private use, are made in much the same way as large ones, but they are generally simpler in construction and contain much less explosive. Small fireworks include firecrackers (paper tubes holding a small amount of explosive), fountains (paper cones filled with chemicals which release colored sparks), and Roman candles (long paper tubes filled with a small amount of explosive and several small stars which shoot out one at a time). Some small fireworks contain no explosive at all and may be as simple as a single chemical wrapped in paper or foil. Examples include smoke balls (filled with a chemical that releases colored smoke) and snakes (filled with ammonium dichromate, which slowly burns and produces a long trail of ash). Sparklers are made by dipping a metal wire in a slurry containing a fuel, an oxidizer, a coloring agent, and aluminum granules, which provide the sparks.

Launching the fireworks

9 Professional fireworks are usually launched by the same companies who make them. If a set piece (a ground-based display that forms a picture or words with colored flares called lances) is to be used, the design to be formed is sketched on graph paper and sent to carpenters who build a wooden frame with thin wooden slats in the shape of the design. If music will accompany the fireworks, the timing of the display is planned to match the tempo of the music.

10 Several hours before the show begins (or a few days in advance, for a very large show), the crew arrives with all the necessary equipment, including fire extinguishers and first aid kits. Mortars to launch the shells are placed in their proper places. Large ones are placed in holes dug in the ground or in steel drums filled with sand. Smaller mortars are placed in wooden racks. The proper shell for each mortar is loaded in place. The frames for set pieces are assembled, lances are attached to the slats, and fuses are attached to the lances. When the display begins, the lances and mortars are lit at the proper times, either with long hand-held flares or with electrical wires attached to a central switchboard. After the show, the crew safely destroys any unexploded duds.

Quality Control

The most important quality control factor in making fireworks is safety. Firework factories are protected from intruders by chain-link fences, **barbed wire**, locked gates, steel doors, and tamper-proof locks. Within these factories, numerous precautions are taken to prevent accidents.

Electricity is the greatest danger. A single small spark can set off a roomful of explosives. All electrical outlets are located outside the building. To avoid generating static electricity, all workers must wear 100% cotton clothing. They touch a copper plate before they enter a building to remove any static electricity they may be carrying. Elastic straps with wires trailing to the graphite floor are worn around the worker's calves, to drain static electricity away to grounding rods buried beneath the building. All work is halted and all workers leave the building if there is any possibility of an electrical storm approaching.

Many other safety measures are used. All work is done by hand, to avoid machines that could produce heat or sparks. In the winter, buildings are heated with hot water rather than hot air, which could cause an explosion. The buildings are small, so no one is more than one or two steps away from an exit. All exits have doors that open wide at the slightest touch. Explosive chemicals are never mixed when wet, because when they dry out they may release gases that could ignite them.

Where To Learn More

Books

Brenner, Martha. *Fireworks Tonight!* Hastings House, 1986.

Plimpton, George. *Fireworks.* Doubleday, 1984.

Periodicals

Begley, Sharon. "Up in the Sky! It's...Hearts! Stars! Bow Ties!" *Newsweek*, July 9, 1990, p. 60.

Conkling, John A. "Pyrotechnics." *Scientific American*, July 1990, pp. 96-102.

Kozlou, Alex. "First Family of Fireworks." *Discover*, July 1990, pp. 40-45.

—Rose Secrest

Fortune Cookie

The history of the fortune cookie is not entirely known, and there are several competing versions of its origins. Despite its association with Chinese restaurants, the fortune cookie was invented in the United States and may have either Chinese or Japanese roots.

Background

A fortune cookie is a crescent-shaped, hollow cookie with a paper inside imprinted with a short saying or "fortune." Fortune cookies are often presented with the bill at the end of a meal in Chinese restaurants. Each diner selects a cookie and breaks it open to read the advice or prediction inside.

The history of the fortune cookie is not entirely known, and there are several competing versions of its origins. Despite its association with Chinese restaurants, the fortune cookie was invented in the United States and may have either Chinese or Japanese roots. Some say the modern fortune cookie has its origins in an ancient Chinese game played by the nobility and members of the upper classes. In the game, the participants were given twisted cakes that contained pieces of **paper** with subjects written on them. The players opened the cakes and made up wise sayings on the topics specified.

A second story claims that David Tsung, a baker who lived in California's San Joaquin Valley, invented fortune cookies in 1818 or 1819 by wrapping written fortunes in egg roll casings. A Presbyterian minister composed messages of goodwill for Tsung's "fortunes." The baker experimented with different kinds of batter until he created the cookie we know today. A century after his invention, in 1922 or 1923, the fortunes themselves evolved into more whimsical words of wisdom. This story is doubtful because California was a Spanish territory in 1818, and few Americans and no Chinese are known to have lived in the San Joaquin Valley during that period.

In her book *Madame Chu's Cooking School*, Grace Chu acknowledges the Chinese parlor game as a possible historical predecessor and notes that birth announcements also have been wrapped in sweet dough and sent. She credits the invention of the modern fortune cookie to George Jung, who started the Hong Kong Noodle Company in 1916 after immigrating to Los Angeles in 1911.

Golden Gate Park in San Francisco also stakes its claim to fortune cookies in another version of their history. Makoto Hagiwara was the caretaker of the Japanese Tea Garden in the park around 1900. San Francisco's Mayor James Phelan reputedly disliked Asian persons and fired Hagiwara. In 1907, Hagiwara was restored to his cherished position, and he thanked those who helped him return to his beloved tea garden by giving them fortune cookies he had invented during his absence.

Raw Materials

Recipes for fortune cookies appear in many cookbooks, and the basic ingredients are flour, sugar, water, and eggs. Other ingredients may vary depending on the recipe, but may include melted butter, salt, vanilla extract, almond extract, and instant tea powder. Commercial manufacturers may also add baking soda, baking powder, turmeric extract, peanut oil, stabilizing agents, and anticaking agents such as silico aluminate.

The fortunes are printed on paper that is treated to be oil- and moisture-resistant. Many manufacturers acquire preprinted for-

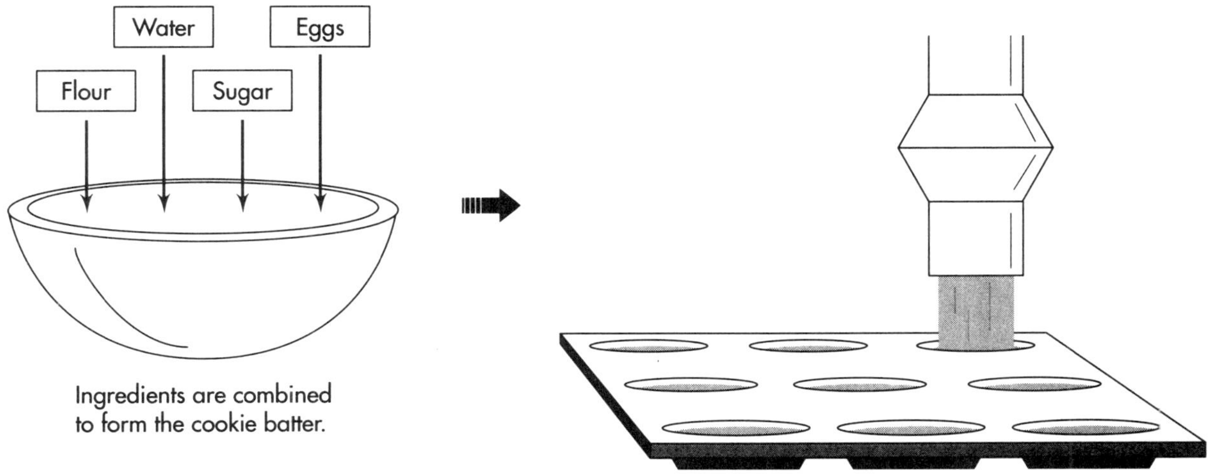

Ingredients are combined to form the cookie batter.

The dough is baked in a specially designed circular oven for about 3–4 minutes.

tunes from the suppliers of their fortune-cookie machines, but some create their own. Also, some suppliers produce custom-made fortunes for weddings and other occasions, or for companies to use in marketing strategies.

The Manufacturing Process

Baking the cookies

1 The ingredients for fortune cookies are mixed together to form the batter. This watery dough is transferred by a pump to the fortune cookie oven, which is circular and contains a number of shallow cups with flat bottoms (about 3 inches [7.6 cm] in diameter) in the shape of the finished but unfolded cookie. When the cups are filled with the correct amount of dough, as regulated by the batter pump, flat metal plates are placed in the cups on top of the dough. The plates flatten the dough and also allow heat to transfer through the metal surfaces against both the top and bottom of the cookie so it is golden brown on both sides. The cookies rotate through the circular oven. One complete orbit takes 3.5 minutes, which is the time it takes the thin dough to bake. In a variation of this method, the dough is poured on a griddle and the round shapes are stamped out by metal forms.

Folding the cookies and fortunes

2 When the cookie has finished baking, it sticks to the plate. As the plates are lifted up, a mechanical arm snatches the cookie from the plate and transfers it to a receiving area where the fortunes are added. The paper strips are sucked by vacuum onto the cookies. The cookies are pushed along to two mechanical fingers that grab the still hot cookie and fold it in half so it resembles a half-moon with the paper fortune inside. The machine then bends the cookie in the middle to its familiar crescent shape (which has been fancifully described as a pair of water wings, a fan, or an extracted molar!) The formed cookie is then cooled by air from a fan and pushed to the packaging area. Cookies are packed in small bags for store sale or large bags for restaurant use, and the bags are placed in cartons to protect their fragile cargo.

Quality Control

As with other food products, exceptional care is taken in meeting government regulations for ingredients and food processing. Ingredients are selected for quality and inspected when they are received. Machines are cleaned regularly, and processes are monitored continuously for product standards and safety. Because the process is

The basic ingredients for fortune cookies are flour, sugar, water, and eggs. Commercial manufacturers may also add baking soda, baking powder, turmeric extract, peanut oil, stabilizing agents, and anticaking agents.

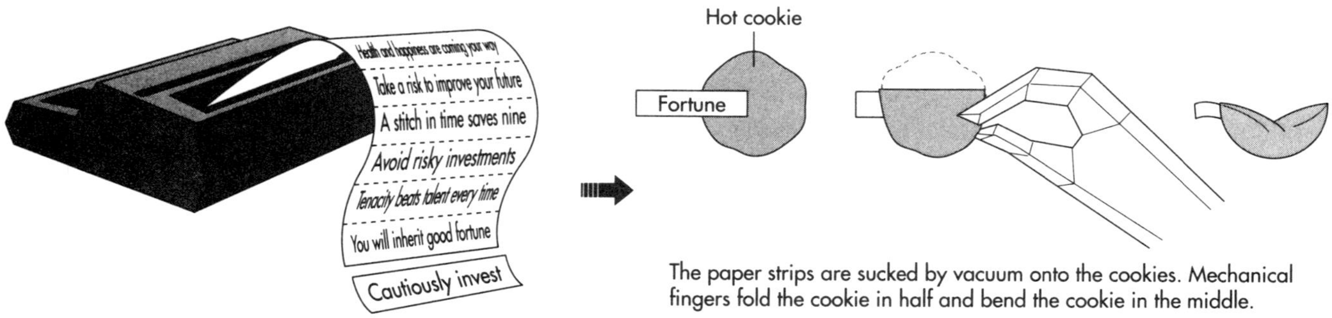

"Fortunes" are printed on paper that is oil and moisture resistant.

The paper strips are sucked by vacuum onto the cookies. Mechanical fingers fold the cookie in half and bend the cookie in the middle.

Many manufacturers acquire preprinted fortunes from the suppliers of their fortune-cookie machines, but some create their own. Also, some suppliers produce custom-made fortunes for weddings and other occasions.

fully and simply mechanized, there is little opportunity for human contact or other contamination. Very little waste results from the cookie-making process.

Another aspect of quality control enters into the preparation of the fortunes. Manufacturers of fortune cookies may sell them in bulk to other food product suppliers or distributors. The company marketing the cookies under its name may review the style and text of the fortunes; for example, the legal department of one large distributor has approved the use of approximately 1,000 fortunes that the manufacturer inserts in cookies bearing the distributor's name.

The Future

The future for fortune cookies is golden for several reasons. Chinese and other Asian cuisines have grown steadily in popularity because of their healthfulness and appealing taste. Fortune cookies are similarly popular because they are light in texture, lightly sweetened, low in calories, and fat-free. Fortune cookies have also joined the "designer foods" bandwagon. Varieties flavored with almond, chocolate, and lemon have been available for some time; but new flavors and colors like blueberry and coffee are now being marketed. The fortunes themselves have also been revamped, with lottery numbers printed on the backs of some producers' fortune papers, and with specialized and modernized sayings or quotations. Packaging has also been tuned to the upscale market with individually wrapped cookies housed in decorator tins. These gourmet versions can even be purchased from home via the shopping channels on television. Finally, "Chinese" fortune cookies have found a new home; the first fortune-cookie plant in China opened in 1993.

Where To Learn More

Periodicals

Heatter, Maida. "Bake Your Holiday Gifts." *New Choices*, December 1994-January 1995, pp. 50-53.

Herman, Jeff. "20 Lucky Hoosier Products Make QVC's Cut." *The Indianapolis Star*, May 20, 1995, p. B3.

Kaufman, Michael T. "Pursuing a New Market: Fortune Cookies for China." *New York Times*, November 7, 1992, p. 1.

"Notes from All Over: China." *Reader's Digest*. June 1993, p. 34.

—*Gillian S. Holmes*

Freeze-Dried Food

Background

Freeze drying is a relatively recent method of preserving food. It involves freezing the food, then removing almost all the moisture in a vacuum chamber, and finally sealing the food in an airtight container. Freeze-dried foods can be easily transported at normal temperatures, stored for a long period of time, and consumed with a minimum of preparation. Once prepared, freeze-dried foods have much the same look and taste as the original, natural products.

The freeze-drying process was developed during World War II as a method of preserving blood plasma for battlefield emergencies without requiring refrigeration or damaging the organic nature of the plasma. The technology was applied to consumer food products after the end of the war. Coffee was one of the first freeze-dried products to be marketed on a large scale. Today, many fruits, vegetables, meats, eggs, and food flavorings are freeze-dried.

Freeze-dried food has many advantages. Because as much as 98% of the water content has been removed, the food is extremely lightweight, which significantly reduces the cost of shipping. This also makes it popular with boaters and hikers who have to carry their food with them. Because it requires no refrigeration, shipping and storage costs are even further reduced. Freeze-dried food is also relatively contamination-free since the dehydration process makes it virtually impossible for yeast and potentially harmful bacteria to survive. Finally, since the physical structure of the food is not altered during the freeze-drying process, the food retains much of its color, shape, texture, and flavor when it is prepared for consumption by reintroducing water. This makes it more attractive to consumers than food preserved by some other methods.

One of the major disadvantages of freeze-dried food is its cost. The equipment required for this process requires a large investment of money, and the process itself is time consuming and labor intensive. These costs are usually passed on to the consumer, which makes freeze-dried food very expensive when compared to other methods of food preservation such as canning or freezing.

Raw Materials

Some foods are extremely well-suited to the freeze-drying process, others do not fare so well. Liquids, thin portions of meat, and small fruits and vegetables can be freeze-dried easily. Coffee is the most common freeze-dried liquid. Chunks or slices of shrimp, crab, lobster, beef, and chicken can be freeze-dried. They are often mixed with vegetables as part of soups or main course entrees. Almost all fruits and vegetables can be freeze-dried, including beans, corn, peas, tomatoes, berries, lemons, oranges, and pineapples. Even items like olives and water chestnuts can be processed this way.

Thick portions of meat and larger, whole vegetables and fruits cannot be freeze dried with any success. With many other foods, it is simply not economical to preserve them by freeze drying.

The Manufacturing Process

A freeze-drying processing facility is usually a large plant with modern equipment. Its

The freeze-drying process was developed during World War II as a method of preserving blood plasma for battlefield emergencies without requiring refrigeration or damaging the organic nature of the plasma.

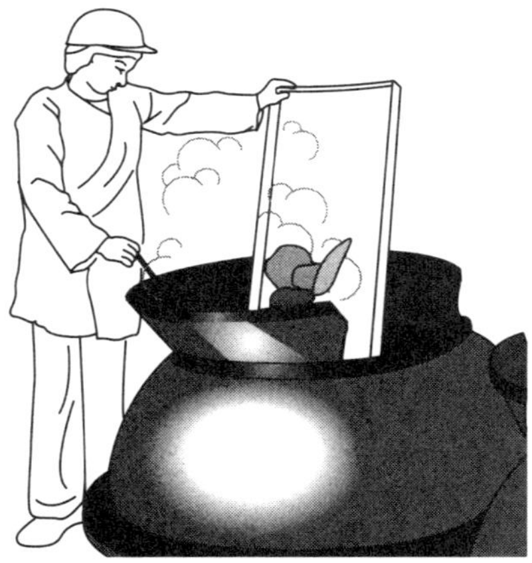

Some foods are cooked before freeze drying; fruits and vegetables are just washed with water.

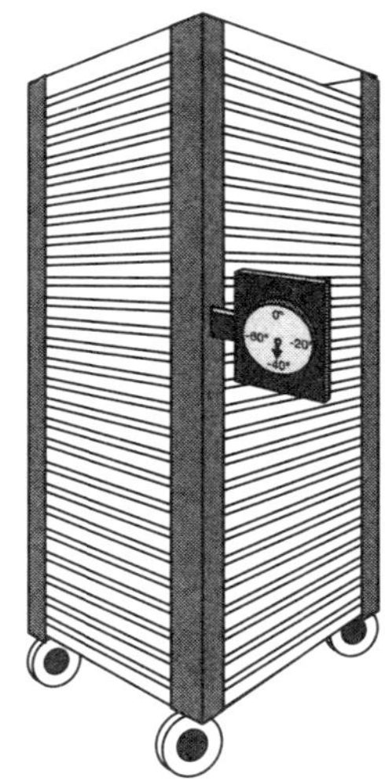

The food pieces are spread on flat metal trays on wheeled carts and placed in a coldroom, where the temperature can be as low as −40°F.

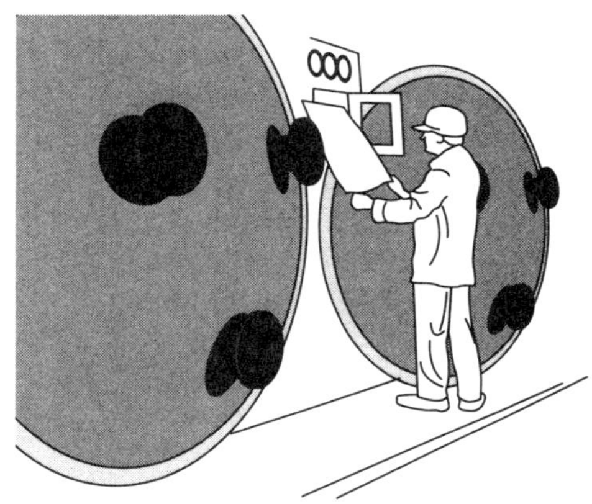

The carts then go to a vacuum drying chamber, where 98% of the water from the food is removed.

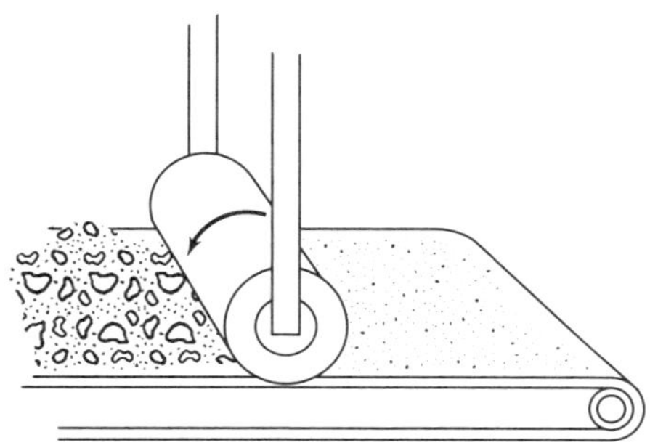

Some food pieces may be ground further or reduced to a powder. Two or more products may be blended.

food-handling areas must be approved by the United States Department of Agriculture, and the company and its employees must adhere to government regulatory procedures. The plant may include a receiving and storage area for raw foods that arrive at the plant in bulk; a food cooking area for those foods that must be cooked before processing; a large area with several large freezing and drying chambers; and a packaging area. The facility may also include a research area where improved methods of freeze-drying foods are developed, and a test kitchen where new preparation techniques to improve the final taste, quality, and texture of the food are tried. Some plants are dedicated to freeze-drying only one product like freeze-dried coffee. Others process a wide range of meats, vegetables, and fruits. Nonfood products such as chemicals and pharmaceuticals are usually processed in separate plants from food products.

The freeze-drying process varies in the details of temperatures, times, pressures, and intermediate steps from one food to another. The following is a generalized description of the process with several specific exceptions noted.

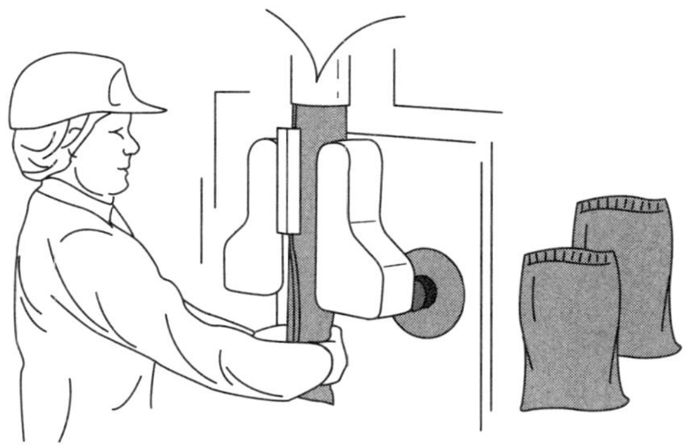

The food is packaged in airtight containers or vacuum packed.

When water is added, the freeze-dried food retains much of its original color, texture, and flavor.

Testing and preparation

1 The food is first checked for contamination and purity. Fruits, meats, and some other edibles are tested for bacterial counts and spoilage. Much of the work of the plant is dependent on the harvest season for each food. In January, for example, the plant would be processing celery, olives, lemons, oranges, and pineapples. In July, it would process green beans, peas, and strawberries, among others.

2 Some kinds of food, like seafood and meats, must be cooked before freeze drying. They are usually purchased already cut into small pieces. If they have not been pre-cooked and frozen, these foods are placed in large, industrial-sized kettles and properly cooked. Fruits and vegetables are usually purchased already cut, pitted, and peeled. These foods are simply washed with sprays of water. Some vegetables, like peas and corn, are quickly scalded, or blanched, before freezing. Coffee is purchased as a pre-brewed concentrated liquid. Because the aroma of coffee is important to consumers, a small amount of coffee bean oil may be added to the liquid. Unlike the water, the oil is not removed during the drying process.

Freezing

3 The food pieces are spread out on flat, metal trays which are stacked 20 to 30 high in slots in a wheeled cart. With food that has been pre-cooked and frozen, the trays are pre-chilled to prevent partial thawing during handling. With liquids like coffee, the pre-brewed coffee is poured into shallow pans. The carts are wheeled into a large, walk-in coldroom where the temperature can be as low as -40°F (-40°C). In this extremely cold temperature, the food is quickly frozen. There are usually a dozen or more coldrooms in operation, and the carts are kept there until it is time to move them into the drying chamber.

Drying

4 The carts are wheeled out of the coldroom and into a vacuum drying chamber. In the case of liquids like coffee, the frozen coffee is first ground up into small particles in a low-temperature grinder. The drying chamber is a large, long, horizontal cylinder with semi-elliptical ends. One end is hinged to open and close. When the trays of frozen food pieces are inside, the chamber is closed and sealed. In a large plant, there may be 20 to 30 drying chambers in operation at any time.

5 The drying procedure involves a process known as sublimation. In sublimation, a solid material is forced to change state into a gaseous material without ever becoming a liquid. In the case of freeze-dried food, the solid ice crystals trapped in the frozen food pieces are forced to change into water vapor without ever becoming liquid water. In the drying chamber, this is accomplished by evacuating the air with a vacuum pump to reduce the pressure to about 0.036 psi (0.0025 bar). The temperature of the food is raised to about 100°F (38°C) by direct conduction through the bottom of the trays, radiation from heat lamps, or microwave heating. When the chamber is evacuated of air, the pressure is below the threshold at which water can simultaneously exist in a solid, liquid, and gaseous (vapor) state. This threshold is known as the triple point of water. Once the pressure falls below this point, the heat causes the ice crystals trapped in the frozen pieces of food to change directly to water vapor. The vapor is drawn off and condensed within the chamber leaving the food behind. The dried food is filled with tiny voids, like a sponge, where the ice crystals were once present. Not only does this make it easier for the food to reabsorb water when it is prepared for consumption, but the dried food retains its original size and shape. The time for this drying process varies. Freeze-dried liquids make take only about four hours to prepare, while semi-solids and solids like soup and sliced meats may take 12 hours or more.

Sizing and blending

6 The dried food pieces are removed from the drying chamber and tested for moisture content and purity.

7 Some food pieces may be ground to a smaller size or may be reduced to a powder. Others may be screened to separate them by size. Two or more different products may also be blended together to meet a customer's specific specifications.

Packaging

8 Freeze-dried foods must be sealed in airtight containers to prevent them from absorbing moisture from the air. Several types of containers may be used: plastic laminated foil pouches, metal and plastic cans, or metal and fiber drums for bulk packaging. Some freeze-dried food is vacuum packed, in which the air is evacuated from the container before sealing. Other food has an inert gas like nitrogen injected into the container before sealing to displace the oxygen in the air and prevent oxidation or spoiling of the food. The packaging is done in the freeze-dry plant almost as soon as the foods come out of the drying chamber. The plant can form, fill, and seal the packages to the desired weight for the end user. Packages that are to be sold directly to the consumer are packed in cartons, stacked on pallets, and transported to the grocery warehouse. Other freeze-dried food is packaged in bulk and sold to a secondary processor for incorporation into other food products. Freeze-dried blueberries, for example, may be sent to a company that makes pancake and muffin mixes.

Quality Control

Each food has different processing, storage, and rehydration requirements. Some of the variables include the sizing of the raw food products before freezing, the cooking or blanching time and temperature, the rate of freezing and final freezing temperature, the rate of application of vacuum and the final vacuum pressure during drying, the rate and method of application of heat and the final dried product temperature, the allowable residual moisture content after drying, the storage temperature and atmosphere (vacuum, nitrogen, etc) after drying, and the rehydration procedures. At large freeze-drying facilities, electronic microprocessors regulate the times, temperatures, and pressures throughout each step of the process. A central computer collects this data, analyzes it using statistical quality control methods, and stores it for later reference. This assures that the food sent out to the public for consumption has been through a strictly controlled process that meets government guidelines and varies only slightly from batch to batch. The computer also collects data on the bacterial and moisture levels of the raw, bulk food products coming into the plant as well as the final freeze-dried products. Special equipment may include computerized gas chromatographs and oxygen analyzers. Even the packaging materials are tested for their ability to prevent water vapor and oxygen transmission.

The Future

Food is not the only material that is freeze-dried. Pharmaceutical products such as antibiotics and **vaccine**s are often preserved this way. Specialty chemicals, pigments, and ceramics powders are also produced using freeze-drying. Currently, there is development work on freeze-drying various aerosol sprays. One of the most interesting applications is freeze-drying flowers to produce bouquets that can be stored for many months before being reactivated to make "fresh" flowers. This would be especially beneficial for those who want flowers that bloom only during a short season.

Where To Learn More

Books

Considine, Douglas M., Ed. *Food and Food Production Encyclopedia.* Van Nostrand Reinhold Company, 1982.

Hui, Y.H., Ed. *Encyclopedia of Food Science and Technology, Volume 2.* John Wiley and Sons, 1992.

—Carol Brennan

Frozen Yogurt

When frozen yogurt was first introduced in the 1970s, it was a distinct failure. Today, it has become one of the most popular low-calorie desserts with sales of over 135 million gallons.

Background

Frozen yogurt is a refreshing, tangy dessert that combines the flavors and textures of ice cream and sherbet. Frozen yogurt is a relative new-comer in the dessert market. The history of frozen desserts dates back thousands of years to Asia where water ices were first made. Although Roman literature describes how the Emperor Nero was treated to exotic fruit juices and wines chilled with mountain snow, it was not until the 13th century that Marco Polo introduced Asian water ices to Italy. The popularity of these frozen desserts spread throughout Europe and within a few centuries, European colonists introduced ice cream in the U.S.

Technological improvements throughout the 1800s simplified the process of making frozen desserts. The first hand-freezer was patented in 1848. Shortly thereafter the first wholesale ice cream manufacturing company in the U.S. was created by Jacob Fussell of Baltimore. By the turn of the century pasteurization machines and homogenizers were developed, which improved the healthfulness and texture of ice cream. The manufacturing process was simplified further with the invention of the direct expansion freezer and the continuous freezing process. Low-temperature refrigerators developed in the 1940s expanded the frozen dessert industry into new markets, leading to the creation of carry-home packages. Finally, in the late 1960s and 1970s, high-tech, high-volume processing machinery allowed the industry to flourish.

Dessert makers had long experimented with a variety of ice cream flavors and styles. In the 1970s, frozen yogurt's entry into the dessert market was a distinct failure—consumers complained that it tasted too much like yogurt. Despite the initial reaction, manufacturers reformulated and refined their frozen yogurt recipes, and the increasingly health-conscious populace of the 1980s finally took to the low-calorie dessert with a vengeance. Frozen yogurt was soon available in a variety of flavors throughout the U.S. It proved to be just as versatile as ice cream, served in cones and cups, with toppings, on crepes, waffles, and banana splits. Frozen yogurt offered a tangier flavor than ice cream and more depth in flavor and texture than sherbet.

During the 1980s the frozen yogurt market reached sales of $25 million in 1986 with triple-digit growth rates. Major ice cream manufacturers quickly jumped on the bandwagon and started producing their own brands of frozen yogurt, recognizing that the low-calorie dessert was here to stay. By the early 1990s, frozen yogurt captured about 10% of the total frozen dessert market with sales of $330 million on 135 million gallons.

Raw Materials

Frozen yogurt gets its unique flavor from strains of *Lactobacillus bulgaricus* and *Streptococcus thermophilous*. The yogurt culture includes all the strains of bacteria in the product and makes up about 1% of the ingredients.

Frozen yogurt is made in much the same way as ice cream and, with the exception of yogurt culture, they are made from similar ingredients. Two of the most important elements in frozen yogurt are water and air. Air is incorporated into the mixture to add volume. Water exists in the liquid elements of the mix and is considered the continuous phase, which goes from liquid to a partially solid state. Frozen yogurt is never completely frozen; it simply contains ice crystals.

The primary ingredient in frozen yogurt is milk and milk products. Milkfat generally makes up between 0.5-6% of the ingredients depending on whether the frozen yogurt is non-fat, low-fat, or regular. Milkfat lends richness to the yogurt and is the synergist for other flavorings.

"Milk solids, not fat" (MSNF) makes up between 8-14% of frozen yogurt. MSNF must be balanced in inverse proportion to fat for the best body of the product. MSNF consists of about 55% milk sugar, or lactose, 37% protein, and the remaining 8% are various minerals. The protein element increases the smoothness, viscosity, and compactness of the frozen dessert and makes it more resistant to melting.

Sugar makes up between 15-17% of the ingredients. Sucrose, in the form of cane or beet sugar, is generally the primary sweetener, though other sweeteners are often combined. Sugar not only adds sweetness to the yogurt but also improves the body and viscosity and increases the concentration of total solids (TS) in the product. Total solids add body and texture as well as food value, since solids take the place of water in the mixture. Egg solids may be used for solids with the added benefit of decreasing the amount of time necessary for freezing the mixture.

Stabilizers, in the form of animal and vegetable gelatins, are added to the frozen yogurt so that it maintains a smooth consistency in retail outlets, where temperature changes can coarsen the texture. Stabilizers reduce crystallization, hinder melting, and improve the handling properties of the frozen yogurt. Emulsifiers are used to help blend liquids that are generally immiscible by creating smaller air cells throughout the mixture. Generally in the form of fatty acids, emulsifiers also add firmness to the body and reduce the time needed to whip the mix. Although stabilizers and emulsifiers occur naturally in milk products, small amounts are usually added, making up only 0.5-0.6% of the entire mixture.

Other ingredients added in small quantities include egg solids, color, mineral salts, and caseinate derivative, such as citrates and phosphates. Additional flavors include fruit, fruit extracts, nuts, cocoa, vanilla, sugars, and spices, such as allspice, cinnamon, cloves, nutmeg, and ginger.

The Manufacturing Process

Processing the mix

1 The ingredients are selected for freshness and quality. They are measured in precise quantities according to the particular recipe. Liquid and dry ingredients are combined separately.

2 The liquids are poured into a vat, mixed together, and heated. Next, the dry ingredients are added to the liquids in a particular order. Meanwhile the batch is stirred and the temperature gradually increased. Most ingredients must be incorporated before the mix is heated to 120°F (49°C) so that the mix does not become lumpy. The mixture must be heated to dissolve and blend the ingredients.

Pasteurizing the mix

3 Pasteurizing the batch is necessary to destroy pathogenic bacteria and to help preserve the finished product. It is also required by law in most regions. Pasteurization is a simple process that involves quickly bringing the mix to a high temperature for a specified time and then quickly reducing the temperature to less than 40°F (4°C). The trend in the industry has been toward increasing the pasteurization temperature to about 175°F (79°C) for about 25-40 seconds. For greater results, batches can be pasteurized at temperatures as high as 210°F (99°C) to 220°F (104°C). These high temperatures also improve the flavor and help blend the ingredients more effectively.

Homogenizing the mix

4 Homogenizing the batch makes it smoother, primarily by decreasing the size of fat globules to less than two micro-

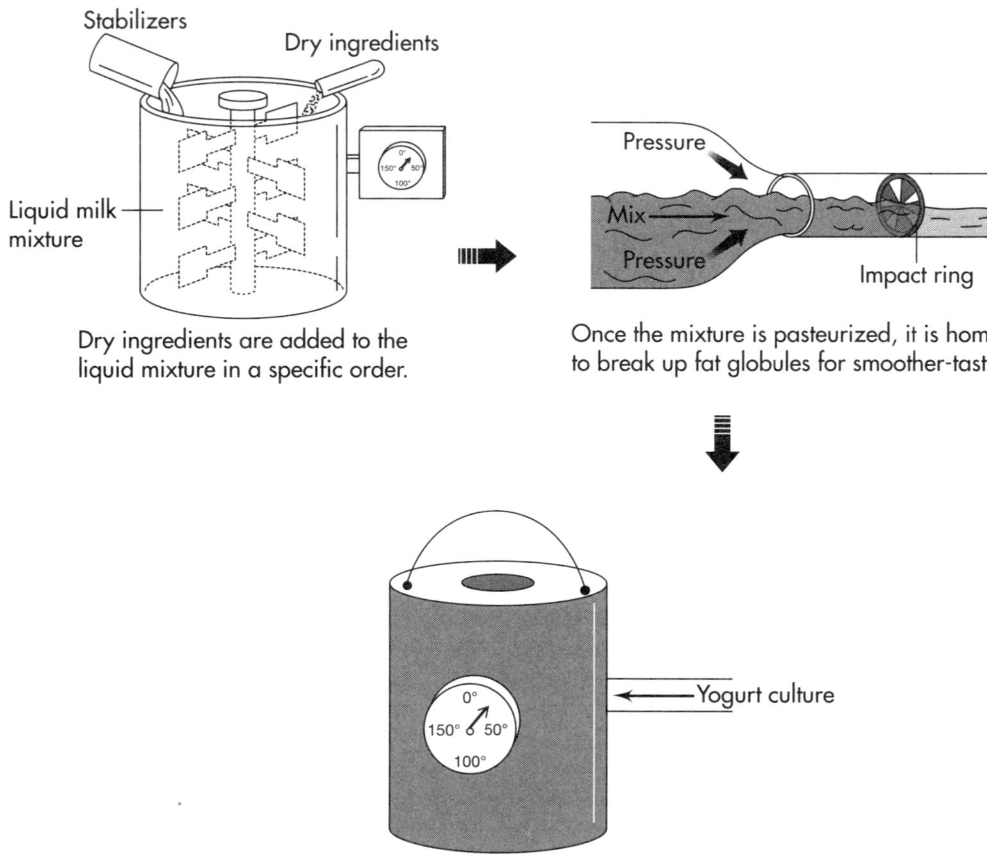

Dry ingredients are added to the liquid mixture in a specific order.

Once the mixture is pasteurized, it is homogenized to break up fat globules for smoother-tasting yogurt.

Yogurt culture is added, and the mixture is cooled and aged for up to four hours.

meters. Without homogenization fat could rise to the top of the mixture and create a layer of cream. Homogenization consists of pumping the batch through a small valve and against an impact ring. Three forces are at work. As the mix passes at a high velocity of about 30,000 fpm (feet per minute) through the valve, shear forces begin to break up the fat particles. The impact ring ruptures the fat further. Completing the process is cavitation, in which vapor bubbles are created by a sudden discharge of pressure. Within the bubbles the fat droplets crash against the vapor walls and disintegrate; thus, the more fat, the more homogenization required.

Inoculating with yogurt culture

5 While the temperature of the mix is 90°F (32°C), it is inoculated with 1% yogurt culture. The mix remains at this temperature until it sets and is ready for cooling.

Cooling and aging

6 After homogenization, the mixture must be cooled. If it is cooled slowly from about 90°F (32°C) to about 40°F (4°C), the mix will become more viscous. Once the temperature falls between 32°F (0°C) and 40°F (4°C), the batch is stored in aging tanks inside coolers. The mix is aged for up to four hours.

Flavoring, coloring, and freezing

7 The final ingredients are mixed together in a flavor vat. These include sweeteners, flavorings, and coloring. This mixture is then pumped into the freezer with the rest of the mixture which is about 20°F (-6°C) to 28°F (-2°C).

8 While the mix is hardening, it is agitated to incorporate air and create over-

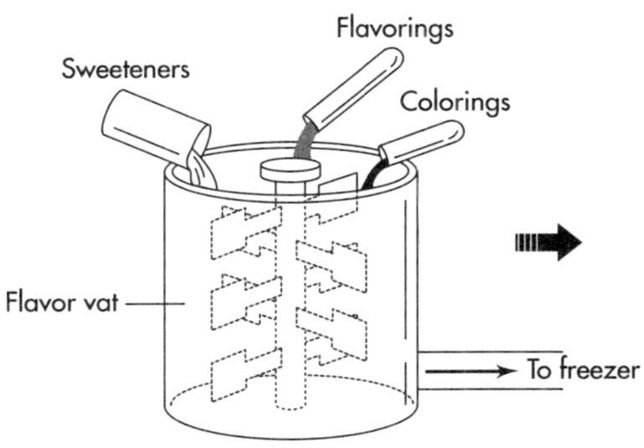

Final ingredients are mixed together in a flavor vat.

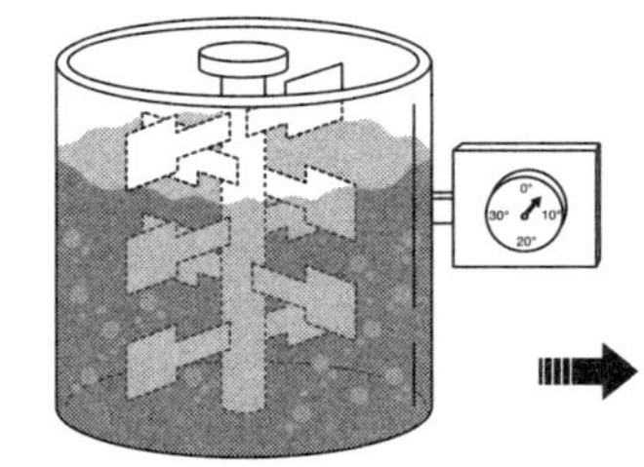

The mix is agitated during the hardening and freezing process to create overrun, or excess volume.

Depending on the size and shape of the package, the final hardening of yogurt takes 30 minutes to 12 hours.

run, or excess volume. The addition of air also smoothes the consistency and creates a more palatable product. In about three minutes the mix begins to freeze and within a few more minutes, the desired overrun, about 50%, is achieved. About one- to two-thirds of the water freezes during this stage.

Packaging and hardening

9 After the desired overrun is reached, the mixture is packaged and placed in freezers where the freezing process continues. The temperature falls quickly, within one or two minutes, to at least 0°F (-17°C) but ideally -15°F (-26°C). For best results, the freezing process should occur rapidly so that the mixture does not form large, coarse ice crystals but small, smooth ones. The frozen yogurt may be stored in continuous or batch freezers. In the former there is a constant flow of product into the freezer, while in the batch method, batches are prepared individually. Depending on the type of freezer and the size and shape of the package, the final hardening takes between 30 minutes to 12 hours.

Shipping

10 The containers of frozen yogurt are piled closely together inside the delivery trucks to minimize the temperature change during shipping. The vehicles are generally mechanically refrigerated at the same temperatures as in the storage facilities in the factory, about -15°F (-26°C), and not above the temperature at the retail outlet. Dry ice may be used as a refrigerant, though it risks heat-shock to the yogurt, which occurs if the temperature falls too low; the freezing point for dry ice is -109°F (-78°C). The frozen yogurt is shipped to retail outlets and food service establishments or to other manufacturers for further processing into novelties.

Quality Control

All aspects of production, packaging, and distribution of the frozen yogurt should be performed with appropriate hygiene to minimize risk of contaminating the food. An automated CIP (clean-in-place) system quickly, easily, and efficiently cleans all the pumps and tanks for maximum protection against pathogenic bacteria.

Manufacturers must adhere to the pasteurization temperatures and minimum times required by the Public Health Service and other regulatory agencies. Tests are regularly conducted for standard plate count of bacteria and coliform and other microorganisms, such as molds and yeast, which could contaminate the product. Also care must be taken to avoid fermentation, which could damage the product by making it more acid, altering color, and curdling.

Finally, frozen yogurt manufacturers must provide accurate information regarding the ingredients and the caloric composition of the product as prescribed by law.

The Future

The future bodes well for frozen yogurt as it expands into new markets with new variations. People's interest in low-fat foods seems unabated and will, no doubt, continue to encourage the food industry to provide low-fat, tasty food. Technology will also continue to improve efficiency in manufacturing frozen yogurt as well the quality of the product.

Where To Learn More

Book

Arbuckle, Wendell Sherwood. *Ice Cream.* 4th ed. AVI Publishing, 1986.

Periodicals

Dexheimer, Ellen. "On The Fat Track...Dairy Processors Race to Develop Better Low Fat Dairy Products." *Dairy Foods Magazine,* May 1992, p. 38.

Mason-Draffen, Carrie. "The Very Fat Facts." *Newsday,* November 22, 1994, p. 13.

Rogers, Paul. "Dairy Experts Forecast the Issues Most Likely to Influence the Industry's Course This Year." *Dairy Foods Magazine,* January 1989, p. 13.

—Audra Avizienis

Gasoline

Background

Gasoline is a volatile, flammable liquid obtained from the refinement of petroleum, or crude oil. It was originally discarded as a byproduct of kerosene production, but its ability to vaporize at low temperatures made it a useful fuel for many machines. The first oil well in the United States was struck by Edwin L. Drake near Titusville, Pennsylvania, in 1859 at a depth of almost 70 feet (21 m). With the development of the four-stroke internal combustion engine by Nikolaus Otto in 1876, gasoline became essential to the automotive industry. Today, almost all gasoline is used to fuel automobiles, with a very small percentage used to power agricultural equipment and aircraft.

Petroleum, a fossil fuel, supplies more energy to the world today than any other source. The United States is the world's leading consumer of petroleum; in 1994, Americans used 7,587,000 barrels of oil per day. Petroleum is formed from the remains of plants and animals that have been held under tremendous pressure for millions of years. Ordinarily, this organic matter would decompose completely with the help of scavengers and aerobic bacteria, but petroleum is created in an anaerobic environment, without the presence of oxygen. Over half of the world's known crude oil is concentrated in the Persian Gulf basin. Other major areas include the coasts of Alaska and the Gulf of Mexico.

Petroleum products, including gasoline, are primarily a mixture of hydrocarbons (molecules containing hydrogen and carbon molecules) with small amounts of other substances. Crude oil is comprised of different lengths of hydrocarbon chains, with some short chains and some very long chains. Depending on how much the oil is broken down, or refined, it may become any number of products. In general, the smaller the molecule, the lower the boiling point. Therefore, gas, with very small chains of one to five carbons, boils at a very low temperature. Gasoline, with 6-10 carbons, boils at a slightly higher temperature. The heaviest oils may contain up to 25 carbon atoms and not reach their boiling point until 761°F (405°C).

Raw Materials

Gasoline is one of the products derived from distilling and refining petroleum. Compounds of organic **lead** were added to gasoline in the past to reduce knocking in engines, but due to environmental concerns this is no longer common. Other chemicals are also added to gasoline to further stabilize it and improve its color and smell in a process called "sweetening."

The Manufacturing Process

Exploration

1 The first step in the manufacture of gasoline is to find its parent ingredient, petroleum. Crude oil is trapped in areas of porous rock, or reservoir rock, after it has migrated there from the area of its origin. Possible areas of oil concentration may be pinpointed by looking for rock types that are commonly found in those areas. Explorers may examine the surface features of the land, analyze how sound waves bounce off

Petroleum, a fossil fuel, supplies more energy to the world today than any other source. The United States is the world's leading consumer of petroleum; in 1994, Americans used 7,587,000 barrels of oil per day.

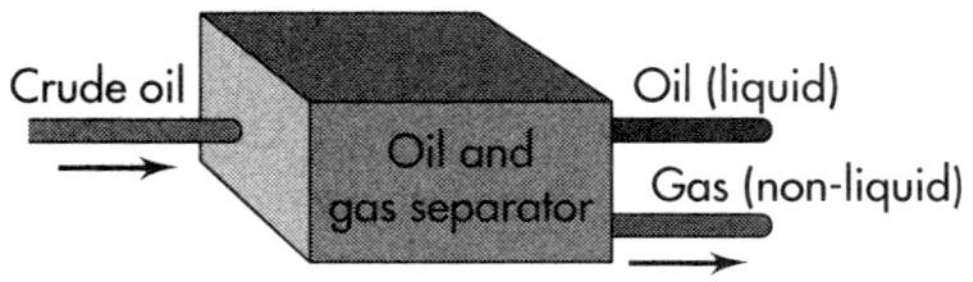

Once crude oil is discovered, a series of pipes transport the oil to the refinery via separators.

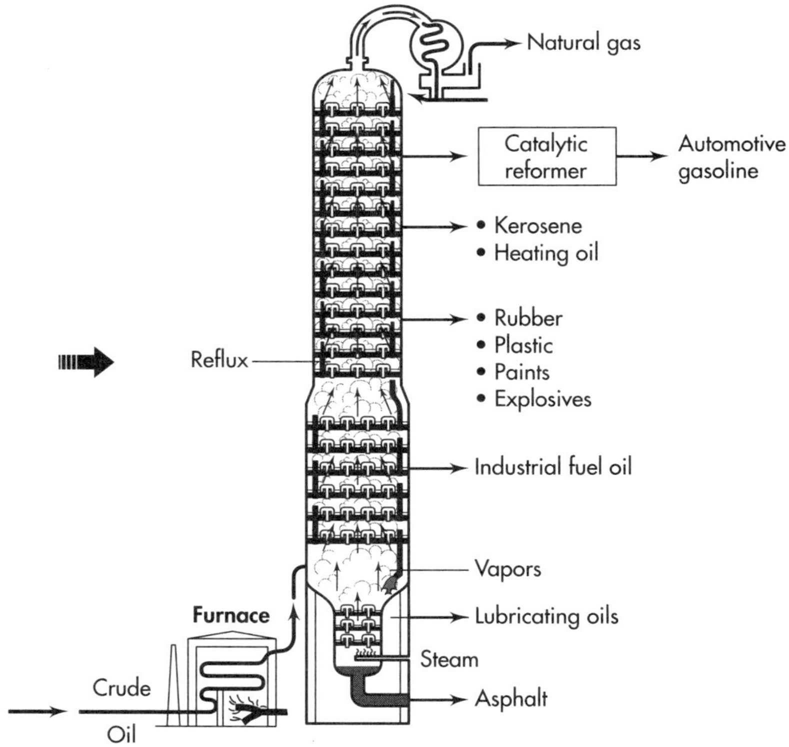

A fractional distillation tower is used to separate different compounds in petroleum. The fuels condense at different temperatures and produce different products.

Gasoline is a volatile, flammable liquid obtained from the refinement of petroleum, or crude oil.

the rock, or use a gravity meter to detect slight differences in rock formations.

2 After a possible oil reservoir is found, the area must be test drilled. Core samples are taken from test wells to confirm rock formations, and the samples are chemically analyzed in order to determine if more drilling is justified. Although the methods used today are more advanced than any of the past, there is still no certainty in oil exploration.

Drilling

3 Crude oil is recovered through wells that can reach over 1,000 feet (305 m) into the rock. The holes are made by rotary drillers, which use a bit to bore a hole in the ground as water is added. The water and soil create a thick mud that helps hold back the oil and prevent it from "gushing" due to the internal pressure contained in the reservoir rock. When the reservoir is reached, the mud continues to hold back the oil while the drill is removed and a pipe is inserted.

Recovery

4 To recover the oil, a complicated system of pipes and valving is installed directly into the drilling well. The natural pressure of the reservoir rock brings the oil out of the well and into the pipes. These are connected to a recovery system, which consists of a series of larger pipes taking the crude oil to the refinery via an oil (liquid) and gas (non-liquid) separator. This method allows the oil to be recovered with a minimum of waste.

5 Eventually, the natural pressure of the well is expended, though great quantities of oil may still remain in the rock. Secondary recovery methods are now required to obtain a greater percentage of the oil. The pressure is restored by either injecting gas into the pocket above the oil or by flooding water into the well, which is far

more common. In this process, four holes are drilled around the perimeter of the well and water is added. The petroleum will float on the water and come to the surface.

Fractional distillation

6 Crude oil is not a good fuel, since it is not fluid and requires a very high temperature to burn. The long chains of molecules in crude oil must be separated from the smaller chains of refined fuels, including gasoline, in a petroleum refinery. This process is called *fractional distillation.*

A fractional distillation tower is a huge unit that may hold up to 200,000 barrels of crude oil. The oil is first pumped into a furnace and heated to over 600°F (316°C), causing all but the largest molecules to evaporate. The vapors rise into the fractionating column, which may be as tall as 150 feet (46 m). The vapors cool as they rise through the column. Since the boiling points of all the compounds differ, the larger, heavier molecules will condense first lower in the tower and the shorter, lighter molecules will condense higher in the tower. Natural gases, gasoline, and kerosene are released near the top. Heavier compounds used in the manufacture of plastics and lubricants are removed lower in the tower.

Fractional distillation itself does not produce gasoline from crude oil, it just removes the gasoline from other compounds in crude oil. Further refining processes are now used to improve the quality of the fuel.

Refining petroleum

7 Catalytic cracking is one of the most important processes in oil refining. This process uses a catalyst, high temperature, and increased pressure to affect chemical changes in petroleum. Catalysts such as aluminum, platinum, processed clay, and acids are added to petroleum to break down larger molecules so that it will possess the desired compounds of gasoline.

Another refining process is polymerization. This is the opposite of cracking in that it combines the smaller molecules of lighter gases into larger ones that can be used as liquid fuels.

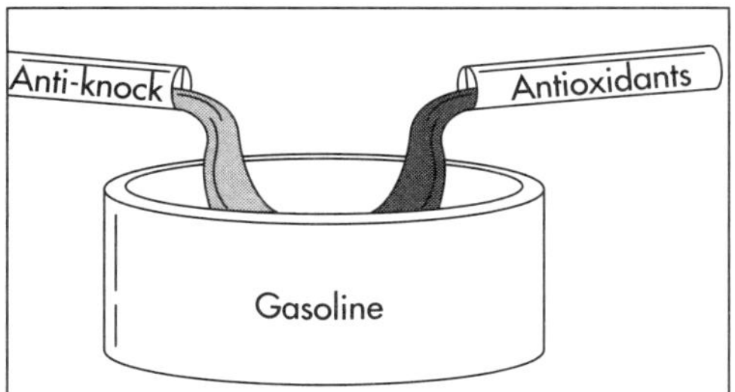

Once gasoline is refined, chemicals are added to reduce knocking and prevent the formation of gum in the engine.

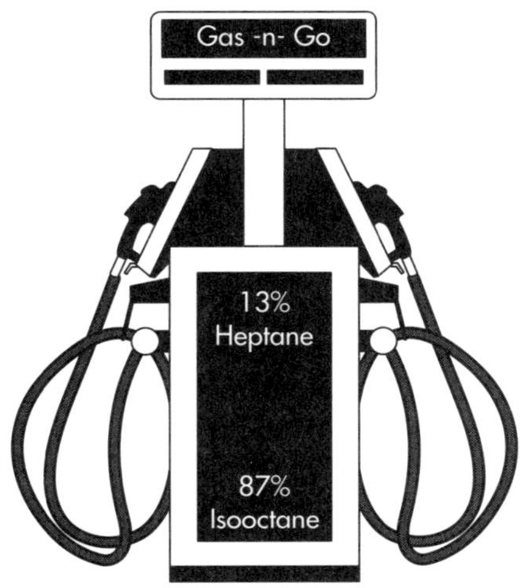

Gasoline is then rated according to the ratio of heptane to isooctane.

Additives

8 Once gasoline is refined, chemicals are added. Some are anti-knock compounds, which react with the chemicals in gasoline that burn too quickly, to prevent "engine knock." In leaded gasoline, tetraethyl lead is the anti-knock additive. (Unleaded gasoline is refined further so the need for anti-knock additives is minimal.) Other additives (antioxidants) are added to prevent the formation of gum in the engine. Gum is a resin formed in gasoline that can coat the internal parts of the engine and increase wear.

Rating gasoline

9 Gasoline is primarily a mixture of two volatile liquids, heptane and isooctane. Pure heptane, a lighter fuel, burns so

quickly that it produces a great amount of knocking in an engine. Pure isooctane evaporates slowly and produces virtually no knocking. The ratio of heptane to isooctane is measured by the octane rating. The greater the percentage of isooctane, the less knocking and the higher the octane rating. For example, an octane rating of 87 is comparable to a mixture of 87% isooctane and 13% heptane.

Byproducts/Waste

On average, 44.4% of petroleum becomes gasoline. There really are no waste products from petroleum. The lighter chemicals are natural gas, liquified petroleum gas (LPG), jet fuel, and kerosene. The heavier products are used for the manufacture of lubricants, plastics, and **asphalt**. In addition, many less valuable products can be chemically converted into more saleable compounds.

The Future

Gasoline, though widely used in many applications today, is destined to become a fuel of the past because petroleum is a nonrenewable resource. Current technology centers on making the most of the remaining petroleum reservoirs and exploring alternative energy sources. New methods to accurately determine the extent of oil reservoirs, automated systems to control oil recovery, and ways of enabling workers to recover more oil from known reservoirs are all being investigated to fully utilize the oil stores available today.

The newest methods in oil field exploration measure the physical size of the reservoir and its volume of oil. Frequently, the pressure inside the well is measured over a period of time as the oil is recovered. Using this data, scientists can determine the size of the reservoir and its permeability. An echo meter, which bounces sound waves off the sides of the reservoir, can also be used to discover the well's characteristics.

Modern oil recovery methods are most often controlled, at least in part, by computerized systems. SCADA (Systems for Supervisory Control of Data Acquisitions) use specialized software to monitor operations through one or more master terminals and several remote terminals. These systems increase efficiency, help prevent mishaps that could harm the environment, and reduce the number of laborers with increased safety.

Enhanced oil recovery methods increase the percentage of oil that can be obtained from a reservoir. In the past, workers were able to extract less than half of the oil contained in a reservoir. New methods involve injecting gases or foams into the well to force out the oil, drilling horizontally into the well, and using more geophysical information to accurately predict the characteristics of the reservoir.

Because gasoline is produced from a limited supply of petroleum, scientists are looking for clean, renewable sources of energy to power machines of the future. Steam power, used in the steamboats of the past, is an energy source that is receiving renewed attention. Electric vehicles have been developed, and solar and wind energies are also powering cars and homes.

Where To Learn More

Book

Shilstone, Beatrice. *The First Book of Oil.* Franklin Watts, Inc., 1969.

Periodicals

Gibbs, L.M. "The Changing Nature of Gasoline." *Automotive Engineering*, January 1994, pp. 99-102.

Langreth, Robert. "Less Smog, More Buildup?" *Popular Science*, April 1995, p. 36.

"Getting the Lead Out." *Motor Trend*, April 1992, pp. 106-107.

—Barry Marton/Kristine M. Krapp

Hay Baler

Background

The term "hay baler" refers to a particular piece of agricultural equipment used to harvest hay. Hay includes grasses and other plants called legumes. Timothy Grass, alfalfa, and clover are common hay crops. These crops, used as animal feed, are cut with a mowing machine when they are about 18 inches (46 cm) tall and still green. The mowing machine cuts and crimps the forage and discharges it into a 4-feet (1.2 m) wide windrow. This crimping process breaks the stems and makes the hay more palatable to cattle. A hayrake is used to turn the hay over so that the windrow can dry completely. The baler gathers the cut hay from the windrow and compresses it into square- or round-shaped bales for easy transportation and storage.

Most hay is stored as bales, with small square bales weighing 40-70 pounds (18-32 kg) and large round bales weighing 750-2,000 pounds (341-908 kg). Small bales must be protected from rain and snow in a dry place such as a barn or hay shed. The large round bales can be left outside because the rain will run off the sides, instead of soaking through and rotting the hay. Small bales can be fed by hand into feed bunks or hay feeders. The larger round bales are handled with a tractor equipped with a "bale mover," a spear-like attachment that pierces the bale and allows the hydraulic loader to lift and transport it to the feeder.

History

During the late 1800s and early 1900s, farming was changing dramatically with the introduction of many new machines. Until that time, hay had been stored loose in the upper story of the barn, or "haymow," where it took up considerable space. By compressing the hay, or baling it, more hay could be stored in the same amount of space. One of the first balers was powered by horses walking on an inclined treadmill. As the leather and wood treadmill belt moved with each step of the horse, it turned a shaft that operated a chain drive. The chain drive, through a variety of sprockets, drove a plunger into the baler, which compressed the hay. Hay was hauled to the baler from the field in wagons, and then forked into the bale chamber by hand. Wooden blocks were dropped into the chamber when the bale reached the right size. Wire or twine was then threaded around the bale and tied by hand. As technology improved, the steam traction engine replaced the horse, and the internal combustion tractor eventually replaced the steam engine. By the 1930s, balers were attached to tractors, and they automatically picked up hay from the ground. Improvements in hydraulics allowed the introduction of the large round baler in the late 1960s. Companies such as John Deere, New Holland, and Hesston have continually refined the baler into a modern farm implement.

How It Works

Hay balers are pulled behind and powered by the tractor in the field. The baler has flotation tires, which reduce the damage to the hay stubble by distributing its weight over a larger area. Also connected to the tractor is the Power Takeoff Shaft (PTO), which transmits rotary power from the trac-

Most hay is stored as bales, with small square bales weighing 40-70 pounds and large round bales weighing 750-2,000 pounds.

tor to the baling mechanism. Along the sides of the hitch are the hydraulic hoses that operate the various controls on the baler using hydraulic pressure from the tractor. When operating the baler, the tractor pulls the baler pickup, a horizontal spool of moving steel teeth, in line with the windrow and engages the PTO drive.

Operation of a Small Square Baler

The hay enters the baler through the pickup, and the teeth gently rake the hay from the ground to prevent the loss of leaves and ingestion of rocks or debris into the baler. Directly behind the pickup is the compressor bar, which holds the hay in place so the auger can feed it into the bale chamber. The bale chamber contains a plunger that drives in and out, each time packing and compressing hay into the desired shape. The plunger also cuts the ends of the hay to make the bale an uniform size. The chamber feeds into a spring tension section that keeps the bale tightly compressed until enough hay has been processed to complete the bale. When the correct length of bale is achieved, a mechanism wraps the bale with two lengths of twine or wire and ties it securely. The twine is carried on spools and fed through two curved needles that are timed to miss the cycle of the plunger. After the twine is in place, a gear mechanism called a knotter ties the knot and cuts the twine free of the supply spool. All of this motion occurs in less than two seconds, and must be carefully timed to prevent interference with the continued operation of the rest of the baler. After it is tied, the bale is pushed down the bale chute and falls to the ground. Some balers have "kickers," or bale ejectors, which throw the bale onto a hay rack pulled behind the baler.

Operation of a Large Round Baler

Like the small baler, the large round baler uses a pickup to load the hay from the ground into the bale chamber. Here, however, the hay is wrapped onto itself by six to eight long rubber belts that are 7 inches (18 cm) wide. As the hay is drawn into the machine, the bale becomes round and fills the bale chamber to capacity. The hay exerts force upon the belts, which is in turn monitored by the hydraulic system. Once a predetermined pressure is reached, a signal is transmitted to the tractor operator. The operator stops the forward movement of the baler, and the bale is automatically wrapped with twine or protective sheeting. After wrapping, the tension on the belts is released and the entire rear portion of the baler is opened by hydraulic cylinders. The bale then simply rolls out onto the ground. The baler is pulled ahead, the rear closes, and baling resumes. The entire process can be operated from the tractor, and bale ejection takes from 15 to 45 seconds.

Raw Materials

Balers are constructed primarily of structural and sheet steel. This steel can be in bar stock, sheets, or in rolls. Hydraulic cylinders, pumps, wheels and tires, belts, and other components are purchased from suppliers and shipped to the baler assembly plant. The belts are made of nylon and **polyester**, a material similar to tires. All raw materials are carefully specified by the engineering staff after extensive testing and research. Many portions of the baler arrive as subassemblies, put together by divisions of the baler manufacturer or by independent suppliers.

The Manufacturing Process

Cutting the sheet metal

1 The outer skin, covers, and shields of the baler are punched or blanked out of sheet steel in a large punch press or by laser cutter. The sheet metal can be taken directly from the roll, or precut into flat sheets as needed. Punch presses operate by forcing a hardened steel punch through the material into a hardened die with up to 200 tons of force. This shears the metal to size, and the rapid stroke of the punch press allows many parts to be produced per hour. More intricate shapes and low-volume parts are cut using an industrial laser to burn through the metal in a preprogrammed pattern. In spite of being slower than the punch press, the laser reduces material waste by arranging the part shapes to most effectively utilize the sheet size. Another advantage of the laser cutter is that it requires practically no set up, which means it can create different-

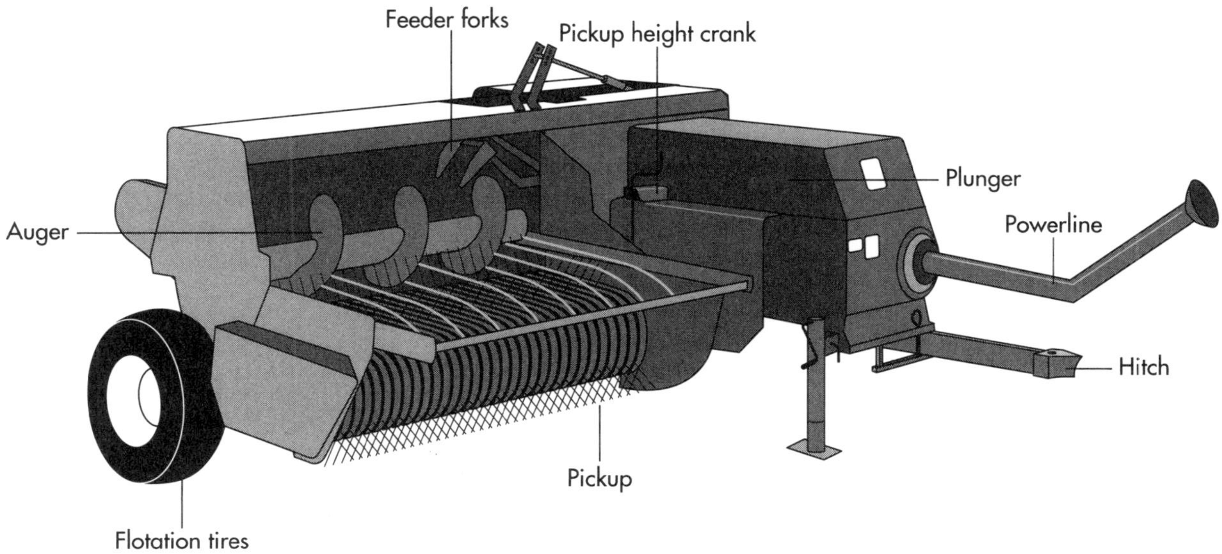

sized parts without physically adapting the machine. This is important, as one machine can then produce hundreds of different parts, in any random order, and provide them to the rest of the manufacturing process as needed.

Cutting the bar stock

2 Bar stock steel that is used for frames, shafts, arms, and other structural parts is cut using a band saw. These saws have many toothed cutting blades driven over two large wheels. The wheels rotate and move the blade continuously to cut through the bar. Bar stock can be round, square, rectangular, solid or hollow, plus many other shapes specific to the desired application. After cutting to length, holes and slots are drilled, punched, or milled into the parts as needed. Structural parts are usually fabricated close to the welding area to minimize time between operations.

Welding

3 Welding is performed primarily by robotic welders. These machines are operated by a computer that has been programmed to move the welding electrode wire through a specific range of motion. The robotic welder is very accurate and makes high-quality, repeatable welds. After the various components are clamped into a large welding fixture by the operator, the robot extends a short piece of welding wire from the welding gun. With the welding current turned off, it gently touches a computer sensor that accurately defines the exact position of the wire tip. Then, the robot touches the tip of the wire to each of the components in two or three places. All of this touching is actually transmitting data to the computer about the exact location of the components in reference to the desired location of the weld. Once complete, the computer program compensates slightly for any deviation from the exact position. The welding current is turned on and the wire is fed into the arc while the robot moves the gun along the joint. This insures that the welds will be exactly at the proper place and achieve the highest possible strength. It also prevents welding any misplaced or incorrect parts, reducing scrap and possible machine failures. In many critical applications, robotic welders can be more accurate than a human welder because of this type of programming.

Cleaning and painting

4 After welding, fabricated parts as well as shields and covers are cleaned and painted. Cleaning is accomplished by hot steam or solvents inside an enclosure. The paint is applied by dipping the parts into an electrostatic paint tank. These tanks are large enough to submerge an entire baler frame, typically 15 feet (4.5 m) long and 10

Balers are constructed primarily of structural and sheet steel. This steel can be in bar stock, sheets, or in rolls. Hydraulic cylinders, pumps, wheels and tires, belts, and other components are purchased from suppliers and shipped to the baler assembly plant.

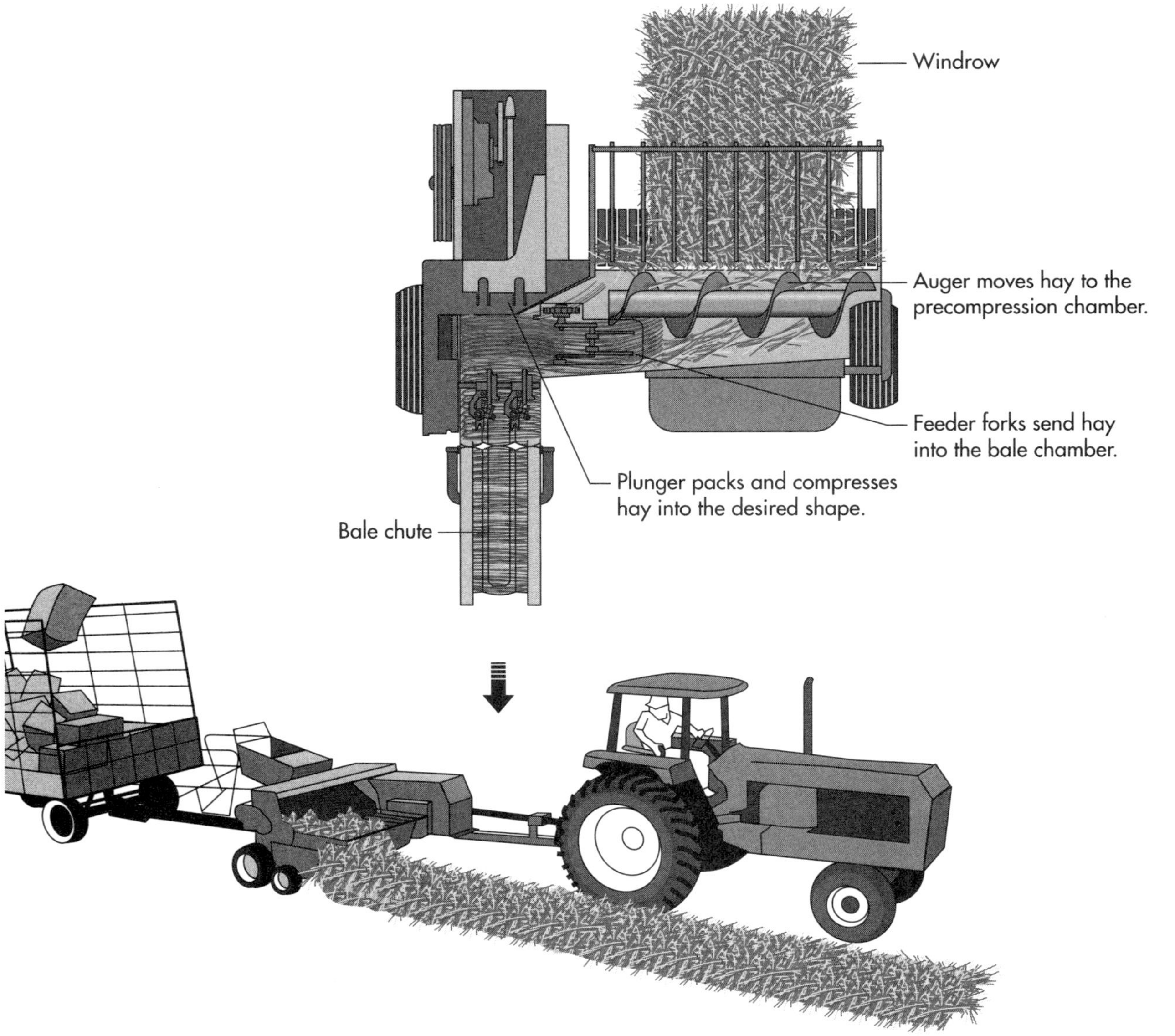

The hay enters the baler through the pickup. Directly behind the pickup is the compressor bar which holds the hay in place so the auger can feed it into the bale chamber. The bale chamber contains a plunger that drive in and out, each time packing and compressing hay into the desired shape.

feet (3 m) wide. They contain 20,000-40,000 gallons (75,700-151,400 l) of paint. The paint is suspended in a water solution and charged with a negative electric charge. The parts are charged positive, and when the parts submerge into the solution, the electrical attraction of the two charges sticks the paint tightly to the metal. The parts are hoisted out of the tank and allowed to drip before they are placed into a drying oven. This drying oven heats the parts to approximately 365°F (185°C), which bakes the paint into a hard finish. Because balers are used and sometimes stored outside, a good finish is needed to protect the steel from rusting.

Assembly

5 After painting, the frame of the baler is placed onto a wheeled cart. This allows the baler to be moved and assembled at a variety of work stations, each adding specific components and performing quality checks during production. The hydraulics, pickup tines, bale tensioners, knotters, and subassemblies are bolted into place. Many design features create special hole patterns, tabs, pins, notches, and other forms to prevent the parts from being assembled in the wrong place or position. After assembly, the hydraulic systems are filled with oil, belt and chain tensions are adjusted, and the

bearings greased. The baler then has the tire and wheel assemblies mounted and identifying decals and stickers attached. Since a baler can be a very dangerous machine, many warning labels are applied to caution operators to keep their hands away from the moving parts.

Inspection and adjustments

6 Finished balers are finally inspected and operated without hay to check the functionality of all the parts. Most are shipped by truck or railcar to dealers all over the world. A technician may be sent out to the field with the new baler to instruct the operator and make final adjustments. Often, changes in the hay crop will require occasional adjustment to the baler to produce quality bales.

Quality Control

During a production run, balers are closely inspected at various stages for proper function and durability. Whenever possible, features are designed into parts and processes to prevent incorrect parts and subassemblies from leaving the assembly line. Occasionally, a completed baler is taken from production for testing either by special test equipment or by actual field trials. Subassemblies can be tested and even destroyed without sacrificing the entire baler. Data acquisition computers are used to record the test performance and, using Computer Aided Design (CAD) programs, offer engineering changes to improve performance. Like automobiles, balers have local service dealers that can notify customers of correct maintenance procedures and/or new features that may be retrofitted to existing balers.

The Future

Processing hay for animal feed has improved greatly in the past 20 years, primarily due to the advent of the large round baler. Even larger balers are on the market, but the round baler seems most popular at present. As the size and complexity of the machine increase, so does the cost. Custom balers, people who bale for resale, are a large portion of the new baler market, along with large corporate farms. These operators usually have the larger tractors to power big balers, and with improvements in hydraulics and bale wrapping, the large baler will probably be the standard for the next several years.

Where To Learn More

Bell, Brian. *Farm Machinery*, 3rd ed. Diamond Farm Book Publishers, 1989.

—Douglas E. Betts

Hearing Aid

The earliest hearing aids were ear trumpets invented sometime in the 17th century. They were long horns with one large opening at one end and a smaller opening at the other end, which was placed in the ear. Today, tiny aids placed in the ear canal are barely visible to others.

Background

The hearing aid is an instrument that amplifies sounds, particularly speech, for people with hearing impairments. It may be worn comfortably behind the ear, in the outer ear, within the ear canal, in the frames of eyeglasses, or against the body or in the clothing. The main elements of the aid are a microphone, an electronic amplifier to make the sound louder, an earphone or receiver, and an ear mold or plastic shell that serves to couple acoustic energy (sound) from the earphone to the eardrum either directly or through plastic tubes. The sound is converted to an electrical signal, amplified, then reconverted to acoustic energy in the inner ear. A battery, the typical power source, can also be contained in the shell.

The microphone and earphone together form a transducer and determine the performance of the aid over a range of frequencies. The adjustment of tone (low and high frequencies) and gain (volume) can be either manual or automatic so that the user can hear enhanced sounds within a comfortable tolerance level.

History

The earliest hearing aids were ear trumpets invented sometime in the 17th century. They were long horns with one large opening at one end and a smaller opening at the other end, which was placed in the ear. The principle behind this instrument being that sound pressure waves entering the large end are condensed into smaller volume, thereby increasing the audible sound pressure.

The first electronic aid was a telephone invented by Alexander Graham Bell in 1876 that converted sound pressure waves to a current and then back to waves. By the 1920s, a more sophisticated telephone-type aid was developed resembling the modern hearing aid with a microphone, electrical circuit, diaphragm, and battery. With the invention of the transistor in 1948, the size and weight of the aid was further greatly reduced. Today, tiny aids placed in the ear canal are barely visible to others, offering great cosmetic appeal to the user. The miniaturization of the hearing aid continues to be an area of research and experimentation. Unfortunately, the smaller the hearing aids become, the greater the manual dexterity required of a user to work the controls.

The Manufacturing Process

Fabricating a hearing aid takes about two hours. Making hearing aids is a customized process requiring skilled technicians to operate magnification glasses and microscopes in a micro-miniature manufacturing environment. The tools are generally handheld and the tasks demand precision movements. The assembler must pay close attention to the wiring diagram and assembly prints so that he or she wires it to produce exact results.

Before fabrication begins, the user is screened by a trained professional. The screening includes a hearing test, and the results are used to create an audiogram covering a variety of parameters. At the screening stage, an impression or mold of the user's outer ear is also taken. The audiogram and the impression are integral to the manufacturing process.

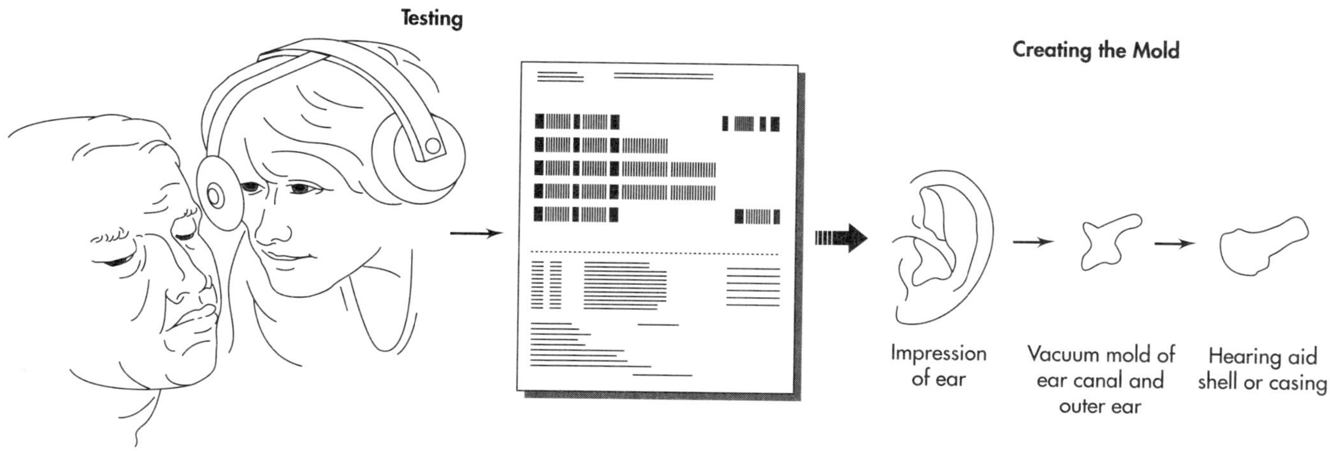

First the user must be screened and given a hearing test; the results are entered on an order card. An impression of the ear is also taken.

The impression of the ear is passed through several stages to create a mold of the ear canal. This mold is then used to create the outer shell of the hearing aid.

Data entry

1 All order data—desired product features and the results of the audiogram—are entered in a computer to determine the operating range for the hearing aid, specifically which levels of amplification are required for the user. For some manufacturers, the computer also selects the electronic circuitry to be used. Typically an order card will be prepared and sent to the production line along with the ear impression. A parts card is also printed and sent to the stock room, where the various components are stored.

Vacuum form of impression

2 In this step, a form or reverse copy of the impression of the outer ear and ear canal is produced. A sheet of clear vinyl is placed over the impression; then heat is applied. When cooled, the impression is removed from the vinyl form and trimmed. Next the impression is dipped in hot wax, giving it a smooth, paper-thin coating for the casting step.

Cast of finished impression

3 Here a technician will place the impression on a metal plate and place rings around it. A clear liquid colloid or particle suspension is poured into the rings, immersing the impression. The liquid is allowed to cure into a rubbery silicone consistency. Lastly, the impression is pulled out of the gel.

Creating the shell

4 The shell or outer casing of the hearing aid is made from this mold. The mold is first heated in 110°F (43°C) water, and air is blown through it to clear away any impurities. At this point, a technician will mix liquid acrylic of equal parts monomer and polymer (for a structurally sound shell) and add the desired pigment to give the shell a pink, tan, or brown color. The technician pours the liquid into the heated colloid mold and after ten seconds, pours off any excess, leaving a thin acrylic shell inside the mold. After 10 minutes of cooling, the technician pulls from the mold a shell that is a perfect replica of the raw impression of the ear canal.

Building the hearing aid into the shell

5 After a technician grinds off the excess flanges from the shell, he will add a vent or opening. A small piece of silicon wire shaped to the vent size is run through the inside of the shell and pulled out. The technician drills holes into the canal end of the shell for the receiver tube. After that, the outside of the shell will be buffed to a

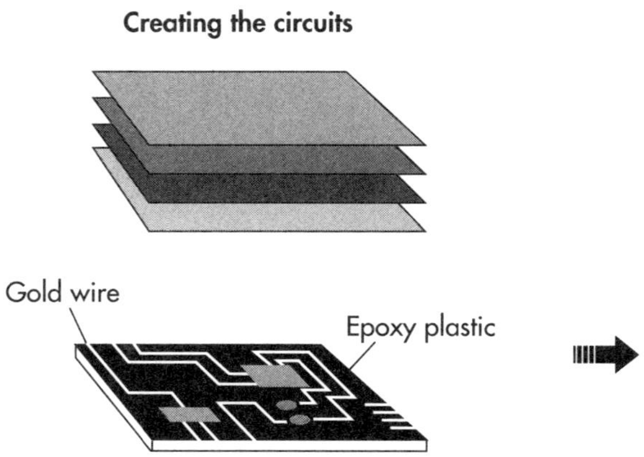

Microminiature circuits are fabricated on a ceramic substrate of alternating layers of conductive and insulating materials.

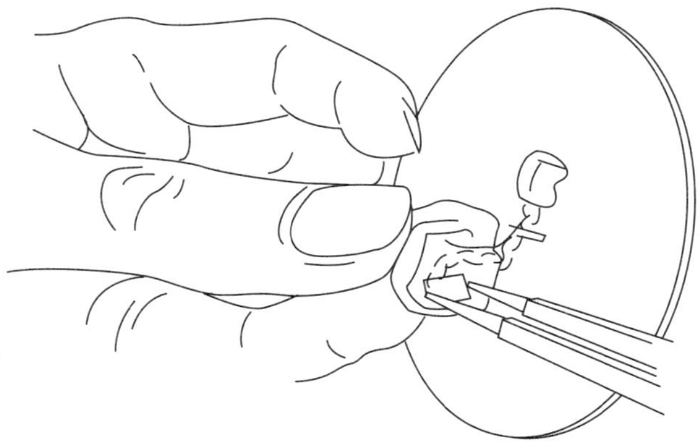

A technician assembles the electronic components and places the microphone, battery compartment, and volume control onto the face plate.

smooth, shiny finish. A technician will size the shell for a face plate or flange—the area that will be exposed outside the ear canal—using the vacuum form from the original impression. The plate will be carefully set at the correct angle for the user's ear.

Creating the microminiature circuits

6 The components and circuits are run on a ceramic substrate base of various designs. The substrate is made by a screen printing technique that alternates layers of conductive and insulating materials, depending on the engineered design. The conducting layer contains gold and silver, and the insulating layers contain silicon compounds. Between the printing of each layer, the substrate is passed by a conveyor through a furnace, where it bakes for two hours at 850°F (454°C). This seals the layers and creates the color patterns characteristic of **printed circuit boards**, only on a smaller scale. The various electronic components are bonded by hand to the gold and silver parts on both sides of the substrate. A technician will interconnect the devices using gold wire of .001 inch (.025 mm) thickness. Lastly, the components are sealed in an epoxy paste and heat-hardened.

Assembling the electronic hardware

7 Working from the parts card or bill of materials determined at the outset of production, a technician assembles the electronic hardware on the face plate where holes have been drilled for mounting the microphone, amplifier, battery compartment, and volume control, all hand-wired with colors for traceability. The wires are soldered into place. After this, the receiver is mounted into the shell and a preliminary hearing check made on the instrument.

Sealing and finishing

8 In the final production steps, the parts are carefully packaged to avoid interferences. A plastic cement is brushed on while the technician performs a listening check to make sure there is no oscillation in the sound quality. Once the cement dries, excess face plate is cut away and the remaining edge ground off with a hand lathe. Finally the aid is buffed and shined to a high gloss and manufacturing is complete. The aid is now ready for final testing.

Quality Control and Testing

Quality control measures are checked throughout production, some of which have been discussed in the process description above. In addition, the shell is given a serial number after it is constructed for tracking purposes. Appearance is important, and a cosmetic check is made as well as a final function check.

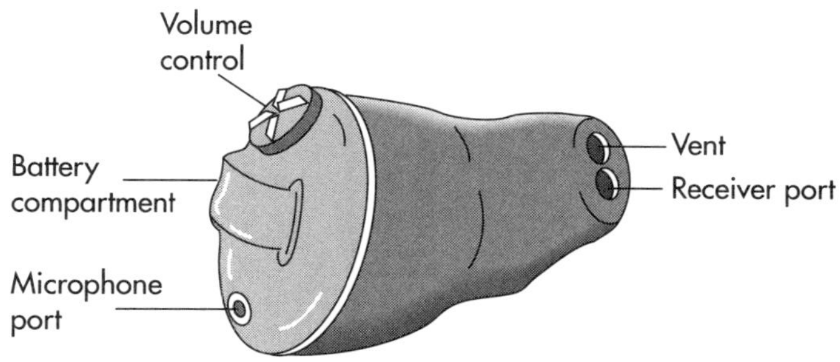

The hearing aid is buffed to a shiny finish and ready for testing.

Hearing aids are tested using a computerized ANSI (American National Standards Institute) program that analyzes the production parameters and produces a performance chart. A technician reviews the chart on-screen, checking tolerance levels and other specifications. He or she will print a copy of the results and include it with the finished hearing aid.

The Future

The future of hearing aids seems to lie in miniaturization. Today's technology can produce aids the size of a fingertip. Also a recent development, customized digitally programmable aids using microchips found in computers allow users to rapidly switch settings to accommodate different situations. Outdoor events, crowded restaurants, and intimate meetings, each with different sound patterns, can be programmed in the chips. This minimizes the quick adjustments some users must make when they move into a new environment. These custom aids can cost $2,000 each.

Where To Learn More

Books

Levitt, Harry, ed. *Recent Advances in Sensory Aids for Hearing Impairment*. Diane Publishing Company, 1994.

Neuman, Arlene C. *Hearing Aids: Recent Developments*. York Press, Inc., 1993.

Pascoe, David P. *Hearing Aids—Who Needs Them?: What They Can Do for You, Where to Buy Them, How to Use Them*. Big Bend Books, 1991.

Staab, Wayne J. *Hearing Aid Handbook*. Tab Books, 1978.

Periodicals

Aldersey-Williams, Hugh. "Invisible Hearing Aid Gives Eardrums Good Vibrations." *New Scientist*, August 1, 1992, p. 18.

Brietzke, C.E. "Listen Up!" *Saturday Evening Post*, September-October 1993, p. 36, 92-93.

Pierson, John. "Form Plus Function: In-the-Ear Hearing Aids with In-Your-Face Colors." *The Wall Street Journal*, October 28, 1994, B1.

—Peter Toeg

Hunting Knife

Belonging to the generation of tools or weapons after stones and clubs, the hunting knife has survived essentially unchanged to the present day.

Background

The hunting knife is an ancient tool that has survived essentially unchanged to the present day. Belonging to the generation of tools or weapons after stones and clubs, the hunting knife gave early hunters the ability to butcher animals for meat and skins. It was also important to their defense from predators and warring tribes.

The first metal hunting knives were made of bronze around 2000 B.C., a time generally referred to as the Bronze Age. During 1500 to 1100 B.C., knives served as patterns for larger swords in Crete and Britain. With the coming of the Iron Age in the period 1000 to 800 B.C., the use of **iron** for the manufacture of knives developed. Iron allowed a sharper, more durable cutting edge than bronze. The major problem with iron knives was that they bent easily. The Vikings solved this problem by adding carbon, or "carbonizing" the iron. This hardened the iron blade and improved its rigidity and sharpness.

In early American history, one particular design of the hunting knife became a standard by which others were compared. This favorite of the early frontiersmen was the Bowie knife, named after the legendary pioneer James Bowie (1796-1836). This single-edged knife was 10-15 inches (25-38 cm) long. The steel blade was straight for most of its length with a concave tapered point.

The modern hunting knife usually has a rigid, single-edged blade with a handle large enough to be grasped firmly. The unsharpened portion of the blade, or tang, extends through the grip area for strength. Folding, lock-back knives and specialized skinning and gutting knives are also used for hunting and dressing game, but the basic, straight hunting knife is the outdoor enthusiast's standard.

Raw Materials

The most important part of the modern hunting knife, the blade, is made of steel or stainless steel. Of particular importance is the carbon content of the metal, which determines the hardness of the blade. Other substances, such as molybdenum, produce other specific improvements in sharpness, toughness, finish, and corrosion resistance.

The guard and pommel may be steel, brass, or aluminum. The stocks, or grips, can be made of many materials. Some knifemakers use exotic woods or animal bone and horn for the stock. Stocks may also be wrapped with leather, or the leather may be in washer-like segments stacked on the tang and contoured to form the entire handle. The materials chosen for the stocks depend greatly on the desired use. More durable, utilitarian materials such as nylon and leather are chosen for knives intended for actual field use, while fancy materials are used for show knives.

The Manufacturing Process

Forming the blade

1 The blade begins as a "blank," cut from flat material that is in a "soft" condition, which means it has not been heat treated. Mass-produced knives are stamped in a

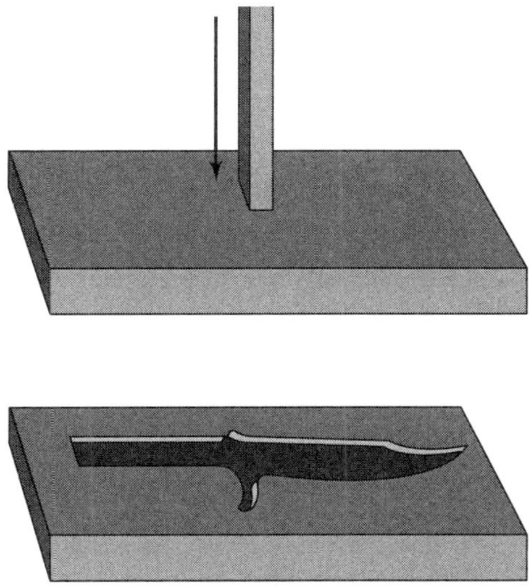

A blank is cut using a punch press. Custom knives may be cut with a saw or an industrial laser.

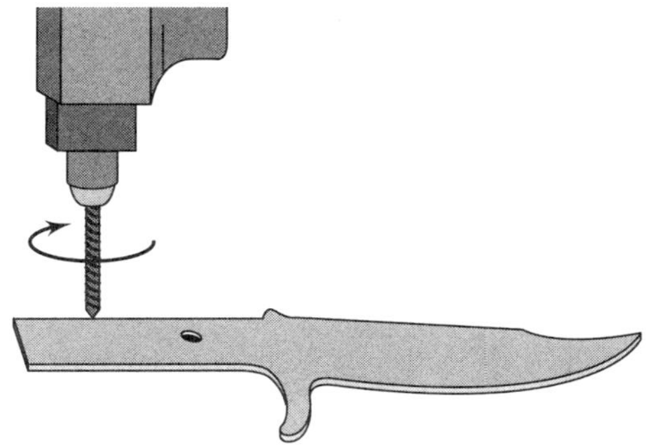

A drill forms the holes for the handle stocks.

punch press, using a hardened punch and die shaped to match the outline of the blade. The punch is forced through the material into the die, blanking the rough blade into shape. Custom knives may be cut out with a saw, or an industrial laser may be used to cut especially intricate blades.

After the blank is cut, holes for the handle stocks are drilled into the tang. Rough shaping is also performed by grinding or machining. This shaping forms the thickness of the blade at various points and reduces the amount of finish grinding on the sharp edge. Any identifying markings or decorative details are stamped while the blade is in a soft condition.

Hardening

2 The blade is hardened to preserve the sharpness of the knife edge. Each blade material may require different hardening and heat-treating methods; however, common steel blades are generally heat treated in the following manner.

To harden the blade, several blades are placed into a ceramic tray in a heat-treating oven. They are laid flat so the weight of the blades does not cause warping. Depending upon the specific requirements of the alloy, the temperature in the oven is raised to approximately 1600°F (871°C). After heating the blades for approximately two hours, the entire tray is removed from the oven and the blades are submerged in oil or water. This rapidly cools the blades and is called quenching. The quenching locks the metal crystals into an intricate microscopic pattern. This process also results in the metal becoming very brittle. After quenching, the blades are reheated to approximately 500°F (260°C). At this point, the metal has a slight dark-reddish color, and the crystals change their alignment slightly. Then the blades are allowed to cool slowly in a process called tempering. This toughens the metal while retaining some of the brittleness needed for fine sharpening. Further heating and cooling cycles may be used to harden other specialized alloys.

Polishing and sharpening

3 After the heat-treated blades are cool, they are polished and sharpened. Polishing is performed by machine or by hand. A flat belt sander is used to produce a smooth, even, "brushed" finish to the sides of the blade. This also polishes out any marks from the punch press operation and removes the surface residue from the heat-treat operation. Next the blade is placed into a grinding fixture that passes it through a series of grinding wheels. Each rotating wheel removes the correct amount of metal

The most important part of the modern hunting knife, the blade, is made of steel or stainless steel. Of particular importance is the carbon content of the metal, which determines the hardness of the blade.

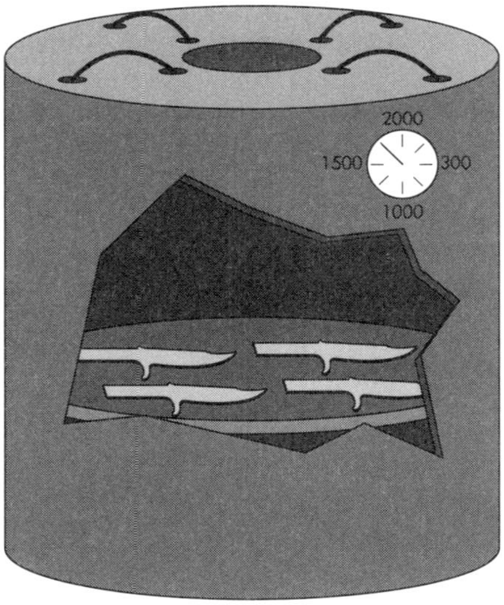

To harden the blade, several blades are placed in a heat-treating oven. Next, they are "quenched" by submerging them in oil or water.

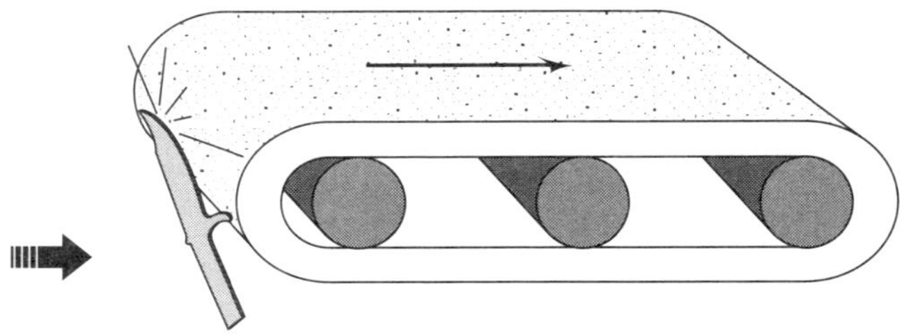

The blade is now polished and sharpened. A belt sander produces a smooth finish and removes surface residue. Next a series of grinding wheels forms the specific features of the blade.

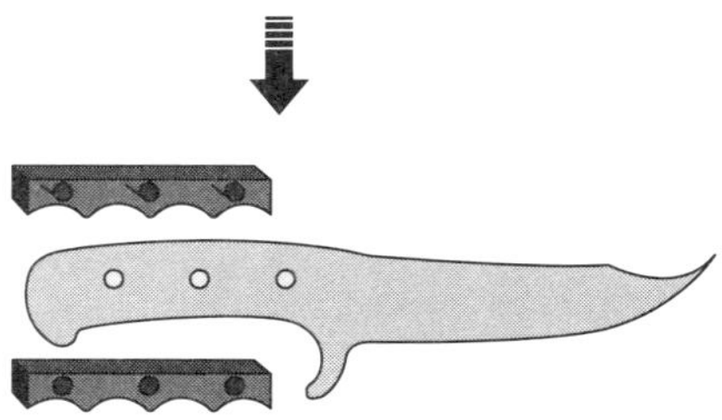

After the blade is honed, the stocks are pressed onto the tang. A buffing wheel smooths the stocks and blends the rivet heads.

to form the edge relief, point, rough edge angle, and other features of the working portion of the blade. When utility knives are made, many of these features are left without further finishing since they are for functional rather than cosmetic purposes. In the case of fancier knives, these features may be hand polished using a rotating buffing wheel and jeweler's rouge. The rouge is a soft, abrasive paste which, when spread onto the cotton buffing wheel and then buffed against the blade surface, polishes the metal to a high luster. When the blade is finished with these steps, it will be quite sharp and may appear ready to use. However, the final sharpening steps are required to produce a long-lasting edge.

Honing

4 To the casual observer, the knife edge appears as simply the result of two intersecting surfaces. Upon closer examination, the true form of the edge is seen to consist of two distinctly separate sets of angles. The first is left by the rough grinding as explained above, while the second is the fine angle that actually forms the cutting edge. This edge is produced by a fine grinding operation called honing. The angle of the hone may be between 17 and 30 degrees to the axis of the blade, depending upon the blade application. A smaller angle will produce a sharper edge, but the edge will wear and become dull more quickly. A fine grinding hone, or "stone," is oiled and gently rubbed on the knife edge. This action produces the finest sharpened edge and is the only true method of properly sharpening a knife blade. When viewed under a microscope, the rough ground edge appears as a series of jagged points. While sharp enough to cut adequately in this condition, the points wear easily and soon the sharp-

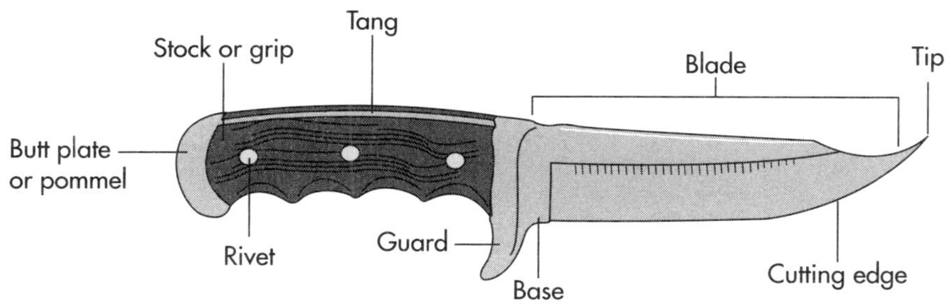

ness is gone. By honing, these points are blended into a consistent edge of equal sharpness.

Assembly of the stocks, guard, and pommel

5 The stocks are riveted or pressed onto the tang of the knife blade. A buffing wheel is used to smooth the stocks and blend the rivet heads. In the case of a leather handle, the washer segments are stacked onto the tang starting at the guard. After the handle length is filled with leather, the pommel is pressed onto the end of the tang and secured with a pin or rivet. This squeezes the leather together, making a tight, easily gripped handle. The leather may be shaped using a grinding wheel and then sealed with a penetrating sealer. The guard and pommel can then be finished by polishing with the buffing wheel. Great care must be taken during the assembly and finishing process as not to damage the knife blade or ruin the sharp edge.

Quality Control

To insure a sharp edge and long life after sharpening, the heat treatment of the blade must be monitored. This is performed by measuring the Rockwell Hardness, a procedure in which the blade or a sample specimen is placed under a hardened point. A heavy weight is exerted upon the point, and the amount of penetration is measured. The dent left by this test is barely visible to the naked eye, and can be done under the handle where it will be hidden. Using various conversion scales, the hardness can then be compared with the desired standard.

Sharpness and finish of the blade are also important. Skilled technicians visually inspect the blade, including using a microscope to closely view the sharpened edge. Any defective blades are returned to the final finishing and honing operation for rework.

The finish of the handle and other portions of the knife are also visually inspected. In particular, show knives are meticulously examined for the slightest defect. Since these knives are a form of art, they will be closely scrutinized by the buyer; any visual defects would lessen the knife's value. Also, if the knife is a duplicate of an historical piece, specific design details and markings are important to the collector.

The Future

The design and construction of the hunting knife has changed little in the last 200 years, and little change is expected in the future. Advances in metallurgy will continue to offer knife manufacturers new steel alloys, which they will use to improve the hardness, durability, and finish of their product. Likewise, new plastics with improved impact resistance, formability, and surface finish will find applications as knife stocks, or grips.

These will be minor changes, though, and will not change the overall design. Today's hunting knife will continue to be a useful and valued tool for the outdoor enthusiast, and will be kept proudly to be handed down to future generations.

Where To Learn More

Books

Boye, David. *Step-by-Step Knifemaking: You Can Do It!* Rodale Press, 1977.

Latham, Sid. *Knifecraft: A Comprehensive Step-by-Step Guide to the Art of Knifemaking*. Stackpole Books, 1978.

Peterson, Harold L. *American Knives: The First History and Collectors' Guide*. Scribner, 1980.

Periodical

Cooper, D. "Points of Interest," *Popular Mechanics*, October 1993, pp. 65-69.

—Douglas E. Betts

Ice Skates

History

Ice skating, in one form or another, has existed for thousands of years. Evidence suggests that as long ago as 1000 B.C. Scandinavians were fashioning crude blades from the shank or rib bones of elk, oxen, and reindeer and strapping them onto boots. A game played on ice between teams has been recorded as early as the second century A.D. In the Netherlands, both men and women skated on the canals during the Middle Ages. Scottish history recounts tales of armies crossing frozen marshes on skates to attack enemy territories. Ice skating became so popular in Scotland that the first skating club was established in Edinburgh in 1742. In 1848, E.W. Bushnell invented the first all-**iron** ice skate that could be clipped onto a boot.

During the 1800s, the popularity of ice skating skyrocketed. Skating clubs opened in London, Vienna, and New York. Rinks were built in Toronto, Canada, and in Davos, Switzerland. In 1876, the first artificially frozen ice rink, called the Glaciarium, opened in London. During the 19th century, the sport of speed skating was introduced and classical dance theory was applied to create the sport of figure skating.

There are three basic types of ice skates: hockey skates, figure skates, and speed skates. Speed skates are designed for optimum swiftness in one direction, with the skater moving right foot over left. The speed skate features a straight blade up to 18 inches (46 cm) long and 0.03-0.06 inch (0.08-0.15 cm) wide. The blade is reinforced with hollow steel tubing. The boot is constructed of very light, thin leather.

Hockey skates are constructed to allow the skater to move both right foot over left and left foot over right. The blade, usually 0.06 inch (0.15 cm) wide, is also reinforced with hollow tubing. The boot is short, measuring 4-5 inches (10-13 cm) from the sole, and reinforced with plastic caps and extra layers of leather at the toe. This protects the skaters' feet from the blades of other skates. The original hockey skate was made of leather with a plasticized sole, a safety tip at the rear, and a hard toe. A ballistic-proof nylon was then introduced that provided even greater protection against cutting. The newest innovation features a plastic molded boot with plastic stanchions and plastic tubing. A heavily padded, removable liner helps to control the fit.

Figure skates are fitted with a 0.125-inch (0.32 cm) steel blade designed for spinning. The blade is hollow on the bottom so that only the outer edges touch the ice. A series of sharp angles at the front of the blade called toe picks facilitate landing from toe jumps. The figure skate has a high boot, measuring 7-8 inches (18-20 cm) from the sole to the top, completely covering the ankle.

Raw Materials

Ice skates are constructed of leather, nylon, plastic, steel, and various other synthetic materials. In most cases, the raw materials are purchased from outside vendors. The ice-skate manufacturer inspects the leather hides closely to insure that the skins have been cleaned and tanned to the company's specifications. Kangaroo leather is one of the popular skins used for figure skates.

Ice skating has existed for thousands of years. Evidence suggests that as long ago as 1000 B.C. Scandinavians were fashioning crude blades from the shank or rib bones of elk, oxen, and reindeer and strapping them onto boots.

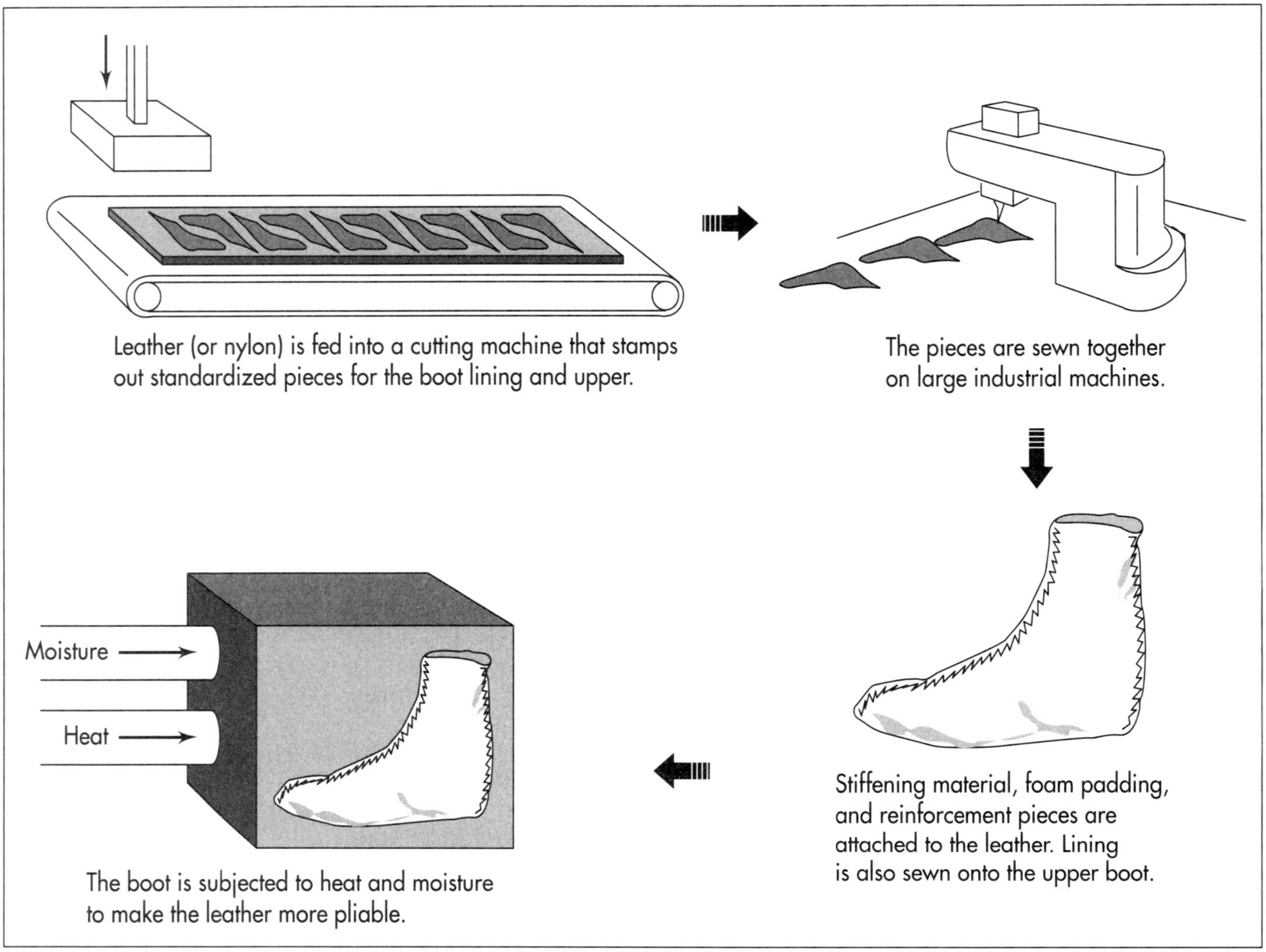

Leather (or nylon) is fed into a cutting machine that stamps out standardized pieces for the boot lining and upper.

The pieces are sewn together on large industrial machines.

Stiffening material, foam padding, and reinforcement pieces are attached to the leather. Lining is also sewn onto the upper boot.

The boot is subjected to heat and moisture to make the leather more pliable.

Knit nylon and molded plastic are commonly used for hockey and speed skates. The leather and nylon are specially treated for water-resistance.

Blades are generally made of tempered steel and coated with a high-quality chrome. Some blade manufacturers may add titanium to the metal. The ice-skate manufacturer contracts with outside manufacturers to supply them with blades in various styles and sizes. Competitive skaters (as opposed to recreational ones) usually have their blades mounted by a specialist.

The cements, stitching threads, and other synthetic materials are also purchased from outside vendors and stocked at the skate manufacturing plant.

The Manufacturing Process

Neither the Olympic Committee nor the U.S. Ice Skating Federation has requirements regarding the manufacture of ice skates. However, the manufacturers pay close attention to the needs and suggestions of professional skaters and coaches.

Cutting the boot segments

1 The leather or nylon is fed into a cutting machine that stamps out eight standardized pieces: four for the boot lining and four for the boot upper. A similar machine cuts pieces for the tongue, sole, heel, and reinforcement sections. The pieces are mechanically punched with a date code, size, order number, and eyelet holes.

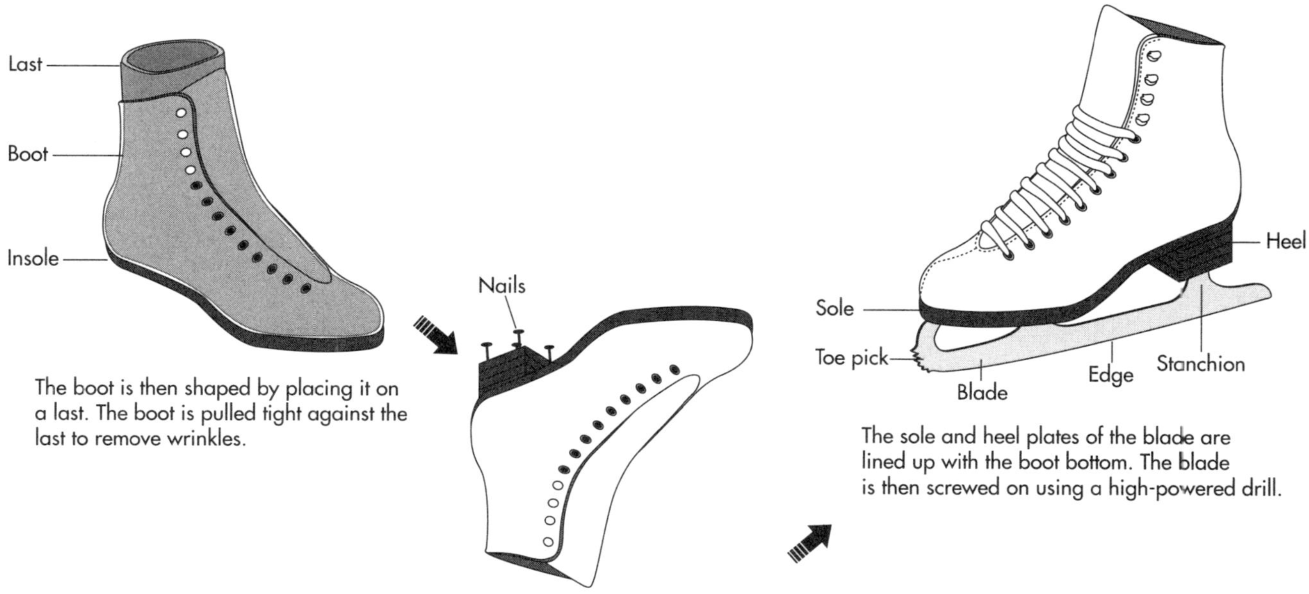

The boot is then shaped by placing it on a last. The boot is pulled tight against the last to remove wrinkles.

An electric grinder removes the leather finish from the boot bottom. The arch is packed with synthetic filler and a steel arch support shank. Square stacks of leather or fiberboard are glued together to form the heel.

The sole and heel plates of the blade are lined up with the boot bottom. The blade is then screwed on using a high-powered drill.

Constructing the upper

2 The boot is sewn together on industrial sewing machines that the plant workers regulate with foot pedals or control panels. The lining pieces are stitched together at the back and toe seams. The upper pieces are stitched together in the same manner. Stiffening material, called a counter and usually made of leather or plastic, is inserted from the ball of the boot around the heel to the middle of the arch to provide support for the ankle. Reinforcement pieces are cemented onto the underside of the upper with latex or neoprene base cement. Foam padding is layered on top of the reinforcement. The lining is then sewn onto the upper boot through all three layers. Excess edges are trimmed. A line of top-stitching provides additional support and decoration. The eyelet placket is also reinforced by means of a strip of leather. The tongue is lined with sponge rubber, then tacked first to the lining, and then to the boot upper, using a cross-butterfly stitch.

3 The partially constructed boot is then subjected to a heat and moisture process called mulling that leaves the leather more pliable.

Shaping the boot

4 The boot is shaped by placing it on a last, which is a plastic foot form corresponding to a particular shoe size and width. The lasts resemble shoe trees. A skate manufacturer can have as many as 102 sizes and widths for women and as many as 57 sizes for men. This large difference in number of lasts may be due to several factors: there are more women skaters (particularly in figure skating) than men, thus increasing the variety demanded. In addition, women may be more particular about fit, as evidenced by the wider variety of width and sizes for women's street shoes than men's.

First an insole is tacked onto the last. Then the boot is pulled tight, by hand, over the bottom of the last. The worker must make sure that all wrinkles are eliminated, working from the arch to the heel then from the arch to the toe. Tacks or cement adhere the arch, heel, and toe to the insole.

Drying and setting the boot to the last

5 The tacks holding the insole to the last are removed. The boot, still stretched over the last, is placed into a drying chamber. Heat is applied to set the boot to the length and width of the last. A worker then removes the last from the interior of the skate boot.

Preparing and attaching the sole

6 The boot is placed upside down on a peg under an electric grinder to remove the leather finish from the boot bottom. The grinding process causes a crevice to form across the middle of the boot bottom where the foot's natural arch occurs. This crevice is packed with a synthetic filler and an 18-gauge spring-steel arch support shank. A leather or rubber sole is then attached to the boot with a urethane base cement.

Attaching the heel

7 The heel is constructed by gluing square stacks of leather or water-repellent fiberboard on top of each other. The height and width of the heel varies with the type of skate. The heel is glued onto the boot and then six to ten long nails are driven through the heel, outsole, and insole. Finally, the heel is trimmed and smoothed.

Attaching the blade

8 The sole and heel plates of the blade are lined up with the sole and heel area of the boot bottom. The worker "eyes" the placement, making sure that the blade is centered on the sole. An equal amount of boot bottom should be exposed at the toe and heel. The worker then screws the blade onto the boot using a high-speed drill on a foot-powered press. Only a few of the screws are inserted into the boot; the remainder are packaged with the ice skates. This allows the skater to adjust the blade as necessary before the remaining screws are inserted and tightened. However, the blades of skates fitted with rubber soles are permanently attached with rivets.

Finishing and polishing

9 The completed skate is hand-polished and sprayed with a solution to make it shine. Laces are threaded through the eyelets. The skates are then boxed for shipping.

Quality Control

The manufacturing process includes several inspection points. At each position, the inspector checks the alignment of the various pieces. Seams and eyelets are checked for straightness and evenness. Structural and visual imperfections such as loose threads and wrinkles are weeded out.

Most ice skate manufacturers have professional skaters on staff who are involved in the design and testing of the product.

Where To Learn More

Book

Jonland, Einar and Jim Fitzgerald. *Inside Ice Skating*. Contemporary Books, 1978.

Periodical

Faiad, Andrea. "The Making of a Boot the Riedell Way." *Skating*, March 1995, pp. 40-41.

—Mary F. McNulty

In-Line Skates

Background

In-line skates were developed in the mid-1980s, but the basic concept of rolling wheels attached to a boot is much older. Earlier roller skates had wheels of wood, plastic, or steel, arranged in pairs. Modern-day in-line skates have wheels made out of polyurethane plastic arranged in a line so that the gliding action is much like that of an ice skating blade. Sometimes this type of skate is called a rollerblade, although this is a trademark name and refers only to a specific brand of skates.

History

People in Scandinavia used **ice skates** as early as 1000 B.C. These skates were made of bone runners tied to boots, and they were used to travel across frozen lakes. The first roller skate is credited to Belgian inventor Joseph Merlin, who in 1760 demonstrated a pair of skates with small metal wheels. A French inventor named Petitbled patented a roller skate in France in 1819 that was something like today's in-line skates. Petitbled's invention had a straight line of wooden, metal, or ivory rollers attached to a wooden sole. Soon after, Robert John Tyers of London came up with what he called the "rolito," a similar skate with five wheels placed in a row on the bottom of a shoe. Roller skates steadily gained in popularity, and the 1849 French opera "La Prophete" even featured a simulated ice skating scene where the performers used wooden-wheeled in-line skates.

These early skates became less common, however, with the 1863 invention of the "rocking skate" by American James Plimpton. Plimpton's skate had a pair of wheels in front and back, mounted on a rubber-padded plate. Roller skaters were able to skate in curves with the Plimpton skate, quite an advantage over previous models. This style of skate soon became the standard. Other innovations were incorporated: high-topped stiff leather boots were attached to metal plates; moveable parts were added to plates to allow skaters to turn corners; and a pair of wheels was attached to an axle on each assembly. Plastic wheels eventually replaced steel or wooden wheels, and basic design improvements over the years have included use of a rubber toe stop, similar to a front brake, and sturdier assembly.

In the 1980s, hockey players Scott and Brennan Olson of Minneapolis, Minnesota, were looking for a way to modify hockey boots so they could cross-train year round. While rummaging through a sporting goods store, they discovered an in-line skate and decided to improve on the design. They began assembling what would be the first "Rollerblade" in-line skate in the basement of their parents' house. The Olsons' first in-line skate had steel frames and skateboard rubber wheels that were riveted to hockey boots. Their prototype was extremely clunky and heavy, but the basic design prevailed.

Hockey players were the first to seriously use in-line skates. During the summer months, they played roller hockey games in the gym and outside on pavement. Nordic and alpine skiers also began to use in-line skates in their training. Soon the popularity of the skates spread to non-athletes as well. Scott Olson created the company that became Rollerblade, Inc. and improved on several of the in-line skate patents. But com-

The first in-line skates were invented by hockey players Scott and Brennan Olson in the basement of their parents' house. This model had steel frames and skateboard rubber wheels riveted to hockey boots.

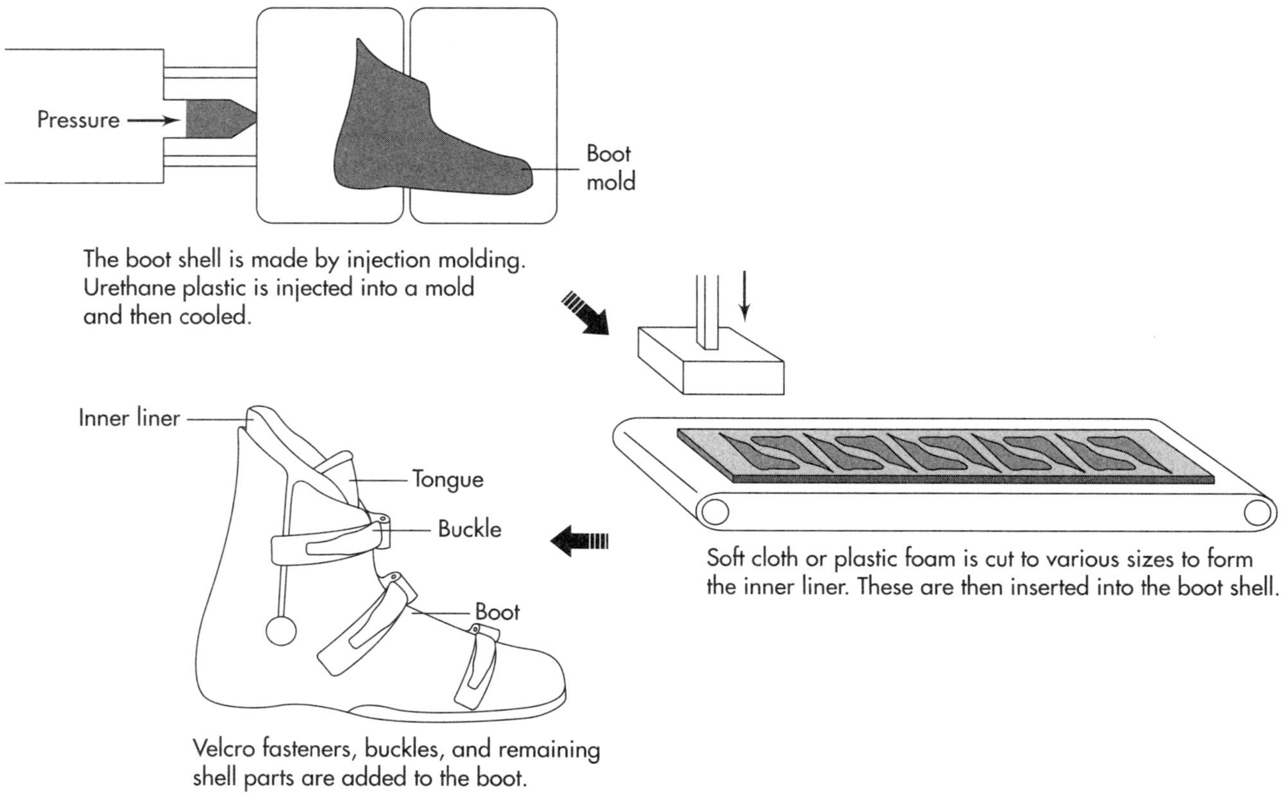

petition from other in-line skate manufacturers forced the company to adopt even more sophisticated manufacturing and marketing efforts. In 1994, Rollerblade patented the Active Brake Technology (ABT) braking system, a cuff-activated braking system that made stopping easier for beginning in-line skaters. The new brake also provides greater speed control. Rollerblade currently holds 16 U.S. patents and 200 pending patents for in-line skating products.

Parts of In-Line Skates

The modern in-line skate has polyurethane wheels aligned on a strip or truck (chassis) made out of plastic or light aluminum. The truck holds the wheels in place with bolts, and on some models the wheels can be detached and changed. The truck is attached to a molded boot, also made of polyurethane, which holds the buckles (made out of plastic, velcro straps, or cloth laces). Some in-line boot shells are leather or a combination of leather and polyurethane plastic. The brakes are made of hard rubber or polyurethane and are located on the back of the skate.

The boots are designed to provide good ankle support by extending 3-4 inches (7-10 cm) above the ankle joint. The upper and lower portions of the boot are connected on each side of the ankle by a hinge system that allows the ankle to flex. For additional comfort, boots are lined with material that absorbs sweat, usually high-density foam covered with nylon clots. Wheels vary in hardness (durometer) and size (diameter). Generally, the harder the wheels, the faster the skates. Wheels have two ball bearings per wheel that are separated by a spacer of either plastic or metal. Bearings help the wheels spin faster and more freely. The spacer helps prevent the bearings from making contact with each other.

Raw Materials

The in-line skate is a product of modern technology. High-grade polyurethane, co-polymer plastics, or carbon fiber Kevlar (material used to make bulletproof vests) are used to make the majority of in-line skating shells or boots. Wheels are also made of

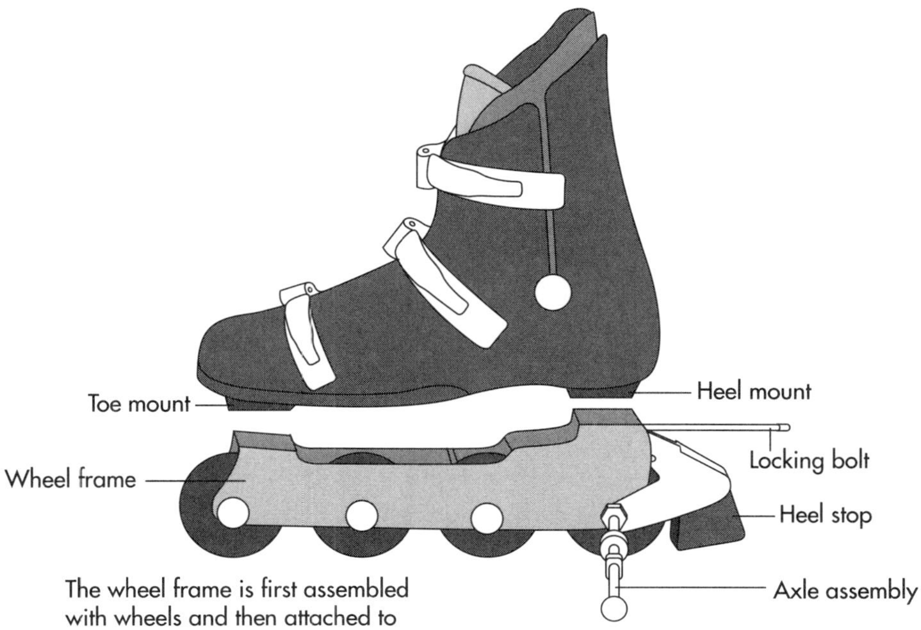

The wheel frame is first assembled with wheels and then attached to the boot shell.

polyurethane. Harder nylon or urethane plastic or steel is used for the chassis and brake components. Frames that hold the wheels can be made of aluminum, carbon fiber, or titanium. Cloth or plastic foam is used to line the inside of the boot. Steel is also used for ball bearings, buckles, and other attachments on the in-line skate itself.

The Manufacturing Process

Separate companies worldwide manufacture the individual in-line skate components. For example, manufacturers of skateboard wheels also make in-line skate wheels. Another company makes the boot shell and chassis; another manufacturer provides the braking system. In-line skates are manufactured using an assembly line process whereby all components move down a conveyor belt and are assembled piece by piece.

Making the boot shell

1 In-line boot shells are made by a process called injection molding. Urethane plastic is injected into a mold to form the boot shell, much like a ski boot. Dye may be added for color. When the mold has cooled, the injection-molding machine ejects the finished boot.

Inserting the liner

2 Soft cloth or plastic foam is cut to various sizes and inserted into the boot shell. The dimensions of the liner should conform to the skater's foot.

Adding the attachments

3 The remaining shell cuff (as well as the tongue), metal buckles, and plastic or velcro fasteners are attached to the shell. Most of these pieces are bought prefabricated.

Assembling the wheels

4 The frame, including wheels and bearings, is attached to the boot shell by rivets or by a coupling system. The frame is often assembled ahead of time. The wheels are placed in a row and attached to the frame. On inexpensive models, the boot and frame are injection molded as one unit.

5 Once the in-line skate has been assembled, it is boxed and delivered to stores.

The Future

In 1990, Rollerblade reported retail sales of more than $100 million and in 1991 controlled about 70% of the in-line skating market. Ultra-Wheels claims about 20% of the market. Since its heyday, however, Rollerblade has lost market share to many competitors, an indication of the widespread popularity of in-line skates. Industry analysts have predicted that the market for in-line skates could reach $1 billion in 1995, compared to $200 million in 1992. The future for the product is highly favorable and newer models that are faster, lighter, and more comfortable are being developed every year.

Where To Learn More

Books

Powell, Mark and John Svensson. *In-Line Skating*. Human Kinetics Publishers, 1993.

Rappelfeld, Joel. *The Complete Blader*. St. Martin's Press, 1992.

Periodicals

Fassett, Wayne. "Blade Runners." *Popular Mechanics*, November 1991, p. 49.

Macnow, Glen. "New Ideas Get a Sporting Chance." *Nation's Business*, December 1992, pp. 62-63.

"Synergy Standouts." *Sporting Goods Business*, February 1994, p. 74.

—Evelyn S. Dorman

Industrial Robot

Background

Industrial robots are mechanical devices which, to a certain degree, replicate human motions. They are used whenever there is a need to reduce the danger to a human, provide more strength or accuracy than a human, or when continuous operation is required. Most robots are stationary, but some move throughout the workplace delivering materials and supplies.

Many people think of robots as the humanoid-type monsters that are seen in science fiction and fantasy movies. While we may someday have the technical ability to produce such a machine, today's robots are actually quite simple devices. Motions that we take for granted—picking up a coin from the table, for instance—are considerably more difficult for a robot. Our brain processes thousands of variable bits of data from our eyes to instruct our arm, wrist, hand, and fingers to reach, grasp, and pick up the **coin**. Even the tactile feel of the coin constantly updates our brain to provide just enough finger pressure to grip the coin securely. Any variations in position are effortlessly compensated for in our brain. To easily and economically program an industrial robot to perform the same task, many of these variables must be restricted or eliminated. Position, reach, weight, and grasp should remain as consistent as possible so that variations do not result in missing or dropping the object. The computer that controls the robot must be programmed by a technician, to "teach" the machine to complete the motion. The areas where robots perform better than humans are in accuracy and repeatability. While some people could pick up the coin with similar motions each time, the robot can perform the operation with exactly the same motions without tiring. Many robots can repeat motions with an accuracy of a few thousandths of an inch and operate 24 hours a day. Because of this tireless, accurate work, robots are a growing segment of industrial equipment purchases. Most are used for repetitive painting and welding operations, while others, known as pick-and-place robots, are used to lift and place products into machines and packages.

History

Robots, or "robotics," are a segment of the broader science of automation. Automation uses machines and computers which can learn or compensate for varying conditions of operation. The term robot can be traced to the Czech word *robota*, which means compulsory labor. The term first appeared in 1921 in the play "R.U.R." (Rossum's Universal Robots) by Czech dramatist Karel Capek. The play described humanoid robots that destroyed their human makers—much the same plot of some modern science fiction thrillers.

Practical robots were first attempted after the development of the computer. In the late 1960s, the Stanford Research Institute designed and built an experimental robot called "SHAKEY." Using a television camera and a computer, this machine was capable of moving and arranging blocks into stacks. General Motors financed a program at the Massachusetts Institute of Technology in the mid-1970s to develop an automated robot for assembly purposes. Here, researcher Victor Scheinman invented the

The term robot can be traced to the Czech word robota, *which means compulsory labor. The term first appeared in 1921 in a play by Czech dramatist Karel Capek, in which he described humanoid robots that destroyed their human makers.*

The first robot installed in American industry, this Unimate "pick-and-place" unit first removed hot metal parts from a die-casting machine at a GM plant in Trenton, New Jersey, in 1961. (From the collections of Henry Ford Museum & Greenfield Village.)

Robots, like any tool, are only as good as the people wielding them. They cannot do jobs they were not designed or programmed to do. They are most effective when the overall system and processes are carefully planned. In addition, workers responsible for them have to be fully educated and trained as well.

In the 1980s, the General Motors Corporation spent upwards of $40 billion on new technologies, many hundreds of millions on robots. Unfortunately, the company did not spend nearly enough on understanding the systems and processes that the robots were supposed to revolutionize or on the people who were to maintain and operate them. The GM plant in Hamtramck, Michigan, was supposed to be a showcase for the company. Instead, by 1988 it was the site of some of the worst in technological utopianism. Robots on the line sometimes painted each other rather than the car bodies passing by; robots occasionally went out of control and smashed into the passing vehicles; a robot designed to install windshields was found systematically smashing them. Once, when a robot ceased working, technicians did not know how to fix it. A hurried call to the manufacturer brought a technician on the next plane. He looked at the robot, pushed the "Reset" button, and the machine was once again operational.

William S. Pretzer

PUMA (programmable universal manipulator for assembly), and the entry of robots into American industry began.

Raw Materials

Robots are mostly built of common materials. Some specialized robots for clean room applications, the space program, or other "high tech" projects may use titanium metal and structural composites of carbon fibers. The operating environment and strength required are major factors in material selection.

Steel, cast **iron**, and aluminum are most often used for the arms and bases of robots. If the robot is mobile, they usually equip them with rubber tires for quiet operation and a positive grip on the floor. Robots contain a significant amount of electronics and wiring, and some are radio or laser controlled. The cylinders and other motion-generating mechanisms contain hydraulic oil or pressurized air. Hoses of silicone, rubber, and braided stainless steel connect these mechanisms to their control valves. To protect the robot from the environment, some exposed areas are covered with flexible neoprene shields and collapsible bellows. Electric motors and linear drives are purchased from automation suppliers along with the controller, or "brain." Controllers are housed in steel electrical cabinets located near the robot's work area or carried on board the robot itself.

The Manufacturing Process

Design

1 Every robot begins with the design phase. These and other factors must be accounted for in the design: job to be performed; speed of operation; environment of operation; hazardous materials involved; length of reach; path of travel; process variables; human involvement; controller capability; and result of failures.

Most manufacturers have a basic machine design to which they incorporate modifications and accessories to meet the specific requirements of the application.

Fabrication

2 Once designed, the base, arms, column, and supports are fabricated. The base is

Once the eventual job of the robot is decided, a design is created. Most manufacturers have a basic machine design, and modifications are added.

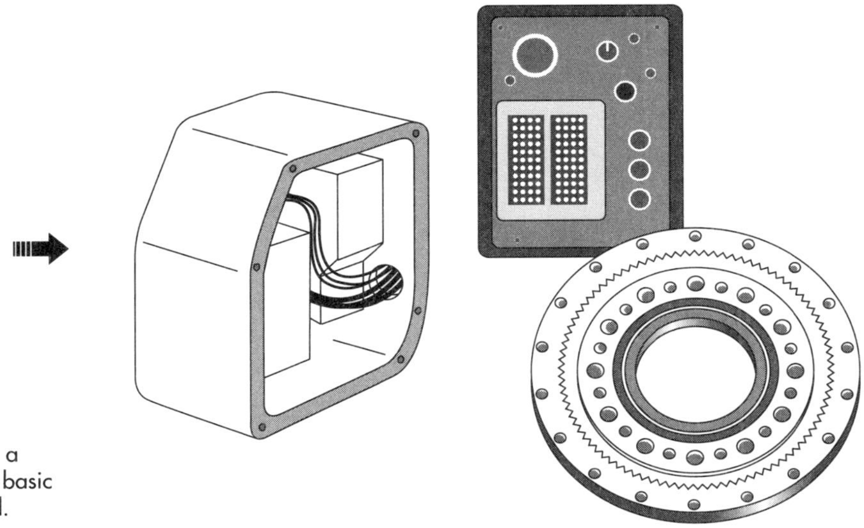

Robots are assembled using many prefabricated, purchased components.

usually heavy, to prevent the robot from tipping over. It is made by casting or by welding, then machined. Many robot manufacturers use robots to weld parts for new ones.

Those areas that mate with the rest of the robot are machined with close dimensional control to assure proper fit and operation of the attaching components. Likewise, the main column and arms are constructed to fit accurately into the final assembly.

Assembly

Robots are assembled using a substantial amount of purchased components such as electric motors, hydraulic cylinders, bearings, wiring, controllers, and other important parts. An industrial robot can contain 2,000 individual parts and is assembled by teams. These teams begin with the base, and assemble components into the robot until it is complete and ready for testing and finishing.

To begin the assembly process, mobile robots first have the traction motors, batteries, axles, wheels, and tires mounted. Stationary robots do not require these items. They are temporarily bolted to the floor for stability during assembly. The moving columns and arms are subassembled with their respective drive motors and then attached to the base. The base contains a ring gear that is motor driven to provide the turning motion. It must mate closely with the drive gear contained in the column. Thrust bearings support the weight of column and arms on the base. A magnetic scale surrounds the bearing and provides electronic position feedback to the controller.

Link

3 The next joint is the link. It acts like an elbow, and connects the arm to the base. A stabilizer support provides positional control to the link, allowing it to move in a predetermined path. These components contain bearing mounts into which pivot shafts are bolted. Each bearing is prelubricated or provided with a lubrication line or fitting. The link contains a position sensor which provides another position signal to the controller.

Arm

4 The arm is assembled onto the upright portion of the link. It provides the most "reach" to the robot and supports the wrist. The arm contains the drive shafts that operate the wrist. Three motors, or a combination of motors and hydraulic cylinders, are

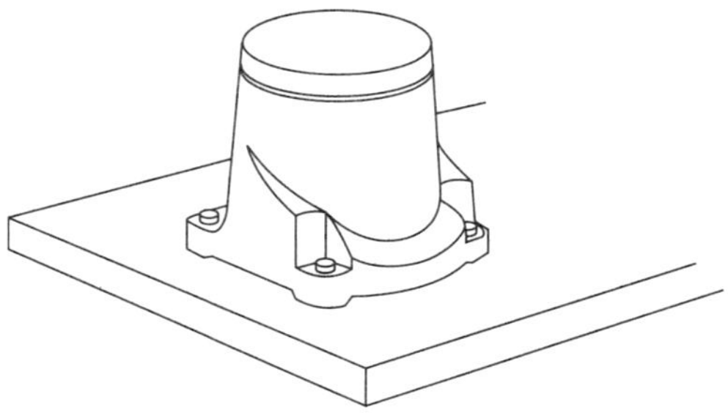

The base is secured to a work area and motors, batteries, axles, and wheels are added.

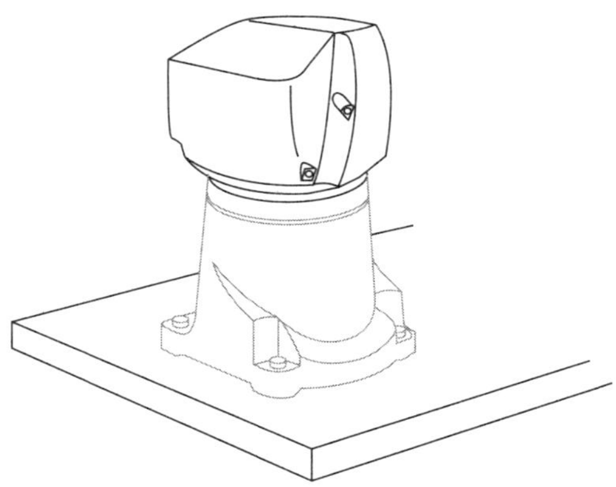

The first piece attached to the base is the link. This acts like a human elbow, connecting the arm to the base and allowing movement.

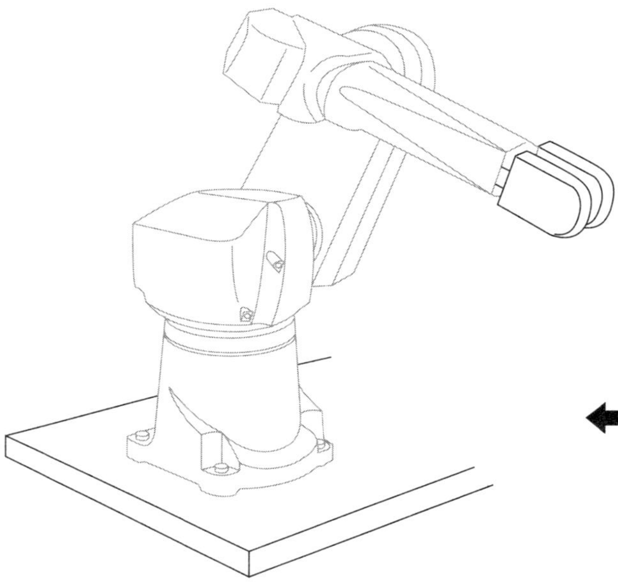

The wrist, perhaps the most critical part of the robot, is attached. This mechanism replicates human wrist motion by turning vertically and horizontally.

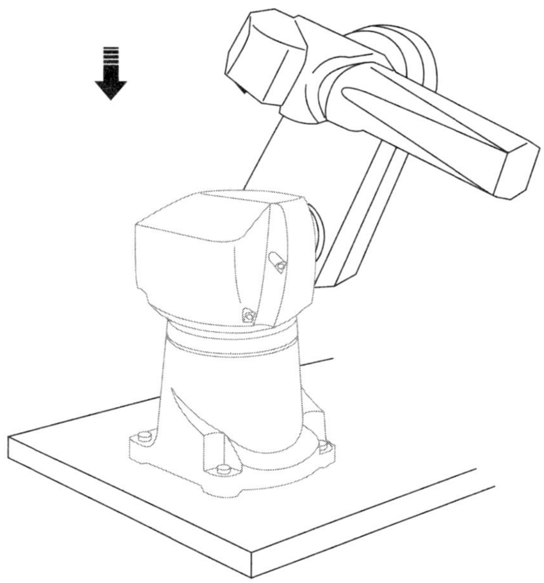

Once the arm is attached, the robot has the ability to reach and have a wide range of movement. This arm also supports the wrist.

An industrial robot can contain 2,000 individual parts and is assembled by teams. These teams begin with the base, and assemble components into the robot until it is complete and ready for testing and finishing.

connected to the drive shafts. Since the arm and link joint must withstand the entire load of the wrist, this is accomplished with large bearings and a pivot pin.

Wrist

5 The wrist is the critical mechanism of the robot. It is the wrist that most replicates human motion by twisting and turning to place the paint gun, welder, or other tool in the correct position. Many robots also have load-sensing electronics in the wrist to signal when an obstruction has been hit, or when a load is too heavy to safely pick up. Additional position sensors and tool control electronics are also assembled into the arm and wrist.

Wiring to the controller

6 Once the mechanical assembly has been completed, the wiring and plumbing of the robot can be finished. All of the

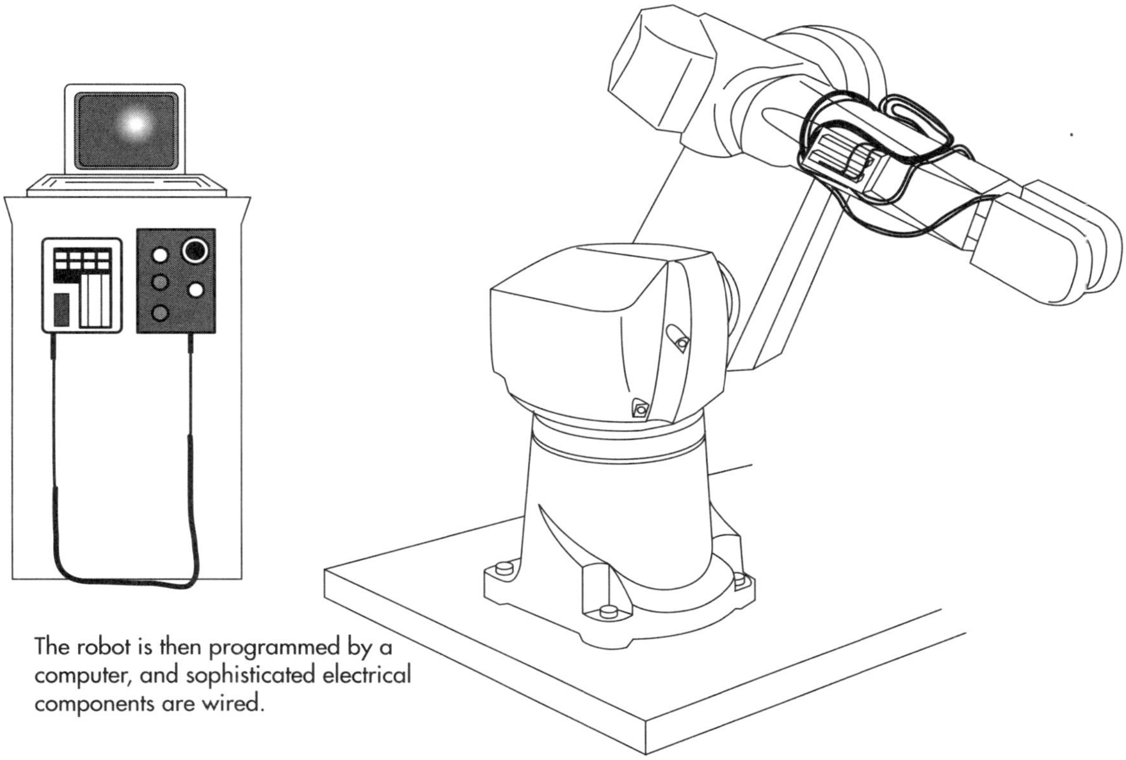

The robot is then programmed by a computer, and sophisticated electrical components are wired.

motor's sensors and electrical components must have wires for power and to carry information back to the control computer. Occasionally, unused space in the arms and base provides a handy place to mount some of the controller electronics, shortening the wiring paths. Hydraulic and air cylinders have hoses that carry pressure to operate them, controlled from the valves in the base. Most of these wires and hoses are routed back to the controller cabinet which, for mobile robots, is attached to the base. If the robot is stationary, this controller is usually mounted several feet away and is connected by an umbilical cord. After assembly, the arms and column of the robot are sometimes covered with guards and shields to protect them from paint spray, welding sparks, or other hazards in the environment.

Installation

7 Installation occurs at the user's site. If stationary, the robot is secured to the floor with bolts. If moving, a guide wire is buried into the floor for the robot to follow from task to task. It follows the wire by radio signals and also uses the wire to communicate with the central controller. Recently, lasers have been used to eliminate the wire. The robot is guided through its path by a laser beam reflected off the walls. Some designs also incorporate video cameras. Stationary applications usually require that fences be constructed around the robot so an unsuspecting human doesn't wander into the robot's work area and be injured. After installation, the robot manufacturer usually provides operation and maintenance training to the customer.

Quality Control

Testing consists of two parts: functional accuracy and a process known as "burn-in." Once the assembled robot is energized with power, a computer program instructs the controller to move the robot arm through a series of motions. Accurate recordings of these motions are made, any problems corrected. Then the robot is placed into operation continuously for several hours. This is called burn-in, and it serves two functions. First, any loss of accuracy can be detected using the data from the functional test. Such an instance would indicate a design problem, loose assembly, defective bearing, or the like. Second, the trial run brings the electronics and hydraulics up to operating temperature. This is important because the controller is programmed with correction factors called offsets. These offsets compensate the feedback from the position sensors to allow for temperature variation of the components. With the machine warmed up, the programmer can place the correction factors into the program to provide optimum performance.

The Future

Robotics is one of the fastest growing segments of the industrial machine market. Driven mainly by advances in computer technology, older robots are quickly made obsolete by new models. Japanese firms are leading the development of robotics, and many of their designs incorporate the new science of artificial intelligence which allows robots to "learn" and "adapt" their operations on their own.

Advances in cameras and electronic vision will also impact the robot in the 1990s. Many robots will enter new areas of use such as medical and food service, which will bring more people into contact with them than previously occurred in the industrial workplace.

Where To Learn More

Books

Berger, Fredericka. *Robots: What They Are, What They Do.* Greenwillow Books, 1992.

Ulrich, Robert A. *The Robotics Primer: The What, Why, and How of Robots in the Workplace.* Prentice-Hall, 1983.

Periodicals

Edelson, Edward. "Robo Surgeons." *Popular Science*, April 1995, pp. 62-65, 90.

Kirschner, Suzanne Kantra and Bob Sillery, eds. "What's New: Dexterous Digits." *Popular Science*, April 1995, p. 11.

—Douglas E. Betts

Integrated Circuit

Background

An integrated circuit, commonly referred to as an IC, is a microscopic array of electronic circuits and components that has been diffused or implanted onto the surface of a single crystal, or chip, of semiconducting material such as silicon. It is called an integrated circuit because the components, circuits, and base material are all made together, or integrated, out of a single piece of silicon, as opposed to a discrete circuit in which the components are made separately from different materials and assembled later. ICs range in complexity from simple logic modules and amplifiers to complete microcomputers containing millions of elements.

The impact of integrated circuits on our lives has been enormous. ICs have become the principal components of almost all electronic devices. These miniature circuits have demonstrated low cost, high reliability, low power requirements, and high processing speeds compared to the vacuum tubes and transistors which preceded them. Integrated circuit microcomputers are now used as controllers in equipment such as machine tools, vehicle operating systems, and other applications where hydraulic, pneumatic, or mechanical controls were previously used. Because IC microcomputers are smaller and more versatile than previous control mechanisms, they allow the equipment to respond to a wider range of input and produce a wider range of output. They can also be reprogrammed without having to redesign the control circuitry. Integrated circuit microcomputers are so inexpensive they are even found in children's electronic toys.

The first integrated circuits were created in the late 1950s in response to a demand from the military for miniaturized electronics to be used in missile control systems. At the time, transistors and **printed circuit board**s were the state-of-the-art electronic technology. Although transistors made many new electronic applications possible, engineers were still unable to make a small enough package for the large number of components and circuits required in complex devices like sophisticated control systems and handheld programmable calculators. Several companies were in competition to produce a breakthrough in miniaturized electronics, and their development efforts were so close that there is some question as to which company actually produced the first IC. In fact, when the integrated circuit was finally patented in 1959, the patent was awarded jointly to two individuals working separately at two different companies.

After the invention of the IC in 1959, the number of components and circuits that could be incorporated into a single chip doubled every year for several years. The first integrated circuits contained only up to a dozen components. The process that produced these early ICs was known as small scale integration, or SSI. By the mid-1960s, medium scale integration, MSI, produced ICs with hundreds of components. This was followed by large scale integration techniques, or LSI, which produced ICs with thousands of components and made the first microcomputers possible.

The first microcomputer chip, often called a microprocessor, was developed by Intel Corporation in 1969. It went into commercial production in 1971 as the Intel 4004.

The impact of integrated circuits on our lives has been enormous. ICs have become the principal components of almost all electronic devices. These miniature circuits have demonstrated low cost, high reliability, low power requirements, and high processing speeds compared to the vacuum tubes and transistors which preceded them.

Intel introduced their 8088 chip in 1979, followed by the Intel 80286, 80386, and 80486. In the late 1980s and early 1990s, the designations 286, 386, and 486 were well known to computer users as reflecting increasing levels of computing power and speed. Intel's Pentium chip is the latest in this series and reflects an even higher level.

How Integrated Circuit Components Are Formed

In an integrated circuit, electronic components such as resistors, capacitors, diodes, and transistors are formed directly onto the surface of a silicon crystal. The process of manufacturing an integrated circuit will make more sense if one first understands some of the basics of how these components are formed.

Even before the first IC was developed, it was known that common electronic components could be made from silicon. The question was how to make them, and the connecting circuits, from the same piece of silicon? The solution was to alter, or dope, the chemical composition of tiny areas on the silicon crystal surface by adding other chemicals, called dopants. Some dopants bond with the silicon to produce regions where the dopant atoms have one electron they can give up. These are called N regions. Other dopants bond with the silicon to produce regions where the dopant atoms have room to take one electron. These are called P regions. When a P region touches an N region, the boundary between them is referred to as a PN junction. This boundary is only 0.000004 inches (0.0001 cm) wide, but is crucial to the operation of integrated circuit components.

Within a PN junction, the atoms of the two regions bond in such a manner as to create a third region, called a depletion region, in which the P dopant atoms capture all the N dopant extra electrons, thus depleting them. One of the phenomena that results is that a positive voltage applied to the P region can cause an electrical current to flow through the junction into the N region, but a similar positive voltage applied to the N region will result in little or no current flowing through the junction back into the P region. This ability of a PN junction to either conduct or insulate depending on which side the voltage is applied can be used to form integrated circuit components that direct and control current flows in the same manner as diodes and transistors. A diode, for example, is simply a single PN junction. By altering the amount and types of dopants and changing the shapes and relative placements of P and N regions, integrated circuit components that emulate the functions of resistors and capacitors can be also be formed.

Design

Some integrated circuits can be considered standard, off-the-shelf items. Once designed, there is no further design work required. Examples of standard ICs would include voltage regulators, amplifiers, analog switches, and analog-to-digital or digital-to-analog converters. These ICs are usually sold to other companies who incorporate them into printed circuit boards for various electronic products.

Other integrated circuits are unique and require extensive design work. An example would be a new microprocessor for computers. This design work may require research and development of new materials and new manufacturing techniques to achieve the final design.

Raw Materials

Pure silicon is the basis for most integrated circuits. It provides the base, or substrate for the entire chip and is chemically doped to provide the N and P regions that make up the integrated circuit components. The silicon must be so pure that only one out of every ten billion atoms can be an impurity. This would be the equivalent of one grain of sugar in ten buckets of sand. Silicon dioxide is used as an insulator and as a dielectric material in IC capacitors.

Typical N-type dopants include phosphorus and arsenic. Boron and gallium are typical P-type dopants. Aluminum is commonly used as a connector between the various IC components. The thin wire leads from the integrated circuit chip to its mounting package may be aluminum or gold. The mounting package itself may be made from ceramic or plastic materials.

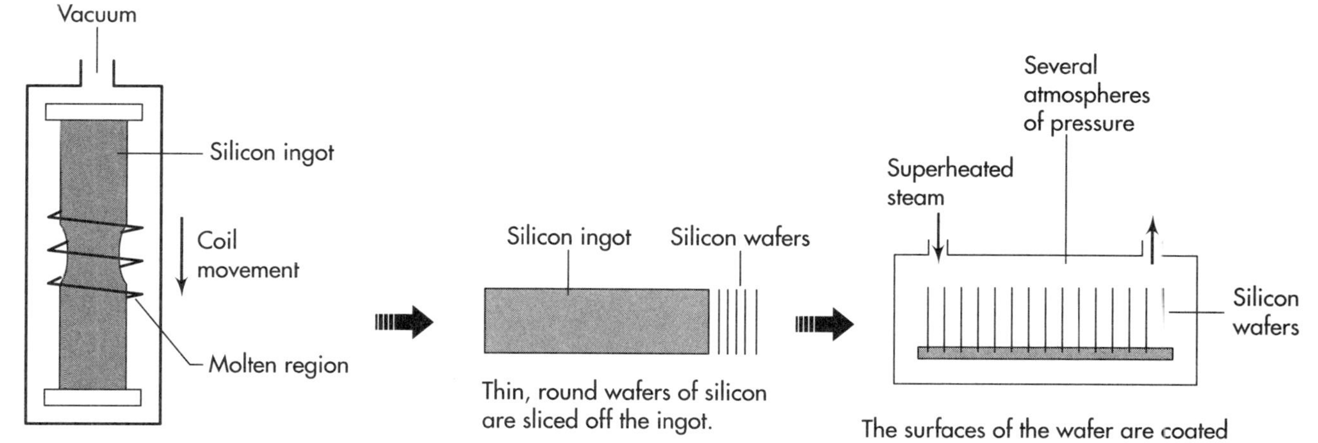

A cylindrical ingot of silicon is held vertically in a vacuum chamber. A high temperature heating coil moves slowly down the length of the ingot.

Thin, round wafers of silicon are sliced off the ingot.

The surfaces of the wafer are coated with a layer of silicon dioxide to form an insulating base. The silicon dioxide is formed by subjecting the wafer to superheated steam under pressure.

The Manufacturing Process

Hundreds of integrated circuits are made at the same time on a single, thin slice of silicon and are then cut apart into individual IC chips. The manufacturing process takes place in a tightly controlled environment known as a clean room where the air is filtered to remove foreign particles. The few equipment operators in the room wear lint-free garments, gloves, and coverings for their heads and feet. Since some IC components are sensitive to certain frequencies of light, even the light sources are filtered. Although manufacturing processes may vary depending on the integrated circuit being made, the following process is typical.

Preparing the silicon wafer

1 A cylindrical ingot of silicon about 1.5 to 4.0 inches (3.8 to 10.2 cm) in diameter is held vertically inside a vacuum chamber with a high-temperature heating coil encircling it. Starting at the top of the cylinder, the silicon is heated to its melting point of about 2550°F (1400°C). To avoid contamination, the heated region is contained only by the surface tension of the molten silicon. As the region melts, any impurities in the silicon become mobile. The heating coil is slowly moved down the length of the cylinder, and the impurities are carried along with the melted region. When the heating coil reaches the bottom, almost all of the impurities have been swept along and are concentrated there. The bottom is then sliced off, leaving a cylindrical ingot of purified silicon.

2 A thin, round wafer of silicon is cut off the ingot using a precise cutting machine called a wafer slicer. Each slice is about 0.01 to 0.025 inches (0.004 to 0.01 cm) thick. The surface on which the integrated circuits are to be formed is polished.

3 The surfaces of the wafer are coated with a layer of silicon dioxide to form an insulating base and to prevent any oxidation of the silicon which would cause impurities. The silicon dioxide is formed by subjecting the wafer to superheated steam at about 1830°F (1000°C) under several atmospheres of pressure to allow the oxygen in the water vapor to react with the silicon. Controlling the temperature and length of exposure controls the thickness of the silicon dioxide layer.

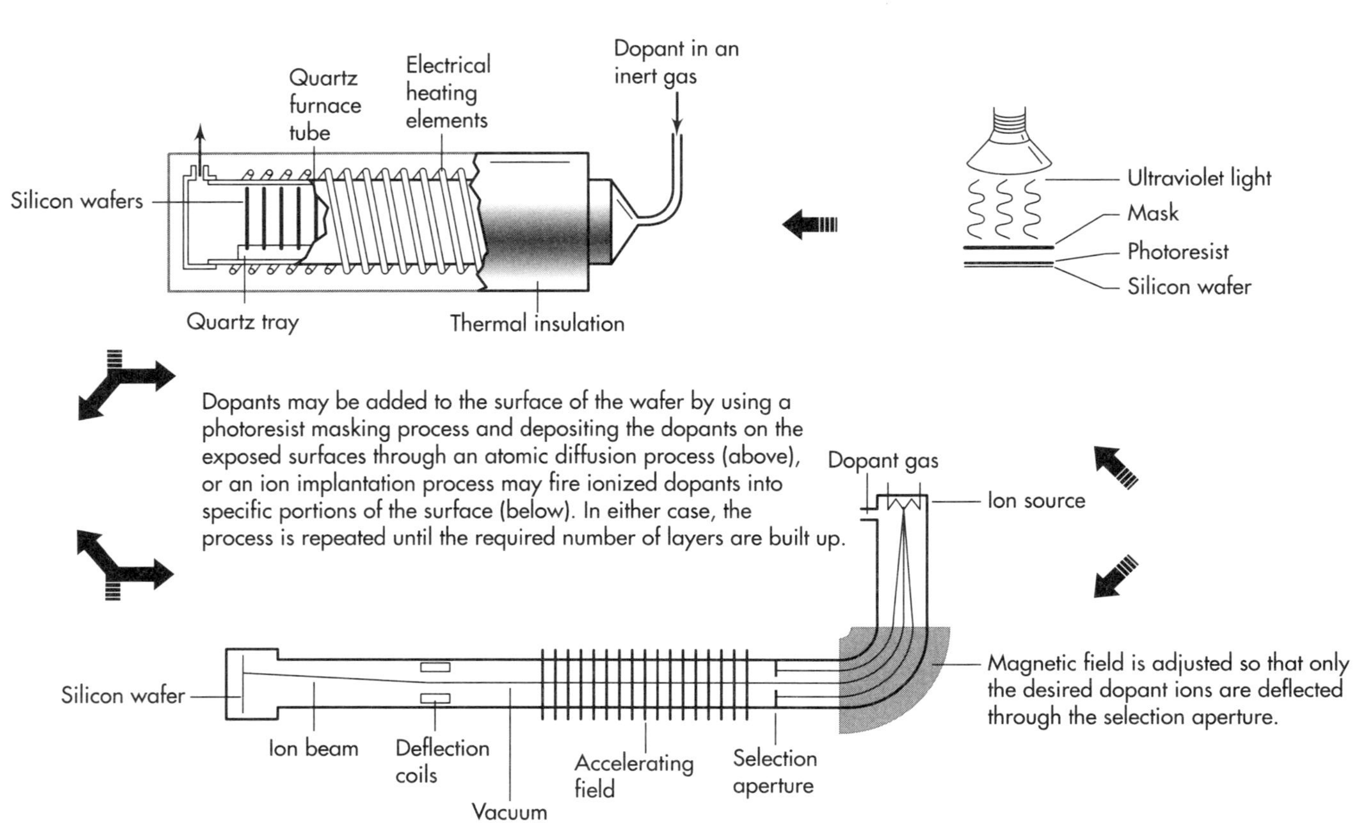

Dopants may be added to the surface of the wafer by using a photoresist masking process and depositing the dopants on the exposed surfaces through an atomic diffusion process (above), or an ion implantation process may fire ionized dopants into specific portions of the surface (below). In either case, the process is repeated until the required number of layers are built up.

Masking

4 The complex and interconnected design of the circuits and components is prepared in a process similar to that used to make printed circuit boards. For ICs, however, the dimensions are much smaller and there are many layers superimposed on top of each other. The design of each layer is prepared on a computer-aided drafting machine, and the image is made into a mask which will be optically reduced and transferred to the surface of the wafer. The mask is opaque in certain areas and clear in others. It has the images for all of the several hundred integrated circuits to be formed on the wafer.

5 A drop of photoresist material is placed in the center of the silicon wafer, and the wafer is spun rapidly to distribute the photoresist over the entire surface. The photoresist is then baked to remove the solvent.

6 The coated wafer is then placed under the first layer mask and irradiated with light. Because the spaces between circuits and components are so small, ultraviolet light with a very short wavelength is used to squeeze through the tiny clear areas on the mask. Beams of electrons or x-rays are also sometimes used to irradiate the photoresist.

7 The mask is removed and portions of the photoresist are dissolved. If a positive photoresist was used, then the areas that were irradiated will be dissolved. If a negative photoresist was used, then the areas that were irradiated will remain. The uncovered areas are then either chemically etched to open up a layer or are subjected to chemical doping to create a layer of P or N regions.

Doping—Atomic diffusion

8 One method of adding dopants to create a layer of P or N regions is atomic diffusion. In this method a batch of wafers is placed in an oven made of a quartz tube surrounded by a heating element. The wafers are heated to an operating temperature of

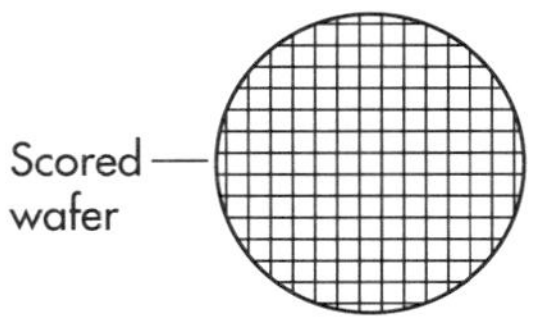

When the integrated circuits on the wafer are complete, the surface of the wafer is scored with a fine diamond cutter, and the individual ICs are separated.

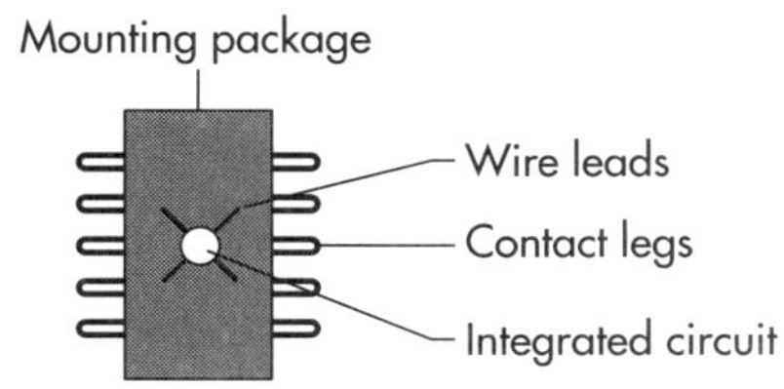

Each individual integrated circuit is bonded to a mounting package, and thin wire leads are connected. The mounting package has short legs that connect to a printed circuit board.

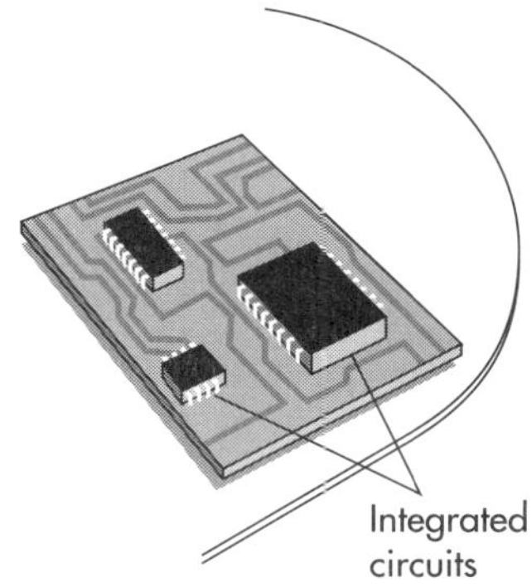

about 1500-2200°F (816-1205°C), and the dopant chemical is carried in on an inert gas. As the dopant and gas pass over the wafers, the dopant is deposited on the hot surfaces left exposed by the masking process. This method is good for doping relatively large areas, but is not accurate for smaller areas. There are also some problems with the repeated use of high temperatures as successive layers are added.

Doping—Ion implantation

9 The second method to add dopants is ion implantation. In this method a dopant gas, like phosphine or boron trichloride, is ionized to provide a beam of high-energy dopant ions which are fired at specific regions of the wafer. The ions penetrate the wafer and remain implanted. The depth of penetration can be controlled by altering the beam energy, and the amount of dopant can be controlled by altering the beam current and time of exposure. Schematically, the whole process resembles firing a beam in a bent **cathode-ray tube**. This method is so precise, it does not require masking—it just points and shoots the dopant where it is needed. However it is much slower than the atomic diffusion process.

Making successive layers

10 The process of masking and etching or doping is repeated for each successive layer depending on the doping process used until all of the integrated circuit chips are complete. Sometimes a layer of silicon dioxide is laid down to provide an insulator between layers or components. This is done through a process known as chemical vapor deposition, in which the wafer's surface is heated to about 752°F (400°C), and a reaction between the gases silane and oxygen deposits a layer of silicon dioxide. A final silicon dioxide layer seals the surface, a final etching opens up contact points, and a layer of aluminum is deposited to make the contact pads. At this point, the individual ICs are tested for electrical function.

Making individual ICs

11 The thin wafer is like a piece of glass. The hundreds of individual chips are separated by scoring a crosshatch of lines with a fine **diamond** cutter and then putting the wafer under stress to cause each chip to separate. Those ICs that failed the electrical test are discarded. Inspection under a microscope reveals other ICs that were damaged by the separation process, and these are also discarded.

12 The good ICs are individually bonded into their mounting package and the thin wire leads are connected by either ultrasonic bonding or thermocompression. The mounting package is marked with identifying part numbers and other information.

13 The completed integrated circuits are sealed in anti-static plastic bags to be stored or shipped to the end user.

Quality Control

Despite the controlled environment and use of precision tools, a high number of integrated circuit chips are rejected. Although the percentage of reject chips has steadily dropped over the years, the task of making an interwoven lattice of microscopic circuits and components is still difficult, and a certain amount of rejects are inevitable.

Hazardous Materials and Recycling

The dopants gallium and arsenic, among others, are toxic substances and their storage, use, and disposal must be tightly controlled.

Because integrated circuit chips are so versatile, a significant recycling industry has sprung up. Many ICs and other electronic components are removed from otherwise obsolete equipment, tested, and resold for use in other devices.

The Future

It is difficult to tell with any certainty what the future holds for the integrated circuit. Changes in technology since the device's invention have been rapid, but evolutionary. Many changes have been made in the architecture, or circuit layout, on a chip, but the integrated circuit still remains a silicon-based design.

The next major leap in the advancement of electronic devices, if such a leap is to come, may involve an entirely new circuit technology. Better devices than the very best microprocessor have always been known to be possible. The human brain, for example, processes information much more efficiently than any computer, and some futurists have speculated that the next generation of processor circuits will be biological, rather than mineral. At this point, such matters are the stuff of fiction. There are no immediate signs that the integrated circuit is in any danger of extinction.

Where To Learn More

Books

Braithwaite, Nicholas and Graham Weaver, Eds. *Electronic Materials.* Butterworths, 1990.

Gibilisco, Stan. *International Encyclopedia of Integrated Circuits*, 2nd ed. Tab Books, a Division of McGraw-Hill, 1992.

Spinks, Brian. *Introduction to Integrated-Circuit Layout.* Prentice-Hall, 1985.

Periodicals

Faggin, Federico. "The Birth of the Microprocessor." *Byte,* March 1992, pp. 145-50.

Rosch, Winn L. "The Evolution of the PC Microprocessor." *PC Magazine,* January 31, 1989, pp. 96-97.

Smith, Gina. "Field Guide to CPUs." *PC/Computing*, March 1993, pp. 123-30.

—Joel Simon/Chris Cavette

Iron

Background

Iron is one of the most common elements on earth. Nearly every construction of man contains at least a little iron. It is also one of the oldest metals and was first fashioned into useful and ornamental objects at least 3,500 years ago.

Pure iron is a soft, grayish-white metal. Although iron is a common element, pure iron is almost never found in nature. The only pure iron known to exist naturally comes from fallen meteorites. Most iron is found in minerals formed by the combination of iron with other elements. Iron oxides are the most common. Those minerals near the surface of the earth that have the highest iron content are known as iron ores and are mined commercially.

Iron ore is converted into various types of iron through several processes. The most common process is the use of a blast furnace to produce pig iron which is about 92-94% iron and 3-5% carbon with smaller amounts of other elements. Pig iron has only limited uses, and most of this iron goes on to a steel mill where it is converted into various steel alloys by further reducing the carbon content and adding other elements such as manganese and nickel to give the steel specific properties.

History

Historians believe that the Egyptians were the first people to work with small amounts of iron, some five or six thousand years ago. The metal they used was apparently extracted from meteorites. Evidence of what is believed to be the first example of iron mining and smelting points to the ancient Hittite culture in what is now Turkey. Because iron was a far superior material for the manufacture of weapons and tools than any other known metal, its production was a closely guarded secret. However, the basic technique was simple, and the use of iron gradually spread. As useful as it was compared to other materials, iron had disadvantages. The quality of the tools made from it was highly variable, depending on the region from which the iron ore was taken and the method used to extract the iron. The chemical nature of the changes taking place during the extraction were not understood; in particular, the importance of carbon to the metal's hardness. Practices varied widely in different parts of the world. There is evidence, for example, that the Chinese were able to melt and cast iron implements very early, and that the Japanese produced amazing results with steel in small amounts, as evidenced by heirloom swords dating back centuries. Similar breakthroughs were made in the Middle East and India, but the processes never emerged into the rest of the world. For centuries the Europeans lacked methods for heating iron to the melting point at all. To produce iron, they slowly burned iron ore with wood in a clay-lined oven. The iron separated from the surrounding rock but never quite melted. Instead, it formed a crusty slag which was removed by hammering. This repeated heating and hammering process mixed oxygen with the iron oxide to produce iron, and removed the carbon from the metal. The result was nearly pure iron, easily shaped with hammers and tongs but too soft to take and keep a good edge. Because the metal was shaped, or wrought, by hammering, it came to be called wrought iron.

Historians believe that the Egyptians were the first people to work with small amounts of iron, apparently extracted from meteorites, some five or six thousand years ago.

Leaning on his long tongs, this young iron puddler's helper posed for this photograph in the early 1860s, when the Sons of Vulcan was a young union. (From the collections of Henry Ford Museum & Greenfield Village.)

Historically, iron was produced by the hot-blast method, or later, the anthracite furnace. Either way, the fundamental activity in iron making involved a worker stirring small batches of pig iron and cinder until the iron separated from the slag. Called "puddling," this was highly skilled work, but was also hot, strenuous, and dangerous. It required a lot of experience as well as a hearty constitution. Puddlers were proud, independent, and highly paid.

Puddlers founded the first trade union in the iron and steel industry, the Sons of Vulcan, in Pittsburgh in 1858. In 1876, this union merged with three other labor organizations to form the Amalgamated Association of Iron and Steel Workers. This was the union that Andrew Carnegie defeated in the Homestead Strike of 1892, leaving the union in shambles and the industry essentially unorganized until the 1930s.

William S. Pretzer

Tools and weapons brought back to Europe from the East were made of an iron that had been melted and cast into shape. Retaining more carbon, cast iron is harder than wrought iron and will hold a cutting edge. However, it is also more brittle than wrought iron. The European iron workers knew the Easterners had better iron, but not the processes involved in fashioning stronger iron products. Entire nations launched efforts to discover the process.

The first known European breakthrough in the production of cast iron, which led quickly to the first practical steel, did not come until 1740. In that year, Benjamin Huntsman took out a patent for the melting of material for the production of steel springs to be used in clockmaking. Over the next 20 years or so, the procedure became more widely adopted. Huntsman used a blast furnace to melt wrought iron in a clay crucible. He then added carefully measured amounts of pure charcoal to the melted metal. The resulting alloy was both strong and flexible when cast into springs. Since Huntsman was originally interested only in making better clocks, his crucible steel led directly to the development of nautical chronometers, which, in turn, made global navigation possible by allowing mariners to precisely determine their east/west position. The fact that he had also invented modern metallurgy was a side-effect which he apparently failed to notice.

Raw Materials

The raw materials used to produce pig iron in a blast furnace are iron ore, coke, sinter, and limestone. Iron ores are mainly iron oxides and include magnetite, hematite, limonite, and many other rocks. The iron content of these ores ranges from 70% down to 20% or less. Coke is a substance made by heating coal until it becomes almost pure carbon. Sinter is made of lesser grade, finely divided iron ore which is roasted with coke and lime to remove a large amount of the impurities in the ore. Limestone occurs naturally and is a source of calcium carbonate.

Other metals are sometimes mixed with iron in the production of various forms of steel, such as chromium, nickel, manganese, molybdenum, and tungsten.

The Ore Extraction and Refining Process

Before iron ore can be used in a blast furnace, it must be extracted from the ground and partially refined to remove most of the impurities.

Extraction

1 Much of the world's iron ore is extracted through open pit mining in which the

Iron ore is generally extracted through open pit mining. The ground is removed from a very large area to expose the ore beneath.

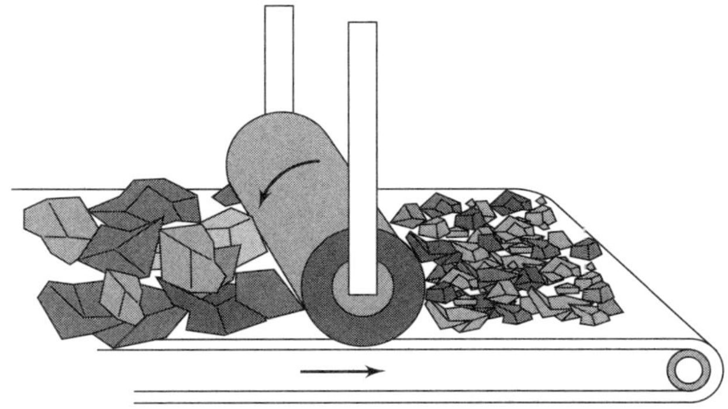

The mined ore is crushed and sorted by grade according to the percentage of iron.

The refined ore is taken from the mining site to the blast furnace where the iron is melted.

surface of the ground is removed by heavy machines, often over a very large area, to expose the ore beneath. In cases where it is not economical to remove the surface, shafts are dug into the earth, with side tunnels to follow the layer of ore.

Refining

2 The mined ore is crushed and sorted. The best grades of ore contain over 60% iron. Lesser grades are treated, or refined, to remove various contaminants before the ore is shipped to the blast furnace. Collectively, these refining methods are called beneficiation and include further crushing, washing with water to float sand and clay away, magnetic separation, pelletizing, and sintering. As more of the world's known supply of high iron content ore is depleted, these refining techniques have become increasingly important.

3 The refined ore is then loaded on trains or ships and transported to the blast furnace site.

The Manufacturing Process

Charging the blast furnace

1 After processing, the ore is blended with other ore and goes to the blast furnace. A blast furnace is a tower-shaped structure, made of steel, and lined with refractory, or heat-resistant bricks. The mixture of raw material, or charge, enters at the top of the blast furnace. At the bottom of the furnace, very hot air is blown, or blasted, in through nozzles called *tuyéres*. The coke burns in the presence of the hot air. The oxygen in the air reacts with the carbon in the coke to form carbon monoxide. The carbon mon-

Pure iron is a scft, grayish-white metal. Although iron is a common element, pure iron is almost never found in nature. Minerals near the surface of the earth that have the highest iron content are known as iron ores and are mined commercially.

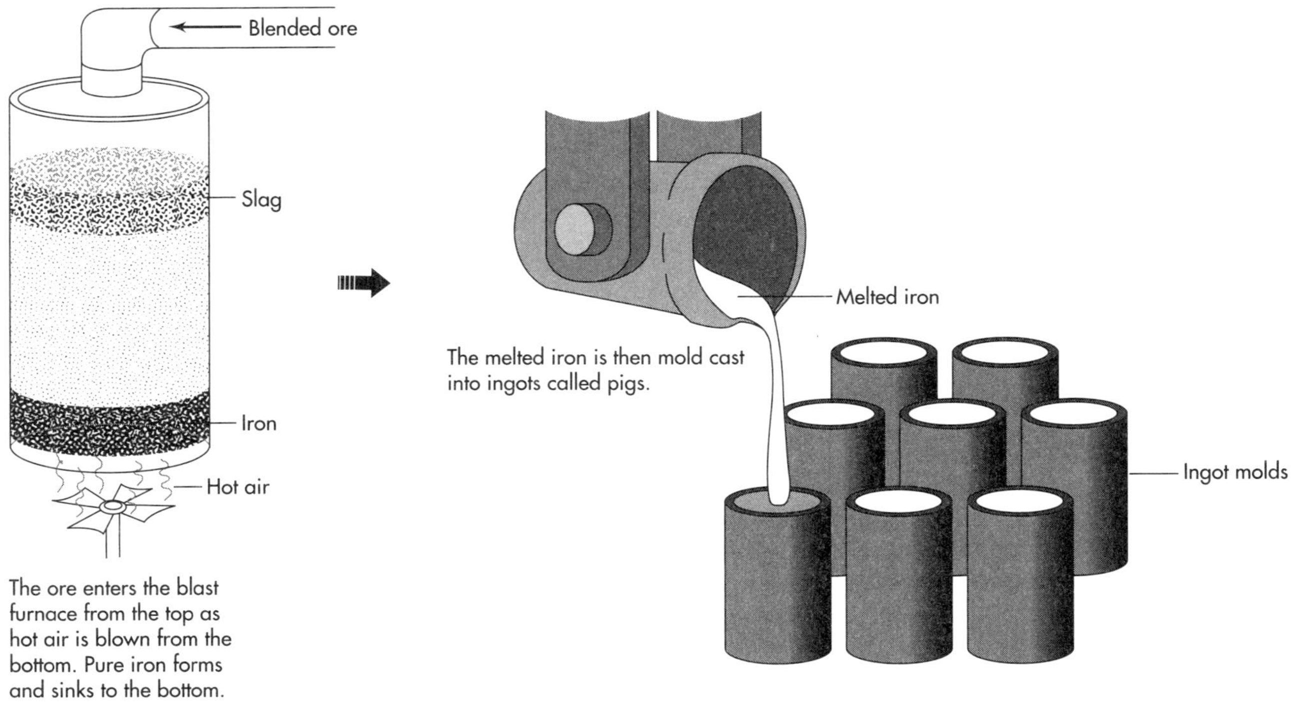

The ore enters the blast furnace from the top as hot air is blown from the bottom. Pure iron forms and sinks to the bottom.

The melted iron is then mold cast into ingots called pigs.

oxide then reacts with the iron ore to form carbon dioxide and pure iron.

Separating the iron from the slag

2 The melted iron sinks to the bottom of the furnace. The limestone combines with the rock and other impurities in the ore to form a slag which is lighter than the iron and floats on top. As the volume of the charge is reduced, more is continually added at the top of the furnace. The iron and slag are drawn off separately from the bottom of the furnace. The melted iron might go to a further alloying process, or might be cast into ingots called pigs. The slag is carried away for disposal.

Treating the gases

3 The hot gases produced in the chemical reactions are drawn off at the top and routed to a gas cleaning plant where they are cleaned, or scrubbed, and sent back into the furnace; the remaining carbon monoxide, in particular, is useful to the chemical reactions going on within the furnace.

A blast furnace normally runs day and night for several years. Eventually the brick lining begins to crumble, and the furnace is then shut down for maintenance.

Quality Control

The blast furnace operation is highly instrumented and is monitored continuously. Times and temperatures are checked and recorded. The chemical content of the iron ores received from the various mines are checked, and the ore is blended with other iron ore to achieve the desired charge. Samples are taken from each pour and checked for chemical content and mechanical properties such as strength and hardness.

Byproducts/Waste

There are a great many possible environmental effects from the iron industry. The first and most obvious is the process of open pit mining. Huge tracts of land are stripped to bare rock. Today, depleted mining sites are commonly used as landfills, then covered over and landscaped. Some of these landfills themselves become environmental problems, since in the recent past, some were used for the disposal of highly toxic substances which leached into soil and water.

The process of extracting iron from ore produces great quantities of poisonous and corrosive gases. In practice, these gases are scrubbed and recycled. Inevitably, however, some small amounts of toxic gases escape to the atmosphere.

A byproduct of iron purification is slag, which is produced in huge amounts. This material is largely inert, but must still be disposed of in landfills.

Ironmaking uses up huge amounts of coal. The coal is not used directly, but is first reduced to coke which consists of almost pure carbon. The many chemical byproducts of coking are almost all toxic, but they are also commercially useful. These products include ammonia, which is used in a vast number of products; phenol, which is used to make plastics, cutting oils, and antiseptics; cresols, which go into herbicides, pesticides, pharmaceuticals, and photographic chemicals; and toluene, which is an ingredient in many complex chemical products such as solvents and explosives.

Scrap iron and steel—in the form of old cars, appliances and even entire steel-girdered buildings—are also an environmental concern. Most of this material is recycled, however, since steel scrap is an essential resource in steelmaking. Scrap which isn't recycled eventually turns into iron oxide, or rust, and returns to the ground.

The Future

On the surface, the future of iron production—especially in the United States—appears troubled. Reserves of high-quality ore have become considerably depleted in areas where it can be economically extracted. Many long-time steel mills have closed.

However, these appearances are deceiving. New ore-enrichment techniques have made the use of lower-grade ore much more attractive, and there is a vast supply of that ore. Many steel plants have closed in recent decades, but this is largely because fewer are needed. The efficiency of blast furnaces alone has improved remarkably. At the beginning of this century, the largest blast furnace in the United States produced 644 tons of pig iron a day. It is believed that soon the possible production of a single furnace will reach 4,000 tons per day. Since many of these more modern plants have been built overseas, it has actually become more economical in some cases to ship steel across the ocean than to produce it in older U.S. plants.

Where To Learn More

Books

Lambert, Mark. *Spotlight on Iron and Steel*. Rourke Enterprises, 1988.

Hartley, Edward N. *Iron and Steel Works of the World*. International Publication, 1987.

Lewis, W. David. *Iron and Steel in America*. Hagley Museum, 1986.

Walker, R. D. *Modern Ironmaking Methods*. Gower Publication, 1986.

—Joel Simon

Jelly Bean

Jelly beans appear to have been introduced around 1900 with other shaped candies. As former President Ronald Reagan's favorite candy, the jelly bean had something of a resurgence in the 1980s, and many "designer" or "gourmet" flavors were introduced.

Background

The jelly bean is a semi-soft candy, shaped like a bean and generally fruit flavored. Long considered a traditional Easter candy, jelly beans are also produced in seasonal colors for other holidays such as Halloween and Independence Day. Basic jelly beans (sometimes also called "pectin beans" because their gel-like centers are flavored with fruit pectin) come in nine colors—red, black, white, green, yellow, brown, orange, pink, and purple. Typically, the bean has the same flavor and color in both the candy center and the sugar shell.

As former President Ronald Reagan's favorite candy, the jelly bean experienced something of a resurgence in the 1980s, and many "designer" or "gourmet" flavors were introduced. These newer incarnations include more exotic fruit flavors like blueberry, pear, cantaloupe, peach, and watermelon; beverage-based flavors such as root beer, champagne, mai tai, and daiquiri; and dessert or other sweet flavors such as bubble gum, marshmallow, mint, cheesecake, and cinnamon. The names of the flavors vary with the manufacturer, and the processing may be varied as well so that the particular jelly bean flavor resembles its "real world" counterpart. For example, the watermelon-flavored bean has a red candy center and a green hard shell like a real watermelon, and a mixed fruit or "tutti-frutti" bean may have a pink center and a speckled exterior to suggest its mix of flavors.

The exact origins of the jelly bean are not known, but it seems to have appeared around 1900 with other shaped candies. The jelly bean has a longer shelf life than many other confections, and its size and durability make it portable. Like other small treats, it was sold as "penny candy" through the first half of the century, including during the Depression. By segregating beans by color, retailers were able to sell jelly beans for particular holidays. In 1976, the gourmet jelly bean was invented by the Herman Goelitz Candy Co., Inc., and the candy assumed a new life as a delicacy. Jelly beans were a fixture of the Reagan White House, and they have flown on the space shuttle as well. New flavors are developed in keeping with taste trends, so the future of the humble bean in both traditional and new guises seems assured.

Raw Materials

The basic ingredients of jelly beans include sugar, corn syrup, and food starch. Relatively minor amounts of lecithin (an emulsifier), anti-foaming agents, beeswax or carnauba wax, salt, and confectioner's glaze are also added. The ingredients that give each bean its character are also relatively small in proportion and may vary depending on the flavor. These include natural and artificial flavors and colors, and, depending on the bean flavor, may include chocolate, coconut, fruit as puree or juice, peanuts, vanilla, oils, cream, or freeze-dried egg, milk, or fruit powders.

Design

The "design" of the jelly bean was time-honored until the mid-1970s when the gourmet or designer jelly bean was developed. Although the shape remained fairly standard, gourmet-type beans are typically smaller and softer than traditional jelly beans. The colors and flavors also are more

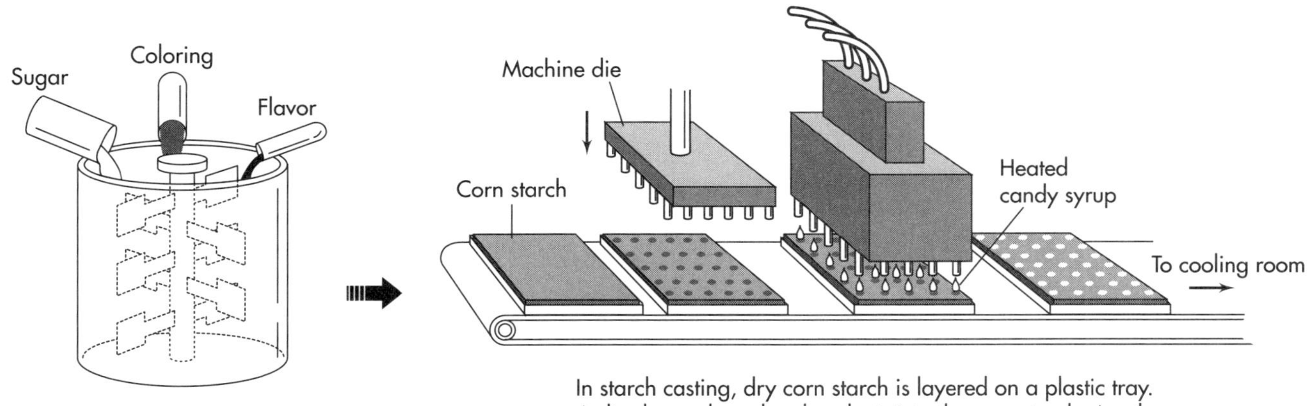

To create the chewy jelly bean center, ingredients are mixed together to a syrupy consistency.

In starch casting, dry corn starch is layered on a plastic tray. A die descends and makes dents into the corn starch. Another machine drops a squirt of candy syrup. The trays are then transferred to a cooling room.

varied, and flavors that decrease in popularity are phased out, while new ones are added in keeping with other candies popular with children and other food fads and trends. Intentional in its design or not, the smaller jelly bean is touted as a low-calorie treat because jelly beans contain little or no fat, and there are about 150 calories in 2 tablespoons of small jelly beans.

Also, some manufacturers make a slightly larger jelly bean for holidays like Easter, Halloween, and Christmas. Forming jelly beans and many other candies does require design and development of the molds used in casting the shapes.

The Manufacturing Process

Cooking and chemistry

1 Each manufacturer's jelly bean and each new flavor/color combination begin in a chemistry laboratory, where the right balance of ingredients is mixed and developed in test batches. New designer flavors are suggested based on marketing studies, and the flavors are tested for taste and visual appeal in the laboratory. Subsequently, a new flavor will be manufactured over a trial time period and test marketed. If the flavor proves popular, it will become a new product. The chemists also develop new flavors with an eye toward the consumer's interest in natural products, and they evaluate its "nutrition facts," including ingredients and caloric content.

2 After the chemists have fine-tuned their recipe, the candy kitchen creates the syrup that forms the jelly bean center by dissolving the sugar and other ingredients in large boilers where the syrup is cooked to the proper temperature and consistency. Flavor and color for the bean center are added to the syrup, which is then piped to the starch casting area. The kitchens also mix and cook the flavor and color syrups for the panning process described below. These syrups are carefully transported to the panning room and are added in measured doses during panning.

Starch casting

3 Formation or shaping of a single jelly bean begins with a process called starch casting. Dry corn starch is a fine, white powder that retains impressions or shapes well. A machine called a mogul deposits a layer of corn starch in a plastic tray and moves the tray to a machine die, which presses dents into the corn starch. Each tray may contain several hundred to over 1,200 of these impressions or dents, each of which is the size and shape of the center of a jelly bean. The mogul moves the trays to a depositor or "filling station" where heated candy syrup is squirted into the tiny molds. From the mogul, conveyors carefully move the trays to cooling rooms in which temperature and humidity are controlled and

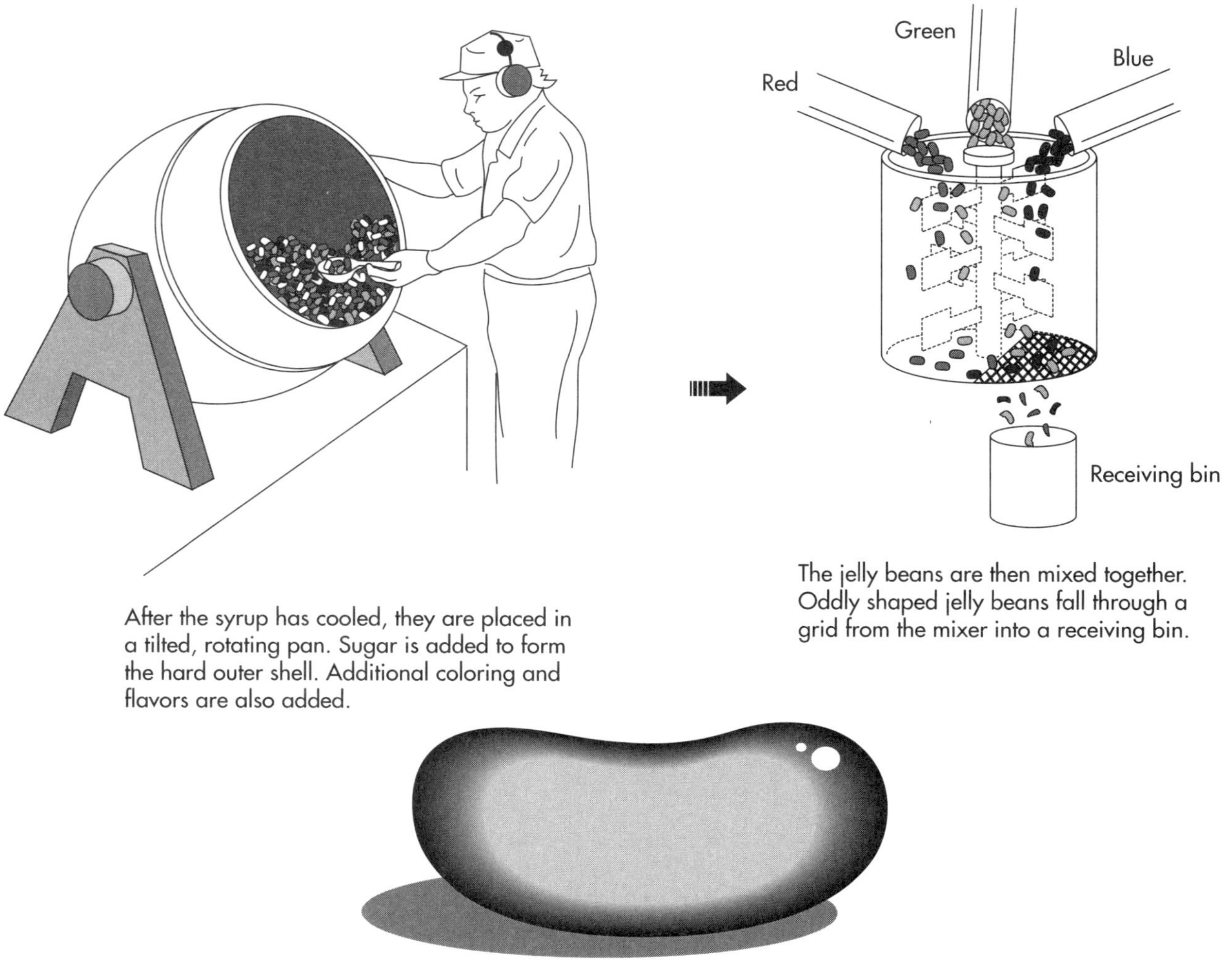

After the syrup has cooled, they are placed in a tilted, rotating pan. Sugar is added to form the hard outer shell. Additional coloring and flavors are also added.

The jelly beans are then mixed together. Oddly shaped jelly beans fall through a grid from the mixer into a receiving bin.

where the liquid candy cools and sets up to form the gummy center of the jelly bean.

The panning process

4 The panning process gives the jelly beans their outer color and flavor, protective sugar shells, and shiny glaze. The trays of candy centers are dumped out. The corn starch absorbs moisture from them during the cooling process; but it is removed, dried, reprocessed, and recycled to create molds for more candies. The centers, which are all the same flavor and color, are placed in stainless steel vessels called "pans" that are globe-shaped and hollow with an opening at one "pole" of the globe. Just like globes, the pans are tilted on their axes so the candies can be placed in them easily and so workers can add other ingredients through the openings. At the bottom "pole" or axis end, the vessel is linked to a rotating power source. The pans rotate the jelly bean centers several hundred times per minute.

5 Sugar is added through the opening, which gradually builds up on the soft center to form a harder, sugar shell. Workers add colors and flavors during the panning process by pouring beakers of syrup supplied by the candy kitchen through the opening in the vessel. They can also observe the jelly bean shells as they form and become colored throughout the process. The beans are essentially finished at this point but are rather dull-looking. To give them their glossy coats, a glaze of confectioner's sugar is added while the beans are still revolving in the pans.

Packaging

6 The process of making the jelly bean takes 6 to 10 days, depending on the kind of bean and the manufacturer. Packaging is the final step before sending the jelly beans to distributors. Jelly beans are placed in trays after panning and are still segregated by color or flavor. The trays of candies are taken to a large bin where they are dumped in and mixed to the desired combination of colors and flavors. The mixing bin is a large, rotating cylinder. On one side, a grid is set in the wall of the bin. Beans that are too small fall through the openings in the grid and into a receiving bin, while beans that are too large stick in the mesh and are removed later. The beans that continue rotating are therefore only the desired size and shape. They fall from the mixing bin onto a conveyor, where workers inspect them and remove any candies that look imperfect. The beans that pass inspection move on the conveyor to a packaging machine, where the candy is weighed and bagged in any of several sizes of bags either for bulk sale or purchase by individual consumers. The packaging machine can package and seal about 80,000 bags of jelly beans a day.

7 Exceptions to the sorting and mixing process occur when jelly beans (usually the gourmet type) are packaged by single flavor, or when the flavors are separated in small compartments in gift or "sampler" boxes that let the taster experience the unique flavors of designer beans. The candies are still sized and inspected, but individual flavors are then placed in funnel-like bins. The small openings fit the compartments in plastic trays in the gift boxes, and a controlled quantity of each flavor is dropped into its specific tray compartment.

Although the candies are thoroughly mixed to try to get an equal distribution of colors, the randomness of conveying and sorting may cause some variations in the mix. The consumer who purchases the larger bag has a better chance, statistically, of getting a near-equal distribution of colors and flavors. Slight variations in size and shape account for one bag of jelly beans containing more beans than the next, even though the contents are weighed. Some manufacturers put more than the stated weight in each package, so the customer may actually get more beans than paid for in each bag.

Quality Control

Jelly beans, like any food product, must meet many regulatory requirements for safety and quality. All ingredients are supplied by vendors and inspected for correct quantities, quality, integrity of packaging, and other criteria. Equipment and materials that contact the food ingredients and product are inspected and cleaned daily or between batches as necessary. Packing materials that contact the jelly beans are formed and handled by machines that are also cleaned daily.

There are a number of product quality assurances among the manufacturing steps, starting with laboratory testing, tasting, observation of color quality, and both machine sorting and inspection to identify and oust imperfect candies.

Factory workers wear special clothing required for food handlers. Because they are working with equipment that generates high heat, has revolving parts, requires electrical supply, and imposes other safety hazards, workers are also protected by a myriad of safety requirements. Some jelly bean factories allow visitors to tour. They are kept at controlled distances from food processing both to protect the visitors and to isolate the candy from possible contamination.

Byproducts/Waste

The jelly bean making process generates very little waste. Sometimes the candy centers are malformed, or the molds collapse, forcing several candies to congeal. These are melted and reused or recycled to salvage the sweeteners. Some manufacturers package and sell imperfectly shaped but edible beans selected during final sorting and inspection.

The Future

New developments are most likely to include changing flavors among gourmet beans as the taste of the consumer follows the latest fashion. Other "revolutions" in jelly beans are less likely, and the future of the jelly bean as an icon among candies seems secure.

Where To Learn More

Brach & Brock Confections. 401 N. Cicero Ave., Chicago, IL 60644. (312) 626-1200.

Goelitz Confectionery. PO Box 1050, 1501 Morrow Ave., North Chicago, IL 60064. (708) 689-2225.

—Gillian S. Holmes

Kayak

Background

Kayaks have become an increasingly popular means of enjoying sporting and leisure boating activities. Kayaks resemble canoes in that both are long, narrow, lightweight paddle boats which are pointed at both ends. Like canoes, kayaks have a hull, which is the hollow bottom shell of the boat. However, several features distinguish kayaks from canoes. Kayaks are water displacement vessels, which means they float just below the surface, forcing the water to be pushed aside as they move. Canoes, on the other hand, float on the water's surface. Because the kayak sits lower in the water, it requires a deck. The deck is the semi-enclosed covering on the kayak's top. The purpose of the deck is to keep waves from washing over the edges of the boat and filling the hull with water. The cockpit is the opening placed in the center of the deck where the kayaker sits. The cockpit is surrounded by a raised lip called the coaming. Other kayak features include interior braces and bulkheads, interior flotation devices, and a cord called the grab loop attached to each end for towing and mooring. Some kayaks include elastic strapping or **bungee cord**s across the deck for carrying lightweight equipment. Other kayaks have waterproof hatches that give access to the forward and aft compartments for additional storage. Some sea kayaks have a rudder controlled by foot pedals to assist steering.

Kayaking is an excellent way to exercise and relax at the same time. A person sits upright in a kayak with legs extended to the front into the hollow hull. A double-bladed **paddle** is used for propulsion. Kayakers pull their crafts through the water by paddling with left and right strokes. A kayaker wears a **life vest** and, sometimes, a protective helmet. Waterproof spray skirts may also be worn. Spray skirts fit around the upper torso of the kayaker and attach to the deck around the cockpit opening to prevent water from splashing into the cockpit when navigating rough water.

The two most common uses of kayaks are racing and recreational boating. Racing kayaks are built for speed and maneuvering on river rapids, while recreational kayaks are built for comfort and longer distance paddling on oceans and lakes. Other design features distinguish racing from recreational kayaks. For example, racing hulls are narrow and v-shaped, whereas recreational hulls are more rounded. The number of seating areas for racing would typically be not more than two, while up to four seats might be provided for recreational purposes. The length, width, and weight would vary as well. Racing models tend to be shorter in length, 11-13 feet (3-4 m) long, and much lighter than recreational models. Some highly specialized racing kayaks weigh less than 25 pounds (11 kg). A typical recreational model will be 13-20 feet (4-6 m) long and may weigh as much as 75 pounds (34 kg). Most designs are no wider than 35 inches (89 cm) across.

History

The design and manufacture of kayaks have gone through many stages over the centuries, from primitive, handmade crafts used for survival to mass-produced sporting boats. Archeological evidence shows that kayaks were used at least 2,000 years ago by Eskimos for transportation, hunting, and

Unlike canoes, which float on the water's surface, kayaks are water displacement vessels that float just below the surface, forcing the water to be pushed aside as they move.

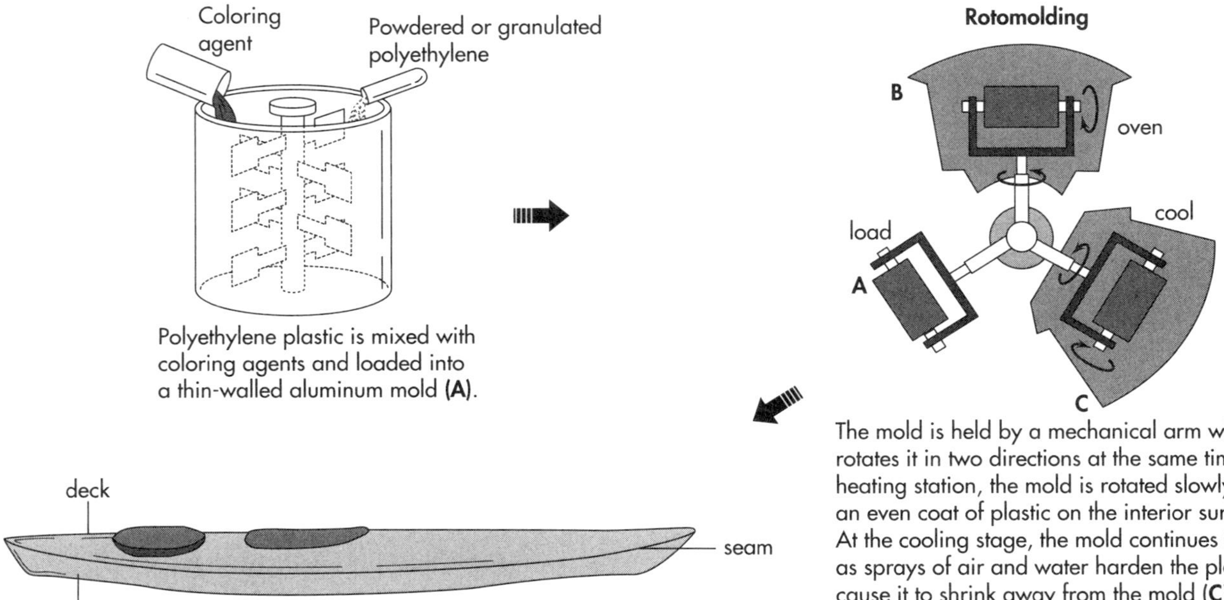

Polyethylene plastic is mixed with coloring agents and loaded into a thin-walled aluminum mold **(A)**.

The mold is held by a mechanical arm which rotates it in two directions at the same time. At the heating station, the mold is rotated slowly to form an even coat of plastic on the interior surface **(B)**. At the cooling stage, the mold continues to rotate as sprays of air and water harden the plastic and cause it to shrink away from the mold **(C)**.

The molded hull and deck are designed to fit together with a single horizontal seam.

fishing. Eskimo kayaks typically weighed about 26 pounds (12 kg), were 18-20 feet (5.5-6 m) long and 20 inches (51 cm) wide. The Eskimos lashed bone or driftwood into frames with seal sinew or gut. Seal or caribou skins were stripped of hair, tied together, and soaked in water before being tightly stretched over the frame. The skins stretched taut as they dried. Seams were waterproofed with boiled seal oil or caribou fat. Limitations such as the availability, shape, and size of natural materials did not hinder the grace and durability of the Eskimo kayak. Modern kayak designers and manufacturers are indebted to these early engineers both for the concept of the kayak as a low, covered boat as well as for specific features which make the boat so seaworthy.

By A.D. 900, kayaks were being used in Europe. New designs in frames and coverings addressed the need to easily transport the kayak over land. One of these designs was a revolutionary collapsible kayak model called a foldboat which was invented in Germany in the 1800s. The foldboat used a rubberized canvas outer layer stretched over a folding tubular frame. The foldboat could be disassembled and carried in just two suitcases.

More recently, the primary use of kayaks has shifted from hunting and transportation to recreation and competitive sport. Kayaking for recreation began on rivers and lakes in the late 1800s. Sea kayaking was widely popularized when Percy Blanfold designed and built thousands of canvas-covered plywood and lumber kayaks in the 1950s and 1960s. Kayaking as a competitive sport began during the Summer Olympic Games in Germany in 1936 and has since gained international appeal.

Modern kayaks are built from covered wooden frames or from shells of **fiberglass** or plastic. Wooden kayaks most closely resemble the ancient Eskimo rib and cross frame construction. They are considered the classic design, and can be built from scratch in a very short time or assembled from kits. Durable marine or exterior-grade plywood is used for the frame. Waterproof glue and mechanical fasteners join the wooden pieces. Copper tacks or stainless steel staples are used for attaching the cotton canvas or cotton duck fabric to the frame. These coverings are treated and finished with airplane dope or exterior paint to waterproof and strengthen the fabric.

In the 1950s, fiberglass-reinforced resins

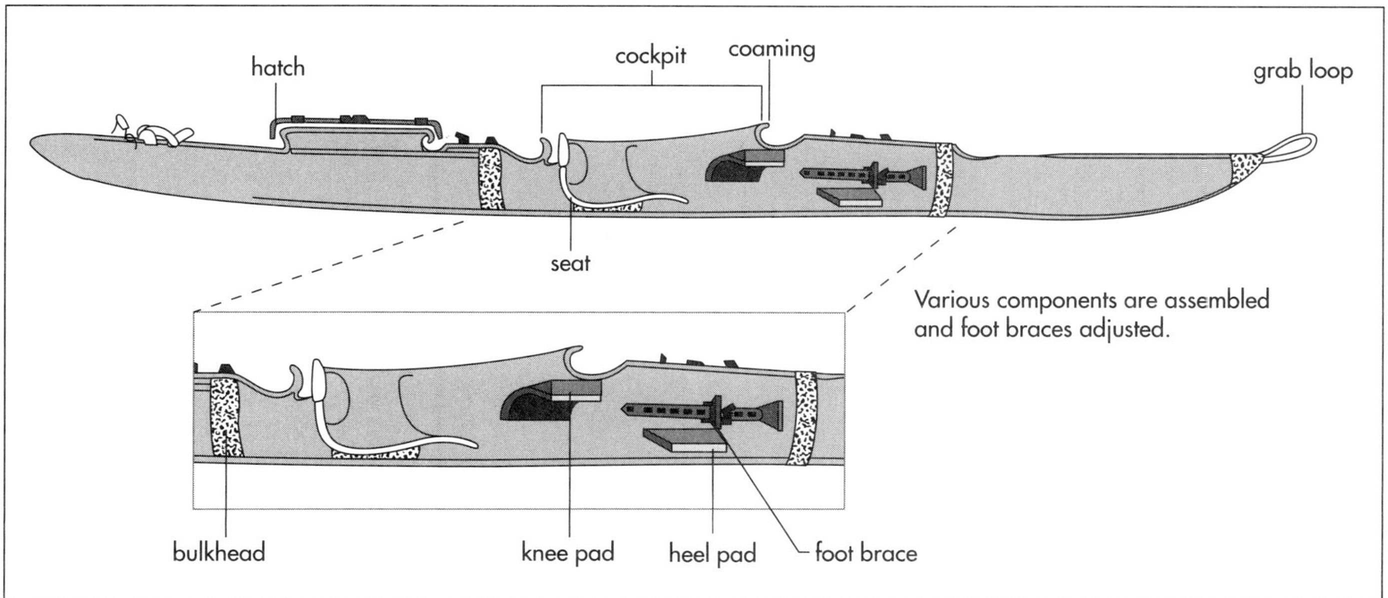

allowed a method of kayak construction that did not require a frame. The shape of the kayak was molded instead. Two molds were needed for this method: a bottom mold which was shaped like the kayak's hull, and an upper mold shaped like the deck. The mold was protected with a release agent to ensure that the pieces would not adhere to it. Using a layering process called hand lay-up, the builder draped resin-saturated cloth over the mold to form the hull and deck pieces. Once the resin hardened, the boat pieces were clamped together along the gunwale where the deck meets the hull. Fiberglass tape was used to seal this seam, inside and out.

Polyethylene and the use of recycled plastics revolutionized kayak construction again in the early 1980s. The raw materials and manufacturing process for this latest advancement are described next.

Raw Materials

The idea that a kayak shell can be made from recycled plastics is very appealing to many environmentally-minded boaters. The primary ingredient of a plastic kayak is polyethylene. Polyethylene is a tough, waxy-textured material that is unaffected by water and many chemicals. It can be repeatedly softened by heating and hardened by cooling. These characteristics make it excellent for the manufacture of kayaks. Adequate supplies of polyethylene recycled from plastic beverage bottles are now available with sufficient durability and strength for use in making kayaks.

Polyethylene is also used to make the seats in a kayak. The flotation devices, bulkheads, and padding for the kayaker's hips, knees, and heels are often made of a closed-cell foam like Ethafoam. The rudder pedals or foot braces are usually made from a lightweight, corrosion-resistant metal like aluminum, as is the rudder itself. The grab loop may be nylon rope with a plastic or wooden toggle, or handle.

The Manufacturing Process

The body of a polyethylene kayak is a long, hollow shell of uniform thickness made by a process called rotational molding. After the shell is made, the seat and other components are added in a manual assembly process. The shell may be molded in one piece, or it may be made from two separate pieces which are joined later. Rotational molding requires fairly rounded contours, so one-piece shells cannot have sharp intersections between surfaces. If the design requires a sharper intersection—such as between the hull and the deck, for example—then a two-piece shell is used. The process for making the two-piece shell is described here.

Loading the mold

1 Measured amounts of powdered or granulated polyethylene plastic are mixed with coloring agents, and an exact amount is loaded into the bottom of a two-piece, thin-walled aluminum mold for either the hull or the deck. The mold is then closed and the two halves secured tightly.

Molding the hull or deck

2 The mold is held by a mechanical arm which can rotate the mold in two directions at the same time. The arm can also move, or index, the mold from one process station to another. After the mold is loaded, it is moved to the heating station where an oven heats it to 480-840°F (250-450°C). As the heat from the mold melts the plastic, the mold is slowly rotated in two directions. Gravitational force causes the melted plastic to flow to the front and rear and up the sides of the rotating mold, eventually covering the entire surface with a uniformly thick skin of plastic. Continued heating then fuses the plastic into a solid layer that attaches itself to the mold.

3 Next the mold is indexed to the cooling chamber. By continuing to rotate the mold and gradually cooling it with carefully directed sprays of air and then water, the kayak hull or deck further hardens and shrinks away from the mold. Repeated cycles of heating and cooling may be required to form the kayak properly. After the final cooling, the hull or deck is removed from the mold.

Assembling the shell

4 The hull and deck are designed so that they fit together at a single seam running horizontally along the length of the boat. Careful sealing ensures that the shell acts as a single structural unit. After sealing, a decorative strip can be added to conceal the seam. On some designs, a keel stiffener is bonded along the length of the shell to give it added strength.

Final assembly

5 Closed-cell foam flotation aids are inserted into each end of the kayak and held in place with adhesive. Some kayaks use inflatable flotation bags instead. These flotation aids keep the kayak level and floating on the surface should the craft capsize. If the kayak design uses bulkheads, then those pieces are bonded in place and sealed with a polyurethane sealant. The seat is either suspended from straps attached to the sides or supported by a piece of foam bonded to the inside of the bottom of the hull. Adjustable foot braces or rudder pedals are attached to the inside of the shell, forward of the cockpit, using non-corrosive metal fasteners. The hatch covers have waterproof gaskets installed before they are put in place and the grab loops are installed. Finally, a drain plug is inserted in a molded hole in the shell.

Shipping

6 Kayaks are relatively large, but lightweight. Packaging to protect the shell from abrasion is more important than providing a strong support. Traditional methods of preparing a kayak for shipping include shrink-wrapping in plastic or placing the kayak in a cardboard box. One environmentally sensitive approach uses heavy, tarp-like cloth bags.

Quality Control

Most of the quality control checks for a rotationally molded kayak are in the molding process. Maintaining the proper rotation speed and oven temperature are critical to producing shells with uniform thickness. Care is required during cooling to avoid warping the shell. The duration of the heating and cooling cycles, and the intervals between those stages, must also be controlled precisely.

The Future

Kayaking continues to grow as a recreational and competitive sport. Improved performance in competitive kayaking will depend on modern technology for new design and construction methods, and may result in the use of new materials. At the same time, many kayak enthusiasts will demand that these materials be environmentally benign, such as recycled plastics.

Where To Learn More

Books

Kulczycki, Chris. *The Kayak Shop.* McGraw Hill, 1993.

Freeman, Zu. *Building a Jawbone Kayak.* Tamal Vista Publications, 1989.

Periodicals

Noble, Peter, Michael Wadden, Timothy Bourke, David Williams, and Knut Nordbo. "An Introduction to Ethnotechnology for Naval Architects: Sea Kayak Design of Yesterday and Tomorrow." *Marine Technology,* October 1994, pp. 305-14.

Stark, Peter. "From Icy Northern Waters Came the Ancient Sea Kayak." *Smithsonian,* November 1992, pp. 150-62.

Wells, Edward O. "Virtual Realities." *Inc.,* August 1993, pp. 51-58.

Other

American Boat and Yacht Council, Inc. *Publication H.29 (2).* July 1, 1986.

—David N. Ford

Ketchup

Ketchup originated in ancient China as a brine of pickled fish or shellfish called "ke-tsiap."

Background

Ketchup, a tangy, seasoned tomato sauce, is one of America's favorite condiments. Although ketchup, also spelled catsup, is used primarily as a relish for hamburgers, hot dogs, and french fries, it is also a common ingredient for sauces, meatloaf, beans, and stews. During the mid-1990s the sales of ketchup exceeded $400 million annually.

The tangy sauce originated in ancient China as a brine of pickled fish or shellfish called "ke-tsiap." Neighboring countries adopted their own variations of "kechap" consisting of fish brine, herbs, and spices. In the late 1600s, English sailors visiting Malaysia and Singapore were so impressed with the sauce that they took samples home. English cooks attempted to duplicate the spicy sauce, but without access to some of the exotic Asian ingredients, they improvised with cucumbers, mushrooms, nuts, oysters, and other variants.

One hundred years later, New Englanders created the definitive tomato ketchup when Maine seamen returned from Mexico and the Spanish West Indies with seeds of an exotic New World fruit called tomato. The tangy tomato ketchup quickly became a popular sauce for codfish cakes, meat, and other foods.

Making ketchup at home was a tedious, day-long process. The tomato mixture, cooked in heavy iron kettles at wood-burning stoves, required constant stirring to prevent it from burning. Scouring the preserving kettles meticulously was also no easy task. To the relief of many homemakers, ketchup became commercially available in the second half of the 1800s.

H.J. Heinz Co. developed one of the first leading brands of mass-marketed ketchup. The classic narrow-neck design of the Heinz ketchup bottle established the norm for the industry. The narrow-neck bottle simplified pouring the ketchup and minimized contact with air, which could darken the sauce. Glass was an ideal container because it was inert and did not react with the ketchup, and the clear glass allowed the consumer to see the product. Initially, the bottles were sealed with cork, dipped by hand into wax to prevent aeration, and topped with foil to further protect it from contamination. By the turn of the century, screw caps provided a more convenient closure. In the 1980s, plastic squeezable containers revolutionized ketchup packaging and soon outsold glass containers. Plastic was not only more convenient than glass for pouring the thick sauce, but also safer. Ten years later, in response to environmental concerns, recyclable plastic containers were also developed.

Raw Materials

The main ingredients of ketchup are tomatoes, sweeteners, vinegar, **salt**, spices, flavorings, onion, and/or garlic. The types of sweetener used are usually granulated cane sugar or beet sugar. Other sweeteners include dextrose or liquid sugar in the form of corn or glucose syrup. The white vinegar, commonly 100-grain distilled, helps to preserve the ketchup. The spices commonly used to enhance the flavor of the tomatoes are allspice, cassia, cinnamon, cayenne, cloves, pepper, ginger, mustard, and paprika. Some manufacturers believe that whole spices produce a superior, more mild flavor than ground spices or spice oils. More modern processes use premixed or encapsulated spices, which are easier to use but more expensive. Whatever the form, spices must be of a high quality.

The various brands of ketchup have slightly different formulas, which vary primarily in the amounts of spices or flavorings. Thicker consistencies require a greater ratio of sugar and spices relative to the tomato juice. Occasionally formulas must be slightly adjusted according to variations in the acid and sugar content of tomatoes, which occurs with changes in growing conditions and types of tomatoes.

The Manufacturing Process

Developing quality tomatoes

1 Ketchup manufacturers must seek out the best quality tomatoes for their product. Tomato varieties are developed which are superior in color, flavor, texture, and yield. Consistency is an important factor, as slight variations in tomato characteristics could alter the flavor and color of the finished product.

Preparing tomatoes

2 Tomatoes are harvested mechanically between June and July. The fruit is commonly conveyed by water from the trucks into a flume, or an inclined channel. The water method washes the tomatoes and protects them from bruising while they pass from the truck to the factory. The U.S. Department of Agriculture or state inspectors approve and grade tomatoes to meet initial requirements. The tomatoes are sorted, washed, and chopped. Next, precooking, or scaling, in stainless steel vats preserves the tomatoes and destroys bacteria.

Pulping

3 The chopped and precooked tomatoes are pumped into pulping machines, or cyclones, which separate seeds, skins, and stems from the pulp. The pulp and juice are filtered through screens and processed further into ketchup, though some may be stored in a paste for use later in the year.

By the 1920s, when this photo was taken, ketchup operations were highly mechanized. (From the collections of Henry Ford Museum & Greenfield Village.)

The history of ketchup and the history of advertising are inextricably intertwined. This is especially true in the case of the H.J. Heinz Company, a firm that pioneered many elements of the prepared food business and the modern advertising industry.

Born in 1844, Henry John Heinz began helping his mother with her gardens along the Allegheny River, just east of Pittsburgh, when he was nine years old. He learned business practices while working as a bookkeeper for his father's brickyard and at night school. By his teens he was employing three women to help process garden products and bottling his mother's horseradish for distribution. Heinz distinguished his horseradish from his competitors by using clear glass bottles to emphasize the product's purity.

Twenty years later, Heinz was operating another family food processing firm. Riding the New York elevator one day in 1892, he saw a sign advertising 21 varieties of shoes. He took the concept, came up with a figure of 57 because he thought it was a memorable number, and created the catch phrase "Heinz 57 Varieties."

In 1893, seeking to bolster attendance at the World's Columbian Exposition in Chicago, Heinz distributed thousands of small tokens throughout the fair grounds. The tokens were redeemable for a free Heinz souvenir, a watch charm in the shape of a pickle, at the food pavilion, which was soon overrun with visitors. The "pickle pin" went on to become one of the best-known corporate souvenirs in history, with over 100 million distributed.

In 1898, Heinz bought the Iron Pier in Atlantic City, New Jersey, renamed it the Heinz Ocean Pier, and operated it until 1945 as a free public attraction with antique displays, lectures, concerts, and motion pictures amid the displays of Heinz products and souvenirs.

William S. Pretzer

Adding ingredients and cooking

4 The pulp is pumped into cooking tanks or kettles and heated to boiling. Foaming may occur if fresh tomato pulp is used, but can be corrected with anti-foaming compounds or compressed air. Precise amounts of sweeteners, vinegar, salt, spices, and flavorings are added to the tomato pulp. Most spices are added early in the cooking process. To avoid excessive evaporation, volatile spice oils and vinegar must be mixed in later. Onions and garlic can be mixed in with the spices, placed in a separate bag, or chopped and added to the pulp. Salt and sugar may be added at any stage of cooking though it is better to add sugar later to prevent burning. The mixture cooks for 30-45 minutes and is circulated by rotating blades installed in the cookers. The temperature must be carefully regulated to insure absorption of the ingredients without overcooking, which creates a flat body.

Finishing

5 Once the cooking is complete, the ketchup mixture passes through a finishing machine. Finishers remove excess fiber and particles through screens, creating a smoother consistency. The ketchup passes to a holding tank before further processing.

6 The ketchup may be milled at higher temperatures and pressures to achieve a smoother consistency.

Removing air

7 The ketchup must be de-aerated to prevent discoloration and growth of bacteria. Excess air might also create unattractive air pockets and impede the closure process.

Filling

8 To prevent contamination, the ketchup passes from the receiving tanks to the filling machines at a temperature not lower than 190°F (88°C). The containers are filled with the ketchup and immediately sealed to retain the freshness of the product. Ketchup containers come in various sizes and shapes, including 14-oz. bottles, No. 10 cans, pouch packs, room-service sizes, and single-serve packets.

Cooling

9 The containers must be cooled to prevent flavor loss through stack burning, which occurs when ketchup stays at high temperatures after cooking is complete. Containers of ketchup may be cooled in cold air or cold water.

Labeling and packing

10 Finally, the ketchup containers are labeled and coded with product information, including ingredients, date and location of manufacture, and shelf-life. The bottled ketchup may be inspected again before shipping. The entire process of ketchup manufacturing generally takes two to three hours.

Quality Control

Some of the commonly used preservatives during the 19th century included benzoate of soda, borax salicylic acid, benzoic, and formaldehyde, all of which could pose health risks when consumed in large quantities. A series of Pure Food Laws beginning in 1906 banned the use of the harmful preservatives.

In 1940, the U.S. government established a "Standard of Identity" for ketchup as tomato-based. Thus consumers could tell from the label that the product was made of tomatoes, since ketchup could also be made from other foods, including bananas, beets, or mangoes.

The quality of ketchup is insured by taking samples of the product during various stages of production. Tomato growers must comply with regulations set by the Environmental Protection Agency and the Food and Drug Administration regarding the use of fertilizers and pesticides. Increasing concern in the closing decades of the 20th century led to increased use of natural fertilizers and pesticides. Inspection is necessary of the tomatoes, ingredients, and of all processing equipment which comes into contact with the product.

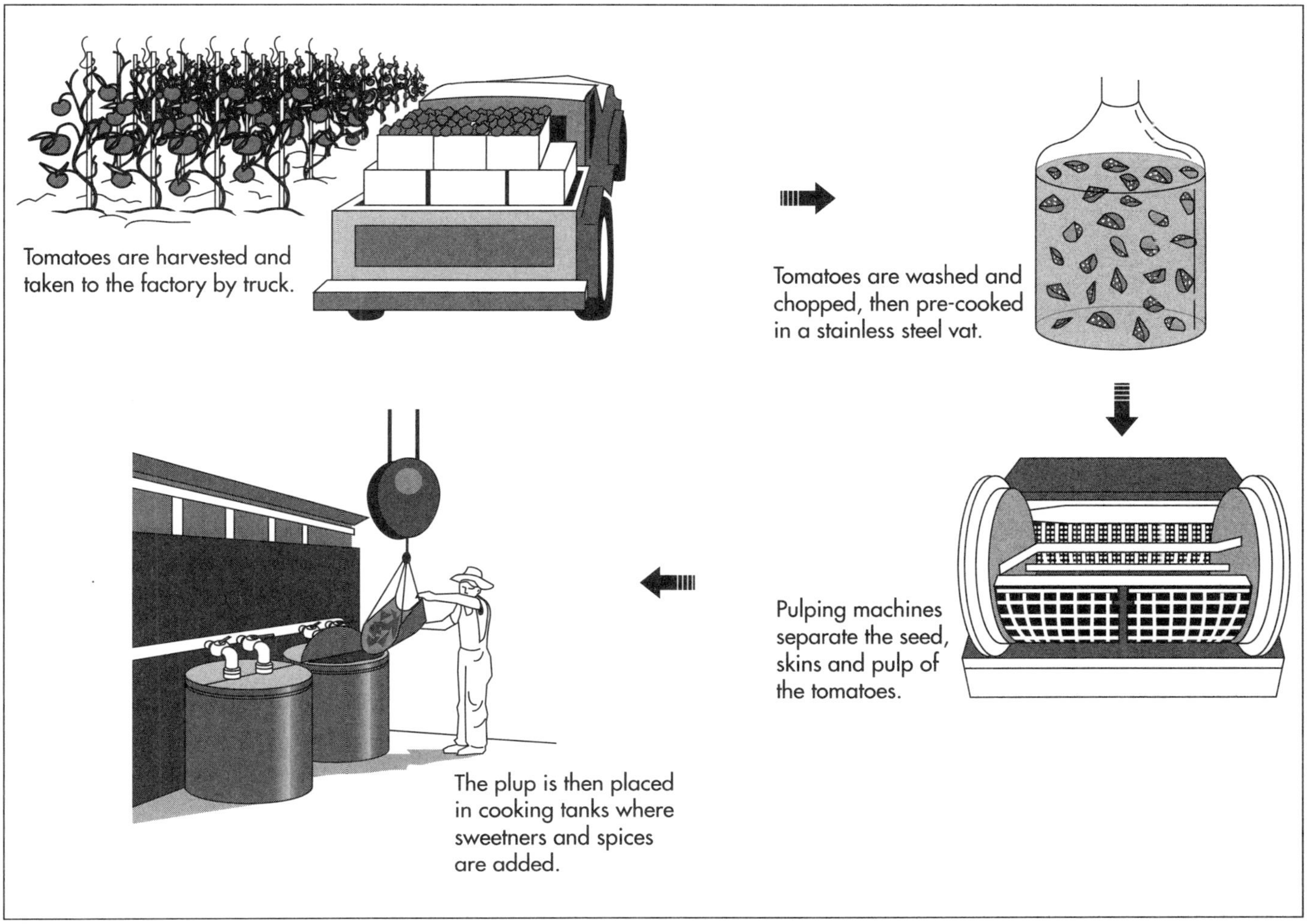

Oxidation of ketchup can darken the color of ketchup, but de-aeration of the sauce during manufacture can prevent this problem. However, once the containers are opened, oxidation may still occur. Although the acidity of ketchup preserves the sauce, manufacturers recommend that once containers are opened they should be refrigerated to prevent deterioration of the ketchup color, flavor, and quality.

To maintain consistency in color and flavor, manufacturers determine the concentration of tomato solids in the mixture, since about one-third of the ketchup's acidity and sugar content depends on the amount of solids. The ketchup Grades A through C must conform to specific concentrations. The quality of the ketchup can be measured by its physical consistency, or body, which refers to the ability of the ketchup to retain its liquid in suspension. The slower the rate, the higher the grade of the ketchup. For instance, the Bostwick Consistometer, recommended by the USDA, set Grades A and B at flow rates at less than 4 inches (10 cm) in 30 seconds at 68°F (20°C).

The Future

Ketchup manufacturers continue to improve the quality of ketchup by developing tomato strains that are superior in color, flavor, and firmness. Tomato hybrids are also engineered to improve resistance to disease and rot, thus decreasing the reliance on chemical pesticides.

In the 1990s, in response to consumer demand for more healthful foods, ketchup manufacturers created low-calorie, low-salt ketchup alternatives. The increasing popularity of Spanish salsas and marinades also influenced manufacturers to develop salsa-style ketchups which were lower in sugar content. Packaging technology continues to

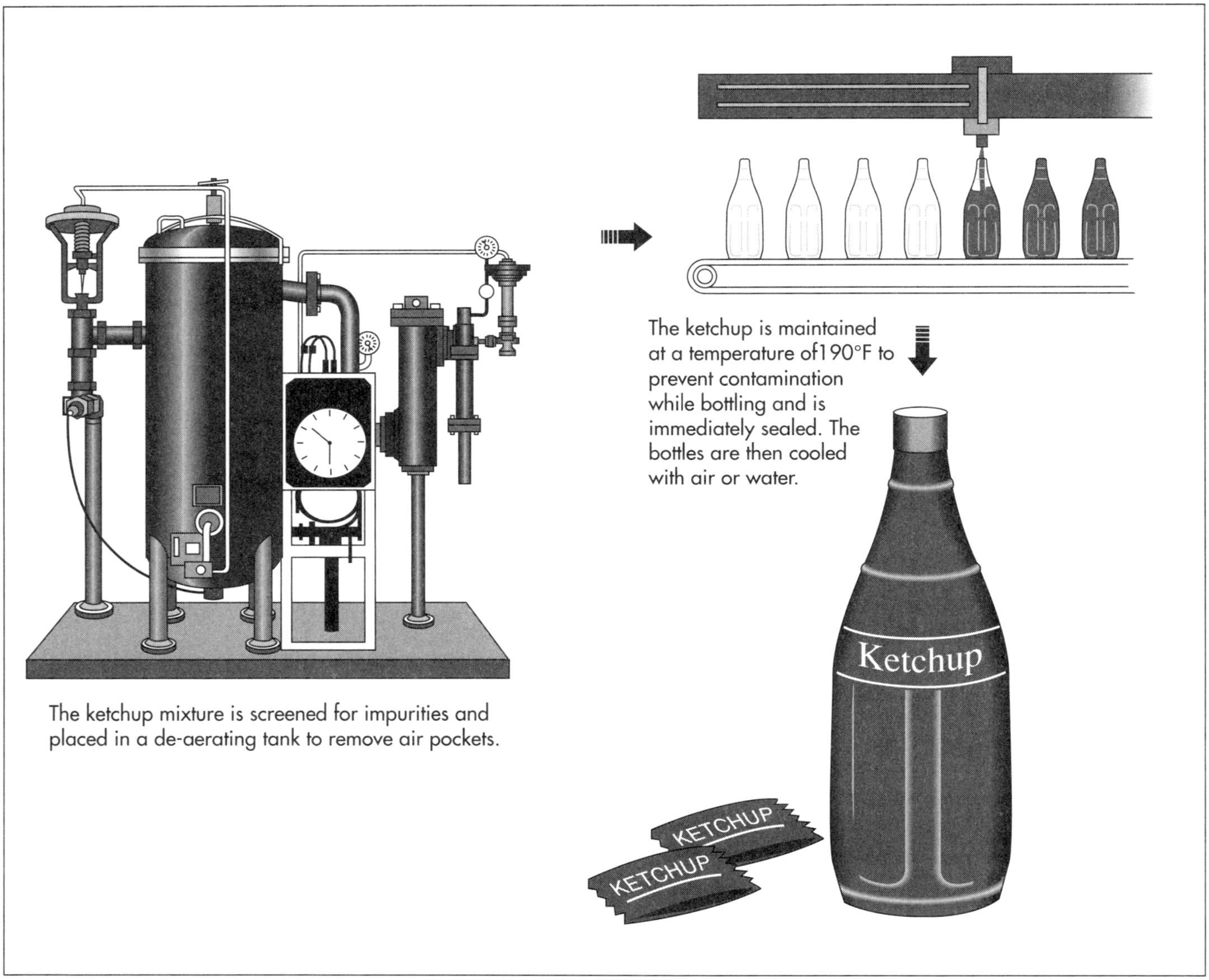

improve as consumers demand safer, more convenient, and recyclable containers.

Where To Learn More

Book

Gould, Wilbur A. *Tomato Production, Processing, and Quality Evaluation*, 2nd ed. AVI Publishing Company, Inc., 1983.

Periodicals

McDermott, Michael J. "Salsamania! Mexican Sauce Marketing." *Food & Beverage Marketing,* August 1993, p. 8.

Strenk, Thomas. "Ketchup." *Restaurant Business Magazine.* May 20, 1993, p. 99.

Wagner, Jim. "Building the Best New Products in America." *Food Processing,* November 1993, p. 16.

Other

All About Heinz Ketchup. H.J. Heinz Co., 1991. P.O. Box 57, Pittsburg, PA, 15230-0057.

—Audra Avizienis

Lead

Background

Lead is a dense, soft, low-melting metal. It is an important component of batteries, and about 75% of the world's lead production is consumed by the battery industry. Lead is the densest common metal except for gold, and this quality makes it effective in sound barriers and as a shield against X-rays. Lead resists corrosion by water, so it has long been used in the plumbing industry. It is also added to paints, and it makes a long-lasting roofing material.

Lead is a health hazard to humans if it is inhaled or ingested, interfering with the production of red blood cells. Its use must be carefully controlled, and several formerly common uses of lead are now restricted by the U.S. government. Lead paint is found in many older buildings, but it is now mostly used on outdoor steel structures such as bridges, to improve their weatherability. A lead compound called tetraethyl lead was added to **gasoline** as early as 1921 because it prevented the "knocking" problem of high-compression automobile engines. However, most gasoline now contains no lead, because lead from car exhaust was a major source of air pollution.

Lead is also commonly used in glass and enamel. In television picture tubes and computer video display terminals, lead helps block radiation, and the inner, though not the outer, portion of the common light bulb is made of leaded glass. Lead also increases the strength and brilliance of crystal glassware. Lead is used to make bearings and solder, and it is important in rubber production and oil refining.

Lead production dates back at least 8,000 years. Lead was used in Egypt as early as 5,000 B.C., and in the time of the Pharaohs it was used in pottery glazes and as solder. It was also cast into ornamental objects. A white lead paint was also used in ancient Egypt, Greece, and Rome. Ancient Rome used lead pipes for its extensive water works. Some of the toxic effects of lead were also noted as early as the Roman era, though lead was also thought to have positive medical qualities. In the 15th and 16th centuries, builders used lead as a roofing material for cathedrals, and lead was also used to hold together the different panels of colored glass in **stained glass** windows. The first lead battery is credited to a French physicist, Gaston Plante, who invented it in 1859. By 1889, so-called lead-acid storage batteries of the modern type were being commercially produced.

Modern lead mines produce about 3 million metric tons of lead annually. This is only about half the lead used worldwide; the remainder is obtained by recycling. The top producer of lead is Australia, followed by the United States, China, and Canada. Other countries with major lead deposits are Mexico, Peru, Russia, and Kazakhstan.

Raw Materials

Lead is extracted from ores dug from underground mines. More than 60 minerals contain some form of lead, but only three are usually mined for lead production. The most common is called galena. The pure form of galena contains only lead and sulfur, but it is usually found with traces of other metals in it, including silver, copper, **zinc**, cadmium, and antimony as well as arsenic. Two other

Lead is the densest common metal except for gold, and this quality makes it effective in sound barriers and as a shield against X-rays.

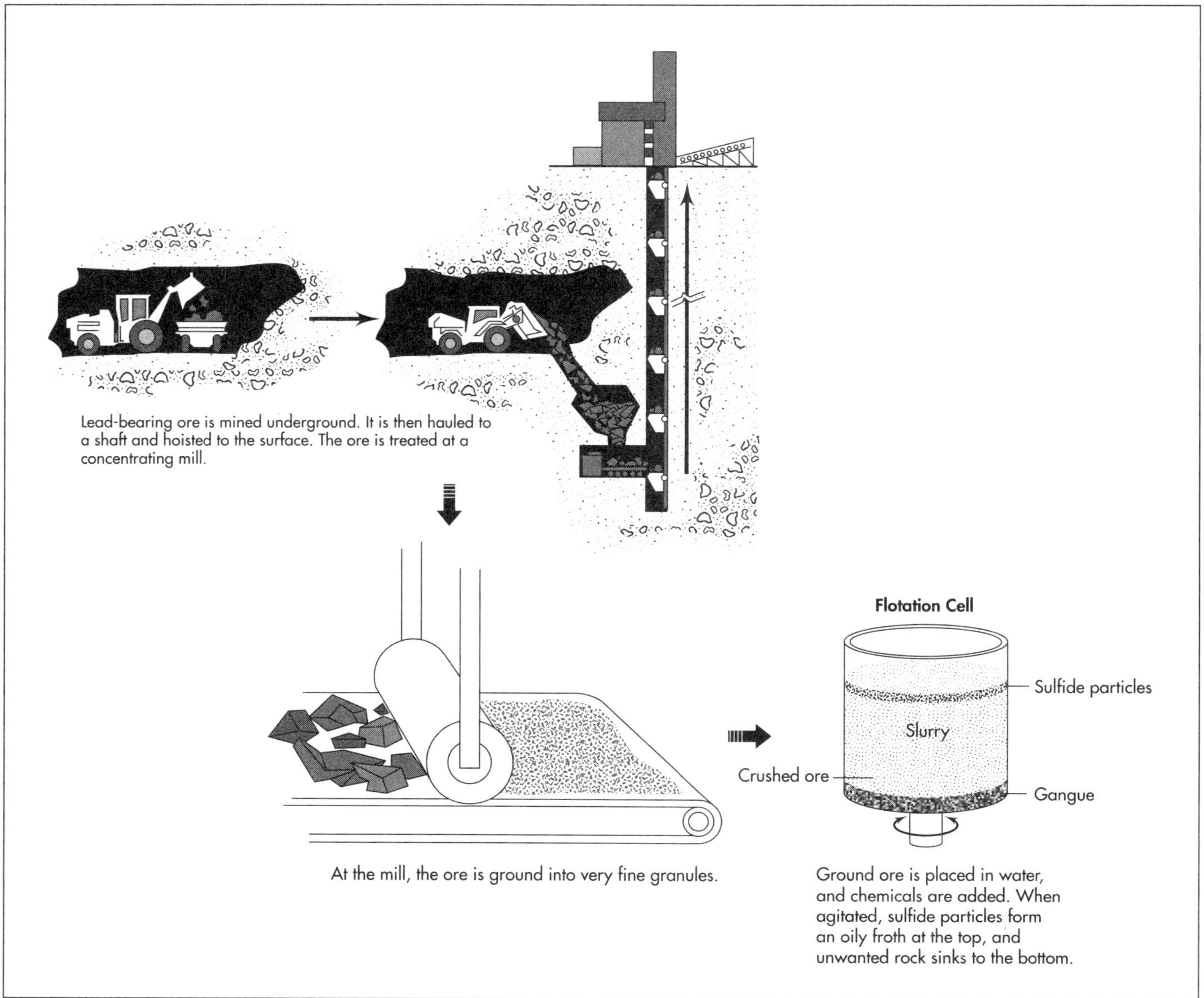

minerals commercially mined for lead are cerussite and anglesite. Over 95% of all lead mined is derived from one of these three minerals. However, most deposits of these ores are not found alone but mixed with other minerals such as pyrite, marcasite, and zinc blende. Therefore much lead ore is obtained as a byproduct of other metal mining, usually zinc or silver. Only half of all lead used yearly derives from mining, as half is recovered through recycling, mostly of automobile batteries.

Besides the ore itself, only a few raw materials are necessary for the refining of lead. The ore concentrating process requires pine oil, alum, lime, and xanthate. Limestone or iron ore is added to the lead ore during the roasting process. Coke, a coal distillate, is used to further heat the ore.

The Manufacturing Process

Mining the ore

1 The first step in retrieving lead-bearing ore is to mine it underground. Workers using heavy machinery drill the rock from deep tunnels with heavy machinery or blast it with **dynamite**, leaving the ore in pieces. Then they shovel the ore onto loaders and trucks, and haul it to a shaft. The shaft at a large mine may be a mile or more from the

drill or blast site. The miners dump the ore down the shaft, and from there it is hoisted to the surface.

Concentrating the ore

2 After the ore is removed from the mine, it is treated at a concentrating mill. Concentrating means to remove the waste rock from the lead. To begin, the ore must be crushed into very small pieces. The ore is ground at the mill, leaving it in particles with diameters of 0.1 millimeter or less. This means the individual granules are finer than table salt. The texture is something like granulated sugar.

Flotation

3 The principal lead ore, galena, is properly known as lead sulfide, and sulfur makes up a substantial portion of the mineral. The flotation process collects the sulfur-bearing portions of the ore, which also contains the valuable metal. First, the finely crushed ore is diluted with water and then poured into a tank called a flotation cell. The ground ore and water mixture is called slurry. One percent pine oil or a similar chemical is then added to the slurry in the tank. The tank then agitates, shaking the mixture violently. The pine oil attracts the sulfide particles. Then air is bubbled through the mixture. This causes the sulfide particles to form an oily froth at the top of the tank. The waste rock, which is called gangue, sinks to the bottom. The flotation process is controlled by means of X-ray analyzers. A flotation monitor in the control room can check the metal content of the slurry using the X-ray analysis. Then, with the aid of a computer, the monitor may adjust the proportion of the chemical additive to optimize recovery of the metal. Other chemicals are also added to the flotation cell to help concentrate the minerals. Alum and lime aggregate the metal, or make the particles larger. Xanthate is also added to the slurry, in order to help the metal particles float to the surface. At the end of the flotation process, the lead has been separated from the rock, and other minerals too, such as zinc and copper, have been separated out.

Filtering

4 After the ore is concentrated in the flotation cells, it flows to a filter, which removes up to 90% of the water. The concentrate at this point contains from 40-80% lead, with large amounts of other impurities, mostly sulfur and zinc. It is ready at this stage to be shipped to the smelter. The gangue, or rock that was not mineral-bearing, must be pumped out of the flotation tank. It may be dumped into a pond resembling a natural lake, and when the pond eventually fills, the land can be replanted.

Roasting the ore

5 The lead concentrate fresh from the filter needs to be further refined to remove the sulfur. After the concentrate is unloaded at what is called the sinter plant, it is mixed with other lead-bearing materials and with sand and limestone. Then the mixture is spread on a moving grate. Air which has been heated to 2,550°F (1,400°C) blows through the grate. Coke is added as fuel, and the sulfur in the ore concentrate combusts to sulfur dioxide gas. This sulfur dioxide is an important byproduct of the lead refining process. It is captured at a separate acid plant and converted to sulfuric acid, which has many uses. After the ore has been roasted in this way, it fuses into a brittle material called sinter. The sinter is mostly lead oxide, but it can also contain oxides of zinc, iron, and silicon, some lime, and sulfur. As the sinter passes off the moving grate, it is broken into lumps. The lumps are then loaded into the blast furnace.

Blasting

6 The sinter falls into the top of the blast furnace, along with coke fuel. A blast of air comes through the lower part of the furnace, combusting the coke. The burning coke generates a temperature of about 2,200°F (1,200°C) and produces carbon monoxide. The carbon monoxide reacts with the lead and other metal oxides, producing molten lead, nonmetallic waste slag, and carbon dioxide. Then the molten metal is drawn off into drossing kettles or molds.

Refining

7 The molten lead as it comes from the blast furnace is from 95-99% pure. It is called at this point *base bullion*. It must be further refined to remove impurities, because commercial lead must be from 99-99.999% pure. To refine the bullion, it is kept in the

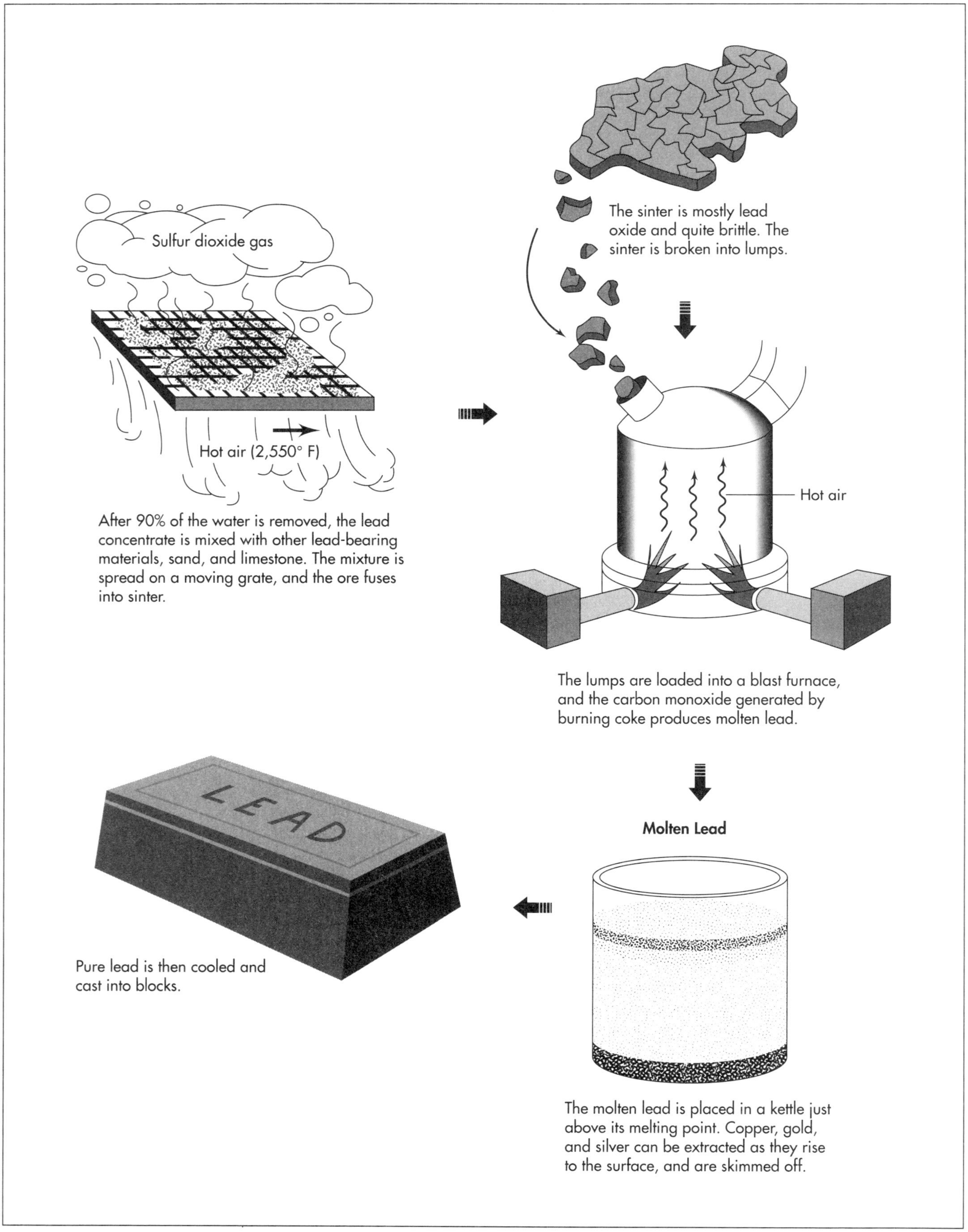
Sulfur dioxide gas
Hot air (2,550° F)
After 90% of the water is removed, the lead concentrate is mixed with other lead-bearing materials, sand, and limestone. The mixture is spread on a moving grate, and the ore fuses into sinter.
The sinter is mostly lead oxide and quite brittle. The sinter is broken into lumps.
Hot air
The lumps are loaded into a blast furnace, and the carbon monoxide generated by burning coke produces molten lead.
Molten Lead
The molten lead is placed in a kettle just above its melting point. Copper, gold, and silver can be extracted as they rise to the surface, and are skimmed off.
LEAD
Pure lead is then cooled and cast into blocks.

drossing kettle at a temperature just above its melting point, about 626°F (330°C). At this temperature, any copper left in the bullion rises to the top of the kettle and forms a scum or dross which can be skimmed off. Gold and silver can be removed from the bullion by adding to it a small quantity of zinc. The gold and silver dissolves more easily in zinc than in lead, and when the bullion is cooled slightly, a zinc dross rises to the top, bringing the other metals with it.

Casting

8 When the lead has been sufficiently refined, it is cooled and cast into blocks which may weigh as much as a ton. This is the finished product. Lead alloys may also be produced at the smelter plant. In this case metals are added to the molten lead in precise proportions to produce a lead material for specific industrial uses. For example the lead commonly used in car batteries, and also for pipe, sheet, cable sheathing, and **ammunition**, is alloyed with antimony because this increases the metal's strength.

Byproducts/Waste

Lead refining produces several byproducts. The gangue, or waste rock, accumulates as the ore is concentrated. Most of the minerals have been removed from the rock, so this waste is not considered by the industry to be an environmental hazard. It can be pumped into a disposal pond, which resembles a natural lake. Sulfuric acid is the major byproduct of the smelting process. Sulfur dioxide gas is released when the ore is roasted at the sinter plant. To protect the atmosphere, fumes and smoke are captured, and the air released by the plant is first cleaned. The sulfur dioxide is collected at a separate acid plant, and converted to sulfuric acid. The refinery can sell this acid as well as its primary product, the lead itself.

Air pollution can result from lead processing as well. The smelter requires a "bag house," that is, a separate facility to filter and vacuum the fumes so that lead is not released into the atmosphere. Nevertheless, lead particles do reach the atmosphere, and in the United States, federal regulations attempt to control how much is allowable. Most of the solid waste product produced by the smelting process is a dense, glassy substance called slag. This contains traces of lead as well as zinc and copper. The slag is more toxic than the gangue, and it must be stored securely and monitored so that it does not escape into the environment or come in contact with populations.

The Future

New developments in the lead industry seem aimed less at improvements in the manufacturing process than towards finding new uses for the lead itself. Since a large proportion of the lead mined and recycled is sold to the automotive industry for batteries, lead producers are quite dependent on the health of the auto industry. But lead producers are interested in finding new applications for lead to give them more market stability.

One recent new application for lead is a lead-**fiberglass** laminate. Lead sheeting can be laminated between gypsum and fiberglass, forming a superior duct material that helps isolate noise. If this is used in an air conditioning unit, for example, it effectively dampens the din of the machine. Another prospective market for lead is in nuclear waste containment. Safely storing radioactive material is a growing concern around the world. The lead industry is researching canisters made of titanium with an inner layer of lead or lead and plastic, contending that a one-inch layer of lead could add 880 years to the life of a properly buried container. And looking to the cars of the future, researchers in the U.S. and several other countries have been studying ways of improving lead-acid battery technology in order to power electric cars.

Where To Learn More

Periodicals

Goodwin, Frank E. and Dodd S. Carr. "Brilliant Performer." *Natural Science*, July 1989, pp. 317-23.

King, Angela. "Producers Hunt New Lead Uses." *American Metal Market*. April 11, 1988, pp. 10-13.

Knights, Mikell. "Higher Auto Output Boosts Lead, Zinc Use." *American Metal Market*, August 6, 1993, pp. 5-12.

Schmitt, Bill. "Lead, Zinc Vie for Place in Future Electric Cars." *American Metal Market*, August 6, 1993, p.6.

—Angela Woodward

Leather Jacket

History

Leather, a material made from tanned animal hides, has been used as clothing since the earliest days of human existence. Prehistoric people wrapped animal skins around their bodies for warmth and to absorb the magical powers that they believed the skins imparted to them. Phoenician sailors often brought brightly embroidered leather garments from Babylonia to the countries they visited. Leather shoes were found in the tombs of Egyptian pharaohs.

Soldiers of the Roman Empire invaded the lands of northern Europe and discovered Teutonic nomads wearing leather garments as protection against the harsh elements. When these soldiers returned to Rome with leather pants, the toga-garbed officials of the city tried to ban their use, but to no avail. Romans were soon using leather for shoes and tunics as well as for breastplates and shields. In fact, the first recorded tanning guild was formed in the Roman Empire.

In the Middle Ages, the Moors introduced the European world to softer Cordovan leather which they made from goatskin. By the Renaissance, tanners' guilds had been organized all over Europe. The Mayan, Incan, and Aztec cultures in Central and South America also used leather, as did the American Indians, who sewed garments from buckskin, doeskin, and buffalo hide.

During the Stone Age, garments were held together with leather straps that had been threaded through holes punched in the hide with crude implements made of stone. The hides were stiff and did not last long before putrefaction made them unwearable. Later, people learned to soften the hides by rubbing them with animal fat and used stones to clean the animal cells from the skins. It is also possible that in trying to dye the skins with various substances, early people discovered further preservation methods.

Preservation methods have varied over the centuries. Hides have been smoked, salted, soaked in urine, rubbed with animal dung, beaten, and dragged over sharp sticks. Prehistoric people and some modern Eskimos have even resorted to chewing the skins to remove hair, pieces of flesh, and ultimately, to soften the leather. In some cultures, the skins were sprinkled with talc and flour to replace the natural oils. Women's leather garments were often doused with perfumes.

Although leather was a valuable trade commodity, the tanning work was dirty and odorous, and leather workers were usually relegated to the outskirts of town.

Ancient Hebrews are credited with inventing the first tanning process using oak bark. The American Indians used fish oil for the same purpose. American colonists found that plants such as the hemlock and chestnut trees could also be used for tanning. In the 19th century, machines were developed to perform these processes and an American chemist developed a tanning method using chromium salts that cut the processing time from weeks or months to just a few hours.

Raw Materials

Antelope, buckskin, lambskin, sheepskin, and cowhide are the hides most commonly

Antelope, buckskin, lambskin, sheepskin, and cowhide are the hides most commonly used to make leather jackets.

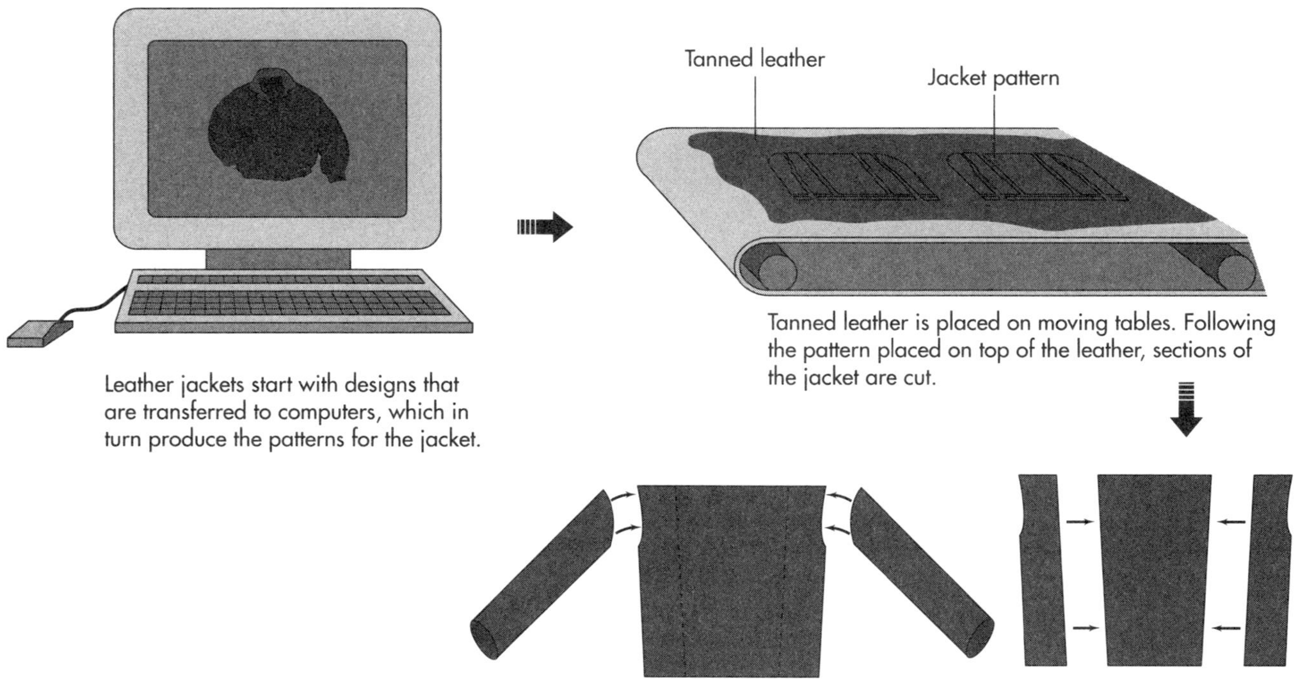

Leather jackets start with designs that are transferred to computers, which in turn produce the patterns for the jacket.

Tanned leather is placed on moving tables. Following the pattern placed on top of the leather, sections of the jacket are cut.

Once all the pattern pieces are cut, the major sections are assembled together.

used to make leather jackets. As soon as the skin is removed from the animal at the meat processing plant, it is refrigerated, salted, or packed in barrels of brine. It is then sent to the tannery where the skins undergo a series of processes designed to preserve and soften the hides. The work performed at the tannery is of utmost importance to insure that the resulting garment is of high quality.

Sewing materials such as thread, lining, seam tape, **button**s, snaps, and zippers are generally purchased from outside vendors and stored in the garment factory.

The Preparation Process

Trimming and cleaning

1 The skins are trimmed and sorted according to size, weight, and thickness. It is necessary to remove any remaining proteins that could stimulate the growth of bacteria. To this end, the skins are soaked in revolving drums filled with water, bactericides, and detergents. Hair is removed with the application of chemical sprays or lime solutions. A scudding machine fitted with dull blades scrapes off any excess hair.

Another round of washing (de-liming) removes these chemicals before the hides are soaked in an acid solution and then bated, an enzyme treatment to further remove the skins' collagens. Finally, the hides are pickled with salt and sulfuric acid.

Tanning

2 The hides undergo one of three types of tanning. Vegetable tanning requires the most time with the skins soaked in progressively stronger tannic acid solutions for several weeks. However, some soft leathers, such as lambskin, may be soaked for as little as 12 hours.

Mineral tanning is significantly faster but can change the color of the leather. For this method, the skins are placed in alum salt-filled drums fitted with paddles that provide a constant agitating motion.

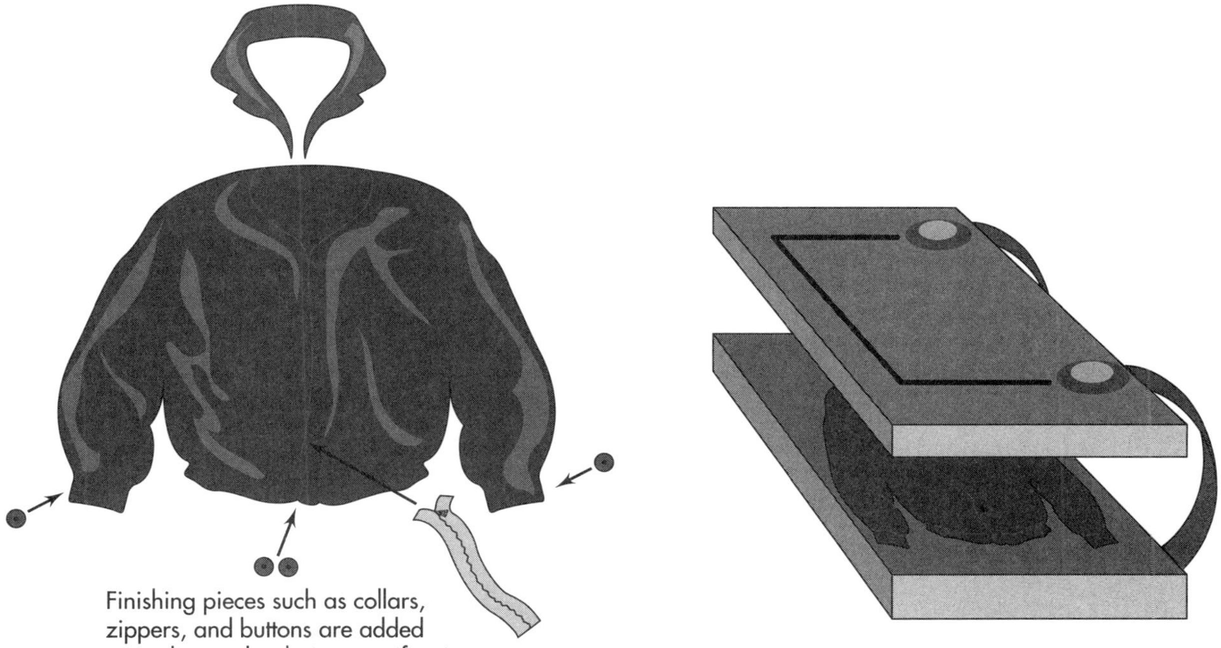

Finishing pieces such as collars, zippers, and buttons are added according to the design specifications.

Next, the jacket is steamed and pressed to give it a distinctive shape.

The third method, oil tanning, is the one that resembles the ancient methods most closely. Here, fish oil is sprayed onto the skins.

Washing and drying

3 After the skins have been tanned, they are washed once again and wrung out thoroughly. Then the skins are passed under a band knife, which cuts the skins to a uniform thickness, and placed on conveyer belts which carry them to drying tunnels. Usually, the skins are stretched on frames to prevent shrinkage during the drying stage. To combat the stiffness that drying produces, the hides are sprayed with water and soap and allowed to hang for a period of time.

After the skins have been conditioned, they are placed in machines designed to rhythmically manipulate the leather so that the fibers are further loosened and made more flexible. In the final drying period, the hides are hung in vacuum-drying cabinets.

When the skins are thoroughly dried, they are buffed with revolving steel cylinders covered with abrasive **paper**. Suede finishes are produced by passing the hides under high-speed emery wheels. At this point, glazes, dyes, and lacquers are applied. The skins are now ready to be sent to the garment factory.

The Manufacturing Process

The development of high-speed sewing machinery changed the face of traditional sewing factories where one person may have worked on a single garment from start to finish. Because leather garments are considered luxury items, hand-construction by highly skilled artisans is still sought by many consumers. However, the following steps are those used in factory mass production.

Jacket design

1 Garment manufacturers typically employ designers to create patterns from which the clothing is made. Computerized machines grade the designs according to government anthropometric tables which

assign sizes based on body height and weight. The computer then produces patterns in a range of sizes from the original design.

Cutting

2 The tanned leather is placed on moving tables called spreaders. Although modern technology allows several layers of fabric to be cut simultaneously, leather is usually cut one layer at a time. The pattern is placed on top of the leather. This is accomplished in one of two ways; tissue-paper patterns may be pinned onto the leather, or the pattern may be marked with tailor's chalk. The spreading table works on the conveyer system, moving the fabric to the cutting machine, which is fitted with either rotary blades or band-knives. The table is either guided by a human operator or run automatically. The most recent technological advance is the computerized laser beam system in which the fabric seams are vaporized rather than cut.

Lining material for the jacket is cut in the same manner. Because it is of a much thinner weight, lining can be placed on the spreaders in multiple layers.

Jacket assembly

3 The jacket is assembled in roughly this order: the sides are stitched to the back portion, sleeve underseams are stitched together, and the sleeves are attached to the armholes. The attachment of finishing pieces such as collars, cuffs, buttonholes, buttons, zippers, and pockets varies according to the design of the jacket. Patch pockets are sewn onto the side pieces before they are stitched to the back portion, and side pockets are sewn in at the same time that the sides are attached to the back. Generally, lining material is attached to each piece before it is sewn onto the jacket.

In mass production, the pieces are moved along a highly sophisticated production line using integrated automatic sewing machines that are capable of sewing as many as 8,000 stitches per minute. In a sequential system, one sewing machine stitches a particular section of the jacket and then moves the garment to another sewing machine which performs the next step. For example, after one machine stitches a cuff to the sleeve, the sleeve moves to another machine where it is attached to the jacket armhole.

A tandem sewing system calls for two or more machines to work on the same garment simultaneously. In this instance, one machine attaches buttons to the front of the jacket while another machine applies the collar.

Each step, from setting thread and needle positions to aligning the fabric to extracting the sewed materials, is pre-programmed. Each sewing machine is equipped with under-bed trimming devices that automatically knot and cut threads after each seam is sewn. Excess threads are carried away to waste receptacles by streams of compressed air.

Operators regulate the work at each station with a modified presser-foot or from a control panel. A stop-motion device allows the operator to halt production to make adjustments such as replacing broken threads or needles.

Molding and pressing

4 A number of pressing processes incorporating heat application, steaming, and blocking are employed to complete the transformation of the animal skins into a jacket. Buck presses equipped with controls and gauges to regulate the amount of steam and pressure are used to give the jacket its distinctive shape, whether a bomber- or blazer-styled jacket. Curved blocks are placed around the collars and cuffs and then heat is applied. The blocks are removed, leaving the collars and cuffs curved.

Final inspection

5 Each jacket is inspected by hand before it leaves the factory floor. The completed jackets are then sheathed in plastic bags, packed into cartons, and shipped to the retailer.

Quality Control

The thoroughness of the tanning process is designed to produce skins that are supple and free of bacteria-causing proteins. Garment manufacturers inspect each shipment of skins for marks, tears, stains, and imperfections.

Today's automated sewing systems are self-correcting. Sophisticated lubricating systems composed of pumps, reservoirs, fluidic controls, and electronic controls insure that the garments are manufactured at a consistent level of quality.

The Future

Although the leather goods industry suffered slightly during the most recent recession, and merchandisers were compelled to implement discount strategies, the Leather Apparel Association predicts a strong comeback during the last decade of the 20th century as many companies expand their clothing lines. New technologies such as the laser cutter are also increasing the rate of production.

As with the fur industry, the leather industry has been the target of some animal rights groups who denounce the killing of animals for human benefit, particularly to create "luxury" items. In an effort to address these concerns, some clothing manufacturers have increased production of artificial leather, a material made from synthetic fibers, which does not use any animals and is less expensive. However, jackets and coats made from real leather remain fashionably popular.

Where To Learn More

Book

Parker, Xenia Ley. *Working with Leather.* Charles Scribner's Sons, 1972.

—Mary F. McNulty

Life Vest

Life vests and life jackets are technically known as PFDs, short for personal flotation devices. The main function of a PFD is to keep a person on the water's surface in a relatively upright position.

Background

Recreational and professional activities bring people in contact with water everyday. Safety on and near the water is an issue for the weekend sail boater as well as for military and rescue personnel. Flotation devices are an important part of water safety. Full-body, industrial survival suits and simple water ski belts are two extremes in the family of personal flotation devices available. Although they differ tremendously in the amount of protection, the goal of all safety apparel is survival.

Life vests and life jackets are technically known as PFDs, short for personal flotation devices. They are designed to keep an individual afloat in the water in the event of an emergency and are considered life-saving equipment. The main function of a PFD is to keep a person on the water's surface in a relatively upright position to allow the person to breathe and not have to tread water to stay afloat. U.S. Coast Guard regulations require a boat to carry at least one Coast Guard approved PFD per person, including one for each person water skiing.

There are five categories of Coast Guard-approved personal flotation devices. Each of the five categories provide different flotation and body positioning specifications. Types I and II PFDs are full-and half-length vests designed to turn an unconscious person from a face-down position to a vertical or slightly head-back position floating in the water. These vests are usually big and bulky. A type III PFD, most commonly used in recreational activities, is also a buoyant vest or jacket. This type is designed to keep a conscious person afloat in a vertical or slightly head-back position. This type comes in many styles and is the most comfortable. A further difference in the I, II, and III types is that the specific degree of buoyancy required increases from type III to type I. Throwable devices such as a ring buoy or the buoyant cushion typically used by boaters as a cushion for sitting are considered Type IV. Type V PFDs are special-purpose devices for aircraft pilots who fly over water, rafters, and ferryboat pilots.

History

Natural materials were first used to create floatation devices. Before 1900, life jackets were made from cork and balsa wood. A material called kapok was later used as the fill material in life vests. Kapok is a vegetable fiber found in tropical tree pods, resembling milkweed. The waxy coating which covers the kapok fiber provided the necessary buoyancy. The kapok fiber was sealed in vinyl plastic packets to prevent exposure to the water. One problem with the vinyl-sealed kapok fiber life jacket was that the packets could be punctured, causing the jacket to lose its buoyancy. Kapok is now prohibited for use in life preservers in most of Europe and in Canada.

The Coast Guard made a significant change in life vest requirements after the 1953 sinking of the ore carrier Carl D. Bradley, in which 33 persons died. Many crewmen were found floating among their life vests, having slipped out of them after the ship had sunk. Thereafter, the Coast Guard required that life jackets be designed so that unconscious persons could not accidentally slip out of them if immersed in water.

In the 1960s, France introduced a life jacket called the flotherchoc. The flotherchoc was a light and flexible body-fitting vest. This design replaced the then popular but awkward horse-collar design. The advantage of this vest was that it was less confining and, therefore, more likely to actually be worn. The flotherchoc was made up of small, air-filled vinyl packets which were placed inside nylon chambers. However the flotherchoc had the same problem as the PFDs which used kapok: over time, the flotherchoc's vinyl packets could lose their buoyancy if punctured.

Plastics are now being used in the manufacture of life vests. Some vests are made from closed-cell foam or foamed plastics which are encased in nylon. Closed-cell foam has been around since the 1940s, but it was not until the 1970s that its use in survival wear was introduced. A closed-cell foam insert is made of tiny, individual air-filled pockets within the foam itself. The air-filled pockets are called cells. This foam structure is similar to a sponge, except that in a sponge the individual cells are connected by tunnels which run throughout the material. Closed-cell foam cells are not connected at all. It is the isolated air-filled pockets which provide the flotation. Closed-cell foam can be punctured over and over again with only minimal effect on its buoyancy. Some of the better closed-cell foam structures will not deteriorate even under tremendous compression. The air-filled pockets also provide some thermal insulation protection against hypothermia.

Described below is the process for the manufacture of a standard Type III personal flotation vest containing closed-cell foam encased in nylon with various finishes, including reflective tape, zippers, snaps, and labels.

Raw Materials

Most of the materials required to manufacture life vests are purchased in bulk from sources outside the manufacturer. Some materials come from custom fabricators and are specifically made to meet certain standards. Threads for stitching seams and for embroidery of logos are purchased from one source. Nylon fabric is purchased in bulk which is typically 60 inches (152 cm) wide by 20 or 30 feet (6 or 9 m) long. The width of the fabric generally corresponds to the standard width of cutting machines. The same width specifications apply to the closed-cell foam which is purchased in thick pieces. Non-corrosive plastic zippers and snaps are purchased from still another outside source as are materials like strapping and reflective tape. Finally, labels specifying Coast Guard Approval and other information regarding the classification of the individual product are obtained from testing organizations such as United Laboratories.

The Manufacturing Process

The steps in the manufacture of life vests are similar to those for any automated garment manufacturing process, differing in specific features such as raw materials and, more importantly, safety specifications. The operations necessary to complete a garment from scratch are known in the industry as "cut-fit-trim." As many as 100 life vests may be manufactured simultaneously in an automated manufacturing process such as the one described here.

Creating markers

1 Pattern designs are digitally fed from a computer into a machine called a plotter, which draws a stencil of the pattern design on long sheets of white paper. The drawn sheets are called markers.

Preparing the nylon

2 A machine called a spreader unrolls a bolt of nylon fabric along a table, usually 66-72 inches (168-183 cm) wide and up to 100 feet (31 m) long. Thin fabrics such as nylon may be layered 25 deep for cutting. Wrinkles are smoothed by the spreader or by hand, and the marker is laid on top of the nylon.

Cutting the pattern

3 Some manufacturers use an automated cutting machine to cut the pattern pieces out of the nylon. Other manufacturers may cut the pieces by hand using a

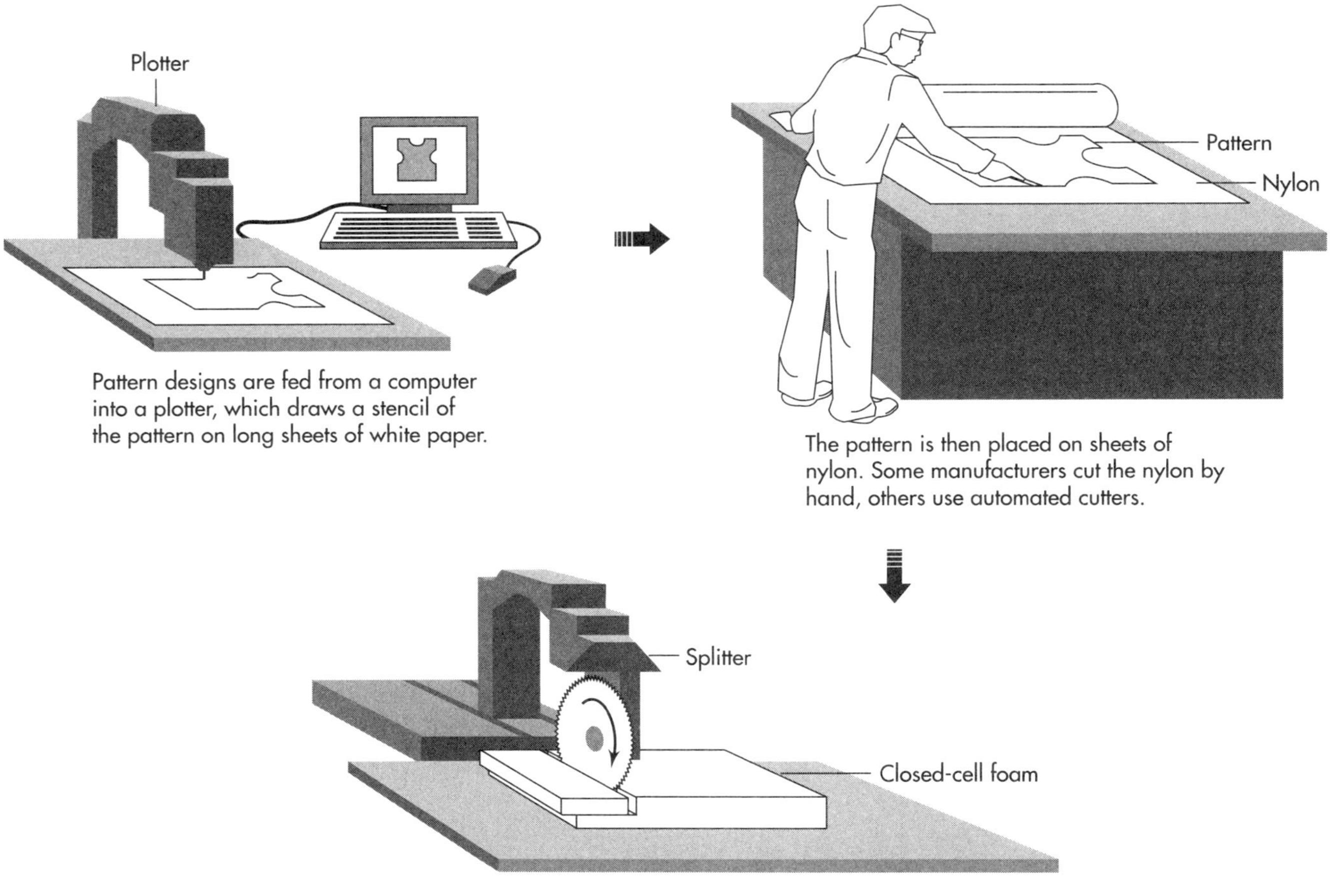

Pattern designs are fed from a computer into a plotter, which draws a stencil of the pattern on long sheets of white paper.

The pattern is then placed on sheets of nylon. Some manufacturers cut the nylon by hand, others use automated cutters.

The closed cell foam is sliced to the desired thickness with a splitter. The foam is then sent through the same pattern cutting process as the nylon.

portable, motor-driven straight knife which resembles a jig saw. In automated cutting, the digital pattern is fed into the cutting machine. A sheet of cellophane, wider than the fabric, is drawn over the top of the marker and fabric layers. A vacuum pulls the cellophane down tight against the table, holding the marker and nylon layers in place. A knife cuts the pattern out of cellophane, marker, and fabric layers at the same time. The cut pattern pieces are then bundled for transfer to assembly areas. The marker, which can be read through the cellophane, identifies the pattern pieces.

Cutting the foam

4 The closed-cell foam is sliced to the desired thickness with a type of band saw called a splitter. A band saw has a long thin blade welded in a continuous loop that travels over a driven wheel up and over one or more idler wheels, and then down through the material being cut. The foam then goes through the pattern cutting process in the same way as the nylon.

5 Reflective tape and small, attachable pieces such as straps are cut from rolls with a smaller cutting machine called a die cutter.

Assembling pattern pieces

6 Sewing professionals, each stationed at an industrial sewing machine, match the pieces and sew them together. The pieces are sewn inside out and then reversed. The

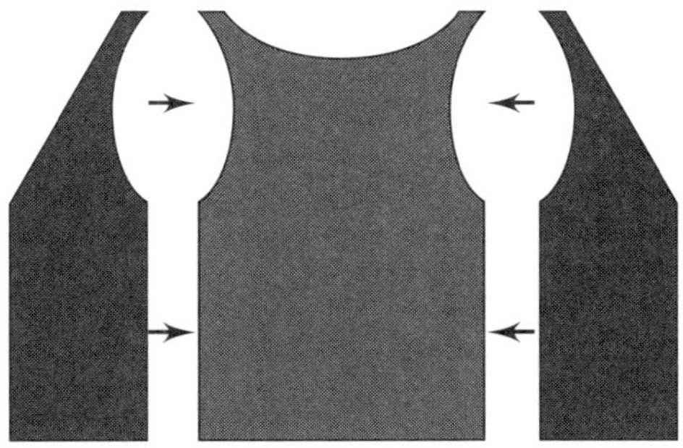

The pieces of the vest are matched together and sewn with industrial sewing machines. Next, foam pieces are inserted.

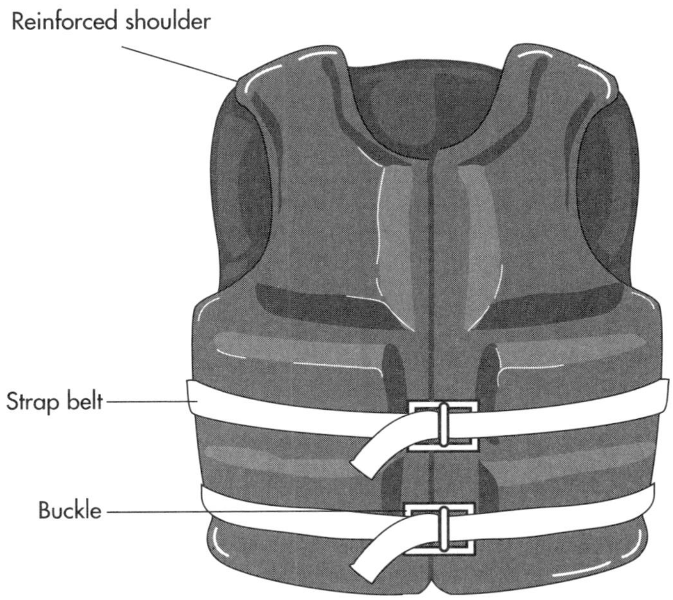

Straps, reflective tape, and labels are added. Snaps are attached by a riveting machine. A computer-aided embroidering machine handles brand names and logos.

cut foam pieces are inserted through an open seam which is then sewn shut.

Finishing

7 Straps, reflective tape, and labels are sewn on last. Snaps are attached by an eyelet or riveting machine much like that used by a shoemaker for shoelace holes. A computer-aided embroidery machine—much like those used by department stores to personalize towels—is used to embroider brand names and logos.

8 Individual finished vests are placed in plastic bags for protection. The bags are packed into corrugated boxes and sent to distribution centers.

Quality Control

Quality assurance in any safety product is very important because a person's life may depend upon the manufacturing quality. The United States Coast Guard and Underwriters Laboratories have been monitoring the development and manufacturing of life jackets for many years. Federal law sets very high manufacturing and performance standards that must be met before the Coast Guard will approve a PFD. The PFDs which are approved are recognized by a stamp on the PFD itself or on an attached tag. Some manufacturers insure that all materials that go into the manufacture of PFDs are checked and meet, or exceed, Coast Guard Standards before any cutting or assembling is done. Defects are also monitored by manufacturer station supervisors. Public response to product performance may also be solicited. Typically, samples from batches are lifted and sent to international standard keepers such as the International Standards Organization (ISO) for comparison with ISO 9001, the highest level of international recognition of quality of design and manufacturing.

Some manufacturers have even the most incidental materials designed and made to exacting specifications for their particular product. For instance, threads for stitching seams and embroidery may be tested for disintegration tolerances. Nylon fabric made from high-tenacity yarns is specifically tested under ultraviolet light for up to 600 hours to check for premature aging due to exposure to the sun. Closed-cell foam may be scientifically developed especially for a single manufacturer with specific floatation needs.

The Future

New developments in life vests and other personal flotation devices will continue to address the comfort of life-saving apparel while not being used for flotation. Flotation devices which inflate only when needed are the newest products to address this issue. Inflatables lie flat on the body and pose no bulky restrictions until inflation occurs. Inflation may be spontaneous, as in the event of water immersion, or manual. Automatic inflation works by the controlled release of carbon dioxide. Some of the newest devices available are inflatable vests, collars, and pillows built into full-body, insulated survival suits. Although currently not Coast Guard approved, some inflatables exceed Type I specifications. The BOAT/US Foundation reports that participants preferred the wearability of the inflatable devices because of their non-restrictive features. Life vests will continue to evolve as designers, manufacturers, and testers overcome challenges such as comfort, controlled inflation, and loss of buoyancy.

Where To Learn More

Periodicals

Baldwin, Mark. "The Survival Suit Saga." *Oceans,* September-October 1986, pp. 40-45.

Sutphen, Hal. "Will It Work When I Need It?" *Cruising World*, March 1994, pp. 19-22.

Sutphen, Hal. "How to Equip for Safety." *Cruising World*, December 1993, pp. 75-80.

Other

Department of Transportation. *The Skippers Course*. United States Coast Guard Office of Boating Safety, Washington DC, 1973.

—David N. Ford

Magnet

Background

A magnet is a material that can exert a noticeable force on other materials without actually contacting them. This force is known as a magnetic force and may either attract or repel. While all known materials exert some sort of magnetic force, it is so small in most materials that it is not readily noticeable. With other materials, the magnetic force is much larger, and these are referred to as magnets. The Earth itself is a huge magnet.

Some magnets, known as permanent magnets, exert a force on objects without any outside influence. The **iron** ore magnetite, also known as lodestone, is a natural permanent magnet. Other permanent magnets can be made by subjecting certain materials to a magnetic force. When the force is removed, these materials retain their own magnetic properties. Although the magnetic properties may change over time or at elevated temperatures, these materials are generally considered to be permanently magnetized, hence the name.

Other magnets are known as electromagnets. They are made by surrounding certain materials with a coil of wire. When an electric current is passed through the coil, these materials exert a magnetic force. When the current is shut off, the magnetic force of these materials drops to nearly zero. Electromagnet materials retain little, if any, magnetic properties without a flow of electric current in the coil.

All magnets have two points where the magnetic force is greatest. These two points are known as the poles. For a rectangular or cylindrical bar magnet, these poles would be at opposite ends. One pole is called the north-seeking pole, or north pole, and the other pole is called the south-seeking, or south pole. This terminology reflects one of the earliest uses of magnetic materials such as lodestone. When suspended from a string, the north pole of these first crude compasses would always "seek" or point towards the north. This aided sailors in judging the direction to steer to reach distant lands and return home.

In our present technology, magnet applications include compasses, electric motors, microwave ovens, coin-operated vending machines, light meters for photography, automobile horns, televisions, loudspeakers, and tape recorders. A simple refrigerator note holder and a complex medical magnetic resonance imaging device both utilize magnets.

The two basic types of magnets are permanent magnets, which exert a force on objects without any outside influence, and electromagnets, which exert a magnetic force only when electric current is applied.

History

Naturally occurring magnetic lodestone was studied and used by the Greeks as early as 500 B.C. Other civilizations may have known of it earlier than that. The word magnet is derived from the Greek name *magnetis lithos*, the stone of Magnesia, referring to the region on the Aegean coast in present-day Turkey where these magnetic stones were found.

The first use of a lodestone as a compass is generally believed to have occurred in Europe in about A.D. 1100 to A.D. 1200. The term lodestone comes from the Anglo-Saxon meaning "leading stone," or literally, "the stone that leads." The Icelandic word is *leider-stein*, and was used in writings of that period in reference to the navigation of ships.

In 1600, English scientist William Gilbert confirmed earlier observations regarding magnetic poles and concluded that the Earth was a magnet. In 1820, the Dutch scientist Hans Christian Oersted discovered the relationship between electricity and magnetism, and French physicist Andre Ampere further expanded upon this discovery in 1821.

In the early 1900s, scientists began studying magnetic materials other than those based on iron and steel. By the 1930s, researchers had produced the first powerful Alnico alloy permanent magnets. Even more powerful ceramic magnets using rare earth elements were successfully formulated in the 1970s with further advances in this area in the 1980s.

Today, magnetic materials can be made to meet many different performance requirements depending on the final application.

Raw Materials

When making magnets, the raw materials are often more important than the manufacturing process. The materials used in permanent magnets (sometimes known as hard materials, reflecting the early use of alloy steels for these magnets) are different than the materials used in electromagnets (sometimes known as soft materials, reflecting the use of soft, malleable iron in this application).

Permanent Magnet Materials

Permanent magnet lodestones contain magnetite, a hard, crystalline iron ferrite mineral that derives its magnetism from the effect the earth's magnetic field has on it. Various steel alloys can also be magnetized. The first big step in developing more effective permanent magnet materials came in the 1930s with the development of Alnico alloy magnets. These magnets take their name from the chemical symbols for the aluminum-nickel-cobalt elements used to make the alloy. Once magnetized, Alnico magnets have between 5 and 17 times the magnetic force of magnetite.

Ceramic permanent magnets are made from finely powdered barium ferrite or strontium ferrite formed under heat and pressure. Their magnetic strength is enhanced by aligning the powder particles with a strong magnetic field during forming. Ceramic magnets are comparable to Alnico magnets in terms of magnetic force and have the advantage of being able to be pressed into various shapes without significant machining.

Flexible permanent magnets are made from powdered barium ferrite or strontium ferrite mixed in a binding material like rubber or a flexible plastic like polyvinyl chloride.

In the 1970s, researchers developed permanent magnets made from powdered samarium cobalt fused under heat. These magnets take advantage of the fact that the arrangement of the groups of atoms, called magnetic domains, in the hexagonal crystals of this material tend to be magnetically aligned. Because of this natural alignment, samarium-cobalt magnets can be made to produce magnetic forces 50 times stronger than magnetite. Headphones for small, personal stereo systems use samarium-cobalt permanent magnets. Samarium-cobalt magnets also have the advantage of being able to operate in higher temperatures than other permanent magnets without losing their magnetic strength.

Similar permanent magnets were made in the 1980s using powdered neodymium iron boron which produces magnetic forces almost 75 times stronger than magnetite. These are the most powerful permanent magnets commercially available today.

Electromagnet Materials

Pure iron and iron alloys are most commonly used in electromagnets. Silicon iron and specially treated iron-cobalt alloys are used in low-frequency power transformers.

A special iron oxide, called a gamma iron oxide, is often used in the manufacture of magnetic tapes for sound and data recording. Other materials for this application include

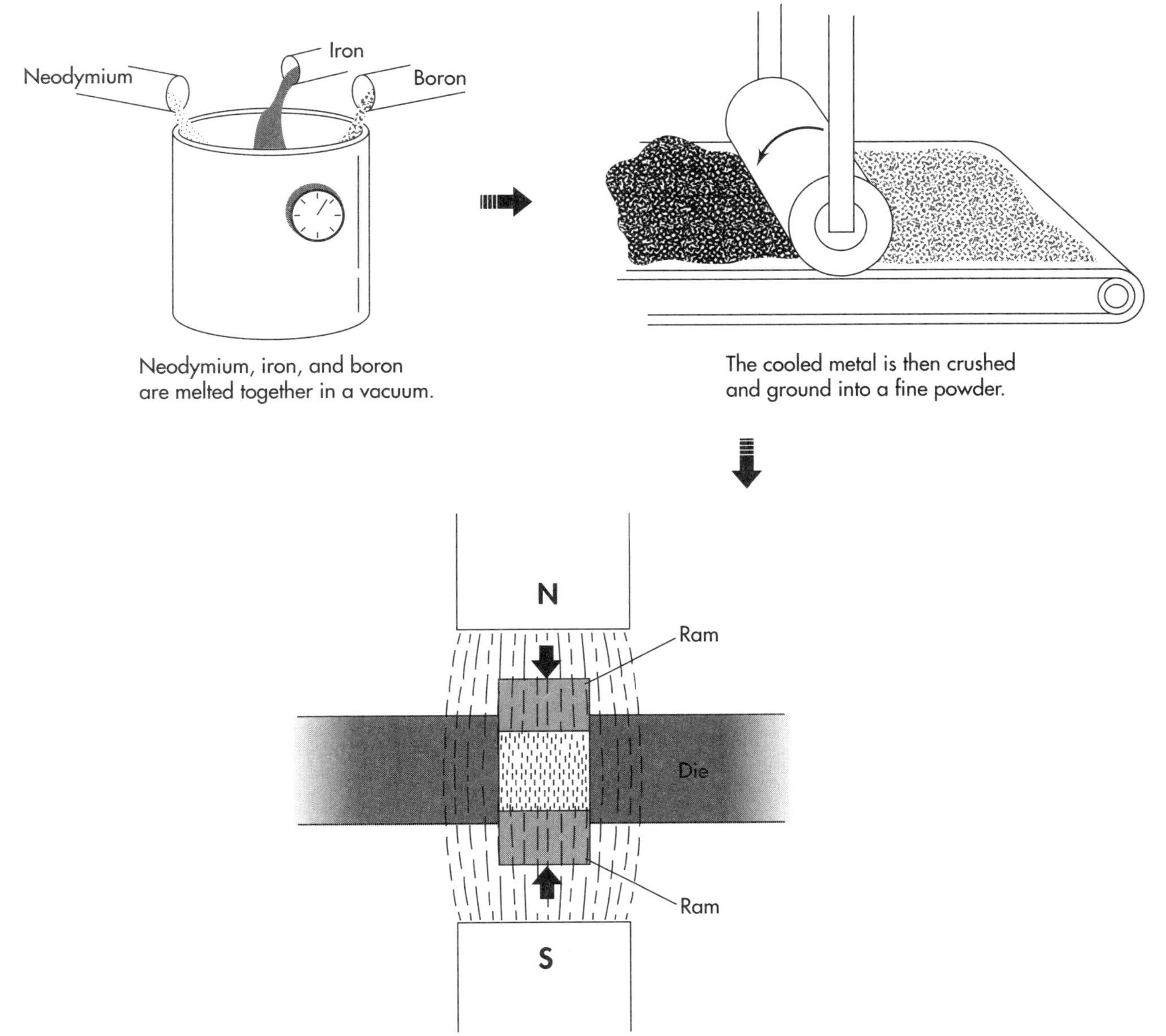

Neodymium, iron, and boron are melted together in a vacuum.

The cooled metal is then crushed and ground into a fine powder.

The powdered metal is placed in a mold. A magnetic force is applied to the material to line up the powder particles. At the same time, hydraulic rams on top and bottom compress the powder.

cobalt-modified iron oxides and chromium dioxide. The material is finely ground and coated on a thin polyester plastic film.

Other Magnetic Materials

Magnetic fluids can be made by encapsulating powdered barium ferrite particles in a single layer of molecules of a long-chain polymer plastic. The particles are then held in suspension in a liquid like water or oil. Because of the plastic encapsulation, the magnetic particles slide over each other with almost no friction. The particles are so small that normal thermal agitation in the liquid keeps the particles from settling. Magnetic fluids are used in several applica-

The above illustrations show a typical powdered metallurgy process used to produce powerful neodymium-iron-boron permanent magnets.

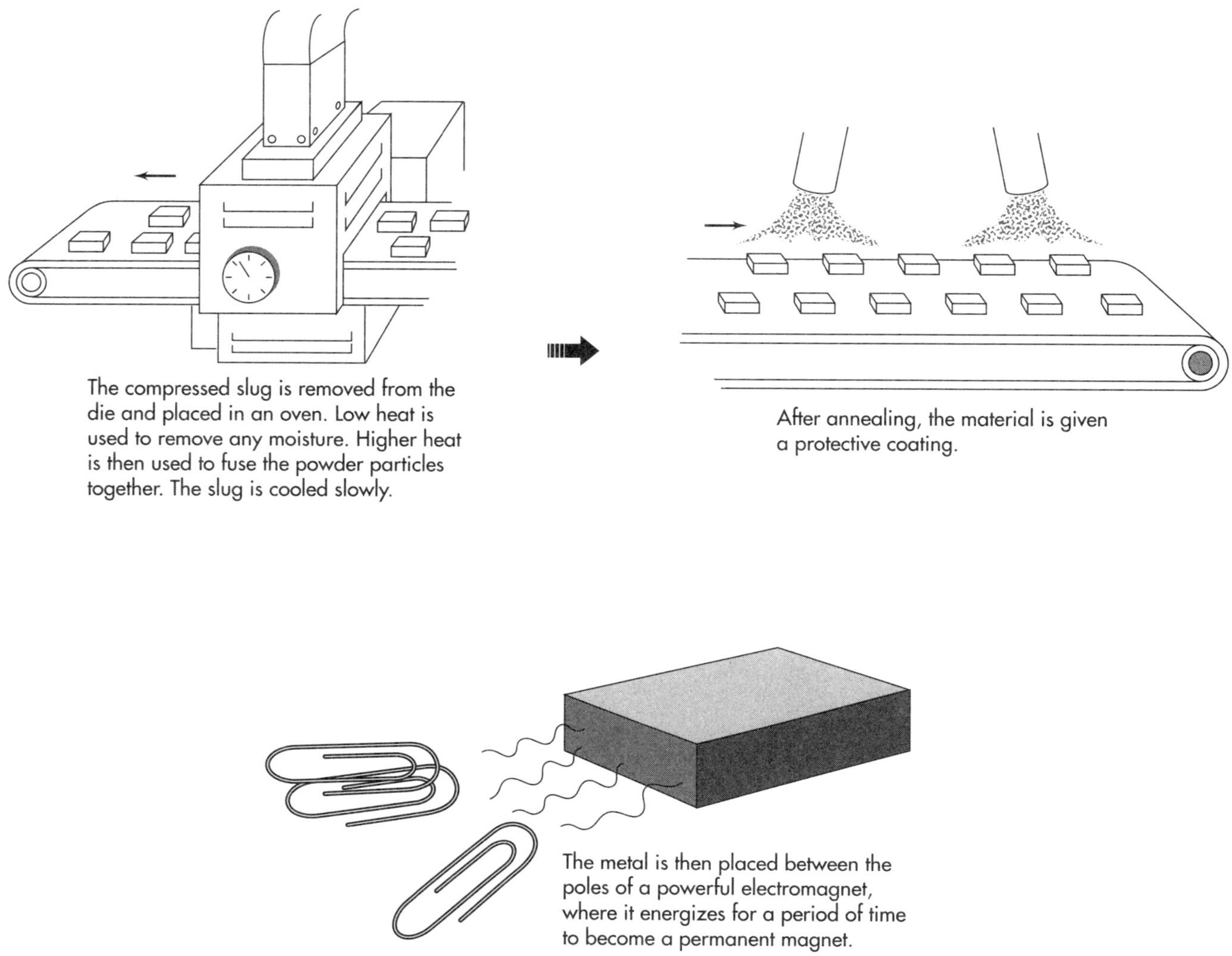

The compressed slug is removed from the die and placed in an oven. Low heat is used to remove any moisture. Higher heat is then used to fuse the powder particles together. The slug is cooled slowly.

After annealing, the material is given a protective coating.

The metal is then placed between the poles of a powerful electromagnet, where it energizes for a period of time to become a permanent magnet.

tions as sealants, lubricants, or vibration damping materials.

The Manufacturing Process

Just as the materials are different for different kinds of magnets, the manufacturing processes are also different. Many electromagnets are cast using standard metal casting techniques. Flexible permanent magnets are formed in a plastic extrusion process in which the materials are mixed, heated, and forced through a shaped opening under pressure.

Some magnets are formed using a modified powdered metallurgy process in which finely powdered metal is subjected to pressure, heat, and magnetic forces to form the final magnet. Here is a typical powdered metallurgy process used to produce powerful neodymium-iron-boron permanent magnets with cross-sectional areas of about 3-10 square inches (20-65 sq cm):

Preparing the powdered metal

1 The appropriate amounts of neodymium, iron, and boron are heated to melting in a vacuum. The vacuum prevents any chemical reaction between air and the melting materials that might contaminate the final metal alloy.

2 Once the metal has cooled and solidified, it is broken up and crushed into small pieces. The small pieces are then ground into a fine powder in a ball mill.

Pressing

3 The powdered metal is placed in a mold, called a die, that is the same length and width (or diameter, for round magnets) as the finished magnet. A magnetic force is applied to the powdered material to line up the powder particles. While the magnetic force is being applied, the powder is pressed from the top and bottom with hydraulic or mechanical rams to compress it to within about 0.125 inches (0.32 cm) of its final intended thickness. Typical pressures are about 10,000 psi to 15,000 psi (70 MPa to 100 MPa). Some shapes are made by placing the powdered material in a flexible, air-tight, evacuated container and pressing it into shape with liquid or gas pressure. This is known as isostatic compaction.

Heating

4 The compressed "slug" of powdered metal is removed from the die and placed in an oven. The process of heating compressed powdered metals to transform them into fused, solid metal pieces is called sintering. The process usually consists of three stages. In the first stage, the compressed material is heated at a low temperature to slowly drive off any moisture or other contaminants that may have become entrapped during the pressing process. In the second stage, the temperature is raised to about 70-90% of the melting point of the metal alloy and held there for a period of several hours or several days to allow the small particles to fuse together. Finally, the material is cooled down slowly in controlled, step-by-step temperature increments.

Annealing

5 The sintered material then undergoes a second controlled heating and cooling process known as annealing. This process removes any residual stresses within the material and strengthens it.

Finishing

6 The annealed material is very close to the finished shape and dimensions desired. This condition is known as "near-net" shape. A final machining process removes any excess material and produces a smooth surface where needed. The material is then given a protective coating to seal the surfaces.

Magnetizing

7 Up to this point, the material is just a piece of compressed and fused metal. Even though it was subjected to a magnetic force during pressing, that force didn't magnetize the material, it simply lined up the loose powder particles. To turn it into a magnet, the piece is placed between the poles of a very powerful electromagnet and oriented in the desired direction of magnetization. The electromagnet is then energized for a period of time. The magnetic force aligns the groups of atoms, or magnetic domains, within the material to make the piece into a strong permanent magnet.

Quality Control

Each step of the manufacturing process is monitored and controlled. The sintering and annealing processes are especially critical to the final mechanical and magnetic properties of the magnet, and the variables of time and temperature must be closely controlled.

Hazardous Materials, Byproducts, and Recycling

Barium and the barium compounds used to make barium ferrite permanent magnets are poisonous and are considered toxic materials. Companies making barium ferrite magnets must take special precautions in the storage, handling, and waste disposal of the barium products.

Electromagnets can usually be recycled by salvaging the component iron cores and copper wiring in the coil. Partial recycling of permanent magnets may be achieved by removing them from obsolete equipment and using them again in similar new equipment. This is not always possible, however, and a more comprehensive approach to recycling permanent magnets needs to be developed.

The Future

Researchers continue to search for even more powerful magnets than those available today. One of the applications of more powerful permanent magnets would be the development of small, high-torque electric motors for battery-powered **industrial robot**s and laptop computer disk drives. More powerful electromagnets could be used for the levitation and propulsion of high-speed trains using pulsed magnetic fields. Such trains, sometimes called maglev trains, would be supported and guided by a central, magnetic "rail." They would move without ever contacting the rail, thus eliminating mechanical friction and noise. Pulsed magnetic fields could also be used to launch satellites into space without relying on expensive and heavy booster rockets.

More powerful magnets could also be used as research tools to develop other new materials and processes. Intense, pulsed magnet fields are currently being used in nuclear fusion research to contain the hot, reacting nuclear plasma that would otherwise melt any solid material vessel. Magnetic fields can also be used in materials research to study the behavior of semiconductors used in electronics to determine the effects of making micro-sized **integrated circuit**s.

Where To Learn More

Books

Brady, George S. and Henry R. Clauser. *Materials Handbook*, 12th Ed. McGraw-Hill, 1986.

Braithwaite, Nicholas and Graham Weaver, eds. *Electronic Materials.* Butterworths, 1990.

Campbell, Peter. *Permanent Magnet Materials and Their Design.* Cambridge University Press, 1994.

Verschuur, Gerrit L. *Hidden Attraction: The History and Mystery of Magnetism.* Oxford University Press, 1993.

Periodicals

Boebinger, Greg, Al Passner, and Joze Bevk. "Building World-Record Magnets." *Scientific American,* June 1995, pp. 58-66.

Duplessis, John. "An Attractive Proposition." *Machine Design*, June 11, 1993, p. 46.

—Chris Cavette

Marbles

Background

Marbles are small, round, spherical objects made from glass or stone and most commonly used in children's games. They are usually less than an inch (2.54 cm) in diameter and often brightly colored or otherwise decorated. Their origins as recreational objects appear to date back several thousand years, and it is also believed that the primitive games played with marbles eventually evolved into the sports that we now know as bowling, billiards, and pinball. Marbles have numerous industrial uses as well—they are the noise inside a can of spray paint and the translucent letters and numbers on a **road sign**. Marbles are also melted down to make **fiberglass**, used in automotive bodies and draperies.

History

Small, round objects made from stone have been unearthed in the excavations of ancient cultures all around the globe. The antecedent of the marble was probably the nut, polished by youngsters in ancient times into a smooth surface for playing games. Both Greek and Roman youths played games with small balls made from clay, and marbles were discovered in the tomb of the young Egyptian pharaoh Tutankhamen. In North America, objects of stone and clay that appear to be marbles have been unearthed from several sites. One of the most famous is the Hopewell burial sites in Ohio. Marbles and marble games for children continued to be a popular form of entertainment well through the Middle Ages. Unfortunately, youngsters playing marble games came to be seen as delinquents, and efforts were made to restrict their marble-playing activities.

Most of the marbles used in medieval and Elizabethan times were made from clay. Around 1600, water-powered stone mills in Germany began producing more polished versions from the marble and alabaster quarries nearby, especially in the regions near Coburg and Oberstein. The word marble is derived from the German term "for the rock," and has come to mean any small, round sphere used as such. Soon the mills began grinding out versions from agate, limestone, brass, and gemstone, and these large operations could grind a marble into shape at the rate of about 800 an hour. This made Germany the center of marble manufacturing for several centuries.

Glass marbles, the most common version of the object today, only came into existence relatively recently in the history of the object. It is debated whether they originated in Venice, where glassblowing had become a well-developed industry since the ninth century, or back in Germany. Historians point to 1846 as the invention of *marbelschere* (marble scissors) by a glass factory employee in Germany. This tool resembled a pair of tongs with a small cup on one end and a slicing device on the other. A molten glass rod would be forcefully inserted into the cup, and the worker would then twist the cup, which would help form the sphere of the marble. Squeezing the tongs shut sliced off the rest of the glass. Such marbles can be identified by their pontil marks, the two tiny tags at each end of the sphere where the cooling glass was severed from the rest of the rod. The objects were further cooled inside a wooden barrel and then taken up with an **iron** spoon and inserted into an annealing oven, a process which yielded a tougher piece of

Because youths who played with marbles were seen as delinquents, the city of Nuremberg, Germany, passed an ordinance in 1503 confining marble games to a meadow outside the city limits. In the English village of St. Gall, church officials were permitted to flay marble-playing youths who had refused to heed previous warnings with cat-o'-nine tails.

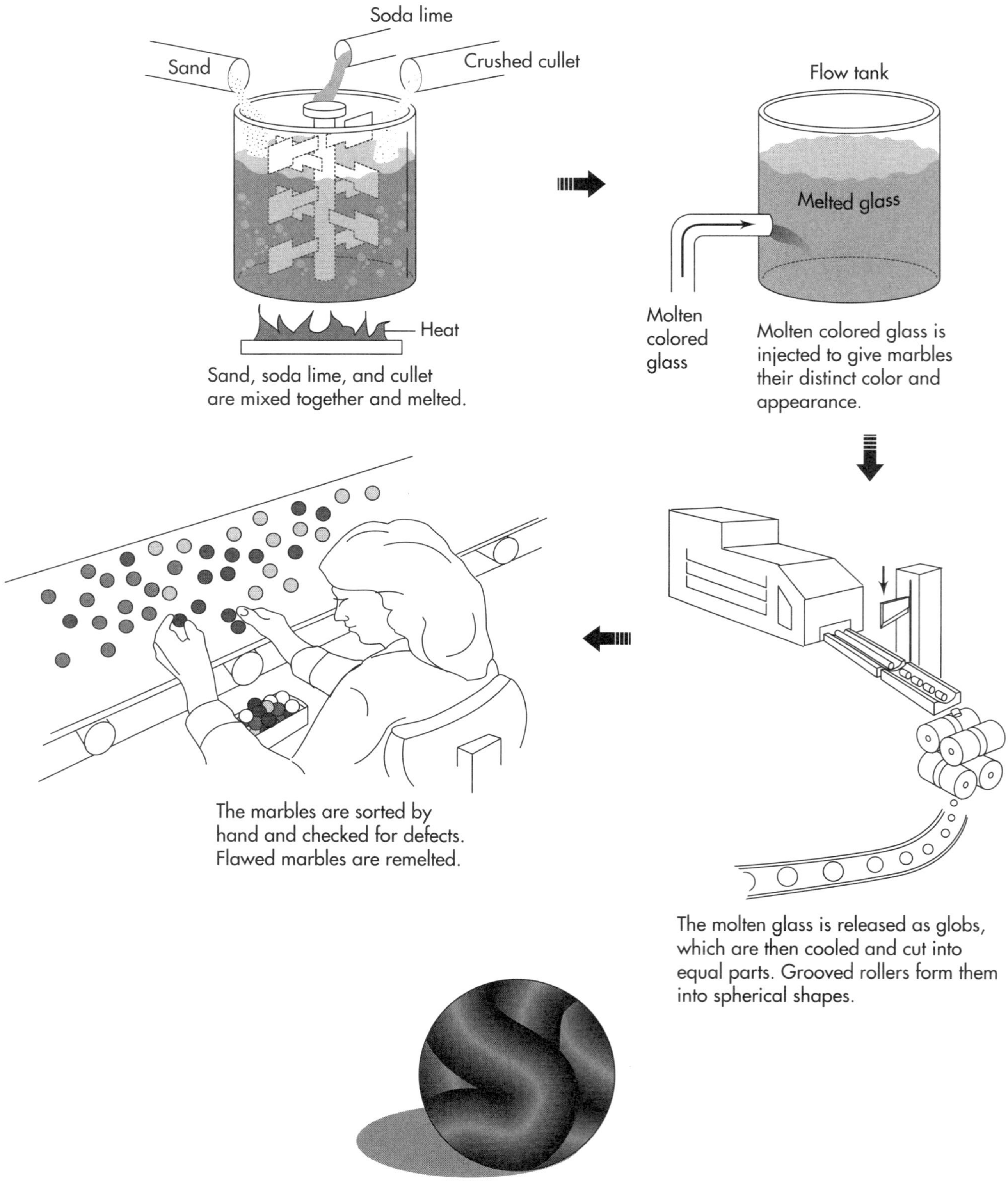
Soda lime
Sand
Crushed cullet
Heat
Sand, soda lime, and cullet are mixed together and melted.
Flow tank
Melted glass
Molten colored glass
Molten colored glass is injected to give marbles their distinct color and appearance.
The molten glass is released as globs, which are then cooled and cut into equal parts. Grooved rollers form them into spherical shapes.
The marbles are sorted by hand and checked for defects. Flawed marbles are remelted.

process which yielded a tougher piece of glass not likely to break or become brittle.

Marble manufacturing migrated to American shores in the later decades of the 19th century. In 1900 Martin Frederick Christensen received a patent for a machine that made near-perfect spheres of steel ball bearings. The first machine-made marbles were manufactured in a barn behind Christensen's house in Ohio, which eventually led to a prosperous manufacturing facility. By 1910 the 33 workers at the M.F. Christensen and Son plant were producing 10,000 marbles a day. The furnaces were fired by natural gas, however, and the onset of World War I brought rationing of the resource and spelled the fiscal end of Christensen's operation.

Akro Agate Company, founded in 1911 and originally based in Akron, Ohio, became the next major marble manufacturer. Further refinements in marble-making machinery came during the 1920s, and Akro Agate enjoyed a position as the major marble manufacturer during the subsequent decades. But the popularity of marbles as toys waned as more sophisticated gadgets entered the children's toy market. Many of the American marble manufacturing firms countered this by diversifying operations into industrial glass production, such as making automobile windshields.

Today, marbles are still produced in record numbers, but most are made in Third World factories. One such operation, Vacor de Mexico, located in Guadalajara, produces about 12 million marbles a day, which are then shipped to 35 different countries.

Raw Materials

Modern marbles are made from a combination of sand, soda lime, silica, and several other ingredients added for pigment or decoration. These other additives range from aluminum hydrate to zinc oxide. The primary component, sand, is essentially loose, granular particles of disintegrated rock. Soda lime is the chemical term for the mixture of calcium hydroxide and sodium or potassium hydroxide. It is a drying agent and absorbs carbon dioxide. Another compound used in marble manufacturing is silica, a white or colorless crystalline found in agate, flint, quartz, and other rocks. Some marbles are also made from cullet, or scrap glass.

The Manufacturing Process

Meltdown

1 Sand, soda lime, and crushed cullet are fed into a large, furnace-driven tank. In the tank, the mixture is heated to 2300°F (1260°C) to melt the raw materials. This can take as long as 28 hours.

Injection

2 Next, the molten mixture moves out of the tank through an opening into another vat known as the flow tank. There an opening in the tank injects molten colored glass. This hot, pigmented glass gives the marbles their distinct appearance. A green marble has been injected with glass containing iron oxide; cobalt results in a blue marble; and manganese will yield a purple one. The use of uranium oxide gives marbles an eerie, greenish-yellow cast. The speed and force of the injection determines the final design of the marble. A grooved feeder device, patented by the Akro Agate Company, was able to produce multicolored marbles known as corkscrews.

Cutting and cooling

3 Next, the still-molten glass is released from the flow tank as globs of glass. Automatic cutting devices slice the mixture into equal parts. The globs travel down metal ramps that simultaneously cool them and perfect their spherical shape. Next, they travel down a second metal slide and are sorted by hand. These grooved rollers were the invention of Horace Hill, a former employee of Christensen and Son and later founder of Akro Agate. This device produced marbles much more quickly and reduced the labor necessary by nearly two-thirds. Marbles with flaws are sent back to another area of the factory for re-melting. The marbles cool off in 5-gallon (19 l) containers that house 5,000 marbles at a time.

Where To Learn More

Books

Barrett, Marilyn. *Aggies, Immies, Shooters, and Swirls: The Magical World of Marbles*. Bulfinch Press, 1994.

Ferretti, Fred. *The Great American Marble Book*. Workman Publishing, 1973.

Periodicals

Block, Stanley A. "The Akro Agate Company." *Antiques and Collecting*, April 1988, pp. 38-40.

"Marbles: An American Industry Still Involved in a Shoot-Out." *Compressed Air*, July-August 1994, pp. 22-29.

—*Carol Brennan*

Milking Machine

Background

The milking machine is a nearly automatic machine installation for milking cows. It is not a single unit, but rather an assembly of components designed to handle as many as 200 cows an hour. The system consists of the cluster (the assembly that is manually attached to the cow), a milk tube, a pulse tube and pulsator, a vacuum pump or blower, and perhaps a recorder jar or milk meter that measures yield. Together, the system allows milk to flow into a pipeline in preparation for shipping to a processing plant.

The cluster consists of teatcups, a shell and liner device that actually performs the milking action, and a claw or manifold that spaces the teatcups and connects them to the milk and pulse tubes. The milk tube carries the milk and air mixture away from the cow's udder to receiving tanks. The pulse tube, or airline, carries the varying air pressure from the pulsator device to the tanks, drawing the milk and fluids out of the cows as well.

In operation, milk is drawn from the cow's teats because a vacuum is created within the cup device, forcing the milk through the teat canal. The pulsator alternates the pressure, first creating a vacuum (milk phase), and then applying air, which causes the flexible liner in the cup to collapse and massage the teat (rest phase). The alternating process of milk-and-rest is continued in a rhythmic pattern for the cows' health and good milk productivity.

History

Early attempts at milking cows involved a variety of methods. Around 380 B.C., Egyptians, along with traditional milking-by-hand, inserted wheat straws into cows' teats. Suction was first used as a basis for the mechanized harvesting of milk in 1851, although the attempts were not altogether successful, drawing too much blood and body fluid congestion within the teat. To encourage further innovations, the Royal Agricultural Society of England offered money for a safe, working milking machine. Around the 1890s Alexander Shiels of Glasgow, Scotland, developed a pulsator that alternated suction levels to successfully massage the blood and fluids out of the teat for proper blood circulation. That device, along with the development of a double-chambered teatcup in 1892, led to milking machines replacing hand milking. After the 1920s machine milking became firmly established in the dairy industry. Today, the majority of all milking is processed by machine.

Early attempts at milking cows involved a variety of methods, including inserting wheat straws into cows' teats. Suction was first used as a basis for the mechanized harvesting of milk in 1851, but it was not until the 1920s that machine milking became firmly established in the dairy industry.

The Manufacturing Process

The milking machine components are created and assembled in several major manufacturing plants throughout the world using traditional processes and procedures. Stainless steel and plastic are used for containers and liners and cast **iron** and steel for vacuum pumps, controls, and metering devices.

Receiving

1 Here stainless steel is received in large sheet or tube form. Stainless steel is used to fabricate components that will come in contact with milk. The sheets are protected from scratches by a vinyl lining, which will be removed later after forming and machin-

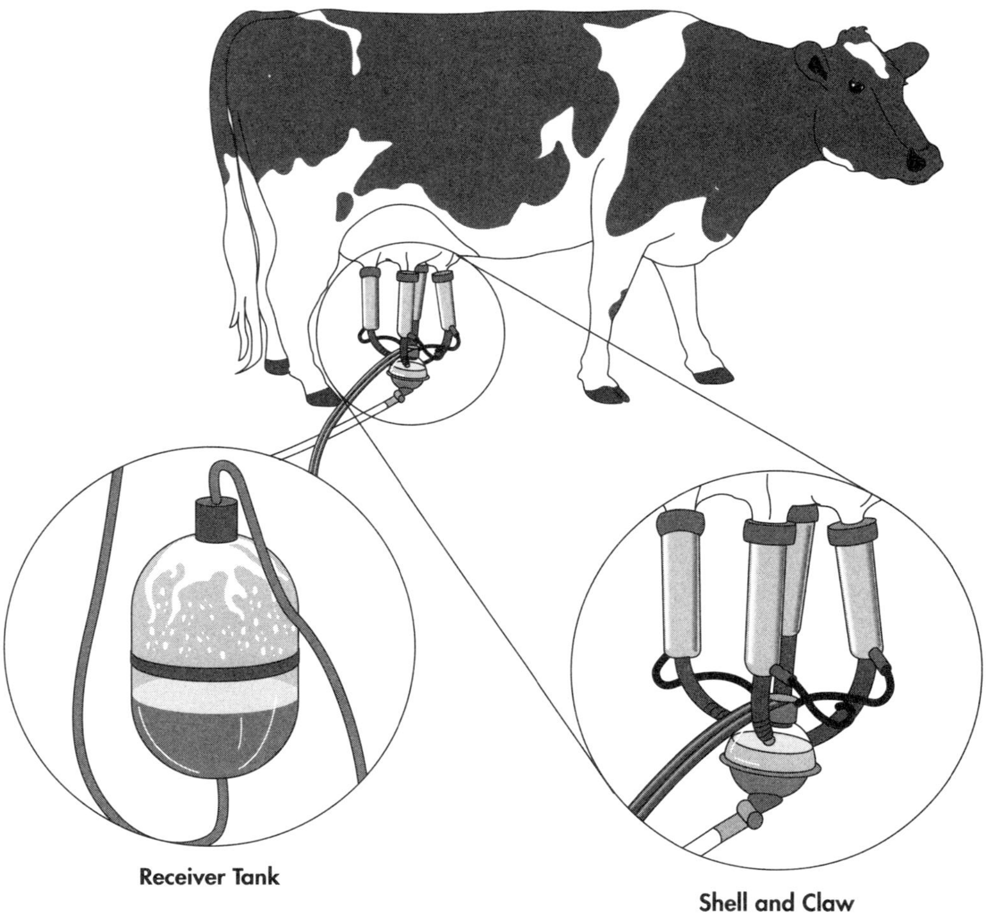

Milking machine components are created using stainless steel and plastic for containers and liners, and cast iron and steel for vacuum pumps, controls, and metering devices.

ing. At this point, the stainless steel from the foundry has a dull finish.

Cutting, machining, and forming

2 The steel sheets may be sent directly to a polishing station where large machines create a smooth finish. Once the initial finish is achieved, the sheets may move on rollers to numerically controlled punch machines, where they are cut into shapes for various parts. From there, some parts are formed or bent into shape on large brakes. Mounting panels for controls, structural components, and small detail parts for the claws and pumps are made at this stage.

Creating the shell and claws

3 The shell that contains the liner is fabricated as an extrusion. This process involves forcing steel tubing over a die under heat and pressure to form the elongated rigid piece. The claw is made as a series of smaller extruded pieces of tubing that form the manifold, which spaces the teatcups in a cluster formation. The extrusions are then manually welded together in fixtures according to the desired size. Both the completed claws and shell have a dull finish that will require polishing.

Making the receivers

4 Receiver tanks are stainless steel tanks that receive milk from the milk lines, generally holding from 15-26 gallons (60-100 l). Although some are customized plastic, most are steel with the heads or ends spun on specialty machines. This production technique shapes a thin steel disc as it is being turned in a lathe. The disc is shaped

as it is forced over a steel shape or mandrel. Once the operator forms the ends of the receiver as cups, another technician will weld the body to the head, leaving orifices or openings for milk input and output. Individually manufactured, the receivers will also be polished by hand. Some receivers will have translucent plastic panel inserts so that dairy farmers can visually gauge the cow's milk production.

Polishing

5 All major stainless steel components are polished to the familiar finish associated with food-handling equipment. The manufacturers use a variety of mechanized belts, cloths, and spinning wheels in what becomes a very labor intensive process to meet government and industry sanitary standards. This polishing is in addition to the polishing the large stainless sheets undergo following receiving. Workers wield an assortment of hand polishers and attachments to shine all contact and protective surfaces, from the claws to receivers to pipelines.

Making the vacuum pumps or blowers

6 These "drivers" of air and fluids through the system are manufactured by select vendors. The unit is basically a sealed iron case with a set of timing gears inside and an impeller, resembling two blades or paddles on a shaft, that spins at over 3,000 revolutions per minute to create a vacuum that will draw fluids in the line. The making of the pumps requires raw castings to be poured for the various parts. The rough parts are hand-machined on a bench by technicians and then placed on a CNC rotating cutter table where the cutting stone makes two precise passes along the pre-determined edges to cut away any rough surfaces. The assembly of the major components—along with seals, bearings, and shafts—is critical because a variation of just 0.002 inch can change the balance and affect the pump. The components are placed on a coordinate measuring machine where the profile is checked so that all parts fit properly. The technicians, using feeler gauges, will set proper clearances before locking the critical gear parts in place and bolting the unit together. Then they will run the pumps in a test booth before releasing them for distribution.

Assembly

7 The manufactured and purchased parts are placed in several different areas where the milking machine subassemblies come together. Technicians will select the parts according to the system design specifications, often customized to meet certain operations. It is at this stage that pressures and loads will gauge equipment performance. Final assembly of all machine components will not occur until after shipment to the farm and installation in barns, where often elaborate stalls and stations will be used to maximize the milk harvest.

Workers at assembly will also make initial line connections from valves to pumps to meters, checking for tolerances and poor fits. Vacuum pumps or blowers, the heart of the milking system, are tested so that both milk and accompanying air can be efficiently removed. The pumps will be tested under simulated loads.

Vacuum controllers, which admit the necessary air to maintain the proper vacuum level, are selected. The air lines and milk lines are selected to size and connection hardware grouped. Controls, also purchased, are prepared for installation.

Standards

Milking machine manufacturers are subject to a variety of standards, some self-imposed. In addition to inspections throughout the manufacturing process, all installations are set up by trained dealers and electrical contractors. Equipment designers follow Association of Agricultural Engineer standards and sanitary guidelines established by a dairy industry council.

The Future

Advances in technology have introduced several new innovations to milking machines. Automatic detacher units that connect loosely to the milking claw allow cows to move and shift freely during milking. Based on the rate of milk flow, the detacher can also detect the end of milking, shutting the vacuum and actually removing the claw from the cow.

Automatic backflushing units are also gaining popularity. These units and systems send chemical and rinse solutions through pipelines and clusters to reduce the risk of infection and mastitis (udder inflammation).

The use of automatic identification systems, such as electronic transponder cow necktags, have enabled dairy farmers to keep track of milk production by individual cows.

Robotics are at the forefront of milking technology, especially in Europe. Automatic attachment devices have been created but not quite perfected. This new innovation will require little manual labor, and the machines will oversee much of the milking process from the time a cow enters a milking center until it leaves to graze.

Where To Learn More

Books

Machine Milking and Lactation. Insight Books, 1991.

Maximizing the Milk Harvest. Milking Machine Manufacturers Council, 1993.

Periodicals

Blankesteijn, Herbert and Daniel Clery. "Milking Automation for All It's Worth." *New Scientist*, February 1, 1992, p. 27.

Spahr, Sid. "Coming: New Dairy Ideas from Europe and Israel." *Successful Farming*, October 1987.

Wittenberg, Gunter. "A Robot to Milk Cows." *The Industrial Robot*, 1993, pp. 22-25.

—Peter Toeg

Nail

Background

A nail consists of a metal rod or shank, pointed at one end and usually having a formed head at the other, that can be hammered into pieces of wood or other materials to fasten them together. A nail is usually made of steel, although it can be made of aluminum, brass, or many other metals. The surface can be coated or plated to improve its corrosion resistance, gripping strength, or decorative appearance. The head, shank, and point may have several shapes based on the intended function of the nail. Of the nearly 300 types of nails made in the United States today, most are used in residential housing construction. The average wood frame house uses between 20,000 and 30,000 nails of various types and sizes.

Nails are divided into three broad categories based on their length. In general nails under 1 inch (2.5 cm) in length are called tacks or brads. Nails 1-4 inches (2.5-10.2 cm) in length are called nails, while those over 4 inches (10.2 cm) are sometimes called spikes. These categories are roughly defined, and there is considerable crossover between them.

The length of a nail is measured in a unit called the penny. This term comes from the use of nails in England in the late 1700s when it referred to the price of one hundred nails of that size. For example, a "ten penny nail" would have cost ten pennies per hundred. The symbol for penny is "d," as in 10d. This designation is believed to go back to the time of the Roman Empire when a similar form of measurement for hand-forged nails involved a common Roman coin known as the *denarius*. Today the term penny only defines the length of a nail and has nothing to do with the price. The shortest nail is 2d which is 1 inch (2.5 cm) long. A 10d nail is 3 inches (7.6 cm) long, and a 16d nail is 3.5 inches (8.9 cm) long. Between 2d and 10d the nail length increases 0.25 inch (0.64 cm) for each penny designation. Beyond 10d there is no logical progression to the lengths and designations.

Nails may have been used in Mesopotamia as early as 3500 B.C. and were probably made of copper or bronze. Later, **iron** was used to make nails. Early nails were shaped, or forged, with hammers. They were usually made one at a time, and were consequently scarce and expensive. By the 1500s a machine was developed which produced long, flattened strips of iron, called nail rods. These strips could then be cut into lengths, pointed, and headed. Nails were so valuable in the early American settlements that in 1646 the Virginia legislature had to pass a measure to prevent colonists from burning down their old houses to reclaim the nails when they moved. Two early nail-making machines were patented by Ezekial Reed of the United States in 1786 and Thomas Clifford of England in 1790. These machines cut tapered pieces from flat iron sheet, then flattened the head. In rural areas, blacksmiths continued to make nails from wrought iron right into the 20th century. The first machine to make nails from metal wire was introduced in the United States in about 1850, and this technique is now used to make most of the nails today.

Design

Most of the 300 different types of nails produced in the United States today require no

There are nearly 300 different types of nails made in the United States today. The average wood frame house uses 20,000 to 30,000 nails of various types and sizes.

This massive, mid-19th-century nail cutting machine used a shearing action to cut nails from bar iron. (From the collections of Henry Ford Museum & Greenfield Village.)

Nails are essential to the construction of wood-framed buildings. This, however, was not always the case. Until the late 18th century, Americans built wooden buildings using heavy timber frames. At places where these massive timbers had to hold together, one end of a post or beam would be cut down to form a tongue ("tenon") and fitted into a hole ("mortise") cut in the adjoining beam. Additional strength could be added by driving wooden pegs through auger holes in the joined timbers. The skill and labor involved in such construction was considerable; carpenters had to be highly skilled individuals and, as such, commanded high prices in colonial America.

Until the end of the 18th century, nails were imported from England or made by local blacksmiths. The smithy, or often his apprentice, took a piece of bar iron maybe 5 feet long and 0.06-0.25 inch in diameter. Holding one end he heated the other, laid it on the anvil and, using the flat face of his hammer, tapered all four sides to about an inch from the end. He then used the peen, or sharpened end of his hammer, or a hardy, a wedge-shaped attachment to his anvil, to cut a notch in the rod. He thrust the sharpened end of the rod into a tapered hole in his anvil and snapped off the short nail. Then he flattened the end of the nail with four or five quick strikes of the hammer and popped it out of the anvil hole with a quick, upward strike at the point.

Between 1790 and 1830, several mechanical devices were developed in Europe and the United States to speed the production and lower the cost of nails. It is not entirely coincidental, therefore, that the balloon-framed house, which relied on two-by-fours held together by nails, was invented in the early 1830s in Chicago. The balloon-frame system required much less skill and labor in carpentry and made use of mass-produced nails.

William S. Pretzer

new design work. Once a nail has been designed, forming dies and processes are developed for its manufacture, and the nail is produced in quantity.

Most nails have a broad, circular head. Finishing nails have a narrow, tapered head which allows them to be countersunk below the surface of the material and covered over to produce a smooth finish. Upholstery nails have decorative heads. Double-headed nails are used to fasten wood forms used in concrete pouring. The nail is driven in up to the first head, leaving the second head protruding. The protruding head allows the nails to be easily removed and the forms quickly dismantled once the concrete has hardened.

The shank is usually designed to be round and smooth. Shanks with serrations, annular grooves, spiral flutes, or helical threads are used when a stronger, more permanent grip is required. Thermoplastic coatings may also be added to the shaft. These coatings heat up through friction while the nail is being driven, then quickly cool and set to lock the nail in place. The diameter of the shank is determined by the type of nail. Most nails, called common nails, have a relatively large diameter. Box nails, originally used to make thin-walled boxes, have a smaller diameter shank than common nails. Finishing nails have a very small diameter shank in order to make the smallest hole possible.

The most typical nail point is a four-sided tapered cut called a diamond point. Other nails may have a blunter point to prevent splitting certain woods. Chisel points, barbed points, needle points, and many others are sometimes used on specialty nails.

As new building materials become available, nail manufacturers work to develop new nails. There are special nails for tile roofing, hardwood flooring, shingles, rain gutters, wall board, sheet metal, and concrete. Some new nails are designed to be driven by air-powered nail guns rather than by a hammer. There have even been new nails designed for specific applications in the aerospace industry.

Raw Materials

Most nails are made of steel. Aluminum, copper, brass, bronze, stainless steel, nickel

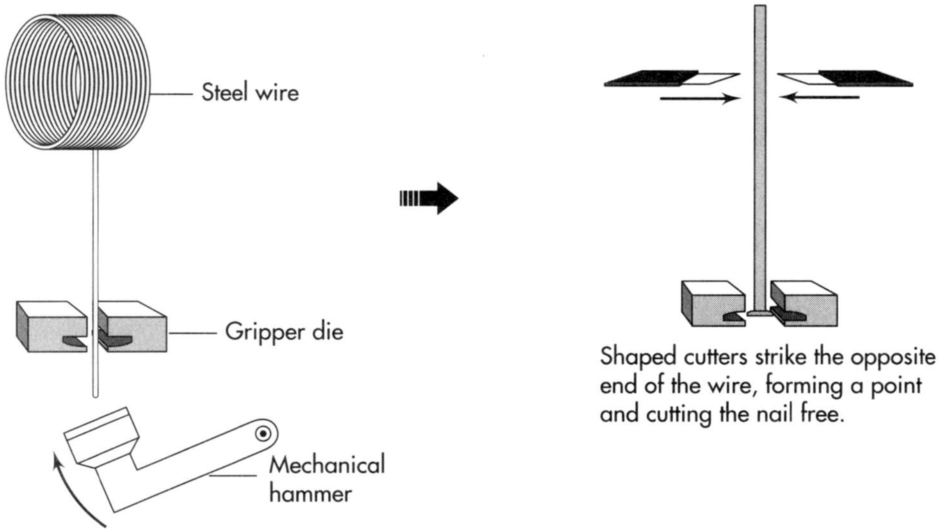

Steel wire is drawn from a coil and fed into a nail-making machine where it is held by a gripper die, which has the shape of the nail head in its mold. A mechanical hammer strikes the wire.

Shaped cutters strike the opposite end of the wire, forming a point and cutting the nail free.

silver, monel, **zinc**, and iron are also used. Galvanized nails are coated with zinc to give them added corrosion resistance. Blued steel nails are subjected to a flame to give them a bluish oxide finish that provides a certain amount of corrosion resistance. So-called cement-coated nails are actually coated with a plastic resin to improve their grip. Some brads are given a colored enamel coating to blend in with the color of the material they are fastening.

The Manufacturing Process

Most nails are made from coils of metal wire. The wire is fed into a nail-making machine which can produce up to 700 nails per minute. The nails may then be further twisted or formed, cleaned, finished, and packaged.

Forming

1 Wire is drawn from a coil and fed into the nail-making machine where it is gripped by a pair of gripper dies. The shape of the head of the nail has been machined into the end of the dies.

2 While the dies clamp the wire in place, the free end of the wire is struck by a mechanical hammer. This deforms the end of the wire into the die cavity to form the head of the nail.

3 With the wire still clamped in the dies, a set of shaped cutters strike the opposite end of the nail, forming the point and cutting the nail free from the rest of the wire coming off the coil.

4 The dies open and an expelling mechanism knocks the nail into a collection pan below the machine. The free end of the wire is drawn from the coil and fed into the machine. The cycle then begins again.

Additional forming

5 Nails with helical twists, serrations, or other surface configurations are fed into other machines that roll, twist, stamp, or cut the required forms. This may be a purely mechanical process or may require heating the material before forming.

Finishing

6 The nails are cleaned in a rotating barrel filled with hot caustic soda. This

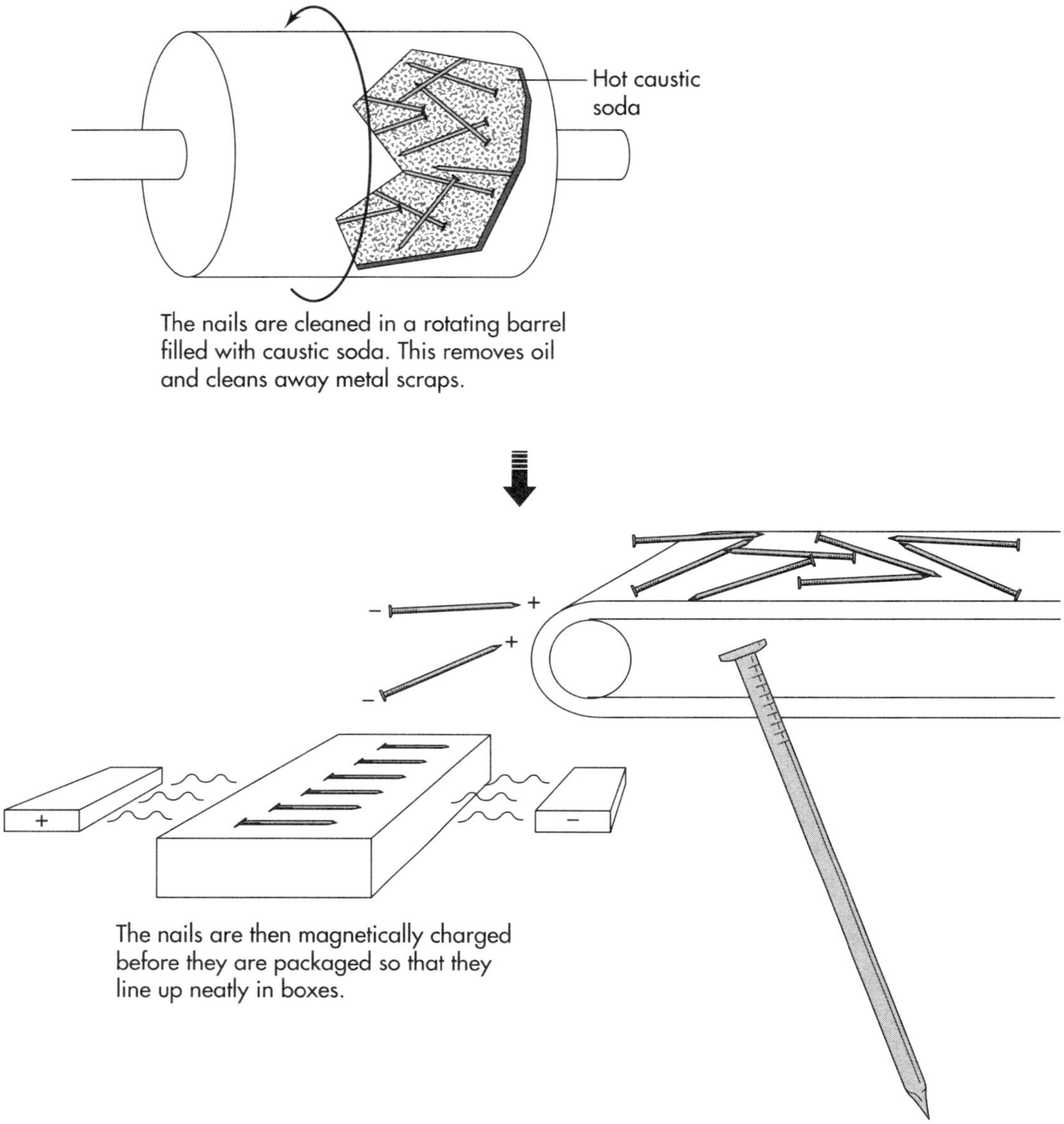

The nails are cleaned in a rotating barrel filled with caustic soda. This removes oil and cleans away metal scraps.

The nails are then magnetically charged before they are packaged so that they line up neatly in boxes.

removes any oil from the forming machine and cleans up any small metal scraps, or nippings, that might be clinging to the nails.

7 Many nails are given a final bright finish before being packaged. This is accomplished by placing the nails in a rotating drum of hot sawdust to lightly polish the surface of the nails. Other nails may be passed through an open flame in an oven to give them a blued finish. Galvanized nails are dipped into a tank of molten zinc in a process called hot-dip galvanizing. A zinc coating may also be applied by heating the nails to about 570°F (300°C) in a closed container filed with a powder composed of zinc dust and zinc oxide. Other coated nails are either dipped or sprayed to obtain their final finish.

8 Depending on the tolerances desired, some specialty nails may also require an additional heat treating step.

Packaging

9 Magnetic elevators convey the finished nails to weighing machines which drop them into open cardboard boxes. As they are

dropped in, a magnetic field aligns them so they stack in neat rows. After they are packaged, the nails are demagnetized. Nails are usually sold in boxes of 1, 5, 10, 25, and 50 pounds. Smaller nails, such as brads, are sold in 2-ounce or 4-ounce boxes and are packaged without being magnetically aligned.

Quality Control

Raw materials must meet certain standards for chemical composition, yield strength, hardness, corrosion resistance and other properties. These are usually certified by the company supplying the wire, and may be independently checked by the nail manufacturer.

During manufacture, nails must also meet certain specifications regarding dimensions and properties. These are achieved using a method known as statistical process control, which periodically samples the dimensions and properties of the nails being produced and evaluates any changes through statistical analysis techniques.

The Future

The demand for mass-produced commodity nails is dependent on the fluctuations in the housing market, which varies with the economy. Demand for these nails is also subject to competition from foreign manufacturers, further reducing profits.

The demand for specialty nails, on the other hand, is expected to continue to grow and be profitable. New building materials, such as composite wood-fiber and cement-based siding and roofing, require new specialty nails. New corrosion-resistant coatings for nails are also being developed.

One unique new nail market is the result of the increase in building restoration and preservation efforts throughout the country. One nail factory in Massachusetts makes old-fashioned cut nails. They estimate that 20% of their work is in producing a variety of these nails for use in authentic building restoration projects.

Where To Learn More

Books

Loveday Jr., Amos J. *The Rise and Decline of the American Cut Nail Industry.* Greenwood Press, 1993.

Vila, Bob. *This Old House Guide to Building and Remodeling Materials.* Warner Books, 1986.

Periodicals

Johnson, Duane. "Nails Get a Better Grip." *The Family Handyman,* September 1994, p. 16.

Whorf, Amy. "A Thumbnail History of Nails." *Country Living,* June 1993, p. 72.

—Laurel M. Sheppard

Neon Sign

Once chiefly used in business establishments, neon displays have found their way into consumer products such as telephones and automobile license plate frames. There are even neon displays that cover portions of the exterior of cars for the ultimate in a "flashy" vehicle.

Background

A neon sign is a lighting display made of glass tubes that have been filled with a gas and bent into the shape of letters or decorative designs. When a high-voltage electrical current is passed through the gas, the tubes emit light. Although neon gas was originally used in these signs, several other gases are also used. These gases, along with different tints and phosphor coatings for the glass tubes, produce a spectrum of over 50 brilliant colors. Neon signs can be as simple as a small advertising sign for beer, or as complex as a multi-story facade on a Las Vegas casino.

Neon signs evolved from scientific experiments in which various gases were subjected to high-voltage currents. In 1856, Heinrich Geissler produced a light source by passing a high-voltage alternating current through a low-pressure gas sealed in a glass tube. Subsequent experiments showed that almost all gases would conduct an electric current, and that many would produce light. The problem was that most of the common gases, like carbon dioxide, would react with the current-carrying electrodes within the sealed tube. This quickly reduced the efficiency of the electrodes until the light sputtered and died. In 1898, Sir William Ramsay and Morris William Travers developed a method for the fractional distillation of liquid air. In the process, they discovered the rare gas elements neon, argon, krypton, and xenon. Using these gases in sealed glass tubes, they produced colored light sources ranging from a bright reddish-orange for neon to an intense grayish-blue or violet for argon. Not only did these gases produce colored light, but they were chemically inert and did not react with the electrodes.

Fractional distillation of liquid air remained an expensive process until 1907 when Georges Claude of France and Karl von Linde of Germany developed a more economical method. Georges Claude's original interest was to produce quantities of oxygen for use in hospitals and industries. The rare gases that were also produced by this distillation process had no ready market, which prompted Claude to seek potential applications. Utilizing the previous experimental work of Ramsay and Travers, he began promoting illuminated signs using tubes filled with neon gas. He displayed his first neon sign at an exposition in Paris in 1910, and made his first commercial installation in 1912. By 1915, business was so promising that he formed the Claude Neon sign company and began selling franchises.

Neon signs came to the United States in 1923 when a Los Angeles car dealer, Earle C. Anthony, bought two of Claude's signs for his Packard dealership. Throughout the 1920s and 1930s, neon tubes were used for signage as well as decorative displays, and they became an integral part of the architecture of many buildings. By 1947, several casinos in Las Vegas began to draw attention with their elaborate neon lights.

During the 1950s and 1960s, neon signs were slowly displaced by plastic signs illuminated from the inside with fluorescent tubes. Recently, neon has made a comeback in both commercial signage and as an artistic medium. In Los Angeles, the Museum of Neon Art features historical and contemporary neon works. It also conducts

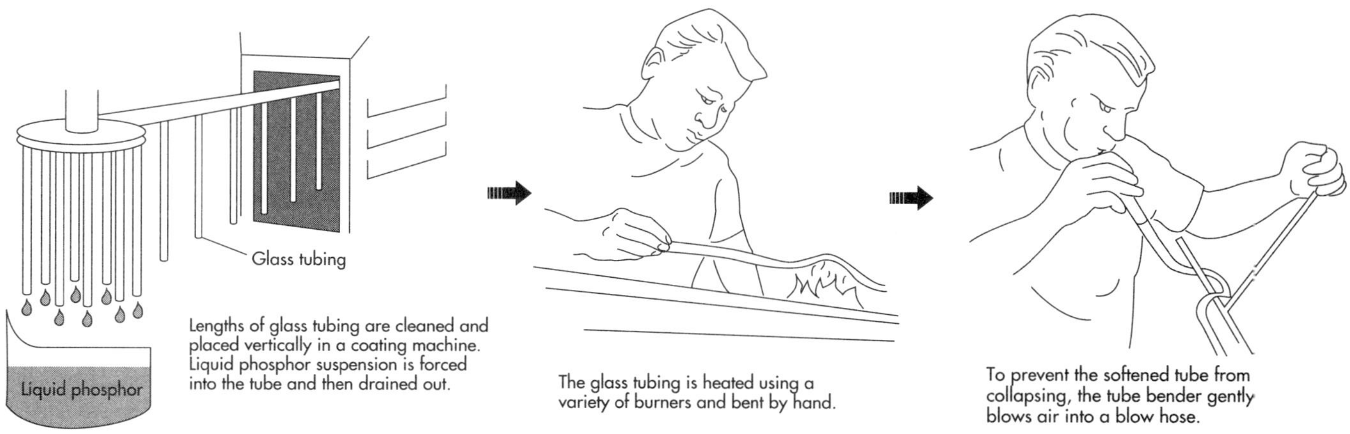

Lengths of glass tubing are cleaned and placed vertically in a coating machine. Liquid phosphor suspension is forced into the tube and then drained out.

The glass tubing is heated using a variety of burners and bent by hand.

To prevent the softened tube from collapsing, the tube bender gently blows air into a blow hose.

monthly tours of notable examples of neon displays throughout the city.

Raw Materials

Although neon gas was originally used in neon signs, it is now only used to produce reds and oranges. Argon, or an argon-neon mixture, is used in most signs. To improve the intensity of the light, a small amount of mercury is added to the argon to produce an intense blue light. This light impinges a variety of light-emitting phosphorescent materials coated on the inside of the glass tube to produce various colors. Optical tints in various colors may also be used, or the glass may be left clear if a strong blue light is desired. Xenon, krypton, and helium gases are sometimes used for special color effects.

The glass tubing used in neon signs is made from soft lead glass that is easily bent and formed. It ranges from 0.3 inches (8 mm) to 1.0 inches (25 mm) in diameter and comes in lengths of 4-5 feet (1.2-1.5 m).

The electrodes in each end of a section of illuminated tubing are usually made from very pure **iron** surrounded by a cylindrical glass jacket or envelope with one end open. A wire is attached to the metal electrode and passes through the closed end of the glass envelope. The closed end is sealed into the end of the sign tubing with the open end protruding into the tube.

The high-voltage electricity to power the sign is provided by a transformer which converts the 120 volts from the electrical lines to as much as 15,000 volts for the sign. Typical current ratings for neon signs are 30-60 milliamperes, although the transformers are usually sized at twice that amount. The transformers are connected to the electrodes in the sign using special wire, called GTO wire, that is insulated for at least 7,500 volts. This wire is also used to connect the individual sections of illuminated tubing in series. The wire is connected to the transformer through an insulated housing made of borosilicate glass with a spring connection on one end. The transformer and wires are purchased from a separate manufacturer and installed by the sign maker.

The sign tubing is supported in several ways. Small indoor signs usually have a thin steel skeleton framework that supports both the tubing and the power transformer. The framework is painted black so it will be less visible, making the sign seem to float in space. Large outdoor signs may be supported by wood, steel, or aluminum structures. The glass tubing is held by glass supports with metal bases. The transformer is placed inside a cabinet to protect it from the weather.

Design

Manufacturing neon signs is as much an art as it is a mechanical process. With only a

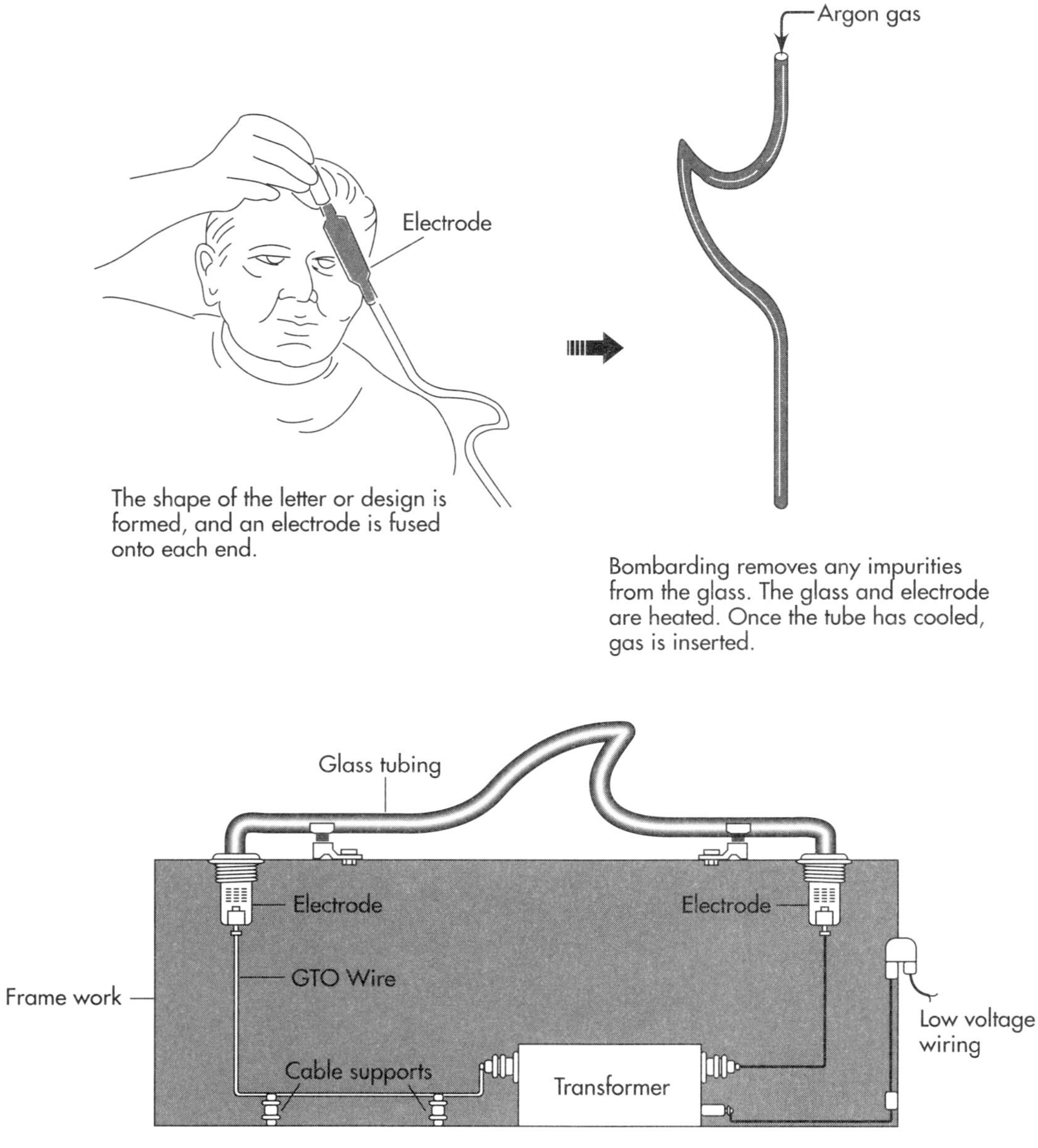

The shape of the letter or design is formed, and an electrode is fused onto each end.

Bombarding removes any impurities from the glass. The glass and electrode are heated. Once the tube has cooled, gas is inserted.

The neon sign is then attached to a transformer and mounted inside a framework to protect the wiring and elements from the weather.

few exceptions each sign is unique and must be designed to fit the desired display within the confines of the available space. Considerations of the diameter of the tubing, the minimum radius the tubing can be bent, and the overall length of tubing the transformer can power all limit the final design. For example the smaller the diameter of the tubing, the brighter the light. Conversely a smaller diameter tubing requires more power, thus limiting the overall length of tubing one transformer can handle.

The Manufacturing Process

Manufacturing neon signs is largely a manual process. It consists of bending the tubing and attaching the electrodes, removing any impurities from within the tubing, then evacuating the air and adding the gas. The following process is typical.

Preparing the tubing

1 Lengths of glass tubing are cleaned and placed vertically in a coating machine. The machine blows a liquid phosphor suspension upwards into the tube and then lets it drain back out the bottom. The tubes are placed vertically in an oven which dries the coating. Color tints are applied in a similar manner. Tubes that are to be filled with neon to form a red or orange light or argon to form a blue light are left clear.

Bending the tubing

2 The design of the sign is laid out in full size on a heat-resistant sheet of asbestos. The glass tubing is carefully heated and softened using a variety of burners. Gas-fired ribbon burners 24 inches (61 cm) or longer are used to make curves in round letters and the sweeping curves of script. Smaller hand torches are used to heat shorter lengths. Using the asbestos template as a guide, the tubing is bent by hand. The tube benders do not wear protective gloves because they must be able to feel the heat transfer and the degree of softening in the glass to determine the right moment to make the bend. To prevent the softened tubing from collapsing, the tube bender attaches a short length of flexible hose, called a blow hose, to one end. While the glass is still soft, the tube bender gently blows into the hose to force the tubing back to its original diameter. Tubes with restricted diameters will not operate properly.

3 Most large neon signs are made of several sections of glass tubing. A length of 8-10 feet (2.4-3.1 m) for each section is considered a practical limit. To make each section, the ends of two lengths of tubing are heated and spliced together. When the shape of the lettering or design has been formed for a section, an electrode is heated and fused onto each end. A small port, called a tubulation, is added to allow the tubing to be evacuated with a vacuum pump. This tubulation port may be part of one of the electrodes or may be a separate piece joined into the tubing.

Bombarding the tubing

4 A process known as bombarding is used to remove any impurities from the glass, phosphors, and electrodes. First the air inside the tubing is evacuated. After the vacuum reaches a certain level, dry air is allowed back into the tubing until the pressure is in the range of 0.02-0.04 inches (0.5-1.0 mm) of mercury. The longer the tubing, the lower the pressure may have to be. A very high-current transformer is connected to the electrodes. For a length of tubing that may normally run on 30 milliamperes, 400-750 milliamperes may used for the bombarding process. The high current heats the glass to about 420°F (216°C), and the metal electrode is heated to about 1400°F (760°C). This heating forces the impurities out of the materials, and the vacuum pump carries the impurities out of the system.

Filling the tube

5 Once the tube has cooled, the gas is inserted under low pressure. The gas must be free from impurities in order for the sign to operate properly and have a long life. The normal fill pressure for a tube 0.6 inches (15 mm) in diameter is about 0.5 inches (12 mm) of mercury. The tubulation port is then heated and sealed off.

Aging the tube

6 The finished gas-filled tubing is put through an aging process. Sometimes this process is referred to as "burning in the tube." The purpose is to allow the gas in the tube to stabilize and operate properly. A transformer, often rated slightly higher than the normal operating current, is attached to the electrodes. The tube should come to full illumination within 15 minutes if neon is used. It may take up to a few hours for argon. If a small amount of mercury is to be added to an argon tube, a droplet was first placed into the tubulation port before it was sealed. The droplet is then rolled from one end to the other to coat the electrodes after the aging process. Any problems such as a flicker in the gas or a hot spot on the tube indicate the tubing must be opened and the bombarding and filling processes repeated.

Installation and mounting

7 Small neon signs are mounted on their framework and wired in the shop. Larger signs may be mounted in pieces and put into place on the building or other support structure where they are interconnected and wired. Very large installations may require months to install.

Quality Control

Pure materials and careful manufacturing processes are required in order to produce a properly operating neon sign. A well-built neon sign should have a life of over 30,000 hours. As a comparison, the average 100-watt light bulb has a rated life of 750-1,000 hours.

Neon signs must meet the requirements of the Underwriters Laboratories to obtain a UL listing. This requires a series of tests by independent testing agencies. Neon signs must also meet the requirements of the National Electrical Code. Outdoor signs must comply with local building codes in their construction and electrical wiring.

The Future

Recent developments in neon sign design include small, electronic transformers that make the audible hum of older neon signs a thing of the past. Neon signs that blink or appear to move are now controlled by programmable electronic controls that replaced the older electromechanical cam-and-switch controls. Neon displays have also found their way into consumer products such as telephones and automobile license plate frames. There are even neon displays that cover portions of the exterior of cars for the ultimate in a "flashy" vehicle.

Neon signs are expected to continue to enjoy a resurgence of interest and applications. Some Japanese companies have expanded the palette of neon lights far beyond the 50 or so colors now commonly used. Neon displays that appear to move are also getting more complex and flamboyant with the aid of computer controls.

Where To Learn More

Book

Stern, Rudi. *Let There Be Neon.* Harry N. Abrams, Inc., 1979.

Periodicals

Gromer, Cliff. "Neon Cowboy: Computer Magic Makes Vegas Lights Come to Life." *Popular Mechanics,* September 1994, pp. 56-57.

Resnick, Rosalind. "Hot Lights on Wheels." *Nation's Business,* June 1994, pp. 14-15.

Woods, Wilton. "Neon Comes Home." *Fortune,* March 20, 1995, p. 22.

—Chris Cavette/Paul Davis

Newspaper

Background

A newspaper is a printed periodical whose purpose is to deliver news and other information in an up-to-date, factual manner. Newspapers appear most commonly in daily editions, but may also be issued twice a day or weekly. While the content of a newspaper varies, it generally consists of a predetermined combination of news, opinion, and advertising. The editorial section is written by reporters and other journalists at the direction of editors and may also be compiled from wire service reports. The advertising content of a newspaper can be divided into two parts, classified and display. Classified ads are small, text-only items obtained via telephone and set into the format by the classified advertising representative. Display ads are obtained by sales representatives employed by the newspaper who actively solicit local businesses for this larger, more visually oriented ad space.

A newspaper is printed on thin **paper** made from a combination of recycled matter and wood pulp, and is not intended to last very long. Large printing presses, usually located at a plant separate from the editorial and advertising headquarters, print the editions, and a network of delivery trucks bring them to the newsstands and geographical distribution centers for subscribers.

History

Public officials in ancient Rome posted news of the day in a public space, but it was not until the invention of the printing press in the late Middle Ages that mass-produced printed matter became possible. One hundred fifty years after the invention of printing from movable type by Johann Gutenberg in 1447, the first regular newspaper, *Avisa Relation oder Zeitung*, appeared in Germany in the early 17th century. The first English-language newspaper, the *Weekly Newes*, began publishing in England in 1622. Over the next few generations, small pamphlets and broadsheets were the primary source of printed information in both England and the colonies of North America, although they were generally geared toward business matters. One of the first newspapers in the U.S. was *Publick Occurrences Both Foreign and Domestick*, which began appearing in Boston in 1690.

These early prototypes of the newspaper eventually developed into publications that appeared on a more regular basis in localized geographic areas. At the time of the American Revolution, 35 newspapers were published in the 13 colonies. Many of these papers and their successors over the next few generations were concerned with political issues of the day and were rather expensive. This changed during the 1830s, however, when technology and publicity popularized "penny papers." The *New York Sun* was one of the first of these to gain widespread readership.

The development of quicker, more efficient printing methods led to a rapid growth of newspapers in the U.S. during the 19th century. As the country expanded and new metropolitan centers sprang up, so did newspapers that served the interests of the region. A growing literacy rate among the populace also helped make such printed matter more popular and profitable. In the latter decades of the 20th century, papers

Many of the earlier newspapers were concerned with political and business issues and were rather expensive. During the 1830s, however, technology and publicity popularized the "penny paper."

Holding a sign stating "We're Printers, by gravy," these young men had their photograph taken in the mid-1890s, possibly to commemorate the end of their apprenticeships. (From the collections of Henry Ford Museum & Greenfield Village.)

Timeliness is of the essence in the newspaper business. Even 150 years ago, New York City publishers would have messengers waiting to meet ships coming from Europe. The messengers would grab the latest dispatches, newspapers, and even novels and race to the printing office. There, rows of compositors would be poised to work all night setting type so that the next afternoon's newspaper could contain European news only two weeks old or the first chapters of a novel published months ago.

With the coming of the telegraph to the western parts of the U.S. in the mid-19th century, editors commonly kept one or two compositors late into the night ready to set stories that came in from the East by telegraph. The dots and dashes of the telegraph message, often consisting of just key words and phrases, were hastily transcribed by the telegrapher and given directly to the typesetters. Compositors were skilled enough to decipher the telegrapher's scribbles, compose full sentences while setting type (letter by letter) by hand, and complete the entire story by deadline.

The Linotype machine, developed in the 1880s, combined the processes of composing text, casting type, and redistributing the type molds. By working a keyboard, the Linotype operator assembled molds, or matrices, of letters, numbers, or punctuation marks in sequence. The matrices were then mechanically held in place while molten type metal was forced into them, creating a line of type ("lin' o' type"). The individual matrices were automatically replaced in the machine's magazine for reuse.

The Linotype increased the speed of a typesetter fourfold. This allowed editors to cut labor costs while getting all the latest news. The machine cost hundreds of compositors their jobs and added to the intensity and pace of the work.

William S. Pretzer

such as the *New York Times* and the *Wall Street Journal* have become esteemed sources of news in the U.S. and have wide distribution outside of the cities where they are produced.

Until the 1980s, many cities had more than one newspaper, and it was not uncommon for a large city to have three or four competing dailies. By the 1990s, many papers had disappeared or merged so that only one or two noncompeting papers coexisted in major cities. Smaller regional newspapers provide a mix of local news with national and international items. Such papers usually have correspondents in New York, Washington, D.C., and the major cities of the world. Tabloid newspapers, presenting more sensational news and features such as detailed crime stories, first appeared in the U.S. in the 1920s. The word tabloid refers to the size of the printed page, which is generally half the size of a standard newspaper.

The Editorial Process

The process of producing a daily edition of a large city newspaper begins with a meeting of the paper's editors, who determine the amount of editorial copy in an issue based on the advertising space that has already been sold. A specific number of pages is agreed upon, and the editorial assignments are made to the various departments. The section of national and international news, generally the first part of the paper, is compiled from correspondents who send in their stories electronically, usually via computer modem, to their editor's computer. There, the editor checks the stories, sometimes rewriting them or increasing or decreasing their length. Additional stories of importance are compiled from wire services such as United Press International, Associated Press, and Reuters. These are organizations that employ reporters in various cities of the globe to compile stories and items quickly for dissemination over telephone wires.

For a typical, newsbreaking story of local origin, the process begins with a correspondent submitting a report, either in person or via computer modem, to the "rewrite" desk person. The rewrite journalist fine-tunes the wording of the story and makes sure it answers the six important questions: who, what, where, when, why, and how. He or she then sends it over to the computer at the city desk. The city desk editor, who is responsible for the paper's local content, looks over the story, makes additional changes if necessary, and sends it over to the news desk. The news editor, who makes the final call about which stories to run in the upcoming edition based on their relevance, may make further changes before submitting the piece to the copydesk. The story arrives there with guidelines for length as well as headline instructions regarding size and type.

From this point, the story is set to be inserted on a certain page that has already been roughly laid out by both the news editor and a makeup editor. A mock-up of the page, essentially a blank form showing where the stories will run and where pictures and advertising will be inserted, is called the "dummy." The makeup editor has already met with the advertising department to determine how such pages will be laid out with ad space. The dummy has rough notes for headlines, story insertions, and graphic elements such as photos and tables of statistics. It also shows the date of the edition as well as a page and section number. After the news editor has determined the placement of the story on the page in question—as well as the other items set to run there—the dummy is sent on to a composing room.

The Manufacturing Process

Typesetting

1 The composing room receives the story in an electronic format, with the computer text file already translated with typeset codes. In a typeset file, the characters are of the same "type"—style, size, and width—as they appear on the pages of the newspaper. The setting of stories into the type that a reader sees went unchanged for several decades until the latter years of the 20th century. Well into the 1800s, type was set by hand, letter by letter. A typesetter dropped small metal letters into a hand-held tray called a "stick." The invention of the Linotype machine in 1884 made possible a quicker, more efficient method of typesetting. Invented by German immigrant Ottmar Mergenthaler of Baltimore, Maryland, this large, cumbersome machine worked by casting hot lead into a line of type with the assistance of an operator who typed in the copy on a keyboard. Individual lines of type were then placed by hand onto a page form. When a page was completed, it was then sent to a stereotyping room where a curved metal plate was made from the page form. The page form was then placed on the printing press.

Modern technology has replaced the Linotype process through a method called phototypesetting. The first step in this process is the transfer of the dummy to the page layout section of the newspaper. There, an operator transfers the instructions on the dummy into a rough page prototype. A printed version may be looked over and adjusted several times by one of the reporters whose story is featured as well as by the copy editor. If another breaking story comes in, this page layout can be altered in a matter of minutes.

Image transference

2 The final version of the page is then approved by the editor on duty—sometimes a night editor in the case of a paper that is slated for a morning edition—and sent over to a process department. There, the page is taken in its computer format and transferred via laser beams onto film in an image setter apparatus. The operator then takes the film to a processor in another section of the paper, who develops it and adjusts it for its final look. Photographs are scanned into another computer terminal and inserted into the page layout. The pages that are set to be printed together are then taped down onto a device called a "stripper," and an editor checks them over once more for errors. The strippers are then put into frames on light-sensitive film, and the image of each page is burned onto the film. The film of each page is inserted into a laser reader, a large facsimile machine that scans the page and digitally transfers the images to the printing center of the newspaper.

At the printing center, typically a large plant separate from the newspaper's editorial offices and centrally located to facilitate

The composing room receives the story in electronic format. It is then transferred onto film and developed.

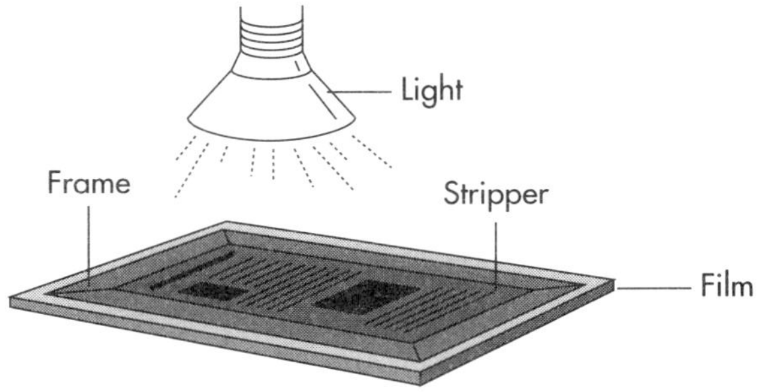

The page layout and photographs are taped onto a device called a stripper and placed over light-sensitive film. The image of each page is burned onto the film.

The film is sent through a laserwriter which makes a negative.

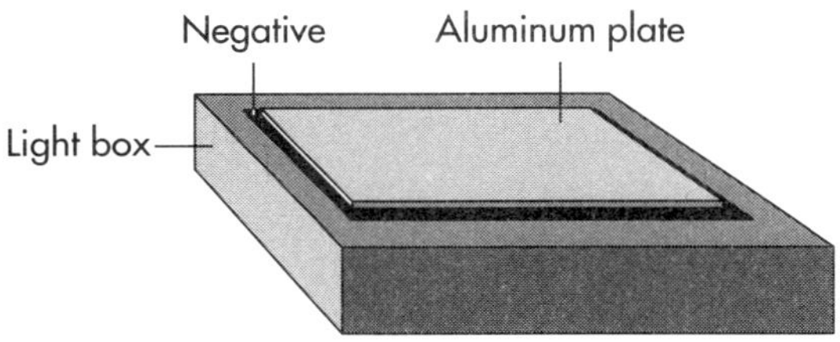

The negative film is placed on a lighted box, and an aluminum plate is placed on top. The ultraviolet light in the box develops the image of the page onto the aluminum plate. These plates are then attached to cylinders.

citywide distribution, the pages arrive at the laser room and are put through a laser writer, another scanning device that makes a negative image of them. In the negative image of the page, the text is white while the blank spaces are black. The final images of each page are further adjusted. This last-minute adjustment may involve fine-tuning of the colored sections and retouching photographs.

Platemaking

3 From these negatives, the forms from which the paper will be printed are composed in a platemaking room. The film of the page, usually done two pages at a time, is then placed on a lighted box. Next, an aluminum plate containing a light-sensitive coating is placed on top of the image of the pages. The light box is then switched on, and ultraviolet light develops the image of the pages onto the aluminum plate. The aluminum plate is then bent at the edges so that it will fit into a press, and is fitted onto plate cylinders.

Printing

4 The aluminum plates of each page next move on to the actual printing press, an enormous machine often two stories high. When the press is running, the noise in the building is deafening and employees must wear earplugs. The most common method of printing newspapers is called web offset. The "web" refers to the large sheets of blank newsprint that are inserted in rolls, sometimes weighing over a ton, into the actual printing press. The reels of newsprint are loaded in at the bottom floor of the press. The rolls are inserted onto a reel

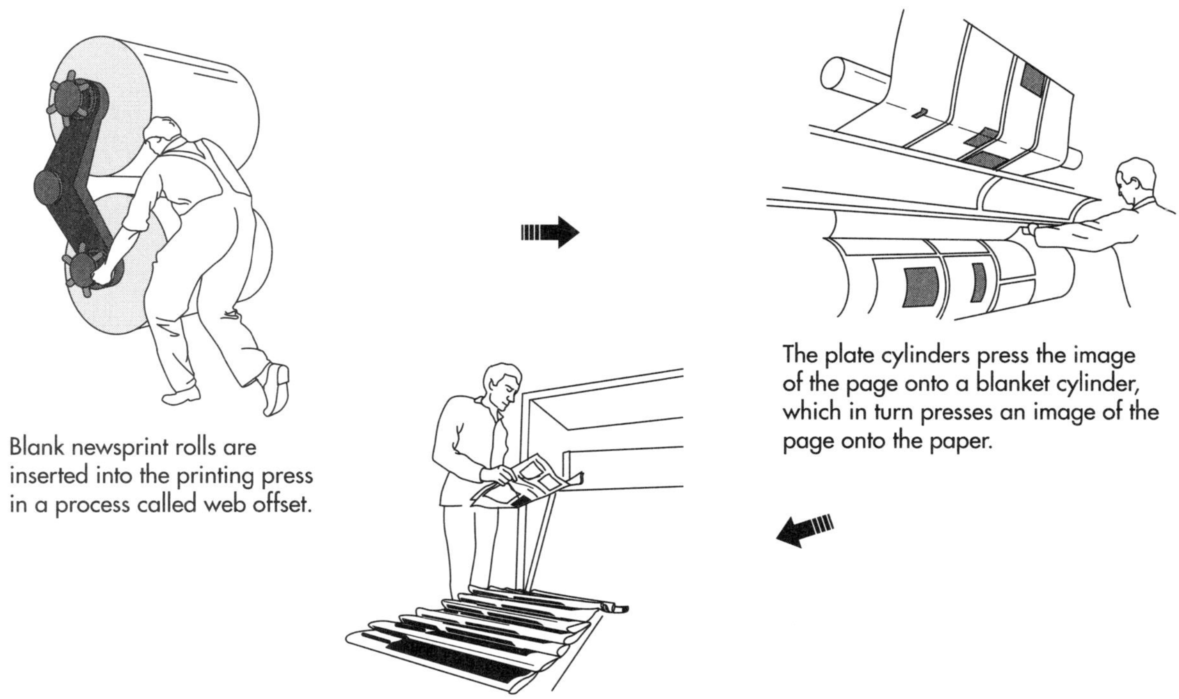

Blank newsprint rolls are inserted into the printing press in a process called web offset.

The plate cylinders press the image of the page onto a blanket cylinder, which in turn presses an image of the page onto the paper.

The newsprint is cut by a large machine and sent through a folder.

stand, which has three components: the first reel brings a roll of paper up to the press, a second is loaded and ready to replace the first roll when it runs out, and a third reel stays empty and ready to be fed with another when the first reel is almost finished. Each roll of blank newsprint has double-sided tape at its edges, so that when one roll runs out in the press, another smoothly takes up where the other left off without interrupting the printing process.

The plate cylinders then press the image of the page onto a blanket cylinder, leaving a version of the page's image on the cylinder's soft material. When the paper runs through the press, the blanket cylinder presses the image onto it. The chemical reaction of the ink, which contains oil, and the squirting of jets of water into the process result in the actual newspaper page of black or colored images on a white background. Since oil and water do not mix, the areas where ink should adhere to the page are black or colored, and water washes away the parts where ink is not needed. This is why this printing process is referred to as "offset."

Next, the large sheets of printed newsprint move on to another large piece of machinery called a folder. There, the pages are cut individually and folded in order. This entire printing process can move as fast as 60,000 copies per hour. Quality control technicians and supervisors take random copies and scan them for printing malfunctions in color, order, and readability. Next, a conveyer belt moves the papers into a mail room section of the plant, where they are stacked into quires, or bundles of 24. The quires then move to another section where a machine wraps them in plastic. The bundles are now ready to be loaded onto delivery trucks for distribution.

The Future

The demise of the printed word, especially in the form of a daily newspaper, is periodically predicted to be imminent by industry analysts. The growth of other news sources

—such as radio and 24-hour television news stations—has helped diminish the impact of newspapers, but the competition between dailies in many cities has forced many of the weaker, less financially-viable newspapers out of business. In many cities, joint-operating agreements—by which two competing papers share business, advertising, and printing departments—has helped to keep two editorially distinct papers afloat.

Bypassing the printed newspaper altogether, on-line computer technology has enabled consumers to pick and choose news from among their own specific interests on the information superhighway. One site on the Internet, one of the most popular providers of access to on-line information, allows a person to create his or her own newspaper. A menu appears onscreen, and the user selects stories from wire services, as well as entertainment features and cartoons, and inserts them onto a template. This template can be generated on a daily basis with a few keystrokes, producing an edition of a customized newspaper almost instantly.

Where To Learn More

Books

Miller, Margaret. *Hot Off the Press! A Day at the Daily News*. Crown, 1985.

Waters, Sarah. *How Newspapers Are Made*. Facts on File, 1989.

Periodicals

Booker, Ellis. "Extra! Extra! Newspapers Paperless." *Computerworld*, July 20, 1992, p. 30.

Kowet, Don. "Laying Out a Futuristic Newspaper." *Insight on the News*, May 14, 1990, p. 48.

—Carol Brennan

Paddle

Background

A paddle is an implement for manually moving and guiding a small boat. A paddle consists of a shaft with a broad flat surface, called a blade, on one or both ends. The area where the blade joins or tapers into the shaft is called the throat. Paddles differ from oars in that they are used without oar locks, the attachments that clamp the oars to a boat. Paddles are used to propel and steer a canoe or **kayak** by pulling or pushing the paddle blade against the water along the sides of the boat. For forward motion the paddler puts the blade in the water and pulls the shaft to the rear, first along one side and then along the other. Controlling direction and steering the boat are accomplished either by repeatedly or strongly paddling on one side of the boat or by altering the fundamental linear stroke pattern.

Paddle shafts can vary in shape and diameter. Round cross sections of 1-1.25 inches (2.5-3.2 cm) in diameter are typical, but elliptical shapes are also used. Some designs taper from fully round at the shaft's midpoint to elliptical at the throat. Other shafts are elliptical over their entire length. Shaft designs can include a bend near the paddle throat to increase the power of each stroke. Materials to protect and cushion the paddler's hands typically cover part of the shaft.

Paddle blades vary in size and shape depending on their intended use and the strength of the paddler. A typical paddle blade is 8 inches (20 cm) wide and 18 inches (46 cm) long from its throat to the tip. Optimal blade size depends on the shoulder power of the individual paddler. Round blades called "pizza type" blades provide an example. Pizza blades are 12-14 inches (30-36 cm) wide and the same distance from throat to tip. They are used for racing and require a strong paddler to be most efficient because their larger area pushes more water with each stroke. Proper blade size is also important to recreational paddlers because a blade that is too small will cause a strong paddler to waste energy and cause unnecessary fatigue. The shape of the blade also affects a paddler's performance. The side of the blade used to push the water is called the power side. The power side of some blades are spoon shaped to scoop the water. This increases the resistance as the paddle moves through the water, and thereby increases the effect of the paddler's stroke.

Paddles can be single or double bladed. Single-bladed paddles have a blade at one end of the shaft and a grip at the other end. Grips can be T-shaped, flared to form a triangular shape, or have a simple rounded end. Canoe paddlers use a single-bladed paddle that is typically about 4-5 feet (1.2-1.5 m) long. They switch their grasp on the paddle's throat and grip from one hand to the other as they alternate strokes on the left and right sides of the canoe. In contrast kayakers use a double-bladed paddle about 7-9 feet (2-2.7 m) long. This enables kayakers to alternate left and right strokes without needing to change their grasp on the paddle. Double-bladed paddles typically have a grip area at the mid point of the shaft. They can also have elliptically shaped shafts with blades set at right angles to each other. This design is said to have feathered blades. Feather-bladed paddles offer an advantage over traditionally oriented blades in that the angle between the feathered blades allows the paddler to pull one blade through the water

Native Americans introduced paddling to fur traders as a faster mode of propulsion than using long, non-bladed poles to push their boats through the water by planting the pole into the river bed.

while the other blade slices horizontally through the air. This reduces air resistance on the blade out of the water and increases the paddler's efficiency. The elliptical shape of the shaft lets the paddler know the orientation of the blades.

History

Ancient paddles were made primarily of wood, and most specimens have disintegrated. However some ancient paddles have been found with features such as ivory fittings on the blade tips to prevent splitting and damage. Native Americans introduced paddling to fur traders as a faster mode of propulsion than using long, non-bladed poles to push their boats through the water by planting the pole into the river bed.

Native peoples of the Arctic regions and Greenland used both single-bladed and double-bladed paddles. Primitive single-bladed paddles were about 63 inches (160 cm) long with a blade about 5 inches (13 cm) wide. These paddles were used for larger and deeper boats where the use of double-bladed paddles would be difficult. Single-bladed paddles were particularly useful when sneaking close to sea mammals, which would dive upon hearing the slightest noise. Hunters held the paddle in the hand away from the game, allowing the hunter to hold his weapon in the hand closest to the game. Double-bladed paddles used in narrower, smaller boats had very narrow blades usually of about 3 inches (7.6 cm) wide and much longer shafts, usually about 94 inches (239 cm) long. The blades were typically positioned on the shaft in the same plane. Some of these primitive double-bladed paddles had convex blades.

Today many people consider wooden paddles to be the most aesthetically pleasing. Wooden paddles are relatively lightweight, about 2.5-3.5 pounds (1.1-1.6 kg). Shafts can be custom made from 1.25-inch (3.2 cm) diameter pole or by piecing together separate halves of a softwood outer layer with a hardwood center. The separate halves are fastened in the middle with a scarf joint, a joining technique that notches two pieces so that they overlap into one continuous piece. The hardwood center extends into the blade as a reinforcement and also to provide a means of securely attaching the blade to the shaft. The wooden blades are constructed by layering and bonding thin wood pieces together with glue or resin. The blades can be faced with more attractive wood veneers. Caps of metal or **fiberglass** fit on the tip of the blade for protection against abrasion. Wooden paddles can also be finished with fiberglass and epoxy resin for improved durability.

Raw Materials

Paddles are made from a variety of materials. Wood, fiberglass, aluminum, and plastic can be used alone or in combination with other materials to make the entire paddle.

The manufacturing process described below is for a paddle with a combination of a cork-filled or foam-filled tempered aluminum shaft with plastic blades and grips. Hollow aluminum tubing of different thicknesses is used as the structural portion of the paddle's shaft. The tubing is purchased in long pieces from a source outside of the fabricator. Cork or foam materials are used to fill the hollow tubing to prevent water infiltration and to help the paddle float. These are also purchased from outside sources. Cork may be purchased in dowel form or as stoppers. Thermoplastic powder and color pigment are bought for making blades and grips at the fabricator's shop. Plastic tubing or other material will wrap around the shaft for comfort and protection of the paddler's hands from the bare aluminum, as well as for protection against corrosion.

The Manufacturing Process

The following paddle manufacturing process is an assembly line operation. It can be used to manufacture several types of paddles. It combines custom-made pieces, alterations to stock materials, and assembly.

Molding the blades and grips

1 Blades and grips are formed using a standard manufacturing process called injection molding. Materials needed in this process are granulated or powdered thermoplastics and color pigment, which are mixed together. Injection molding is a process in which the raw materials are funneled through a hopper at the top of the injection molding

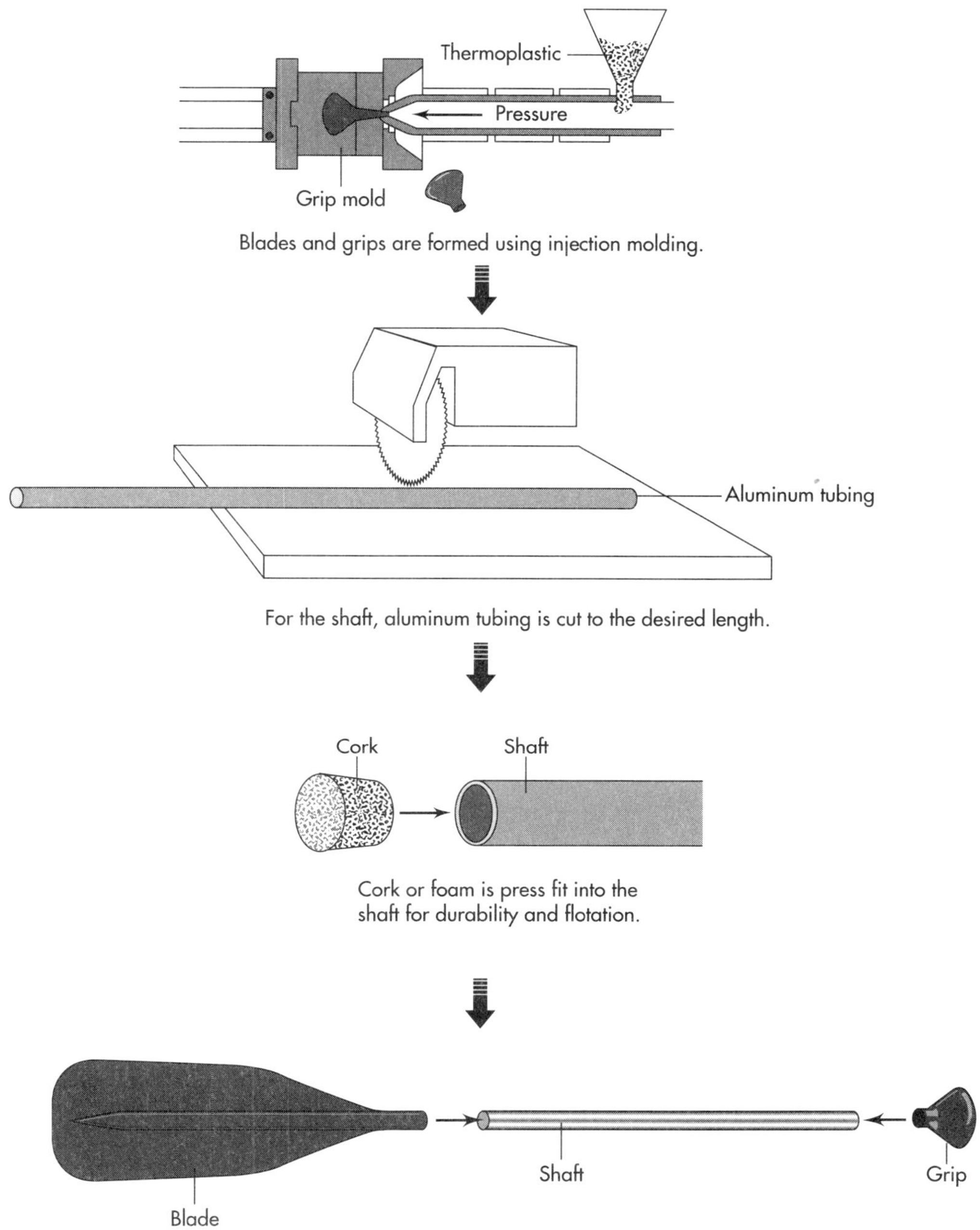

Blades and grips are formed using injection molding.

For the shaft, aluminum tubing is cut to the desired length.

Cork or foam is press fit into the shaft for durability and flotation.

Plastic tubing is slipped over the shaft. The blade and grip are attached.

machine into a heated barrel that mixes and softens them. The softened material is then forced, or injected, through the barrel by a retractable screw or plunger into a relatively cool mold cavity for hardening. In paddle manufacturing the mold cavity is shaped like the blade or grip. Once the blade or grip is fully cured, the detachable mold cavity is opened, thereby releasing the molded plastic piece. The injection molding process can operate continuously since the screw can retract to fill the barrel with fresh raw materials while the piece in the cavity hardens.

2 Blades and grips must be inspected for flash. Flash is excess plastic that seeped into the seams of the mold cavity and hardened. Blades and grips are cleaned by hand

to remove the excess plastic and smooth the edges.

Cutting the shaft

3 Aluminum tubing is cut to length using a table saw. The plastic tubing or other protective material used to cover the shaft is also cut by this method.

4 Once the shafts are cut to size, the cork or foam can be press fit into the shaft. A press fit is the name for a method of attaching pieces together using the tightness of the bond that results from forcing the two pieces together. Excess cork or foam is trimmed by hand.

Assembly

5 The exact order of assembly can vary. In a typical operation plastic tubing or other protective material is slipped over the aluminum shaft. The plastic tubing is heated with a hot air gun, which causes it to shrink tightly onto the shaft. In some assembly operations the grip is pushed onto the aluminum shaft before the plastic tubing. In this case the plastic tubing shrinks over the neck of the grip, therefore securing the grip to the shaft as well.

6 The neck of the blade is press fit on to the plastic covered shaft at the paddle's throat. This helps prevent the blade from becoming loose and spinning on the shaft.

7 In another variation, the grips and blades are shaped so that they are pushed inside the ends of the plastic covered aluminum tubing by machine. Hydraulic or pneumatic presses squeeze and slightly deform the ends of the aluminum tubing into the neck of the blade or grip to securely fasten the plastic pieces.

Packaging

8 Once the paddle is fully assembled, it is wrapped in a clear plastic bag. The paddles are then packed in corrugated boxes, about 30 to a box, for shipping to distributors.

Byproducts/Waste

The primary byproducts of the process described above are scraps of aluminum from the cutting of the shafts and plastic scraps and shavings from forming the blades and grips. The plastic can be melted and reused in other injection molded products. The aluminum is recycled in a manner similar to the recycling of **aluminum beverage can**s.

The Future

Recent research has led to lighter weight paddles made from various materials. For example Olympic kayak racer Greg Barton researched the design, construction, and testing of lightweight composite kayak paddles. His new paddles weigh less than half the weight of the best wooden paddle today. Other research at the University of California's Livermore Laboratories has developed instrumentation designed to instantaneously measure the power output and synchrony of rowers in team boats as well as physiological parameters such as heart rate. This instrumentation may lead to improvements in the design and manufacture of paddles.

Where To Learn More

Books

Gordon, I. Herbert. *Canoeing Made Easy.* The Globe Peguot Press, 1992.

Johannaber, F. *Injection Molding Machines, A User's Guide.* Macmillan Publishing Co., 1979.

Kulczycki, Chris. *The Kayak Shop.* McGraw Hill, 1993.

Seidman, David. *The Essential Sea Kayaker.* International Marine Publishing, 1992.

Periodicals

"Bent Shaft Canoe Paddle to Speed up Your Trip." *Materials Engineering,* November 1984, p. 41.

Noble, Peter, Michael Wadden, Timothy Bourke, David Williams, and Knut Nordbo. "An Introduction to Ethnotechnology for Naval Architects: Sea Kayak Design of Yesterday and Tomorrow." *Marine Technology,* October 1994, pp. 305-14.

—David N. Ford

Paper

Background

Formed from wood pulp or plant fiber, paper is chiefly used for written communication. The earliest paper was papyrus, made from reeds by the ancient Egyptians. Paper was made by the Chinese in the second century, probably by a Chinese court official named Cai Lun. His paper was made from such things as tree bark and old fish netting. Recognized almost immediately as a valuable secret, it was 500 years before the Japanese acquired knowledge of the method. Papermaking was known in the Islamic world from the end of the eighth century A.D.

Knowledge of papermaking eventually moved westward, and the first European paper mill was built at Jativa, in the province of Valencia, Spain, in about 1150. By the end of the 15th century, paper mills existed in Italy, France, Germany, and England, and by the end of the 16th century, paper was being made throughout Europe.

Paper, whether produced in the modern factory or by the most careful, delicate hand methods, is made up of connected fibers. The fibers can come from a number of sources including cloth rags, cellulose fibers from plants, and, most notably, trees. The use of cloth in the process has always produced high-quality paper. Today, a large proportion of cotton and linen fibers in the mix create many excellent papers for special uses, from wedding invitation paper stock to special paper for pen and ink drawings.

The method of making paper is essentially a simple one—mix up vegetable fibers, and cook them in hot water until the fibers are soft but not dissolved. The hot water also contains a base chemical such as lye, which softens the fibers as they are cooking. Then, pass a screen-like material through the mixture, let the water drip off and/or evaporate, and then squeeze or blot out additional water. A layer of paper is left behind. Essential to the process are the fibers, which are never totally destroyed, and, when mixed and softened, form an interlaced pattern within the paper itself. Modern papermaking methods, although significantly more complicated than the older ways, are developmental improvements rather than entirely new methods of making paper.

Raw Materials

Probably half of the fiber used for paper today comes from wood that has been purposely harvested. The remaining material comes from wood fiber from sawmills, recycled **newspaper**, some vegetable matter, and recycled cloth. Coniferous trees, such as spruce and fir, used to be preferred for papermaking because the cellulose fibers in the pulp of these species are longer, therefore making for stronger paper. These trees are called "softwood" by the paper industry. Deciduous trees (leafy trees such as poplar and elm) are called "hardwood." Because of increasing demand for paper, and improvements in pulp processing technology, almost any species of tree can now be harvested for paper.

Some plants other than trees are suitable for paper-making. In areas without significant forests, bamboo has been used for paper pulp, as has straw and sugarcane. Flax,

Paper was made by the Chinese in the second century, probably by a Chinese court official named Cai Lun. His paper was made from such things as tree bark and old fish netting.

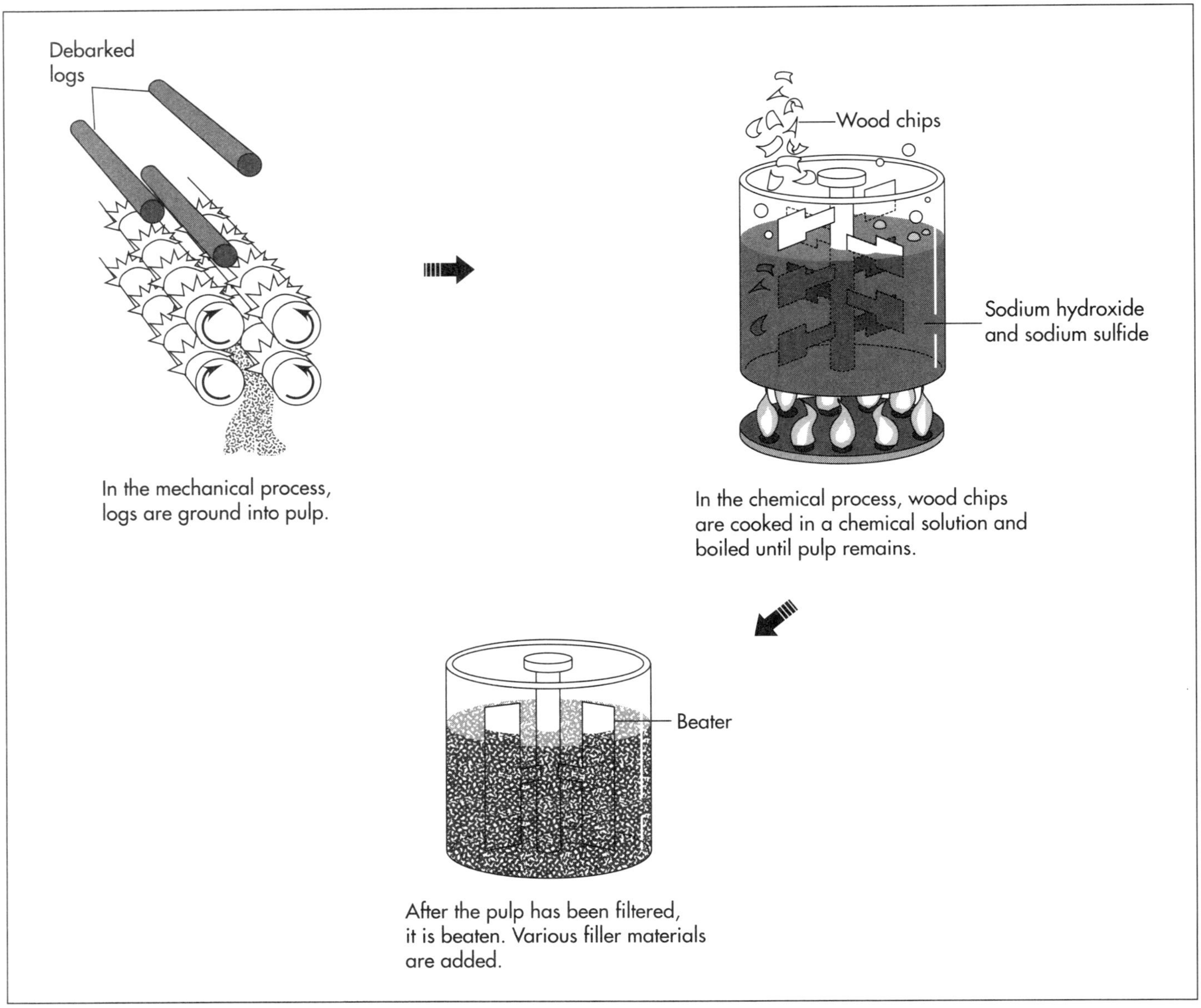

Most paper is made by a mechanical or chemical process.

hemp, and jute fibers are commonly used for textiles and **rope** making, but they can also be used for paper. Some high-grade **cigarette** paper is made from flax.

Cotton and linen rags are used in fine-grade papers such as letterhead and resume paper, and for bank notes and security certificates. The rags are usually cuttings and waste from textile and garment mills. The rags must be cut and cleaned, boiled, and beaten before they can be used by the paper mill.

Other materials used in paper manufacture include bleaches and dyes, fillers such as chalk, clay, or titanium oxide, and sizings such as rosin, gum, and starch.

The Manufacturing Process

Making pulp

1 Several processes are commonly used to convert logs to wood pulp. In the mechanical process, logs are first tumbled in drums to remove the bark. The logs are then sent to grinders, which break the wood down into pulp by pressing it between huge revolving slabs. The pulp is filtered to remove foreign objects. In the chemical process, wood chips from de-barked logs are cooked in a chemical solution. This is done in huge vats called digesters. The chips are fed into the digester, and then boiled at high pressure in a solution of

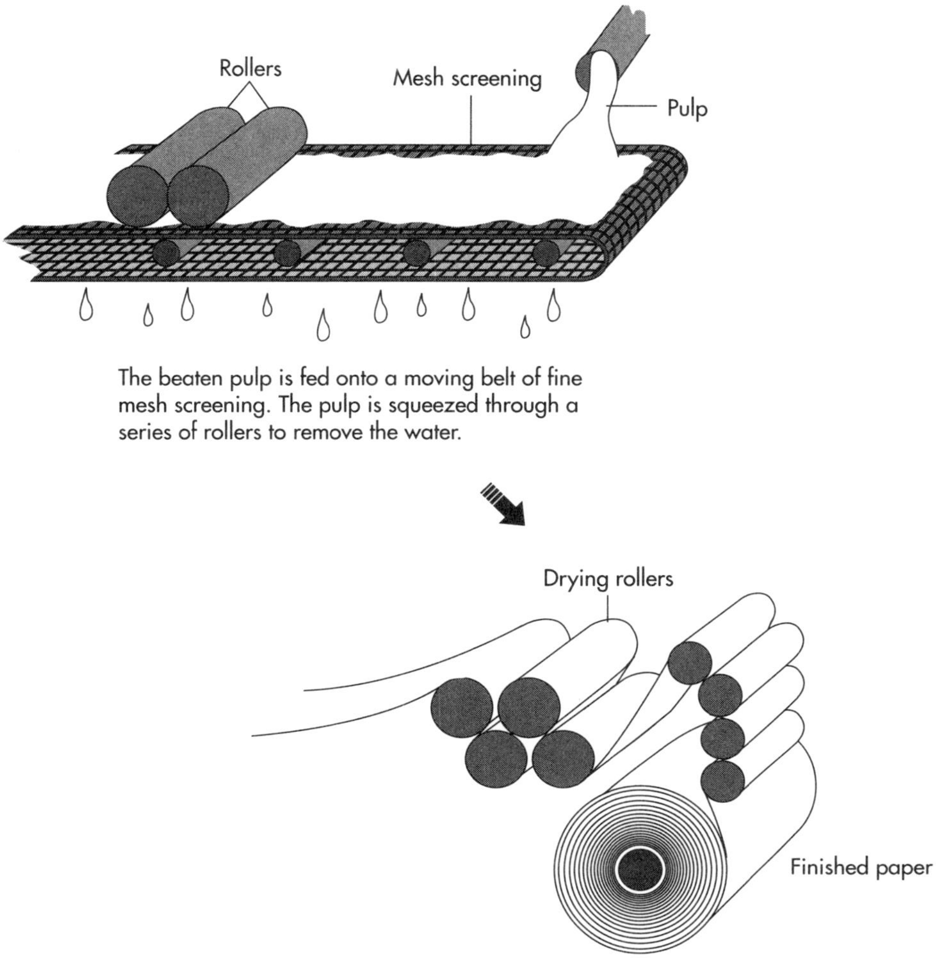

The beaten pulp is fed onto a moving belt of fine mesh screening. The pulp is squeezed through a series of rollers to remove the water.

The paper passes over a series of steam-heated cylinders to remove the remaining water. The dried paper is then wound onto large reels.

sodium hydroxide and sodium sulfide. The chips dissolve into pulp in the solution. Next the pulp is sent through filters. **Bleach** may be added at this stage, or colorings. The pulp is sent to the paper plant.

Beating

2 The pulp is next put through a pounding and squeezing process called, appropriately enough, beating. Inside a large tub, the pulp is subjected to the effect of machine beaters. At this point, various filler materials can be added such as chalks, clays, or chemicals such as titanium oxide. These additives will influence the opacity and other qualities of the final product. Sizings are also added at this point. Sizing affects the way the paper will react with various inks. Without any sizing at all, a paper will be too absorbent for most uses except as a desk blotter. A sizing such as starch makes the paper resistant to water-based ink (inks actually sit on top of a sheet of paper, rather than sinking in). A variety of sizings, generally rosins and gums, is available depending on the eventual use of the paper. Paper that will receive a printed design, such as gift wrapping, requires a particular formula of sizing that will make the paper accept the printing properly.

Pulp to paper

3 In order to finally turn the pulp into paper, the pulp is fed or pumped into

giant, automated machines. One common type is called the Fourdrinier machine, which was invented in England in 1807. Pulp is fed into the Fourdrinier machine on a moving belt of fine mesh screening. The pulp is squeezed through a series of rollers, while suction devices below the belt drain off water. If the paper is to receive a watermark, a device called a dandy moves across the sheet of pulp and presses a design into it.

The paper then moves onto the press section of the machine, where it is pressed between rollers of wool felt. The paper then passes over a series of steam-heated cylinders to remove the remaining water. A large machine may have from 40 to 70 drying cylinders.

Finishing

4 Finally, the dried paper is wound onto large reels, where it will be further processed depending on its ultimate use. Paper is smoothed and compacted further by passing through metal rollers called calendars. A particular finish, whether soft and dull or hard and shiny, can be imparted by the calendars.

The paper may be further finished by passing through a vat of sizing material. It may also receive a coating, which is either brushed on or rolled on. Coating adds chemicals or pigments to the paper's surface, supplementing the sizings and fillers from earlier in the process. Fine clay is often used as a coating. The paper may next be supercalendered, that is, run through extremely smooth calendar rollers, for a final time. Then the paper is cut to the desired size.

Environmental Concerns

The number of trees and other vegetation cut down in order to make paper is enormous. Paper companies insist that they plant as many new trees as they cut down. Environmentalists contend that the new growth trees, so much younger and smaller than what was removed, cannot replace the value of older trees. Efforts to recycle used paper (especially newspapers) have been effective in at least partially mitigating the need for destruction of woodlands, and recycled paper is now an important ingredient in many types of paper production.

The chemicals used in paper manufacture, including dyes, inks, bleach, and sizing, can also be harmful to the environment when they are released into water supplies and nearby land after use. The industry has, sometimes with government prompting, cleared up a large amount of pollution, and federal requirements now demand pollution-free paper production. The cost of such clean-up efforts is passed on to the consumer.

Where To Learn More

Books

Biermann, Christopher J. *Essentials of Pulping & Papermaking*. Academic Press, 1993.

Bell, Lilian A. *Plant Fibers for Papermaking*. Liliaceae Press, 1992.

Ferguson, Kelly, ed. *New Trends and Developments in Papermaking*. Miller Freeman, Inc., 1994.

Munsell, Joel. *Chronology and Process of Papermaking, 1876-1990*. Albert Saifer Publisher, 1992.

Periodicals

deGrassi, Jennifer. "Primitive Papermaking." *Schools Arts*, February 1981, pp. 32-33.

Kleiner, Art. "Making Paper." *Co-Evolution Quarterly*, Winter 1980, p. 138.

Lamb, Lynette. "Tree-Free Paper." *Utne Reader*, March-April 1994, p. 40.

Saddington, Marrianne. "How to Make Homemade Paper." *Mother Earth News*, December-January 1993, p. 30+.

Sessions, Larry. "Making Paper." *Family Explorer*, October 1994.

—*Lawrence H. Berlow*

Pasta

Background

Pasta is a universally enjoyed food, and almost every country serves a type of noodle. In China, it is *mein*; Japan, *udon*; Poland, *pierogi*; Germany, *spaetzle*. The popularity of pasta can be attributed to several factors: it is easily manufactured, it takes up little storage space, it is easy to cook, and it is rich in complex carbohydrates.

Ancient Etruscan meals of gruel and porridge were eventually replaced with more appetizing unleavened **bread** cakes. Food historians believe these cakes may have been the precursor to pasta. Opinions about where the noodle originated vary. The Italian explorer Marco Polo has been commonly credited with bringing the noodle back to Italy from his travels in the Orient during the 1300s. However, some contend that a close examination of Polo's papers reveals that he reported enjoying a certain *type* of noodle in China, comparing it favorably to the pasta he was accustomed to eating in Italy.

Nevertheless, it is true that Chinese noodles have been around for centuries. The vermicelli-like transparent noodles are made from the paste of germinated mung beans and are usually soaked in water before they are boiled or fried. (Pasta has not always been prepared by boiling. In fact, boiled noodles were once considered a relatively bland meal. Frying or grilling were the preferred preparations.) Koreans claim to have taught the Japanese how to make *soba* noodles in the 12th century, using Chinese buckwheat grown in the northern regions where rice paddies could not survive.

Early French writers also mention a dish called *pastillum*, essentially a ravioli-like pouch filled with meat. However, the Italians have staked the claim so vehemently that today we generally think of pasta dishes as Italian in origin. In fact, the word "pasta" comes from the Italian phrase "paste (dough) alimentari (relating to nourishment)."

The first industrial production of pasta occurred in Naples in the early 15th century. The site was chosen for its naturally fluctuating temperatures, sometimes as much as four times a day, which provided the hot and cold temperatures necessary for drying. Mechanical drying was not invented until 1800.

Raw Materials

Pasta is made from a mixture of water and semolina flour. Semolina is a coarse-ground flour from the heart, or endosperm, of durum wheat, an amber-colored high protein hard wheat that is grown specifically for the manufacture of pasta. With a lower starch content and a higher protein content than all-purpose flours, semolina flour is easily digested. Farina, rougher granulations of other high-quality hard wheat, is also used to make some pastas. The semolina and farina flour are enriched with B-vitamins and iron before they are shipped to pasta plants.

Eggs are sometimes added to the mixture for color or richness. Federal guidelines stipulate that egg noodles contain a minimum of 5.5% egg solids. Vegetable juices, such as spinach, beet, tomato, and carrot, can also be added for color and taste. In recent years, the addition of herbs and spices such as garlic, basil, and thyme has become popular.

The word "pasta" comes from the Italian phrase "paste (dough) alimentari (relating to nourishment)."

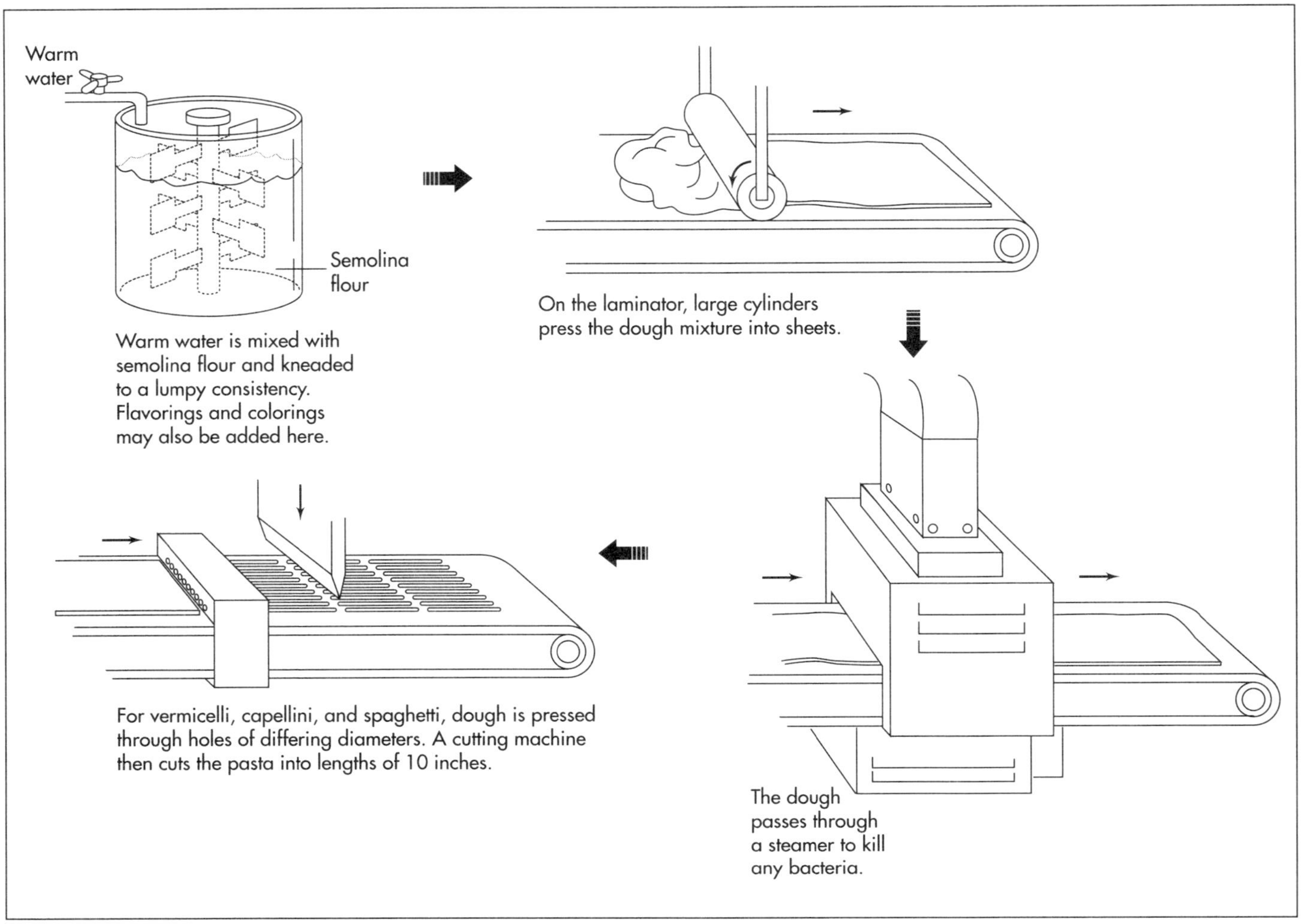

The Manufacturing Process

Mixing and kneading

1 The semolina is stored in giant silos that can hold up to 150,000 pounds (68,100 kg). Pipes move the flour to a mixing machine equipped with rotating blades. Warm water is also piped into the mixing machine. The mixture is kneaded to a lumpy consistency.

Flavoring and coloring

2 Eggs are added to the mixture if the product is an egg noodle. If pasta is to be a flavored variety, vegetable juices are added here. A tomato or beet mixture is added for red pasta, spinach for green pasta, carrots for orange pasta. Herbs and spices can also be folded in for additional flavoring.

Rolling

3 The mixture moves to a laminator where it is pressed into sheets by large cylinders. A vacuum mixer-machine further flattens the dough while pressing air bubbles and excess water from the dough to reach the optimum water content of 12%.

Pasteurization

4 The roll of dough moves through a steamer, which heats the dough to 220°F (104°C) in order to kill any existing bacteria.

Cutting

5 Depending on the type of noodle to be produced, the dough is either cut or pushed through dies. Ribbon and string-style pasta—such as fettucine, linguine, spaghetti, and capellini (angel hair)—are cut by rotating blades. To make tube or shell-shaped pasta such as rigatoni, ziti, elbow

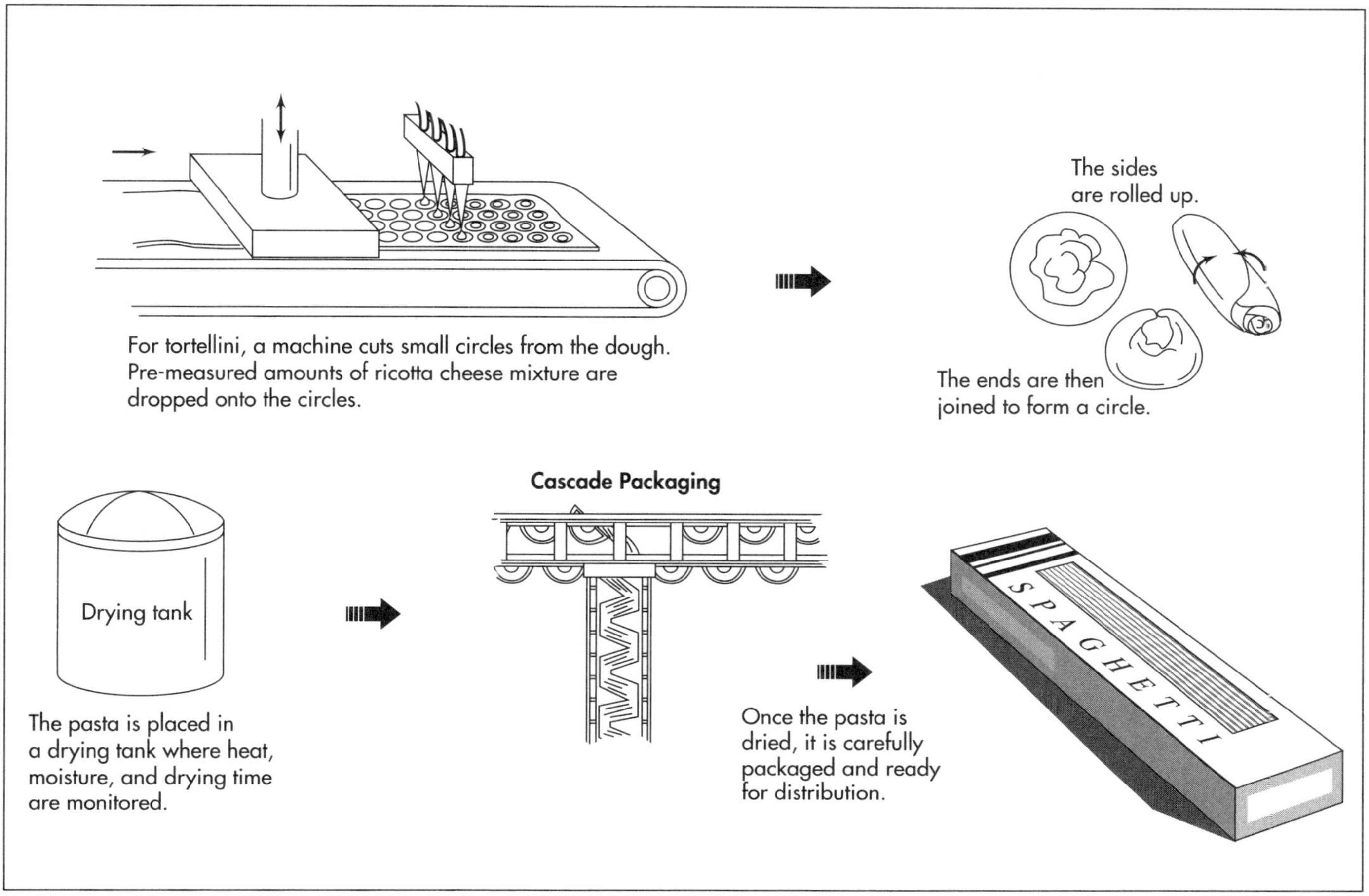

macaroni, and fusilli, the dough is fed into an extruder which then pushes it through metal dies. The size and shape of the holes in the die determine the type of pasta.

To make vermicelli and capellini, the pasta dough is pushed through holes between 0.8-0.5 mm in diameter. The cutting machine then cuts the pasta into lengths of 10 inches (250 mm) and twists it into curls. Spaghetti ranges from 1.5-2.5 mm in diameter and is left straight.

Tortellini (filled pasta rings) are made on a separate machine. The machine cuts small circles from a roll of dough. A bucket of ricotta cheese mixture drops a pre-measured amount of cheese onto the circle of dough. The dough is then folded over and the two ends are joined to form a circle.

To make ravioli (filled pasta squares), pre-measured quantities of cheese filling are dropped by machine at pre-measured intervals on a sheet of pasta. Another sheet of pasta is placed over this sheet as it moves along a conveyer belt. The two layers then pass under a cutting machine that perforates the pasta into pre-measured squares.

Drying

6 The pasta is placed in a drying tank in which heat, moisture, and drying time are strictly regulated. The drying period differs for the various types of pasta. It can range from three hours for elbow macaroni and egg noodles to as much as 12 hours for spaghetti. The drying time is critical because if the pasta is dried too quickly it will break and if it is dried too slowly, the chance for spoilage increases. The oxygen level in the tank is also regulated, and lab technicians test frequently for salmonella and other bacteria.

Careful handling of the pasta during the drying period is also crucial. Spaghetti is the most fragile of the noodles and is therefore hung high above the floor.

Packaging

7 Fresh pasta is folded in pre-measured amounts into clear plastic containers. As the containers move along a conveyer belt, a plastic sheet covers each container and is sealed with a hot press. At the same time, a small tube sucks the air of the container and replaces it with a mixture of carbon dioxide and nitrogen to prolong the product's shelf-life. Labels listing the type of noodle, nutritional information, cooking instructions, and expiration date are attached to the top of the containers.

Dried pasta is loaded, either manually or by machine, into stainless steel buckets (usually of heavy gauge type 304) which move along a conveyer belt to the appropriate packaging station. The pasta is measured by machine into pre-printed boxes, which also list the type of noodle, ingredients, preparation, and expiration date. Again, careful handling is important. For example, because lasagna noodles are particularly fragile, workers place them on metal slides that ease the pasta into boxes. The boxes are then sealed by machine.

Conveying system can be constructed in "S," "C," or "Z" configurations, or as horizontal conveyer belts. These systems move the pasta up and down and across the plant at heights up to 10 feet (3 m). Workers at the floor-level stations monitor the packaging process. The mechanism allows for workers to package the pasta manually if necessary.

Quality Control

The manufacturing of pasta is subject to strict federal regulations for food production. Federal inspectors schedule regular visits to insure that the company is adhering to government laws. In addition, each company sets its own standards for quality, some of which are set in practice before the pasta reaches the plant. Lab technicians test the semolina flour for color, texture, and purity before it is removed from rail cars. Protein and moisture content are measured and monitored on sophisticated quality control computer software.

In the plant, technicians constantly test the pasta for elasticity, texture, taste, and tolerance to overcooking. Plant workers are required to wear hairnets and plastic gloves. Mixing machines are scrupulously cleaned after each batch of pasta passes through them. The drying process is strictly monitored to guard against spoilage.

Homemade Pasta

The popularity of pasta has spread to the home-cooking arena. Pasta-rolling machines and pasta cookbooks are available at housewares stores and in cooks' catalogs. The recipe for homemade pasta is similar to the industrial process with the exception that eggs are generally used in all home pasta recipes. Sometimes oil is added to the mixture, particularly if a lesser grade of flour is used.

The flour is measured out onto a wooden or marble surface and formed into a mound with a well in the center. Eggs, water, oil and any other desired ingredients are poured into the well and mixed lightly with a fork. Then, beginning from the outside of the mound, the flour is incorporated into the center.

The dough is kneaded for approximately five minutes until a smooth, elastic ball is achieved. Rolling the dough into sheets is done with a long Italian-style rolling pin or with a rolling machine. Most rolling machines have attachments for cutting the dough into various forms of pasta such as spaghetti, fettucine, lasagna, or ravioli. The dough can also be cut by hand using a sharp knife or rolling blade. Specially marked rolling pins that imprint squares on the dough or ravioli trays can be used for making stuffed pasta. Extrusion machines for making tube-style pasta such as rigatoni or fusilli can also be purchased for home use.

The Future

Pasta continues to increase in popularity. The National Pasta Foods Association estimates that the average American will eat more than 29 pounds (13 kg) of pasta each year by the turn of the century. Highly rated for its nutritional value, pasta is an ideal meal for people who are paying more attention to their dietary intake. In addition, people are finding less time to prepare meals, and pasta is easily made.

Pasta manufacturers are responding to this demand by introducing a wide variety of dried and fresh pastas. One recent innovation is no-boil pasta that is partially cooked at the plant, making this already easy-to-prepare food even simpler to bring to the table at mealtime. New lines of fat- and cholesterol-free ravioli are on the market as well as organically-grown pasta products. Two new grains, South American quinoa and Egyptian kamut, are being used to make wheat-free pasta.

Where To Learn More

Books

Bugialli, Guiliano. *Bugialli on Pasta*. Simon and Schuster, 1988.

Toussaint-Samat, Maguelonne. *A History of Food*. (Translated from the French by Anthea Bell). Blackwell Publishers, 1992.

Periodicals

Bannon, Lisa. "Italians Do Still Eat Oodles of Noodles, But Trend Is Limp." *Wall Street Journal*, May 10, 1994, p. A1.

"What Is Pasta?" Borden, Inc., 1994.

"Custom-Manufactured Pasta." *Food Engineering*, January 1991, p. 71.

Giese, James. "Pasta: New Twists on an Old Product." *Food Technology*, February 1992, p. 118-26.

McMath, Robert. "Pasta's New World Order." *Adweek's Marketing Week*, November 25, 1991, p. 26

—*Mary F. McNulty*

Perfume

True perfumes are defined as extracts or essences and contain a percentage of oil distilled in alcohol. The United States is the world's largest perfume market with annual sales totalling several billions of dollars.

Background

Since the beginning of recorded history, humans have attempted to mask or enhance their own odor by using perfume, which emulates nature's pleasant smells. Many natural and man-made materials have been used to make perfume to apply to the skin and clothing, to put in cleaners and cosmetics, or to scent the air. Because of differences in body chemistry, temperature, and body odors, no perfume will smell exactly the same on any two people.

Perfume comes from the Latin "per" meaning "through" and "fumum," or "smoke." Many ancient perfumes were made by extracting natural oils from plants through pressing and steaming. The oil was then burned to scent the air. Today, most perfume is used to scent bar **soap**s. Some products are even perfumed with industrial odorants to mask unpleasant smells or to appear "unscented."

While fragrant liquids used for the body are often considered perfume, true perfumes are defined as extracts or essences and contain a percentage of oil distilled in alcohol. Water is also used. The United States is the world's largest perfume market with annual sales totalling several billions of dollars.

History

According to the Bible, Three Wise Men visited the baby Jesus carrying myrrh and frankincense. Ancient Egyptians burned incense called *kyphi*—made of henna, myrrh, cinnamon, and juniper—as religious offerings. They soaked aromatic wood, gum, and resins in water and oil and used the liquid as a fragrant body lotion. The early Egyptians also perfumed their dead and often assigned specific fragrances to deities. Their word for perfume has been translated as "fragrance of the gods." It is said that the Moslem prophet Mohammed wrote, "Perfumes are foods that reawaken the spirit."

Eventually Egyptian perfumery influenced the Greeks and the Romans. For hundreds of years after the fall of Rome, perfume was primarily an Oriental art. It spread to Europe when 13th century Crusaders brought back samples from Palestine to England, France, and Italy. Europeans discovered the healing properties of fragrance during the 17th century. Doctors treating plague victims covered their mouths and noses with leather pouches holding pungent cloves, cinnamon, and spices which they thought would protect them from disease.

Perfume then came into widespread use among the monarchy. France's King Louis XIV used it so much that he was called the "perfume king." His court contained a floral pavilion filled with fragrances, and dried flowers were placed in bowls throughout the palace to freshen the air. Royal guests bathed in goat's milk and rose petals. Visitors were often doused with perfume, which also was sprayed on clothing, furniture, walls, and tableware. It was at this time that Grasse, a region of southern France where many flowering plant varieties grow, became a leading producer of perfumes.

Meanwhile, in England, aromatics were contained in lockets and the hollow heads of canes to be sniffed by the owner. It was not until the late 1800s, when synthetic chemicals were used, that perfumes could be mass marketed. The first synthetic perfume was nitrobenzene, made from nitric acid and benzene. This synthetic mixture gave off an almond smell and was often used to scent soaps. In 1868, Englishman William Perkin synthesized coumarin from the South American tonka bean to create a fragrance that smelled like freshly sown hay. Ferdinand Tiemann of the University of Berlin created synthetic violet and vanilla. In the United States, Francis Despard Dodge created citronellol—an alcohol with rose-like odor—by experimenting with citronella, which is derived from citronella oil and has a lemon-like odor. In different variations, this synthetic compound gives off the scents of sweet pea, lily of the valley, narcissus, and hyacinth.

Just as the art of perfumery progressed through the centuries, so did the art of the perfume bottle. Perfume bottles were often as elaborate and exotic as the oils they contained. The earliest specimens date back to about 1000 B.C. In ancient Egypt, newly invented glass bottles were made largely to hold perfumes. The crafting of perfume bottles spread into Europe and reached its peak in Venice in the 18th century, when glass containers assumed the shape of small animals or had pastoral scenes painted on them. Today perfume bottles are designed by the manufacturer to reflect the character of the fragrance inside, whether light and flowery or dark and musky.

Raw Materials

Natural ingredients—flowers, grasses, spices, fruit, wood, roots, resins, balsams, leaves, gums, and animal secretions—as well as resources like alcohol, petrochemicals, coal, and coal tars are used in the manufacture of perfumes. Some plants, such as lily of the valley, do not produce oils naturally. In fact, only about 2,000 of the 250,000 known flowering plant species contain these essential oils. Therefore, synthetic chemicals must be used to re-create the smells of non-oily substances. Synthetics also create original scents not found in nature.

Some perfume ingredients are animal products. For example, castor comes from beavers, musk from male deer, and ambergris from the sperm whale. Animal substances are often used as fixatives that enable perfume to evaporate slowly and emit odors longer. Other fixatives include coal tar, mosses, resins, or synthetic chemicals. Alcohol and sometimes water are used to dilute ingredients in perfumes. It is the ratio of alcohol to scent that determines whether the perfume is "eau de toilette" (toilet water) or cologne.

The Manufacturing Process

Collection

1 Before the manufacturing process begins, the initial ingredients must be brought to the manufacturing center. Plant substances are harvested from around the world, often hand-picked for their fragrance. Animal products are obtained by extracting the fatty substances directly from the animal. Aromatic chemicals used in synthetic perfumes are created in the laboratory by perfume chemists.

Extraction

Oils are extracted from plant substances by several methods: steam distillation, solvent extraction, enfleurage, maceration, and expression.

2 In *steam distillation*, steam is passed through plant material held in a still, whereby the essential oil turns to gas. This gas is then passed through tubes, cooled, and liquified. Oils can also be extracted by boiling plant substances like flower petals in water instead of steaming them.

3 Under *solvent extraction*, flowers are put into large rotating tanks or drums and benzene or a petroleum ether is poured over the flowers, extracting the essential oils. The flower parts dissolve in the solvents and leave a waxy material that contains the oil, which is then placed in ethyl alcohol. The oil dissolves in the alcohol and rises. Heat is used to evaporate the alcohol, which once fully burned off, leaves a higher concentration of the perfume oil on the bottom.

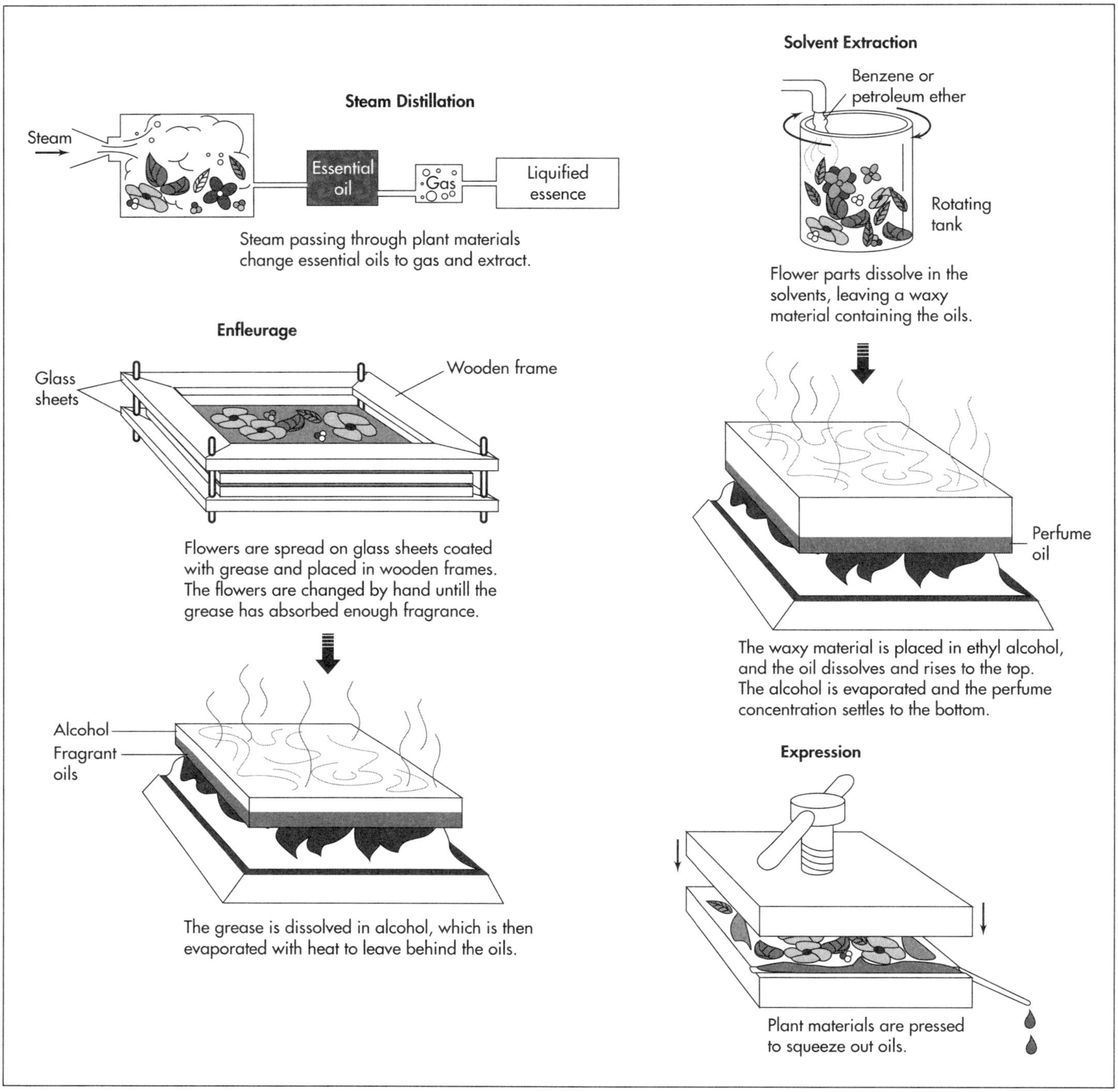

Oils are extracted from plant substances by steam distillation, solvent extraction, enfleurage, maceration, or expression.

4 During *enfleurage*, flowers are spread on glass sheets coated with grease. The glass sheets are placed between wooden frames in tiers. Then the flowers are removed by hand and changed until the grease has absorbed their fragrance.

5 *Maceration* is similar to *enfleurage* except that warmed fats are used to soak up the flower smell. As in solvent extraction, the grease and fats are dissolved in alcohol to obtain the essential oils.

6 *Expression* is the oldest and least complex method of extraction. By this process, now used in obtaining citrus oils from the rind, the fruit or plant is manually or mechanically pressed until all the oil is squeezed out.

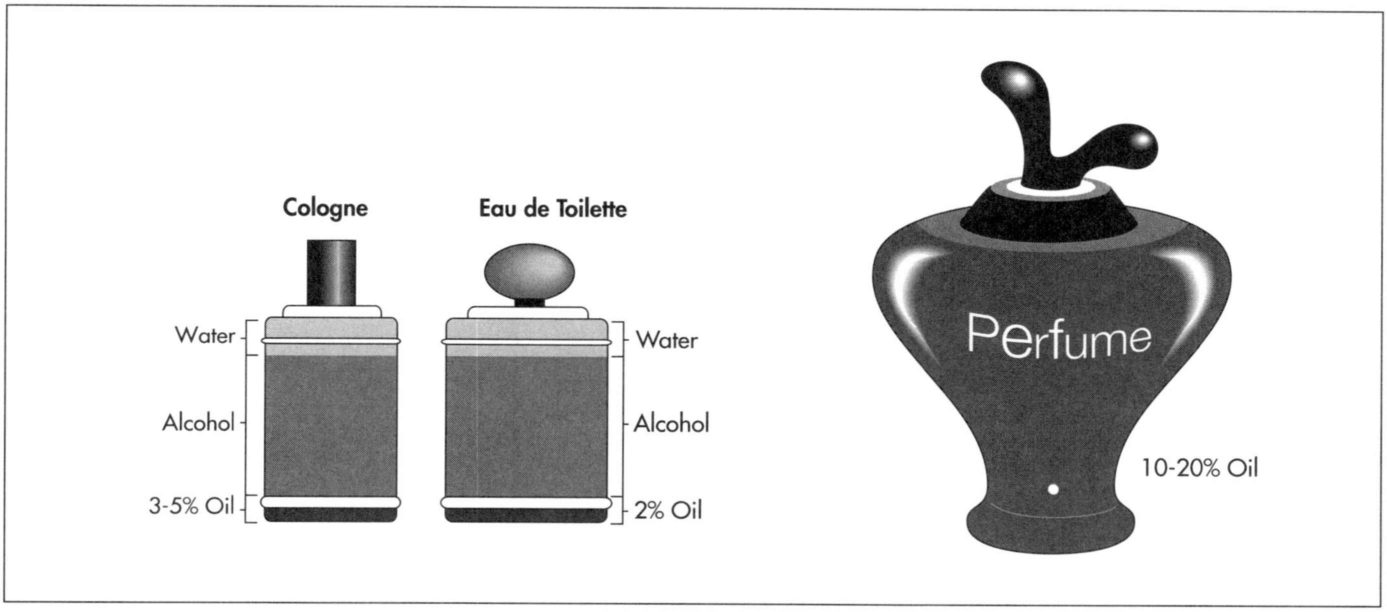

It is the ratio of alcohol to scent that determines perfume, eau de toilette, and cologne.

Blending

7 Once the perfume oils are collected, they are ready to be blended together according to a formula determined by a master in the field, known as a "nose." It may take as many as 800 different ingredients and several years to develop the special formula for a scent.

After the scent has been created, it is mixed with alcohol. The amount of alcohol in a scent can vary greatly. Most full perfumes are made of about 10-20% perfume oils dissolved in alcohol and a trace of water. Colognes contain approximately 3-5% oil diluted in 80-90% alcohol, with water making up about 10%. Toilet water has the least amount—2% oil in 60-80% alcohol and 20% water.

Aging

8 Fine perfume is often aged for several months or even years after it is blended. Following this, a "nose" will once again test the perfume to ensure that the correct scent has been achieved. Each essential oil and perfume has three notes: "Notes de tete," or top notes, "notes de coeur," central or heart notes, and "notes de fond," base notes. Top notes have tangy or citrus-like smells; central notes (aromatic flowers like rose and jasmine) provide body, and base notes (woody fragrances) provide an enduring fragrance. More "notes," of various smells, may be further blended.

Quality Control

Because perfumes depend heavily on harvests of plant substances and the availability of animal products, perfumery can often turn risky. Thousands of flowers are needed to obtain just one pound of essential oils, and if the season's crop is destroyed by disease or adverse weather, perfumeries could be in jeopardy. In addition, consistency is hard to maintain in natural oils. The same species of plant raised in several different areas with slightly different growing conditions may not yield oils with exactly the same scent.

Problems are also encountered in collecting natural animal oils. Many animals once killed for the value of their oils are on the endangered species list and now cannot be hunted. For example, sperm whale products like ambergris have been outlawed since 1977. Also, most animal oils in general are difficult and expensive to extract. Deer musk must come from deer found in Tibet and China; civet cats, bred in Ethiopia, are kept for their fatty gland secretions; beavers from Canada and the former Soviet Union are harvested for their castor.

Synthetic perfumes have allowed perfumers more freedom and stability in their craft, even though natural ingredients are considered more desirable in the very finest perfumes. The use of synthetic perfumes and oils eliminates the need to extract oils from animals and removes the risk of a bad plant harvest, saving much expense and the lives of many animals.

The Future

Perfumes today are being made and used in different ways than in previous centuries. Perfumes are being manufactured more and more frequently with synthetic chemicals rather than natural oils. Less concentrated forms of perfume are also becoming increasingly popular. Combined, these factors decrease the cost of the scents, encouraging more widespread and frequent, often daily, use.

Using perfume to heal, make people feel good, and improve relationships between the sexes are the new frontiers being explored by the industry. The sense of smell is considered a right brain activity, which rules emotions, memory, and creativity. Aromatherapy—smelling oils and fragrances to cure physical and emotional problems—is being revived to help balance hormonal and body energy. The theory behind aromatherapy states that using essential oils helps bolster the immune system when inhaled or applied topically. Smelling sweet smells also affects one's mood and can be used as a form of psychotherapy.

Like aromatherapy, more research is being conducted to synthesize human perfume—that is, the body scents we produce to attract or repel other humans. Humans, like other mammals, release pheromones to attract the opposite sex. New perfumes are being created to duplicate the effect of pheromones and stimulate sexual arousal receptors in the brain. Not only may the perfumes of the future help people cover up "bad" smells, they could improve their physical and emotional well-being as well as their sex lives.

Where To Learn More

Periodicals

Bylinsky, Gene. "Finally, A Good Aphrodisiac?" *Fortune*, October 21, 1991, p. 18.

Green, Timothy. "Making Scents Is More Complicated Than You Think." *Smithsonian*, June 1991, pp. 52-60.

Iverson, Annemarie. "Ozone." *Harper's Bazaar*, November 1993, pp. 208-40.

Lord, Shirley. "Message In a Bottle." *Vogue*, May 1992, p. 220.

Raphael, Anna. "Ahh! Aromatherapy." *Delicious!*, December 1994, pp. 47-48.

—Evelyn S. Dorman

Pet Food

Background

Pet food is a specialty food for domesticated animals that is formulated according to their nutritional needs. Pet food generally consists of meat, meat byproducts, cereals, grain, vitamins, and minerals. In the U.S. about 300 manufacturers produce more than 7 million tons of pet food each year, one of the largest categories of any packaged food. Pet owners can choose from more than 3,000 different pet food products, including the dry, canned, and semi-moist types, as well as snacks such as biscuits, kibbles, and treats. In the 1990s, this $8-billion industry feeds America's 52 million dogs and 63 million cats.

Commercially produced pet food has its origins in a dry, biscuit-style dog food developed in England in 1860. Shortly afterwards, manufacturers produced more sophisticated formulas, which included nutrients considered essential for dogs at the time. At the beginning of the 20th century, pre-packaged pet foods were also available in the U.S. Initially they consisted primarily of dry cereals, but after World War I, dog food made of canned horse meat was available. The 1930s ushered in canned cat food and a dry, meat-meal type of dog food. Some innovations by the 1960s were dry cat food, dry expanded-type dog food, and semi-moist pet food.

Beginning in the 1980s, trends in the pet food market included greater demand for dry foods and less for canned foods. Research suggested that a soft diet of canned dog food led to gum disease more quickly than did dry food. In general, the growing health-consciousness of the public led to an increased interest in more nutritious and scientific formulas for pet foods, such as life-cycle products for younger and aging pets, and therapeutic foods for special health conditions of the pet, such as weight loss and urinary problems. Pet food producers were also more inclined to use less fatty tissue and tallow and more protein-rich tissue. Finally, the pet snack category grew in popularity with products like jerky snacks, sausage-shaped pieces, biscuits, and biscuit pieces called kibbles.

In the U.S. about 300 manufacturers produce more than 7 million tons of pet food each year, one of the largest categories of any packaged food.

Raw Materials

The primary ingredients in pet food are byproducts of meat, poultry, and seafood, feed grains, and soybean meal. Among the animals used in rendering are livestock, horses, and house pets which have been put to sleep. The National Animal Control Association estimated that each year about 5 million pets were shipped to rendering plants and recycled into pet food during the 1990s. They are generally listed as meat or bone meal in the ingredient lists.

The animal parts used for pet food may include damaged carcass parts, bones, and cheek meat, and organs such as intestines, kidneys, liver, lungs, udders, spleen, and stomach tissue. Cereal grains, such as soybean meal, corn meal, cracked wheat, and barley, are often used to improve the consistency of the product as well as to reduce the cost of raw materials. Liquid ingredients may include water, meat broth, or blood. **Salt**, preservatives, stabilizers, and gelling agents are often necessary. Gelling agents allow greater homogeneity during processing and also control the moisture. They

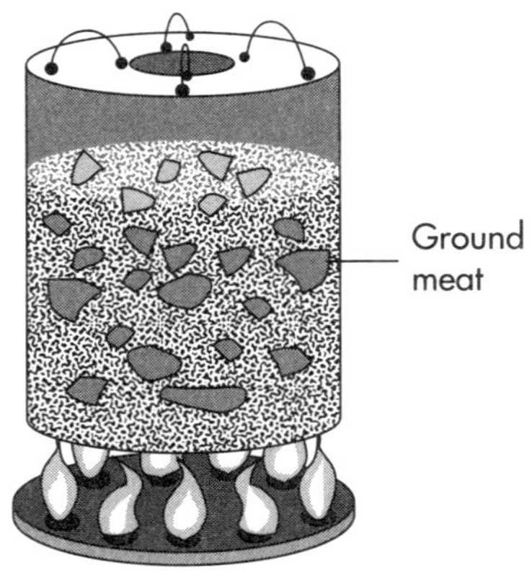

Rendered meat arrives at the factory where it is ground to the desired texture. The meat is then cooked in a continous cooker with steam.

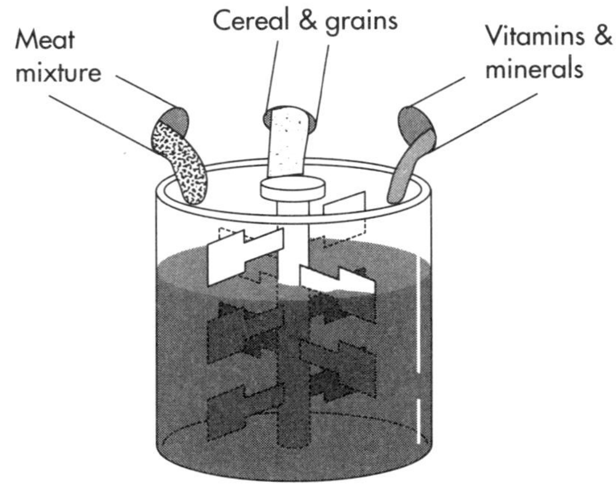

The meat mixture is blended with other ingredients such as cereals, grains, vitamins, and minerals.

include bean and guar gums, cellulose, carrageenan, and other starches and thickeners. Palatability can be enhanced with yeast, protein, fat, fish solubles, sweeteners, or concentrated flavors called "digests." Generally, artificial flavors are not used, though smoke or bacon flavors may be added to some treats. Most manufacturers supplement pet foods with vitamins and minerals, since some may be lost during processing.

Ingredients vary somewhat depending on the type of pet food. The basic difference between canned and dry pet foods is the amount of moisture. Canned food contains between 70 and 80% moisture, since these are generally made from fresh meat products, while dry pet food contains no more than 10%. Additional ingredients used for dry foods include corn gluten feed, meat and bone meal, animal fats, and oils. For a meat-like texture, dry foods require more amylaceous, or starch ingredients; proteinaceous adhesives, such as collagen, albumens, and casein; and plasticizing agents. Semi-moist pet foods usually require binders, which come from a variety of sources, such as gels, cereal flours, sulfur-containing amino acids, lower alkyl mercaptans, lower alkyl sulfides and disulfides, salts, and thiamin. Semi-moist products may also incorporate soybean flakes, bran flakes, soluble carbohydrates, emulsifiers, stabilizers, and dried skim milk and dried whey.

Antioxidants are often used to retard oxidation and rancidity of fats. These include butylated hydroxy anisole (BHA), butylated hydroxy toluene (BHT), and tocopherol. To prevent mold and bacterial growth, producers use either sucrose, propylene glycol, sorbic acid, or potassium and calcium sorbates.

The Manufacturing Process

Except for the ingredients, the general manufacturing process for pet food is similar to that for processed food. The flesh products used in pet foods must first be rendered, or processed, to separate the water, fat, and protein components, including soft offals (viscera) and hard offals (e.g. bones and hoofs). Generally, meat is rendered by outside companies and shipped to pet food manufacturers. The meat products intended for canned food must be delivered fresh and used within three days. Frozen meat products may be used for dry foods.

The manufacturing process entails grinding and cooking the flesh and flesh byproducts. Next, the meat is mixed with the other

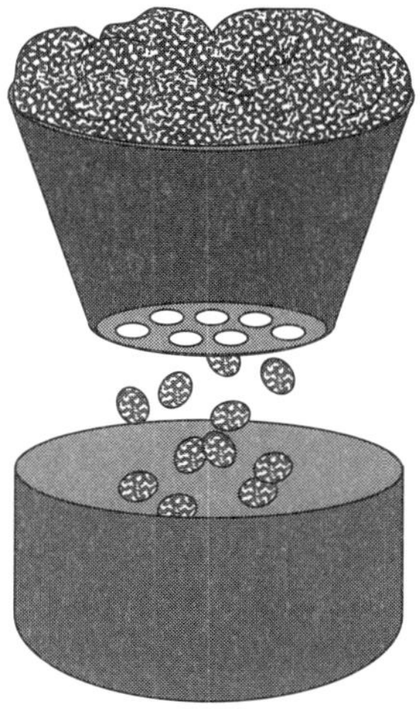

For dry and semi-moist foods, the mixture is extruded through orificed plates to obtain the proper shape.

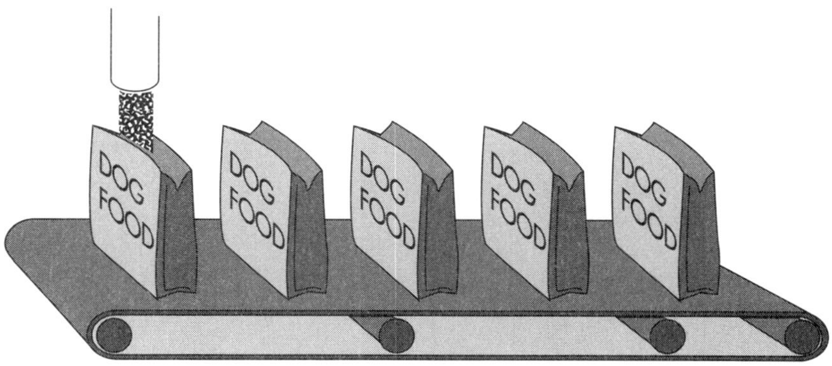

Measured amounts of the product are packaged into pre-printed bags.

ingredients, and if the recipe requires, the mixture is shaped into the appropriate forms. The finished product is filled into containers and shipped to distributors.

Innovations in pet food processing and packaging have led to better quality products with longer shelf life. Canned dog foods that are vacuum packed have a shelf life of three to five years and are very stable with little or no loss in nutritional value. Dry dog food, on the other hand, has a shelf life of only 10 to 12 months and requires the addition of preservatives, though some manufacturers are using natural preservatives such as vitamins E and C.

Rendering the meat

1 Generally, rendering is performed by meat processors. Rendering entails rupturing fat cells, either by heat or enzymatic- and solvent-extraction, and then drying the residue.

Grinding and pre-cooking the meat

2 The meat products are coarsely ground to the desired texture.

3 To facilitate further processing, the ground meat is cooked in a continuous cooker with live steam at the appropriate temperature.

4 The flesh products are reground after initial cooking to produce a more uniform consistency. For semi-moist or chunky foods, the batches are deliberately cooked unevenly to create the desired chunky texture.

Blending and shaping

5 The meat mixture is blended with other ingredients such as cereal grains, vitamins, and minerals.

6 Dry and semi-moist foods are usually heated so the mixture will partially dextrinize, or thicken, the starch. To achieve the marbled-look of real meat, the meat mixture may be cooked unevenly and half of the batch colored red and the other white. Semi-moist foods must be stabilized to retain the proper amount of moisture in the dry and semi-moist parts of the food.

7 Dry and semi-moist foods may be extruded under high pressure through a device with orificed plates to obtain the shape and size of the specific product, for instance, the form of biscuits, kibbles, meatballs, patties, pellets, or slices. An alternative to extrusion is to gelatinize and expand the mixture. For marbled meat, the mixture of red and white meat is extruded together and broken into chunks.

Packaging and labeling

8 Measured amounts of the product are packaged into appropriate containers. Dry foods are poured into pre-printed containers. Moist canned foods are vacuum sealed to reduce the oxygen content and prevent spoilage of fats in the food.

Sterilizing

9 Cans of pet food are sterilized by passing them through a retort, or heating chamber. The retort may be either a batch or continuous hydrostatic type. The cans are heated to about 250°F (121°C) for 80 minutes, though the cooking temperatures and times depend on the contents, steam pressure, and can size.

10 The cans are quickly cooled to about 100°F (38°C). Next, the cans are dried and labeled.

11 The containers are packaged into corrugated cardboard boxes or shrink-wrapped with plastic in corrugated cardboard trays. The pet food is ready for shipping to distributors.

Quality Control

Pet food manufacturers must conform to the rules and regulations set by several agencies at the federal and state levels, including the Food and Drug Administration (FDA), Federal Trade Commission (FTC), and U.S. Department of Agriculture (USDA). The USDA controls meat quality and determines which animals can be used in pet foods. The FDA regulates ingredients by setting maximum and minimum limits on certain nutrients and by banning the use of medications or antibiotics in foods, since pet food is sometimes accidentally eaten by children. The job of the Association of American Feed Control Officials (AAFCO), a non-governmental advisory group with representatives in each state, is to register the 3,000 brands and sizes of pet food.

The "guaranteed analysis" statement found on pet food labels was created nearly a century ago when some manufacturers used undesirable ingredients like sand or limestone to add weight to their pet food. The guaranteed analysis ensures minimum percentages of crude protein and crude fat and maximum percentages of crude fiber and moisture. The term "crude" refers to a method of testing the elements. Other guarantees may include minimum amounts of calcium, phosphorus, sodium, and linoleic acid in dog food, and ash, taurine, and magnesium in cat food. The maximum allowable moisture for canned food is 78%, while dry foods may contain as much as 12% moisture.

Proper labeling of pet foods is required to provide accurate information to the purchaser. Guidelines are set by the FDA's Center for Veterinary Medicine and the AAFCO. Six basic elements should be on the label: the product name, net weight, name and address of the manufacturer or work area that may be 200 feet (61 m) wide and five stories high. The area must be immaculately clean and dust-free, and the operations of the roll-coating machines are controlled by arrays of control panels in the fully automated process. Machines coat precise amounts of emulsion in micro-thin layers on the wide strips of plastic base; a single, dried layer of emulsion may be six one-hundred-thousandths of an inch thick. Successive layers of three emulsions are applied to the base to make color film, and each emulsion layer has its own color-forming chemicals called linked dyes. The three emulsion layers in color film respond to blue, green, and red light, so each photograph is a triple latent image with the sandwiched color range reproduced by processing. The strips of emulsion-coated base (now film) are cut into progressively narrower widths, perforated so the film can be advanced in the camera, and spooled, except for instant film and sheet film that are packed flat.

Packaging

5 Film is packed in cartridges, cassettes, rolls, instant packs, or sheets. Cartridges are used in certain types of cameras and include a take-up spool that is built in so the exposed film and cartridge are removed as a unit. Cassettes are made for cameras that use film in the 35-millimeter format. They consist of a spool enclosed in a metal jacket. The tongue of the film is drawn over the pressure plate at the back of the camera to a take-up spool that is built into the camera. When the film is finished, it is rewound onto the spool in the cassette, and the unit is removed. Rollfilms consist of paper-backed film that is packed on a spool like the one in the camera. The film is wound onto the spool in the camera, and that spool and film are removed. The spool on which the film was packed originally can then be moved to the receiving side of the camera, and a new roll inserted. The packs for instant cameras contain 8 to 12 sheets that are ejected individually after each shot. Sheet film is used for specialized applications like x-ray film.

Plastic cartridges for cartridge-type film are made by injection molding, in which fluid-like plastic is squirted mechanically into forms or molds. These are hardened, removed from the molds, and trimmed and smoothed. The spooled film is then placed in the cartridges and sealed. The metal canisters are printed on the outside, cut to shape and size, trimmed and smoothed, and edged with protective plastic. The metal is shaped around the spools of film. Plastic canisters and caps are also made for the film canisters, as are other types of outer packaging such as foil-lined paper pouches, and the outer cartons. The packaging is dated, shrink-wrapped in plastic in quantities appropriate for sale, packed in cardboard containers for shipping, and stored in air-conditioned rooms to await shipment.

Quality Control

In all phases of manufacture, photographic film is extremely sensitive to light, heat, dust, and impurities. Air flow into the film-manufacturing rooms is washed and filtered. Temperature and humidity are carefully regulated. Production rooms are scrubbed clean daily, and plant workers wear protective clothing and enter sensitive work areas through air showers that clean personnel of dust and contaminants. Each step of manufacture is carefully inspected and controlled. For example, the chromium-plated wheel on which the base is formed is inspected to maintain a mirror-like finish because tiny imperfections will affect the quality of the film. Finally, samples of film are removed from completed batches and subjected to many tests, including the taking of photographs with the samples.

Byproducts/Waste

Factory workers and the environment must also be protected from the hazardous chemicals, fumes, and wastes that can be generated during the process. Protective clothing keeps the product clean and insulates the workers from possible contaminants. Air released to the outside is also filtered and monitored. Extensive recycling is done, not only to protect the environment but also to salvage valuable materials such as silver for purifying and reuse. The photographic film industry was also among the first to use incineration successfully to burn wastes efficiently and control emissions.

Plastic Wrap

Background

Plastic wrap is a form of food packaging consisting of a thin film of flexible, transparent polymer that clings to itself and to food containers to form a tight seal. The plastic keeps the food fresh by protecting it from air and by preventing dry foods from absorbing moisture and wet foods from losing moisture. It also seals in odors to prevent them from spreading to other foods stored nearby.

Plastics are artificial polymers; that is, they consist of gigantic molecules formed by combining thousands of small molecules of the same kind into a long chain. These small molecules are known as monomers, and the process of combining them is known as polymerization. Natural polymers include such familiar substances as **silk**, rubber, and cotton.

The first plastic was made by the British chemist Alexander Parkes in 1862, who produced a substance he called Parkesine from cotton, nitric acid, sulfuric acid, castor oil, and camphor. Two years later in the United States John Wesley Hyatt improved this product and named it celluloid. Celluloid was a tremendous success and was used to make many different products, but it was highly flammable.

The first completely artificial polymer (unlike celluloid, which was a derivative of the natural polymer cellulose) was Bakelite, which was produced from phenol and formaldehyde by the Belgian chemist Leo Baekeland in 1908. Many other polymers were developed during the 20th century, including such important products as artificial rubber and artificial fibers such as nylon.

The first plastic used for wrapping was cellophane, another derivative of cellulose invented by the Swiss chemist Jacques Brandenberger in 1911. It had the advantage of being transparent, and was used for packaging as early as 1924. Cellophane was the most common form of plastic film made until 1963, when it was overtaken by polyethylene.

Polyethylene was discovered by accident by research workers at the British company Imperial Chemicals Industries in 1933, when they mixed benzene and ethylene at high temperature and pressure. Polyethylene was first used chiefly for electrical insulating material. It was first made into a film in 1945 by the Visking Corporation in the United States, and has grown in popularity ever since.

Polyvinyl chloride (PVC) was produced before World War II and was originally used as an inferior substitute for rubber, but films of this substance were not made in any quantity until the 1950s. PVC is used today in many different products such as pipes, flooring, electric cables, shoes, and clothing, as well as plastic wrap.

Polyvinylidene chloride (PVDC) film was developed by the Dow Chemical Company during World War II for military use. It offered a high degree of protection from moisture and resistance to oils, greases, and corrosive chemicals, so it was used to package sensitive equipment such as optical devices and aircraft engine components. In

Polyvinylidene chloride (PVDC) film was developed by the Dow Chemical Company during World War II for military use to package sensitive equipment such as optical devices and aircraft engine components. In 1952 it was offered to the public under the familiar trade name Saran Wrap.

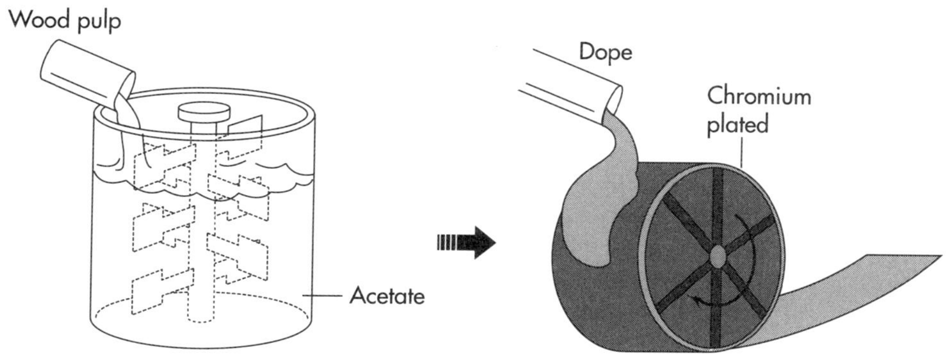

To make the base, wood pulp is mixed with acetate to form a syrup. Solid pellets precipitate and are dissolved in solvents to form a dope.

The dope is spread on a large rotating wheel in thin layers. The solvents evaporate, leaving behind a thin sheet of plastic.

made from animal bones and hides was used to coat glass plates, and silver iodide was precipitated inside the gelatin layer. The plates and their dried jelly could be exposed, then the photograph could be developed later by rewetting the gelatin. The complicated procedure of manufacturing the plate, exposing it, and processing it into the finished photograph was broken into parts that made the photographer's work easier and made photography and photo processing a manufacturing industry.

George Eastman combined the paper base of Talbot's calotype with the gelatinous silver nitrate emulsion from Maddox's process to invent flexible roll film in 1884. Eastman quickly made the transition to an emulsion-bearing plastic, transparent film by 1889, which was a year after his company introduced the first Kodak camera. These developments made photography a simple, compact, portable practice that is now the most popular hobby in the United States.

Raw Materials

A roll of film consists of the emulsion and base that compose the film itself, the cassette or cartridge, and outer protective packaging. The materials used to make the emulsion are silver, nitric acid, and gelatin. The base consists of cellulose and solvents that are mixed to form a thick fluid called dope. Film that is packed in a cassette (35-millimeter film is typically packed this way) requires a metal spool, the protective metal canister, and plastic strips at the canister opening where the film emerges. Other sizes of film including Polaroid film are protected from light and air by plastic cartridges or packs. Outer packaging, which varies among film products, is made from foil-lined paper, plastic, and thin cardboard cartons. The outer packaging is also insulating and protects the film from exposure to light, heat, and air.

The Manufacturing Process

Base

1 For most films, the base to which the light-sensitive emulsion is fixed consists of cellulose acetate, which is wood pulp or cotton linters (short cottonseed fibers) mixed with acetate to form a syrup. Solid pellets of cellulose acetate precipitate or separate out of the syrup and are washed and dried. The pellets are dissolved in solvents to form the transparent, honey-like dope. The dope is spread in a thin, even sheet on a wheel that is two stories in diameter. The wheel is plated with chromium for a smooth finish, and it turns slowly. The solvents in the dope volatilize or evaporate as the wheel turns. The process is much like the applying and drying of nail polish. The remaining base is a thin sheet of plastic that is of a uniform thickness measured in ten-thousandths of an inch. When it is dry, the base is

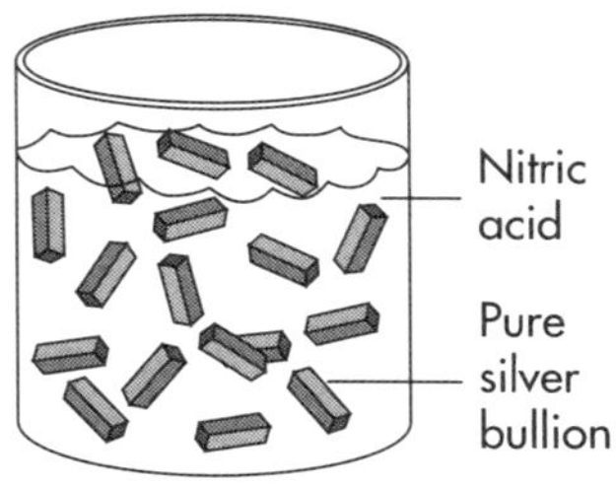

To make the coating emulsion, pure silver bullion is dissolved in nitric acid. As the solution cools, silver nitrate crystals form. These crystals are mixed with gelatin to form the coating.

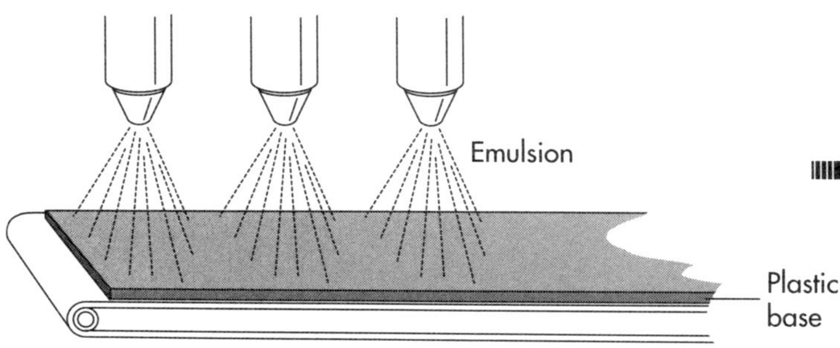

The emulsion is layered onto the plastic base in three uniform coats. Each emulsion layer has its own color-forming chemicals which respond to blue, green, and red light.

The emulsion-coated base is then cut into strips and perforated so the film can be advanced in the camera.

removed from the wheel and wound on 54-inch (137 cm) diameter reels.

Emulsion

2 Silver is the main ingredient of the emulsion. Pure silver bullion is received at the manufacturing plant in bars that are checked by weight and serial number. The bars are dissolved in a strong solution of nitric acid, and the process releases heat. After the acid has completely dissolved the silver, the solution is stirred constantly and cooled. Cooling causes crystals of silver nitrate to grow, much like salt crystals in water. The crystals are wet with water that also separates out. The crystals are removed from the solution and whirled in centrifuges with sieve-like openings to remove the water and keep the crystals pure. At this point in the process, the chemical solutions are light-sensitive, so further manufacturing processes are completed in darkness.

3 Meanwhile, gelatin has been made using distilled water and treated with chemicals including potassium iodide and potassium bromide. The gelatin serves as a binding agent to hold the silver nitrate crystals, and also to fix them to the base. The gelatin and chemicals are mixed in cookers that are lined with silver so the emulsion remains pure. As the mixture cools, silver halide salts (chemical combinations of the silver, iodide, and bromide) form as fine crystals that remain suspended in the gelatin to make the emulsion.

Coating process

4 The emulsion is pumped through a piping system to "coating alley," a huge

The Future

Film manufacturers are continually improving the quality of film so that photographs are sharper, color is truer, graininess is reduced, and film speed is improved. Several new camera films use "T-grain" emulsion technology, in which the molecular structure of the silver halide crystals is modified to create silver grains shaped like tiny tablets. The flat shape helps them collect light efficiently, so sharper photographs are produced from higher-speed films. This technology also benefits the environment because fewer chemicals are needed for processing film, and the opportunity for chemicals to enter the environment is reduced.

The next advance in photography does not require film at all; the film-free camera stores photographs digitally without any film. Digital cameras electronically transfer images to computers which can then print the images.

Where To Learn More

Books

Bailey, Adrian and Adrian Holloway. *The Book of Color Photography*. Alfred A. Knopf, 1979.

Collins, Douglas. *The Story of Kodak*. Harry N. Abrams, Inc., Publishers, 1990.

Periodical

Antonoff, Michael. "Digital Snapshots from My Vacation." *Popular Science*, June 1995, pp. 72-76.

Other

From Glass Plates to Digital Images. Eastman Kodak Company, 1994. 343 State St., Rochester, NY 14650. (716)724-4000.

—Gillian S. Holmes

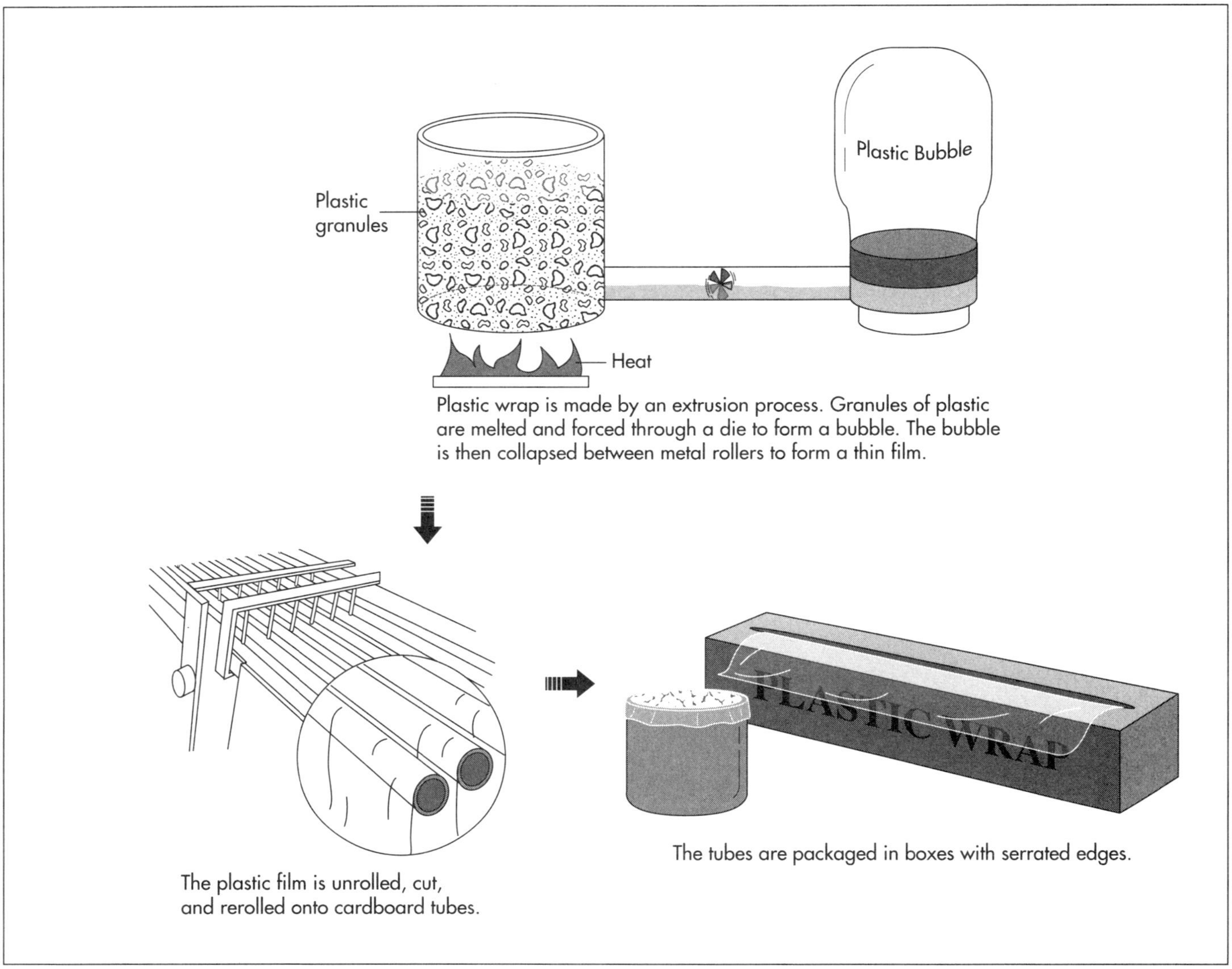

1952 it was offered to the public under the familiar trade name Saran Wrap.

Raw Materials

Most household plastic wrap is made from polyethylene, PVC, or PVDC. These polymers are all derived from simple hydrocarbons such as methane or ethylene, which are produced from natural gas or petroleum. Polyethylene is made directly from ethylene. PVC is made from vinyl chloride, derived from ethylene, or from acetylene, derived from methane. PVDC is made from vinyl chloride and vinylidene chloride, a derivative of 1,1,2-trichloroethane, which in turn is derived from ethylene or acetylene.

Some plastic films, including cellophane, are derived from cellulose, which is obtained from wood pulp or from linters, tiny fibers that cling to cotton seeds after the longer cotton fibers have been removed by a cotton gin. At least one form of plastic film (Pliofilm, a trademark of Goodyear) is derived from rubber.

The Manufacturing Process

Processing the raw materials

1 The chemicals needed to synthesize polymers are usually obtained from petroleum. Crude oil is heated in a furnace to about 752°F (400°C). Vapors from the oil move into a fractionating column, a tall tower containing a series of chambers. The chambers are warmer at the bottom of the

Photographic Film

Film was "discovered" in a chemistry laboratory by Johann Henrich Schulze, a German doctor, in 1727 when he mixed chalk, silver, and nitric acid in a flask to make silver nitrate.

Background

Photographic film is a chemically reactive material that records a fixed or still image when the film is exposed to light. Typically, film is placed in a camera, and light from the image being photographed is allowed to enter and is focused and sometimes made larger or smaller by the **camera lens**. The film is exposed to the image by opening a shutter in the camera body, and the combination of the speed of the shutter and the film speed (which is the chemical reactivity of the film) controls the amount of light that strikes the film. The image is recorded on the film, but it is a latent or invisible image. When the film is removed from the camera, it is developed by chemical processes into a visible image. This visible image is negative or the reverse in brightness of the way our eyes see light; the brightest parts of the photographed object appear the darkest on the negative where the film received the most exposure to light. The negative image is made positive, or as our eyes see it, by another type of processing whereby the negative is printed on sensitive paper. Color-reversal films are positives and are used for making slides. All of the elements of the process—the parts of the camera, the type and parts of the lens, the type of film, including its chemistry, the developing process, the printing process, and the type of paper—contribute to the sharpness or trueness of the finished photograph.

History

Film was "discovered" in a chemistry laboratory. In 1727, Johann Henrich Schulze, a German doctor, mixed chalk, silver, and nitric acid in a flask to make silver nitrate. When the solution was exposed to sunlight, it changed color from white to purple. When Schulze pasted cutouts of letters and numbers on the outside of a flask of freshly made solution and exposed it to the light, the cutouts appeared to have been printed on the solution. Although the discovery marked the birth of photography, it was not used for over 100 years. In 1839, Louis Daguerre, a French painter, created a photographic process in which liquid iodine was placed on a silvered copper plate, and the plate was exposed to light. The liquid iodine was the emulsion, or light-reactive chemical, and the copper plate was the base for these photographs called "daguerreotypes." The American inventor Samuel F.B. Morse learned the art of daguerreotypy and taught it to Matthew Brady, who made images of the Civil War that are treasured both as historical records and artistic landmarks in photography.

Daguerreotypy was cumbersome to use; the "wet plate" process was awkward, the box-type cameras had to hold the large plates, and the finished photographs were the size of the plates. While Daguerre was developing his process, William Henry Fox Talbot, an English archaeologist, created his own process called "calotype," meaning "beautiful picture" in 1841. Talbot coated a paper base with an emulsion of silver iodide and produced a negative by a developing process. The calotype is more like today's film and photographic process, and the intermediate step resulting in a negative permitted more than one print to be made.

The flexibility of photography was improved further in 1871 when R.L. Maddox invented the "dry plate" process. Gelatin distributor, guaranteed analysis, ingredient list, and nutritional information. The product name should accurately describe the contents and adhere to the "percentage" rules. The "95%" rule requires that if the product name suggests that meat, poultry, or fish is the primary ingredient, as in "Barbara's Beef Dog Food," it must contain 95% or more of that ingredient, excluding water used in processing. If two meat ingredients are listed as the primary ingredients, the two together must equal 95%.

The "25%" rule, or "dinner" rule, applies to items such as "chicken dinner," "meat entree," and terms like platter, formula, nuggets, and so on. It requires that the food listed must make up between 25 and 95% of all ingredients by weight. If more than two ingredients are in the name, each must be at least three percent in weight and the primary ingredient must be listed first, as all the ingredients on the label must be listed in predominance by weight.

A third rule is the "three percent" rule, or the "with" rule, which applies to minor ingredients listed on the label. For example, "Charlie's Chicken Cat Food with Cheese" should contain at least three percent cheese. Finally, the "flavor rule" requires that if a flavor ingredient, such as meat meal, is included in the name it must be detectable. To prevent misleading customers, the word "flavor" must be in the same size and style as the corresponding ingredient. Any pictures on the label must not be misleading either.

All the ingredients should correspond to the specific names listed in the Official Publication of the AAFCO. Any preservatives, stabilizers, colors, and flavorings must conform to the GRAS rule, "Generally Recognized as Safe." The term "natural" should not be applied to products containing artificial flavors, colors, or preservatives.

Calories per serving and per container should be listed in much the same manner as foods for human consumption, in kilocalories per kilogram. Package codes must be printed on all containers.

Other associations also monitor pet foods and evaluate their effects on pets, such as the American Animal Hospital Association (AAHA), the American Veterinary Medical Association (AVMA), and the Pet Food Institute (PFI).

Where To Learn More

Book

Ockerman, H.W. and C.L. Hansen. *Animal By-Product Processing.* VCH Publishers, 1988.

Periodicals

Corbin, James. "Promote Product Acceptance with a Lesson on Pet Food Labels." *Pet Product News,* July 1993, p. 40.

Ducommun, Debbie. "The Dog Food Debate." *Pet Product News*, May 1994, p. 45.

Dzanis, David A. "Understanding Pet Food Labels." *FDA Consumer,* October 1994, p. 10.

Eckhouse, John. "How Dogs and Cats Get Recycled into Pet Food," Part One. *The San Francisco Chronicle,* February 19, 1990, p. C1.

———. "Pet Food Is Big Business," Part Two. *The San Francisco Chronicle,* February 20, 1990, p. B1.

McDermott, Michael J. "Pedigreed Pet Foods." *Food & Beverage Marketing,* May 1991, p. 20.

Presley Noble, Barbara. "All about Dog and Cat Food: Will the American Pet Go for Haute Cuisine?" *The New York Times,* December 16, 1990, p. 5.

Other

"Pet Food Institute Fact Sheet, 1994." Pet Food Institute, 1200 19th St., NW, Ste 300, Washington, DC 20036. (202) 857-1120.

—Audra Avizienis

tower and cooler at the top. The various substances that make up petroleum rise through the chambers as gases until they reach the temperature at which they become liquids. Since each substance has a different boiling point, they liquefy in different chambers and can be separated and collected.

2 Most polymers begin with very simple hydrocarbons that have low boiling points. These substances do not liquefy in the fractionating column, but instead remain in the form of gases which can be removed from the top of the tower. They may also be obtained from natural gas, which is mostly methane. Another source for these chemicals is naphtha, a mixture of liquid hydrocarbons, obtained from a fractionating column, which are heavier than **gasoline** but lighter than heavy oil. Naphtha is heated under pressure to break down the liquid hydrocarbons into smaller molecules, a process known as cracking. A catalyst is added to enable cracking to take place at a lower temperature and pressure than it would without it. The catalyst may be a natural or artificial clay (a mixture of alumina and silica or a zeolite (any of various minerals containing aluminum, silicon, oxygen, and other elements in combination with water.) Catalytic cracking usually takes place at a temperature of about 932°F (500°C) under a pressure of about 100 kilopascals. The cracked naphtha is then distilled in a manner similar to that in a fractionating column to separate its components.

Polymerization

3 Polyethylene is polymerized from ethylene, which is obtained from cracking. Ethylene is heated in a pressure chamber to about 338°F (170°C) at a pressure of about 200,000 kilopascals in the presence of a small amount of oxygen. The oxygen breaks the ethylene down into free radicals, which combine with each other to form chains of polyethylene. About one percent of a nonreacting gas such as propane is added to prevent the chains from becoming too long.

4 PVC is polymerized from vinyl chloride, which can be obtained either by mixing acetylene with hydrochloric acid or ethylene with chlorine. Ethylene is more commonly used because it is efficiently obtained from the cracking of naphtha. If acetylene is used it must first be synthesized by heating methane to about 2732°F (1500°C) or through various other chemical reactions. Vinyl chloride is mixed with water and agitated to form a suspension, much as oil and vinegar are mixed to form salad dressing. Various suspending agents such as starch and gelatin are added to keep the mixture from separating. The temperature of the suspension is raised to about 104°F (40°C) or 122°F (50°C) and an initiator, usually an organic peroxide, is added to start the reaction. The vinyl chloride molecules react with each other to form chains of PVC. The mixture is cooled and particles of PVC are separated from the water in a centrifuge and dried in an oven.

5 PVDC is polymerized from a mixture of about 15% vinyl chloride and about 85% vinylidene chloride. To produce vinylidene chloride, first 1,1,2-trichloroethane is made by mixing acetylene, hydrochloric acid, and chlorine, or by mixing ethylene and chlorine. The 1,1,2-trichloroethane then reacts with calcium hydroxide or sodium hydroxide to produce vinylidene chloride. Polymerization of PVDC proceeds in much the same way as PVC.

6 Polyethylene is naturally flexible, but PVC and PVDC must have plasticizers added or they will be hard and rigid. Various organic and inorganic esters can be used as plasticizers. Generally the liquid plasticizer is slowly sprayed into dry polymer powder and heated to about 302°F (150°C) to form a homogeneous mixture.

Making plastic wrap

7 Plastic wrap is made by extrusion. In this process granules of plastic are heated until they melt at about 212°F (100°C) for polyethylene and about 392°F (200°C) for PVC and PVDC. The liquid is then forced through a die to form a tube of warm, stretchable plastic. At regular intervals compressed air is blown into the side of the moving tube to form large bubbles. This stretches the plastic to the desired thinness. The thin plastic cools rapidly, and the bubble is collapsed between metal rollers to form a film. The film is wound around a large metal roller to form a roll that may hold several

kilometers of plastic wrap. The plastic film on these rolls is then unrolled, cut to the proper length (usually about 49 feet [15 m]) and width (about 1 foot [0.33 m]), and rerolled onto small cardboard tubes. (This rolling, unrolling, and rerolling tends to give the plastic wrap a slight negative charge of static electricity, that helps it cling.) The cardboard tubes of plastic wrap are placed in cardboard boxes that have a serrated edge at the opening so that the consumer can tear off the desired length. Some also have a sticky spot on the box to catch the edge of the plastic wrap so it doesn't stick to the tube. The boxes of plastic wrap are then stacked in cartons and shipped to retailers.

Quality Control

A variety of standard tests exist to ensure that plastic wrap is effective. The most important are tests for permeability, impact resistance, and tear strength.

Water vapor permeability is measured by filling a dish with calcium chloride, a highly water-absorbent substance. It is covered with a sample of plastic wrap and weighed. The dish is then placed in a chamber with a controlled temperature and humidity. After a measured amount of time the dish is weighed again. The increase in weight shows how much water vapor has passed through the plastic. This test can also be done by filling the dish with water instead of calcium chloride and measuring the decrease in weight to see how much water vapor has escaped. These tests are performed at 73°F (23°C) with a relative humidity of 50%, at 90°F (32°C) with a relative humidity of 50%, and at 100°F (38°C) with a relative humidity of 90%.

Gas permeability is measured by placing a sample of plastic wrap between two chambers. The upper chamber contains a pressure of 100 kilopascals, and the lower chamber contains a vacuum connected to a tube containing liquid mercury. As the air in the upper chamber passes through the plastic wrap it increases the pressure in the lower chamber and forces the level of mercury to drop. The change in the level reveals how much air has penetrated the plastic.

Impact resistance is measured by dropping weights of increasing size on test samples until half of them break, at which point the weight is recorded. It can also be measured by filling bags made from the plastic wrap that is being tested with sand and dropping them on a hard surface from increasing heights until they burst. The height at which this occurs is then recorded. Impact resistance is also measured by shooting a small steel ball propelled by pressurized air through a sheet of plastic wrap and measuring how much the plastic slows it down.

Tear strength consists of tear initiation strength (the force required to start a tear) and tear propagation strength (the force needed to continue a tear). To measure tear initiation strength a sample shaped like a shallow V is pulled between two jaws until it begins to tear. This unusual shape is selected to provide a 90 degree angle that provides a controlled starting point for the tear. Tear propagation strength is measured by pulling apart a sample containing a precut slit.

In general, PVDC is stronger and less permeable than polyethylene, which is less permeable than PVC.

Environmental Concerns

Since plastic wrap is difficult, if not impossible, to recycle and is rarely reused, it does contribute to waste. One consumer group, considering such factors as the energy and raw materials needed for manufacture, the wastes released during manufacturing and disposal, the ability to be recycled, and the typical amounts used, has rated plastic wrap as "Good." By comparison, reusable plastic containers were rated as "Excellent," plastic bags as "Very Good," aluminum foil and freezer bags as "Good," and freezer papers as "Poor." Another concern is the possibility that exposure to certain plasticizers in plastic wrap could be harmful. These chemicals are absorbed from plastic wrap into hot and fatty foods. Although they have never been shown to cause harm in humans, plasticizers have been proved to cause cancer when fed in large amounts to lab animals. PVC wrap can consist of as much as one-third plasticizers, PVDC wrap consists of about 10%

plasticizers, and polyethylene wrap usually contains no plasticizers.

Where To Learn More

Book

Gait, A. J. and E. G. Hancock. *Plastics and Synthetic Rubbers*. Pergamon Press, 1970.

Periodical

"Keeping Food Fresh." *Consumer Reports*, March 1994, pp. 143-47.

—*Rose Secrest*

Player Piano

Replacing the traditional roll-player piano, the disc-driven player piano has a player action that uses electronically-driven solenoids as strikers to reproduce music as recorded by a musician.

Background

The player piano begins as a standard, or non-player, acoustic piano. It contains a keyboard, and, when the keys are struck, felt hammers strike tuned metal strings to create musical tones. The intricate collection of felt hammers and connecting parts is called the "piano action." The piano action can be manipulated with the musician's fingers, or, if the piano is equipped with a player system, the piano action can be operated by the "player action" of the player piano itself.

The traditional, roll-operated player piano has a pneumatic (air-driven) player action that operates the piano action. Associated with each piano key is a striker, which resembles a miniature bellows. As each striker is pneumatically collapsed, its corresponding key is pushed to play. The full set of strikers comprises the player action. The player action "reads music" by detecting coded perforations in paper rolls. This detection is done by a system of pneumatic tubes and valves, and, as the perforations are noted, the pneumatic system tells the strikers when to collapse. An electric vacuum motor or pedals that are pumped by foot supply the pneumatic power.

The traditional roll-player piano has several disadvantages. The player action must be built in the piano during assembly; it can't be added later. Also, the pneumatic strikers push their keys at one dynamic level only, so the notes of the music are heard, but without soft or loud variations.

The modern player piano has a player action that uses electronically-driven solenoids as strikers. A solenoid is a tubular coil that acts like a magnet when traversed by an electric current. The solenoids respond to signals from a Musical Instrument Digital Interface or MIDI, which is the universal electronic language spoken by keyboards, music synthesizers, and other electronic instruments. The MIDI is considerably more sophisticated than the roll-player system, and plays both the notes and dynamics as recorded by the musician. Libraries of recorded music are stored on floppy or compact disks so the piano can play virtually any piece of music. If equipped with appropriate software and electronics, the player piano can also play recorded vocalists, instruments, or a full symphony orchestra. Thanks to the common MIDI language, it can also be linked to sequencers, drum machines, and synthesizers as well as computers.

History

When King Henry VIII of England died in 1547, his musical instruments were inventoried and, among them, was "an instrument that goethe with a whele without playing upon." Early player pianos were known to have been built by piano builder Samuel Bidermann of Augsburg, Germany. He equipped three spinets (small, upright pianos) with pinned barrels similar to those in music boxes during his lifetime from 1540 to 1622.

Interest in inventing a self-playing piano resurfaced in the late 1890s. Several different mechanisms were developed and sold, but the paper rolls were not interchangeable. Melville Clark, an inventor and piano designer, developed an 88-note, standardized roll size for the player industry, and built his Apollo player piano to this standard in 1901. By 1908, other manufacturers had adopted his standard. The paper rolls were punched by duplicating machines called perforators from a master roll used as a pattern. The master rolls were tediously hand-punched by skilled workers directly from sheet music. To avoid this production difficulty, Clark invented the "marking piano" in 1912. His marking piano punched the master roll data as the musician performed a piece of music. The marking piano was used from 1912 to 1931.

Historic performances by artists of the day were preserved in live recordings, and the marking piano made the Roaring Twenties the heyday of the player piano until the popularity of the phonograph and radio surpassed it. In 1926, its peak sales year, over 10 million piano rolls were sold. The marking piano was retired in 1931, restored in 1971 to record the performances of other outstanding artists, and designated a National Historic Mechanical Engineering Landmark in 1992. Following all-time low sales in the 1950s, roll-player pianos experienced a revival in the 1970s, and traditional player pianos and paper rolls are still manufactured and sold.

In the 1920s, experiments began with electric or electromagnetic devices in place of the piano soundboard. Electronic pianos, keyboards, and music synthesizers use electronic circuits or tuned metal pieces instead of strings to produce sound. Some generate sounds approaching those of a conventional piano, but they are most valued for producing effects that are electronically altered from the acoustic piano voice. Electronic applications began to merge with the acoustic piano in the 1970s; acoustic pianos (as opposed to electronic pianos) were equipped with various components such as digital cassette drives or computers to enable the acoustic piano to be played using electronic "brains" and power. Several major manufacturers now produce disc-driven player pianos, and systems and software are sold in kits to convert acoustic pianos to players. The instrument that has been idle since grade-school piano lessons can thus be converted to an entertainment center. The remainder of this article will focus on the manufacture of the disc-driven piano.

Raw Materials

Essentially no raw materials are used directly in the manufacture of disc-driven player pianos. Manufacturers provide specifications to vendors and then purchase required parts from them. These parts may include factory-produced acoustic pianos in which the player systems are installed.

Design

Engineers design the player piano system and develop new products. The manufacturer's legal staff, meanwhile, performs a thorough patent search so that new approaches do not infringe on existing patents, and the firm's own original concepts can be patented promptly. Prototypes are hand-assembled, tested, and evaluated for marketability and feasibility of construction. The engineers provide design details and specifications for every component of the product, including electrical, electronic, and mechanical parts; tooling and special machining or assembly; and requirements for operation, installation, testing, maintenance, and repair. Because few parts may be made on site, the search for parts or suppliers capable of making them may be worldwide.

Software engineers must also provide the programming so that computer chips can convert notes and data representing musical expressions into electronic signals that drive the solenoids. More than 1,000 pages of computer code may be required for this complicated communication.

To appeal to a broad market, the player piano must be able to play a wide range of music, so a large music library in software form has also been developed and updated. The "codes" from traditional player piano rolls don't translate easily to the MIDI language, but actual performances are ideally suited to the digital system because they convey interpretive elements. Manufacturers maintain their own recording studios and recruit name performers to play their piano stylings. Music editors review the recorded performances, correct any errors or inconsistencies, or modify performances in accordance with musical scores, as appropriate. The legal staff secures contracts, releases, and licenses to the music and performances.

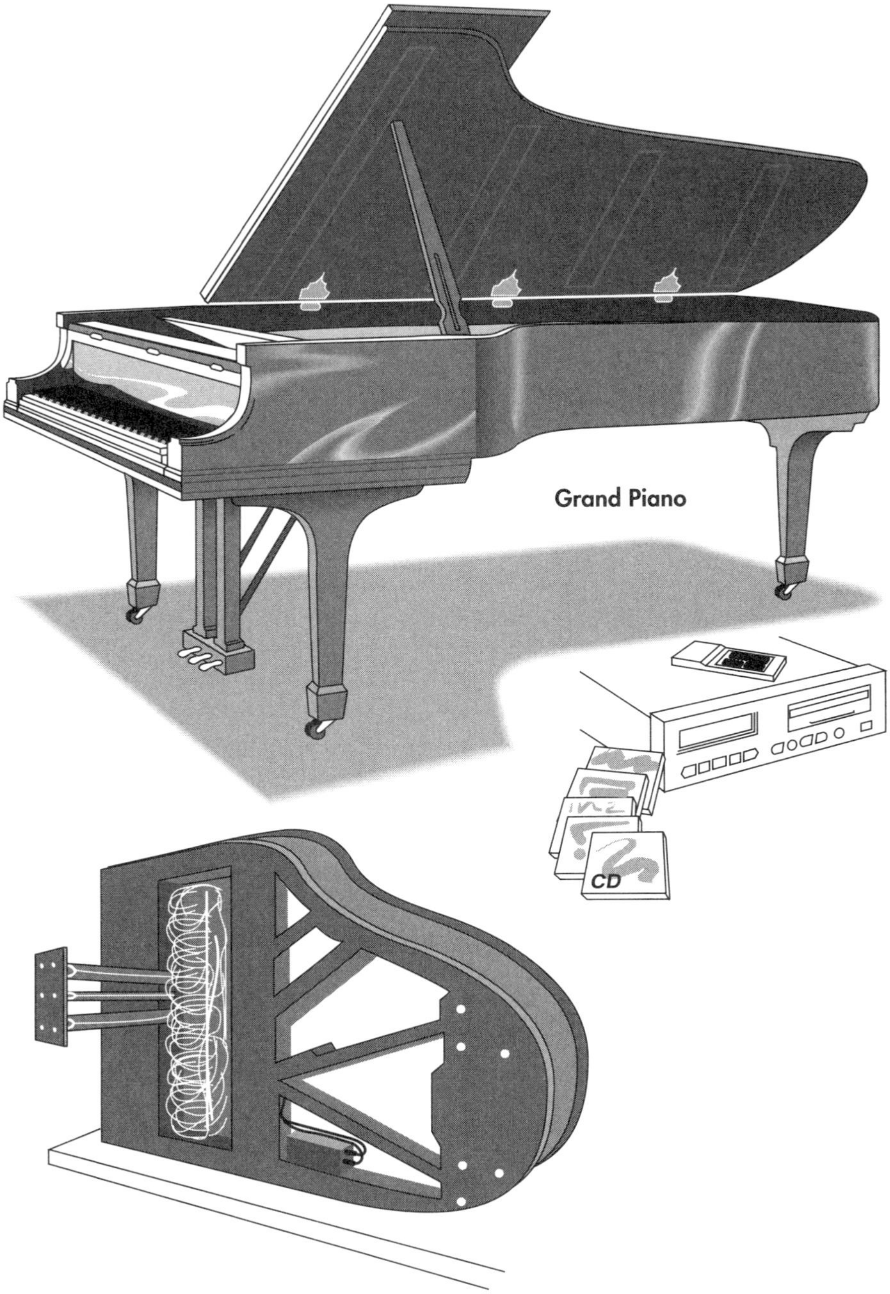

Disk-driven player pianos can be purchased outright, or a player system can be bought separately as a kit and installed on a regular piano.

The Manufacturing Process

Player subassembly

1 The player subassembly is the player action that consists of 88 solenoids and plungers that are computer-activated to move the keys. It is made of preexisting parts that are put together along an assembly line using a combination of hand-assembly and automated processes. The electronics, including the power supply, control box, key activation devices, and interconnections, are controlled by the components on a

printed circuit board. Circuit boards in panel form (multiple boards of the same kind made on one large panel and separated after components are inserted) are used to facilitate automated assembly because components can be added easily. Components of a single kind are supplied in long rolls that are fed through an automated sequencer. The sequencer checks to make sure each part is correct, then organizes an array of parts into the proper sequence for insertion in the circuit board. Some parts are formed, shaped, and inserted in the board by an automated dip inserter that can also check the appropriateness of the part. Other parts are added by hand.

2 The parts are soldered in place by passing the circuit board through a wave soldering machine. It washes the board with flux to remove contaminants, heats the board and components by infrared heat to reduce thermal shock as the board is soldered, then passes the underside of the board over a wave of molten solder. The solder flows up each component and board perforation by capillary action to the upper surface of the board. It cools and hardens, fixing the components in place.

3 The completed circuit boards are installed in control boxes and tested as a quality assurance step. Software is loaded and the control boxes are activated over a two-day-long test period, or "burn-in" process, that simulates the actual working of the player system. Any failures during that time are detected and corrected.

4 On another part of the player subassembly line, the solenoids and plungers are made and installed on a rail or support. Magnetic wire is wrapped by machine on the solenoid bases and sealed in place. Wire connections are also attached to each solenoid, and the plunger assemblies are paired with solenoids. The solenoids and plungers are assembled on aluminum rails. Depending on the manufacturer, the solenoid/plungers vary in configuration, and the rail may be one piece or several sections. Rail mounting brackets will be used later to mount the solenoids under the piano keyboard. A similar assembly is made to activate the piano pedals.

Recording strip

5 The player piano can also record music played on the piano if a recording strip is installed; this allows students or performers to play back their work and aids composers in creating and revising their compositions. The recording strip consists of a mounting rail and tiny electronic film switches that are cased in plastic. There is one switch or finger per key; and the recording strip is sensitive to the key or keys that are struck, the duration they are depressed, and the velocity or intensity (soft or loud) per key strike. The recording strip is linked electronically to the piano's computer, where signals are processed into MIDI messages. This data can be stored by analog media like a cassette player or VCR, or directed to a floppy disk drive. It can be played back by the piano or other MIDI-compatible instrument.

Disc-player

6 The disc-player and other devices are connected to the player piano via the control box. It may be a floppy disk drive, compact disc (CD) player, or both; player pianos can also be attached to VCRs in combination with televisions or computer monitors so lyrics can be synchronized for singalongs or other programmed entertainment. Most manufacturers purchase these components and remote controls for them from outside suppliers.

Accompaniment and speakers

7 As another added feature, the piano can play along with all manner of recorded accompaniments including vocals, other individual instruments, or an orchestra. The "soundboxes" for these additions are electronic speakers that can be furnished as accessories to the piano and that are supplied by vendors. Software is added to supply the other sounds via the MIDI system.

Piano installation

8 On another assembly line, the player subassembly and recording strip are installed in a piano. The pianos are supplied by the manufacturer's own piano division or an outside vendor. A slot is cut in the wood under the piano keyboard, but the structural integrity of the piano is not affected. The

solenoid rail or set of rails is mounted under the piano so that the plungers are under the tails of the keys. The recording strip is similarly installed. The solenoid links with the pedals are also connected, the control box is mounted under the keyboard, and the power supply is mounted to the underside of the piano. The piano action must be adjusted at this point. The keys rest on felt pads that compress with use, so the pianos are programmed to play overnight in another burn-in process. After a 12-hour "concert," the felt has compressed. Then the components are checked to see that they have worked properly, and the piano action is adjusted to suit the felt compression. The pianos are packed and shipped when installation is complete.

Kit assembly

9 Some manufacturers provide their disc-driven systems in kits that can be used to retrofit standard pianos to players. As the completed subassemblies, control boxes and power supplies, and recording strips come off the assembly lines, they are packaged and stored. Kits are then packed with these units as well as disc-players and accessories. The kits are shipped to installers who have been specially trained to install the particular manufacturer's system. Installation typically takes up to two days, including an overnight burn-in session after which the piano action is adjusted and component workings are checked as they are in the factory.

Quality Control

Quality control is essential in every production step. As parts are received from vendors around the world, they are checked for compliance with specifications and other engineering requirements. Along the player subassembly line, workers perform quality checks as part of their responsibility. For example, as circuit boards emerge from the wave soldering machine, the operator inspects the board to see that parts have been soldered in place, no bridging of solder has occurred from one component to another, and the board itself has not warped or cracked. The operator also monitors the machine for temperature of the solder and overall operation of the machine. Electronic assemblies must also meet the requirements of Underwriters' Laboratories (UL). UL representatives visit the manufacturers frequently and randomly to check for compliance with their materials, operations, and safety codes. Other quality checks, such as the burn-ins of control boxes and piano actions, are integrated into the manufacturing process, as described.

The Future

Acoustic pianos are instruments long-favored by listeners, and their future is assured by the disc-driven player piano which provides the sounds and entertainment benefits without requiring the listener to become a virtuoso. Disc-driven player pianos have initiated substantial increases in the sales of pianos, and are seen as the growth sector of the industry. In the disc-driven player piano's future, manufacturers are striving to reproduce performance nuances. Continued merging of technologies also seems likely as professional recording studios are used to input music and as methods are found of linking the player piano to other electronic media.

Where To Learn More

Book

Dolge, Alfred. *Pianos and Their Makers.* Dover Publications, Inc., 1972, pp. 131-62.

Periodicals

Berkman, Robert J. "The Q-R-S Marking Piano." *ASME Book No. HH 9201.* The American Society of Mechanical Engineers, 1992, 8 pp.

Matzer, Marla. "Play It Again." *Forbes*, February 27, 1995, pp. 138-39.

Slutsker, G. "The Player Piano Makes a Comeback." *Forbes*, September 4, 1989. p. 124.

Other

"Story & Clark Player Pianos and QRS Pianomation: Elegance and Technology." Story & Clark Pianos, Div. QRS Music, Inc. Catalog 9501, 1995, 36 pp. (716) 885-4600.

"The PianoDisc Story," The Music Trades, November 1993. (916) 973-8710.

—Gillian S. Holmes

Polyester

Background

Polyester is a synthetic fiber derived from coal, air, water, and petroleum. Developed in a 20th-century laboratory, polyester fibers are formed from a chemical reaction between an acid and alcohol. In this reaction, two or more molecules combine to make a large molecule whose structure repeats throughout its length. Polyester fibers can form very long molecules that are very stable and strong.

Polyester is used in the manufacture of many products, including clothing, home furnishings, industrial fabrics, computer and recording tapes, and electrical insulation. Polyester has several advantages over traditional fabrics such as cotton. It does not absorb moisture, but does absorb oil; this quality makes polyester the perfect fabric for the application of water-, soil-, and fire-resistant finishes. Its low absorbency also makes it naturally resistant to stains. Polyester clothing can be preshrunk in the finishing process, and thereafter the fabric resists shrinking and will not stretch out of shape. The fabric is easily dyeable, and not damaged by mildew. Textured polyester fibers are an effective, nonallergenic insulator, so the material is used for filling pillows, quilting, outerwear, and sleeping bags.

History

In 1926, United States-based E.I. du Pont de Nemours and Co. began research into very large molecules and synthetic fibers. This early research, headed by W.H. Carothers, centered on what became nylon, the first synthetic fiber. Soon after, in the years 1939-41, British research chemists took interest in the du Pont studies and conducted their own research in the laboratories of Calico Printers Association, Ltd. This work resulted in the creation of the polyester fiber known in England as Terylene.

In 1946, du Pont purchased the right to produce this polyester fiber in the United States. The company conducted some further developmental work, and in 1951, began to market the fiber under the name Dacron. During the ensuing years, several companies became interested in polyester fibers and produced their own versions of the product for different uses. Today, there are two primary types of polyester, PET (polyethylene terephthalate) and PCDT (poly-1, 4-cyclohexylene-dimethylene terephthalate). PET, the more popular type, is applicable to a wider variety of uses. It is stronger than PCDT, though PCDT is more elastic and resilient. PCDT is suited to the heavier consumer uses, such as draperies and furniture coverings. PET can be used alone or blended with other fabrics to make clothing that is wrinkle and stain resistant and retains its shape.

Raw Materials

Polyester is a chemical term which can be broken into *poly*, meaning many, and *ester*, a basic organic chemical compound. The principle ingredient used in the manufacture of polyester is ethylene, which is derived from petroleum. In this process, ethylene is the polymer, the chemical building block of polyester, and the chemical process that produces the finished polyester is called polymerization.

Polyester is used in the manufacture of many products, including clothing, home furnishings, industrial fabrics, computer and recording tapes, and electrical insulation.

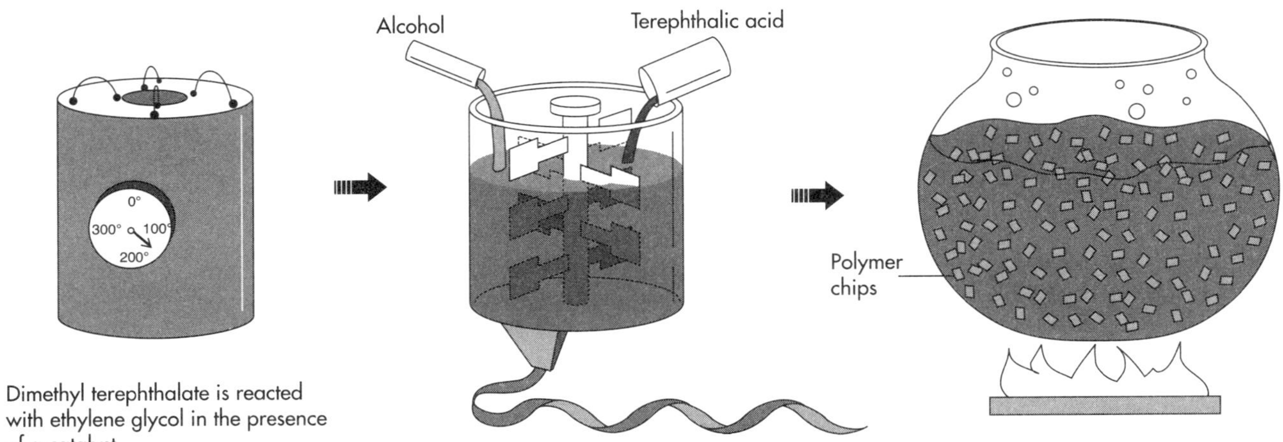

Dimethyl terephthalate is reacted with ethylene glycol in the presence of a catalyst.

The resulting monomer alcohol is combined with terephthalic acid. The newly formed polyester is extruded through a slot to form long ribbons. The ribbons are cooled and cut into tiny chips.

The chips are melted to form a syrup-like solution.

The Manufacturing Process

Polyester is manufactured by one of several methods. The one used depends on the form the finished polyester will take. The four basic forms are filament, staple, tow, and fiberfill. In the filament form, each individual strand of polyester fiber is continuous in length, producing smooth-surfaced fabrics. In staple form, filaments are cut to short, predetermined lengths. In this form polyester is easier to blend with other fibers. Tow is a form in which continuous filaments are drawn loosely together. Fiberfill is the voluminous form used in the manufacture of quilts, pillows, and outerwear. The two forms used most frequently are filament and staple.

Manufacturing Filament Yarn

Polymerization

1 To form polyester, dimethyl terephthalate is first reacted with ethylene glycol in the presence of a catalyst at a temperature of 302-410°F (150-210°C).

2 The resulting chemical, a monomer (single, non-repeating molecule) alcohol, is combined with terephthalic acid and raised to a temperature of 472°F (280°C). Newly-formed polyester, which is clear and molten, is extruded through a slot to form long ribbons.

Drying

3 After the polyester emerges from polymerization, the long molten ribbons are allowed to cool until they become brittle. The material is cut into tiny chips and completely dried to prevent irregularities in consistency.

Melt spinning

4 Polymer chips are melted at 500-518°F (260-270°C) to form a syrup-like solution. The solution is put in a metal container called a spinneret and forced through its tiny holes, which are usually round, but may be pentagonal or any other shape to produce special fibers. The number of holes in the spinneret determines the size of the yarn, as the emerging fibers are brought together to form a single strand.

5 At the spinning stage, other chemicals may be added to the solution to make the resulting material flame retardant, antistatic, or easier to dye.

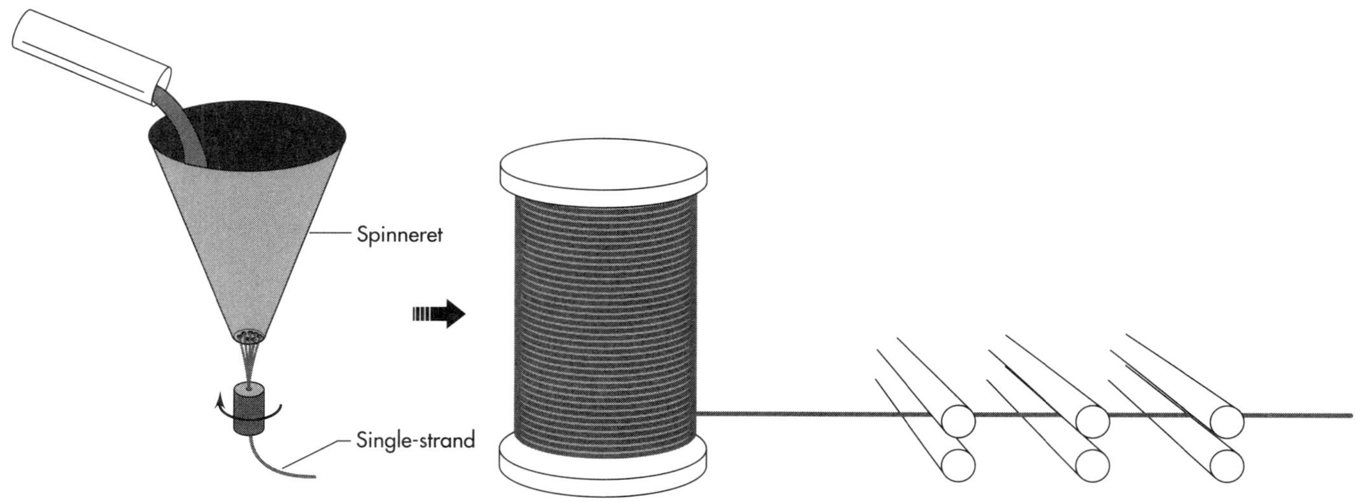

The solution is forced through tiny holes in a spinneret to form a single strand.

Stretching the fiber forces the random polyester molecules to align in a parallel formation.

Drawing the fiber

6 When polyester emerges from the spinneret, it is soft and easily elongated up to five times its original length. The stretching forces the random polyester molecules to align in a parallel formation. This increases the strength, tenacity, and resilience of the fiber. This time, when the filaments dry, the fibers become solid and strong instead of brittle.

7 Drawn fibers may vary greatly in diameter and length, depending on the characteristics desired of the finished material. Also, as the fibers are drawn, they may be textured or twisted to create softer or duller fabrics.

Winding

8 After the polyester yarn is drawn, it is wound on large bobbins or flat-wound packages, ready to be woven into material.

Manufacturing Staple Fiber

In making polyester staple fiber, polymerization, drying, and melt spinning (steps 1-4 above) are much the same as in the manufacture of filament yarn. However, in the melt spinning process, the spinneret has many more holes when the product is staple fiber. The rope-like bundles of polyester that emerge are called tow.

Drawing tow

1 Newly-formed tow is quickly cooled in cans that gather the thick fibers. Several lengths of tow are gathered and then drawn on heated rollers to three or four times their original length.

Crimping

2 Drawn tow is then fed into compression boxes, which force the fibers to fold like an accordion, at a rate of 9-15 crimps per inch (3-6 per cm). This process helps the fiber hold together during the later manufacturing stages.

Setting

3 After the tow is crimped, it is heated at 212-302°F (100-150°C) to completely dry the fibers and set the crimp. Some of the crimp will unavoidably be pulled out of the fibers during the following processes.

Cutting

4 Following heat setting, tow is cut into shorter lengths. Polyester that will be

blended with cotton is cut in 1.25-1.50 inch (3.2-3.8 cm) pieces; for rayon blends, 2 inch (5 cm) lengths are cut. For heavier fabrics, such as **carpet**, polyester filaments are cut into 6 inch (15 cm) lengths.

The Future

Following its introduction to the United States in 1951, polyester quickly became the country's fastest-growing fiber. Easy care of the permanent press fabric made polyester doubleknits extremely popular in the late 1960s. However, polyester has suffered an "image problem" since that time, and clothes made out of polyester were often devalued and even ridiculed. Several new forms of polyester introduced in the early 1990s may help revitalize the image of polyester. A new form of polyester fiber, called microfiber, was introduced to the public in 1991. More luxurious and versatile than traditional polyester, microfiber fabrics are difficult to tell apart from silk fabrics. Clothing designers such as Mary McFadden have created a line of clothes using this new form of polyester. Textile researchers at North Carolina State University are developing a form of polyester that may be as strong as Kevlar, a superfiber material used to make bulletproof vests. This type of polyester may eventually be used as composite materials for cars and airplanes.

Where To Learn More

Books

Corbman, Bernard P. *Textiles: Fiber to Fabric*. 6th ed. Gregg Division, McGraw-Hill, 1983, pp. 374-92.

Encyclopedia of Textiles. 3rd ed. Prentice-Hall, Inc., 1980, pp. 28-33.

Polyester: Fifty Years of Achievement. State Mutual Book & Periodical Service, 1993.

Periodicals

Fellingham, Christine. "Will You Learn to Love Polyester?" *Glamour*, April 1992, p. 204.

Templeton, Fleur. "Show Me a Bulletproof Leisure Suit, In Pink." *Business Week*, July 6, 1992, p. 65.

Thomas, Marita. "At 50 Years, Polyester Gains New Fashion Vitality." *Textile World*, December 1993, p. 62+.

—Kristine M. Krapp

Printed Circuit Board

Background

A printed circuit board, or PCB, is a self-contained module of interconnected electronic components found in devices ranging from common **beeper**s, or pagers, and radios to sophisticated radar and computer systems. The circuits are formed by a thin layer of conducting material deposited, or "printed," on the surface of an insulating board known as the substrate. Individual electronic components are placed on the surface of the substrate and soldered to the interconnecting circuits. Contact fingers along one or more edges of the substrate act as connectors to other PCBs or to external electrical devices such as on-off switches. A printed circuit board may have circuits that perform a single function, such as a signal amplifier, or multiple functions.

There are three major types of printed circuit board construction: single-sided, double-sided, and multi-layered. Single-sided boards have the components on one side of the substrate. When the number of components becomes too much for a single-sided board, a double-sided board may be used. Electrical connections between the circuits on each side are made by drilling holes through the substrate in appropriate locations and plating the inside of the holes with a conducting material. The third type, a multi-layered board, has a substrate made up of layers of printed circuits separated by layers of insulation. The components on the surface connect through plated holes drilled down to the appropriate circuit layer. This greatly simplifies the circuit pattern.

Components on a printed circuit board are electrically connected to the circuits by two different methods: the older "through hole technology" and the newer "surface mount technology." With through hole technology, each component has thin wires, or leads, which are pushed through small holes in the substrate and soldered to connection pads in the circuits on the opposite side. Gravity and friction between the leads and the sides of the holes keeps the components in place until they are soldered. With surface mount technology, stubby J-shaped or L-shaped legs on each component contact the printed circuits directly. A solder paste consisting of glue, flux, and solder are applied at the point of contact to hold the components in place until the solder is melted, or "reflowed," in an oven to make the final connection. Although surface mount technology requires greater care in the placement of the components, it eliminates the time-consuming drilling process and the space-consuming connection pads inherent with through hole technology. Both technologies are used today.

Two other types of circuit assemblies are related to the printed circuit board. An **integrated circuit**, sometimes called an IC or microchip, performs similar functions to a printed circuit board except the IC contains many more circuits and components that are electrochemically "grown" in place on the surface of a very small chip of silicon. A hybrid circuit, as the name implies, looks like a printed circuit board, but contains some components that are grown onto the surface of the substrate rather than being placed on the surface and soldered.

History

Printed circuit boards evolved from electrical connection systems that were developed

There is no such thing as a standard printed circuit board. Each board has a unique function for a particular product and must be designed to perform that function in the space allotted.

in the 1850s. Metal strips or rods were originally used to connect large electric components mounted on wooden bases. In time the metal strips were replaced by wires connected to screw terminals, and wooden bases were replaced by metal chassis. But smaller and more compact designs were needed due to the increased operating needs of the products that used circuit boards. In 1925, Charles Ducas of the United States submitted a patent application for a method of creating an electrical path directly on an insulated surface by printing through a stencil with electrically conductive inks. This method gave birth to the name "printed wiring" or "printed circuit."

In the 1943, Paul Eisler of the United Kingdom patented a method of etching the conductive pattern, or circuits, on a layer of copper foil bonded to a glass-reinforced, non-conductive base. Widespread use of Eisler's technique did not come until the 1950s when the transistor was introduced for commercial use. Up to that point, the size of vacuum tubes and other components were so large that the traditional mounting and wiring methods were all that was needed. With the advent of transistors, however, the components became very small, and manufacturers turned to printed circuit boards to reduce the overall size of the electronic package.

Through hole technology and its use in multi-layer PCBs was patented by the U.S. firm Hazeltyne in 1961. The resulting increase in component density and closely spaced electrical paths started a new era in PCB design. Integrated circuit chips were introduced in the 1970s, and these components were quickly incorporated into printed circuit board design and manufacturing techniques.

Design

There is no such thing as a standard printed circuit board. Each board has a unique function for a particular product and must be designed to perform that function in the space allotted. Board designers use computer-aided design systems with special software to layout the circuit pattern on the board. The spaces between electrical conducting paths are often 0.04 inches (1.0 mm) or smaller. The location of the holes for component leads or contact points are also laid out, and this information is translated into instructions for a computer numerical controlled drilling machine or for the automatic solder paster used in the manufacturing process.

Once the circuit pattern is laid out, a negative image, or mask, is printed out at exact size on a clear plastic sheet. With a negative image, the areas that are not part of the circuit pattern are shown in black and the circuit pattern is shown as clear.

Raw Materials

The substrate most commonly used in printed circuit boards is a glass fiber reinforced (**fiberglass**) epoxy resin with a copper foil bonded on to one or both sides. PCBs made from paper reinforced phenolic resin with a bonded copper foil are less expensive and are often used in household electrical devices.

The printed circuits are made of copper, which is either plated or etched away on the surface of the substrate to leave the pattern desired. (See "additive" and "subtractive" processes described in step 3 under The Manufacturing Process). The copper circuits are coated with a layer of tin-lead to prevent oxidation. Contact fingers are plated with tin-lead, then nickel, and finally gold for excellent conductivity.

Purchased components include resistors, capacitors, transistors, diodes, integrated circuit chips, and others.

The Manufacturing Process

Printed circuit board processing and assembly are done in an extremely clean environment where the air and components can be kept free of contamination. Most electronic manufacturers have their own proprietary processes, but the following steps might typically be used to make a two-sided printed circuit board.

Making the substrate

1 Woven glass fiber is unwound from a roll and fed through a process station

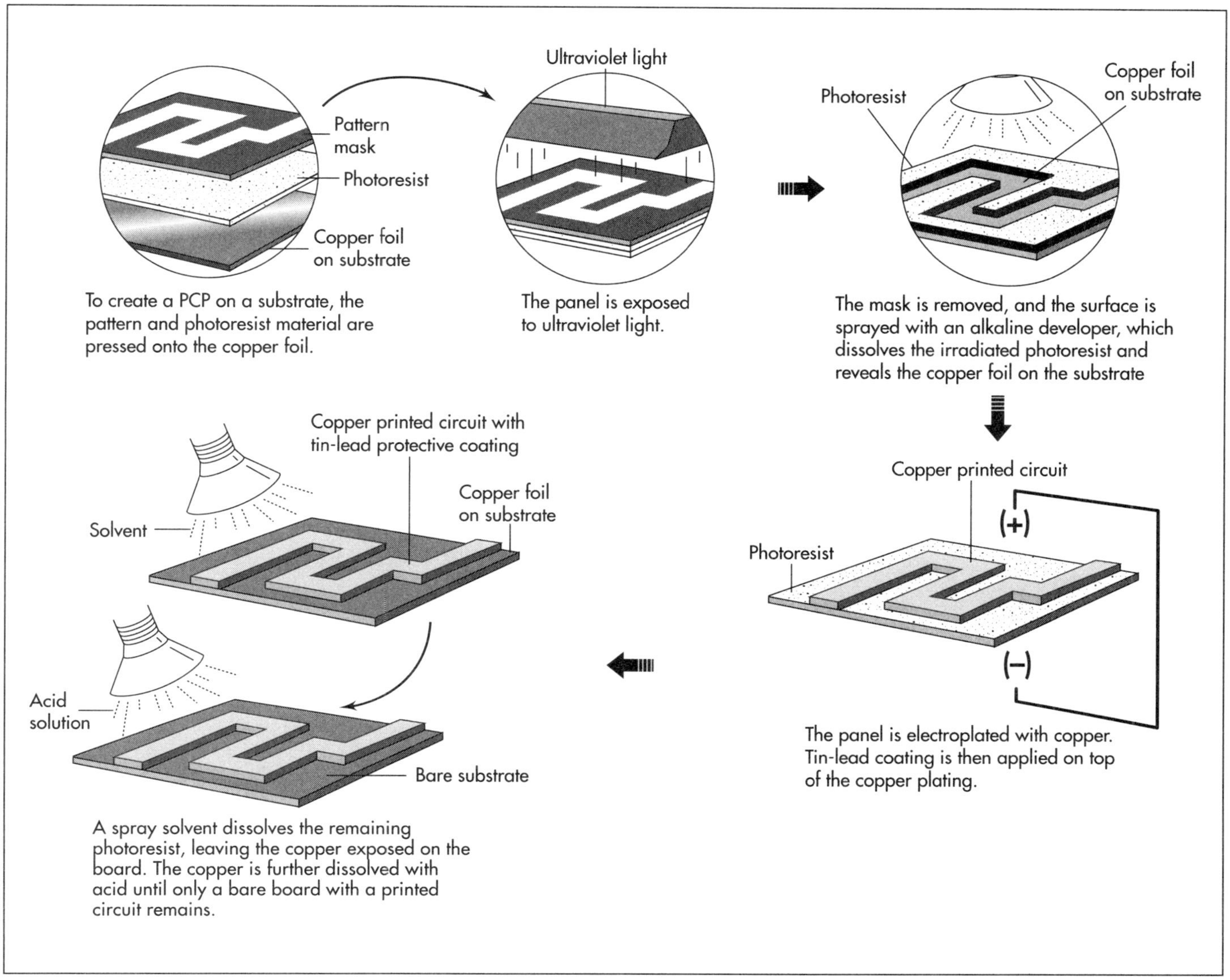

To create a PCP on a substrate, the pattern and photoresist material are pressed onto the copper foil.

The panel is exposed to ultraviolet light.

The mask is removed, and the surface is sprayed with an alkaline developer, which dissolves the irradiated photoresist and reveals the copper foil on the substrate

The panel is electroplated with copper. Tin-lead coating is then applied on top of the copper plating.

A spray solvent dissolves the remaining photoresist, leaving the copper exposed on the board. The copper is further dissolved with acid until only a bare board with a printed circuit remains.

The above illustrations show an enlarged section of a PCB.

where it is impregnated with epoxy resin either by dipping or spraying. The impregnated glass fiber then passes through rollers which roll the material to the desired thickness for the finished substrate and also remove any excess resin.

2 The substrate material passes through an oven where it is semicured. After the oven, the material is cut into large panels.

3 The panels are stacked in layers, alternating with layers of adhesive-backed copper foil. The stacks are placed in a press where they are subjected to temperatures of about 340°F (170°C) and pressures of 1500 psi for an hour or more. This fully cures the resin and tightly bonds the copper foil to the surface of the substrate material.

Drilling and plating the holes

4 Several panels of substrate, each large enough to make several printed circuit boards, are stacked on top of each other and pinned together to keep them from moving. The stacked panels are placed in a **CNC** machine, and the holes are drilled according to the pattern determined when the boards were laid out. The holes are deburred to remove any excess material clinging to the edges of the holes.

5 The inside surfaces of the holes designed to provide a conductive circuit from one side of the board to the other are plated with copper. Non-conducting holes are plugged to keep them from being plated

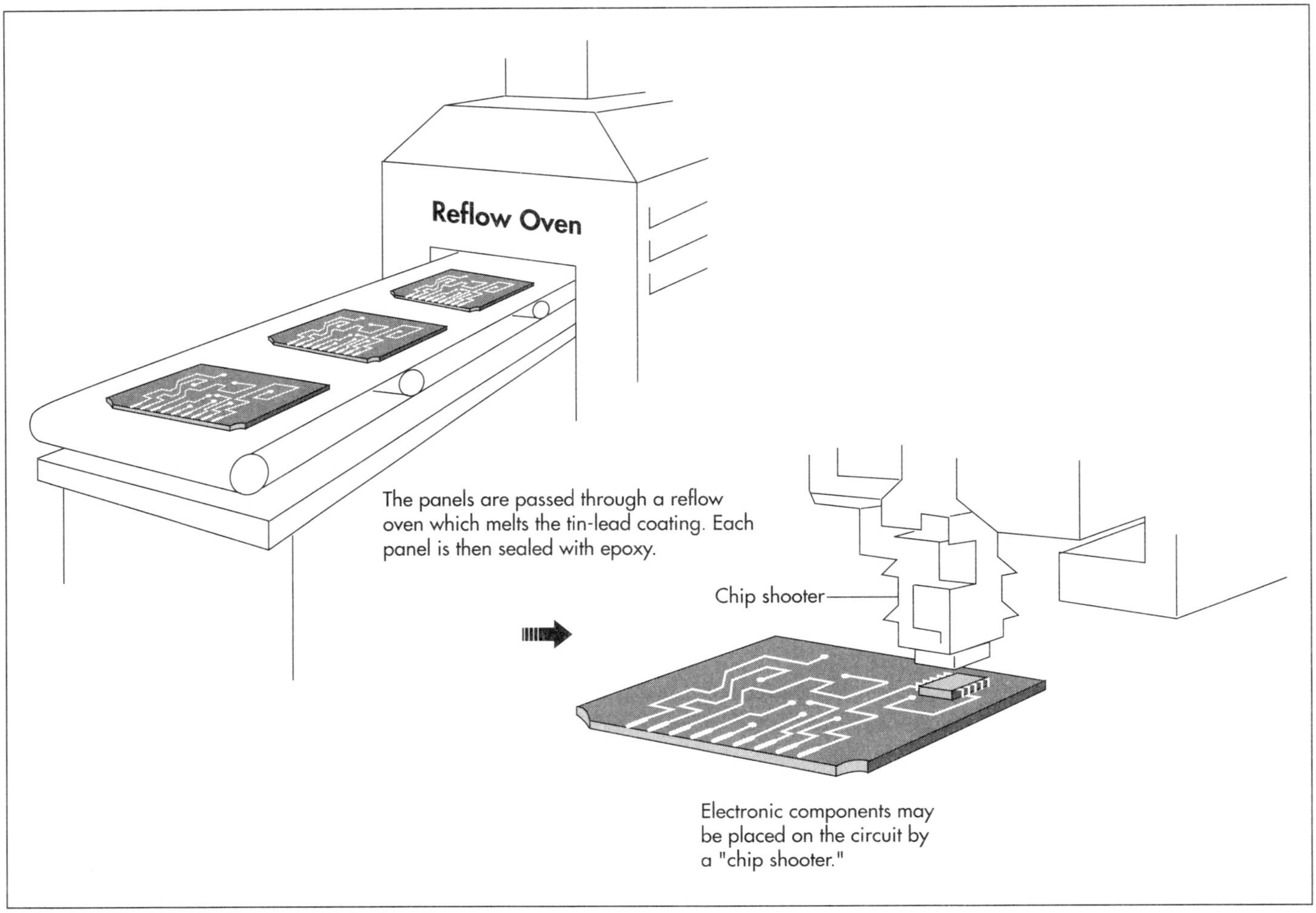

or are drilled after the individual boards are cut from the larger panel.

Creating the printed circuit pattern on the substrate

The printed circuit pattern may be created by an "additive" process or a "subtractive" process. In the additive process, copper is plated, or added, onto the surface of the substrate in the desired pattern, leaving the rest of the surface unplated. In the subtractive process, the entire surface of the substrate is first plated, and then the areas that are not part of the desired pattern are etched away, or subtracted. We shall describe the additive process.

6 The foil surface of the substrate is degreased. The panels pass through a vacuum chamber where a layer of positive photoresist material is pressed firmly onto the entire surface of the foil. A positive photoresist material is a polymer that has the property of becoming more soluble when exposed to ultraviolet light. The vacuum ensures that no air bubbles are trapped between the foil and the photoresist. The printed circuit pattern mask is laid on top of the photoresist and the panels are exposed to an intense ultraviolet light. Because the mask is clear in the areas of the printed circuit pattern, the photoresist in those areas is irradiated and becomes very soluble.

7 The mask is removed, and the surface of the panels is sprayed with an alkaline developer that dissolves the irradiated photoresist in the areas of the printed circuit pattern, leaving the copper foil exposed on the surface of the substrate.

8 The panels are then electroplated with copper. The foil on the surface of the substrate acts as the cathode in this process, and the copper is plated in the exposed foil areas to a thickness of about 0.001-0.002 inches (0.025-0.050 mm). The areas still

covered with photoresist cannot act as a cathode and are not plated. Tin-lead or another protective coating is plated on top of the copper plating to prevent the copper from oxidizing and as a resist for the next manufacturing step.

9 The photoresist is stripped from the boards with a solvent to expose the substrate's copper foil between the plated printed circuit pattern. The boards are sprayed with an acid solution which eats away the copper foil. The copper plating on the printed circuit pattern is protected by the tin-lead coating and is unaffected by the acid.

Attaching the contact fingers

10 The contact fingers are attached to the edge of the substrate to connect with the printed circuit. The contact fingers are masked off from the rest of the board and then plated. Plating is done with three metals: first tin-lead, next nickel, then gold.

Fusing the tin-lead coating

11 The tin-lead coating on the surface of the copper printed circuit pattern is very porous and is easily oxidized. To protect it, the panels are passed through a "reflow" oven or hot oil bath which causes the tin-lead to melt, or reflow, into a shiny surface.

Sealing, stenciling, and cutting the panels

12 Each panel is sealed with epoxy to protect the circuits from being damaged while components are being attached. Instructions and other markings are stenciled onto the boards.

13 The panels are then cut into individual boards and the edges are smoothed.

Mounting the components

14 Individual boards pass through several machines which place the electronic components in their proper location in the circuit. If surface mount technology is going to be used to mount the components, the boards first pass through an automatic solder paster, which places a dab of solder paste at each component contact point. Very small components may be placed by a "chip shooter" which rapidly places, or shoots, the components onto the board. Larger components may be robotically placed. Some components may be too large or odd-sized for robotic placement and must be manually placed and soldered later.

15 The components are then soldered to the circuits. With surface mount technology, the soldering is done by passing the boards through another reflow process, which causes the solder paste to melt and make the connection.

16 The flux residue from the solder is cleaned with water or solvents depending on the type of solder used.

Packaging

17 Unless the printed circuit boards are going to be used immediately, they are individually packaged in protective plastic bags for storage or shipping.

Quality Control

Visual and electrical inspections are made throughout the manufacturing process to detect flaws. Some of these flaws are generated by the automated machines. For example, components are sometimes misplaced on the board or shifted before final soldering. Other flaws are caused by the application of too much solder paste, which can cause excess solder to flow, or bridge, across two adjacent printed circuit paths. Heating the solder too quickly in the final reflow process can cause a "tombstone" effect where one end of a component lifts up off the board and doesn't make contact.

Completed boards are also tested for functional performance to ensure their output is within the desired limits. Some boards are subjected to environmental tests to determine their performance under extremes of heat, humidity, vibration, and impact.

Toxic Materials and Safety Considerations

The solder used to make electrical connections on a PCB contains **lead**, which is considered a toxic material. The fumes from the solder are considered a health hazard, and the soldering operations must be carried out in a closed environment. The fumes must be given appropriate extraction and cleaning before being discharged to the atmosphere.

Many electronic products containing PCBs are becoming obsolete within 12-18 months. The potential for these obsolete products entering the wastestream and ending up in landfills has many environmentalists concerned. Recycling efforts for electronic products include refurbishing older products and reselling them to customers that don't need, or have access to, newer, state-of-the-art electronics. Other electronics are disassembled and the computer parts are salvaged for resale and reuse in other products.

In many countries in Europe, legislation requires manufacturers to buy back their used products and render them safe for the environment before disposal. For manufacturers of electronics, this means they must remove and reclaim the toxic solder from their PCBs. This is an expensive process and has spurred research into the development of non-toxic means of making electrical connections. One promising approach involves the use of water-soluble, electrically conductive molded plastics to replace the wires and solder.

The Future

The miniaturization of electronic products continues to drive printed circuit board manufacturing towards smaller and more densely packed boards with increased electronic capabilities. Advancements beyond the boards described here include three-dimensional molded plastic boards and the increased use of integrated circuit chips. These and other advancements will keep the manufacture of printed circuit boards a dynamic field for many years.

Where To Learn More

Books

Braithwaite, Nicholas and Graham Weaver, eds. *Electronic Materials.* Butterworths, 1990.

Koshel, Dal., ed. *Manufacturing Engineer's Reference Book.* Butterworth-Heinemann, 1993.

Lotter, Bruno. *Manufacturing Assembly Handbook.* Butterworths, 1986.

Periodicals

Alford, William. "Screen Printing PC Boards." *Electronics Now,* September 1993, pp. 38-41.

Fernando, James R. "Successful Implementation of a CIM Strategy for a PCB Manufacturing Facility." *Electronic Manufacturing,* March 1990.

Kirkland, Carl. "What Ever Happened to Molded 3D Circuit Boards?" *Plastics World,* February 1993, pp. 32-36.

Nishioka, Alan. "Iron-On PC Board Patterns." *Electronics Now*, September 1993, pp. 42-45.

Yam, Philip. "Plastics Get Wired." *Scientific American*, July 1995, pp. 82-87.

—David N. Ford/Chris Cavette

Road Sign

Background

Road signs use shapes, colors, words, and symbols to communicate a message to drivers. Without such signs, the movement of traffic would be disorderly and unpredictable. Virtually all traffic signs use retroreflective sheeting, which is designed to reflect some of the light from vehicle headlights back to the driver so that the sign will be visible at night. Color and shape can also provide cues to motorists even when the words or symbols on the sign are unintelligible. Regulatory signs, such as speed limit signs, are usually rectangular and use a white background. Stop signs, on the other hand, have a distinct octagonal shape and a red background in order to catch the driver's eye.

Designers must utilize elements like shape and consider material properties in creating signs that drivers can see and understand in time to react appropriately. Contrast, which is a measure of the brightness of the message in relation to its background, is an important property of any sign. The environmental backdrop—usually green vegetation and blue sky—must also be considered in the design process. A border is placed around all signs to distinguish them as geometric shapes in contrast to nature.

In order to maintain similar appearances among traffic signs, the federally approved *Manual on Uniform Traffic Control Devices* (MUTCD) provides specifications for sign dimensions and the use of symbols. In addition, the MUTCD prescribes that all signs be either reflectorized or illuminated.

History

Citizens in America began forming automobile clubs in the early 1900s. These groups took it upon themselves to mark their local sections of highway with signs to warn and direct drivers. This scattered effort resulted in a wide variety of sign designs and messages in different sections of highway, which caused confusion among motorists. In 1924, the first steps toward national uniformity in road signs were taken by the Bureau of Public Roads. Designers were faced with developing signs to guide a largely illiterate population of motorists. As a result, color and shape were especially important components of signs from the beginning.

Early signs lacked the reflectorized technology prevalent in signs today. In 1924 bright yellow was chosen as the background color for all warning signs, and white was the background color for all remaining signs. These lighter colors provided the greatest contrast with black lettering, especially when the signs were seen with the aid of headlights at night. Later signs used glass beads to produce a reflectorized effect at night. Beads—approximately 0.75 inch (20 mm) in diameter—were glued on the signs in the shape of numbers (such as the speed limit) or symbols to inform and warn nighttime drivers.

The development of retroreflective sheeting by the 3M company in the 1940s changed the face of traffic signs forever. This material, with reflective elements like glass beads on or under a transparent plastic film, enabled better visibility of signs at night. Unlike diffuse reflection and mirror, or specular, reflection, retroreflection allows a surface to return a portion of light to the original source. In diffuse reflection, the reflected light is scattered in all directions, as when sunlight bounces off a car. In mir-

The first traffic sign using reflective sheeting was installed on the outskirts of Minneapolis in 1939. The surface of the sheeting was covered with tiny glass beads that produced the desired retroreflectivity.

In diffuse reflection, the reflected light is scattered in all directions. In mirror reflection, light bounces and reflects off the surface at an angle opposite to the source. Retroreflection allows light beams to "bend" and return toward the original light source.

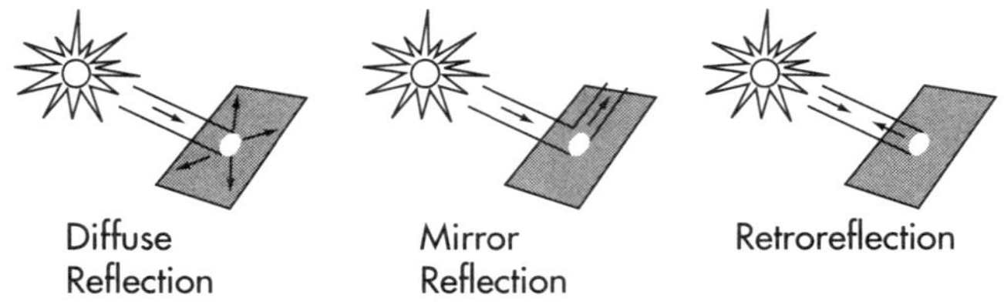

Types of Reflection

ror reflection, light bounces and reflects off the surface at an angle opposite to the source; this is similar to a pool ball striking the table cushion at a shallow angle and bouncing to the other end of the table. Retroreflective material, on the other hand, allows light beams to "bend" and return toward the original light source.

The first traffic sign using reflective sheeting was installed on the outskirts of Minneapolis in 1939. The surface of the sheeting was covered with tiny glass beads that produced the desired retroreflectivity. However, dirt tended to accumulate on the grainy surface and during wet weather, the water would coat the surface and diminish the reflective effects of the beads.

These problems were solved within a couple of years. An enclosed lens system was developed, essentially by covering the beaded sheeting with a transparent film that maintained the surface's retroreflective qualities. This type of sheeting, referred to at the time as "flat-top sheeting," is now known as engineering grade sheeting. It is the most economical grade and can be used on signs in areas with light traffic and slow speeds.

The next major development came in the late 1960s when encapsulated lens sheeting was invented, basically by adding a resin base and an additional reflector coat behind the glass beads. This high intensity material is three to four times as bright as engineering grade, and it retains its reflectivity longer; it is now the most commonly used type of reflective sheeting.

Another significant innovation came in 1989 with the substitution of microscopic prismatic reflectors for the traditional glass beads. There are about 7,000 microprisms per square inch (about 10 per sq mm) of this type of sheeting, producing about three times the brightness of the encapsulated lens variety. This is the most durable and most costly type of high performance sheeting currently available.

Raw Materials

Traffic signs consist of three basic components: a blank, background sheeting, and sign copy. Blanks, usually constructed of plywood, aluminum, or steel, serve as the framework of the sign. Plywood is the least expensive blank material. It is fairly strong but is susceptible to weather damage since it is porous; plywood blanks must be overlaid with a thin layer of plastic. Aluminum will not rust, but it is very lightweight and must be reinforced with metal braces along the back. It is the most expensive blank option. Steel is a more economical alternative to aluminum; it is also more sturdy and does not need reinforcement. Rusting can be prevented by applying a coat of **zinc** to the steel blank.

Background sheeting and the letter and symbols for the sign copy are cut from retroreflective sheeting. This sheeting consists of tiny glass beads or microprisms embedded in a flexible plastic surface; this construction allows light from car headlights to be reflected off the sign and back to the driver. Colored light is reflected from the sign if the sheeting is dyed with a pigment. For instance, to make "STOP" signs, red dye can be added to the sheeting mixture when it is in a liquid form.

The Manufacturing Process

The production of signs can involve many different processes, depending on whether the

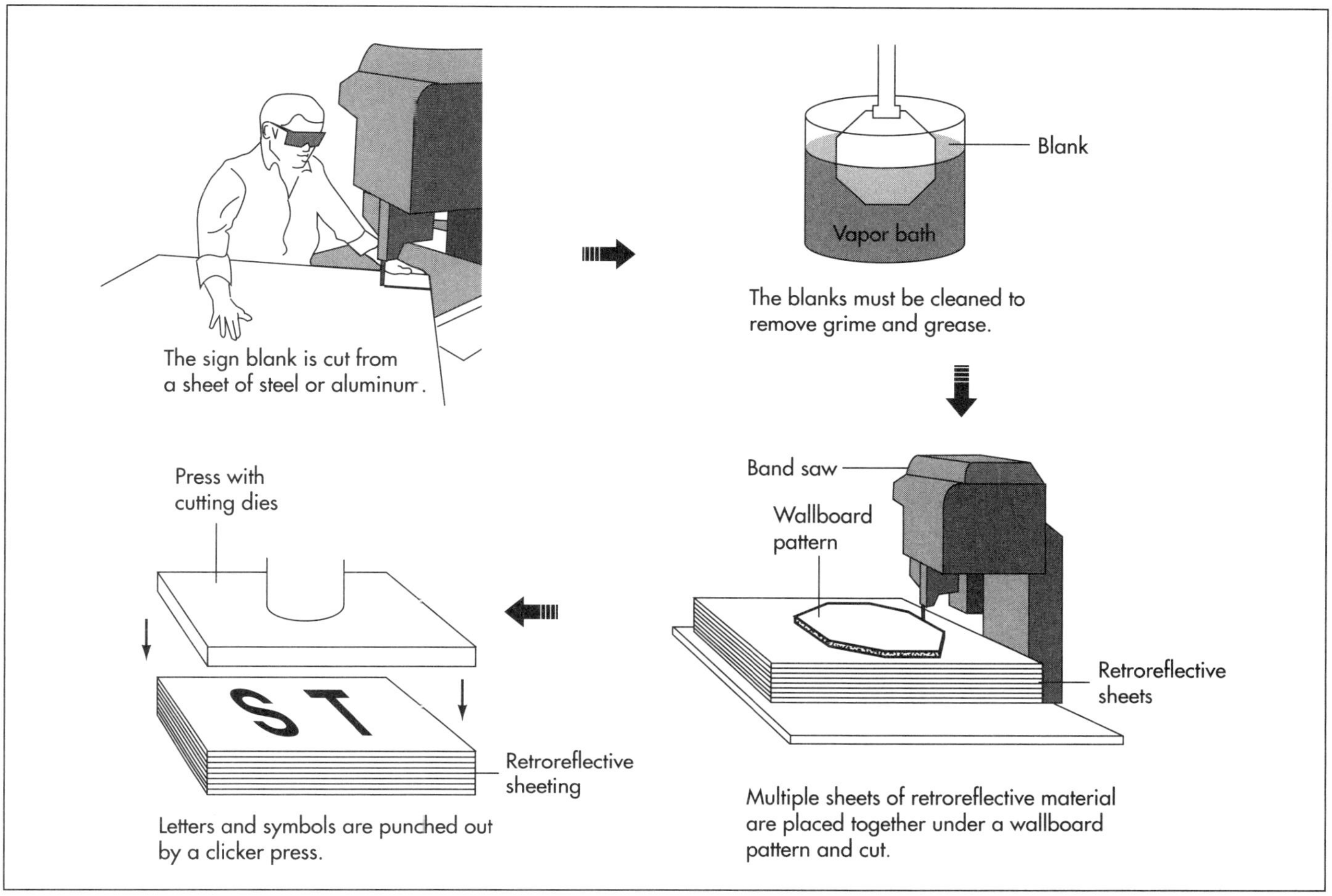

retroreflective sheeting uses a heat-sensitive or pressure-sensitive adhesive and whether silk-screening, etching, or other coloring processes are used. Many traffic signs, however, undergo the following process using heat-sensitive adhesives.

Cutting the blank

1 The sign blank is cut, usually from a sheet of steel or aluminum, by a metal shear machine or a band saw. Corners are rounded using the rounding-selection mode on a punch machine. Holes for mounting the sign are punched or drilled.

Checking the blanks

2 The blanks are checked for any defects or contamination. Blanks must be free of grime in order for background sheeting to adhere properly. The "Tape Snap" test checks for the presence of dirt. A piece of transparent cellophane tape is applied to the dry blank surface and "snapped" up at a right angle. The presence of color or particles on the tape indicates contamination. Any trace of oil or wax is tested by the "Water Break" exercise. Water poured over the blank surface should flow evenly and completely; beading action denotes contamination.

Degreasing the blanks

3 The blank surface is wiped with mineral spirits or naphtha to remove greasy fingerprints. The surface is dried with a clean, lint-free cloth before the solution evaporates. The blank is then degreased by immersion in a bath of trichloroethylene or percholorethylene vapor. Certain alkaline solutions can be used instead of the vapors in the bath. A water rinse afterward is not necessary.

Cutting the retroreflective sheets

4 Using scissors, razor blades, a knife, or a paper cutter, individual background

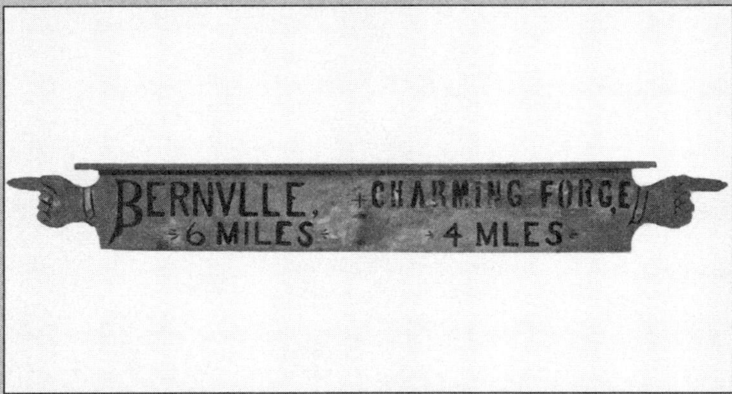

This 1875 carved and painted road sign from Berks County, Pennsylvania, sufficed when passing vehicles were going only three or four miles an hour. (From the collections of Henry Ford Museum & Greenfield Village.)

Beginning in the 1880s, well before the automobile age, the League of American Wheelmen, a national bicycle club, lobbied for better roads and signs. Their pioneering efforts were only later expanded by automobile clubs and enthusiasts. Road signs had to fill some very demanding functions. They had to be comprehensible from a distance on the darkest night and the brightest afternoon against diverse backgrounds. They had to provide succinct information at a sufficient distance so that drivers had time to make decisions and safely maneuver their vehicles in the appropriate direction. As travel far from home became more common, standardization became more desirable.

Interestingly, as more and more Americans took to the road for longer distances, companies responded by displaying distinctive signs on the road. In the 1930s, specifically, companies looked for design features that would allow motorists to make timely decisions even when travelling at the breakneck speed of 35 mph! Texaco took the lead among big oil companies, replacing its regionally styled gasoline stations with a standardized design in 1936. Designed by Walter Dorwin Teague, the new ice-box shaped building was white with three horizontal green stripes and the red Texaco star. A distinctive banjo-shaped sign carrying the Texaco emblem stood in front. From a quarter mile away, drivers knew they were approaching a Texaco station even without reading a word.

William S. Pretzer

retroreflective sheets are cut by hand. Multiple sheets, on the other hand, are cut using a band saw. In this process, the shape of the sign is traced on to a 0.125-inch (3.2 mm) wallboard. This wallboard is laid on top of about 50 sheets, secured and nailed to a hardboard cutting base. The band saw follows the pattern and cuts the sheets.

5 Letters and symbols are punched out from white or black retroreflective sheeting either by hand or by using a "clicker" press. Up to 29 sheets can be placed in the press at once; cutting dies, which function much like cookie-dough cutters, are placed in the machine to produce the desired characters.

Applying sheet to sign blank

6 The adhesive liner on the back of the background sheeting is removed in one motion, and the sheeting is applied to the dry blank surface. The sign is cranked through a large squeeze-roller applicator to remove air bubbles trapped between the sheeting and the blank. Edges are then trimmed.

Heating the sign

7 The sign is placed in a heat lamp vacuum applicator for one minute, removed, and allowed to cool before the sign copy and border are placed on the sign. The squeeze-roller applicator or a hand roller is used over the copy to eliminate air bubbles. The sign is then covered with a plastic slipsheet and placed in the heat lamp vacuum applicator for another minute.

The Future

Although exotic possibilities like projecting holographic traffic signs above the roadway have been suggested for the future, it is more likely that drivers will see conventional refinements of signs to make them more visible, particularly at night. Internally illuminated street signs are already in use in parts of Nevada and California. Because urban areas—the Las Vegas casino "Strip" being an extreme case—are often brightly lit throughout the night, retroreflective signs are harder for drivers to read. Electrically powered signs in which internal light bulbs illuminate translucent copy make the signs significantly more visible.

Interplex Solar has developed self-illuminated traffic signs that use light-emitting diodes (LEDs) powered by solar-charged batteries. The LEDs, similar to those used in some calculator readout displays, outline the sign's letters, symbols, and border. A light-sensitive photocell turns the sign on and off

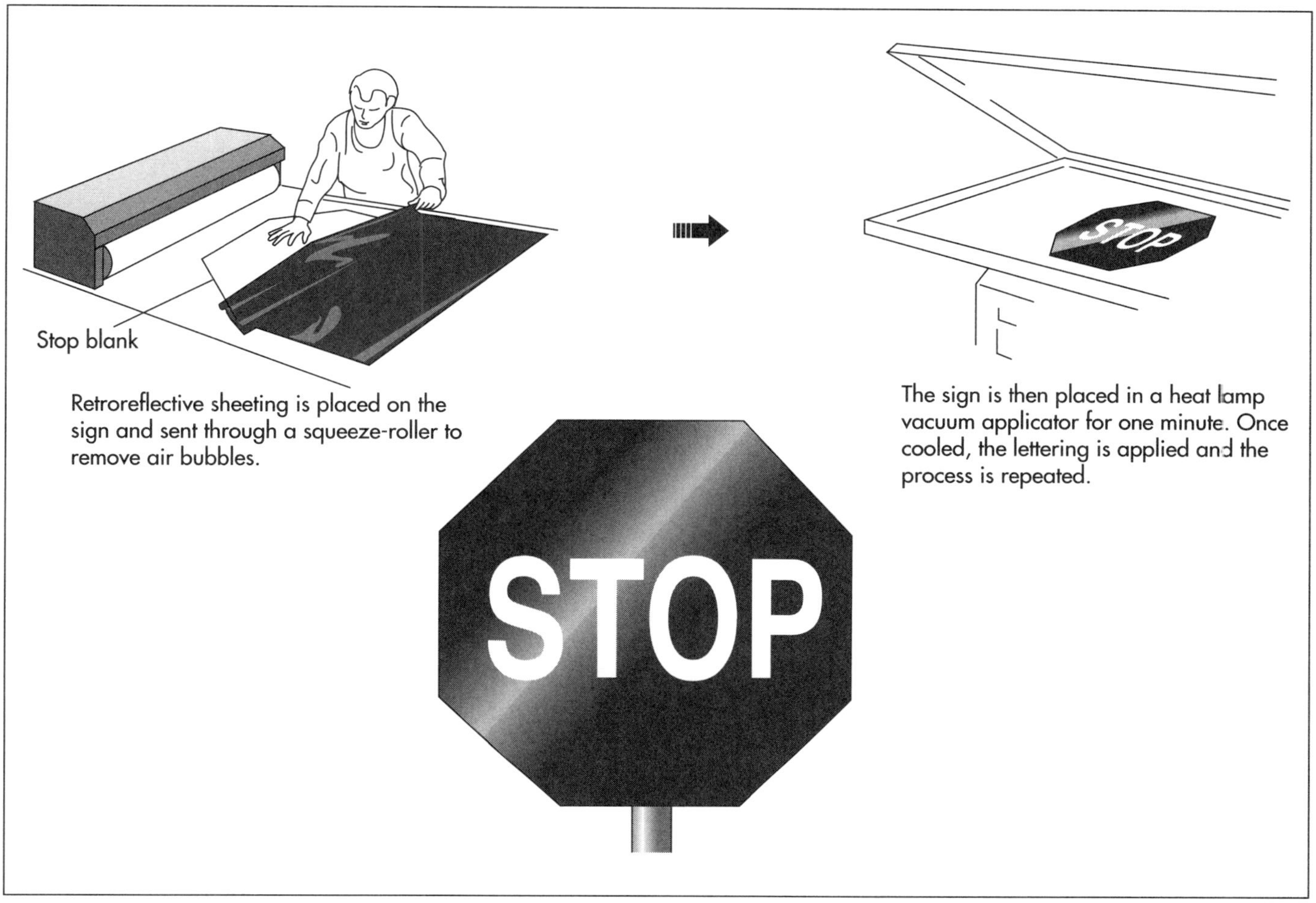

as needed. Already the highway departments of five states have either tested or ordered some of these highly visible signs. Because these signs cost $450 each (roughly five times the cost of a traditional traffic sign), their use may be restricted to more dangerous road passages and high-accident intersections.

Where To Learn More

Books

Outdoor Testing of Reflective Sign Materials. U.S. Department of Agriculture. Forest Service. Government Printing Office (GPO), 1985.

Standard Highway Signs. U.S. Department of Transportation. Federal Highway Administration. GPO, 1979.

Manual on Uniform Traffic Control Devices. U.S. Department of Transportation. Federal Highway Administration. GPO, 1983.

Traffic Control Devices Handbook. U.S. Department of Transportation. Federal Highway Administration. GPO, 1983.

Other

Sign Shop Practices Manual. 3M Corporation, 1978. 1-800-553-1380.

—Bridget K. Hall

Rope

Although the origin of rope is unknown, the Egyptians were the first people to develop special tools to make rope. Egyptian rope dates back to 4000 to 3500 B.C. and was made of water reed fibers. The use of ropes pulled by thousands of slaves allowed the Egyptians to move the heavy stones required to build the pyramids.

Background

A rope is a bundle of flexible fibers twisted or braided together to increase its overall length and tensile strength. The use of ropes for hunting, carrying, lifting, and climbing dates back to prehistoric times. Ropes were originally made by hand using natural fibers. Modern ropes are made by machines and utilize many newer synthetic materials to give them improved strength, lighter weight, and better resistance to rotting. More than half of the rope manufactured today is used in the fishing and maritime industries.

Although the origin of rope is unknown, the Egyptians were the first people to develop special tools to make rope. Egyptian rope dates back to 4000 to 3500 B.C. and was generally made of water reed fibers. Other Egyptian rope was made from the fibers of date palms, flax, grass, papyrus, leather, or camel hair. The use of such ropes pulled by thousands of slaves allowed the Egyptians to move the heavy stones required to build the pyramids. By about 2800 B.C., rope made of hemp fibers was in use in China. Rope and the craft of rope making spread throughout Asia, India, and Europe over the next several thousand years. By the fourth century, rope making in India had become so specialized that some makers produced rope intended only for use with elephants. Leonardo da Vinci (1452-1519) drew sketches of a concept for a rope-making machine, and by the late 1700s several working machines had been built and patented. Rope continued to be made from natural fibers until the 1950s when synthetic materials such as nylon became popular. Despite the changes in materials and technology, rope making today remains little changed since the time of the ancient Egyptians.

Rope is sometimes generally referred to as cordage and can be divided into four categories based on its diameter. Cordage under 0.1875 inches (0.5 cm) in diameter includes twine, clothesline, sash cord, and a tar-covered hemp line called marline. These are not considered to be true rope. Cordage with a diameter of 0.1875 to 0.5 inches (0.5-1.3 cm) is a light-duty rope and is sometimes referred to as "small stuff." Cordage with a diameter of 0.5 to about 1.5 inches (1.3-3.8 cm) is considered to be true rope. Cordage over about 1.5 inches (3.8 cm) in diameter is generally called a hawser and is used for mooring large ships.

Rope construction involves twisting fibers together to form yarn. For twisted rope, the yarn is then twisted into strands, and the strands twisted into rope. Three-strand twisted rope is the most common construction. For braided rope, the yarn is braided rather than being twisted into strands. Double-braided rope has a braided core with a braided cover. Plaited rope is made by braiding twisted strands. Other rope construction includes combinations of these three techniques such as a three-strand twisted core with a braided cover. The concept of forming fibers or filaments into yarn and yarn into strands or braids is fundamental to the rope-making process.

Raw Materials

Rope may be made either from natural fibers, which have been processed to allow them to be easily formed into yarn, or from synthetic materials, which have been spun into fibers or extruded into long filaments.

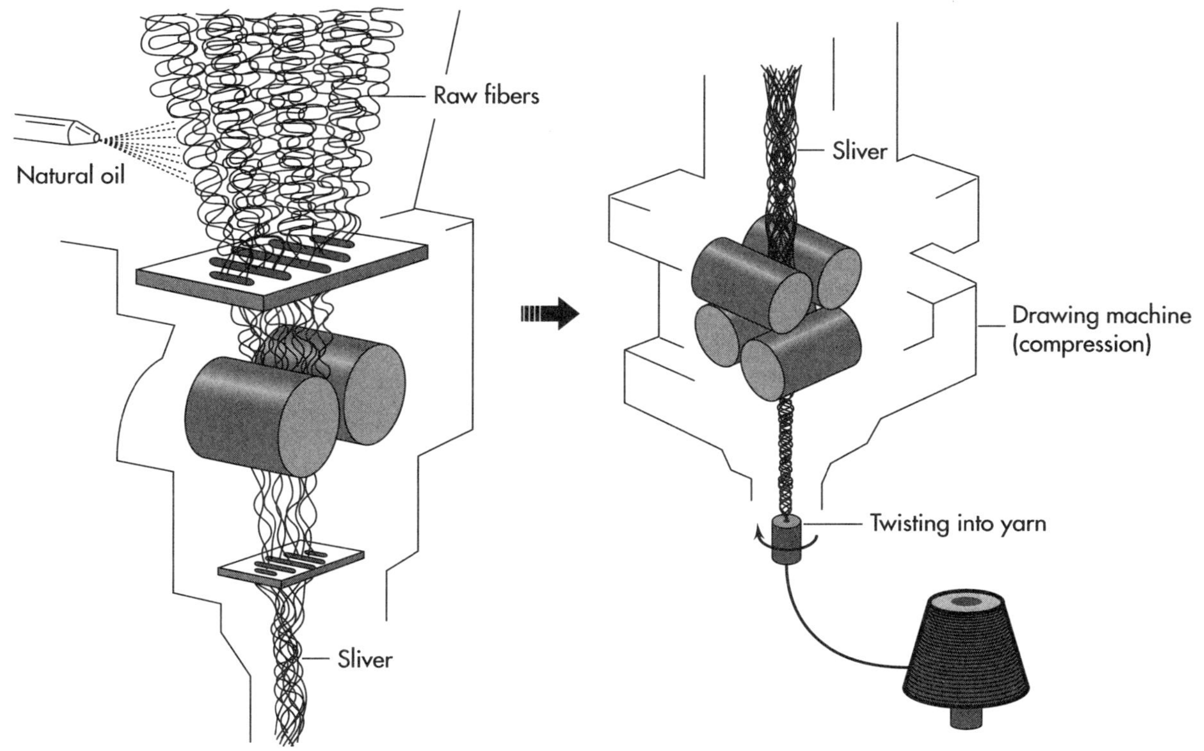

Natural fibers are first lubricated with oil. They are then sent through a series of machines that remove dirt and straighten the fibers.

The sliver is sent through drawing machines to make yarn.

Natural fibers include hemp, sisal, cotton, flax, and jute. Another natural material is called manila hemp, but it is actually the fibers from a banana plant. Sisal was used extensively to make twine, but synthetic materials are replacing it. Manila rope is still used by traditionalists, but it can rot from the inside, thus losing its strength without giving any outward indication.

Synthetic fibers include nylon, **polyester**, polypropylene and aramid. Polypropylene costs the least, floats on water, and does not stretch appreciably. For these reasons it makes a good water ski tow rope. Nylon is moderately expensive, fairly strong, and has quite a bit of stretch. It makes a good mooring and docking line for boats because of its ability to give slightly, yet hold. Aramid is the strongest, but is also very expensive. Nylon and polyester may be spun into fibers about 4-10 inches (10-25 cm) long. Ropes made from spun synthetic fibers feel fuzzy and are not as strong as ropes made from long, continuous filaments. Some ropes use two different synthetic materials to achieve a combination of high strength and low cost or high strength and smooth surface finish.

Wire rope may be made from iron or steel wires. This is commonly referred to as cable and is used in bridges, **elevator**s, and cranes. It is made by a different process than fiber or filament ropes.

The Manufacturing Process

Fibers and filaments are first formed into yarn. The yarn is then twisted, braided, or plaited according to the type of rope being made. The diameter of the rope is determined by the diameter of the yarn, the num-

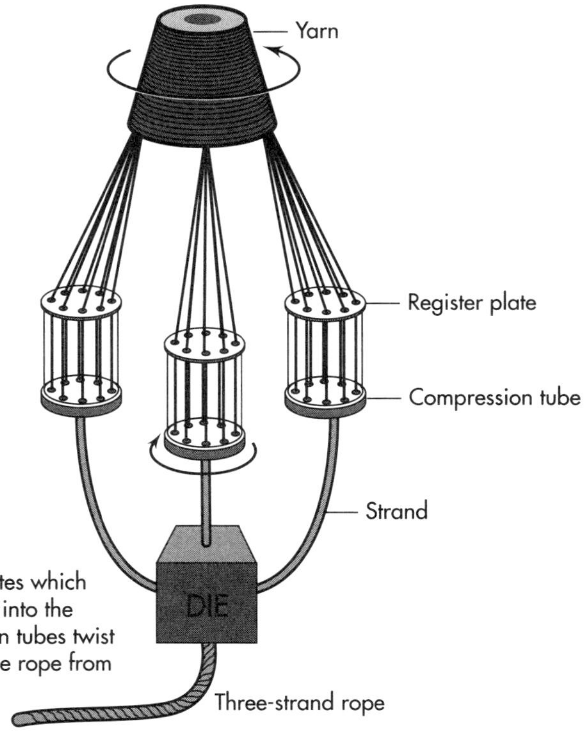

The yarn is fed through register plates which control the amount of yarn passing into the compression tubes. The compression tubes twist in the desired direction and form the rope from the three strands of yarn.

ber of yarns per strand, and the number of strands or braids in the finished rope.

Processing the fibers and filaments

1 If the rope is to be made from raw natural fibers, the fibers are first lubricated with a natural oil. They are then fed into a series of machines that remove any dirt, straighten the fibers, spread them apart, and comb them with several sets of steel-toothed combs. Each set of combs has the teeth set closer together as the fibers proceed through the process. This produces a loose, continuous ribbon of fibers called a sliver. The fibers in the sliver have been aligned along the long axis of the ribbon. Synthetic fibers follow a similar process, but tend to align more easily.

If the rope is to be made from long filaments of synthetic material, several filaments are grouped together in a process called doubling or throwing. This produces a sliver of multiple plies of filaments.

2 The sliver is run through the rollers of a drawing machine to compress it before it is twisted into yarn. Yarn that has a right-hand twist (to the right and up) when viewed from the end is said to have a "Z" twist, and yarn that has a left-handed twist (to the left and up) is said to have an "S" twist. Sometimes this is referred to as right-hand laid yarn and left-hand laid yarn. The finished yarn is wound on spools called bobbins. At this point, the yarn may be dyed various colors to produce a strand, or an entire rope, of a particular color. This is especially helpful in finding a specific line in a maze of rigging on a sailboat.

Forming twisted rope

3 The bobbins of yarn are set on a frame known as a creel. For three-strand, right-hand twist rope, Z-twist yarns would be used to make each strand. The ends of the yarns are fed through a hole in a register plate which keeps the yarns in the proper relation to each other. The ends of the yarns are then fed into a compression tube. As the yarn is pulled through the compression tube, the tube twists it in the S-twist direction, opposite of the yarn twist, to produce a tight strand.

4 The strands are either transferred to strand bobbins or fed directly into the closing machine. For common three-strand rope, three S-twist strands would be used. The closing machine holds the strands firmly with a tube-like clamp called a laying top. The end of each strand is then passed through a rotating die which twists the strands in the Z-twist direction, locking them together. This process is called closing the rope.

5 The finished rope is wound onto a reel. When the end of the strands has been reached, the finished coil of rope is removed from the reel and tied together with bands of smaller rope. The ends are either taped or, if the rope is a synthetic material, melted with heat to prevent them from unraveling.

Forming braided rope

6 Braided ropes are commonly made from synthetic materials. The bobbins of yarn are set up on several moving pendants on a braiding machine. Each pendant travels in an oscillating pattern, weaving the yarn into a tight braid. A set of rollers pulls the braid through a guide to lock, or set, the braid and keep tension on the rope. In some machines the braiding process is accomplished by feeding the yarns through separate counter-rotating register plates. One yarn is woven in one direction followed by another in the opposite direction, and so on, to form an interlocked braid.

7 If a double-braided rope is being formed, the first braid becomes the core, and the second braid is immediately woven on top of it to form the outer covering, called the coat.

8 As the rope emerges from the rollers, it is taken up on a reel. The finished coil is then removed and banded, and the ends are taped or melted.

Forming plated rope

9 Eight-plaited rope consists of four S-twist strands and four Z-twist strands. The strands are paired together with one S-twist and one Z-twist in each pair. These pairs are then held together and braided with the other pairs. The manufacturing process first follows the twisted rope process to make the strands, then the braided rope process to form the final rope.

Quality Control

The level of quality control depends on the intended use of the rope. Ropes intended for general purpose use are sold by diameter and tensile strength. Tensile strength is determined by breaking a sample piece under load. Basic raw material specification and a visual inspection are the only quality control measures used for these ropes. Ropes intended for high-risk applications—such as rappelling, rescue work, and lifting objects over people—are more closely inspected and tested. These ropes have a finite service life and may also have a color code or other coding to indicate the date of manufacture. Some ropes incorporate some type of wear tracer formed into the rope. These tracers are usually a single yarn of contrasting color placed just under the outer wrap of yarn. Should any abrasion or overextension of the rope occur, this filament would be exposed, indicating an unsafe condition and requiring that the rope be replaced.

The Future

The future of rope making is directly linked to improvements in materials. Over the years, almost every conceivable type of rope configuration has been attempted. In the past, new materials have allowed rope makers to reduce the diameter of the rope while maintaining the tensile strength and improving the resistance to weathering and abrasion. It is expected that a new generation of very strong, very light fibers and forming techniques will produce even further improvements in ropes.

Where To Learn More

Book

Merry, Barbara. *The Splicing Handbook.* International Marine, 1987.

Periodical

Foster, G.P. "New Fiber Rope Technologies Drive Increased Applications." *Sea Technology*, July 1989, pp. 15-16.

—Douglas E. Betts/Chris Cavette

Rough Terrain Forklift

Industrial forklifts are commonly used in warehouses and around truck and train loading docks. Rough terrain forklifts, as the name implies, are designed to run on rough, unpaved surfaces. They are commonly used around construction sites or in military applications.

Background

A forklift is a mobile machine that uses two prongs, or forks, to lift and place loads into positions which are ordinarily difficult to reach. Forklifts generally fall into two categories: industrial and rough terrain. Industrial forklifts are commonly used in warehouses and around truck and train loading docks. They have small tires designed to run on paved surfaces and are usually powered by an internal combustion engine running on **gasoline**, diesel, or propane fuel. Some smaller industrial forklifts are powered by an electric motor running off an internal battery. Rough terrain forklifts, as the name implies, are designed to run on rough, unpaved surfaces. They are commonly used around construction sites or in military applications. They have large, pneumatic tires and are usually powered by an internal combustion engine running on gasoline, diesel, or propane fuel. Rough terrain forklifts can have a vertical tower, which lifts loads straight up, or a telescoping boom, which lifts loads up and out from the base of the machine.

The rough terrain forklift dates back to about 1946 when a two-pronged lift attachment was placed on a power buggy or tractor chassis. This early machine was used around construction sites and could lift about 1,000 pounds (454 kg) to a height of 30 inches (76 cm). Rapid development of vertical tower forklifts for industrial use was adapted to rough terrain forklifts as well. By the mid-1950s, capacities of 2,500 pounds (1,135 kg) and lift heights of up to 30 feet (9 m) were available.

In 1958, the first four-wheel drive rough terrain forklift was introduced. It had a capacity of 6,000 pounds (2,724 kg) at a lift height of 22.5 feet (7 m), or 3,000 pounds (1,362 kg) at 35 feet (11 m). In 1962, the first telescoping-boom rough terrain forklift came on the market. The telescoping boom allowed loads to be placed out from the base of the machine, both above grade and below grade. This was especially handy in crowded construction areas where open trenches, construction debris, or other construction work prevented a vertical lift forklift from operating close to the area where the material was needed.

Developments during the 1970s and 1980s brought improvements in the telescoping boom design and the introduction of features such as automatic hydraulic frame leveling for increased stability. Requirements of the Occupational Safety and Health Act (OSHA) resulted in improved operator cabs and controls during this period.

Today, rough terrain forklifts are a common sight on construction projects. They handle everything from pallets of concrete block to stacks of plywood to roof beams. The larger models use a telescoping boom with lift capacities up to 10,000 pounds (4,540 kg), vertical reaches up to 40 feet (12 m) and forward reaches of 25 feet (7 m) or more. They are usually a low-profile design and can pass through openings as low as 8 feet (2 m) high to gain access to the interior of a structure. Two-wheel steering, four-wheel steering and four-wheel crab steering (all wheels turned in the same direction) configurations are available.

Raw Materials

The frame, cab, boom, and body of a telescoping-boom rough terrain forklift are usually fabricated by the forklift manufacturer. Steel is the most common material for these subassemblies. Some steel or aluminum castings or forgings may also be used. Nonmetallic materials such as nylon plastic blocks are sometimes used as guides in the boom assembly. The remainder of the parts are usually purchased as finished products and are installed by the forklift manufacturer. Purchased products include the engine, transmission, axles, wheels, tires, brakes, seat, gauges, lights, back-up alarm, hoses, and hydraulic cylinders. The hydraulic fluid, lubricants, and fuel are purchased in bulk quantities and are added as required.

Design

A typical telescoping-boom rough terrain forklift is long and low with a pair of wheels at the extreme front and another pair located towards the rear. The boom is mounted at the rear of the forklift off a pivot that is raised several feet above the level of the frame. The cab is mounted on the left-hand side of the frame structure with the bottom half of the cab low and between the tires. The hydraulic fluid tank and fuel tank are mounted opposite the cab on the right-hand side. The engine and transmission are mounted within the frame along the centerline of the vehicle.

Beyond this basic configuration, various manufacturers have their own unique designs and options. Some forklifts use a single hydraulic cylinder to elevate the boom, while others use two cylinders. Some models have a side-to-side hydraulic frame leveling capability which tilts the frame up to 10 degrees relative to the axles to compensate for extreme axle articulation. This is used, for example, when the tires on one side of the forklift are up on a mound of dirt and the tires on the other side are down in a rut. Other special features include fork attachments that swing up to 45 degrees left and right to allow exact placement of the load.

The Manufacturing Process

The telescoping-boom rough terrain forklift is generally manufactured in separate, functional group sections: hydraulics, powertrain (engine, transmission, etc.), electrical, chassis, and boom. Individual components are either purchased or created from raw materials, and joined into subassemblies. The subassemblies are then brought together in the final assembly area where the forklift is completed. The actual flow of work varies from one manufacturer to another, but the following is a typical process.

Materials preparation

1 The raw steel material—sheet, plate, bars and tubes—are first cut to size and machined. Plate up to 0.75-inch (1.9 cm) thick is cut into shape, or "burned," by oxyacetylene or plasma gas cutting torches controlled by numerical controlled machines. Thinner steel sheet is cut with a shear and bent into shape as required by press brakes. During the cutting and machining, the steel is held in place with large fixtures, or clamping devices, to ensure dimensional accuracy.

Welding

2 The parts which will be welded together are first tack welded in place. These would include components of the chassis, cab, and boom, among others. A tack weld is simply a small weld, or fusion of the two pieces of material, to keep the pieces from shifting during the final welding process. The whole assembly is then welded by numerical-controlled (NC) machines which place the welds in exactly the right areas, with the right welding temperatures, and the right feed rate for the welding rod. This is important to obtain a weld which will provide the required strength and meet the standards of the American Welding Society. As with the machining step, a variety of fixtures are used to ensure dimensional accuracy.

Shot blasting

3 At this station, steel parts are placed on a rotating table or a conveyor belt in a large chamber. When the chamber doors close, the parts are blasted with thousands of BB-sized metal pellets that are shot at high speed from dozens of openings in the walls of the chamber. This process cleans away the rough scale that naturally forms on the surface of steel when it comes from the steel

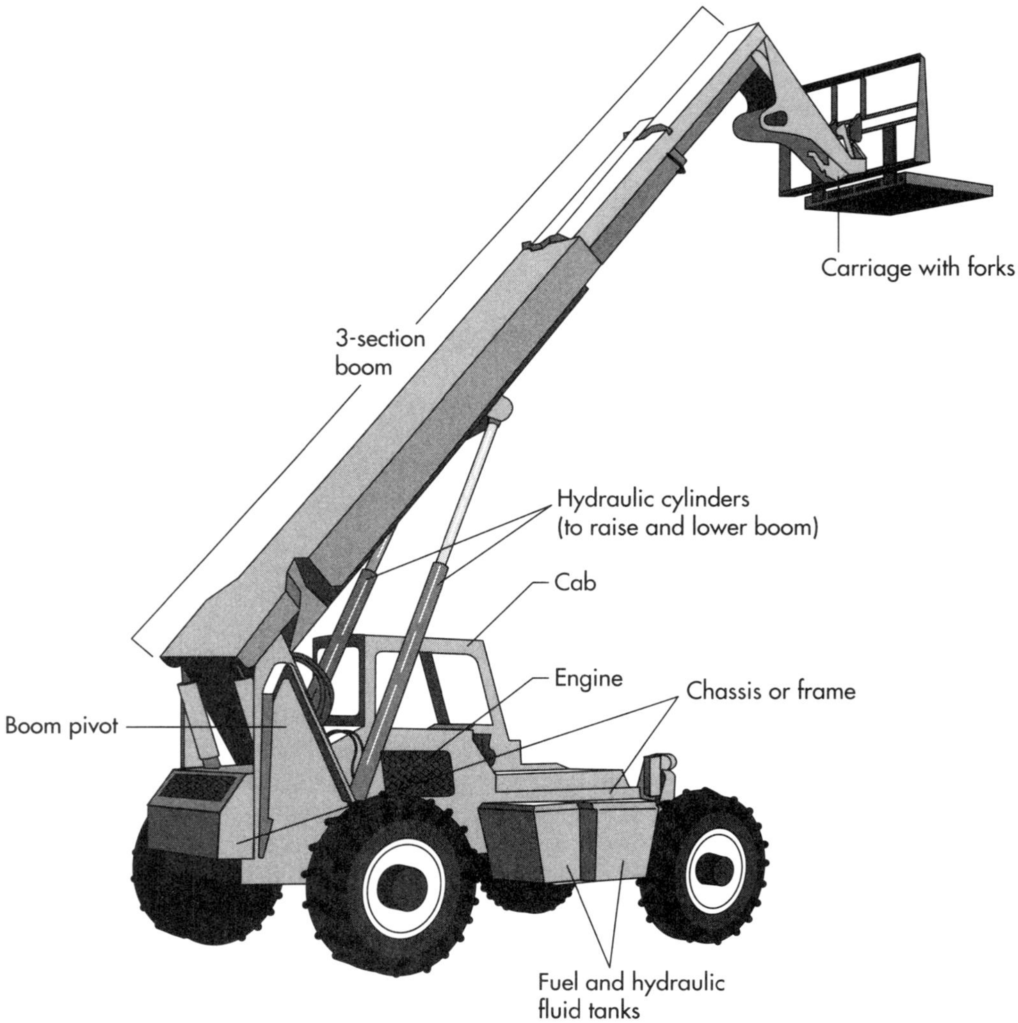

The frame, cab, boom, and body of a telescoping-boom rough terrain forklift are usually fabricated by the forklift manufacturer. The remainder of the parts are usually purchased as finished products and are installed by the forklift manufacturer. Purchased products include the engine, transmission, axles, wheels, tires, brakes, seat, gauges, lights, back-up alarm, hoses, and hydraulic cylinders.

mill. It also cleans away the small welding splatter commonly found in the welded areas. This shot blasting is the first step in preparing the parts for painting.

Painting

4 All exposed parts, except the boom, are now painted to protect the surfaces. The boom is painted after the telescoping sections have been manually assembled in step 5. In preparation, all parts are thoroughly washed in a detergent bath and then rinsed. A second acid wash and rinse cleans the metal further and also applies a thin phosphorus coating to improve paint adherence. In the paint booth, fine paint particles are sprayed from a spraygun that also imparts an electrostatic charge to each particle. The part being painted is electrically charged to the opposite polarity of the paint. This causes the paint to be drawn to the part and results in an even coat of paint over the entire surface. After painting, the parts are baked in ovens to produce a hard coating.

Subassembly

5 The parts are now sent to several functional group work stations. The boom is built at one station, the cab at another, the chassis at another, and so on. The boom is made of two to four rectangular sections of long, hollow steel tube. The size of each

section is smaller than the previous one and the sections slide, or telescope, into each other. Inside each section, a hydraulic cylinder and chain device cause the boom sections to extend or retract when maneuvering loads. Nylon guides prevent the steel sections from rubbing on each other, and stops are installed to prevent the sections from sliding out of each other when the boom is operating below grade level at a downward angle.

The chassis work group installs electrical wiring and hoses and bolts the engine supports in place. The cab group installs the instrument panel, controls, wiring, and seat. The powertrain group joins the transmission to the engine, mounts the engine accessories and hydraulic pumps, and connects electrical wiring to various sensors on the engine.

Final assembly

6 All of the subassemblies are now brought to the final assembly area. The tires, wheels, hubs, and brakes are installed on the axles, and the axles are installed on the underside of the chassis. The engine and transmission are lowered into the chassis and secured to their mounts. The driveshaft(s) connecting the transmission and the drive axle(s) are connected. The cab, fuel tank, and hydraulic fluid tank are installed. The boom assembly is lowered onto its pivot point and the hydraulic cylinders that raise and lower the boom are installed. Hose and electrical connections are made between all the subassemblies. Fluids (oil, hydraulic fluid, fuel) are added as required. Instruction and warning decals are applied in the cab and on the boom.

Start-up and testing

7 Each unit is started and run through a series of functional tests with actual loads for up to 1.5 hours. Any final adjustments or settings are made at this time.

Shipping

8 Finished forklifts are shipped to the customer or distributor by truck or by rail. Two or three forklifts are usually shipped on the same load to minimize the freight charges.

Quality Control

Inspections and tests are essential to the manufacturing process to ensure the product meets all standards and safety requirements. Critical components are placed on a coordinate measuring machine that optically checks dimensions, alignment, and geometry following fabrication. Welders, and even the NC welding machines, must have American Welding Society certification. Other parts are visually inspected during their fabrication and assembly.

In addition to the part-by-part inspection, the entire forklift design is tested for proper function. One of the critical tests is the American Society of Mechanical Engineers (ASME) stability test. This test determines how much weight can be safely handled at various distances, or reaches, from the forklift. For example, a forklift with a 10,000-pound (4,540 kg) lift capacity is limited to a maximum lift height of 20 feet (6 m) and a maximum forward reach of 8 feet (2 m) when lifting a full 10,000-pound load. For a full 25-foot (7.6 m) forward reach, the load capacity of this forklift is reduced to 2,000 pounds (908 kg) without outriggers, or stabilizing legs, and 3,250 pounds (148 kg) with outriggers. Warning labels and charts in the cab caution the operator of these limitations.

The Future

A wide variety of attachments have been developed for rough terrain forklifts to improve their utility. Winches, booms, and rotating fork carriages allow the forklift to place materials more accurately. Articulating booms, or booms with two separate extendible arms, can reach up and over structures to place loads on interior roof slopes or in the center of upper floors. Other attachments and enhancements can be expected in the future.

Additional built-in safety features are also expected. Load-reach management devices can automatically restrict the reach of the forklift based on the load being handled rather than relying on the operator. These devices would determine the weight of the load using pressure sensors and feed this information to a small electronic memory device which had all the load-reach limitations programmed into it. As the load is being maneuvered into position, the memory would compare the angle and extension of the boom with the safety limits. A warning device or a lock-up mechanism would prevent the operator from over-reaching and possibly causing the boom to fail or the forklift to tip over.

Where To Learn More

Periodicals

Petersen, Julie. "Cat's Big Secret: Killer Forklifts." *Mother Jones,* November-December 1993, p. 13.

Schwind, G.F. "What's New in the Lift Truck Marketplace." *Material Handling Engineering,* February 1993, pp. 49-56.

—Peter Toeg/Chris Cavette

Safety Pin

Background

A coiled piece of steel wire, sharpened at one end with a catch plate at the other, the modern safety pin is a device that can be traced back to New York City in about 1825. An indebted inventor named Walter Hunt sought a way to repay a $15 debt. He designed a "safe pin" for securing clothing, which protected fingers from the sharp end. He patented the design in 1849 and sold his idea for $400. The design has remained virtually the same ever since. However, the manufacture of its most common material, steel, has greatly improved, as has the mechanized process used in forming the pins. In 1864, E.J. Manville invented an automatic fourslide machine, which was the forerunner of machines used today.

Even though the current design is a modern one, the safety pin is an ancient fastener. Coiled bronze pins, embellished with gold and several inches long, have been found in Egyptian tombs. The Greeks and Romans called them *fibulae* (Latin for brooch) and used some to fasten garments, while others were mainly ornamental. Dating from the seventh century B.C., elaborately decorated fibulae often had rows of lions or sphinxes along the catch plate, either carved in relief or soldered. An Iranian pin from this period was shaped like a human hand and embellished with two lions placed head to tail, while an Etruscan fibula from the eighth century was decorated with ducks. The fibula became widely used throughout the ancient world as the Roman Empire expanded. In the Middle Ages, the design reverted to one resembling a straight pin. These were fashioned out of skewers of wood for common people, or out of bone, ivory, silver, gold, or brass for those of wealth and high position. In the 15th century, pins were manufactured from drawn wire, a process that still exists in the manufacture of modern safety pins. Today the largest user of safety pins is the retail sewing notions market, while the largest commercial user is the laundry and cleaning industry.

The safety pin was patented in 1849 by an indebted inventor who wanted to repay a $15 debt.

Raw Materials

The most common metals used in safety pins are spring steel, brass, and stainless steel. Most fasteners without critical strength requirements are made from spring steel, an alloy of **iron** that has a high carbon content (more than 0.5%). By varying the proportion of carbon added to iron, spring steel can have high tensile strength and resiliency that allows it to flex and return to its original shape.

Brass is an alloy of approximately two-thirds copper and one-third **zinc**. More expensive than spring steel, this metal will polish up to a high luster. It is nonmagnetic and easily worked into shape. The strength of brass is adequate for most safety pin applications.

More expensive than brass, stainless steel is an alloy that includes chromium and nickel (manganese is sometimes used instead of nickel). With a mirror-like finish, stainless steel is used when corrosion, temperature, and strength are considerations. However, for some safety pin applications, a type of low-carbon stainless steel—ferritic, which costs less than stainless steel—is used.

Other raw materials include a variety of coatings and platings. The most common

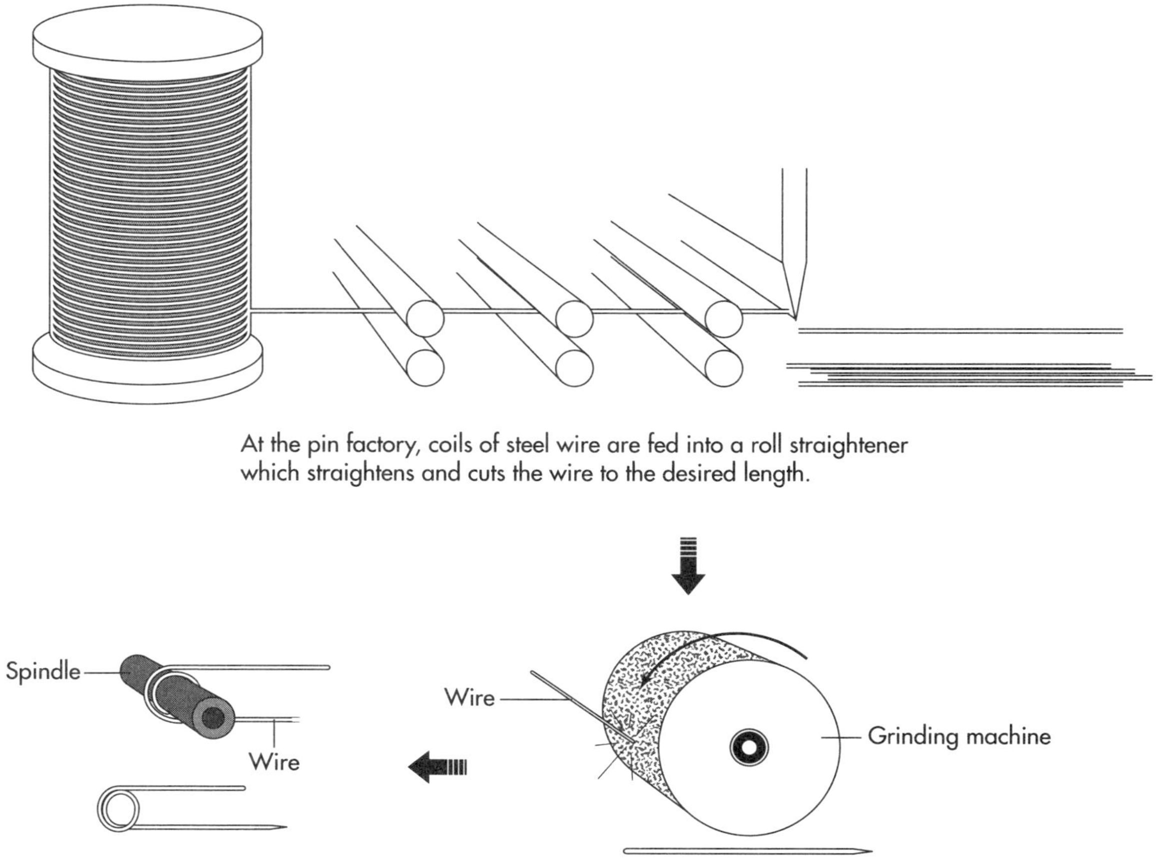

At the pin factory, coils of steel wire are fed into a roll straightener which straightens and cuts the wire to the desired length.

On a fourslide machine, the wire end is coiled around a spindle.

A grinding machine is used to remove nicks and create a sharp point.

finish on spring-steel pins is chrome, followed by a final wax coat. If the pin is to be subjected to damp environments or to a piece of cloth for a long period of time, other treatments are used to prevent stains from corrosion of the steel. One of these methods is to coat the pin with the chemical chromate, particularly on spring-steel safety pins. For brass safety pins, nickel coatings are used, as they resist some chemicals and remain attractive for a long period. A more expensive treatment for brass pins, primarily for decorative purposes, is gilting. By tumbling the pins in an acid bath, the top layer of brass is removed, revealing a shiny, gold-like surface.

The Manufacturing Process

The modern manufacture of safety pins is completely automatic. Specialized machines have been developed to perform multiple steps, replacing many workers and increasing the efficiency of the manufacturing process. Over 3 million safety pins can be made by one factory in a day, almost 1 billion a year. There are only two U.S. companies that make safety pins, although there are many more companies abroad.

Making the wire

1 Steel wire is refined from pig iron, an early stage in iron refining. The ore at this point still contains many impurities, including carbon, which makes iron very brittle. In a converter (a high-temperature oven) oxygen is pumped at high speed into the molten pig iron to burn out the carbon. The molten steel is then formed into slabs by a method known as continuous casting. The metal is poured into a mold. Water jets cool the metal, while straightening rollers

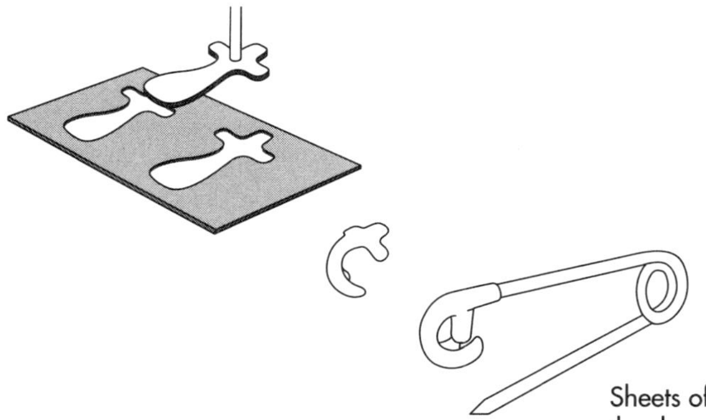

Sheets of metal are stamped to form the doughnut-shaped cap. The cap is then crimped around the pin.

form the mass into a bar called a billet. It is then pickled in acid to remove oxide scale, passed through rollers, heat treated, and then cooled slowly, a process known as annealing. The cold metal is then shaped into wire by drawing it through several dies of decreasing size. During this stage, the metal undergoes frequent annealing to prevent brittleness. As it passes through increasingly narrower dies, the wire achieves the correct gauge (thickness).

Cutting the wire

2 At the pin factory, coils of high carbon spring-steel wire are loaded onto spools and fed into the roll straightener. This machine straightens and cuts the wire to the correct length, from an inch up to a foot long. Most cutting machines can be adjusted to accommodate various diameters of wire.

Forming

3 The cut wire pieces are carried by conveyors or by cart to the grinding machines. The wire pieces are hopper-fed and pressed against grinding wheels. One end of the wire is ground to a point. It is also in the grinder that the metal pieces are polished and nicks and burrs removed from the tapered edges.

4 The forming process is done by a fourslide machine, which allows for a range of motions in four directions. The machine is custom tooled to form the safety pin. The wire pins are hopper-fed into the machine, where they are picked up on a chain. They are then coiled around an arbor, or spindle. The unsharpened end is bent into a hook that will hold the cap.

5 Meanwhile, steel sheets are fed into the machine from the other side. The sheets of metal are stamped on a die, producing oddly-shaped pieces that will form the doughnut-shaped cap.

6 A stamping operation forms the two-dimensional die-cut steel pieces into three-dimensional caps.

7 Inside the machine, the sharpened, hooked, and coiled wire meets up with the cap. The cap is then crimped around the hooked end of the wire. At this point, the safety pin is fully formed and the clasp is closed.

8 For those users who purchase the safety pins with the point open and ready to use, the cost is higher. A mechanical device at the end of the chains opens the pins, and finishing must be done in smaller batches, as the pins nest and tangle together.

Finishing and packaging

9 The pins are loaded into perforated plastic baskets. If plating is required, steel pins are generally chromed, while brass pins are nickel plated. In electroplating, the formed pins are placed into a tank of the

coating solution. The tank is electrically charged, forcing the chemicals in the solution to adhere to the pins in a fine layer. The electroplated pins are shaken by the rotating tank to ensure an even coating.

10 The pins are rinsed off with water and then placed in another tank, where a chemical bath gives them their final polish.

11 The pins are washed in a detergent solution and given a wax finish.

12 The formed and finished pins are then packaged. Lots of 10 gross per box is the standard bulk, or industrial pack. Most consumers purchase safety pins in retail polybags or blister packs.

The Future

Although more modern fasteners like velcro have been introduced in the 20th century, they have not replaced the safety pin. Its simplicity and usefulness seem to guarantee its future into the next century. There is still a large market for safety pins in the less industrialized nations. In India, for example, pins and sewing needles are kept and used for generations, passed on from mother to daughter. In an economy such as this, people do not have easy access to alternative fasteners, and great value is placed on what many of us perceive as a commodity item.

Where To Learn More

Book

Petroski, Henry. *The Evolution of Useful Things.* Knopf, 1993.

Periodical

Fastening, Joining & Assembly Reference Issue. *Machine Design,* November 13, 1986.

—Renee M. Rottner

Salt

Background

Salt is the common name for the substance sodium chloride (NaCl), which occurs in the form of transparent cubic crystals. Although salt is most familiar as a food supplement, less than 5% of the salt produced in the United States is used for that purpose. About 70% is used in the chemical industry, mostly as a source of chlorine. Salt is also used for countless other purposes, such as removing snow and ice from roads, softening water, preserving food, and stabilizing soils for construction.

The earliest humans obtained their salt from natural salt concentrations, called licks, and from meat. Those people who lived near the ocean may have also obtained it by chewing seaweed or from the natural evaporation of small pools of seawater. Meat became a more important source of salt as hunting was developed, as did milk when sheep, goats, horses, camels, reindeer, and cattle were domesticated. Even today, certain peoples—such as the Inuit of the far north, the Bedouin of the Middle Eastern deserts, and the Masai of east Africa—use no other form of salt.

As agriculture developed, leading to an increased population and a diet consisting mostly of plants, it became necessary to devise ways of obtaining salt in greater amounts. The earliest method of salt production was the evaporation of seawater by the heat of the sun. This method was particularly suited to hot, arid regions near the ocean or near salty lakes and is still used in those areas. Solar evaporation was soon followed by the quarrying of exposed masses of rock salt, which quickly developed into the mining of underground deposits of salt. Two thousand years ago the Chinese began using wells to reach underground pools of salt water, some of which were more than 0.6 miles (1.0 km) deep.

In areas where the climate did not allow solar evaporation, salt water was poured on burning wood or heated rocks to boil it. The salt left behind was then scraped off. During the time of the Roman empire, shallow lead pans were used to boil salt water over open fires. In the Middle Ages these were replaced with **iron** pans which were heated with coal. In the 1860s a procedure known as the Michigan process or the grainer process was invented, in which salt water was heated by steam running through pipes immersed in the water. This process is still used to produce certain types of salt. By the late 1880s open pans were replaced by a series of closed pans, in a device known as a multiple-effect vacuum evaporator, which had been used in the sugar industry for about 50 years.

Today the United States is the world's largest producer of salt, followed by China, Russia, Germany, the United Kingdom, India, and France.

Raw Materials

Salt is obtained from two sources: rock salt and brine. Rock salt is simply crystallized salt, also known as halite. It is the result of the evaporation of ancient oceans millions of years ago. Large deposits of rock salt are found in the United States, Canada, Germany, eastern Europe, and China. Sometimes pressure from deep inside the Earth forces up large masses of rock salt to form salt domes. In the United States, salt domes

Although salt is most familiar as a food supplement, less than 5% of the salt produced in the United States is used for that purpose. The majority, about 70%, is used in the chemical industry.

are found along the Gulf Coast of Texas and Louisiana.

Brine is water containing a high concentration of salt. The most obvious source of brine is the ocean, but it can also be obtained from salty lakes such as the Dead Sea and from underground pools of salt water. Large deposits of brine are found in Austria, France, Germany, India, the United States, and the United Kingdom. Brine may also be artificially produced by dissolving mined rock salt or by pumping water into wells drilled into rock salt.

Natural brines always contain other substances dissolved along with salt. The most common of these are magnesium chloride, magnesium sulfate, calcium sulfate, potassium chloride, magnesium bromide, and calcium carbonate. These substances may be as commercially valuable as the salt itself. Rock salt may be quite pure, or it may contain various amounts of these substances along with rocky impurities such as shale and quartz.

For table salt, however, additives are usually mixed in. Most table salt is iodized in order to provide the trace element iodine to the diet. This helps to prevent goiter, a disease of the thyroid gland. To supply iodine, a small amount of potassium iodide is added. Table salt also contains a small amount of various chemicals used to keep the salt from absorbing water and caking. These chemicals include magnesium carbonate, calcium silicate, calcium phosphate, magnesium silicate, and calcium carbonate.

The Manufacturing Process

Processing rock salt

1 Underground salt deposits are usually discovered by prospectors searching for water or oil. When salt is detected, a **diamond**-tipped, hollow drill is used to take several regularly spaced core samples throughout the area. These samples are analyzed to determine if salt mining would be profitable.

2 When a site is selected for mining, shafts are sunk into the center of the salt deposit. Then a machine that looks like a gigantic chain saw is used to cut a slot about 6.0 inches (15 cm) high, about 66 feet (20 m) wide, and about 10 feet (3 m) deep into the salt at floor level. This process is known as undercutting. A series of holes are drilled into the undercut salt with an electric drill containing a tungsten carbide bit. These holes are filled with an explosive such as **dynamite** or ammonium nitrate. Electric blasting caps connected to long wires are attached, and the explosive is detonated from a safe distance. Cutting and blasting are repeated in a pattern that leaves pillars of salt standing to support the roof of the mining area. This is known as the room-and-pillar method and is also used in coal mines.

3 Chunks of blasted rock salt are transported to an underground crushing area. Here they are passed over a grating known as a grizzly which collects pieces smaller than about 9 inches (23 cm). Larger pieces are crushed in a rotating cylinder between metal jaws with spiked teeth. The salt is then transported outside the mine to a secondary crushing area where a smaller grizzly and a smaller crusher reduce the particle size to about 3.2 inches (8 cm). At this point foreign matter is removed from the salt, a process known as picking. Metal is removed by **magnet**s and other material by hand. Rocky material may also be removed in a Bradford breaker, a rotating metal drum with small holes in the bottom. Salt is dumped into the drum, breaks when it hits the bottom, and passes through the holes. Rocky matter is generally harder than salt, so it does not break and does not go through. The picked salt then goes to a tertiary crushing area, where an even smaller grizzly and crusher produce particles about 1.0 inch (2.5 cm) in size. If smaller particles are needed, the salt is passed through a grinder consisting of two metal cylinders rolling against each other. If purer salt is needed, rock salt is dissolved in water to form brine for further processing. Otherwise the crushed or ground salt is passed through screens to sort it by size, poured into bags, and shipped to the consumer.

Processing brine

4 The simplest method of evaporating brine is solar evaporation, but it can only

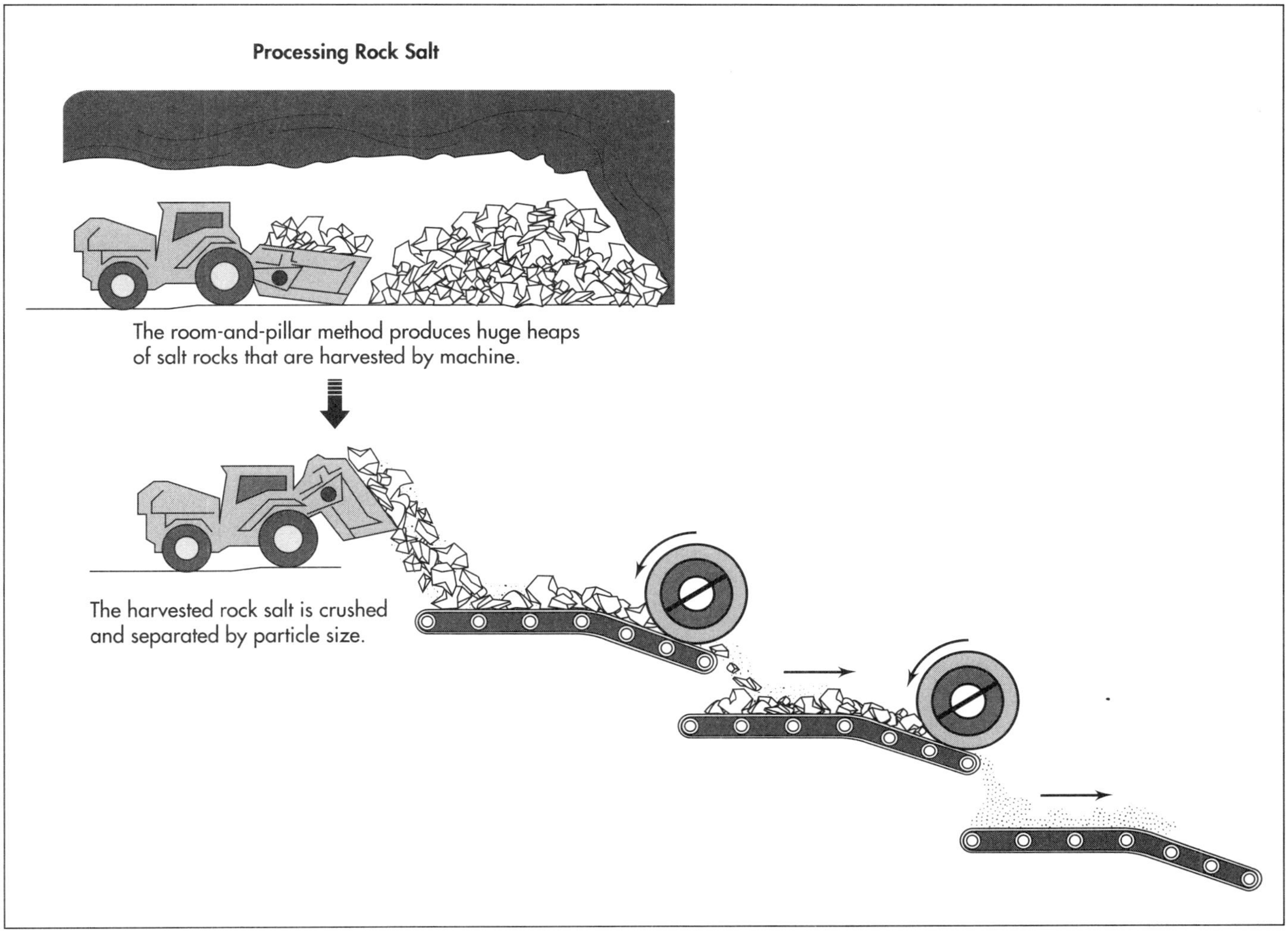

be used in hot, dry, sunny places. The brine is collected into shallow ponds and allowed to evaporate in the sun. Insoluble impurities such as sand and clay and slightly soluble impurities such as calcium carbonate settle to the bottom as evaporation begins. The brine is pumped or moved by gravity flow to another pond where calcium sulfate settles out as evaporation continues. The remaining brine is moved to yet another pond where the salt settles out as evaporation proceeds. The brine is moved one more time before evaporation is complete to prevent highly soluble impurities such as magnesium chloride, magnesium sulfate, potassium chloride, and magnesium bromide from settling out with the salt. These substances may be collected separately for commercial use.

5 The salt is scooped up by machines running on temporary railroad tracks laid on top of the layer of salt. It is then washed with highly concentrated salt water. This water contains so much salt that it cannot hold any more, so the salt is washed free of any trace impurities without dissolving. The washed salt is removed from the salt water, rinsed with a small amount of fresh water, and piled into huge stacks to drain for two or three months. At this point the salt is about 99.4% pure and can be used for many industrial purposes. If purer salt is needed, it is rewashed in salt water and fresh water, allowed to drain for one or two days, then dried in a hot air oven at about 365°F (185°C). This salt is about 99.8% pure and can be used for food processing.

6 Most brine is processed by a multiple-effect vacuum evaporator. This device consists of three or more closed metal cylinders with conical bottoms. Brine is first treated chemically to remove calcium and magnesium compounds. It then fills the bot-

Rock salt is simply crystallized salt. It is the result of the evaporation of ancient oceans millions of years ago. Large deposits of rock salt are found in the United States, Canada, Germany, eastern Europe, and China.

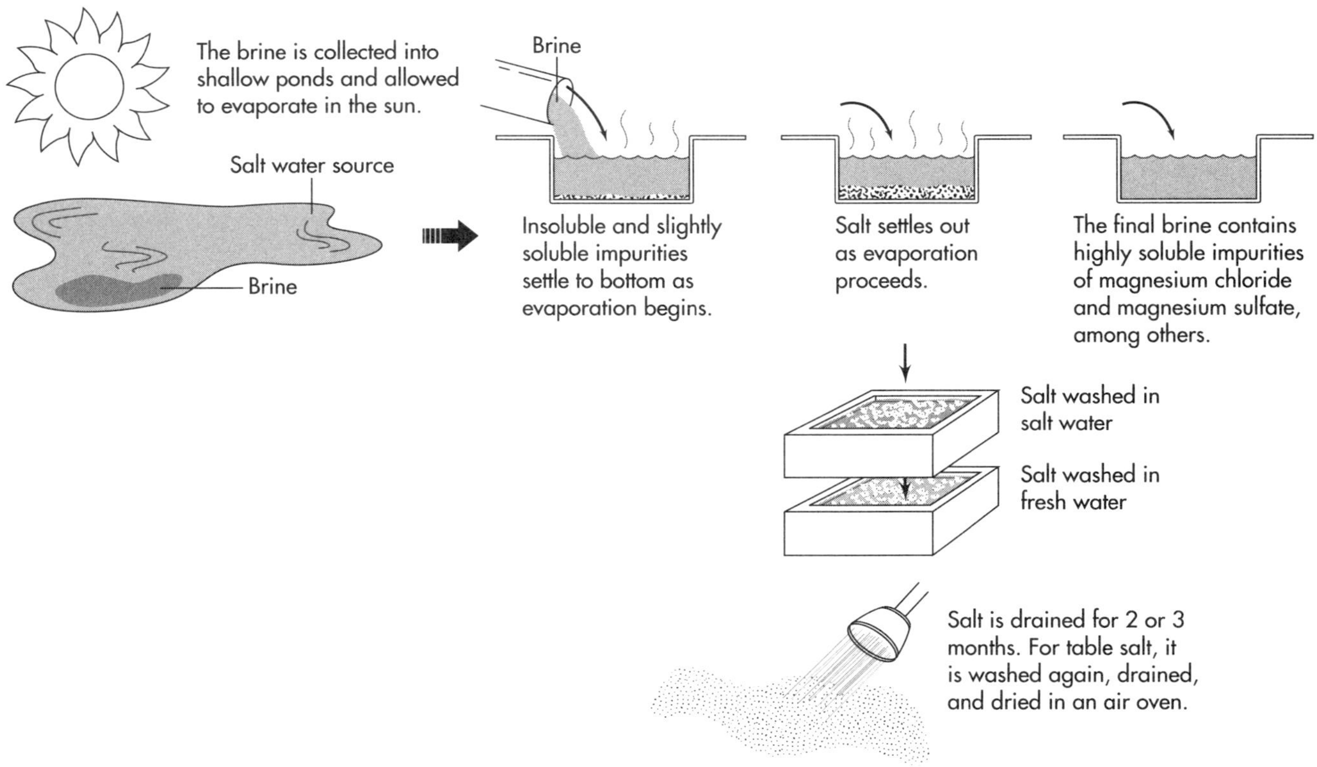

Brine is water containing a high concentration of salt. The most obvious source of brine is the ocean, but it can also be obtained from salty lakes and underground pools of salt water.

tom of the cylinders. The brine in the first cylinder passes through tubes heated by steam. The brine boils and its steam enters the next cylinder, where it heats the brine there. The steam from this brine heats the brine in the next cylinder, and so on. In each cylinder the condensation of steam causes the pressure inside to drop, allowing the brine to boil at a lower temperature. Salt is removed from the bottom of the cylinders as a thick slurry. It is filtered to remove excess brine, dried, and passed through screens to sort the particles by size. Salt made this way is known as vacuum pan salt and consists of small cubic crystals.

7 Brine may also be processed in a grainer. The brine is chemically purified and pumped into a long open pan heated by steam running through pipes immersed in the brine. The brine is heated to a temperature slightly below the boiling point and flakes of salt form on its surface as it evaporates. Usually a temperature of about 194°F (90°C) is used. Lower temperatures produce larger flakes and higher temperatures produce smaller flakes. The flakes grow until they sink to the bottom of the pan, where they are collected and dried. Grainer salt consists of small flakes rather than cubes and is preferred for certain uses in food processing. Sometimes the Alberger process is used, in which the brine is first partially evaporated in a vacuum evaporator then moved to a grainer. This process produces a mixture of flakes and cubes.

8 At this point salt used for most purposes is ready to be packaged in bags or boxes and shipped to consumers. To make iodized table salt, however, potassium iodide is added, then magnesium carbonate, calcium silicate, calcium phosphate, magnesium silicate, or calcium carbonate is added to make it free-flowing. The salt is then packaged and shipped to restaurants and grocery stores.

Quality Control

Specifications for salt vary widely according to the intended use. Salt intended for human consumption must be much purer than salt used for melting snow and ice, but salt used for certain scientific purposes may need to be even purer.

For most purposes, rock salt is allowed to have a gray, pink, or brown tinge rather than being pure white. The impurities that cause these colors may make up as much as 4% of a test sample. To test solubility, a 0.7-ounce (20 g) sample is placed in 6.8 fluid ounces (200 ml) of water. It should completely dissolve in no more than 20 minutes.

Evaporated salt intended for food processing is very pure, containing as much as 99.99% sodium chloride before additives are mixed in. This is important not only for safety and good taste, but because certain impurities can cause problems with certain foods. For example, small amounts of calcium tend to toughen vegetables. Traces of copper or iron tend to destroy vitamin C and to increase the rate at which fatty foods become rancid. In addition, calcium and magnesium both tend to make salt absorb more water, causing it to cake.

Health Aspects

Salt intake—or more precisely, sodium intake—is a controversial topic in health care today. Healthy adults can safely consume 0.2-0.4 ounces (6-11 g) of salt daily, which is equivalent to 0.08-0.14 ounces (2400-4400 mg) of sodium. For some people with high blood pressure, salt intake should be reduced. About one-third to one-half of all hypertensive people are salt-sensitive and will benefit from a low-sodium diet. Since there is no way to tell who these people are, most hypertensives under medical care will be placed on such a diet to see if it helps. A low-sodium diet usually aims to reduce sodium intake to less than 0.08 ounces (2400 mg) per day. While some have suggested that everyone should reduce salt intake, others point out that there is no evidence that salt restriction is of any benefit to otherwise healthy individuals.

Where To Learn More

Books

Adshead, Samuel A.M. *Salt and Civilization*. MacMillan, 1992.

Multhauf, Robert P. *Neptune's Gift*. Johns Hopkins, 1978.

Periodicals

Dornberg, John. "A 700-Year-Old Mine in Poland Is a Shrine to Human Ingenuity." *Smithsonian*, March 1994, pp. 96-106.

Vogel, Hans Ulrich. "The Great Well of China." *Scientific American*, June 1993, pp. 116-21.

Young, Gordon. "The Essence of Life: Salt." *National Geographic*, September 1977, pp. 380-401.

—Rose Secrest

Self-Adhesive Note

The self-adhesive, or sticky, note was created by scientists at the 3M company during the 1960s. It wasn't until 1980, however, that "Post-it" notes were sold to American consumers.

Background

Self-adhesive notes, also called "sticky notes," are partially adhesive, detachable note **paper**s that have revolutionized memo making. Throughout the country, it seems almost every surface in an office is amenable to the little yellow notes: they're stuck to desks, computers, files, folders, and rolodexes. Home is not off bounds either; the notes appear frequently on cabinet doors, telephones, and refrigerators. The tremendous appeal of sticky notes is in part because they are so easy and convenient to use. They can be placed exactly where you want them without any fasteners like tacks, paper clips, or staples. They are also reliable; they do not easily float away, nor do they leave glue stains or dents like paper clips do.

As pervasive as sticky notes are, they are a newcomer in the office supply market. The creators of the notes, Minnesota Mining & Manufacturing Company (3M), first distributed their well-known Post-it brand nationally in 1980. Four years later, it was the company's best-selling product. A decade later, sales of Post-it notes exceeded $100 million. By 1995, industry analysts estimated a sales of $500 million, one of the five best-selling office products in the world, along with Scotch tape, liquid paper, copy paper, and file holders. 3M still dominated the market well into the 1990s, keeping a tight rein on the company secret to making the unique note pads.

History

Sticky notes were a long time in the making, involving many people and obstacles. Ten years passed between the discovery of the adhesive and its application. During the mid-1960s, the 3M company was conducting a four-year program dubbed "Polymers for Adhesives." Spencer Silver, one of the participating chemists, became interested in a new family of polymers developed by Archer Daniels Midland, Inc. (ADM). Silver acquired the ADM monomers and performed an experiment in which he mixed an unusually large quantity of the element with the reaction mixture, totally contrary to established scientific principles. Ordinarily, this process, called polymerization catalysis, required mixing very precise ratios of the elements.

By defying scientific rationale, Silver discovered a totally unique phenomenon—a new polymer that was only partly sticky, not "aggressively" adhesive. This polymer would ultimately serve as the "tack" between the sticky note paper and other surfaces, holding them together but allowing easy separation without damaging either surface. Significantly, the polymer was also scientifically reliable, that is, the combination would always lead to the same result. Silver was fascinated by his discovery and investigated potential applications for it, but to no avail. The first product that Silver conceived was a sticky bulletin board that was covered with the adhesive.

After years of little success with the glue and the end of the "Plymers for Adhesives" program, Silver was transferred in 1970 to the System Research Group, where he met biochemist Robert Oliveira. They began investigating an application for this unusual glue. In 1974 Arthur Fry, a 3M corporate scientist, stumbled upon the perfect application for Silver's adhesive. One Sunday at his church, his bookmarks kept slipping out of the choir

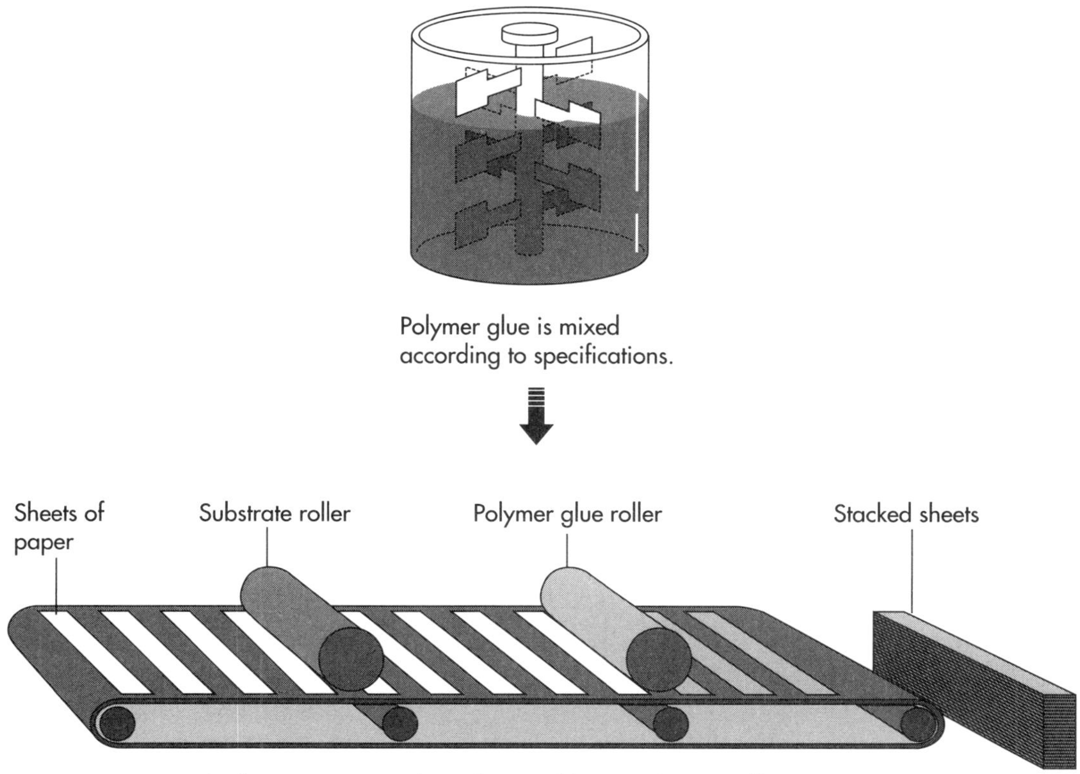

As the paper passes through a machine, continuous rollers prime the back of the paper with a substrate. Polymer glue is applied on top of the substrate. Another device stacks the sheets to form large pads.

hymnal. It occurred to him how convenient it would be to have a "permanently temporary" bookmark that would stick to the page, but could be removed without harming it. The sticky notes would have to have several properties: they would have to be moderately sticky on part of the back side, not sticky on the front, and they would have to be stacked into pads. One problem facing the researchers and engineers was that the polymer molecules tended to stick more to themselves than other objects. Thus, if two objects were joined by the adhesive and then separated, the glue would come off unevenly from the surfaces. The polymer required a substrate onto which it could adhere well. To address this problem, Henry Courtney and Roger Merril created a method of priming the substrate on the back of the paper, so that the glue would permanently stick to that surface.

Another hurdle to overcome was creating the machinery to apply the substrate and glue to the paper and then to stack the sticky paper into tear-off pads. It was not an easy task, since the company had for decades manufactured adhesive products which were dispensable in rolls. Fry faced opposition from many 3M designers and engineers, who were convinced the note pads could not be made. Fry began to work on the project in his home basement and within a few months, he had a machine that churned out the first tear-off pads. The company eagerly planned to move the machinery to the 3M plant, but with all of Fry's adjustments, the machine had grown too big to be carried out the door. Fry's basement wall was knocked out, and production of the pads soon began on a small scale at the 3M plant. They were designed in yellow, to contrast with the standard white documents, and they came in 1.5 by 2.0 inch (3.8 by 5.1 cm) and 3.0 by 5.0 inch (7.6 by 12.7 cm) pads. The "Press'n'Peel" pads were tested within the company, where staffers quickly became addicted to the notes. Jack Wilkins, who was the Commercial Tape marketing director at the time, remarked,

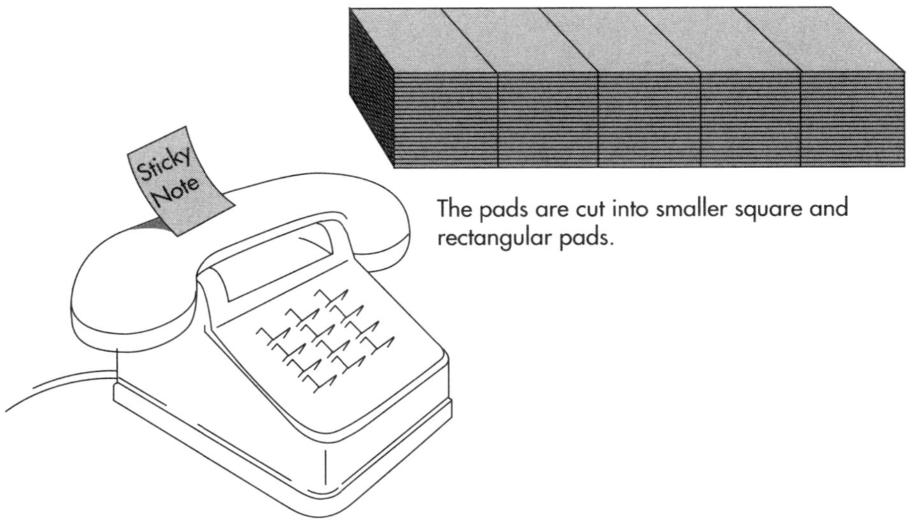

The pads are cut into smaller square and rectangular pads.

"Once people started using them it was like handing them marijuana. Once you start it, you can't stop."

In 1977, the first formal test markets for the note pads in Denver, Tulsa, Tampa, and Richmond were complete failures. The notes were a new concept and people simply did not know how to use the pads. The whole project was almost scrapped. Some people in the company, however, believed that the problem could be solved with personal demonstrations. They concentrated their efforts in Boise in 1978, with heavy advertising and lots of demonstrations. It worked, and orders came in immediately.

The following year, 3M changed the name from Press'n'Peel to Post-it notes. In 1980 Post-it notes went into national distribution, and four years later, the tear-off pads became 3M's most successful new product. A few years later, the note pads were developed in a variety of colors, sizes, and styles, including preprinted notes with headings such as "copy," "rush," and "FYI." Competitors inevitably began to market versions of the sticky notes as its popularity grew.

Raw Materials

The basic element in the self-adhesive note is, of course, paper. Yellow is the standard color, although assorted color pads are also available. The key ingredient is the unique polymer adhesive which makes the note pads "temporarily sticky" so they may be removed from a surface and reused. A machine applies the adhesive, stacks the paper, and cuts the paper pads to the specific sizes.

The Manufacturing Process

Preparing the glue

1 The polymer-based adhesive is mixed and placed into the machine which applies the glue to the paper.

Preparing the paper

2 For preprinted pads, the memo headings are printed onto large sheets of paper. The paper is placed into the machine and prepared to be fed through the priming device.

3 As the paper passes through the machine, continuous rollers prime the back side of the paper with a substrate. Next, the polymer glue is applied on top of the substrate.

Forming the pads

4 As the paper passes from the glue application, another device stacks the sheets of paper on top of each other to form large pads. Attached to the bottom of the pad is a piece of paper listing the company and brand names.

5 The pads are cut into smaller square and rectangular pads in the standard sizes, ranging from the smaller 1.5 by 2.0 inches (3.8 by 5.1 cm) to 8.5 by 11 (22 by 28 cm).

Packaging

6 The pads are labeled with the product name and manufacturer and wrapped in plastic. The pads are packaged together in bulk into larger boxes and shipped to distributors and wholesalers.

The Future

The future of sticky notes in the U.S. looks highly promising beyond the 1990s because of the product's usefulness in the office as well as at home. Although competitors began entering the market, 3M still dominated the category it had created. 3M also applied their polymer adhesive to more than 300 new applications, ranging from medical bandages to reusable interior-decorating kits.

Where To Learn More

Books

Gershman, Michael. *Getting It Right the Second Time.* Addison-Wesley, 1990, pp. 193-98.

Nayak, P. Ranganath and John M. Ketteringham. *Breakthroughs!* Pfeiffer, 1994.

Other

Kahn, Aron. "Post-It Notes Popping Up Everywhere." *Knight-Ridder/Tribune Business News*, September 14, 1993.

"3M Developing Electronic Post-it Notes." *Computergram International,* May 11, 1995, #2661.

—Audra Avizienis

Silk

The popularization of silk is credited to Chinese Empress Si-Ling, also referred to as the Goddess of the Silkworm, who apparently raised silkworms and designed a loom for making silk fabrics.

Background

Silk has set the standard in luxury fabrics for several millennia. The origins of silk date back to Ancient China. Legend has it that a Chinese princess was sipping tea in her garden when a cocoon fell into her cup, and the hot tea loosened the long strand of silk. Ancient literature, however, attributes the popularization of silk to the Chinese Empress Si-Ling, to around 2600 B.C. Called the Goddess of the Silkworm, Si-Ling apparently raised silkworms and designed a loom for making silk fabrics.

The Chinese used silk fabrics for arts and decorations as well as for clothing. Silk became an integral part of the Chinese economy and an important means of exchange for trading with neighboring countries. Caravans traded the prized silk fabrics along the famed Silk Road into the Near East. By the fourth century B.C., Alexander the Great is said to have introduced silk to Europe. The popularity of silk was influenced by Christian prelates who donned the rich fabrics and adorned their altars with them. Gradually the nobility began to have their own clothing fashioned from silk fabrics as well.

Initially, the Chinese were highly protective of their secret to making silk. Indeed, the reigning powers decreed death by torture to anyone who divulged the secret of the silkworm. Eventually, the mystery of the silk-making process was smuggled into neighboring regions, reaching Japan about A.D. 300 and India around A.D. 400. By the eighth century, Spain began producing silk, and 400 years later Italy became quite successful at making silk, with several towns giving their names to particular types of silk.

The first country to apply scientific techniques to raising silkworms was Japan, which produces some of the world's finest silk fabrics. Other countries that also produce quality silks are China, Italy, India, Spain, and France. China was the largest exporter of raw silk in the early 1990s, accounting for about 85% of the world's raw silk, worth about $800 million. Exports of China's finished silk products were about half of the world's total at about $3 billion.

Silk is highly valued because it possesses many excellent properties. Not only does it look lustrous and feel luxurious, but it is also lightweight, resilient, and extremely strong—one filament of silk is stronger then a comparable filament of steel! Although fabric manufacturers have created less costly alternatives to silk, such as nylon and **polyester**, silk is still in a class by itself.

Raw Materials

The secret to silk production is the tiny creature known as the silkworm, which is the caterpillar of the silk moth *Bombyx mori*. It feeds solely on the leaves of mulberry trees. Only one other species of moth, the *Antheraea mylitta,* also produces silk fiber. This is a wild creature, and its silk filament is about three times heavier than that of the cultivated silkworm. Its coarser fiber is called *tussah*.

The life cycle of the *Bombyx mori* begins with eggs laid by the adult moth. The larvae emerge from the eggs and feed on mulberry leaves. In the larval stage, the Bombyx is the caterpillar known as the silkworm. The silkworm spins a protective cocoon around itself so it can safely transform into a

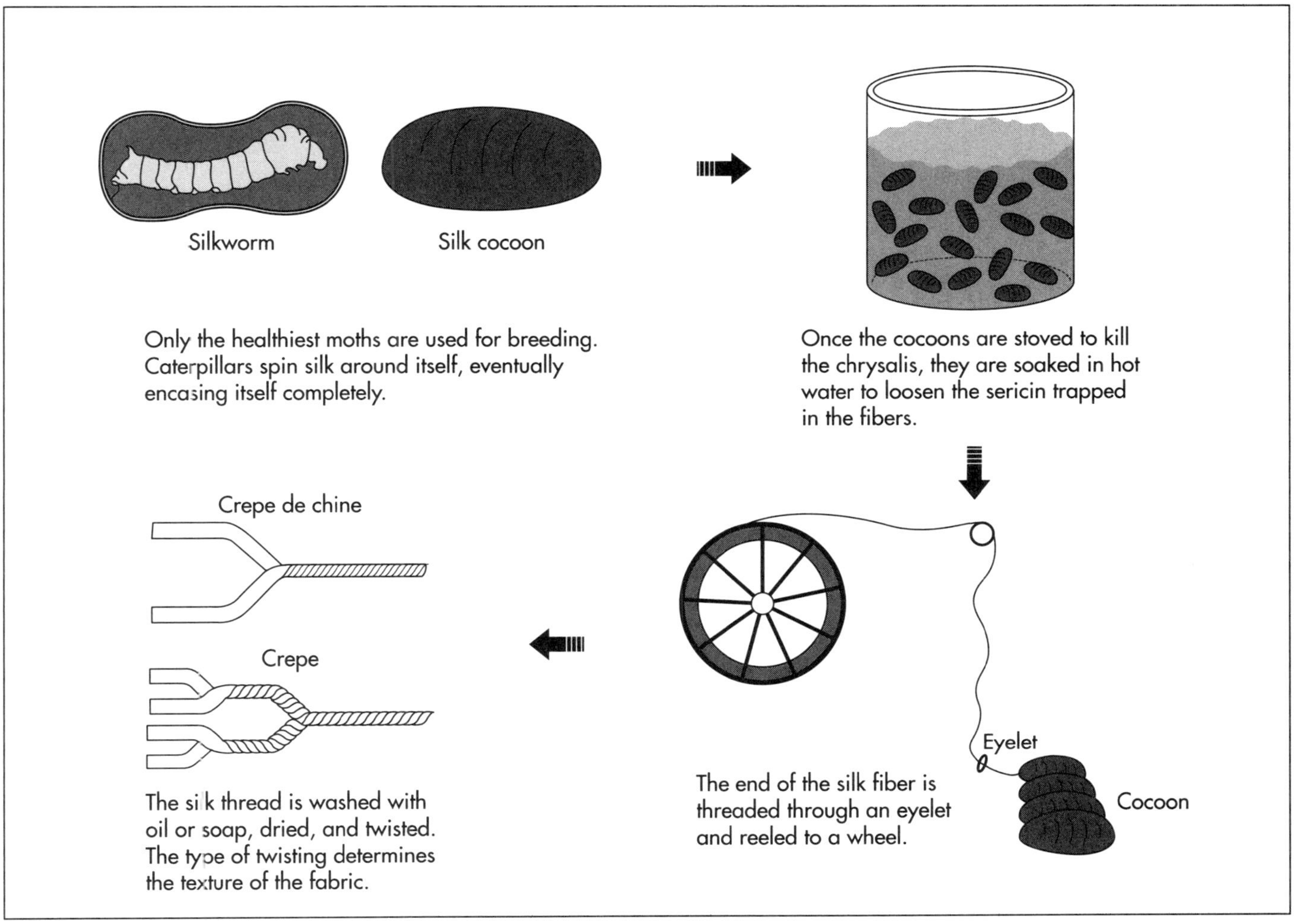

chrysalis. In nature, the chrysalis breaks through the cocoon and emerges as a moth. The moths mate and the female lays 300 to 400 eggs. A few days after emerging from the cocoon, the moths die and the life cycle continues.

The cultivation of silkworms for the purpose of producing silk is called sericulture. Over the centuries, sericulture has been developed and refined to a precise science. Sericulture involves raising healthy eggs through the chrysalis stage when the worm is encased in its silky cocoon. The chrysalis inside is destroyed before it can break out of the cocoon so that the precious silk filament remains intact. The healthiest moths are selected for breeding, and they are allowed to reach maturity, mate, and produce more eggs.

Generally, one cocoon produces between 1,000 and 2,000 feet of silk filament, made essentially of two elements. The fiber, called fibroin, makes up between 75 and 90%, and sericin, the gum secreted by the caterpillar to glue the fiber into a cocoon, comprises about 10-25% of silk. Other elements include fats, salts, and wax. To make one yard of silk material, about 3,000 cocoons are used.

Sericulture

Breeding silkworms

1 Only the healthiest moths are used for breeding. Their eggs are categorized, graded, and meticulously tested for infection. Unhealthy eggs are burned. The healthiest eggs may be placed in cold storage until they are ready to be hatched. Once the eggs are incubated, they usually hatch within seven days. They emerge at a mere one-eighth of an inch (3.2 mm) long and must be maintained in a carefully controlled envi-

The secret to silk production is the tiny creature known as the silkworm, which is the caterpillar of the silk moth *Bombyx mori*.

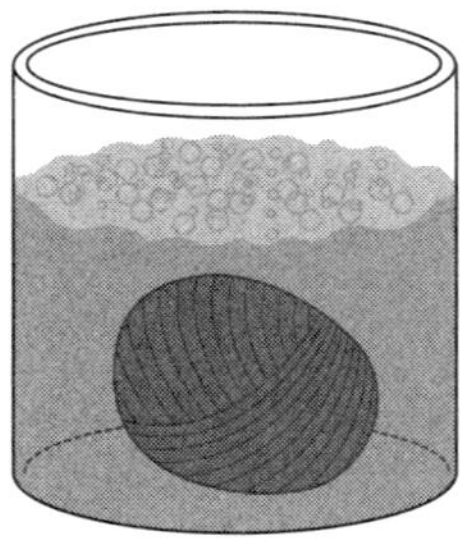

To create softness and shine, the remaining sericin is removed by soaking the yarn in warm, soapy water.

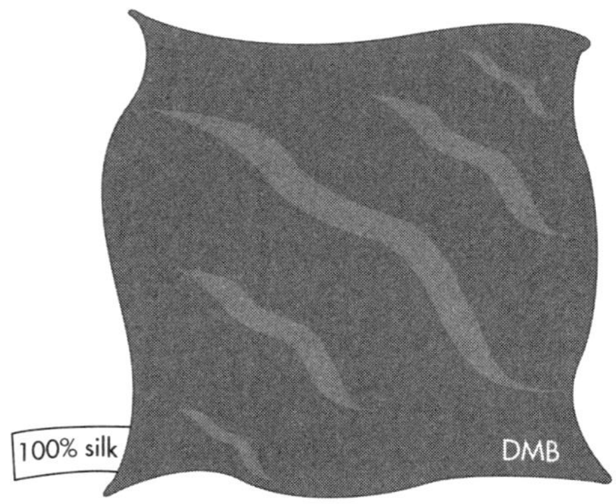

The yarn may be dyed before woven into fabric.

ronment. Under normal conditions, the eggs would hatch once a year in the spring when mulberry trees begin to leaf. But with the intervention of sericulturists, breeding can occur as many as three times per year.

Feeding the larva

2 The silkworms feed only on the leaves of the mulberry tree. The mulberry leaves are finely chopped and fed to the voracious silkworms every few hours for 20 to 35 days. During this period the worms increase in size to about 3.5 inches (8.9 cm). They also shed their skin, or molt, four times and change color from gray to a translucent pinkish color.

Spinning the cocoon

3 When the silkworm starts to fidget and toss its head back and forth, it is preparing to spin its cocoon. The caterpillar attaches itself to either a twig or rack for support. As the worm twists its head, it spins a double strand of fiber in a figure-eight pattern and constructs a symmetrical wall around itself. The filament is secreted from each of two glands called the spinneret located under the jaws of the silkworm. The insoluble protein-like fiber is called fibroin.

4 The fibroin is held together by sericin, a soluble gum secreted by the worm, which hardens as soon as it is exposed to air. The result is the raw silk fiber, called the bave. The caterpillar spins a cocoon encasing itself completely. It can then safely transform into the chrysalis, which is the pupa stage.

Stoving the chrysalis

5 The natural course would be for the chrysalis to break through the protective cocoon and emerge as a moth. However, sericulturists must destroy the chrysalis so that it does not break the silk filament. This is done by stoving, or stifling, the chrysalis with heat.

The Filature

Sorting and softening the cocoons

6 The filature is the factory in which the cocoons are processed into silk thread. In the filature the cocoons are sorted by various characteristics, including color and size, so that the finished product can be of uniform quality. The cocoons must then be soaked in hot water to loosen the sericin. Although the silk is about 20% sericin, only 1% is removed at this stage. This way the gum facilitates the following stage in which the filaments are combined to form silk thread, or yarn.

Reeling the filament

7 Reeling may be achieved manually or automatically. The cocoon is brushed to

locate the end of the fiber. It is threaded through a porcelain eyelet, and the fiber is reeled onto a wheel. Meanwhile, diligent operators check for flaws in the filaments as they are being reeled.

8 As each filament is nearly finished being reeled, a new fiber is twisted onto it, thereby forming one long, continuous thread. Sericin contributes to the adhesion of the fibers to each other.

Packaging the skeins

9 The end product, the raw silk filaments, are reeled into skeins. These skeins are packaged into bundles weighing 5-10 pounds (2-4 kg), called books. The books are further packaged into bales of 133 pounds (60 kg) and transported to manufacturing centers.

Forming silk yarn

10 Silk thread, also called yarn, is formed by throwing, or twisting, the reeled silk. First the skeins of raw silk are categorized by color, size, and quantity. Next they are soaked in warm water mixed with oil or **soap** to soften the sericin. The silk is then dried.

11 As the silk filaments are reeled onto bobbins, they are twisted in a particular manner to achieve a certain texture of yarn. For instance, "singles" consist of several filaments which are twisted together in one direction. They are turned tightly for sheer fabrics and loosely for thicker fabrics. Combinations of singles and untwisted fibers may be twisted together in certain patterns to achieve desired textures of fabrics such as crepe de chine, voile, or tram. Fibers may also be manufactured in different patterns for use in the nap of fabrics, for the outside, or for the inside of the fabric.

12 The silk yarn is put through rollers to make the width more uniform. The yarn is inspected, weighed, and packaged. Finally, the yarn is shipped to fabric manufacturers.

Degumming thrown yarn

13 To achieve the distinctive softness and shine of silk, the remaining sericin must be removed from the yarn by soaking it in warm soapy water. Degumming decreases the weight of the yarn by as much as 25%.

Finishing silk fabrics

14 After degumming, the silk yarn is a creamy white color. It may next be dyed as yarn, or after the yarn has been woven into fabric. The silk industry makes a distinction between pure-dye silk and what is called weighted silk. In the pure-dye process, the silk is colored with dye, and may be finished with water-soluble substances such as starch, glue, sugar, or gelatin. To produce weighted silk, metallic substances are added to the fabric during the dying process. This is done to increase the weight lost during degumming and to add body to the fabric. If weighting is not executed properly, it can decrease the longevity of the fabric, so pure-dye silk is considered the superior product. After dyeing, silk fabric may be finished by additional processes, such as bleaching, embossing, steaming, or stiffening.

Spun Silk

Not all of the silk filament is usable for reeled silk. The leftover silk may include the brushed ends or broken cocoons. This shorter staple silk may be used for spinning silk in a manner of fabrics like cotton and linen. The quality of spun silk is slightly inferior to reeled silk in that it is a bit weaker and it tends to become fuzzy. The waste material from the spun silk can also be used for making "waste silk" or "silk noil." This coarse material is commonly used for draperies and upholstery.

The Future

Sericulture is an ancient science, and the modern age has not brought great changes to silk manufacture. Rather, man-made fibers such as polyester, nylon, and acetate have replaced silk in many instances. But many of the qualities of silk cannot be reproduced. For example, silk is stronger than an equivalent strand of steel. Some recent research has focused on the molecular structure of silk as it emerges from the silkworm, in order to better understand how new, stronger artificial fibers might be constructed. Silk spun by the silkworm starts out as a liquid secretion. The liquid passes through a brief interim state with a semi-ordered molecular structure known as nematic liquid crystal, before it solidifies into a fiber. Materials scientists have been able to manufacture durable fibers using liquid crystal source material, but only at high temperatures or under extreme pressure. Researcher are continuing to study the silkworm to determine how liquid crystal is transformed into fiber at ordinary temperatures and pressures.

Where To Learn More

Books

Corbman, Bernard P. *Textiles: Fiber to Fabric.* 6th ed. McGraw-Hill, 1983.

Deshpande, Chris. *Silk.* Garrett Educational Corporation, 1995.

Parker, Julie. *All About Silk: A Fabric Dictionary & Swatchbook.* Rain City Publishing, 1992.

Scott, Philippa. *The Book of Silk.* Thames & Hudson, 1993.

Periodicals

"Chinese Exports of Silk Textiles." *Daily News Record,* August 23, 1994, p. 9.

Ostroff, Jim. "U.S. Textile, Apparel Firms Commend New China Pact." *Daily News Record,* January 19, 1994, p. 2.

Yanxi, Wang. "The Chinese Nonwovens Industry Marches towards the Year 2000." *Nonwovens Industry,* November 1993, p. 38.

—Audra Avizienis

Ski

Background

Although skiing is a popular sport today, the first skis were used as a quick and efficient means of transportation rather than for recreation. It is believed that skis originated in the Scandinavian nations about 5,000 years ago. Early inhabitants of Sweden, Norway, and Finland probably took the idea of the earlier snowshoe and used long femur bones from animals to make the first skis. The oldest extant pair of skis can be found in the Djugarden Museum in Stockholm, Sweden. Experts estimate the animal-bone skis are at least 4,000 years old. The early models closely resembled snowshoes, but other discovered skis—thought to be around 2,000 years old—appear similar to the ones we know today in their elongated shape and upward-curving front tip. Skis can be seen in pictographic paintings of the prehistoric era, and the first written mention of skis came around A.D. 1000 in the Norse sagas. The word "ski" is a Norwegian term for this early type of conveyance, but the Germanic and Latin root of the word means "to split"—the splitting of the bone into a pair of skis.

History

Bone skis were attached to the wearer via crude leather thongs, and early skiers proceeded without the benefit of poles. The use of skis evolved from an efficient means of winter transportation into a military tool with the Battle of Oslo in 1200, when Norwegian scouts were sent on skis to investigate the Swedish enemy camp. By the late Middle Ages skis were used regularly in battle, and soldiers were routinely issued skis or snowshoes. They were also popular with doctors, clergy, midwives, and others whose livelihood involved long treks through Scandinavian landscapes. Medieval skis were made from wood and were about 7.5 feet (2.3 m) long, 2 inches (5 cm) thick, and 5 inches (13 cm) wide. The prototype of today's ski boot was only a simple leather shoe, and it was attached to the ski itself with a leather or willow branch binding around the toes. The first heel strap was used in the 18th century by Norwegian soldiers in a ski unit, thus enabling them to ski downhill faster without losing control. They were also the first to use a pole.

Great regional differences existed in ski shape and length, with each Scandinavian village or hamlet making its own particular style, but one popular model widely used during the 19th century was the *Osterdal.* The Osterdal consisted of one short ski called the *andor* that was usually from 4-6 feet (1.2-1.8 m) in length. It was used to push off against the snow, and its bottom was sometimes covered with fur. The longer ski of the pair, reaching between 8 and 10 feet (2.4-3 m) in length, was used for gliding and was grooved on the bottom to guide it smoothly across the snow. It was also during the 19th century that skiing developed into a sport in Sweden and Norway, a change preceded by the invention of standard methods of turning and stopping. The sport's popularity eventually spread to Europe and the United States by the beginning of the 20th century. Scandinavian immigrants brought their enthusiasm for skiing to northern states such as Michigan and Minnesota as well as the western frontier. Many of these early skiers were Norwegian miners who are responsible for the first ski competitions and resorts in the U.S.

Skiing as a serious sport began in earnest around World War II. Downhill skiing had

The oldest extant pair of skis can be found in the Djugarden Museum in Stockholm, Sweden. Experts estimate the animal-bone skis are at least 4,000 years old.

While entitled "The Game of Ski," this 1892 article in Harper's Weekly *referred to the sport as "snow skating." (From the collections of Henry Ford Museum & Greenfield Village.)*

High in the Sierra Nevada Mountains between Placerville, California, and Carson City, Nevada, John A. Thomson, a native of Norway, delivered mail in the 1860s and 1870s. Thomson carried 100-pound packs as he made his way through the snow-covered ranges on what were referred to as "snowshoes" in those days. "Snowshoe" Thomson was a legend of the Sierras and one of the pioneers of American skiing. When he died at the age of 49 in 1876, friends erected his granite tombstone with a pair of snow skis carved on it.

Even before Thomson was drudging through the high Sierras, miners were careening down mountain slopes on "wooden wings." Tournaments were held in the Sierras as early as the 1850s with miners making bets of beer, silver dollars, and gold claims. Not long after, popular skiing competitions were being held in Alta, Utah; Aspen, Colorado; Berlin, New Hampshire; and Ishpeming, Michigan. Invented in Scandinavia and Europe, skiing was quickly and eagerly adopted by Americans in places where climate and geography lended themselves to winter sports. However, it was not until the 1890s that skiing became a true participatory sport, drawing people from all walks of life and regions of the country.

The 1930s were an era of rapid growth for the sport. Fueled by interest in the 1928 Winter Olympic Games in St. Moritz, Switzerland, American winter sports enthusiasts were exhilarated by hosting the 1932 Winter Olympics at Lake Placid, New York. Just the year before, the Boston and Main Railroad inaugurated a regular run of its ski train from urban centers to the skiing ranges of New England. In 1930 it is estimated that there were 75 ski clubs with 3,500 skiers; by 1940 there were over 2 million skiers in the United States.

William S. Pretzer

become popular in the Swiss Alps, especially after the "snowplow" method of slowing down was developed by Austrian enthusiast Mathias Zdarsky. Zdarsky also shortened the length of the still-wooden skis to 8 feet (2.4 m) and introduced a second pole for better balance in speedy downhill treks. Ski clubs sprang up in colleges around the northeastern U.S., and the sport was an integral part of the first Winter Olympics held in Chamonix, France, in 1924. Ski resorts also appeared in Vermont, New Hampshire, Colorado, California, and Idaho as well as in the Alps and Scandinavian countries. Their popularity increased even further after the development of the rope tow in 1932 and the chair lift five years later. In the decades after World War II, enthusiasm for skiing took on epic proportions and millions of enthusiasts joined in.

Despite the sport's increasing popularity, little had changed about the ski itself. Still constructed of light wood, usually hickory or ash, steel edges were introduced on the bottom to give them better glide, but a quickening depletion of inexpensive wood led to the development of skis made from more modern materials. Metal skis, especially the Dow Metal Air Ski, became common in the mid-1950s. This brand competed with the Truflex, developed by three men who worked together in the aircraft industry. These skis had no steel edges and often became stuck in wet snow because wax could not be used on their metal undersides. Aluminum was first used in skis by American skier Howard Head. The light metal was sandwiched around a wooden core and fused by glue and heat, but this aluminum underside froze easily. Head next fashioned a ski made from lightweight and flexible plastic, and added steel edges. The first pairs of Head skis were expensive but popular with part-time ski enthusiasts who found them easy to use, and within a matter of years the experts began using them as well. Meanwhile, bindings had evolved from simple leather straps to **iron** devises that attached the toe to the ski, leaving the heel free to move.

Raw Materials

In modern skis, the integral part of the unit is the inner core, which can be made from a variety of materials. When skis were constructed entirely of wood, the core's material was irrelevant. With the advent of metal, however, the core determined the strength and flexibility of the ski. Ski manufacturers and aficionados are split into two camps, one group preferring wood and the other foam as the material of choice. When using wood, manufacturing engineers must be extremely precise in matching the wood of each inner core in the pair. The weight, strength, and character of the wood must correspond precisely so that both right and left skis perform in the same manner at high speeds. Ash, beech, poplar, and okume are the most common types of wood used in skis.

Foam was first introduced as core material in the 1970s and yields a lighter ski than those with wooden cores. Foam core is more easily controlled in the manufacturing process and absorbs vibrations better than wood. It has the added advantage of being cheaper than wood. Most foam cores are made from polyurethane. A third type of material used in the core is aluminum. In skis with aluminum cores, the metal is fashioned into a honeycomb pattern. These cores are light and retain an excellent tensile strength from the aluminum, but are also more flexible than those with wood cores.

The outer part of the ski may be manufactured from a wide array of materials. Most common are **fiberglass**, carbon fibers, or a type of epoxy. The bottom part of the ski, the one designed for contact with the snow, is called the base. Polyethylene is the most popular material used in the bases of modern skis. One of the drawbacks of the polyethylene base is its softness, and with time the ski can become scratched by small stones and ice. A polyethylene candle is used by skiers and ski repair technicians to patch such scratches on the base. Additionally, because of its chemical nature, polyethylene is easily broken down by ultraviolet rays. This is remedied by applying a coat of wax to the skis after each use. Wax manufacturers make several different formulations of wax that are geared toward the type and temperature of the snow. The edges of skis are made of steel, which may be regular strength or hard tempered.

Design

Although a ski appears to be a rather simple piece of lightweight synthetic material, the factors involved in its design are based on complex principles of physics, engineering, and materials science. The first consideration is the weight and strength of the ski. It must be lightweight enough to glide easily across snow, yet strong enough to support the skier in the event of a sudden stop. The ski must also be waterproof and relatively resistant to damage from ice or rocks encountered at high speeds. Finally, the ski itself must have a permanent camber, or bend, to it. There are two types of camber: *bottom camber* refers to the arc of the ski as seen from the side, and its purpose is to evenly distribute the weight of the skier. The ski is narrower at the center in order to distribute the weight more efficiently. *Side camber* refers to the arc on each side of the ski as seen from above, designed to enable the skier to turn while in motion.

There are four basic types of skis. The first are downhill skis, used for speed skiing. They are longer, heavier, and wider than the other types of ski, and have less flexibility. A typical length is 87 inches (2.2 m). The binding is located towards the rear of the ski. The slalom ski is better for agility and quick turns. They are shorter and lighter than downhill skis, but are not recommended for high speeds. A common length of a slalom ski is 81 inches (2 m). The third type of ski, the giant slalom, combines the speed afforded by the downhill and the easy turns afforded by the slalom. The combination, or standard ski, is the generic model geared toward most skiers. All of the above skis are used in Alpine, or downhill skiing. Cross-country skis are called Nordic and are shorter and designed much differently.

Skis are manufactured in a variety of lengths, measured in centimeters. The length of the ski depends on the height and experience of the skier, and the type of terrain on which it will be used. For Alpine skiing, there are several different models—cruising skis, for taking long turns downhill at a high rate of speed; bump skis, designed to travel effortlessly over clumps of snow known as moguls, and powder skis, for cutting through deep snow. The categories are combined in the all-around ski, designed for use in all of the above situations.

Other factors guide ski design. The type of snow on which the ski will be used is a cru-

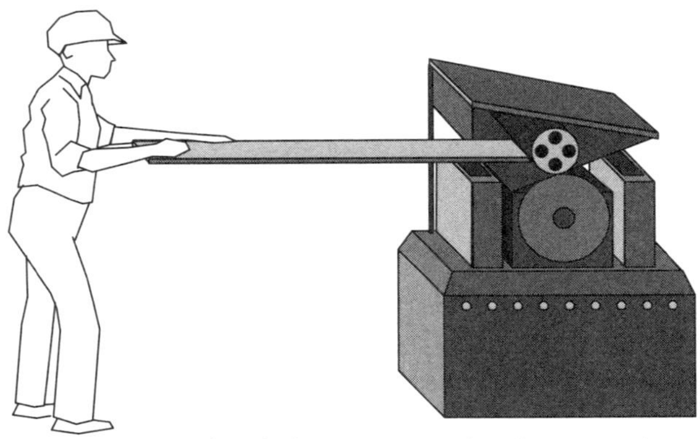

For wood and aluminum core skis, the material is put through a computer-operated mill for precise cutting and finishing.

cial element. Harder, heavy natural snows, characteristic of mountainous regions, require a tougher ski with less flexibility. Man-made snow, common to ski resorts where the climate does not yield a sufficient snowfall, requires less hardy skis. The skill of the skier is also an important factor. A professional skier will have excellent control of the dynamics of the ski and the sport, and looks for a ski that will help him move faster through the snow. A beginning skier requires a ski that is easy to manage.

The design of the ski's core is also important, since the core determines the amount of vibration the skier will feel. The speed at which the ski travels downhill or across terrain results in vibrations that affect the skier. Too much vibration tires the skier out and makes it hard to control the skis. To eliminate this, ski engineers attempt to design an inner construction that absorbs as much of the vibration as possible without sacrificing the life of the ski. If all vibration was eradicated, the ski would perform poorly so a right balance must be achieved. A ski will also be grooved in one or two lines along its base. This enables it to maintain a straight line when gliding through the snow.

The Manufacturing Process

Modern skis can be classified into three different types: laminated, torsion-box, and single-shell. The laminated method is the one most commonly used in ski manufacturing, since the combination of various materials allows for greater array of models. Laminated skis may contain one or all of the following materials: plastic, fiberglass, carbon, wood, steel, aluminum, neoprene, or ceramics. These elements are sandwiched or layered on the top and bottom of the core. In a torsion-box ski, the core material is surrounded by fiberglass or carbon fibers. Instead of being sandwiched around the core, the layers wrap around it in what is known as "wet wrap construction." The core is then sealed with resin and heat. This more complex manufacturing process means that torsion-box skis are more expensive than laminated ones, but they provide better handling and thus are more appealing to serious skiers. Single-shell skis are made from a strong inner material such as wood, but enclosed by a flexible fiberglass or plastic shell and sides. The lighter weight of single-shell skis means that more control is located at the tip of the skis, giving the user better steering control and turning ability.

Milling the core

1 The manufacturing process for most skis originates in a warehouse-type room of the factory. Here, all the raw materials—the roughly-cut blocks of wood for wood cores, the steel edges, sheets of polyethylene for the bases, rolls of fiberglass or aluminum—are sorted and stacked. For skis with wood cores, the pre-laminated wood is

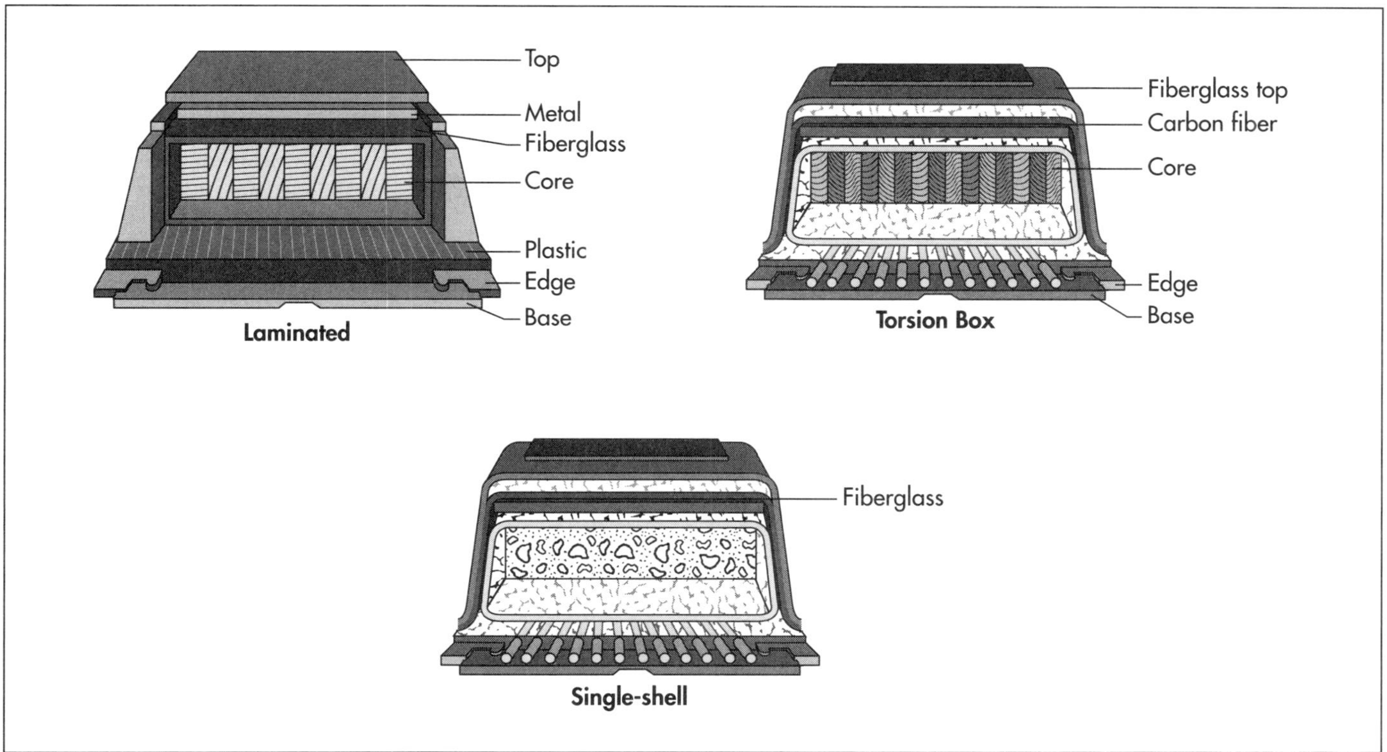

Modern skis can be classified into three different types. In laminated skis, various layers sandwich the core on top and bottom. In torsion-box ski construction, the layers wrap around the core. In single-shell construction, polyurethane foam is sometimes injected into the ski shell. As this foam expands and hardens, the top and sides form one sheet.

put through a mill for precise cutting and finishing. These mills are computer-operated by a technician who sets the parameters through numerical controls on the machine. This process is also used for aluminum cores. The machine cuts and mills the material, including the core's thickness and sidecut, and also collects the excess, which are returned to the suppliers. In skis with polyurethane cores, the top and base layers are put into a mold, then a press, and the foam is injected with a hose. As it expands and hardens, the foam fuses the layers and becomes the core.

Assembling the layers

2 In the next step, the core and the rest of the layers, including the top and base, are placed into a mold and then into a press. Heat and pressure result in a rudimentary ski, then epoxy resin is used to completely seal the layers together. Single-shell, or "cap" skis, require a more complicated assembly process and more precise molding chambers. They still use wood or foam cores, but the top and sides are united in one sheet and this makes the core less integral to the ski's overall performance.

Bonding the base and edges

3 The polyethylene base, the running surface of the ski, is bonded onto the ski. Steel edges are also bonded onto the ski surface.

Applying graphics

4 Another important difference between the laminated and torsion-box types of skis and single-shell ones is the application of graphics. Flashy colors and logos are a trademark of ski design, and manufacturers use graphics to differentiate their models. In single-shell skis, graphics are applied to the top, or cap, before the parts go into the mold. The heat-transfer process is commonly used, a method that can best be compared to the way an emblem is applied to a **t-shirt** with an iron. In the ski industry, this process is called sublimation, and it yields a visually complex but clear image. Another method used in single-shell graphics applications is backprinting, in which graphics are applied in reverse to a clear sheet through silk-screening, then flipped over. During assembly, the freshly applied graphics remain protected during the molding stage.

Graphics may be added using a silkscreen process. Next, a coat of laquer is applied to protect the design.

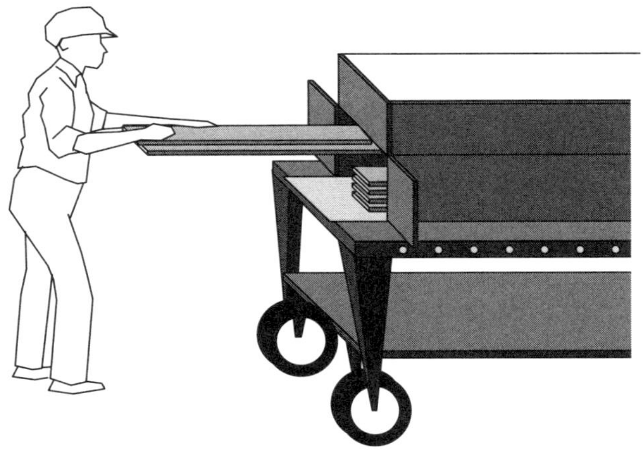

The finished ski is then put through a sanding and grinding machine. The skis are matched into pairs and waxed.

With traditional laminated and torsion-box skis, the graphics are applied in one of the final steps of the manufacturing process. The actual ski goes through a silkscreen process, where the design is first put on a piece of silk or other thin fabric. The areas which are not to be colored are covered with an impermeable substance, and then ink is forced through the fabric. This must be done several times over, one press for each color, and the ski must be allowed to dry between presses. A curtain coat of lacquer is then applied. Because of the complexities of this process, skis made this way often take days to dry, whereas single-shell skis can be finished in a matter of hours.

Finishing

5 The final finishing process is nearly the same among all types of skis. The bases must be put through a machine that grinds and polishes them, and this is done with a combination of belt sanding and stone grinding. Stone grinding is considered superior to the belt method. Next, the skis must be matched up into pairs. The edges are also beveled and polished. This is done with a machine that tests their flexibility and camber. After they are paired, a quality control technician makes certain that they are well-matched. In the final stages, the steel edges are oiled, the skis waxed, wrapped in polyethylene, and boxed for shipping to retail outlets.

Quality Control

Although ski factories employ technicians who check the skis during each stage of the manufacturing process, much of the testing is done after the skis arrive in stores. It often takes some time for all the synthetic materials to set properly, and the surface of the ski may change during shipping. This process by which skis are readied for the slopes is called tuning, and is performed by a ski technician, or "tuner," employed in retail ski outlets or pro shops. The tuner uses files to make the base of the ski as flat as possible. The steel edges are further beveled, because the base of the ski will shrink when it hits the snow. The tips and tails are again sanded and waxed for protection and better glide.

The Future

Experts predict that ski manufacturing will become more and more geared toward the production of the single-shell ski. Its more cost-efficient methods and improved performance are the primary reasons for this forecast—single-shells are lighter and handle vibration more effectively. Although retooling factories is expensive, the single-shell manufacturing process takes only a few hours—compared to days for laminated and torsion-box construction—and requires less workers. In time, these factors will reduce the cost of the ski. Environmental concerns are also reduced by single-shell production. Ski factories both in the United States and in Europe must adhere to increasingly stringent government regulations in regard to the exposure of workers to chemicals and noise, as well as the reduction of waste during the manufacturing process. In the manufacture of single-shell skis, these elements are greatly diminished. In all types of ski factories, new mills and assembly machines now collect excess materials for return to suppliers, and modern grinding machines are quieter and safer. Robotics technology is also increasingly used in the manufacturing process.

Where To Learn More

Books

Masia, Seth. *The Ski Maintenance and Repair Handbook.* Contemporary Books, 1982.

Ski Magazine's Encyclopedia of Skiing. Harper & Row, 1978.

Sports Illustrated Skiing. Time/Harper & Row, 1985.

Periodicals

Chalsma, Jennifer K. and Kenneth J. Korane. "Engineering the Ultimate Ski." *Machine Design*, February 7, 1991, pp. 26-32.

Grout, Dill. "Capped Crusaders." *Skiing*, October 1993, pp. 138-44.

—Carol Brennan

Smoke Detector

There are two basic types of smoke detectors: photoelectric smoke detectors which use an optical beam to search for smoke and the ionization chamber smoke detector (ICSD) which employs a radioactive material to ionize the air in a sensing chamber.

Background

A smoke detector is a device that senses the presence of smoke in a building and warns the occupants, enabling them to escape a fire before succumbing to smoke inhalation or burns. Equipping a home with at least one smoke detector cuts in half the chances that the residents will die in a fire. In 1992 the readers of *R&D Magazine* selected home smoke alarms as one of the "30 Products that Changed Our Lives." Smoke detectors became widely available and affordable in the early 1970s. Prior to that date, fatalities from fires in the home averaged 10,000 per year, but by the early 1990s the figure dropped to fewer than 6,000 per year.

Two basic types of smoke detectors are currently manufactured for residential use. The photoelectric smoke detector uses an optical beam to search for smoke. When smoke particles cloud the beam, a photoelectric cell senses the decrease in light intensity and triggers an alarm. This type of detector reacts most quickly to smoldering fires that release relatively large amounts of smoke.

The second type of smoke detector, known as an ionization chamber smoke detector (ICSD), is quicker at sensing flaming fires that produce little smoke. It employs a radioactive material to ionize the air in a sensing chamber; the presence of smoke affects the flow of the ions between a pair of electrodes, which triggers the alarm. Between 80 and 90% of the smoke detectors in American homes are of this type. Although most residential models are self-contained units that operate on a 9-volt battery, construction codes in some parts of the country now require installations in new homes to be connected to the house wiring, with a battery backup in case of a power failure.

The typical ICSD radiation source emits alpha particles that strip electrons from the air molecules, creating positive oxygen and nitrogen ions. In the process, the electrons attach themselves to other air molecules, forming negative oxygen and nitrogen ions. Two oppositely charged electrodes within the sensing chamber attract the positive and negative ions, setting up a small flow of current in the air space between the electrodes. When smoke particles enter the chamber, they attract some of the ions, disrupting the current flow. A similar reference chamber is constructed so that no smoke particles can enter. The smoke detector constantly compares the current flow in the sensing chamber to the flow in the reference chamber; if a significant difference develops, an alarm is triggered.

History

The development of these life-saving appliances began in 1939 when Ernst Meili, a Swiss physicist, devised an ionization chamber device capable of detecting combustible gases in mines. The real breakthrough was Meili's invention of a cold-cathode tube that could amplify the small electronic signal generated by the detection mechanism to a strength sufficient to activate an alarm.

Although ionization chamber smoke detectors have been available in the United States since 1951, they were initially used only in factories, warehouses, and public buildings because they were expensive. By 1971 residential ICSDs were commercially available;

they cost about $125 per detector and sold at a rate of a few hundred thousand per year.

A flurry of new technological developments occurred over the next five years, reducing the cost of the detectors by 80% and boosting sales to 8 million in 1976 and 12 million in 1977. By this time, solid-state circuitry had replaced the earlier cold-cathode tube, significantly reducing the size of the detectors as well as their cost. Design refinements, including more energy-efficient alarm horns, enabled the use of commonly available sizes of batteries rather than the hard-to-find specialty batteries that had previously been required. Improvements in the circuitry made it possible to monitor both the decrease in voltage and the build-up of internal resistance in the battery, either of which would trigger a signal to replace the power source. The new generation of detectors could also function with smaller amounts of radioactive source material, and the sensing chamber and smoke detector enclosure were redesigned for more effective operation.

Raw Materials

An ICSD smoke detector is composed of a housing made of polyvinylchloride or polystyrene plastic, a small electronic alarm horn, a **printed circuit board** with an assortment of electronic components, and a sensing chamber and reference chamber, each containing a pair of electrodes and the radioactive source material.

Americium 241 (Am-241), a radioactive isotope, has been the preferred source material for ICSDs since the late 1970s. It is very stable and has a half-life of 458 years. It is usually processed with gold and sealed within gold and silver foils.

The Manufacturing Process

The production of a smoke detector consists of two major steps. One is fabrication of the Am-241 into a form (typically a foil) that can be installed into the sensing and reference chambers. The other is assembly of the entire ICSD, beginning either with all of the individual components or with prefabricated sensing and reference chambers obtained from the manufacturer of the radioactive source material. The following description covers all steps, even though some may be done by different manufacturers. Tests and inspections at several stages of the assembly process ensure a reliable product.

Radioactive source

1 The process begins with the compound AmO_2, an oxide of Am-241. This substance is thoroughly mixed with gold, shaped into a briquette, and fused by pressure and heat at over 1470°F (800°C). A backing of silver and a front covering of gold or gold alloy are applied to the briquette and sealed by hot forging. The briquette is then processed through several stages of cold rolling to achieve the desired thickness and levels of radiation emission. The final thickness is about 0.008 inches (0.2 mm), with the gold cover representing about one percent of the thickness. The resulting foil strip, which is about 0.8 inches (20 mm) wide, is cut into sections 39 inches (1 meter) long.

2 Circular ICSD source elements are punched out of the foil strip. Each disc, which is about 0.2 inches (5 mm) in diameter, is mounted in a metal holder. A thin metal rim on the holder is rolled over to completely seal the cut edge around the disc.

The sensing and reference chambers

3 One disc of source material is mounted in the sensing chamber and another is mounted in the adjacent reference chamber. The electrodes are installed in both chambers and connected to external leads which project out of the bottoms of the chambers.

The circuit board

4 Printed circuit boards are prepared from design schematics by punching holes for the component leads and by laying a copper trace on the back to form the paths for electric currents. On the assembly line, the various electronic components (diodes, capacitors, resistors, etc.) are inserted into the proper holes on the board. Leads extending out the back of the board are trimmed.

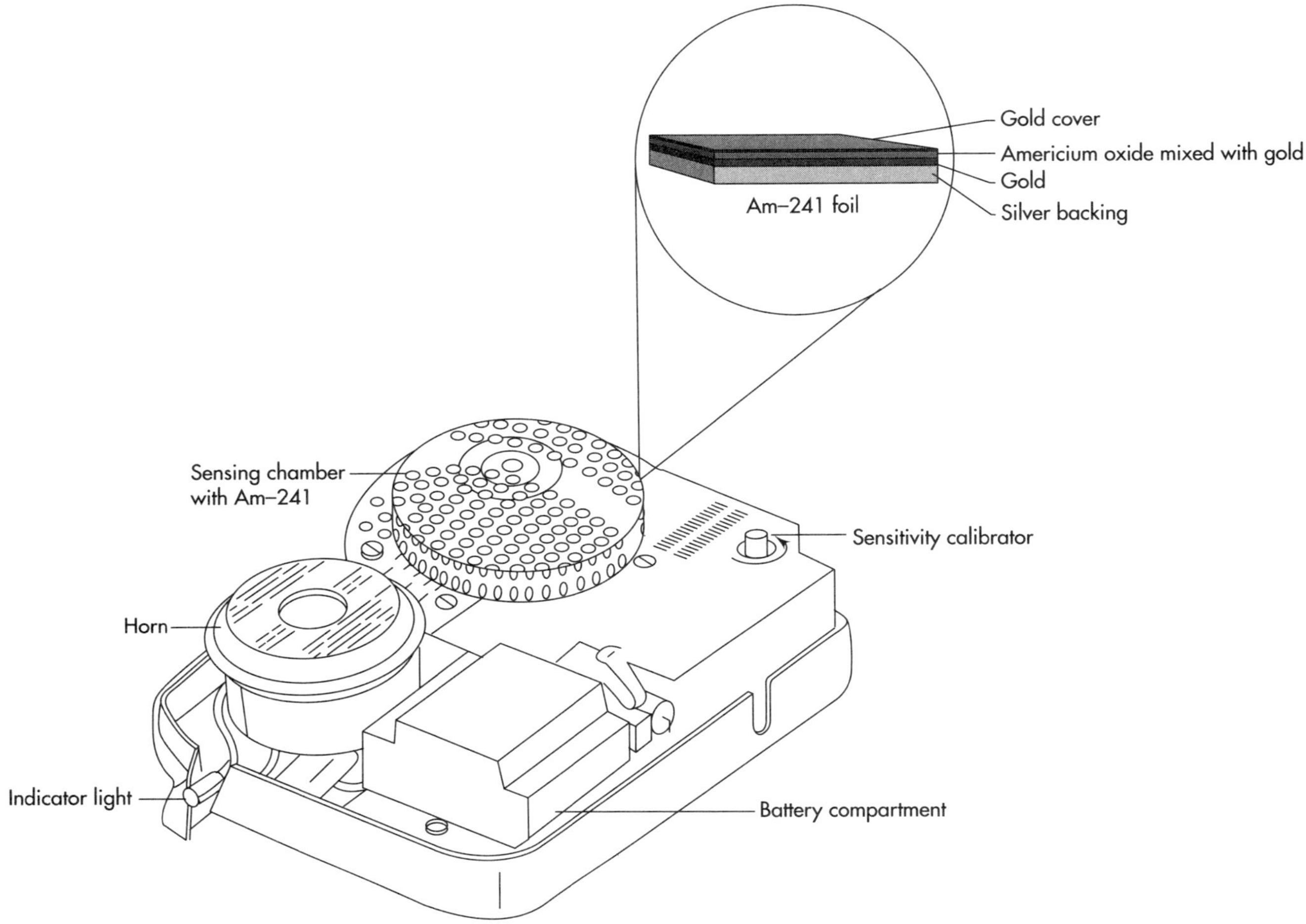

The most common type of smoke detector, ICSD is equipped with an alarm horn, a printed circuit board, and a sensing and reference chamber containing radioactive source material.

5 The sensing chamber, reference chamber, and an alarm horn are installed on the printed circuit board.

6 The board then passes over a wave solder machine, which solders the electronic components into place.

Housing

7 The plastic housing consists of a mounting base and a cover. Both are made by injection molding process in which powdered plastic and molding pigments are mixed, heated, forced into a mold under pressure, then cooled to form the final pieces.

Final assembly

8 The circuit board is seated on the plastic mounting base. A test button is installed so the device can be tested periodically after installation in the home. A mounting bracket is added to the base, and the cover is added to complete the assembly.

9 The smoke detector is packaged in a cardboard box, along with a battery and an owner's manual.

New Developments

Some recent developments may make smoke detectors even more effective. One model, for example, uses a strobe light alarm to alert hearing-impaired people of danger. The remote strobe light can be mounted in a bedroom even though the detector may be located in another room or hallway, giving the same advantage of early warning available to hearing people when an alarm sounds from outside the bedroom.

In 1993 Newtron Products redesigned a traditional smoke detector to fit in the standard air filters of a central heating or air conditioning system in order to examine air that circulates through an entire building. When it detects smoke, the device shuts off the system's blower to prevent the air flow from helping spread the smoke and fire. In addition, it triggers an alarm that resonates through the duct work and is audible anywhere in the building.

Another kind of fire detector may utilize sound. Investigators at the Building and Fire Research Laboratory of the National Institute of Standards and Technology have found that various types of housing materials, such as wood, plastic, and **drywall**, make identifiable sounds as they expand from rapid heating. Piezoelectric transducers can detect those sounds even before the materials actually begin to burn. This would be especially helpful in detecting incipient fires caused by overheated electrical wiring within a building's walls.

Where To Learn More

Books

Belanger, R., D.W. Buckley, and J.B. Swenson. *Environmental Assessment of Ionization Chamber Smoke Detectors Containing Am-241*. Science Applications Inc., November 1979.

Bukowski, Richard W., and G.W. Mulholland. *Smoke Detector Design and Smoke Properties*. National Bureau of Standards, National Engineering Laboratory, Center for Fire Research, 1978.

Periodicals

Andrews, Edmund L. "Central System for Smoke Detection." *The New York Times*, February 1, 1993, p. D2.

"Sounds Like Fire." *Discover*, May 1994, p. 16

"Listening for Hidden Fires." *Science News*, July 24, 1993, p. 63.

"Smoke Detectors: Essential for Safety." *Consumer Reports*, May 1994, pp. 336-39.

—Loretta Hall

Soap

Manufacture of soap began in England around the end of the 12th century. Because soapmakers had to pay a heavy tax on all the soap they produced, soap was a luxury item, and it did not come into common use until after the tax was repealed in 1853.

Background

Soap is a combination of animal fat or plant oil and caustic soda. When dissolved in water, it breaks dirt away from surfaces. Through the ages soap has been used to cleanse, to cure skin sores, to dye hair, and as a salve or skin ointment. But today we generally use soap as a cleanser or **perfume**.

The exact origins of soap are unknown, though Roman sources claim it dates back to at least 600 B.C., when Phoenicians prepared it from goat's tallow and wood ash. Soap was also made by the Celts, ancient inhabitants of Britain. Soap was used widely throughout the Roman empire, primarily as a medicine. Mention of soap as a cleanser does not appear until the second century A.D. By the eighth century, soap was common in France, Italy, and Spain, but it was rarely used in the rest of Europe until as late as the 17th century.

Manufacture of soap began in England around the end of the 12th century. Soapmakers had to pay a heavy tax on all the soap they produced. The tax collector locked the lids on soap boiling pans every night to prevent illegal soap manufacture after hours. Because of the high tax, soap was a luxury item, and it did not come into common use in England until after the tax was repealed in 1853. In the 19th century, soap was affordable and popular throughout Europe.

Early soap manufacturers simply boiled a solution of wood ash and animal fat. A foam substance formed at the top of the pot. When cooled, it hardened into soap. Around 1790, French soapmaker Nicolas Leblanc developed a method of extracting caustic soda (sodium hydroxide) from common table **salt** (sodium chloride), replacing the wood ash element of soap. The French chemist Eugene-Michel Chevreul put the soap-forming process (called in English saponification) into concrete chemical terms in 1823. In saponification, the animal fat, which is chemically neutral, splits into fatty acids, which react with alkali carbonates to form soap, leaving glycerin as a byproduct. Soap was made with industrial processes by the end of the 19th century, though people in rural areas, such as the pioneers in the western United States, continued to make soap at home.

Raw Materials

Soap requires two major raw materials: fat and alkali. The alkali most commonly used today is sodium hydroxide. Potassium hydroxide can also be used. Potassium-based soap creates a more water-soluble product than sodium-based soap, and so it is called "soft soap." Soft soap, alone or in combination with sodium-based soap, is commonly used in shaving products.

Animal fat in the past was obtained directly from a slaughterhouse. Modern soapmakers use fat that has been processed into fatty acids. This eliminates many impurities, and it produces as a byproduct water instead of glycerin. Many vegetable fats, including olive oil, palm kernel oil, and coconut oil, are also used in soap making.

Additives are used to enhance the color, texture, and scent of soap. Fragrances and perfumes are added to the soap mixture to

The above illustrations show the kettle process of making soap.

cover the odor of dirt and to leave behind a fresh-smelling scent. Abrasives to enhance the texture of soap include talc, silica, and marble pumice (volcanic ash). Soap made without dye is a dull grey or brown color, but modern manufacturers color soap to make it more enticing to the consumer.

The Manufacturing Process

The kettle method of making soap is still used today by small soap manufacturing companies. This process takes from four to eleven days to complete, and the quality of each batch is inconsistent due to the variety of oils used. Around 1940, engineers and scientists developed a more efficient manufacturing process, called the continuous process. This procedure is employed by large soap manufacturing companies all around the world today. Exactly as the name states, in the continuous process soap is produced continuously, rather than one batch at a time. Technicians have more control of the production in the continuous process, and the steps are much quicker than in the kettle method—it takes only about six hours to complete a batch of soap.

The Kettle Process

Boiling

1 Fats and alkali are melted in a kettle, which is a steel tank that can stand three stories high and hold several thousand pounds of material. Steam coils within the kettle heat the batch and bring it to a boil. After boiling, the mass thickens as the fat reacts with the alkali, producing soap and glycerin.

Salting

2 The soap and glycerin must now be separated. The mixture is treated with salt, causing the soap to rise to the top and the glycerin to settle to the bottom. The glycerin is extracted from the bottom of the kettle.

Strong change

3 To remove the small amounts of fat that have not saponified, a strong caustic solution is added to the kettle. This step in the process is called "strong change." The mass is brought to a boil again, and the last of the fat turns to soap. The batch may be given another salt treatment at this time, or the manufacturer may proceed to the next step.

Pitching

4 The next step is called "pitching." The soap in the kettle is boiled again with added water. The mass eventually separates into two layers. The top layer is called "neat soap," which is about 70% soap and 30% water. The lower layer, called "nigre," contains most of the impurities in the soap such as dirt and salt, as well as most of the water. The neat soap is taken off the top. The soap is then cooled. The finishing process is the

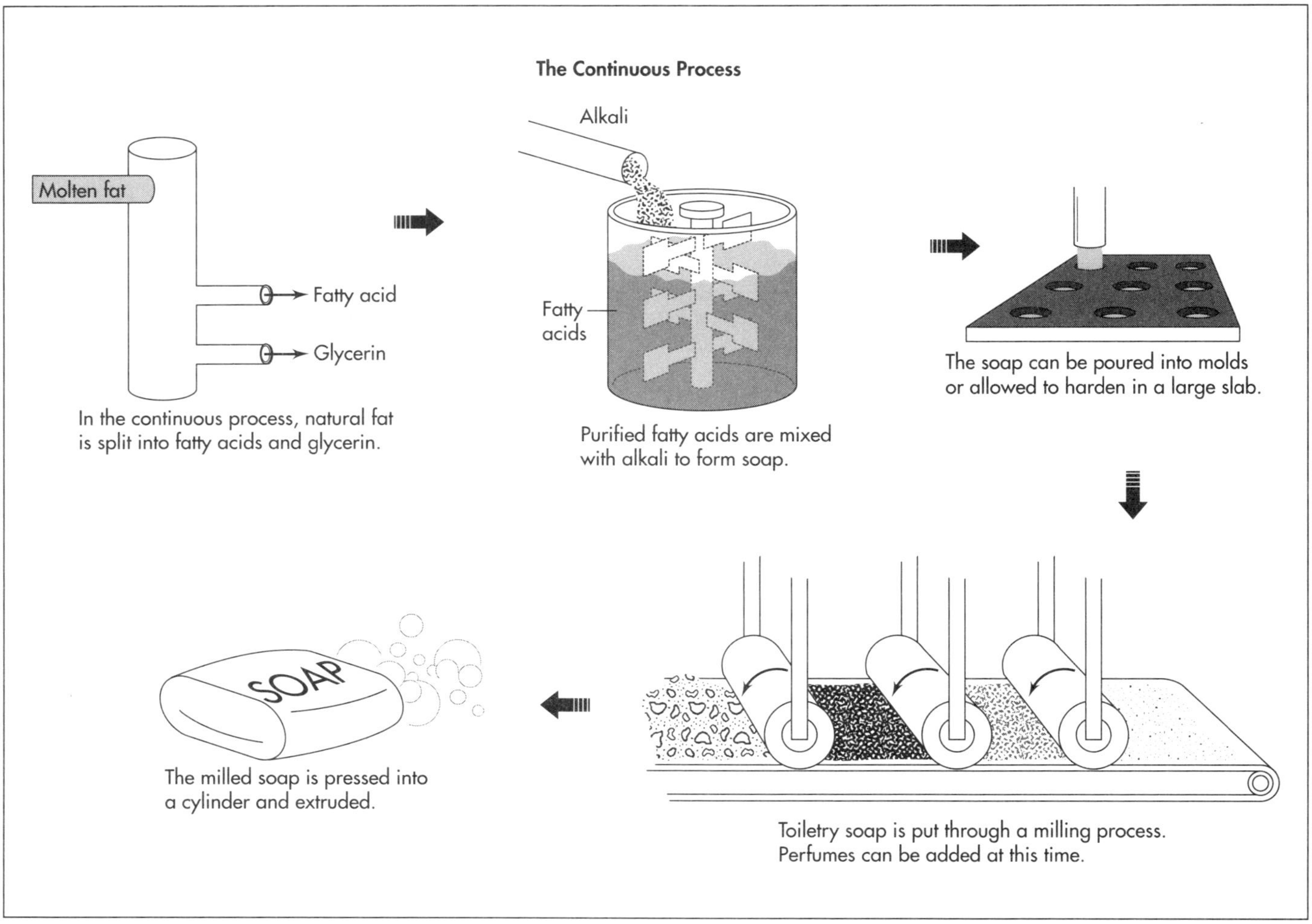

Developed around 1940 and used by today's major soap-making companies, the above illustrations show the continuous process of making soap.

same as for soap made by the continuous process.

The Continuous Process

Splitting

1 The first step of the continuous process splits natural fat into fatty acids and glycerin. The equipment used is a vertical stainless steel column with the diameter of a barrel called a hydrolizer. It may be as tall as 80 feet (24 m). Pumps and meters attached to the column allow precise measurements and control of the process. Molten fat is pumped into one end of the column, while at the other end water at high temperature (266°F [130°C]) and pressure is introduced. This splits the fat into its two components. The fatty acid and glycerin are pumped out continuously as more fat and water enter. The fatty acids are then distilled for purification.

Mixing

2 The purified fatty acids are next mixed with a precise amount of alkali to form soap. Other ingredients such as abrasives and fragrance are also mixed in. The hot liquid soap may be then whipped to incorporate air.

Cooling and finishing

3 The soap may be poured into molds and allowed to harden into a large slab. It may also be cooled in a special freezer. The slab is cut into smaller pieces of bar size, which are then stamped and wrapped. The entire continuous process, from splitting to finishing, can be accomplished in several hours.

Milling

4 Most toiletry soap undergoes additional processing called milling. The milled

bar lathers up better and has a finer consistency than non-milled soap. The cooled soap is fed through several sets of heavy rollers (mills), which crush and knead it. Perfumes can best be incorporated at this time because their volatile oils do not evaporate in the cold mixture. After the soap emerges from the mills, it is pressed into a smooth cylinder and extruded. The extruded soap is cut into bar size, stamped and wrapped.

Byproducts

Glycerin is a very useful byproduct of soap manufacture. It is used to make hand lotion, drugs, and nitroglycerin, the main component of explosives such as **dynamite**.

Where To Learn More

Books

Cavitch, Susan M. *The Natural Soap Book: Making Herbal and Vegetable-Based Soaps.* Storey Communications, 1995.

Maine, Sandy. *The Soap Book: Simple Herbal Recipes.* Interweave Press, 1995.

Spitz, Luis, ed. *Soap Technologies in the 1990s.* American Oil Chemists Society, 1990.

Other

About Soap. Procter & Gamble, 1990. (513) 983-1100.

—Sheila Dow

Soft Drink

The world's most famous soft drink, Coca-Cola was invented by Atlanta pharmacist John Pemberton in 1886.

Background

Soft drinks are enormously popular beverages consisting primarily of carbonated water, sugar, and flavorings. Nearly 200 nations enjoy the sweet, sparkling soda with an annual consumption of more than 34 billion gallons. Soft drinks rank as America's favorite beverage segment, representing 25% of the total beverage market. In the early 1990s per capita consumption of soft drinks in the U.S. was 49 gallons, 15 gallons more than the next most popular beverage, water.

The roots of soft drinks extend to ancient times. Two thousand years ago Greeks and Romans recognized the medicinal value of mineral water and bathed in it for relaxation, a practice that continues to the present. In the late 1700s Europeans and Americans began drinking the sparkling mineral water for its reputed therapeutic benefits. The first imitation mineral water in the U.S. was patented in 1809. It was called "soda water" and consisted of water and sodium bicarbonate mixed with acid to add effervescence. Pharmacists in America and Europe experimented with myriad ingredients in the hope of finding new remedies for various ailments. Already the flavored soda waters were hailed as brain tonics for curing headaches, hangovers, and nervous afflictions.

Pharmacies equipped with "soda fountains" featuring the medicinal soda water soon developed into regular meeting places for local populations. Flavored soda water gained popularity not only for medicinal benefits but for the refreshing taste as well. The market expanded in the 1830s when soda water was first sold in glass bottles. Filling and capping the gaseous liquid in containers was a difficult process until 1850, when a manual filling and corking machine was successfully designed. The term "soda pop" originated in the 1860s from the popping sound of escaping gas as a soda bottle was opened.

New soda flavors constantly appeared on the market. Some of the more popular flavors were ginger ale, sarsaparilla, root beer, lemon, and other fruit flavors. In the early 1880s pharmacists experimented with powerful stimulants to add to soda water, including cola nuts and coca leaves. They were inspired by Bolivian Indian workers who chewed coca leaves to ward off fatigue and by West African workers who chewed cola nuts as a stimulant. In 1886 an Atlanta pharmacist, John Pemberton, took the fateful step of combining coca with cola, thus creating what would become the world's most famous drink, "Coca-Cola". The beverage was advertised as refreshing as well as therapeutic: "French Wine Cola—Ideal Nerve and Tonic Stimulant." A few years later another pharmacist, Caleb Bradham, created "Pepsi-Cola" in North Carolina. Although the name was a derivation of pepsin, an acid that aids digestion, Pepsi did not advertise the beverage as having therapeutic benefits. By the early 20th century, most cola companies focused their advertising on the refreshing aspects of their drinks.

As flavored carbonated beverages gained popularity, manufacturers struggled to find an appropriate name for the drinks. Some suggested "marble water," "syrup water," and "aerated water." The most appealing name, however, was "soft drink," adapted in the hopes that soft drinks would ultimately supplant the "hard liquor" market. Although the idea never stuck, the term soft drink did.

Until the 1890s soft drinks were produced manually, from blowing bottles individually to filling and packaging. During the following two decades automated machinery greatly increased the productivity of soft drink plants. Probably the most important development in bottling technology occurred with the invention of the "crown cap" in 1892, which successfully contained the carbon dioxide gas in glass bottles. The crown cap design endured for 70 years.

The advent of motor vehicles spawned further growth in the soft drink industry. Vending machines, serving soft drinks in cups, became regular fixtures at service stations across the country. In the late 1950s **aluminum beverage can**s were introduced, equipped with convenient pull-ring tabs and later with stay-on tabs. Light-weight and break-resistant plastic bottles came into use in the 1970s, though it was not until 1991 that the soft drink industry used plastic PET (polyethylene terephthalate) on a wide scale.

Soft drink manufacturers have been quick to respond to consumer preferences. In 1962 diet colas were introduced in response to the fashion of thinness for women. In the 1980s the growing health consciousness of the country led to the creation of caffeine-free and low-sodium soft drinks. The 1990s ushered in clear colas that were colorless, caffeine-free, and preservative-free.

Raw Materials

Carbonated water constitutes up to 94% of a soft drink. Carbon dioxide adds that special sparkle and bite to the beverage and also acts as a mild preservative. Carbon dioxide is an uniquely suitable gas for soft drinks because it is inert, non-toxic, and relatively inexpensive and easy to liquefy.

The second main ingredient is sugar, which makes up 7-12% of a soft drink. Used in either dry or liquid form, sugar adds sweetness and body to the beverage, enhancing the "mouth-feel," an important component for consumer enjoyment of a soft drink. Sugar also balances flavors and acids.

Sugar-free soft drinks stemmed from a sugar scarcity during World War II. Soft drink manufacturers turned to high-intensity sweeteners, mainly saccharin, which was phased out in the 1970s when it was declared a potential carcinogen. Other sugar substitutes were introduced more successfully, notably aspartame, or NutraSweet, which was widely used throughout the 1980s and 1990s for diet soft drinks. Because some high-intensity sweeteners do not provide the desired mouth-feel and aftertaste of sugar, they often are combined with sugar and other sweeteners and flavors to improve the beverage.

The overall flavor of a soft drink depends on an intricate balance of sweetness, tartness, and acidity (pH). Acids add a sharpness to the background taste and enhance the thirst-quenching experience by stimulating saliva flow. The most common acid in soft drinks is citric acid, which has a lemony flavor. Acids also reduce pH levels, mildly preserving the beverage.

Very small quantities of other additives enhance taste, mouth-feel, aroma, and appearance of the beverage. There is an endless range of flavorings; they may be natural, natural identical (chemically synthesized imitations), or artificial (chemically unrelated to natural flavors). Emulsions are added to soft drinks primarily to enhance "eye appeal" by serving as clouding agents. Emulsions are mixtures of liquids that are generally incompatible. They consist of water-based elements, such as gums, pectins, and preservatives; and oil-based liquids, such as flavors, colors, and weighing agents. Saponins enhance the foamy head of certain soft drinks, like cream soda and ginger beer.

To impede the growth of microorganisms and prevent deterioration, preservatives are added to soft drinks. Anti-oxidants, such as BHA and ascorbic acid, maintain color and flavor. Beginning in the 1980s, soft drink manufacturers opted for natural additives in response to increasing health concerns of the public.

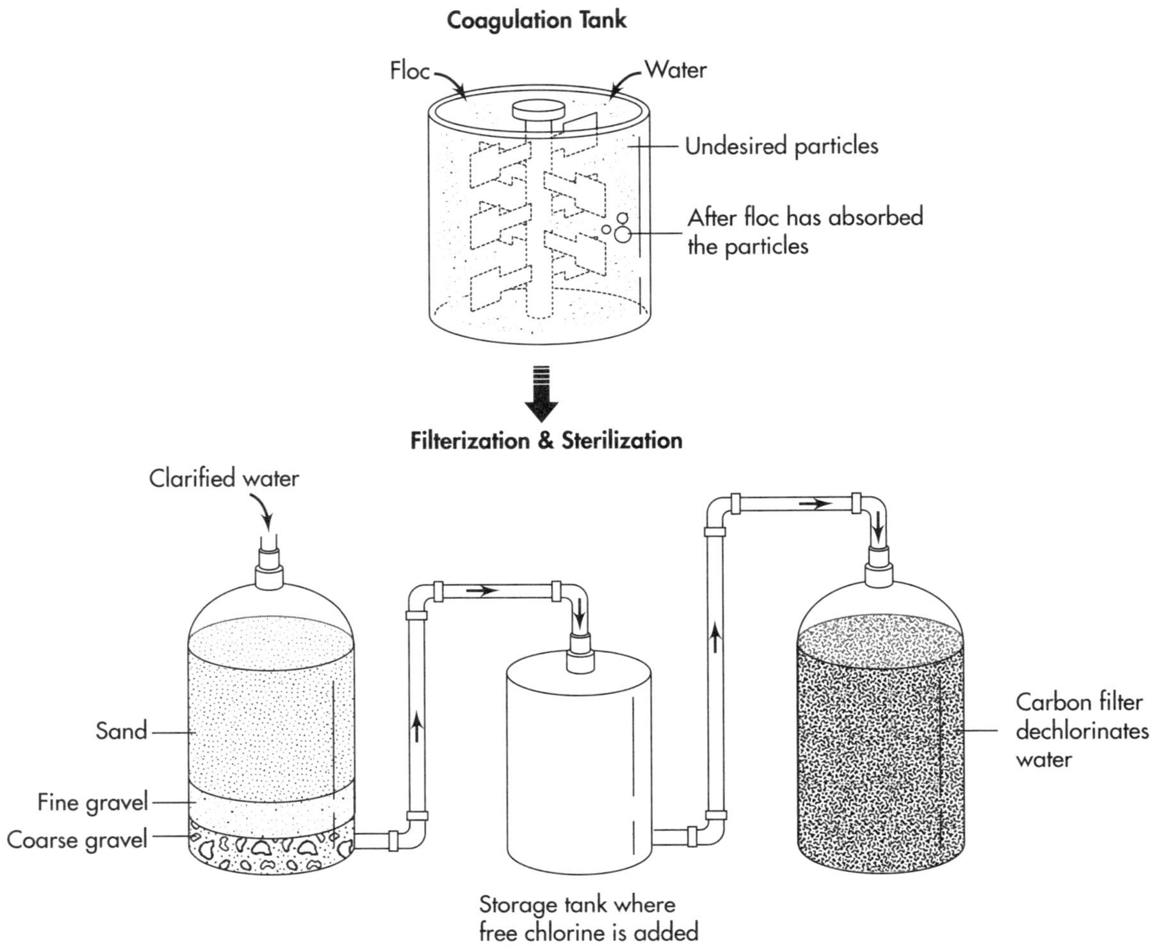

Impurities in the water are removed through a process of coagulation, filtration, and chlorination. Coagulation involves mixing floc into the water to absorb suspended particles. The water is then poured through a sand filter to remove fine particles of floc. To sterilize the water, small amounts of chlorine are added to the water and filtered out.

The Manufacturing Process

Most soft drinks are made at local bottling and canning companies. Brand name franchise companies grant licenses to bottlers to mix the soft drinks in strict accordance to their secret formulas and their required manufacturing procedures.

Clarifying the water

1 The quality of water is crucial to the success of a soft drink. Impurities, such as suspended particles, organic matter, and bacteria, may degrade taste and color. They are generally removed through the traditional process of a series of coagulation, filtration, and chlorination. Coagulation involves mixing a gelatinous precipitate, or floc (ferric sulphate or aluminum sulphate), into the water. The floc absorbs suspended particles, making them larger and more easily trapped by filters. During the clarification process, alkalinity must be adjusted with an addition of lime to reach the desired pH level.

Filtering, sterilizing, and dechlorinating the water

2 The clarified water is poured through a sand filter to remove fine particles of floc. The water passes through a layer of sand and courser beds of gravel to capture the particles.

3 Sterilization is necessary to destroy bacteria and organic compounds that might spoil the water's taste or color. The water is pumped into a storage tank and is dosed with a small amount of free chlorine. The chlorinated water remains in the storage

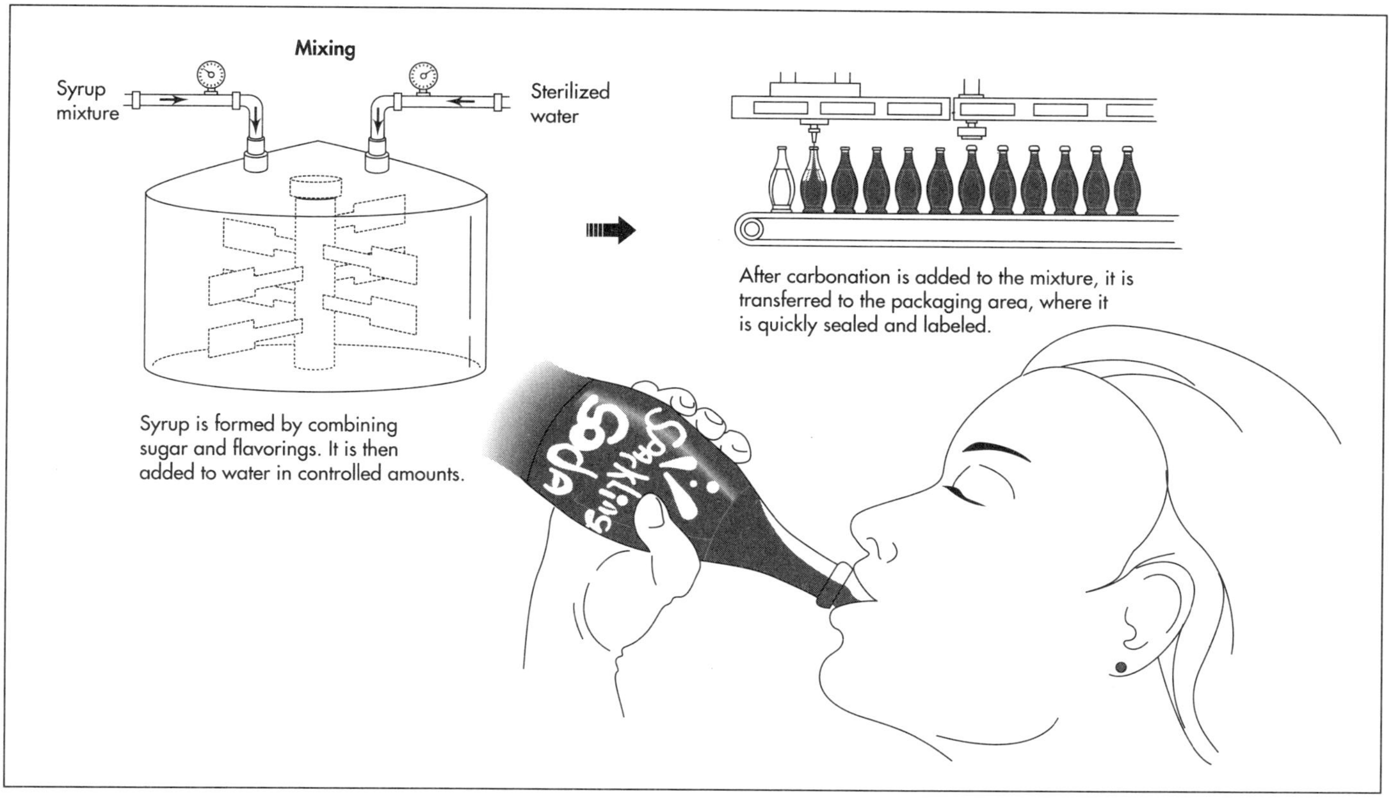

Syrup is formed by combining sugar and flavorings. It is then added to water in controlled amounts.

After carbonation is added to the mixture, it is transferred to the packaging area, where it is quickly sealed and labeled.

tank for about two hours until the reaction is complete.

4 Next, an activated carbon filter dechlorinates the water and removes residual organic matter, much like the sand filter. A vacuum pump de-aerates the water before it passes into a dosing station.

Mixing the ingredients

5 The dissolved sugar and flavor concentrates are pumped into the dosing station in a predetermined sequence according to their compatibility. The ingredients are conveyed into batch tanks where they are carefully mixed; too much agitation can cause unwanted aeration. The syrup may be sterilized while in the tanks, using ultraviolet radiation or flash pasteurization, which involves quickly heating and cooling the mixture. Fruit based syrups generally must be pasteurized.

6 The water and syrup are carefully combined by sophisticated machines, called proportioners, which regulate the flow rates and ratios of the liquids. The vessels are pressurized with carbon dioxide to prevent aeration of the mixture.

Carbonating the beverage

7 Carbonation is generally added to the finished product, though it may be mixed into the water at an earlier stage. The temperature of the liquid must be carefully controlled since carbon dioxide solubility increases as the liquid temperature decreases. Many carbonators are equipped with their own cooling systems. The amount of carbon dioxide pressure used depends on the type of soft drink. For instance, fruit drinks require far less carbonation than mixer drinks, such as tonics, which are meant to be diluted with other liquids. The beverage is slightly over-pressured with carbon dioxide to facilitate the movement into storage tanks and ultimately to the filler machine.

Filling and packaging

8 The finished product is transferred into bottles or cans at extremely high flow rates. The containers are immediately sealed with pressure-resistant closures, either tin-

plate or steel crowns with corrugated edges, twist offs, or pull tabs.

9 Because soft drinks are generally cooled during the manufacturing process, they must be brought to room temperature before labeling to prevent condensation from ruining the labels. This is usually achieved by spraying the containers with warm water and drying them. Labels are then affixed to bottles to provide information about the brand, ingredients, shelf life, and safe use of the product. Most labels are made of **paper** though some are made of a plastic film. Cans are generally pre-printed with product information before the filling stage.

10 Finally, containers are packed into cartons or trays which are then shipped in larger pallets or crates to distributors.

Quality Control

Soft drink manufacturers adhere to strict water quality standards for allowable dissolved solids, alkalinity, chlorides, sulfates, iron, and aluminum. Not only is it in the interest of public health, but clean water also facilitates the production process and maintains consistency in flavor, color, and body. Microbiological and other testing occur regularly. The National Soft Drink Association and other agencies set standards for regulating the quality of sugar and other ingredients. If soft drinks are produced with low-quality sugar, particles in the beverage will spoil it, creating floc. To prevent such spoilage, sugar must be carefully handled in dry, sanitized environments.

It is crucial for soft drink manufacturers to inspect raw materials before they are mixed with other ingredients, because preservatives may not kill all bacteria. All tanks, pumps, and containers are thoroughly sterilized and continuously monitored. Cans, made of aluminum alloy or tin-coated low-carbon steel, are lacquered internally to seal the metal and prevent corrosion from contact with the beverage. Soft drink manufacturers also recommend specific storage conditions to retailers to insure that the beverages do not spoil. The shelf life of soft drinks is generally at least one year.

Recycling

The $27 billion dollar soft drink industry generated about 110 billion containers each year in the early 1990s. About half of soft drink containers were aluminum cans and the other half, about 35 billion, were PET plastic bottles. Nearly 60% of all soft drink containers were recycled, the highest rate for any packaging in the United States. Environmental concerns continued to lead to improvements and innovations in packaging technology, including the development of refillable and reusable containers.

The Future

In the 1990s there were more than 450 types of soft drinks on the market and new flavors and sweeteners are developed all the time to meet market demands. In the future, advanced technology will lead to greater efficiency of soft drink production at all stages. New methods of water clarification, sterilization, and pasteurization will improve production and minimize the need for preservatives in soft drinks. Concerns with consumer health, safety, and the environment will continue to have a positive impact on trends in the soft drink industry.

Where To Learn More

Books

Louis, J.C. *The Cola Wars.* Everest House, 1980.

Mitchell, Alan J., ed. *Formulation and Production of Carbonated Soft Drinks.* AVI, 1990.

Oliver, Thomas. *The Real Coke.* Random House, 1986.

Riley, John J. *A History of the American Soft Drink Industry.* Arno Press, 1972.

—Audra Avizienis

Stained Glass

Background

The technology for making glass dates back at least 5,000 years, and some form of stained glass was used in European Christian churches by the third or fourth century A.D. The art of stained glass flowered in the 12th century with the rise of the Gothic cathedral. Today only 10% of all stained glasses are used in churches and other religious buildings; the rest are used in residential and industrial architecture. Though stained glass has traditionally been used in windows, its use has expanded to lamp shades, Christmas ornaments, and even simple objects a hobbyist can make.

Stained glass has had various levels of popularity throughout history. The 12th and 13th centuries in Europe have been designated as the Golden Age of Stained Glass. However, during the Renaissance period, stained glass was replaced with painted glass, and by the 18th century it was rarely, if ever, used or made according to medieval methods. During the second half of the 19th century, European artists rediscovered how to design and work glass according to medieval principles, and large quantities of stained glass windows were made.

In America, the stained glass movement began with William Jay Bolton, who made his first window for a church in New York in 1843. But he was to be in the business for only six or seven years before returning to his native England. No other American practiced the art professionally until Louis Comfort Tiffany and John La Farge began working with stained glass near the end of the 19th century. In fact, the art of stained glass in the United States languished until the 1870s, and did not undergo a true revival until the turn of the century. At this time, American architects and glassmen journeyed to Europe to study medieval glass windows, returning to create similar art forms and new designs in their own studios.

A leaded stained glass window or other object is made of pieces of glass, held together by **lead**. The pieces of glass are about 1/8-inch (3.2 mm) thick and bound together by strips, called "cames" of grooved lead, soldered at the joints. The entire window is secured in the opening at regular intervals by metal saddle bars tied with wire and soldered to the leads and reinforced at greater intervals by tee-bars fitted into the masonry. A faceted glass panel differs slightly from traditional leaded stained glass in that it is made up of pieces of slab (dalle) glass approximately 8 inches square, or in large rectangular sizes, varying in thickness from 1-2 inches (2.5-5 cm). These slabs are not held together with lead; rather they are embedded in a matrix of concrete, epoxy, or plastic.

Raw Materials

Glass is made by fusing together some form of silica such as sand, an alkali such as potash or soda, and lime or lead oxide. The color is produced by adding a metallic oxide to the raw materials.

Copper oxide, under different conditions, produces ruby, blue, or green colors in glass. Cobalt is usually used to produce most shades of blues. Green shades can also be obtained from the addition of chromium and iron oxide. Golden glass is sometimes colored with uranium, cadmium sulfide, or titanium, and there are fine selenium yellows as well as vermilions. Ruby colored glass is made by adding gold.

Today only 10% of all stained glasses are used in churches and other religious buildings; the rest are used in residential and industrial architecture.

Processing Stained Glass

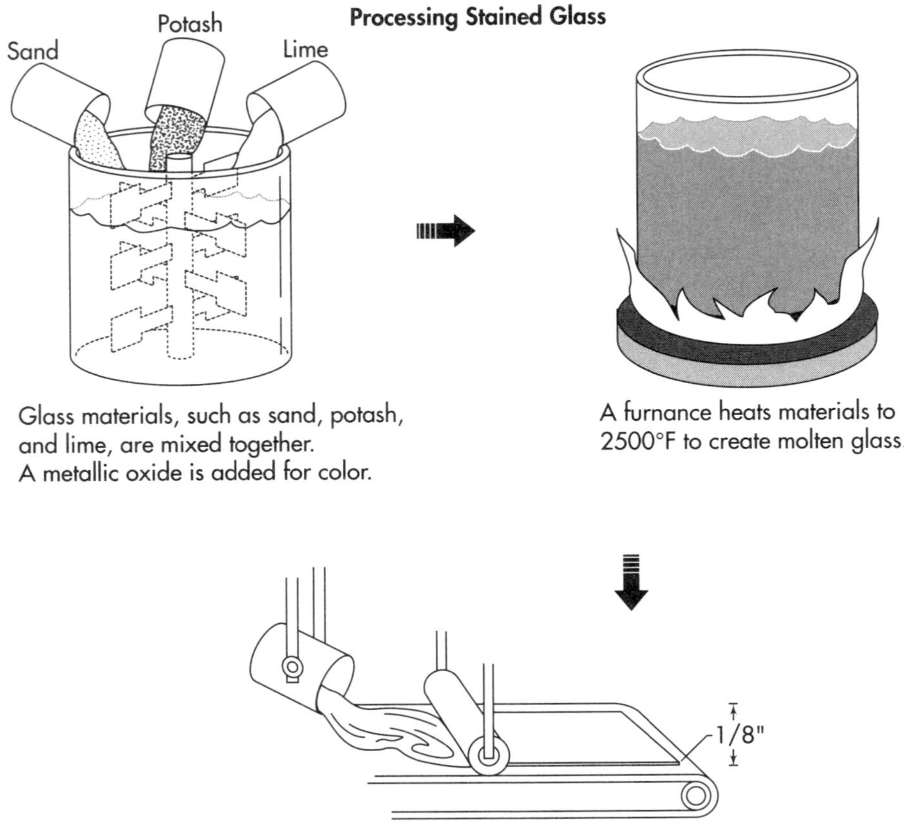

Glass materials, such as sand, potash, and lime, are mixed together. A metallic oxide is added for color.

A furnance heats materials to 2500°F to create molten glass.

The molten glass is then rolled into thin sheets. These sheets are then cooled, trimmed to different sizes, and packaged for craftsmen.

The Manufacturing Process

Stained glass is still made the same way it was back in the Middle Ages and comes in various forms. For the glass used in leaded glass windows, a lump of the molten glass is caught up at one end of a blow pipe, blown into a cylinder, cut, flattened and cooled. Artisans also vary this basic process in order to produce different effects. For example, "flashed glass" is made by dipping a ball of molten white glass into molten colored glass which, when blown and flattened, results in a less intense color because it will be white on one side and colored on the other.

So-called "Norman slabs" are made by blowing the molten glass into a mold in the shape of a four-sided bottle. The sides are cut apart and form slabs, thin at the edges and as much as 0.25 inch (0.6 cm) thick at the center. Another form of glass, known as cathedral glass, is rolled into flat sheets. This results in a somewhat monotonous regularity of texture and thickness. Other similarly made glasses are referred to as marine antique, but have a more bubbly texture.

Processing the stained glass

1 Large manufacturers of stained glass mix the batch of raw materials, including alkaline fluxes and stabilizing agents, in huge mixers. The mix is then melted in a modern furnace at 2500°F (1371°C). Each ingredient must be carefully measured and weighed according to a calculated formula, in order to produce the appropriate color. For cathedral glass, the molten glass is ladled into a machine that rolls the glass into 1/8-inch (3.2

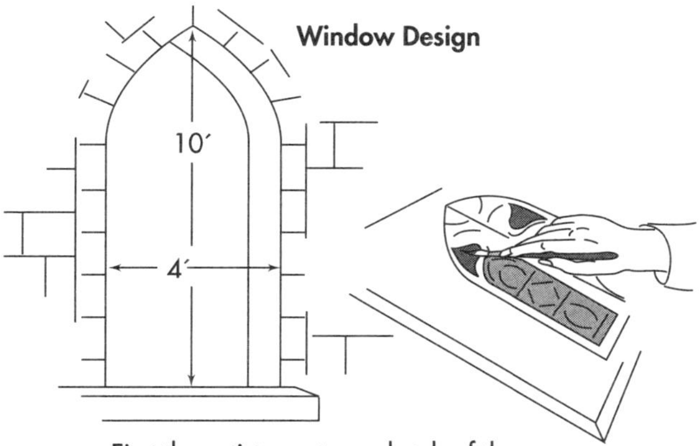

First the artist creates a sketch of the design, which is then used to create a full-size sketch.

The next process is cartooning, in which cutlines and leadlines are outlined on the full-size sketch. Another copy of the sketch, called a pattern drawing, is cut apart to use for modeling the pieces of glass.

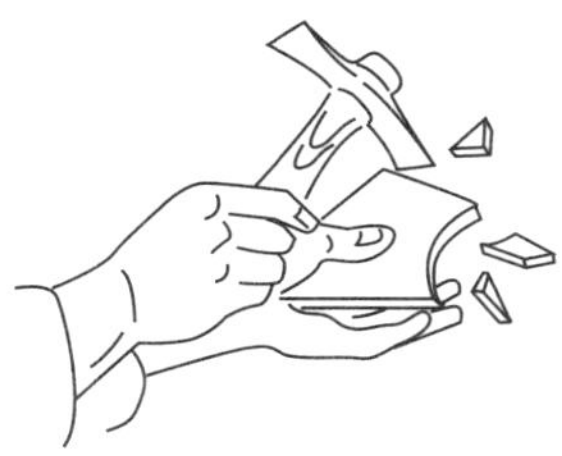

The cut pieces of the pattern drawing are placed over pieces of the desired colored glass and then, using a two-edged sharp hammer or a cutting wheel, the glass is trimmed to the proper shape.

mm) thick sheets. The sheets are then cooled in a special furnace called an annealing lehr. The glass is then inspected, trimmed to standard size, and packed into cases.

At a typical factory, eight to ten different color runs are made per day. Some manufacturers cut a small rectangle of glass from each run in order to provide a sample of each color to their customers. There are hundreds of colors, tints, and patterns available, as well as a number of different textures of cathedral glass. Different textures are produced by changing the roller to one having the desired texture. Glass manufacturers are continuously introducing new colors and types of glass to meet the demands of their customers.

Creating the window pattern

2 Though some of the tools to make stained glass windows have been improved, the windows are still hand crafted as they were centuries ago. The first step of the process involves the artist creating a small scale version of the final design. After the design has been approved, the craftsperson takes measurements or templates of the actual window openings to create a pattern. This pattern is usually drawn on **paper** or cardboard and is the actual size of the spaces to be filled with glass.

Next a full-sized drawing called the cartoon is prepared in black and white. From the cartoon, the cutline and pattern drawings are

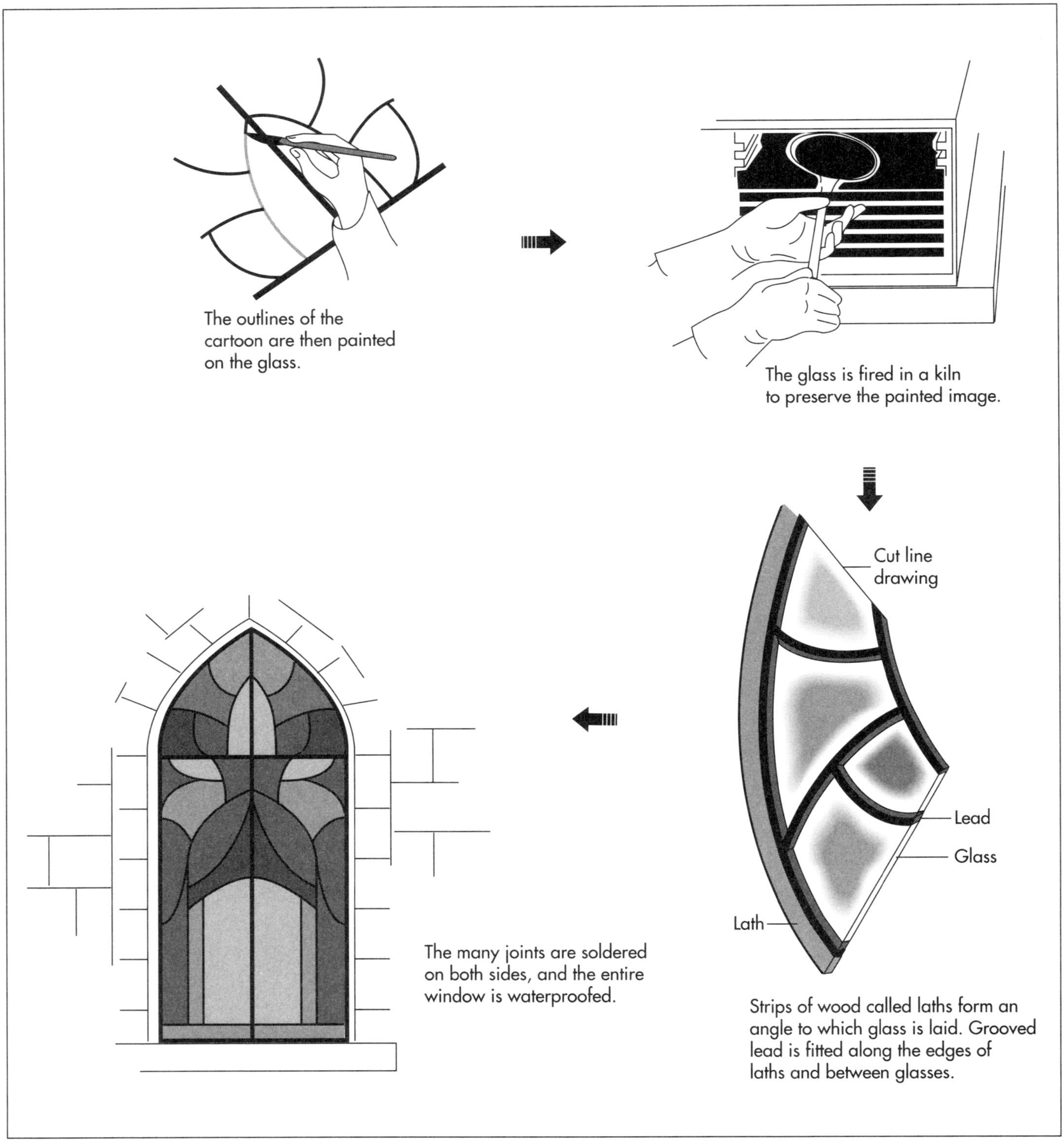

made. The modern cutline drawing is a careful, exact tracing of the leadlines of the cartoon on heavy paper. The leadlines are the outlines of the shapes for patterns to which the glass is to be cut. This drawing serves as the guide for the subsequent placing and binding with lead of the many pieces of glass.

The pattern-drawing is a carbon copy of the cutline drawing. It is cut along the black or lead lines with double-bladed scissors or a knife which, as it passes through the middle of the black lines, simultaneously cuts away a narrow strip of paper, thus allowing sufficient space between the segment of glass for the core of the grooved lead. This core is the supporting wall between the upper and lower flanges of the lead.

Cutting and painting

3 Colored glass is then selected from the supply on hand. The pattern is placed on a piece of the desired color, and with a **diamond** or steel wheel, the glass is cut to the shape of the pattern. After the glass has been cut, the main outlines of the cartoon are painted on each piece of glass with special paint, called "vitrifiable" paint. This becomes glassy when heated. The painter might apply further paint to the glass in order to control the light and bring all the colors into closer harmony. During this painting process, the glass is held up to the light to simulate the same conditions in which the window will be seen. The painted pieces are fired in the kiln at least once to fuse the paint and glass.

Glazing and leading

4 The next step is glazing. The cutline drawing is spread out on a table and narrow strips of wood called laths are nailed down along two edges of the drawing to form a right angle. Long strips of grooved lead are placed along the inside of the laths. The piece of glass belonging in the angle is fitted into the grooves. A strip of narrow lead is fitted around the exposed edge or edges and the next required segment slipped into the groove on the other side of the narrow lead. This is continued until each piece has been inserted into the leads in its proper place according to the outline drawing beneath.

Finishing

5 The many joints formed by the leading are then soldered on both sides and the entire window is waterproofed. After the completed window has been thoroughly inspected in the light, the sections are packed and shipped to their destination where they are installed and secured with reinforcing bars.

Faceted glass

6 For faceted glass windows, the process begins the same way, with the cutline and pattern drawings being made with carbons in a similar manner. The pattern drawing is then cut to the actual size of the piece of glass with ordinary scissors since there is no core of lead to allow for. The thick glass slabs next are cut with a sharp double-edged hammer to the shape of the pattern. To give the slab an interesting texture, the worker then chips round depressions in the glass with the same hammer. This is called faceting.

Instead of glazing with lead, a matrix of concrete or epoxy is poured around the pieces of glass. The glass pieces have first been glued to the outline drawing, which is covered with a heavy coating of transparent grease so that the paper can be removed after the epoxy sets. The whole is enclosed within a wooden form, which is the exact size and shape of the section being made. The worker must wear gloves during this process, since epoxy resin is a toxic material. After hardening, the section is cleaned and cured prior to shipping and installation.

The process for making an entire stained glass window can take anywhere from seven to ten weeks, since everything must be done by hand. Cost can vary widely depending on complexity and size, though some windows can be created for a cost as low as $500. The customer can choose an existing pattern rather than create an entirely new one to minimize costs. In this case, the pattern can be customized by altering shapes or by changing the placement of the central image.

The Future

In the last 20 years there has been an explosion in growth of glass studios in the United States and it appears this growth will continue. For instance, in Ohio alone the number of studios has increased from a mere half a dozen to at least 100. The Stained Glass Association of America membership includes 500 studio owners and 300 manufacturers. The circulation of its quarterly publication totals 6,000. There has been a resurgence in restoration overseas, and the home market continues to grow. The hobby market also appears strong, with one publication serving this market having a circulation of 15,000. It is clear that stained glass is now recognized as a true art form no matter where it is used, and innovative designs using this medium will continue to flourish.

Where To Learn More

Books

Clark, Willene B. *The Stained Glass Art of William Jay Bolton.* Syracuse University Press, 1992.

Clarke, Brian, ed. *Architectural Stained Glass.* McGraw-Hill, 1979.

Plowright, Terrance. *Stained Glass Inspirations and Designs.* Kangaroo Press, Australia, distributed by Seven Hills Book Distributors, 1993.

Other

Achilees, Rolf and Neal A. Vogel. *Stained Glass in Houses of Worship.* Inspired Partnerships Inc. and the National Trust for Historic Preservation, 1785 Massachusetts Avenue NW, Washington, DC 20036.

The Story of Stained Glass, 1984. The Stained Glass Association of America, PO Box 22642, Kansas City, MO 64113. 800-888-7422, 816-333-6690.

—Laurel M. Sheppard

Sunscreen

Background

The image of a healthy person today does not necessarily include a deep, dark tan. Research has linked exposure to the sun's ultraviolet rays to skin cancer, premature wrinkles, and other skin problems. While protective clothing such as hats, pants, and long sleeve shirts are the most effective blocks against these damaging rays, sunscreen lotions also provide remarkable protection for the skin.

The light emitted by the sun consists of three frequency bands of radiation: infrared, visible, and ultraviolet. Of the three, only the ultraviolet is harmful to most humans. Ultraviolet (UV) radiation is further divided into three categories: UVA, UVB, and UVC. UVA radiation penetrates the skin without burning the surface layers. These rays can penetrate to a depth of 0.04 inches (1.0 mm) and cause damage to cell membranes and the immune system. UVA radiation has been linked to skin cancers and premature aging and wrinkling of the skin. UVB radiation is responsible for the painful, red burn people get after prolonged exposure to the sun. UVB rays also cause skin cancer and can damage the cornea and lens of the eye. The third category, UVC radiation, is generally absorbed by the earth's atmosphere and is not considered harmful. Sunscreen lotions that provide protection against both UVA and UVB radiation are known as having a broad spectrum of protection.

There are two basic types of sunscreen lotions on the market: products that penetrate the outermost layer of skin to absorb ultraviolet rays, and products which coat the surface of the skin to act as physical barriers to ultraviolet rays. Both of these types are rated with a sun protection factor (SPF), which lets the consumer know how much protection against UVB rays the product provides. The SPF of a product is the ratio of the time required for a person's protected skin to redden after being exposed to sunlight compared to the time required for the same person's unprotected skin to redden. For example, a product with SPF15 means that a person whose unprotected skin would redden in ten minutes can apply the product and stay in the sun 15 times longer, or 150 minutes, before they get a sunburn.

There are two basic types of sunscreen lotions on the market: products that penetrate the outermost layer of skin to absorb ultraviolet rays, and products which coat the surface of the skin to act as physical barriers to ultraviolet rays.

Researchers believed for a long time that UVB rays—the rays that actually cause a sunburn—were solely responsible for all forms of skin cancer. However, recent studies prove that UVA rays are also responsible. Although many sunscreens now contain UVA protectors, there are currently no standards set by the Food and Drug Administration (FDA) for protection against UVA rays. The SPF rating on a product applies only to protection against UVB rays. The FDA requires strict regulations and testing prior to the manufacture of any new sunscreen lotion. Sunscreen producers go through an expensive and lengthy process to get FDA approval. This approval authorizes the manufacturer to produce the exact formulation applied for and is limited to only one SPF rating and one specific usage.

Development and Testing

Today's target market for sunscreens are highly specialized. Sunscreen products are continually being redeveloped to meet the changing needs of specific consumers. For instance, formulations for athletes may contain ingredients which are more waterproof and sweatproof to provide protection for up to eight hours. Athletes may also desire a lotion that feels dry so as not to affect their grip. Children's skins are more sensitive than adults because the outermost layer is thinner. This supports the observation that most sun damage to the skin occurs during childhood and the early teenage years. Sunscreens developed for the children's market tend to contain natural ingredients such as aloe vera and vitamin E.

In the development phase for a new sunscreen, a team of chemists and lab technicians develop the sunscreen formulation from synthetic and natural ingredients. Initial formulations are made in quantities of 10 gallons (38 l) and are stored in stainless steel vats. These initial formulations are tested and finalized before an application to the FDA for approval is made. FDA approval requires further testing which may be done in-house or by an outside laboratory. Examples of the kind of testing required for FDA approval include tests to measure the effective sun protection factor according to FDA guidelines, tests to determine how safe the product is to use on the skin, and tests to measure the waterproof tolerance of a lotion.

Raw Materials

Many combinations of synthetic and natural ingredients may go into the formulation of a single sunscreen. A formulation is generally geared towards a specific SPF rating or the needs of a specific consumer group. Perhaps the best-known synthetic material used for protection against UVA rays is avobenzone, or Parsol 1789, which is used in products worldwide. Broad spectrum protection is provided by other synthetic ingredients such as benzophenone and oxybenzone, which protect by absorbing UV light. PABA (para-aminobenzoic acid) was once a popular UV-absorbing sunscreen ingredient, but it can cause skin irritation in some people and is now replaced by Padimate-O, a derivative of PABA. Other broad spectrum synthetic ingredients are octyl methoxycinnamate and menthyl anthranilate.

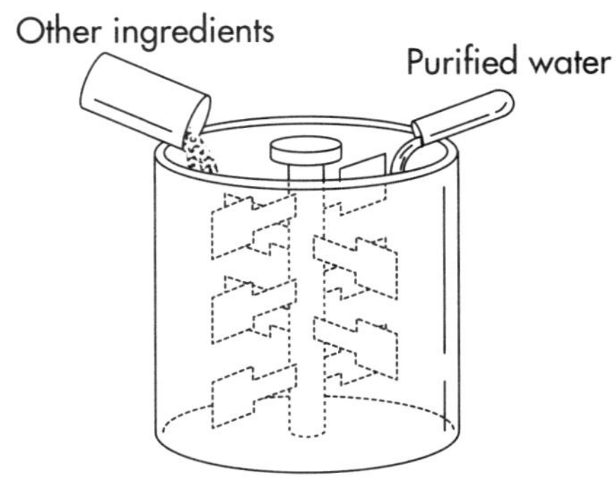

Purified water and other purchased ingredients like avobenzone and aloe are mixed together.

Titanium dioxide is a natural mineral and a popular ingredient for broad spectrum protection. Titanium dioxide works by scattering UV light instead of absorbing it. Although not as opaque as zinc oxide, it has a similar whitening effect in the higher SPF ratings. Antioxidants are often combined with titanium dioxide to slow down the oxidation of oils and thereby delay the deterioration of the lotion. Some examples of natural antioxidants are vitamins E and C, rice bran oil and sesame seed oil. Another popular antioxidant in the natural category is green tea. Many newer sunscreen products also contain skin soothing and moisturizing additives such as aloe and chamomile.

The Manufacturing Process

Sunscreen products may be manufactured, bottled, and shipped from a single facility, or portions of this work may be handled outside of the company. The fully automated manufacturing process described here uses some of both approaches.

Formulating the lotion

1 Water is purified using a method called reverse osmosis. Reverse osmosis extracts pure, fresh water by forcing water under pressure through a semipermeable membrane which separates pure water molecules from salts and other impurities.

Blow Molding

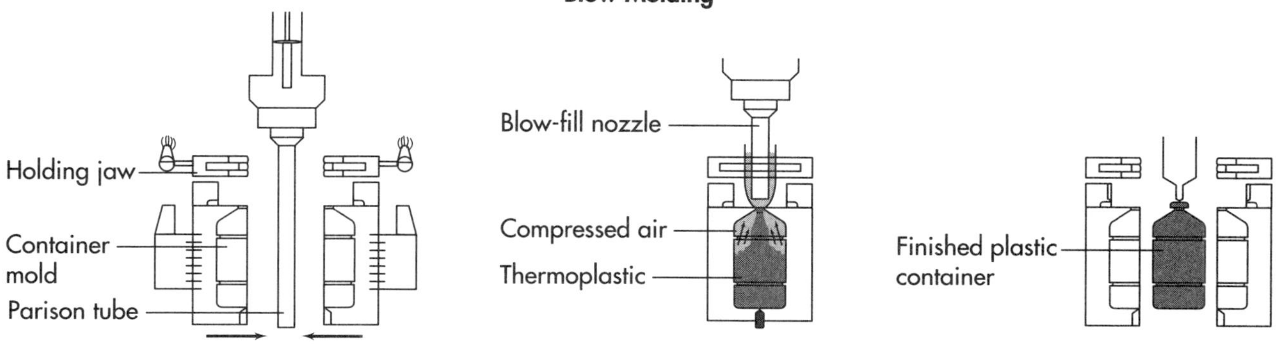

The sunscreen container is made by a blow molding method, in which thermoplastic is extruded into a tube and blown into a mold.

2 Ingredients are purchased from outside sources and mixed with the purified water according to the recipe of the final formulation. The recipe is recorded on a vat sheet which lists the exact measurements for each ingredient. Measurements are converted from the initial 10-gallon (38 l) recipe used in the development stage to larger quantities for commercial use.

Making the containers

3 A blow molding facility manufactures the plastic containers for the sunscreen. In some cases this is done outside of the company. Blow molding is a method in which thermoplastics, plastics which soften when heated and harden when cooled, are extruded into a tube, called a parison, and placed into an open mold. The mold is closed around the heated parison, and the parison is pinched at the bottom to form a seal. Compressed air is blown through the top of the parison which forces the softened plastic to expand to the inside walls of the mold, creating the shape of the container.

4 Containers are moved to a printing facility where logos and product information are printed and, in some cases, stamped onto the containers. Stamping embosses thin metal foil onto the surface of the container in the desired shape, usually a logo. The printed or stamped containers are then stored for use when needed.

Filling the containers

5 Stainless steel tanks with capacities up to 1,000 gallons (3784 l) are used in the filling process. Filling takes place in a separate, sterile room with a conveyor system of many incoming tracks. Machine operators monitor the automated process. Containers and caps enter the fill room on conveyor tracks. The sunscreen lotion flows from the vats through stainless steel piping to a pressure filling machine which inserts a retractable nozzle into each container and fills it with a measured amount of sunscreen lotion.

Capping the containers

6 Most containers are capped automatically along the production line. Some containers include caps with pumps to allow easy dispensing of the sunscreen. These pump caps require manual assembly by operators as the containers leave the fill phase.

Shipping

7 The filled and capped containers are boxed in quantities of 12 to be pelletized and secured to a skid by shrink wrapping for transport to distributors.

Byproducts/Waste

Sanitized plastic scraps from the container molding process are reground and used again in molding. Containers which have

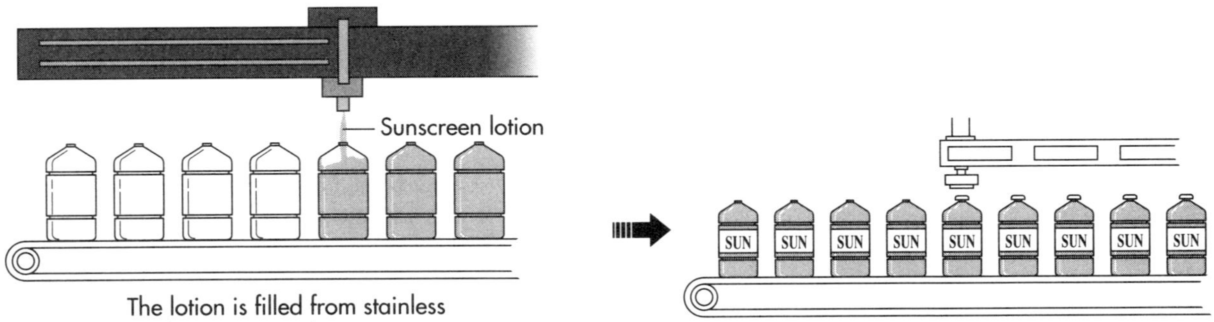

been through the printing process and have flaws are passed on to other companies and made into products such as patio furniture.

The Future

Researchers look to nature in search of the next wave in sunscreen development. Some plants have natural defenses against the damaging rays of the sun. For example the single-cell alga called *Dunaliella Bardawil* that thrives in the Dead Sea and the Sinai desert makes its own sunscreen. Scientists at the Weizmann Institute of Science in Rehovot, Israel, isolated the protein that this plant produces when sunlight gets too intense. The protein acts as a solar deflector by funnelling light down to where photosynthesis takes place. Excess light which could interfere with photosynthesis is shunned by a yellow-orange pigment produced by the algae.

The human body also has a natural defense called melanin. Melanin is the brownish-black pigment found in skin and hair. It reflects and absorbs ultraviolet rays to provide a broad spectrum of protection. Dark-skinned people have a higher concentration of melanin and, as a result, have a lower incidence of skin cancer as well as less physical and medical signs of aging skin. Melanin was once painstakingly collected by extraction from exotic sources such as cuttle fish and cost about $3,000 per ounce ($101 per ml). However it can now be made in an inexpensive procedure using fermentation jars. One technology being used as a method of incorporating melanin into sunscreen lotions is to encapsulate it into microsponges which hold the melanin on the surface of the skin where it is most effective. The microsponges are invisible to the eye, and can only be seen under a microscope. Researchers continue to gain approval for the many uses of natural and synthetic melanin as an ingredient in sunscreen formulations.

Where To Learn More

Books

Conry, Tom, ed. *Consumer's Guide to Cosmetics*. Anchor Press/Doubleday, 1980.

Winter, Ruth. *A Consumer's Dictionary of Cosmetic Ingredients*. Crown Publishers, 1984.

Periodicals

Bylinsky, Gene. "Mass-Producing Nature's Sunscreen." *Fortune*, June 1992, p. 131.

Davey, Tom. "Sunscreen Success Gives Biotech New Healthy Glow." *San Francisco Business Times*, October 1994, p. 8.

Demetrakakes, Pan. "Packaging Patterns Make Mark On Makeup." *Packaging*, March 1994, pp. 20-21.

Groves, Gordon A. "Toward More Effective UVA Protection." *Drug and Cosmetic Industry*, August 1994, pp. 37-38.

Kintish, Lisa. "The Battle of The Sun." *Soap/Cosmetics/Chemical Specialties For May,* May 1993, pp. 42-48.

—David N. Ford

Surfboard

The surfboard, and the sport of surfing, are believed to have originated in Polynesia as early as A.D. 400. The Polynesians brought the sport with them when they settled in Hawaii. Their early wood boards were 12-20 feet long and weighed nearly 200 pounds.

Background

A surfboard is used in the sport of surfing. A typical surfboard is about 18-24 inches (46-61 cm) wide, 72-120 inches (183-305 cm) long, and several inches thick. It has a lightweight, buoyant core covered with a hard shell. In use, the surfer lays face down on the surfboard and paddles out into the ocean to the point where waves are beginning to rise. The surfer turns the board towards shore, paddles rapidly to match the speed of an incoming wave, then quickly stands up and balances on the board as it is propelled by the face of the breaking wave. One variation of the surfboard is the sailboard, which includes a short mast and sail used for wind surfing. Another variation is the body board, which is shorter than a surfboard and is ridden in the prone position.

The surfboard, and the sport of surfing, are believed to have originated in Polynesia as early as A.D. 400. The Polynesians brought the sport with them when they settled in Hawaii. Hawaiian surfboards were made of wood from various trees on the islands. They were carved and shaped by hand, then stained and finished using the natural juices and oils of plants. The largest boards, called *'olos*, were 144-240 inches (3.6-6 m) long and weighed nearly 200 pounds (91 kg). Experimentation with wooden Hawaiian surfboards during the 1920s and 1930s resulted in hollow board designs and the use of redwood and balsa laminates to reduce the weight.

The first **fiberglass** surfboard was built in 1946. It consisted of two hollow, molded halves with a redwood stiffener, or stringer, running down the center. In 1949, Bob Simmons built the first board with a buoyant, styrofoam core sandwiched between two thin, plywood veneers and sealed with resin.

The birth of the modern surfboard came in 1958 when Hobie Alter started producing boards with polyurethane foam cores. Later, he went on to develop fiberglassing techniques using polyester resins to form the outer shell. Today, almost every surfboard uses this construction.

Raw Materials

The typical surfboard has a rigid polyurethane foam core with an outer shell of fiberglass cloth and **polyester** resins. If a stringer is used in the design, it is usually made of wood such as redwood, basswood, or spruce. Colored fiberglass stringers can also be used. The fin, or skeg, is made of wood or laminated layers of fiberglass and resin.

Design

The history of surfboard design has been one of constant experimentation. Except for a period in the 1960s when there was an effort to market standardized, mass-produced boards, most surfboards have been individually designed and hand crafted by talented surfboard builders. Over the last four decades, boards have gotten shorter, then longer, then shorter again. One fin was followed by two fins, then three fins, as builders tried different designs to improve the board's ability to perform maneuvers. Some board builders used channels cut lengthwise along the bottom to improve stability.

Today, surfboard builders continue to experiment with board design as surfers search

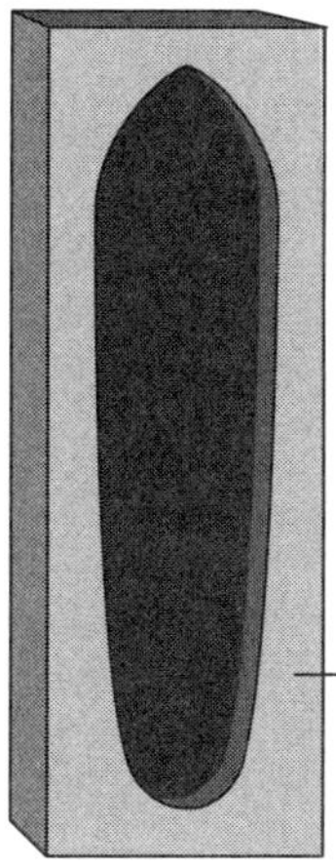

The foam core (or blank) is formed in a large two-piece cement mold. The mold is heated. When liquid polyurethane is poured, the heat triggers a chemical reaction, creating a dense, white foam.

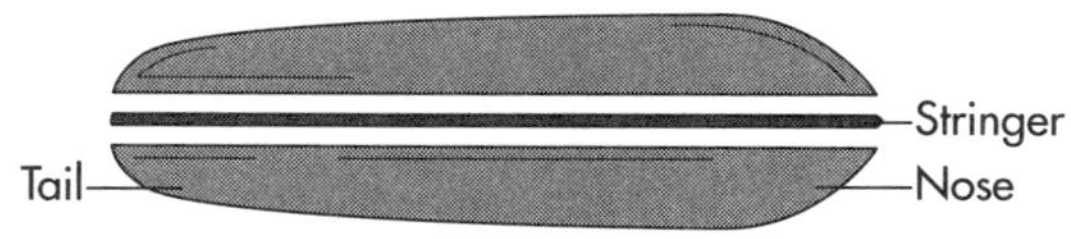

When the core hardens, it is cut in half. A stringer, made of wood or fiberglass, is glued in the middle, and the two pieces are clamped back together.

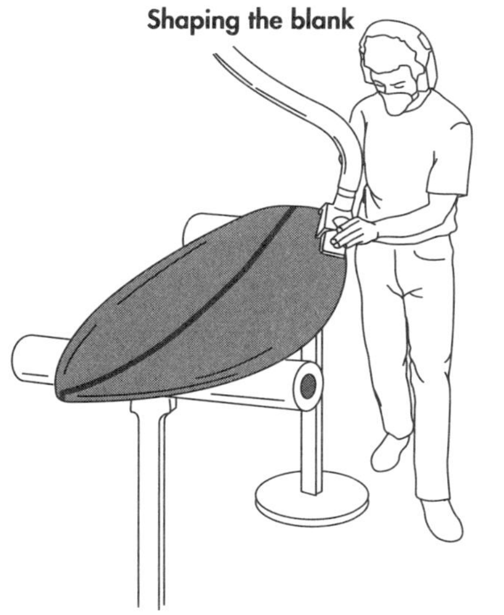

After the outline has been cut with a saber saw, the contours of the blank are shaped with a power planer. Here, the top, or deck, area is shaped. A design may be applied after the board is cleaned.

for that "perfect board." Some serious surfers use as many as five or ten different boards depending on their style of surfing and the surf conditions expected at a particular beach.

The Manufacturing Process

Most surfboards are built one at a time in small surfboard shops. Although techniques and materials vary from one surfboard builder to another, the following is a typical process.

Forming the foam core

1 The foam core, or blank, is formed in a large, cement mold roughly the shape of the surfboard. The mold is constructed in two halves, and the inside is lined with a special paper that keeps the foam from sticking to the mold. The two halves are clamped together and the mold is heated. When the liquid polyurethane chemicals are poured into the mold, the heat triggers a chemical reaction which begins forming a dense, white foam. Surfboard builders call this process "blowing the blank." After 25 minutes, the mold is opened and the foam

Laminating the shell

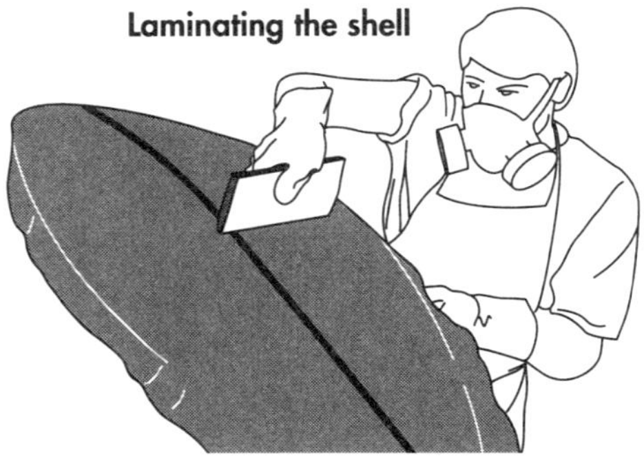

Next, laminating resin is spread over a fiberglass cloth using a rubber squeegee. The resin will harden in 15 minutes.

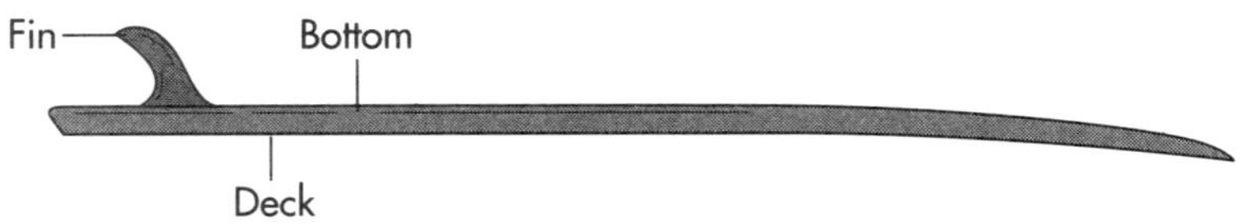

A filler coat of resin is applied. The board is flipped over, and the fin is affixed using fiberglass tape and a laminating resin. A small hole is drilled through the tail for the leg leash.

The board is cleaned and decals added. A final gloss coat is applied before the board is ready for use.

core is taken out and allowed to finish hardening.

Adding the stringer

2 Once the core is hard, it is cut in half vertically from the nose to the tail. A thin stringer is glued between the two halves, and the core is then clamped back together to dry. Stringers provide stiffness and help keep the board from breaking in half.

Shaping the blank

3 The outline of the finished board is traced onto the rough core using a wooden template as a guide. The outline is then cut out with a saber saw. Starting with the bottom of the blank, the surface is smoothed and contoured to its final shape with a power planer. There are no templates or automatic machines to do this job, just the trained eye and experienced touch of the surfboard builder. When the bottom is finished, the board is flipped over and the top is shaped. A power sander removes any ridges left by the planer, and the stringer is contoured with a hand plane. Rough sandpaper is used to shape the sides, or rails. The blank is given a final sanding with fine paper, the position for the fin is marked, and the builder signs the blank with a special design or signature.

Laminating the outer shell

4 The shaped blank is now ready to be covered with fiberglass and resin to form the hard, outer shell of the surfboard. First, the blank is blown clean with compressed air. If the board is to be colored or have a design on it, acrylic paint is applied directly to the foam with a spray gun or airbrush. When the paint is dry, fiberglass cloth is laid over the surface of the blank and cut to fit. The top of the board, or deck, is laminated first. A polyester resin, known

as a laminating resin, is mixed with a second chemical called a catalyst. This starts a chemical reaction which will cause the resin to harden in 15 minutes. The resin is poured over the fiberglass and spread evenly using a rubber squeegee. All of the fiberglass must be covered without leaving too much or too little resin in any spot. This process is known as glassing. When the deck is finished, the board is flipped over and the process repeated on the bottom. The board is then flipped once more, and the deck is given a second layer of fiberglass and resin for added strength and wear resistance. The laminating resin remains slightly tacky and rubbery when dry.

Applying the filler coat and adding the fin

5 A second coat of resin, called the filler coat or sanding resin, is applied next. The filler coat fills any surface imperfections left in the laminating resin. Sometimes, this coat is called a hot coat resin and contains wax. In either case, this resin contains a slightly different mix of chemicals which cause it to harden completely. The deck is coated first and the board is flipped over. The fin is secured with fiberglass tape and a laminating resin. When the fin resin is dry, the bottom of the board and the fin are given a filler coat. When both sides are dry, a small hole is drilled through the tail to attach the leg leash. The leg leash is an elastic cord, sometimes made of surgical rubber tubing, that the surfer attaches to one ankle. The leg leash keeps the board from floating away when the surfer falls, or "wipes out."

Sanding the board

6 Any excess resin must be carefully sanded away. A power sander is used for the broad surfaces, but the rails and other sharply contoured surfaces are hand sanded to avoid gouging into the fiberglass layer.

Final finishing

7 The board is blown clean with compressed air to remove any residual sanding dust. On some boards, decals or color graphics are added at this point. A final coat of gloss resin is then brushed onto the board. Like the other two layers of resin, this final gloss coat is mixed with a catalyst and will harden within 15 minutes. The board is set aside for at least 12 hours to allow the gloss coat to completely harden. As a final step, the board may be wet sanded with very fine sandpaper, then rubbed, buffed, and polished.

Quality Control

A surfboard is visually inspected several times during the manufacturing process. The blank is inspected for voids and other defects after it comes out of the mold. The shaping step, which is critical to the appearance and performance of the board, takes place in a well-illuminated area to allow the builder to spot any imperfections. The board is given a final inspection after the sanding and finishing steps to ensure it meets the craft standards of the builder.

Toxic Materials and Safety Considerations

Some of the materials and processes used in building a surfboard are hazardous. Surfboard builders must use the proper safety equipment and have an understanding of the dangers involved. The polyurethane chemicals used to make the foam core are toxic and flammable. This process requires explosion-proof fume removal equipment and careful control of the room temperature and humidity. The shaping process produces fine foam dust which can be harmful if inhaled. A dust mask is required for the person performing this task. Finally, the laminating resin gives off poisonous fumes which require the use of an appropriate respirator for the person doing the glassing.

The Future

Experimentation with surfboard design, materials, and construction techniques has produced some new approaches to surfboard manufacturing. As with anything new, there are advantages and disadvantages to each approach.

In the area of surfboard design, the use of computers—especially those known as computer aided design, or CAD, systems—has simplified the design process. With CAD, the board builder can create a three-dimensional picture of a new board design, change dimensions and contours, then print out a finished drawing and contour templates. This saves considerable time over the traditional method of building and trying each new design, but many builders still rely on their eyes and hands to judge the look and feel of a new board.

In the area of materials, some builders have tried boards built with a styrofoam core instead of polyurethane and an epoxy resin instead of polyester. The advantages of this combination are lighter weight, greater strength, and better impact resistance. The epoxy resin also produces less toxic fumes. The disadvantages include greater complexity to the resin preparation process, longer time to manufacture, and significantly greater cost. A variation of this approach uses graphite fiber cloth for reinforcement rather than glass fiber (fiberglass). This adds even more cost and produces boards in only one color—black.

New approaches to surfboard construction include a computer-numerical-controlled (**CNC**) shaping machine that can shape and sand a blank in about 25 minutes instead of the several hours required for hand shaping. The disadvantage is that the machine is very expensive and must be reprogrammed every time a new design is required. Another approach uses an existing surfboard as a mold pattern, then produces a duplicate shell which is filled with foam. Total time from start to finish is about 4.5 hours. Once again, however, the machine is very expensive and cannot produce new designs without an existing board to use as a pattern.

For the foreseeable future, surfers are expected to continue to demand custom-built boards at reasonable prices. The majority of this demand will be met by the hundreds of small surfboard crafters who build boards one at a time by hand.

Where To Learn More

Books

Kinstle, James. *Surfboard Design and Construction.* Natural High Express Publishing Company, 1975.

Lueras, Leonard. *Surfing: The Ultimate Pleasure.* Workman Publishing, 1984.

Shaw, Stephen M. *Surfboard.* Transmedia, 1983.

Wardlaw, Lee. *Cowabunga!: The Complete Book of Surfing.* Avon Books, 1991.

Young, Nat. *The History of Surfing.* Palm Beach Press, 1983.

Periodical

Krakauer, Jon. "Their Turf Is the Surf." *Smithsonian Magazine*, June 1989, pp. 106-19.

—*Chris Cavette*

T-Shirt

Background

T-shirts are durable, versatile garments with mass appeal that may be worn as outerwear or underwear. Since their creation in 1920, T-shirts have evolved into a two-billion dollar market. T-shirts are available in a variety of colors, patterns, and styles, such as the standard crew neck and V-neck, as well as tank tops and scoop necks. T-shirt sleeves may be short or long, capped, yoked, or raglan. Additional features include pockets and decorative trim. T-shirts are also popular garments on which to display one's interests, tastes, and affiliations using customized screen prints or heat transfers. Printed shirts may feature political slogans, humor, art, sports, as well as famous people and places. T-shirts are also inexpensive promotional vehicles for products and special events.

T-shirts fit just about anyone in any size, from infants to seniors. Adult sizes are generally small, medium, large, and extra-large, while sizes for toddlers are determined by month and weight. In addition, to compensate for the larger heads of infants relative to their bodies, shirts are specially designed with shoulder openings that may be fastened with buttons or snaps.

Raw Materials

The majority of T-shirts are made of 100% cotton, **polyester**, or a cotton/polyester blend. Environmentally conscious manufacturers may use organically grown cotton and natural dyes. Stretchable T-shirts are made of knit fabrics, especially jerseys, rib knits, and interlock rib knits, which consist of two ribbed fabrics that are joined together. Jerseys are most frequently used since they are versatile, comfortable, and relatively inexpensive. They also are a popular material for applying screen prints and heat transfers. Some jerseys come in tubular form, simplifying the production process by reducing the number of seams. Rib knit fabrics are often used when a snugger fit is desired. Many higher quality T-shirts are made of durable interlock rib knit fabrics.

Neckbands add support to the garment and give the neckline of the T-shirt a more finished look. Neckbands are generally one-by-one inch rib knits, although heavier fabrics or higher quality T-shirts may require two-by-two rib knits. Neckband fabrics may be tubed rib knits of specific widths, or flat fabric that must be seamed. Additional T-shirt materials include tape or seam binding, made of a twill or another stiff fabric. Binding reinforces the neckline and shoulder seams and by covering the seams, it protects them from ripping apart under tension. Alternatively, elastic may be used at the shoulder seams so they remain flexible.

Thread is of course an essential element in sewing any garment. Several types and colors of thread may be used to make a single T-shirt. Some manufacturers use white thread for seams on all their shirts, regardless of color, thus eliminating the extra labor involved in changing the thread. Visible topstitching is done with a color of thread that blends with the fabric. Colorless, or monofilament, thread could be used for hems of any color fabric, again eliminating the need to change thread often, though monofilament thread may irritate the skin somewhat. Finally, optional decorative features may include trim, such as braiding,

Since their creation in 1920, T-shirts have evolved into a $2-billion market. T-shirts are available in a variety of colors, patterns, and styles and fit just about anyone.

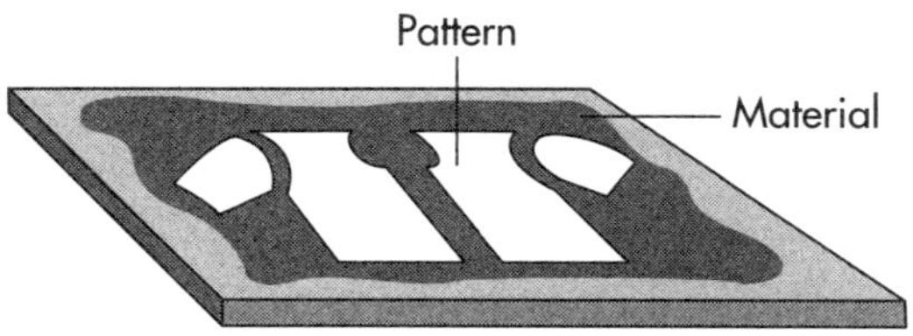

Once a T-shirt style is designed, the fabric sections are cut to the pattern dimensions.

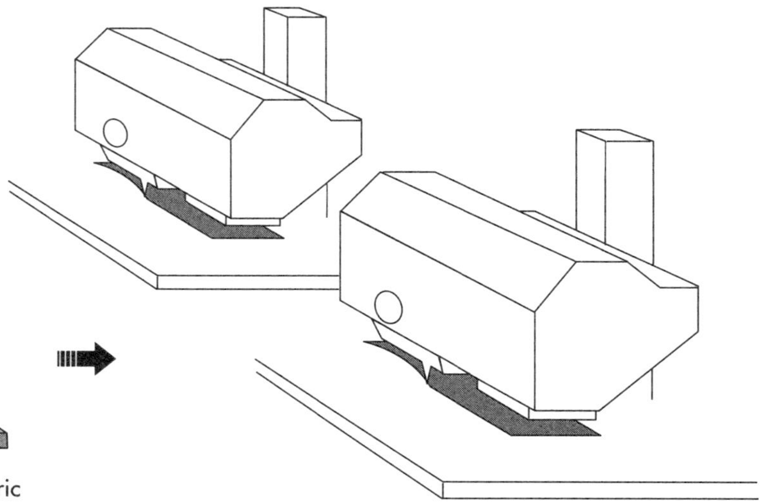

For fabric that is not tubed, the separate pieces for the front and back sections are stitched together on the sides.

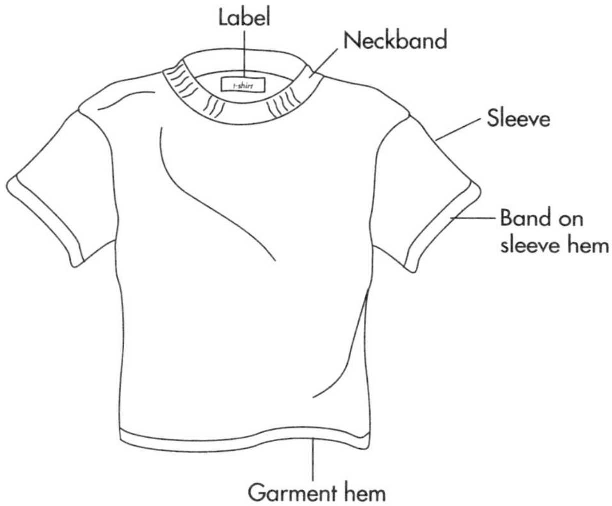

Making T-shirts is a fairly simple and largely automated process. Specially designed machines integrate cutting, assembling, and stitching for the most efficient operations.

contrasting cuffs, appliqués, and heat transfer or screen print designs.

The Manufacturing Process

Making T-shirts is a fairly simple and largely automated process. Specially designed machines integrate cutting, assembling, and stitching for the most efficient operations. The most commonly used seams for T-shirts are narrow, superimposed seams, which are usually made by placing one piece of fabric onto another and lining up the seam edges. These seams are frequently stitched with an overedge stitch, which requires one needle thread from above and two looper threads from below. This particular seam and stitch combination results in a flexible finished seam.

Another type of seam that may be used for T-shirts are bound seams, in which a narrow piece of fabric is folded around a seam, as at the neckline. These seams may be stitched together using a lockstitch, chainstitch, or overedge stitch. Depending on the style of the T-shirt, the order in which the garment is assembled may vary slightly.

Styling

1 The T-shirt style is designed and the dimensions are transferred to patterns. Adjustments are made for size differences and stylistic preferences.

Cutting

2 The T-shirt sections are cut to the dimensions of the patterns. The pieces consist of a tubed body, or separate front and back sections, sleeves, perhaps pockets, and trim.

Assembling the front and back

3 For fabric that is not tubed, the separate pieces for the front and back sections must be stitched together at the sides. They are joined at the seam lines to form a simple, narrow, superimposed seam and stitched together using an overedge stitch. Care must be taken to avoid a needle cutting the yarn of the fabric, which can lead to tears in the garment.

Assembling the sleeves

4 The hems of sleeves are generally finished before they are fitted into the garment, since it is easier to hem the fabric while it is flat. An automated system moves the sleeves to the sewing head by conveyor. The edge may be finished by folding it over, forming the hem and stitching, or by applying a band. The band may be attached as a superimposed seam or folded over the edge as binding.

5 If the T-shirt body is tubular, the sleeve material is first sewn together, and then set into the garment. Alternatively, if the T-shirt is "cut and sewn," the unseamed sleeve is set into place. Later during the final stage of sewing the shirt, the sleeve and side seams are sewn in one action.

Stitching the hem

6 The garment hem is commonly sewn with an overedge stitch, resulting in a flexible hem. The tension of the stitch should be loose enough to allow stretching the garment without tearing the fabric. Alternative hem styles include a combination of edge finishing stitches.

Adding pockets

7 Pockets may be sewn onto T-shirts intended for casual wear. Higher quality T-shirts will insert an interlining into the pocket so that it maintains its shape. The interlining is inserted into the pocket as it is sewn onto the T-shirt front. Pockets may be attached to the garment with automated setters, so the operator only has to arrange the fabric pieces, and the mechanical setter positions the pocket and stitches the seam.

Stitching the shoulder seams

8 Generally, shoulder seams require a simple superimposed seam. Higher quality T-shirt manufacturers may reinforce seams with tape or elastic. Depending on the style of the T-shirt, the seams at the shoulder may be completed before or after the neckband is attached. For instance, if a tubular neckband is to be applied, the shoulder seams must first be closed.

Attaching the neckband

9 For crew neck shirts, the neck edge should be slightly shorter in circumference than the outer edge where it is attached to the garment. Thus, the neckband must be stretched just the right amount to prevent bulging. Tubular neckbands are applied manually. The bands are folded, wrong sides together, stretched slightly, and aligned with the neckline. The superimposed seam is stitched with an overedge stitch.

Bound seams are finished with a cover stitch and are easy to achieve. Bound seams may be used on a variety of neckline styles. The process entails feeding ribbed fabric through machines which fold the fabric and apply tension to it.

Some neckbands on lower-priced shirts are attached separately to the front and back necklines of the garment. Thus when the shoulder seams are stitched, seams are visible on the neckband.

V-necks require the extra step of either lapping or mitering the neckband. In the former process, one side is folded over the other. A mitered seam is more complex, requiring an operator to overlap the band accurately and stitch the band at center front. An easier method for a V-neck look is to attach the band to the neckline and then sew a tuck to form a V.

Finishing the neckline

10 Necklines with superimposed seams may be taped, so that the shirt is stronger and more comfortable. Tape may be extended across the back and over the shoulder seams to reinforce this area as well and to flatten the seam. The seam is then cover stitched or top stitched.

Label setting

11 One or more labels are usually attached at the back of the neckline. Labels provide information about the manufacturer, size, fabric content, and washing instructions.

Optional features

12 Some T-shirts will have trim or screen prints added for decorative purposes. Special T-shirts for infants have larger openings at the head. The shoulder seams are left open near the neck, and buttons or other fasteners are attached.

Finishing operations

13 T-shirts are inspected for flaws in the fabric, stitching, and thread.

14 High-quality T-shirts may be pressed through steam tunnels before they are packaged. Packaging depends on the type of T-shirt and the intended distribution outlet. For underwear, the shirts are folded and packaged in pre-printed bags, usually of clear plastic, that list information about the product. Shirts may be boarded, or folded around a piece of cardboard, so that they maintain their shape during shipping and on the shelf. Finally, they are placed into boxes by the dozen or half-dozen.

Quality Control

Most of the operations in manufacturing clothing are regulated by federal and international guidelines. Manufacturers may also set guidelines for the company. There are standards that apply specifically to the T-shirt industry, which include proper sizing and fit, appropriate needles and seams, types of stitches, and the number of stitches per inch. Stitches must be loose enough to allow the garment to stretch without breaking the seam. Hems must be flat and wide enough to prevent curling. T-shirts must also be inspected for proper application of necklines, which should rest flat against the body. The neckline should also recover properly after being slightly stretched.

The Future

Exposure to sun's harmful rays has become a concern to many people who enjoy outdoor activities. In addition to **sunscreen** and sun glasses, sun-blocking T-shirts are now available. Founded by Harvey Schakowsky, SPF Wear company has introduced a line of clothing, including T-shirts, that blocks out 93-99% of ultraviolet rays. A typical T-shirt blocks out only 50% of the rays. Using a fabric called Solarweave, these new T-shirts are made out of synthetically woven nylon treated with a special chemical substance.

Where To Learn More

Books

Carr, H. and B. Latham. *Technology of Clothing Manufacture.* Oxford BSP Professional Books, 1988.

Glock, Ruth E. and Grace I. Kunz. *Apparel Manufacturing: Sewn Product Analysis.* Macmillan, 1990.

Solinger, J. *Apparel Manufacturing Handbook.* Van Nostrand Reinhold, 1980.

Periodicals

Callahan, Peter. "Sunday Best: Protective Wear for Your Day in the Sun." *Omni*, October 1992, p. 35.

Kopkind, Andrew. "From A to Tee." *Harper's Bazaar*, July 1993, pp. 34-36.

—Audra Avizienis

Tea bag

Background

Tea has existed as a beverage since 2000 B.C. The brewing, serving, and drinking of tea are time-honored rituals throughout the world. While there is general agreement that the tea trade began in China, both China and India lay claim to discovering the dietary properties of tea leaves. The Chinese tell the story of a mythical emperor named Shen Nung who was so particular about his nutrition that he boiled his drinking water before he drank it. One day, the story goes, the wind caught some of the leaves on the tree branches that he had used to build a fire. The leaves floated into his boiling water and, lo, tea was created.

In India, the discovery is attributed to Bodhidharma, an actual person who founded the Ch-an School of Buddhism. In A.D. 527, after four years of a self-imposed nine-year meditation, Bodhidharma grew sleepy. In an attempt to stay awake, he began to chew on the twigs of a nearby tree and suddenly found himself wide awake; he had discovered tea.

The tea bush is a white-flowered evergreen in the Camellia family. Chinese documents record it as indigenous to the Hunan province in southwest China. In modern times, it is generally accepted that the original tea bush grew in India and was brought to China. It thrives in a rocky terrain. In approximately A.D. 350, tea cultivation was also reported in the Szechwan province along the Yangtze River.

During the T'ang Dynasty in the eighth century, tea drinking achieved the status of an art form. Tea merchants hired a man named Lu Yu to compile the first written record of Chinese tea ceremonies. Entitled *Ch'a Ching* (*The Class of Tea*), the three-volume work revolutionized the tea industry. The second volume includes an exhaustive list of the equipment necessary to brew tea correctly; all-told, 24 items are listed.

In the 12th century, monks from the Zen sect of Buddhists brought the tea-brewing process home to Japan from their travels in China. Yeisei, a Buddhist abbot, is credited with writing Japan's first tea book: *Kitcha-Yojoki* (*Book of Tea Sanitation*). Since then, Zen Buddhism and tea have grown inseparable.

Several Europeans mentioned bringing tea back home from their visits to the Orient. One of them was Giambattista Ramusio, an editor of travel books and a diplomatic representative of the Venetian government. Gaspar da Cruz, a Portuguese Jesuit priest and missionary, also brought tea when he returned. In 1520, Ferdinand Magellan discovered the straits that now bear his name around the southern tip of South America and opened the door to a what would become a well-traveled westward trade route between Europe and the Orient.

Holland was the first to record the purchase of tea in 1607. The tea was first sold at apothecary shops, then in stores where spices and sugar were sold. By the 18th century, stores devoted entirely to the sale of tea and coffee had opened. The first tea sold to the English public occurred in 1657 at a coffee house called Exchange Alley. When King Charles II married the Portuguese princess Catherine of Braganza in 1661, her dowry included tea. Fifty-five years later, on October 12, 1712, Thomas Twining opened England's first tea shop.

England's East India Company grew prosperous from its opium trade to China and

Tea bags were invented quite by accident in 1904 when Thomas Sullivan, an enterprising merchant, wrapped samples of tea leaves in silk bags and sent them to prospective customers, some of whom dipped the bags directly into boiling water.

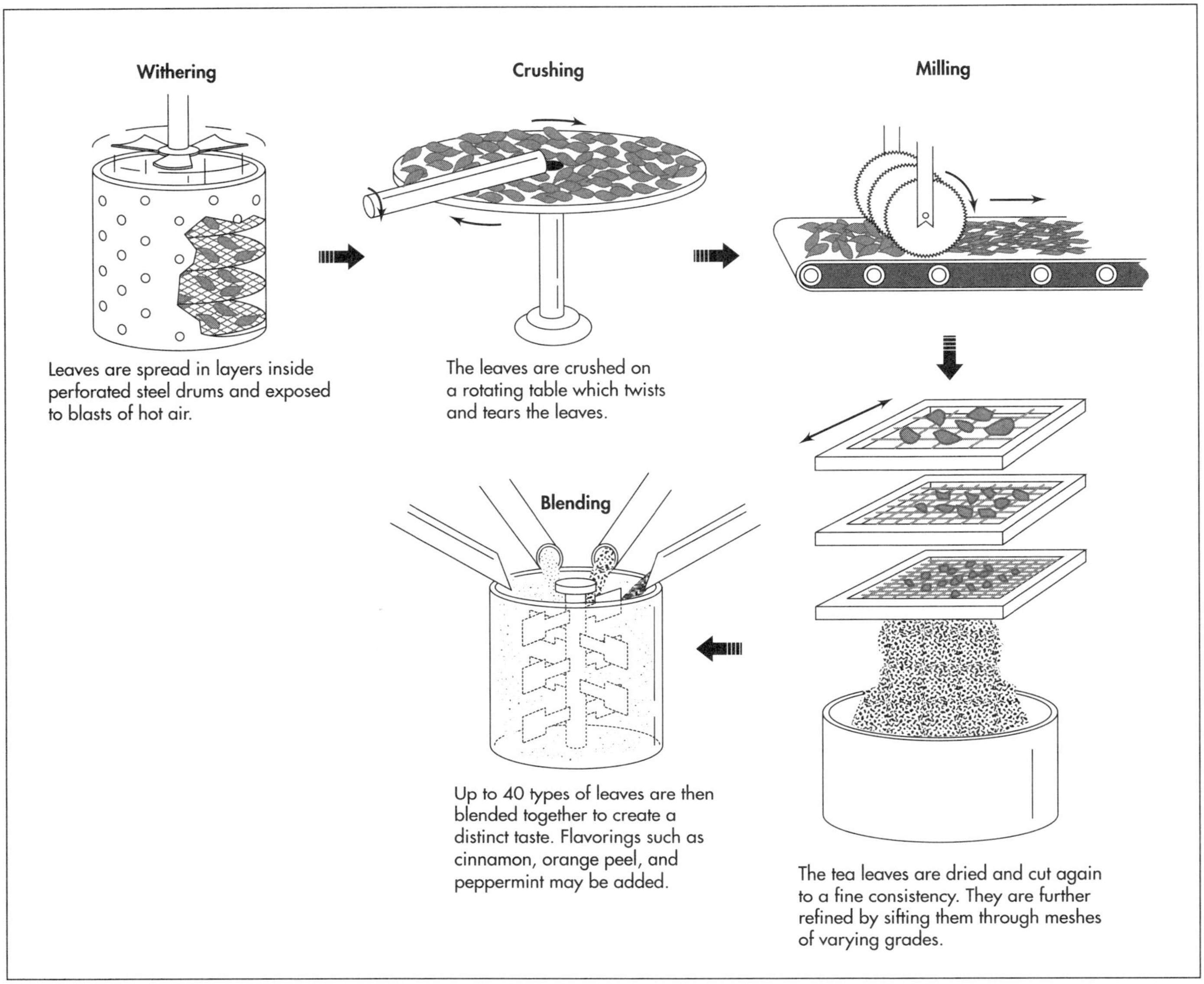

although the English addiction to tea was not as detrimental as the Chinese addiction to opium, the company also profited by the sale of tea it brought back in exchange. The English government ignored the questionable nature of the company's business so that it could levy a high import tax on tea. In spite of the tariff, or perhaps because of it, a great deal of tea was smuggled into the country. The tea tax also figured prominently in the American Revolution, as witnessed by the infamous Boston Tea Party when rebellious colonists pitched a shipment of East India Company tea into the harbor.

Until the early 20th century, tea was sold loose. Tea bags were invented quite by accident in 1904 when Thomas Sullivan, an enterprising merchant, wrapped samples of tea leaves in silk bags and sent them to prospective customers, some of whom dipped the bags directly into boiling water. The silk bags gave way to gauze pouches and eventually to specially treated filter **paper**.

Herb teas, caffeine-free and cultivated from a variety of plant leaves, flowers, roots, bark, and seeds, have become enormously popular over the last 20 years. Although they were probably brewed as early as prehistoric times, herb teas were primarily consumed for medicinal purposes.

Raw Materials

Tea bags are composed of two main ingredients: processed tea leaves and filter-paper

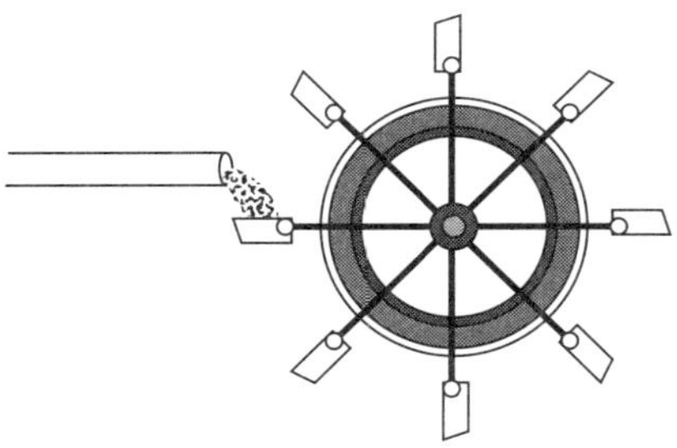

The processed and blended tea leaves are sent through flow tubes to a doser wheel, which separates the tea into pre-measured amounts of two grams.

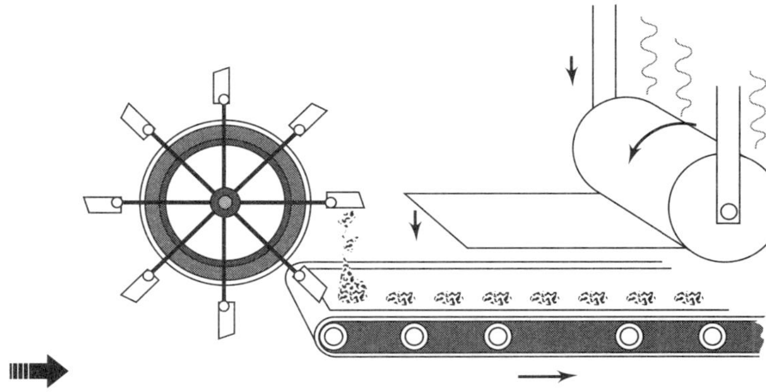

The doser wheel drops the tea onto a layer of paper on a conveyor belt. A second layer of paper is lowered on top of the tea. Next a heat-sealing drum quickly seals the papers together.

The tea is packaged and ready to be sold for consumer enjoyment.

bags. The top tea leaves and leaf buds are hand-picked from the plant. The leaves are then subjected to several processes including withering, rolling, drying, cutting, and blending. The intensity and duration of each process differs according to the type of tea.

The filter paper is made primarily of abaca, the leafstalk of Philippine bananas also known as Manila hemp.

The Manufacturing Process

Withering

1 Traditional withering practices call for manually spreading the leaves in thin layers and exposing them to the open air for 18-20 hours. Modern factory methods place the leaves in troughs, perforated drums, or tunnels and expose them to mechanical blasts of hot air. This process oxidizes the polyphenols, or tannins, the primary active ingredient, and turns the tea leaves a coppery color.

Crushing

2 The leaves are crushed either by hand or on rotating tables called rolling machines. Either method twists the leaves so that they are eventually coated with their own juices and torn into smaller pieces. Some companies employ high-tech machin-

ery similar to tobacco-cutting machines to crush and tear the tea leaves.

Drying

3 Black tea leaves are mechanically dried using a high-temperature method to seal in juices and flavor. This process turns the leaves to their characteristic black color.

Oolong tea leaves are rolled, dried, and rolled again. The drying time is shorter than that for black tea, therefore the fermentation is less natural and half or less of the polyphenols are oxidized.

Green tea leaves are steamed within 24 hours of harvesting, using either moist or dry heat in perforated drums or hot **iron** pans. This process destroys enzymes and prevents fermentation and the oxidation of polyphenols.

Herb tea is simply bundled together and hung upside down to air dry.

Milling

4 After the leaves are dried, they are brought to a mill room, where they are cut with a rotating blade into varying degrees of fineness, depending on the type of tea. The cut leaves are further refined by sifting them through mechanical sieves with meshes of varying grades. The tea used in tea bags are typically broken-grade or small-sized teas because they require a shorter brewing time.

Blending

5 The leaves are blended according to company recipes to achieve a uniform taste and texture. Most teas are a blend of between 20-40 types of tea leaves. The blending process may also include the addition of natural flavorings such as cinnamon, orange peel, nutmeg, cloves, chocolate, licorice root, peppermint, ginger, crushed hibiscus flowers, fennel seeds, and chicory root.

Measuring

6 The processed and blended tea leaves are stored in hoppers that hold up to 800 pounds (363 kg) of tea. Flow tubes connect each hopper to a doser wheel. The doser wheel resembles a Ferris wheel with small chambers in the place of seats. Air pushes the leaves through the flow tube and into the wheel which separates the tea into the chambers in pre-measured amounts, usually two grams.

Tea bag assembly

7 Two large rolls of filter paper are fed over the top and underneath the doser wheel. As each chamber arrives at the bottom of the doser wheel, it releases the tea onto the bottom paper layer of paper as it moves along a conveyer belt. The top layer of paper is lowered onto the lower layer so that each measure of tea is sandwiched between the two layers.

8 A conveyer belt moves the three components to a heat-sealing drum fitted with an indentation pattern. The drum quickly seals the paper along the indentation lines. The timing of this process is closely monitored because too much heat would adversely affect the tea.

9 The sealed paper continues along a conveyer belt until it reaches a perforation blade that is calibrated to cut the paper into precise squares. After a string and tag are stapled to the bag, they are dropped into pre-printed boxes.

Quality Control

Professional tea tasters check each batch of tea before it is inserted into the filter paper. Tea tasting is an art, not unlike wine tasting. Cups of brewed tea are lined up along with bowls of the tea leaves from the same batch. Tasters slurp the tea to the back of their throats, atomizing the tea so that they can taste it and smell it at the same time. The tasters also examine the unbrewed tea leaves to check for cleanliness, purity, and freshness.

The tea must also meet company standards. Each tea is blended to achieve a particular taste and appearance, therefore company recipes are strictly followed for consistency. Consistency is also maintained through computerized control systems that regulate the speed of the manufacturing machinery and heating processes. The systems alert plant workers to breakdowns and jams.

The Future

In spite of coffee's popularity, tea continues to be a fashionable beverage. The Tea Council of the United States estimates that Americans alone consume 122 million cups of tea each day. Worldwide, it is the second most popular beverage, preceded only by water.

Recently, the scientific community has become interested in the potential health benefits of tea, particularly in those properties that could lower blood pressure and blood cholesterol levels, stabilize blood sugar, prevent tooth decay, and inhibit the growth of cancerous tumors.

Where To Learn More

Books

Goodwin, Jason. *A Time for Tea*. Alfred A. Knopf, 1991.

Perry, Sara. *The Tea Book: A Gourmet Guide to Buying, Brewing, & Cooking*. Chronicle Books, 1993.

Woodward, Nancy Hyden. *Teas of the World*. Macmillan Publishing Company, 1980.

Periodicals

Booe, Martin. "Tea Tempest." *Chicago Tribune*, January 16, 1995, Sec. 5, p. 1.

Carton, Barbara. "Tea Companies Get Steamed Up About Who Invented the Wheel." *The Wall Street Journal*, January 12, 1995, p. B1.

Charles, Dan. "A Cup of Green Tea a Day May Keep Cancer Away." *New Scientist*, September 14, 1991, p. 17.

Evans, Mary Anne. "A Perfect Cup of Tea." *Gourmet*, November 1993, pp. 52, 54.

Raloff, Janet. "Add Tea to That Old 'Apple a Day' Adage." *Science News*, October 30, 1993, p. 278.

—Mary F. McNulty

Telescope

The earliest known telescope was a refractor built by the Dutch eyeglass maker Hans Lippershey in 1608 after he accidentally viewed objects through two different eyeglass lenses held a distance apart.

Background

A telescope is a device used to form images of distant objects. The most familiar kind of telescope is an optical telescope, which uses a series of lenses or a curved mirror to focus visible light. An optical telescope which uses lenses is known as a refracting telescope or a refractor; one which uses a mirror is known as a reflecting telescope or a reflector. Besides optical telescopes, astronomers also use telescopes that focus radio waves, X-rays, and other forms of electromagnetic radiation. Telescopes vary in size and sophistication from homemade spyglasses built from cardboard tubes to arrays of house-sized radio telescopes stretching over many miles.

The earliest known telescope was a refractor built by the Dutch eyeglass maker Hans Lippershey in 1608 after he accidentally viewed objects through two different eyeglass lenses held a distance apart. He called his invention a *kijker*, "looker" in Dutch, and intended it for military use. In 1609, the Italian scientist Galileo Galilei built his own telescopes and was the first person to make astronomical observations using them. These early telescopes consisted of two glass lenses set within a hollow **lead** tube and were rather small; Galileo's largest instrument was about 47 inches (120 cm) long and 2 inches (5 cm) in diameter. Astronomers such as Johannes Kepler in Germany and Christian Huygens in Holland built larger, more powerful telescopes throughout the 1600s. Soon these telescopes got too large to be easily controlled by hand and required permanent mounts. Some were more than 197 feet (60 m) long.

The ability to construct enormous telescopes outpaced the ability of glassmakers to manufacture appropriate lenses for them. In particular, the problems caused by chromatic aberration (the tendency for a lens to focus each color of light at a different point, leading to a blurred image) became acute for very large telescopes. Scientists of the time knew of no way to avoid this problem with lenses, so they designed telescopes using curved mirrors instead.

In 1663, the Scottish mathematician James Gregory designed the first reflecting telescope. Alternate designs for reflectors were invented by the English scientist Isaac Newton in 1668 and the French scientist N. Cassegrain in 1672. All three designs are still in use today. In the 1600s, there was no good way to coat glass with a thin reflective film, as is done today to make mirrors, so these early reflectors used mirrors made out of polished metal. Newton used a mixture of copper, tin, and arsenic to produce a mirror which could only reflect 16% of the light it received; today's mirrors reflect nearly 100% of the light that hits them.

It had been known as early as 1730 that chromatic aberration could be minimized by replacing the main lens of the telescope with two properly shaped lenses made from two different kinds of glass, but it was not until the early 1800s that the science of glassmaking was advanced enough to make this technique practical. By the end of the 19th century, refracting telescopes with lenses up to a meter in diameter were constructed, and these are still the largest refracting telescopes in operation.

Reflectors once again dominated refractors in the 20th century, when techniques for constructing very large, very accurate mirrors were developed. The world's largest optical telescopes are all reflectors, with mirrors up to 19 feet (6 m) in diameter.

Raw Materials

A telescope consists of an optical system (the lenses and/or mirrors) and hardware components to hold the optical system in place and allow it to be maneuvered and focused. Lenses must be made from optical glass, a special kind of glass which is much purer and more uniform than ordinary glass. The most important raw material used to make optical glass is silicon dioxide, which must not contain more than one-tenth of one percent (0.1%) of impurities.

Optical glasses are generally divided into crown glasses and flint glasses. Crown glasses contain varying amounts of boron oxide, sodium oxide, potassium oxide, barium oxide, and zinc oxide. Flint glasses contain lead oxide. The antireflective coating on telescope lenses is usually composed of magnesium fluoride.

A telescope mirror can be made from glass that is somewhat less pure than that used to make a lens, since light does not pass through it. Often a strong, temperature-resistant glass such as Pyrex is used. Pyrex is a brand name for glass composed of silicon dioxide, boron oxide, and aluminum oxide. The reflective coating for telescope mirrors is usually made from aluminum, and the protective coating on top of the reflective coating is usually composed of silicon dioxide.

Hardware components that are directly involved with the optical system are usually manufactured from steel or steel and **zinc** alloys. Less critical parts can be made from light, inexpensive materials such as aluminum or acrylonitrile-butadiene-styrene plastic, commonly called ABS.

The Manufacturing Process

Making the hardware components

1 Metal hardware components are manufactured using standard metalworking machines such as lathes and drill presses.

2 Components made from ABS plastics (usually the external body of the telescope) are produced using a technique known as injection molding. In this process the plastic is melted and forced under pressure into a mold in the shape of the final product. The plastic is allowed to cool back into a solid, and the mold is opened to allow the component to be removed.

Making optical glass

3 The glass manufacturer mixes the proper raw materials with waste glass of the same type as the glass to be made. This waste glass, known as cullet, acts as a flux; that is, it causes the raw materials to react together at a lower temperature than they would without it.

4 This mixture is heated in a glass furnace until it has melted into a liquid. The temperature needed to form molten glass varies with the type of glass being made, but it is typically about 2550°F (1400°C).

5 The temperature of the molten glass is raised to about 2820°F (1550°C) to force air bubbles to come to the surface. It is then allowed to cool while being stirred constantly until it has reached about 1830°F (1000°C), at which point it is an extremely thick fluid. This viscous, molten glass is poured into molds with roughly the same shape as the lenses required.

6 After the glass has cooled to about 570°F (300°C), it must be reheated to about 1020°F (550°C) to remove internal stresses that form during the initial cooling period and which weaken the glass. It is then allowed to cool slowly to room temperature. This process is known as annealing. The final lens-shaped chunks of glass are known as blanks.

Making the lenses

The blanks are processed by the telescope manufacturer in three steps: cutting, grinding, and polishing. A mirror is formed in exactly the same way as a lens until the reflective coating is applied.

7 First a high-speed, rotating cylindrical cutter with a round diamond blade,

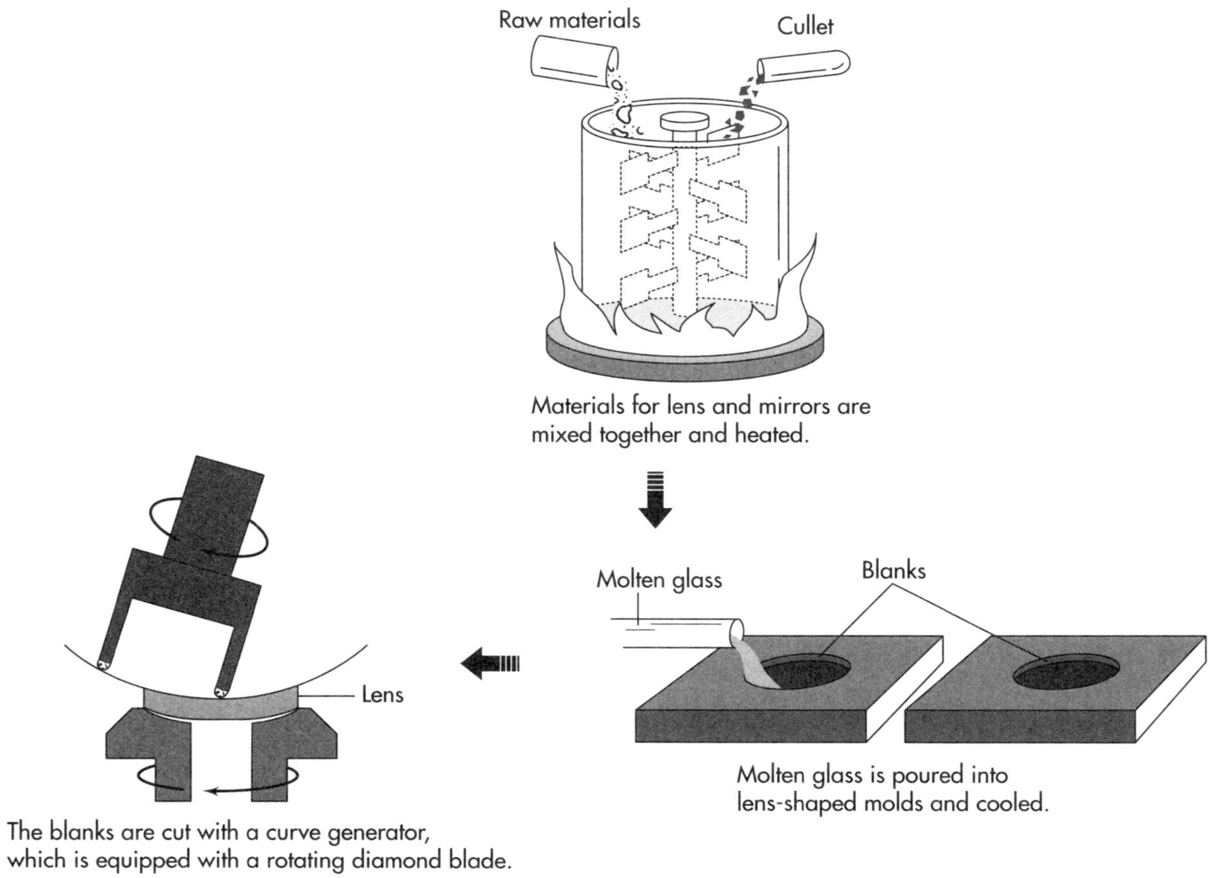

known as a curve generator, shaves the surface of the lens until a close approximation of the desired curve is achieved. The cut lens is inspected with a spherometer to check the curvature and is recut if necessary. The time required for cutting varies greatly with the type of glass being cut and the kind of lens being shaped. A lens may require several cuttings, each of which may take anywhere from a few minutes to more than half an hour.

8 Several cut blanks are placed on a curved block in such a way that their surfaces line up as if they were all part of one large spherical curve. This is necessary so that the grinding machine can grind them all in the same way. A cast iron grinding surface known as a tool is pressed onto them. During grinding, the block of lenses rotates while the tool is free to move at random on top of it. Between the tool and the block flows a slurry containing water, an abrasive to do the grinding (usually silicon carbide), a coolant to prevent the lenses from being damaged by overheating, and a surfactant to keep the abrasive from settling out. The speed at which the block rotates, the force placed on the lenses, the exact contents of the slurry, and other variables are controlled by experienced opticians to produce the exact type of lens desired. Each lens is once again inspected with a spherometer and reground if necessary. The total grinding process may take anywhere from one hour to eight hours. The ground lenses are cleaned and moved to the polishing room.

9 The polishing machine is similar to the grinding machine, but the tool is made from pitch—a thick, soft, resinous substance derived from coal tar or wood tar. A pitch tool is made by placing tape around the circumference of a curved dish, pouring in hot, liquid pitch with other ingredients such as beeswax and jeweler's rouge, and letting it

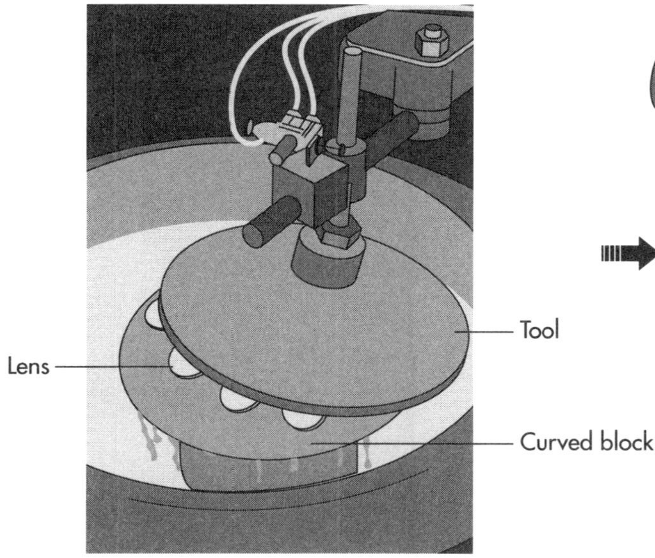

The lenses are then placed on a curved block, which grinds the lenses using a cast iron tool and abrasives.

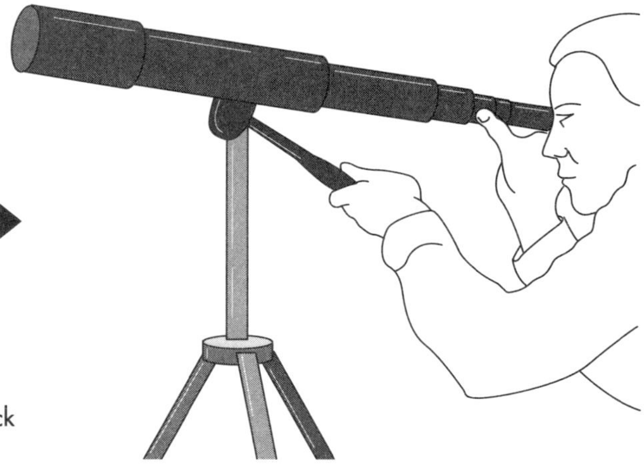

Once the lenses and mirrors are cleaned and polished, they are assembled onto the rest of the telescope.

cool back into a solid. A pitch tool can polish about 50 lenses before it must be reshaped. Polishing proceeds in the same manner as grinding, but instead of an abrasive the slurry contains a polishing substance, usually cerium dioxide, in the form of a very fine pink powder. The polished lenses are optically inspected and repolished if necessary. The polishing procedure may take anywhere from half an hour to four or five hours. The lenses are cleaned and are ready for coating.

Applying coatings

10 To make a lens into a mirror, a very thin, very smooth coating of aluminum is applied. Aluminum is heated in a vacuum to form a vapor. A negative electrostatic charge is applied to the surface of the lens so that the positively charged aluminum ions are attracted to it. Similar procedures are followed to apply a coating of silicon dioxide to protect the fragile surface of a mirror or to apply an antireflective coating of magnesium fluoride to the surface of a lens. The finished lens or mirror is inspected, labeled with a date of manufacture and a serial number, and stored until needed.

Assembling and shipping the telescope

11 The hardware components, lenses, and mirrors required to make a particular model of telescope are assembled by hand in an assembly line process. The completed telescope is packed with close-fitting expanded polystyrene foam to protect it from damage during shipping. The telescope is packed in a cardboard box and shipped to the retailer or consumer.

Quality Control

The most critical aspect of quality control for an optical telescope is the accuracy of the lenses and mirrors. During the cutting and grinding stages, the physical dimensions of the lens are measured very carefully. The thickness and the diameter of the lens are measured with a vernier caliper, an instrument which looks something like a monkey wrench. The outer, fixed jaw of the caliper is placed against one side of the lens and the inner, sliding jaw is gently moved until it meets the other side of the lens. In a classic vernier caliper, the dimensions of the lens are read very accurately using a scale which moves along with the inner jaw and which is compared with a stationary scale attached to the outer jaw. This type of caliper works much like a slide rule. There also exist electronic versions of this instrument, in which the measured dimension automatically appears on a digital display.

The curvature of a lens is measured with a spherometer, a device which resembles a pocket watch with three small pins protruding from its base. The outer two pins are fixed in place while the inner pin is free to move in and out. The spherometer is gently placed on the surface of the lens. Depending on the type of curve, the middle pin will either be higher than the other two pins or lower than the other two pins. The movement of the inner pin moves a needle on a calibrated dial on the face of the spherometer. This value is compared with the standard value that should be obtained for the desired curvature.

Tolerances vary with the type of lens being manufactured, but a typical acceptable variation might be plus or minus 0.0008 inches (20 micrometers). For a flat lens, generally one destined to become a flat mirror, the tolerance is much smaller, usually about plus or minus 0.00004 inches (1.0 micrometer).

During the polishing stage, these instruments are not accurate enough to ensure that the lens will work properly. Optical tests, which measure the way light is affected by the lens, must be used. One common test is known as an autocollimation test. The lens is placed in a dark room and is illuminated with a low intensity pinpoint light source. A diffraction grating (a surface containing thousands of microscopic parallel grooves per inch) is placed at the point where the lens should focus light. The grating causes an interference pattern of dark and light lines to form in front of and behind the focal point. The true focal point can thus be found precisely and compared with the theoretical focal point for the type of lens desired.

In order to test a flat lens, a lens that is known to be flat is placed face down on the lens that is to be tested, which rests on a piece of black felt. The microscopic gaps between the two lenses cause an interference pattern to appear when gentle pressure is applied. The light and dark lines are known as Newton's rings. If the lens being tested is flat, the lines should be straight and regular. If the lens is not flat, the lines will be curved.

The Future

The techniques used to produce excellent lenses and mirrors have been well understood for many years, and major innovations in this area are unlikely. One area of active research is in coating technology. New coating substances may be developed to provide better protection for mirrors and better prevention of loss of light through reflection for lenses.

A more dramatic area of progress is in the electronic accessories that accompany telescopes. Amateur astronomers will soon be able to obtain telescopes with built-in computer guidance systems that will enable them to automatically point the telescope at a selected celestial object and to track it night by night. They will also be able to attach video cameras to their telescopes and film such astronomical phenomena as lunar eclipses and the movements of planets and moons.

Where To Learn More

Books

Asimov, Isaac. *Eyes on the Universe: A History of the Telescope.* Houghton Mifflin, 1975.

Bell, Louis. *The Telescope.* Dover, 1981.

Manly, Peter L. *Unusual Telescopes.* Cambridge University Press, 1991.

Periodicals

Mullins, Mark. "A Truly Economical Telescope." *Sky and Telescope*, December 1993, pp. 91-92.

Nash, J. Madeleine. "Shoot for the Stars." *Time*, April 27, 1992, pp. 56-57.

Nelson, Ray. "Reinventing the Telescope." *Popular Science*, January 1995, pp. 57-59, 85.

—*Rose Secrest*

Tofu

History

Tofu is a highly versatile and nutritious food that is made from soybean curds. Although the word "tofu" is Japanese, the food seems to have originated in ancient China, where the Mandarin term is "doufu." The creation of tofu is generally attributed to the ruler Liu An of Huai-nan during the second century B.C. The creation of tofu was probably accidental. Although soybeans are not technically a grain, the Chinese considered the soybean one of their essential Five Sacred Grains, along with rice, wheat, barley, and millet. It is likely that Liu An prepared soybeans in much the same way as grains, by drying, mashing, and boiling. The addition of sea **salt** would not only have seasoned the puree, it would have also acted as a solidifying agent, forming curds. Another theory suggests that the curding process was simply imported from neighboring regions. Regardless, soybeans appear to have been processed into tofu by the second century B.C. using a sea water precipitate to solidify the tofu, a process still used by many manufacturers.

According to ancient text, soybeans were cultivated in northern China at least as early as the 15th century B.C., during the Chang period. A sixth century Chinese encyclopedia of agriculture, the first of its kind, cites that the explorer Choken brought back soybeans to China from his expeditions to Greece, Rome, and India. However, according to legend, the soya plant was cultivated centuries earlier. In 2838 B.C., the emperor Sheng-nung wrote a treatise on plants which describes the soya plant in detail. Chinese agricultural experts in 2207 B.C. also wrote about soybean cultivation. Clearly, soybeans were an important staple crop in China for quite some time. The soybean was also recognized for its regenerative properties: the roots of soybeans contain nodules, which discharge nitrogen, and thereby enrich the soil. This important quality made its mark on the old ideogram for the soya plant "su," which contained little lines to symbolize roots.

During the eighth century, Chinese Buddhist missionaries introduced the soya plant to Japan and Korea, although they may have been used there much earlier. Buddhist monks believed that a vegetarian diet was healthier for the spirit so they advocated eating the protein-rich tofu as an alternative to meat. First the upper classes of Japan adopted tofu into their diets and by 1400, during the Muromachi period, tofu was popular among all classes in Japan.

Until about World War II most Japanese and Chinese tofu was made in small family-run shops, each of them using the same ingredients, methods, and tools. In the 1960s, the Japanese Food Research Institute made recommendations for modernizing and standardizing tofu production throughout the nation. Their suggestions included using calcium sulfate as the thickening agent, rather than the natural sea water precipitate, *nigari*. They also recommended using pressure cookers to speed the process. Hydraulic presses and centrifuges replaced manual lever presses and hand-turned screw presses. Higher speed grinders and aluminum boxes replaced the original wooden boxes. Despite the improvements in efficiency and productivity, many believed that the new methods compromised the flavor of tofu. Traditionalist manufacturers still retain much of the old-style tofu production.

The United States is the largest soybean producer in the world, providing about two-thirds of the global supply.

The ingredients for making tofu are few. They include the milk of soybeans, water, and coagulating agents. The modern tofu manufacturing process is largely an automated version of the traditional hand method.

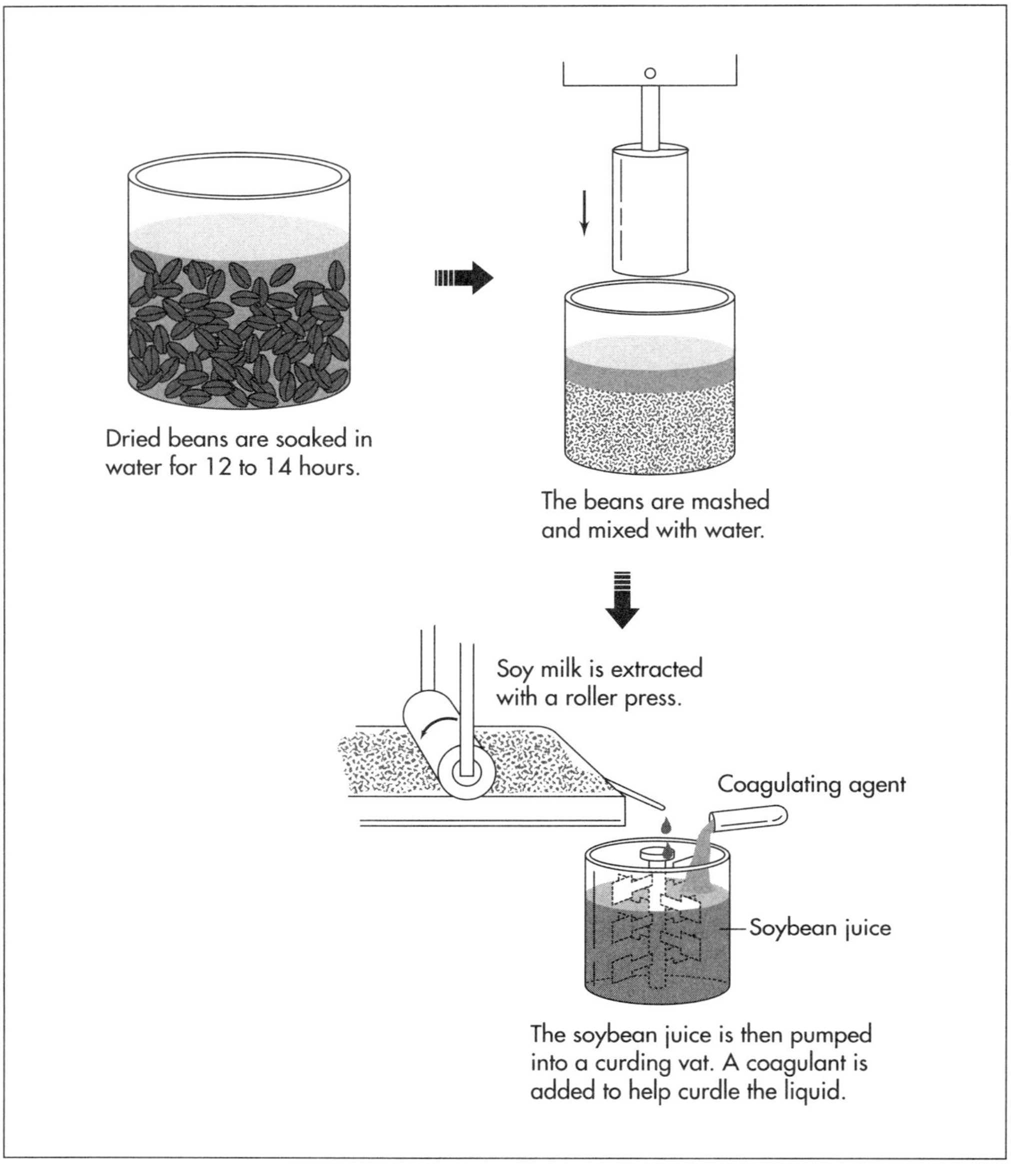

In the U.S., Americans did not readily adopt soybeans into their diet. They were available by the mid-1700s and popularized by Chinese immigrants traveling to California during the Gold Rush of the mid-1800s. As the Chinese immigrants later traveled to other parts of the country, they spread the recipe for tofu. The demand for tofu and other Asian foods also increased after World War II as military personnel returned from Asia, some with Asian spouses. By the 1900s soybeans were grown as a cash crop—primarily for regenerating the soil during crop rotation—as animal feed, and for producing oil and oilcake residue for manufacturing purposes. By the 1950s the U.S. rivaled Asia in its production of soybeans. Forty years later, the U.S. would become the largest soybean producer in the world, providing about two-thirds of the global supply.

During the 1970s, with the growing popularity of ethnic foods in the U.S., mainstream grocers began to offer products like tofu. At the same time, the price of meat soared, and tofu finally caught on as a popular substitute for meat, since tofu is high in protein and low in saturated fat. The increasing health-consciousness of the U.S. populace also contributed to the popularity of tofu. Twenty years later the tofu industry grew to more than a 4-million-pound indus-

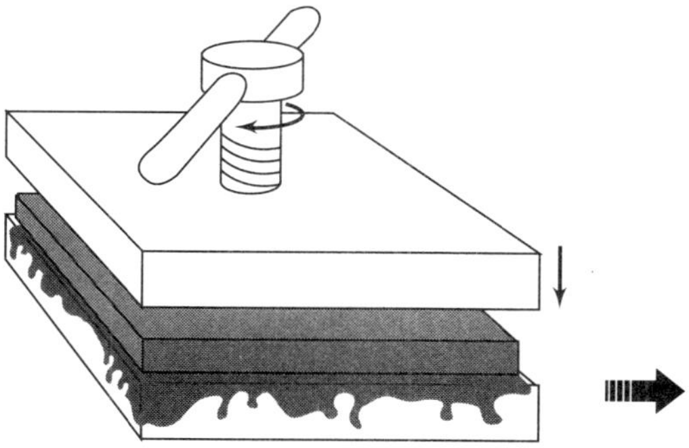

The curds are pressed either with a hand-turned screw press or lever press.

The pressed tofu is cut into blocks and washed in vats of water.

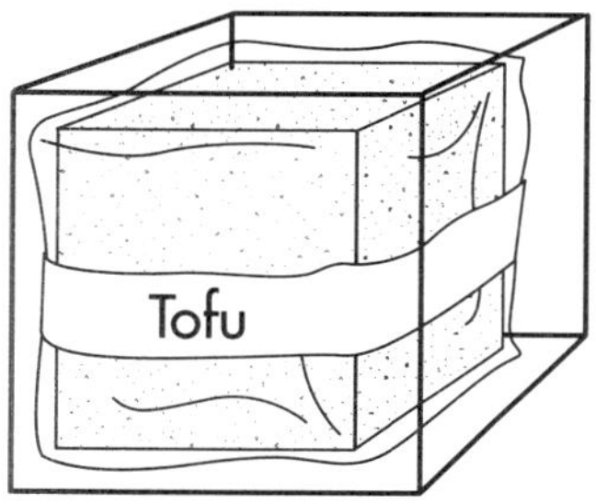

Tofu is packaged and pasteurized before it is sold.

try. Tofu is used in a variety of ways: as a meat substitute, an additive to entrees, a dessert base, and a liquid base for sauces, dips, and healthful shakes.

The Soybean

The soybean is perfectly balanced in the major food categories of protein, fat, and carbohydrates as well as in vitamins, and minerals. Soybeans also contain an ideal amount and combination of amino acids that are necessary for assimilating nutrients by the human body.

The scientific name for the soybean is *Glycine max,* and it is part of the botanical family Leguminosae. The soya plant has a slightly woody stalk and reaches a height of 30-36 inches (76-91 cm). The entire plant is covered in greenish hair. The leaves grow in groups of three and fall off as the beans mature. The soya plant produces papilionaceous (butterfly-shaped) flowers that are either white, red, or purple. The pods grow from 1-2 inches (2.5-5 cm) in length, each holding two or three seeds, which become soybeans. Soya seeds are either round or oval and are similar in size to peas. Their color is usually yellow but they may also be green, purple, brown, or a mixture of colors. Soybeans are pulses, that is, the plant has a symbiotic relationship with the bacteria, called rhizobia, that emit nitrogen through nodules in the soya plant's roots.

The soya plant may grow as far north as 52 degrees latitude, even though it is really a sub-tropical plant. Each climate requires slight alterations for growing soybeans, but in general, the beans are sown in the middle of May with heavy machinery. As the beans ripen, the soya leaves fall off. After the short growing period of 15 weeks, only the stalks and pods remain. The plants are harvested mechanically.

The Manufacturing Process

The ingredients for making tofu are few. They include the milk of soybeans, water, and coagulating agents. The modern tofu manufacturing process is largely an automated version of the traditional method, and much of the modern equipment is made in Japan. While an individual tofu maker might work with 20 gallons (76 l) of beans at a time, a contemporary processing facility can produce about 3.5 tons of tofu per day, using 5.7 tons of soybeans.

The first step in making tofu is soaking the soybeans and extracting the milk. A coagulant is added to curd the milk. Traditionally the coagulant used is *nigari*, which is a sea water precipitate rich in minerals such as magnesium and calcium chlorides. But modern manufacturers use either calcium sulfate or magnesium chloride. The soya curds are then processed into tofu in the desired form, primarily in custard-like blocks. A variety of textures may be produced, depending on the water content. Tofu comes in soft, firm, and extra-firm, as well as silken or in liquid form. A number of tofu flavors, such as Jalapeno and Cheddar, are also available.

Soaking the beans

1 Dried beans, which come in 60-pound (27 kg) sacks, are soaked in water for 12 to 14 hours. The beans soften as they absorb the water and double in size.

Processing the soybeans

2 After soaking, the beans are mashed with special Japanese stone grinders or other pureeing machines and mixed with water into a slurry. The slurry is boiled to neutralize an enzyme that hinders digestion.

3 The soy milk is extracted with a roller press, separating it from the pulp, which consists of the hull and fiber. This process may take about two hours to complete. The remaining pulp can be used to feed livestock.

Solidifying the soy milk

4 Once the whey is extracted, the soybean juice is pumped into curding vats. A coagulating agent is mixed in, such as calcium sulfate, magnesium chloride, or *nigari*. The coagulant alters the pH and curds the milk much like the process for making cottage cheese. This step takes about 20 minutes.

Pressing the tofu

5 Traditionally the curds are pressed with hand-turned screw presses or simple lever presses. The tofu may be pressed in cheesecloth-lined boxes. Modern systems use centrifuges or hydraulic presses. The whey drains off, leaving soft blocks of pressed curds. Tofu can be produced in a variety of textures, from a dense cheese-like texture to a softer or liquid form.

Cutting the tofu

6 Automated cutters slice the cake tofu into one-pound (.45 kg) blocks. The tofu blocks are washed in vats of water where they firm up and are stored until they are ready to be processed further.

Packaging the tofu

7 Tofu may be packaged into shrink-wrapped blocks or continuous thermoform packages. Water may be added to the packages, or tubs, and then they are sealed, weighed, and dated. Some companies process the soy milk directly in its package.

Pasteurizing the tofu

8 The packaged tofu is pasteurized at about 180°F (82°C). Pasteurization extends the shelf life of tofu to about 30 days. The tofu is then chilled in water until it is ready to be placed into boxes and shipped to distributors. Tofu must be refrigerated at below 45°F (7°C) to keep it fresh.

Quality Control

During the 1970s, when tofu was still a fledgling product in the U.S., there were few guidelines regulating the tofu industry. Therefore tofu manufacturers turned to the guidelines set for meat and dairy processors.

One of the early complaints about tofu in the U.S. was that it tasted bland, too beany, and astringent. But genetic engineering greatly improved the flavor of the soybeans used for tofu manufacture. Researchers have discovered that the enzyme lipoxygenase causes the off-flavor, and they have been able to breed soybeans with a lower content of lipoxygenase. Another method is to neutralize the undesirable taste by adding flavors to the tofu.

The Future

Technology will continue to improve the flavor and texture of tofu. Dozens of new tofu products enter the market each year and have expanded that segment to more than $100 million in the 1990s. Demand for soy-based food products will most likely continue to rise as medical research uncovers the health benefits associated with soybean consumption, namely the prevention and treatment of heart disease and cancer.

Where To Learn More

Book

Toussaint-Samat, Maguelonne. *A History of Food,* translated from the French by Anthea Bell, Blackwell Reference, 1987 in French, 1992 translation.

Periodicals

Clifford, Carlsen. "Utilitarian Soybean Curd Goes Exotic." *San Francisco Business Times,* June 28 1991, p. 1.

"Soy: The Bean Most Likely to Succeed in Fending Off Cancer, Heart Disease." *Environmental Nutrition,* May 1994, p.1.

Gerrietts, Marcie, Linda Cooke, and Marcia Wood. "Soy! It's No Ordinary Bean: Part 2." *Agricultural Research,* November 1993, p. 10.

Hewitt, Linda. "An Attractive Alternative: Protein from Fungi and Soya." *Food Manufacture,* May 1994, p. 25.

Kevin, Kitty. "Tofu Comes of Age." *Food Processing,* June 1994, pp. 81-82.

"Tofu Expands as a Healthy Alternative in American Diets." *Supermarket Business Magazine,* February 1992, p. 16A.

Wagner, Bill. "Surprising Sources for New Foods." *FDA Consumer,* November 1993, p. 28.

—Audra Avizienis

Toothbrush

On average each person in the United States purchases three toothbrushes every two years, although the American Dental Association (ADA) recommends that brushes be changed every three to four months.

Background

A toothbrush is a dental instrument used for cleaning teeth, ideally in conjunction with toothpaste or mouthwash. The toothbrush consists of a plastic handle and nylon bristles attached to the head of the brush. Contemporary designs offer a variety of styles and shapes in a market that has swelled to $600 million in the mid-1990s, in part because of price increases, but also because toothbrushes are replaced frequently. On average each person in the United States purchases three toothbrushes every two years, although the American Dental Association (ADA) recommends that brushes be changed every three to four months. The best-selling toothbrushes during the 1990s were the Oral-B brand, produced by Gillette Co.; a range of toothbrushes from the Colgate Palmolive Co.; and the Reach toothbrush made by Johnson & Johnson.

Modern medical research has shown that brushing teeth properly can prevent cavities, gingivitis, and periodontal, or gum, disease, which causes at least one-third of adult tooth loss. Gum disease occurs when plaque builds up, forming a gelatinous film that coats the teeth and gums. Plaque consists of about 75% bacteria, and it grows quickly. If teeth are not brushed correctly and frequently, it could lead to the calcification of saliva minerals, forming tartar.

Brushing one's teeth has long been considered an important part of dental hygiene. As long ago as 3000 B.C. ancient Egyptians constructed crude toothbrushes from twigs and leaves to clean their teeth. Similarly, other cultures such as the Greeks, Romans, and Indians cleaned their teeth with twigs. Some would fray one end of the twig so that it could penetrate between the teeth more effectively. By the 15th century the Chinese had designed a more sophisticated toothbrush complete with a brush attached to a handle. A century later the English nobility were using toothbrushes fashioned out of silver.

Toothbrush design has gone through few substantial changes in its long history. Until the early 1900s, toothbrush bristles were generally made of Siberian hog hair. But in 1938, the soft-bristled Miracle Tuft Toothbrush was invented. Within a decade, Oral-B was mass producing soft-bristled toothbrushes. In 1961, the electric toothbrush was introduced. Beginning in the late 1970s, the toothbrush industry started churning out a variety of new designs. They included variations in bristle shape, size, and texture, as well as unconventional handle styles.

Styles of Toothbrushes

By the 1990s, countless styles of toothbrushes filled the shelves of supermarkets and drug stores, many claiming their superiority over other brands. Consumers could choose from toothbrushes with soft or hard bristles, with natural or synthetic bristles, and in a variety of sizes, colors, and configurations. For instance, the Reach toothbrush by Johnson & Johnson was the first toothbrush designed with an angled handle which was intended to make brushing back teeth easier. A later variation on the Reach brush included bristles in a zigzag design.

Colgate also offered an angled brush. One model, called Rippled Bristles, was designed to reach the plaque trapped between teeth. Gillette designed the Oral-B Indicator

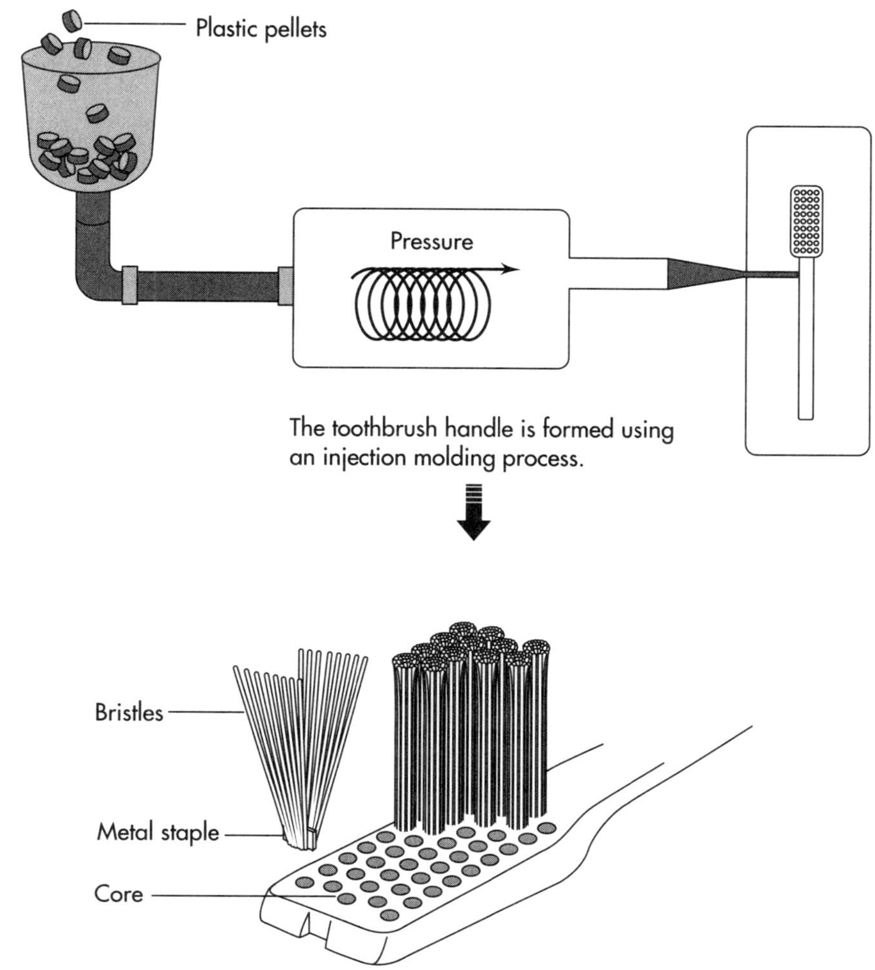

The toothbrush handle is formed using an injection molding process.

The bristles are positioned into the core automatically and fastened with tiny staples.

toothbrush that signaled the user to buy a new brush. The tips of the bristles were coated with a blue dye that would fade to white after about four months of use. The Oral-B Plaque Remover featured taller, contoured bristles that Gillette claimed massaged the gums. SmithKline Beecham designed the AquaFresh Flex, with a flexible, angled handle intended to reduce the pressure put on gums and teeth.

An environmentally sound toothbrush was designed by Jack Hokanson. His Hoke2 brush featured a non-disposable handle with a replaceable bristle head. Some of the most uniquely shaped brushes include the Radius, which featured a wide handle designed for an easier grasp and a large, shoe-brush shaped head packed with bristles. The Collis Curve toothbrush featured bristles which curved inward so that they would hug the teeth, allowing the user to brush the front and back of the teeth at the same time.

Improvements in electric toothbrushes included battery-operated models, rotating heads, and tufts of rotating bristles. A high-tech electronic toothbrush, called the Interplak, featured two rows of bristles with tufts that would spin at 4200 rpm, constantly reversing direction, and pulsating against the teeth. The Interplak, as its name suggests, was designed to remove plaque.

Although many dental experts believe that almost any type of soft-bristled brush is

The injection molding process involves forcing melted plastic pellets into a toothbrush mold and cooling it. The handle, head, and even the small holes are automatically formed.

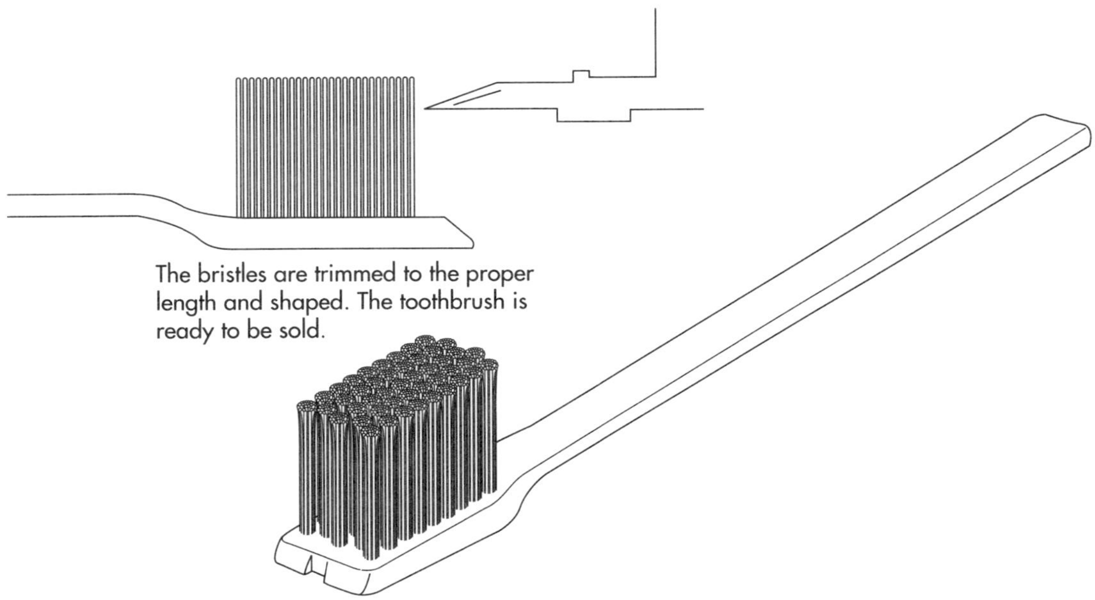

effective if properly used, great strides have been made in specialty toothbrushes. For instance, electric toothbrushes are helpful for individuals with limited mobility or dexterity in their hands, such as people afflicted with arthritis. Other special designs include interproximal toothbrushes, which have small triangular brush heads that can clean under fixed bridges and in between widely spaced teeth. Smaller sizes are also available for people with smaller mouths. For people with highly sensitive teeth, toothbrushes with extra-soft bristles made of polished nylon are available.

The Manufacturing Process

Contemporary toothbrushes are produced mechanically. Generally, toothbrushes consist of plastic handles and nylon or natural boar bristles.

Molding the handles

1 Plastic is mixed and shaped into pellets. The pellets are then placed in an injection molding machine, which heats the plastic until it is melted. A rotating screw or plunger forces the liquid plastic into the handle molds. The molds form the entire handle, including the small holes, called cores, into which the bristles are inserted. The molds are securely clamped, and pressure is applied to the molds while the plastic cools. Once the molds have adequately cooled, the clamps are removed, and small pins push the handles out of the molds.

The filling machine

2 The bristles, which are usually made of nylon, are positioned into the core of the handle automatically. The bristles are then stapled into the core with tiny metal staples.

Trimming the bristles

3 Next, the toothbrush passes through a trimming machine which slices the bristles to the correct length and shape for the particular design.

Packaging the toothbrushes

4 The toothbrushes are packaged into cardboard and/or plastic containers. Labels are attached to the package, providing product information such as bristle hardness, as well as recommendations for usage. If the brand is approved by the American Dental Association, the Seal of Acceptance is also stamped on the container.

5 Finally, the packaged toothbrushes are bundled into larger shipping boxes or crates and transported to distributors.

Quality Control

The American Dental Association tests a number of toothbrushes and other dental products each year. The ADA measures the efficiency and comfort of toothbrushes and those which meet with their standards are awarded the "ADA Seal of Acceptance." By the mid-1990s the ADA approved more than 45 different toothbrush brands.

Where To Learn More

Periodicals

"New Ways to Save Your Teeth?" *Consumer Reports,* August 1989, pp. 504-09.

Deveny, Kathleen. "Toothbrush Makers." *The Wall Street Journal,* October 22, 1991, pp. B1, B10.

———. "Today's Toothbrushes." *The Wall Street Journal,* November 10, 1992, pp. B1, B9.

Higbee, Beth. "The Great Tooth-Brush-Off." *Prevention,* March 1994, pp. 53-55.

Pierson, John, "Form and Functions." *The Wall Street Journal,* June 11, 1991, p. B1.

—Audra Avizienis

Traffic Signal

The first illuminated traffic signal was installed in London, England, in 1868. It was manually turned and consisted of two gas lamps, one red and one green, with semaphore arms atop a pole. Shortly after its inauguration, however, it blew up while the lamps were being lit, killing a policeman.

Background

A traffic signal, or stoplight as it is also known, controls vehicle traffic passing through the intersection of two or more roadways by giving a visual indication to drivers when to proceed, when to slow, and when to stop. In some cases, traffic signals also indicate to drivers when they may make a turn. These signals may be operated manually or by a simple timer which allows traffic to flow on one roadway for a fixed period of time, and then on the other roadway for another fixed period of time before repeating the cycle. Other signals may be operated by sophisticated electronic controllers that sense the time of day and flow of traffic to continually adjust the sequence of operation of the signals. Traffic engineers use signals to avoid traffic congestion and improve safety for both motorists and pedestrians alike.

The first illuminated traffic signal was installed in London, England, in 1868. It was manually turned and consisted of two gas lamps, one red and one green, with semaphore arms atop a pole. Shortly after its inauguration it blew up while the lamps were being lit and killed a policeman. The first electric traffic signal was installed in Cleveland, Ohio, in 1914. It consisted of a green and red light with a warning buzzer to indicate when the light was about to change. The first signal to use the familiar green, yellow, and red lights was installed in New York City in 1918. It was operated manually from an elevated observation post in the middle of the street. In Los Angeles, traffic lights consisted of green and red lights used in conjunction with a warning gong and a pair of semaphore arms lettered "stop" and "go."

A modern traffic signal system consists of three basic subsystems: the signal lights in their housing, the supporting arms or poles, and the electric controller. The signal lights and housing are known as the signal light stack. A single stack usually consists of three lights: a green light on the bottom to indicate the traffic may proceed, a yellow light in the middle to warn traffic to slow and prepare to stop, and a red light on the top to indicate the traffic must stop. Because some people are red-green color blind, there has been an effort to standardize on a vertical stack of lights with red at the top so that these people can perceive the signal condition by the position of the light rather than the color. Each light has a fresnel lens which may be surrounded or hooded by a visor to make it easier to see the light in bright sunlight. A fresnel lens consists of a series of concentric angled ridges on the outer surface of the lens which bend the light to focus it in a parallel beam. The light stack may have a dark-colored backing plate to make the signals more distinguishable by blocking out surrounding lights from buildings and signs. There are one or more signal light stacks for each direction of each roadway. The electric controller is usually mounted in a weatherproof box on one of the corners of the intersection. More elaborate traffic signals may also have electromagnetic sensors buried in the roadway to detect the flow of traffic at various points.

Raw Materials

The housing or body of each signal light stack is usually made of corrosion-resistant aluminum. Some housings are made of molded polypropylene plastic. The lens for each light is made of tinted glass or plastic. The bulb, known as the lamp, is designed for long life. It is purchased from a light bulb manufacturer. The bulb is partially surrounded by a polished metal reflector to direct the light forward. The hood or visor is made from aluminum or molded plastic.

The supporting arms or poles are usually made of galvanized steel for strength and corrosion-resistance. They may also be made of **fiberglass**. The controller is housed in a steel or aluminum enclosure. The electrical components within the controller—switches, relays, and timers—are purchased from various electrical component manufacturers. The wiring between the components is copper with a heavy neoprene rubber or plastic insulation.

The Manufacturing Process

A traffic signal is fabricated in the manufacturer's plant, then installed and wired at the site of the intersection.

Making the signal stack

1 The housing or body of each signal stack is die cast, as are the lens door and the bulb door which attach to the body. They may be cast as individual housings and doors for each light, or as larger units for each stack. The die-casting process uses a large, two-piece, steel mold called a die. Inside the mold, called the cavity, is the reverse image of the part to be cast. The die is placed in a machine which clamps the two halves together with a force of 2,400,000 pounds (1,090,000 kg). Molten aluminum is poured into the "shot end" of the die, and a plunger rams the metal into the cavity under a pressure of approximately 2,000 pounds per square inch (138 bar). The molten metal is forced into every portion of the cavity and cools. After about 15 seconds, the die is opened and the hot part is ejected. The part is then cooled for about 30 minutes. Under normal conditions a die-casting machine like this can produce about 30 parts per hour.

Installed in Detroit in 1920, this was the nation's first three-color, four-way traffic light. (From the collections of Henry Ford Museum & Greenfield Village.)

The introduction of the automobile at the turn of the century was chiefly a matter of novelty, and early streets were not paved, engineered, or controlled for automobiles. The first autos joined horse-drawn carriages, pushcarts, horses, bicycles, and pedestrians on the streets.

Initially traffic control was simply non-existent. People and vehicles negotiated their own way through intersections without control. In urban areas police officers eventually tried to direct traffic and soon employed "Stop-Go" hand signs. Problems constantly arose as officers changed the signals with no advanced warning, and motorists found themselves stopping in the middle of the intersection or coming to a near stop each time they approached an intersection. Either way, traffic did not flow so much as it lurched along.

Morgan Garrett, an African American entrepreneur and inventor, experimented with automatic traffic signals in the early 1920s. His first version, placed in the middle of the intersection, worked like a railroad semaphore, with the arms moving to different positions to signal drivers coming from different directions. Garrett's main innovation was the introduction of an intermediate position, the equivalent of yellow on a modern traffic light. This allowed motorists to anticipate a change and slow down only when necessary. Garrett later sold his invention to the General Electric Co., which later made electric, three-light, four-way traffic signals.

William S. Pretzer

2 Once the cast part is cooled, it is trimmed. The trimming process uses a stamping die to shear off any excess metal. The part is then visually inspected, and a hand file is used to remove any sharp burrs. The points where the doors are to mount to

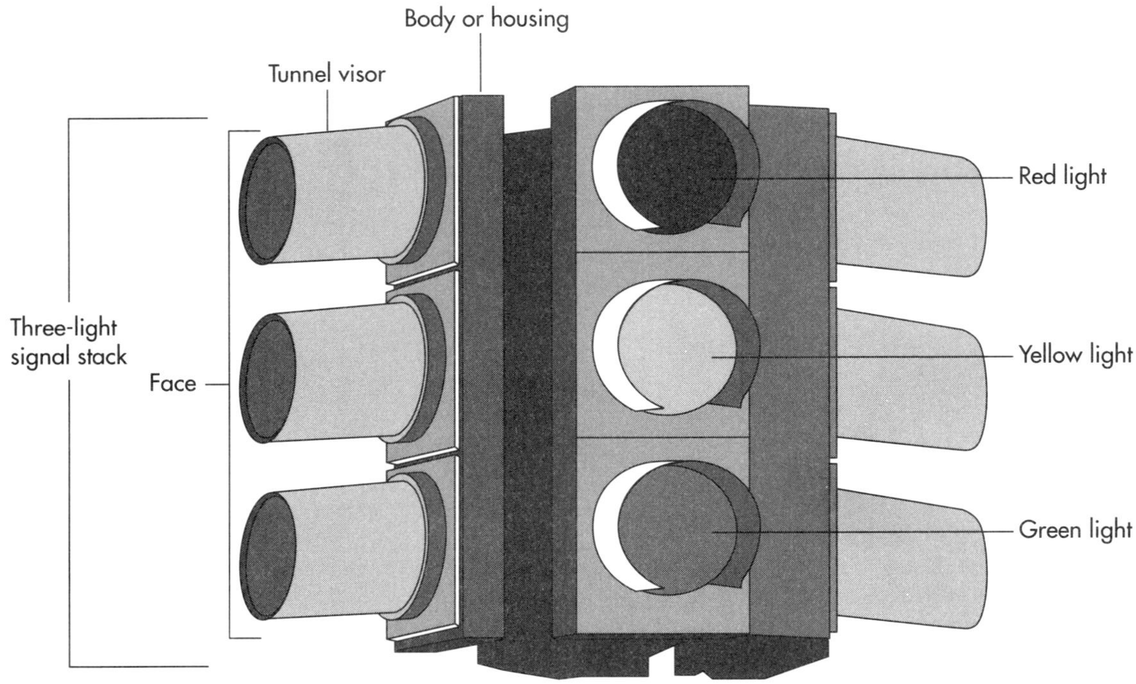

Major components of a traffic signal.

the housing are machined to ensure they will fit properly. The doors are attached to the housing with hinge pins and retained by spring latches. If the housings and doors have been cast individually, they are assembled to form the stack. The holes used for fastening the stack to the support structure are drilled. The stack is cleaned, painted, and placed in a drying oven.

3 The painted stack is transferred to a final assembly area where the lamps, lampholders, reflectors, and lenses are installed. Stainless steel screws and fasteners are used. The lenses are sealed against the lampholder assembly with weatherproof gaskets. The sheet metal visors, which were fabricated in another operation, are attached to each light. Wiring from each light is routed through the hollow stack housing to the stack mounting point.

Making the controller

4 The housing for the electrical controller may be cast or fabricated. It is trimmed, machined, and painted in much the same manner as the signal stack. Inside it has mounting points to which the electrical control boards are attached. The traffic signal manufacturer may assemble the electrical components, or may have this done by another company.

Making the supports

5 The supports may be cast, spun, or fabricated. Supports are hollow with an electrical junction box built into the base to connect with the wires coming underground from the controller. Some supports include decorative details to make the signals fit in with the architecture of the surrounding area. In some signal installations the light stacks are hung from heavy steel cables spanning the intersection.

Installation

6 Underground electrical conduits between the controller and each signal support location are put in place and connecting wires are pulled through the conduits. Power for the signal is brought to the controller location. If sensors are to be placed in the road, they are connected to the controller as well. The supports are bolted in place, and wires are pulled through the hollow supports between the base and the

signal stack mounting point. The signal stacks are attached to the supports. The wiring is then connected between all the elements of the system. Each individual light is given a final adjustment to aim it properly, the sequence timers in the controller are set to the specifications determined by the traffic engineers, and the system is cycled several times to test for proper operation of each element.

Quality Control

The manufacturing process for traffic signals is subject to the standard inspections and control practices found in any similar production facility. These include both conventional and statistical methods. The installation on the job site is subject to review by an electrical inspector from the agency placing the signal. Wiring must comply with the National Electrical Code. The location of the light and any other structural considerations must also meet various federal, state, and local ordinances. A registered professional engineer must review and approve the plans to ensure the installation meets the national requirements for traffic control devices.

The Future

With the ever-increasing use of computers, traffic signals in the future will become more sophisticated. Many systems already feature a remote-controlled activation system which allows **fire engine**s and other emergency vehicles to change the signal to green in their direction as they approach an intersection. Some cities are developing networks of traffic signal controllers that interact to keep traffic moving through heavily congested areas or reroute traffic during peak traffic hours. Other advances might include integrating speed warning devices and systems to check for stalled traffic or accidents.

Where To Learn More

Periodicals

Loggins, J.W. "New Traffic Signal Concepts." *Public Works.* March 1993.

Nadis, S.J. "Gridlock Terminator: Neural Nets Predict the Traffic Future." *Omni,* October 1994, p. 22.

—Douglas E. Betts/Chris Cavette

Vaccine

The term vaccination is derived from the Latin for cow, vacca, and vaccinia, the virus drawn from cowpox.

Background

The development of vaccines to protect against viral disease is one of the hallmarks of modern medicine. The first vaccine was produced by Edward Jenner in 1796 in an attempt to provide protection against smallpox. Jenner noticed that milkmaids who had contracted cowpox, a relatively innocuous infection, seemed to be resistant to smallpox, a disease of humans that regularly reached epidemic levels with extremely high mortality rates.

Jenner theorized (correctly) that cowpox, a disease of animals, was similar to smallpox. He concluded that the human reaction to an injection of cowpox virus would somehow teach the human body to respond to both viruses, without causing major illness or death. Today, smallpox is totally eradicated. Only two frozen samples of this virulent virus exist (one in the United States, the other in Russia), and as of mid-1995 there is serious scientific debate about whether to destroy the samples, or keep them for further laboratory study.

A virus is a small bit of RNA (Ribonucleic acid) and/or DNA (deoxyribonucleic acid), the material in all living cells that instructs the cell how to grow and reproduce. Viruses cannot reproduce by themselves, but only by taking over the nucleus of a host cell and instructing the cell to make additional viruses. When a virus successfully invades an organism, it takes over the cell growth process in the host.

Under ordinary circumstances, the human body responds to viral invasion in several different ways. Generalized immunity to a virus can be developed by the cells in the body that are targets of viral invasion. In this situation, viruses are prevented from gaining access to host cells. A more common protection is the body's ability to develop blood and lymph cells that destroy or limit the efficacy of the invading virus. Often, an infected human body will "learn" how to respond to a specific virus in the future, so that a single infection, especially from a relatively benign virus, usually teaches the body how to respond to additional invasions from the same virus. The common cold, for example, is caused by one of several hundred viruses. After recovering from a cold, most people are resistant to the particular virus that caused the particular cold, although similar cold viruses will still cause similar or identical symptoms. For some innocuous viruses, a person might even develop immunity without becoming visibly ill.

Virus Families

There are usually several variations or strains of any particular virus. Depending on the number of varieties, a biologist might group viruses as types or strains. Vaccines frequently are made from more than one group of related viruses; a preventive reaction to the multivalent vaccination will probably cause immunity to almost all of the group's variants, or at least to those variants which a person is likely to encounter. Choice of the specific members of the group to use in a vaccine are made with painstaking care and deliberation.

The Manufacturing Process

Manufacturing an anti-virus vaccine today is a complicated process even after the ardu-

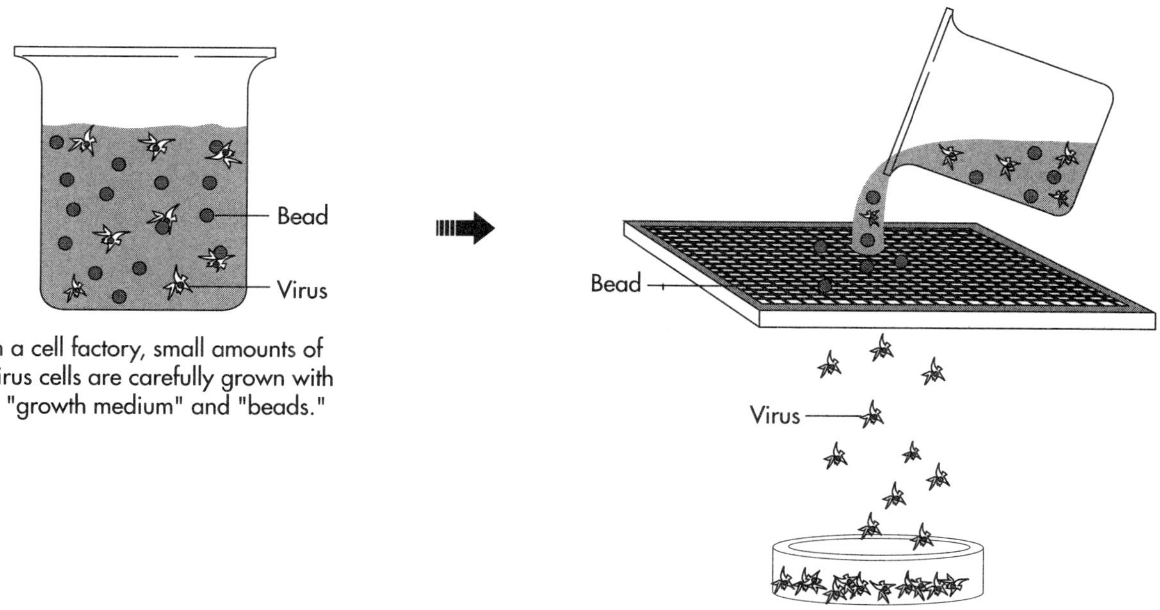

In a cell factory, small amounts of virus cells are carefully grown with a "growth medium" and "beads."

The virus is then separated using a filtering system to remove the beads.

ous task of creating a potential vaccine in the laboratory. The change from manufacturing a potential vaccine in small quantities to manufacturing gallons of safe vaccine in a production situation is dramatic, and simple laboratory procedure may not be amenable to a "scale up" situation.

The Seed

1 Manufacturing begins with small amounts of a specific virus (or seed). The virus must be free of impurities, including other similar viruses and even variations of the same type of virus. Additionally, the seed must be kept under "ideal" conditions, usually frozen, that prevent the virus from becoming either stronger or weaker than desired. Stored in small glass or plastic containers, amounts as small as only 5 or 10 cubic centimeters, but containing thousands if not millions of viruses, will eventually lead to several hundred liters of vaccine. Freezers are maintained at specified temperatures; charts and/or dials outside of the freezer keep a continuous record of the temperature. Sensors will set off audible alarm signals and/or computer alarms if the freezer temperature goes out of range.

Growing the virus

2 After defrosting and warming the seed virus under carefully specified conditions (i.e., at room temperature or in a water bath), the small amount of virus cells is placed into a "cell factory," a small machine that, with the addition of an appropriate medium, allows the virus cells to multiply. Each type of virus grows best in a medium specific to it, established in pre-manufacturing laboratory procedures, but all contain proteins from mammals in one form or another, such as purified protein from cow blood. The medium also contains other proteins and organic compounds that encourage the reproduction of the virus cells. As far as the virus is concerned, the medium in a cell factory is a host for reproduction. Mixed with the appropriate medium, at appropriate temperature, and with a predetermined amount of time, viruses will multiply.

3 In addition to temperature, other factors must be monitored, including the pH of the mixture. pH is a measure of acidity or basicity, measured on a scale from 0 to 14, and the viruses must be kept at a defined pH within the cell factory. Plain water, which is

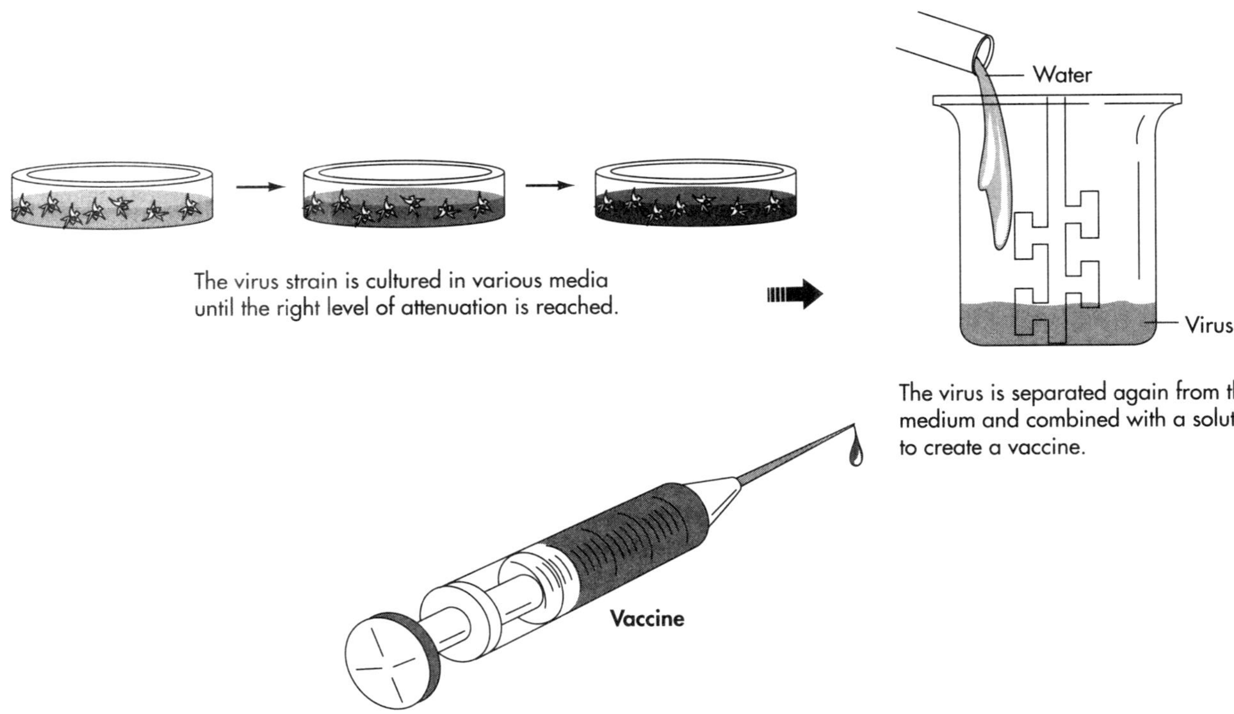

The virus strain is cultured in various media until the right level of attenuation is reached.

The virus is separated again from the medium and combined with a solution to create a vaccine.

neither acidic or basic (neutral) has a pH of 7. Although the container in which the cells are growing is not very large (perhaps the size of a 4-8 quart pot), there are an impressive number of valves, tubes, and sensors connected to it. Sensors monitor pH and temperature, and there are various connections for adding media or chemicals such as oxygen to maintain the pH, places to withdraw samples for microscopic analysis, and sterile arrangements for adding the components of the cell factory and withdrawing the intermediate product when it is ready.

4 The virus from the cell factory is then separated from the medium, and placed in a second medium for additional growth. Early methods of 40 or 50 years ago used a bottle to hold the mixture, and the resulting growth was a single layer of viruses floating on the medium. It was soon discovered that if the bottle was turned while the viruses were growing, even more virus could be produced because a layer of virus grew on all inside surfaces of the bottle. An important discovery in the 1940s was that cell growth is greatly stimulated by the addition of enzymes to a medium, the most commonly used of which is trypsin. An enzyme is a protein that also functions as a catalyst in the feeding and growth of cells.

In current practice, bottles are not used at all. The growing virus is kept in a container larger than but similar to the cell factory, and mixed with "beads," near microscopic particles to which the viruses can attach themselves. The use of the beads provides the virus with a much greater area to attach themselves to, and consequently, a much greater growth of virus. As in the cell factory, temperature and pH are strictly controlled. Time spent in growing virus varies according to the type of virus being produced, and is, in each case, a closely guarded secret of the manufacturer.

Separation

5 When there is a sufficient number of viruses, they are separated from the beads in one or more ways. The broth might be forced through a filter with openings large enough to allow the viruses to pass through, but small enough to prevent the beads from passing. The mixture might be centrifuged several times to separate the virus from the beads in a container from

which they can then be drawn off. Still another alternative might be to flood the bead mixture with another medium which washes the virus off the beads.

Selecting the strain

The eventual vaccine will be either made of attenuated (weakened) virus, or a killed virus. The choice of one or the other depends on a number of factors including the efficacy of the resulting vaccine, and its secondary effects. Flu vaccine, which is developed almost every year in response to new variants of the causative virus, is always an attenuated vaccine. The virulence of a virus can dictate the choice; rabies vaccine, for example, is always a killed vaccine.

6 If the vaccine is attenuated, the virus is usually attenuated before it goes through the production process. Carefully selected strains are cultured (grown) repeatedly in various media. There are strains of viruses that actually become stronger as they grow. These strains are clearly unusable for an attenuated vaccine. Other strains become too weak as they are cultured repeatedly, and these too are unacceptable for vaccine use. Like the porridge, chair, and bed that Goldilocks liked, only some viruses are "just right," reaching a level of attenuation that makes them acceptable for vaccine use, and not changing in strength. Recent molecular technology has made possible the attenuation of live virus by molecular manipulation, but this method is still rare.

7 The virus is then separated from the medium in which it has been grown. Vaccines which are of several types (as most are) are combined before packaging. The actual amount of vaccine given to a patient will be relatively small compared with the medium in which it is given. Decisions about whether to use water, alcohol, or some other solution for an injectable vaccine, for example, are made after repeated tests for safety, sterility, and stability.

Quality Control

To protect both the purity of the vaccine and the safety of the workers who make and package the vaccine, conditions of laboratory cleanliness are observed throughout the procedure. All transfers of virus and media are conducted under sterile conditions, and all instruments used are sterilized in an autoclave (a machine that kills organisms by heat, and which may be as small as a jewel box or as large as an **elevator**) before and after use. Workers performing the procedures wear protective clothing which includes disposable tyvek gowns, gloves, booties, hair nets, and face masks. The manufacturing rooms themselves are specially air conditioned so that there is a minimal number of particles in the air.

The Approval Process

In order for prescription drugs to be sold in the United States, a drug manufacturer must meet strict licensing requirements established by law and enforced by the Food and Drug Administration (FDA).

All prescription drugs must undergo three phases of testing, although data from the second phase can sometimes be used to meet third phase requirements. Phase 1 testing must prove that a medicine is safe, or at least that no untoward or unexpected effects will occur from its administration. If a medicine passes Phase I testing, it must next be tested for efficacy—it must do what it is supposed to do; medicine cannot be sold that is useless, or that makes claims for an effect that it does not have. Finally, Phase III testing is designed to quantify the effectiveness of a medicine or drug. Although vaccines are expected to have effectiveness close to 100%, certain medicines might well be acceptable even if they have limited effectiveness, as long as the prescribing physician knows the odds.

The entire manufacturing process is reviewed carefully by the FDA which examines records of procedures as well as visiting the manufacturing site itself. Each step in the manufacturing process must be documented, and the manufacturer must demonstrate a "state of control" for the manufacturing process. This means that scrupulous records must be kept for every step in the process, and there must be written instructions for each step of the process. Except in cases of grievous error, the FDA does not determine if each step in a process is correct, but only that it is safe and is documented sufficiently to be performed as the manufacturer stipulates.

The Future

Producing a usable, safe antiviral vaccine involves a large number of steps which, unfortunately, cannot always be done for each and every virus. There is still much to be done and learned. The new methods of molecular manipulation have caused more than one scientist to believe that the vaccine technology is only now entering a "golden age." Refinements of existing vaccines are possible in the future. Rabies vaccine, for example, produces side effects which make the vaccine unsatisfactory for mass immunization; in the United States, rabies vaccine is now used only on patients who have contracted the virus from an infected animal and are likely, without immunization, to develop the fatal disease.

The HIV virus, which biologists believe causes AIDS, is not currently amenable to traditional vaccine production methods. The AIDS virus rapidly mutates from one strain to another, and any given strain does not seem to confer immunity against other strains. Additionally, a limited, immunizing effect of either attenuated or killed virus cannot be demonstrated in either the laboratory or in test animals. No HIV vaccine has yet been developed.

Where To Learn More

Books

Dulbecco, Renato and Harold S. Ginsberg. *Virology.* 2nd edition. J.B. Lippincott Company, 1988.

Plotkin, Stanley A. and Edward A. Mortimer, Jr., eds. *Vaccines.* W.B. Saunders Company, 1988.

—Lawrence H. Berlow

Violin

Background

The violin is the most modern embodiment of stringed musical instruments played with a bow. Like the guitar and other plucked string instruments, bowed instruments date from antiquity. Although its precise origins are not completely understood, it is probable that the violin (and its larger siblings the viola and violoncello) evolved during the mid-16th century in Northern Italy. In addition to perhaps being the maker of the first true violins, Andrea Amati (ca. 1500-1577) was the patriarch of the Cremona school of violin making. During the next 150 years, other members of the Amati family and their followers, who included Antonio Stradivari (1644-1737) and Bartolomeo Giuseppe Guarneri (1698-1744), brought the violin to its highest level of perfection both as a musical instrument and as a work of art. During the 17th century, violin making spread to all of the other countries of Europe and, in the 18th and 19th centuries, to the rest of the world. Although violins have been and are being turned out in large numbers by factories in Europe and Asia, most fine violins are handmade by individual craftsmen using essentially the same methods employed by classical Italian makers several hundred years ago.

Tools

Most of the tools required for violin making are the same as those used for most types of hand woodworking and carving: planes, chisels, gouges, knives, saws, and scrapers. In addition, a few specialized tools are needed. These include a thickness caliper, small curved bottom "thumb" planes, purfling groove cutter, peg hole reamer and matching peg shaver, bending iron, clamps of various types, and patterns. Many violin makers take pride in making some of their own tools. Indeed, one of the keys to success as a violin maker is developing the skills associated with making, using, and maintaining sharp edged tools.

Raw Materials

The back, sides (ribs), and neck of the violin are most often made of matching quarter-sawn (cut along the radius of the log) maple. There are many species of maple, growing in different parts of the world, which are suitable. The criteria for selection include the straightness of the grain, the density and the figure of the wood, all of which contribute to the tonal characteristics and visual beauty of the finished instrument. The top of the violin is made of quarter-sawn spruce. The internal parts of the violin—the corner and end blocks and the linings—are usually made of spruce or willow, while purfling can be made of many different woods and/or "fiber" (thick **paper** or cardboard). The fingerboard is made of ebony, the bridge is maple, and the other fittings (pegs, tailpiece, chin rest) are ebony, rosewood, or boxwood. Rather than making these items from scratch, they are usually purchased in a finished or semi-finished form and customized or installed by the maker.

Although violins have been and are being turned out in large numbers by factories, most fine violins are handmade by individual craftsmen using essentially the same methods employed by classical Italian makers several hundred years ago.

The Manufacturing Process

The ribs

1 The first step in making a violin is to make the sides (ribs) from which the out-

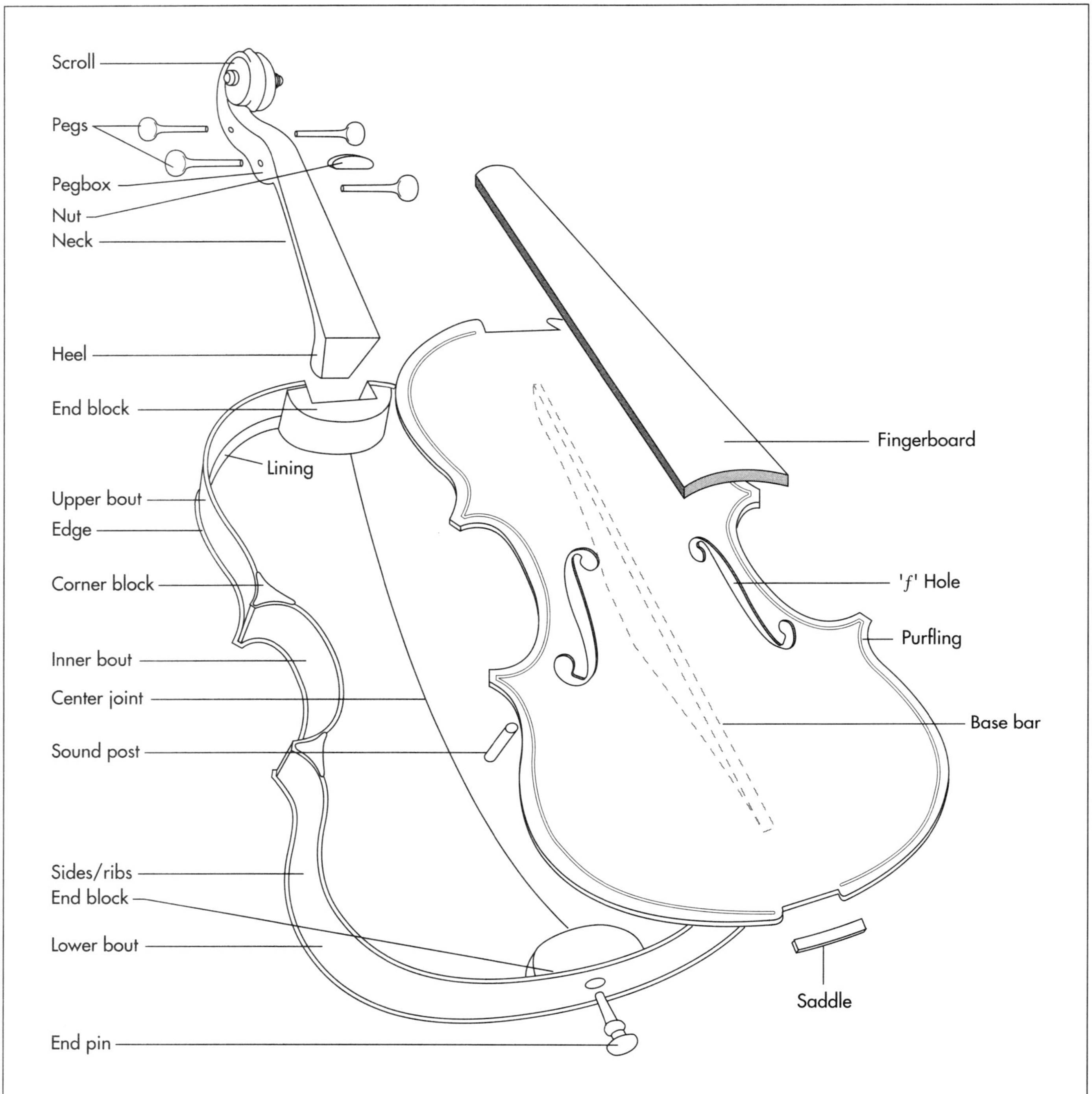

line of the top and back will be taken. The ribs are constructed on an "inside" mold, which is a wooden form about 15 to 18 mm thick cut to the exact outline of the inside of the violin. Pieces of wood for the corner and end blocks are cut to approximate size and temporarily glued to cutouts on the mold at the proper locations. With the aid of a pattern and using gouges and files, the blocks are trimmed to the final shape of the inside contour of the violin. Slices of maple slightly wider than the height of the ribs, which are about 30 to 32 mm for a violin, are cut and planed to a thickness of 1 mm. Pieces are bent to the shape of the mold and blocks using a heated metal form. After trimming to slightly longer than the final length, the ribs are glued. To hold the ribs in place until the glue dries, countermolds shaped to match the outside contours of the ribs at the corners and ends are used. Pressure is applied by clamps or wrapping with string. Care must be taken to avoid gluing the ribs to the mold, which must be removed in a subsequent step. The ribs are glued in sequence starting with the middle bouts, which must be trimmed to final length at the corners before the upper and lower ribs can be added. The linings, strips of willow or spruce about 2 by 8 mm, are bent to shape using the bending iron, cut to length, and glued to the inside of the ribs. The corners are trimmed to their final shape, and the top and bottom surfaces of the ribs, linings, and blocks are planed and filed to be level at the final height.

Top and back

2 The tops of violins are almost always made from a wedge of wood which is cut or split, with the edges of resulting pieces glued together. This joint, for which the pieces must fit with absolute perfection, then becomes the centerline of the top. Maple for the backs of violins is treated the same way to make a two-piece back, however, it is possible to find a piece wide enough to make a one-piece back. Planing the wood to create a perfect center joint is an exacting task. After the pieces have been planed to fit well—as seen by holding the pieces together in front of a light—the edge of one piece is coated with chalk and rubbed against the mating edge. The areas in which chalk is transferred from one piece to the other identify places which must be shaved slightly with the plane to perfect the fit. This chalk fitting procedure is repeated until the fit between the two pieces is perfect, after which they are glued together and clamped. After gluing the center joint, the flat side of the back and top are planed flat. The ribs are placed on this flat side; a tracing around the ribs enlarged by 3 mm establishes the outline of the top or back (a 20 mm square is added at the top of the back for the button). These outlines are cut out using a bow or coping saw; many modern violin makers use a power band or scroll saw.

The outside arching of the top and back are next carved using gouges, thumb planes, and scrapers for the final smoothing. Five transverse and one longitudinal arching guides are consulted frequently as the plates take shape. The arching guides, which are different for the top and back, plus the outline embodied in the mold, determine the design or model of the instrument. Most modern makers follow or copy the designs of great makers such as Stradivari or Guarneri, while some have evolved their own. Next the interior sides of the top and back are carved out. The final thicknesses of the wood has a major influence on the acoustic performance of the finished instrument, and there are many systems in use for arriving at the optimum distribution of thicknesses. In general, most methods involve testing the resonance frequencies of the plates by tapping, flexing, or exciting them with sound, coupled with measurements of the thickness of the plate at many locations using a graduation caliper. Then, depending on the results and on the desired outcome, wood is gradually removed from various locations. Usually, makers seek to establish certain relationships, e.g. octaves, between the various resonances of each plate and between the two plates. Typical thicknesses of a finished back are 4.5 mm at the center decreasing to 2.4 to 2.5 mm in the upper and lower bouts. The thicknesses of the top are more uniform: about 3 mm overall, and perhaps slightly thicker between the soundholes in the area of the soundpost.

Completing the top

3 The outline of the sound holes is transferred to the top, and these are cut out using a sharp knife; some makers use a punch or drill to cut the round holes. The bassbar is made of very straight grained, quarter-cut spruce (much like the top). The area where it fits is outlined on the side of the top, and the rough blank is trimmed to precisely fit the arching. The chalk-fitting method is employed again in this step. The bar is then glued in place and trimmed to its final contour. This again involves testing the resonance of the top, which was altered by the cutting of the sound holes as well as the addition of the bassbar.

Completing the body

4 The mold is now removed from the rib assembly by loosening the temporary glue bonds of the blocks to the mold. The

top and back are then glued to the ribs. The glue holding the back should be full-strength. Thinner, weaker glue is used for the top; this provides for easy removal if service or adjustment is necessary, and will allow the seams to open in extremes of humidity and temperature rather than produce cracks in the wood itself. The groove for the purfling is marked a precise distance from the edges using a purfling cutter. The groove is deepened with a sharp knife and the wood in the groove removed with a purfling pick. The purfling strips, which can be bought ready or made by the violin maker, are bent to fit the groove using the bending iron. The pieces are then cut to the exact length, mitered to fit the corners, and glued in place. The channels which run over the purfling just inside the edges are cut with a gouge and blended into the arching with gouge, planes, and scrapers. Finally, the edge is rounded using knife, file, and perhaps sandpaper. (This is one of the only places in which sandpaper is used in the construction of a violin. All of the other surfaces should be finished with scrapers, which provides a crisp appearance to the workmanship and best reveals the beauty of the wood.)

The neck

5 A block of maple matching the back is squared on the sides and top with a plane. Next, the outline of the side view of the neck and scroll is traced on the quarter-cut side of the block. The wood outside the outline is sawed away. Patterns and outlines for the peg box, top surface of the neck, and the scroll are traced. A razor saw is then used to cut away wood around the scroll and neck outlines. Gouges and scrapers are used to finish the carving of the scroll, the details of which are one of the ways in which the violin maker expresses his individuality. The pegbox is excavated using chisels and gouges. The neck is cut to final dimensions using planes, knives, and scrapers. A mortise (cavity) to receive the neck is cut into the upper ribs, block, and top of the violin's body. The cut of the mortise and the root of the neck must be very precise, since the correct height and angle of the neck are critical to achieving a good tonal result. Chalk fitting is again employed. The neck is then glued into the mortise, and the final shaping of the heel of the neck and the button on the back is done.

Varnishing

6 There is a great deal of lore associated with the varnishing of violins. It has even been asserted that secret recipes are responsible for the extraordinary tonal characteristics of the old Italian violins. Regardless of its possible effects on tone, it is certainly true that the varnish does serve other important purposes of beautifying the appearance and protecting the wood from wear, damage, moisture, and dirt. Thus the selection and application of varnish is vitally important. Because there are many types of varnish and working methods, the following rather general outline of finishing is provided:

- The finished violin is hung up to age for a time (in some cases several months or more), and may be exposed to sunlight. This will cause the wood to darken and bring out its figure. Many makers use less time-consuming alternatives.

- A sealer or pore filler is then applied.

- The varnish is applied in several coats. This may include coats of clear varnish followed by additional coats of colored varnish. Varnish is essentially a coating consisting of resins, which may be natural substances (e.g. copal or seedlac) or man-made. Color is imparted to varnishes by adding pigments or dyes. The color of the individual coats may be varied to produce the desired appearance. Following the colored varnish, an additional coat or two of clear varnish may be applied to protect the layers underneath.

- Since old-looking violins are more appealing to many players, some makers "antique" their instruments. The various methods of antiquing are usually trade secrets, and makers pride themselves on their individual results.

- The surface of the fully dried varnish may be rubbed out using some combination of abrasives (pumice, rottenstone, fine emery paper, etc.) and polishes.

- The part of the neck between the heel and the peg box is not varnished. Rather it is stained, sanded very smooth with fine emery paper and "french polished" (an application of shellac, and/or alcohol, and oil).

Fitting up

7 The top of the neck is planed flat, and the underside of the ebony fingerboard is planed to fit and glued in place. The sides and top are finished with planes, scrapers and emery paper to be smooth and to have exactly the correct curvature. Gauges and straightedge are consulted frequently during this process. The ebony nut is cut to size, lightly glued at the top of the fingerboard, dressed to final shape, and grooves filed for the strings. A mortise is cut at the bottom of the violin into which is glued the ebony saddle. The pegs are shaved to the proper taper and diameter. Peg holes are drilled and reamed to match the pegs. Likewise, a hole at the bottom of the ribs is drilled, reamed, and fitted with the end pin.

The bridge and soundpost are the last parts to be fashioned; their fit and position greatly affect the sound and playing qualities of the violin. Starting with a precut blank, the feet of the bridge are cut to fit the arching of the top at the proper position—between the nicks of the soundholes. The top of the bridge is cut to an arch which matches the curvature of the fingerboard and provides the proper height of the strings. The front side (facing the neck) is planed down to a thickness of about 4.5 mm at the bottom and tapering from the middle to 1.5 mm at the top. Grooves for the strings are cut and filed using a gauge to establish their proper position and spacing. The soundpost transmits the vibrations of the strings to the back of the violin. It is cut from a round piece of straight-grained spruce about 6 mm in diameter. Its length and ends must be cut so that it fits precisely in the proper location inside the violin, about 3 mm behind the treble foot of the bridge. A gauge may be used to measure the approximate length of the soundpost, but the final fit is a trial and error process. The soundpost is inserted and its position adjusted through the soundholes using a special tool. The strings are now fitted into the tailpiece, extended over the bridge and wound on the pegs. Once all four strings are installed, they may be tuned up to pitch and the violin played for the first time. What follows will be a period of adjustment as the violin becomes accustomed to the tension of the strings and their vibration. Numerous adjustments to the position of the soundpost, the bridge, types of strings, and perhaps other factors are usually necessary to optimize the tonal characteristics and playability.

The Future

It is likely that fine violins will continue to be handmade in the manner described above. However, there is a long history of experiments with new designs and materials of construction. Recent products of this are violins made of synthetic materials such as plastic. Some of these have solid bodies, while others are of a traditional design using synthetic materials for some parts. There are also electric violins, in which the vibrations of the strings are converted to an electrical signal by a pick-up or microphone, which is then amplified and output to a speaker or computer interface. There are a number of such "high tech" instruments on the market today; they are mainly used to play jazz and popular music. In the realm of classical music, the traditional violin is by far the dominant choice.

Where To Learn More

Books

Hill, William Henry, et al. *Antonio Stradivari: His Life and Work 1644-1737*. Dover Publications, 1963.

Hill, William Henry, et. al. *The Violin-Makers of the Guarneri Family (1626-1762)*. Holland Press, 1965.

Sacconi, Simone F. *The Secrets of Stradivari*. Cremona, 1979.

Buchanan, George. *Making Stringed Instruments—A Workshop Guide*. N. Y. 1990.

—Andrew M. Sherman

Violin Bow

The practice of using a bow of some sort to make musical sound is so ancient that its origin can only be surmised. The most likely scenario is that the ancient hunting bow, its string treated with mixtures of wax and resin to hold the strands together, served as either instrument or bow in different contexts.

Background

Several types of stringed musical instruments, among them the **violin**, the viola, and cello, cannot be successfully played without a bow, and are therefore referred to as "bowed stringed instruments." Because they are almost always heard while being bowed, the bow is considered an integral part of their tone production, contributing its own individual character and timbre. The use of different bows on the same instrument will produce correspondingly different tonality as a result. Most instrumentalists believe the bow's quality to be as important as the instrument's, and fine bows are therefore manufactured and selected with the utmost care.

History

The practice of using a bow of some sort to make musical sound is so ancient that its origin can only be surmised. The most likely scenario is that the ancient hunting bow, its string treated with mixtures of wax and resin to hold the strands together, served as either instrument or bow in different contexts. From this primitive origin, the bow went through countless stages of evolution. The latest and most important to us today are the so-called "early" bow and the "modern" bow. All the bows of these types have important things in common: they are tapered sticks of special woods that are permanently bent to an arch, and have a flattened length of horsehair, stretched, under some tension, from end to end of the stick. One end is usually pointed, and the other squared off and usually fitted with a small raised portion to fasten and adjust the hair tension. The pointed end of each is called the "tip," and the raised portion of the other end, the "nut," or later, the "frog." (Experts are unclear as to how the latter name evolved.)

The early bow (sometimes referred to as the "baroque" bow) is based on the oldest and most obvious of designs, and has a curve that bows away from the hair. This type of bow was in common use until some time in the early 19th century, when the modern bow came into use. Although their design made these bows agile and responsive, their delicacy was not suitable for the pressure needed for louder and more forceful playing. As the concert halls and orchestras became larger, the violin family instruments received subtle modifications to suit the demands of the great performers. No modification was possible for the early bow however, and it suffered a swift extinction at the hand of the modern bow. After the modern bow's inception, the early bow became almost unheard of until it was revived in the late 1960s by early music enthusiasts seeking to recreate the ambiance of that time period.

The modern bow was a revelation after its introduction in France around the turn of the 19th century. The Tourte family is generally given credit for giving the modern bow its accepted final form, much as Antonio Stradivari contributed to the making of the violin. Modern bow manufacture reached its pinnacle in Paris between the mid-19th to the mid-20th centuries, and bowmakers came from all over Europe to collaborate with the famous French workshops and share their excellent reputation for bowmaking. The biggest changes in the modern bow involved inverting the curve of the stick into the hair, to give it more tension and resistance; shortening the tip to a squat hatchet-

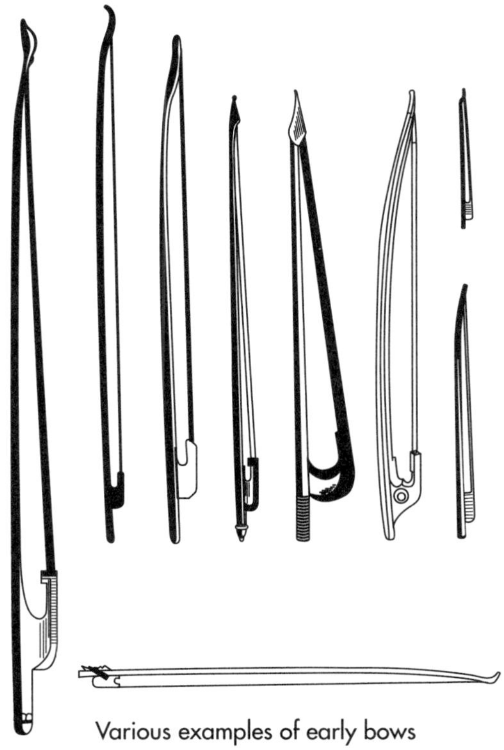

Various examples of early bows

like shape to quicken the flex of the stick; introduction of a screw and eye adjuster for finer adjustment of the hair; and the adoption of Pernambuco wood as the standard wood for the stick. Eventual further improvements included adaptation of a *ferrule* on the frog to hold the hair spread the full width of the frog, at any tension. The makers experimented with many subtle modifications, including building sticks with round or octagonal shafts, using precious metals and materials for the mountings, and incorporating subtle changes in the dimension and curvature of the stick. Today, fine bows are made in much the same, if not exactly the same, manner as they once were by the craftsmen who designed them in France over 150 years ago.

Raw Materials

The making of the bow begins with the selection and rough cutting of the correct woods and raw materials. Pernambuco wood is the accepted type of wood from which the stick of the bow is fashioned. Pernambuco wood grows only in the Amazon delta region in a Brazilian state of the same name. Actually there are several sub-species of this wood, many of which are completely extinct, and others which are rapidly nearing extinction. After harvesting, the logs are sawn into planks, and then into "blanks" which are cut into the rough outline resembling the stick and its tip. The ebony for the frog is split from log cross sections into small wedges which resemble the finished outside dimensions. Sheet silver or gold is prepared to the thickness of the various metal fittings, and a round ebony stick or dowel is prepared to make the adjuster barrel. The decorative pearl slide and pearl eyes are fashioned from specially milled sheets of abalone or mother of pearl shell, sawn and filed to rough size and shape.

The Manufacturing Process

Roughing the stick

1 Roughing the stick refers to the process of carving and planing the stick to its approximate finished dimensions. The squared off blank of Pernambuco wood is either held

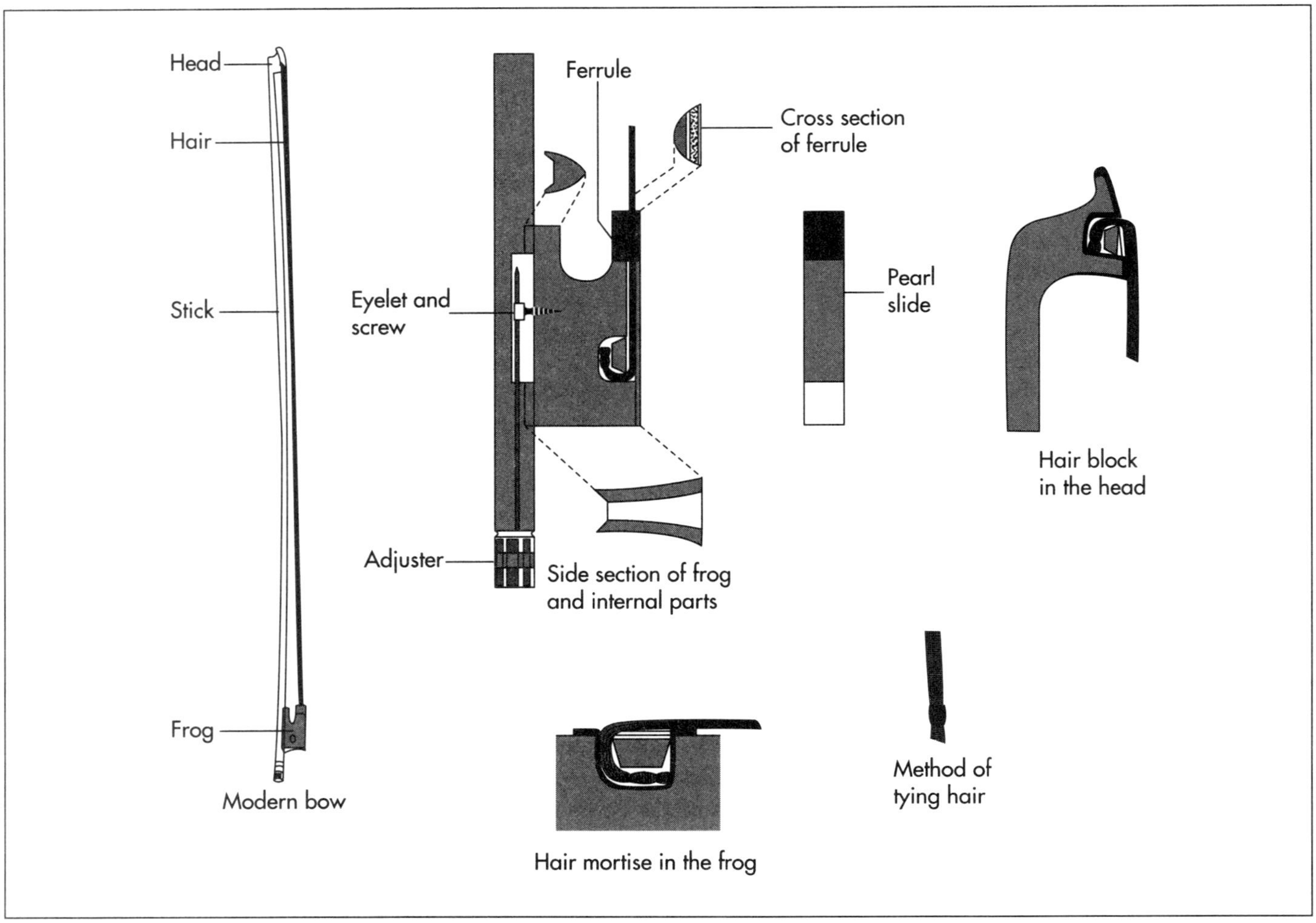

Major parts of a violin bow.

across the corner of a bench, or along the length of a special board, and planed by hand with specially designed planes, fashioning the stick into its characteristic octagon shape. Using a direct heat device such as a spirit lamp or gas burner, the stick is heated slowly until it becomes flexible enough to bend. When ready, the stick is bent into an approximate or rough curve. When cooled, the stick is set aside, and the work on the frog begins.

Roughing out the frog

2 The frog begins with the fashioning of the metal fittings. Several parts require soldering as well as bending and shaping. The ferrule, much like a half round ring, is a semicircular length of sheet silver soldered to a flat one. The button for the adjuster needs one or two silver rings. The other metal parts include a silver liner, which is shaped to three facets of an octagon on a steel die, to conform to the three facets on which it will contact the stick. If the frog is designed with a back plate, the plate is shaped and bent to the 90 degree angle of the back of the frog where it is to be inlaid.

The ebony wedge is trimmed to length and planed true to its center on all sides with a small razor sharp block plane. The various metal fittings are fit onto the frog in their respective places. Although modern commercial manufacture uses milling machinery to accomplish this part, the best modern builders have no problem doing this work by hand.

The fitting of the metal to the frog begins with drilling a 3-mm hole called the "throat" just under the area where the ferrule is located. The ferrule is fit onto the wider part of the throat with a knife and small chisels until it fits back flush and level. The sides are shaped concave with a gouge. The slot for the pearl slide, with its 20 degree undercut sides, is next to be shaved out with the chisels. The cavity for the hair, called the hair mortise, is drilled and carved into the frog with a bow drill and chisel. The liner is then fit to the narrow edge of the frog's length using the chisels. The liner conforms to the top three facets of the stick's octagonal shape and is the bearing surface of the frog against the stick. A tapered silver backplate extending from the back of the pearl slide slot to the center facet of the liner is inlaid to the flat end of the frog. The frog is then shaped using a knife, files, and small scrapers made of thin steel sheet. The decorations, called the eyes, are inlaid into the sides of the frog at this point. Then the ebony dowel for the adjuster button is fashioned separately on a lathe.

Fitting the frog to the stick

3 After the frog is done, the next job is to fit it to the roughed out stick. This is done by chalking the liner of the frog and rubbing it against the facets of the stick at the point where the frog makes contact. Through a process of marking the stick in this way and carefully planing, scraping, and filing the marks away, the frog is brought into the proper contact with the three bottom facets. Then holes are drilled in the stick for the screw and eye assembly which attaches the frog to the brass nut at the end of the stick.

Finishing the stick and frog

4 The first step here is to fit an ivory plate to the head or tip of the bow. A plate of ivory is prepared with a raised section for the right angle of the "beak" with a thin lamination of ebony veneer all along its inside surface. The ivory is glued to the bottom face of the head.

The shaping of the head is done with a knife and files. This work usually follows an established model and is accomplished with the means of a pattern or template, which is alternately traced and compared with the carving as it progresses. The elegant head models of the classic bows are often very beautiful, and have inspired connoisseurs the world over to collect them. All the great bowmakers imprinted their work with their own personal style, and experts are easily able to recognize most of the important styles, each head being akin to the signature of the maker. Once the head is finished, the mortise for the hair is cut into it, and the finishing of the stick can continue.

The stick must be now brought into final dimension, a process called graduation. The stick tapers from 3.5-5.0 mm just behind the head to 6.5-8.5 mm at the button end. Using a gauge or caliper, the craftsman skillfully planes this transition of thickness into the stick. The whole process must be done while preserving the integrity of a perfect octagon. The octagon's transition into the head is most difficult, and ends with the top three facets converging upwards, the two side facets becoming the side of the head, and the bottom three becoming the back of the head and the *chamfers* (a thin finish cut, at a 45 degree angle to the sides). All of this work is accomplished with either the plane in the case of the facets, or the knife and file for the detailing of the head. The stick is simply held by hand across a flat board or the corner of the workbench while planing the facets. The head is simply held in the hand while finishing.

If the stick is to be finished round, as many are, the edges of the octagon are planed away after graduation, and the stick is rounded off in this manner to an area about 1.6-2.4 inches (4-6 cm) in front of the frog. The area where one holds the bow is almost always octagonal.

Treating the stick

5 The bow usually has no real varnish as such because Pernambuco is inherently dark and oily. But the stick may be subjected to a number of chemical treatments to achieve its characteristic chocolate brown color. Bathing the stick with nitric acid, and then following with a neutralizing exposure to ammonia fumes is the most common color treatment. The bow is given additional sheen and protection by a technique known as "French polishing." This involves the application of a dilute solution of shellac, sometimes mixed with other gums or resins, with a lightly oiled rag held wrapped around the fingers.

The roughing and finishing of bow sticks do not vary in technique from hand making to commercial manufacture. Most violin bows are made completely by hand. Only the speed of production, quality of materials, and diligence in finishing distinguish the difference between the mediocre and the sublime.

Lapping and hairing the bow

6 The lapping or winding acts as a grip for the stick and is often called the "grip." It usually covers a 3-inch (7.6 cm) length starting from just in front of the frog and going toward the tip. It consists of some material, usually silver wire, wound in a compact spiral fashion around the stick. Part of the winding nearest the frog is covered with leather to protect the spot where the player's thumb rests.

The hairing of the bow is quite routine, as the hair wears from playing and must be frequently replaced. The horsehair is purchased already selected, drawn, and bundled in uniform lengths. A small amount is separated from this and a small rosined knot is fashioned at one end using very strong thin thread. The knot is made stronger by inserting the end of the hair into the heat of a flame, and expanding the hair behind it. A small wooden plug is carved to fit the mortise in the head, and the hair is turned under and fastened into the head with this plug, which holds the hair spread across the top edge in a neat uniform flat strip. With the frog all the way forward, the hair is measured to length, cut off, and after much combing and arranging, tied in a similar way at the end near the frog. Another wooden plug is fashioned for the mortise of the frog. The ferrule is slid over the hair and after much more combing, the hair is turned over and fastened again with a wooden plug, this time into the frog. The hair is combed again before pushing in the pearl slide, and again after sliding on the ferrule. A wooden wedge is carved to fit the width of the ferrule so as to hold the hair spread in a ribbon-like fashion. After some more combing, the wedge is pushed into the ferrule against the hair and trimmed off with a knife. With the application of some rosin on the hair, for grip, the new bow is ready to play.

Where To Learn More

Books

Henderson, Frank. *How to Make a Violin Bow*. Murray Publishing Co., 1977.

Retford, William. *Bows and Bow Makers.* The Strad, 1964.

Roda, Joseph. *Bows for Musical Instruments of the Violin Family*. W. Lewis, 1959.

—Peter Psarianos

Whiskey

Background

Whiskey (usually spelled whisky in Canada and Scotland) is a spirit produced from fermented grain and aged in wood. A spirit is any alcoholic beverage in which the alcohol content has been increased by distillation. Other spirits include brandy (distilled from wine), rum (distilled from sugarcane juice or molasses), vodka (distilled from grain but not aged), and gin (also distilled from grain and unaged but flavored with juniper berries and other ingredients.)

Undistilled alcoholic beverages such as mead, wine, and **beer** have been produced since at least 7000 B.C. The process of distillation (heating an alcoholic beverage in order to boil off, collect, and concentrate the alcohol) was first used in China no later than 800 B.C. to produce rice spirits. About the same time in other parts of Asia, distillation was used to produce arrack, a beverage similar to rum, made from rice and sugarcane juice or palm juice. The ancient Arabs, Greeks, and Romans all distilled wine to produce beverages similar to modern brandy. The practice of distillation spread to western Europe with the Arabs in the eighth century, particularly in Spain and France.

No one knows where or when the first grain spirits were produced, but they certainly existed in Europe no later than 500 years ago. Some claim that whiskey was invented in Ireland as long as 1,000 years ago and carried to Scotland by monks. In any case, the first written records of Scottish whiskeymaking date as far back as 1494. (The word whiskey comes from the Irish Gaelic *uisge beatha* or the Scottish Gaelic *uisge baugh*, both meaning "water of life.")

Spirits were carried to the New World with the earliest European settlers. Rum was distilled in New England in the early 17th century, and distillation also took place in New York as early as 1640. During the early 18th century whiskeymaking became an important industry in the western part of the American colonies, particularly in western Pennsylvania. Farmers found it difficult to store their perishable grains and to transport them to distant eastern cities. It was much simpler to use them to make whiskey, which could be stored for years and more easily transported.

Whiskey played an important part in the early history of the United States, especially during the so-called Whiskey Rebellion of 1794. Farmers in western Pennsylvania refused to pay an unpopular tax on whiskey and attacked federal officers who tried to collect it. After the home of the local tax inspector was burned by a group of 500 armed rebels, President George Washington sent in 13,000 troops to stop the uprising. The rebellion ended without bloodshed, and the power of the federal government was firmly established. Many whiskeymakers moved farther west, into what was then Indian territory, to escape federal authority. They settled in southern Indiana and Kentucky, areas that are still famous for whiskey.

American whiskeymaking reached a peak in 1911, when about 400 million liters were produced, a figure not exceeded until after Prohibition. On November 16, 1920, the Volstead Act became the Eighteenth Amendment to the Constitution of the United States, and no American whiskey was legally made until the amendment was

The word whiskey comes from the Irish Gaelic uisge beatha *or the Scottish Gaelic* uisge baugh, *both meaning "water of life."*

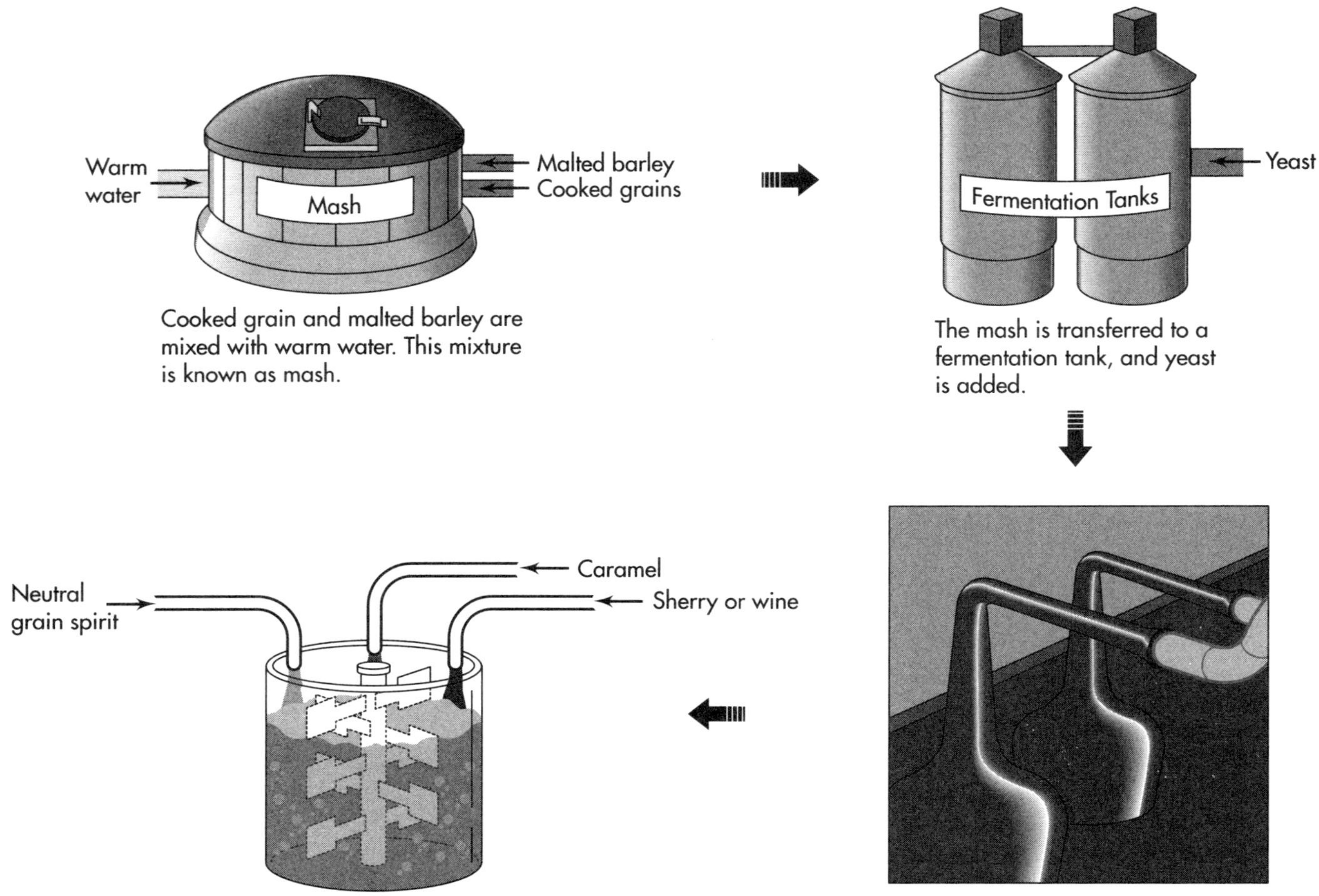

Cooked grain and malted barley are mixed with warm water. This mixture is known as mash.

The mash is transferred to a fermentation tank, and yeast is added.

This liquid, called wash or distiller's beer, is distilled, a process whereby alcohol is evaporated and condensed.

Most whiskeys are blended to give better flavor and color.

repealed on December 5, 1933. Production reached another peak in 1951, when about 800 million liters were made. Today about 400 million liters are produced each year.

The earliest devices for distillation consisted of a closed, heated container, a long tube (known as a condenser) through which the alcohol vapor could cool and turn back into a liquid, and a receptacle to catch the alcohol. These were later refined into pot stills, in which alcohol vapor from a heated copper pot was condensed in a helical, water-cooled copper tube called a worm. Pot stills are still often used to make whiskey in Scotland and Ireland and brandy in France. In Scotland in 1826 Robert Stein invented continuous distillation, in which alcohol could be distilled continually rather than batch by batch. This process was improved by the Irishman Aeneas Coffey in 1831 and is still used to make most mass-produced whiskey today.

Whiskey is popular around the world and is made almost everywhere. The United States makes and consumes more whiskey than any other nation, but the most celebrated whiskey is still Scotch whiskey, often just called Scotch.

Raw Materials

Whiskey is made from water, yeast, and grain. The water used is often considered the most important factor in making good whiskey. It should be clean, clear, and free

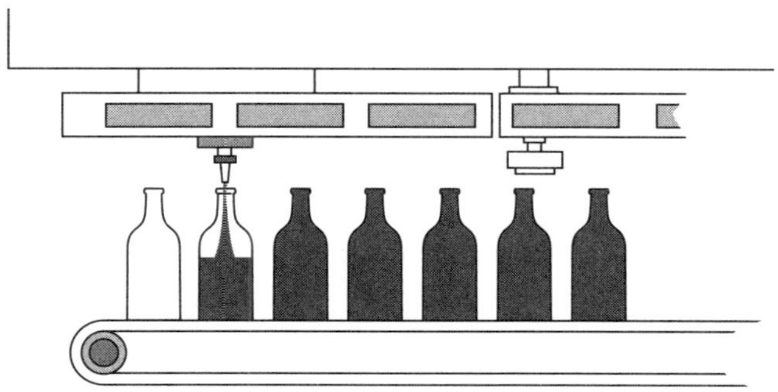

The whiskey is always bottled in glass. Automated machinery can process 400 bottles a minute.

from bad-tasting impurities such as iron. Water that contains carbonates, found in areas that are rich in limestone, is often used in the United States, particularly in Maryland, Pennsylvania, Indiana, and Kentucky. Scottish water is famous for being suited to making fine whiskey, for reasons that are still somewhat mysterious.

Every whiskeymaker keeps a supply of yeast available, grown on barley malt and kept free from bacterial contamination. Some whiskeymakers use several kinds of yeast to control the fermentation process precisely.

The type of grain used varies with the kind of whiskey being made, but all whiskeys contain at least a small amount of malted barley, which is needed to start the fermentation process. Scotch malt whiskey contains only barley. Other whiskeys contain barley in combination with corn, wheat, oats, and/or rye. Corn whiskey must contain at least 80% corn, while Bourbon whiskey and Tennessee whiskey must contain at least 51% corn. Rye whiskey must contain at least 51% rye, and wheat whiskey must contain at least 51% wheat.

Straight whiskeys contain no other ingredients, but blended whiskeys may contain a small amount of additives such as caramel color and sherry.

The Manufacturing Process

Preparing the grain

1 Truckloads of grain are shipped directly from farms to the whiskey manufacturer to be stored in silos until needed. The grain is inspected and cleaned to remove all dust and other foreign particles.

2 All grains except barley are first ground into meal in a gristmill. The meal is then mixed with water and cooked to break down the cellulose walls that contain starch granules. This can be done in a closed pressure cooker at temperatures of up to 311°F (155°C) or more slowly in an open cooker at 212°F (100°C).

3 Instead of being cooked, barley is malted. The first step in malting barley consists of soaking it in water until it is thoroughly saturated. It is then spread out and sprinkled with water for about three weeks, at which time it begins to sprout.

During this germination the enzyme amylase is produced, which converts the starch in the barley into sugars. The sprouting is halted by drying the barley and heating it with hot air from a kiln. For Scotch whiskey, the fuel used in the kiln includes peat, a soft, carbon-rich substance formed when plant matter decomposes in water. The peat gives Scotch whiskey a characteristic smoky taste. The malted barley is then ground like other grains.

Mashing

4 Mashing consists of mixing cooked grain with malted barley and warm water. The amylase in the malted barley converts the starch in the other grains into sugars. After several hours the mixture is converted into a turbid, sugar-rich liquid known as mash. (In making Scotch malt whiskey the mixture consists only of malted barley and water. After mashing the mixture is filtered to produce a sugar-rich liquid known as wort.)

Fermenting

5 The mash or wort is transferred to a fermentation vessel, usually closed in Scotland and open in the United States. These vessels may be made of wood or stainless steel. Yeast is added to begin fermentation, in which the single-celled yeast organisms convert the sugars in the mash or wort to alcohol. The yeast may be added in the form of new, never-used yeast cells (the sweet mash process) or in the form of a portion of a previous batch of fermentation (the sour mash process.) The sour mash method is more often used because it is effective at room temperature and its low pH (high acidity) promotes yeast growth and inhibits the growth of bacteria. The sweet mash method is more difficult to control, and it must be used at temperatures above 80°F (27°C) to speed up the fermentation and to avoid bacterial contamination. After three or four days, the end product of fermentation is a liquid containing about 10% alcohol known as distiller's beer in the United States or wash in Scotland.

Distilling

6 Scottish whiskeymakers often distill their wash in traditional copper pot stills. The wash is heated so that most of the alcohol (which boils at 172°F [78°C]) is transformed into vapor but most of the water (which boils at 212°F [100°C]) is not. This vapor is transferred back into liquid alcohol in a water-cooled condenser and collected. Most modern distilleries use a continuous still. This consists of a tall cylindrical column filled with a series of perforated plates. Steam enters the still from the bottom, and distiller's beer enters from the top. The beer is distilled as it slowly drips through the plates, and the alcohol is condensed back into a liquid. With either method, the product of the initial distillation—known as low wine—is distilled a second time to produce a product known as high wine or new whiskey, which contains about 70% alcohol.

7 The temperature of distillation and other factors determine the proportions of water, alcohol, and other substances (called congeners) in the final product. If it contains more than 95% alcohol it will have no flavor because it has no congeners. This product is known as grain neutral spirit and is often used to add alcohol without adding taste during blending. If the final product has too many congeners of the wrong kind it will taste bad. Distillers remove bad-tasting congeners (usually aldehydes, acids, esters, and higher alcohols) in various ways. Some congeners boil at a lower temperature than alcohol and can be boiled off. Some are lighter than alcohol and will float on top, where they can be poured off.

8 Tennessee whiskey is unique in that the high wine is filtered through charcoal before it is aged. The charcoal is produced by burning wood from sugar maples. This filtration removes unwanted congeners and results in a particularly smooth whiskey. Premium Tennessee whiskey may be filtered through charcoal again after it is aged to produce an even smoother product.

Aging

9 Water is added to the high wine to reduce its alcohol content to about 50% or 60% for American whiskeys and about 65% or higher for Scotch whiskeys. Scotch whiskeys are aged in cool, wet conditions, so they absorb water and become less alcoholic. American whiskeys are aged in warmer, drier conditions so they lose water and become more alcoholic. Whiskey is aged in wooden barrels, usually made from charred white oak. White oak is used because it is one of the few woods that can hold a liquid without leaking but which also allows the water in the whiskey to move back and forth within the pores of the wood, which helps to add flavor. In the United States these barrels are usually new and are only used once. In most other countries it is common to reuse old barrels. New barrels add more flavor than used barrels, resulting in differences in the taste of American and foreign whiskeys.

The aging process is a complex one, still not fully understood, but at least three factors are involved. First, the original mixture of water, alcohol, and congeners react with each other over time. Second, these ingredients react with oxygen in the outside air in oxidation reactions. Third, the water absorbs substances from the wood as it moves within it. (Charring the wood makes these substances more soluble in water.) All these factors change the flavor of the whiskey. Whiskey generally takes at least three or four years to mature, and many whiskeys are aged for ten or fifteen years.

Blending

10 Straight whiskeys and single malt Scotch whiskeys are not blended; that is, they are produced from single batches and are ready to be bottled straight from the barrel. All other whiskeys are blended. Different batches of whiskey are mixed together to produce a better flavor. Often neutral grain spirit is added to lighten the flavor, caramel is added to standardize the color, and a small amount of sherry or port wine is added to help the flavors blend. Blended Scotch whiskey usually consists of several batches of strongly flavored malt whiskeys mixed with less strongly flavored grain whiskeys. A few blends contain only malt whiskeys. Blending is often considered the most difficult and critical process in producing premium Scotch whiskeys. A premium blended Scotch whiskey may contain more than 60 individual malt whiskeys which must be blended in the proper proportions.

Bottling

11 Glass is always used to store mature whiskey because it does not react with it to change the flavor. Modern distilleries use automated machinery to produce as many as 400 bottles of whiskey per minute. The glass bottles move down a conveyor belt as they are cleaned, filled, capped, sealed, labeled, and placed in cardboard boxes. The whiskey is ready to be shipped to liquor stores, bars, and restaurants.

Quality Control

Although the making of good whiskey is still more of an art than an exact science, there are certain basic precautions that all whiskeymakers take to ensure quality. The water used must be taken from an appropriate natural source. It must be filtered so that it is free from organic matter. The grain used must be very clean. It is also passed through screens to eliminate grains that are too small. The yeast is carefully grown to avoid contamination by other microorganisms. The temperature of distillation is monitored with thermometers in the boiling liquid, which are visible through glass windows in the still. During aging, samples of whiskey are evaluated by experienced tasters to determine if it is mature. The blending process is supervised by master blenders to produce a final product with the proper taste.

Byproducts/Waste

Very little of the ingredients used in whiskeymaking are wasted.

The portion of the fermented mash which remains after the distillation can be used for animal feed. The charred white oak barrels used only once in the United States are often sold overseas to age foreign whiskeys. The charcoal used to filter Tennessee whiskey can be pressed into charcoal briquets for barbecues.

Where To Learn More

Books

Jackson, Michael. *The World Guide to Whisky*. Darling Kindersley, 1987.

Grossman, Harold J. and Harriet Lembeck. *Grossman's Guide to Wines, Beers, and Spirits*. Charles Scribner's Sons, 1977.

Lichine, Alexis. *Alexis Lachine's New Encyclopedia of Wines and Spirits*. Knopf, 1976.

Periodicals

Asher, Gerald. "Single Malt Scotch Whiskey." *Gourmet*, December 1989, pp. 94-99.

DeMarco, Dan and Frank Bechard. "New Weigh Scales Smooth Distillery's Production, Improve Inventory Control." *Food Engineering*, October 1986, pp. 95-96.

Johnson, Julie. "Mysteries of the Malt." *New Scientist*, January 26, 1991, pp. 56-59.

Letwin, William. "More Than a Drink." *National Review*, April 18, 1994, p. 14+.

—Rose Secrest

Zinc

Background

Zinc is an elemental metal. It is listed on the Periodic Table as "Zn," with an atomic number of 30 and an atomic weight of 65.37, and it melts at 788°F (420°C). Zinc is usually a gray metallic color, but it can be polished to a shiny silver luster. In nature, it is only found as a chemical compound, not as pure zinc, and can be used as a raw material for castings and coatings.

During the era of the Roman Empire, people used zinc to alloy copper into brass for weapons. In this crude process, the zinc was captured by the copper during the heating of the ores, though little was realized at the time about the importance of zinc in metallurgy. The name zinc may be derived from the German word "zinn," which means tin. The scientific discovery of zinc is credited to Nadreas Sigismund Marggraf, a German chemist who isolated pure zinc in 1746. The first production facility, or smelter, was founded in Bristol, England by William Champion shortly thereafter.

Only about 5% of the world's zinc supply is mined in the United States, with the balance coming primarily from India, Mexico, and Canada. Approximately 6.7 million metric tons of zinc ore are produced worldwide. Roughly two thirds of the zinc used in the United States is imported.

Applications

Zinc is primarily used for galvanizing steel against corrosion, die casting of intricate machine parts, and in batteries and other electrical applications. Zinc is also alloyed with copper to form brass.

Galvanizing steel involves applying a thin coating of zinc to all exposed surfaces of the steel to guard against corrosion. Zinc offers excellent corrosion resistance because it is more easily oxidized by the atmosphere. Oxidation occurs when metal is exposed to air or water, and electrons from the metal transfer to the oxygen. When zinc is tightly bonded to steel, the zinc frees up its electrons more readily than the steel, leaving the stronger metal beneath intact. The application of the zinc coating is accomplished by dipping the steel into molten zinc or by electrolytic plating of the steel with zinc, much like chrome plating.

Die-casting alloys typically contain 96% zinc and 4% aluminum. The die-casting process uses a two-piece steel die and a casting press to hold the die halves together during injection of the molten metal. Inside the steel die is a cavity that has the negative image of the part to be cast. The molten metal is injected into the cavity under pressure, accurately filling the entire void. The metal cools, and the press opens the die halves, revealing the formed part. The zinc cast parts are very close to the desired shape, requiring little machining before they are placed into an assembly. Typical applications include copier, aircraft, and medical instrument parts. Automobile makers use zinc die castings for emblems, moldings, door handles, and brackets. Zinc die castings are easily chrome plated for durability and appearance.

One unique application of zinc takes particular advantage of its ability to transfer its corrosion resistance properties by electrical contact. This application is called a "sacrificial anode." The anodes, made of almost

Zinc is primarily used for galvanizing steel against corrosion, die casting of intricate machine parts, and in batteries and other electrical applications.

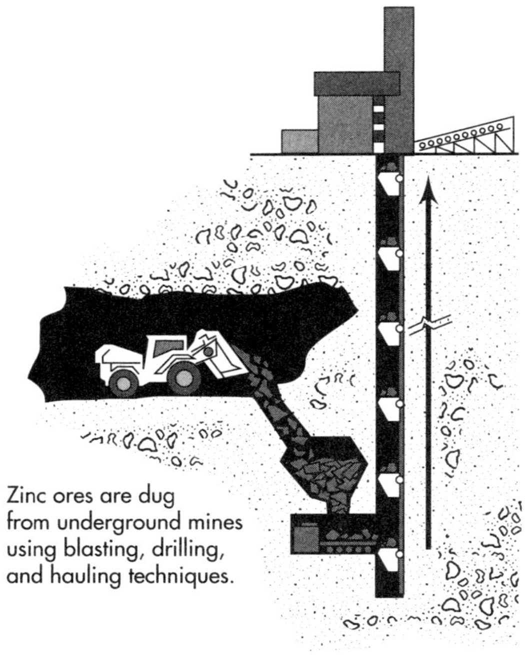

pure zinc, are bolted to aluminum marine engines. During operation in water, especially **salt** water, the oxidation forms a weak electrical current, which may corrode the hull and engine parts. Since zinc is easily oxidized in the presence of this electrical current, it "sacrifices" itself by corroding quickly, consuming all of the electrical imbalance in the ship. The remaining aluminum hull and engine are not corroded as a result. As it is consumed, the anode must be replaced to assure continued protection.

In an application similar to the sacrificial anode, zinc is used as a component in battery production. The dry cell battery creates a chemical reaction with zinc in a metal housing (or "can") that results in a voltage potential between two connections. An electrical device, such as a flashlight or portable radio, can be connected to the battery and powered by the electricity produced. Thus connected, the reaction maintains the electrical current for the duration of the available chemical reactants.

Zinc as a compound is used in pharmaceuticals, rubber, cosmetics, paint, and ceramic glaze. Other compounds use zinc in **cathode-ray tube**s, soldering flux, and wood preservatives.

The Manufacturing Process

Mining

1 Zinc ores are dug from underground mines using conventional blasting, drilling, and hauling techniques. The ores occur as zinc sulfide (also called sphalerite), zinc carbonate (smithsonite), zinc silicate (calimine), and in compounds of manganese and **iron** (franklinite). Zinc ore is sometimes mined in conjunction with silver or **lead** ores. In addition to the ore itself, oil and sulfuric acid are required for the breakdown of the ores; and electricity, coke, or natural gas are needed to provide the heat energy for smelting.

Froth flotation

2 Zinc can be produced by a process called froth flotation, which is also used for reduction of copper and lead ores. This process involves grinding the zinc ore to a fine powder, mixing it with water, pine oil, and flotation chemicals, and then agitating the mixture to "float" the zinc to the surface. A variety of chemicals are used to coat the important zinc particles and prevent them from becoming wetted by the water. Then air is injected, and the coated minerals

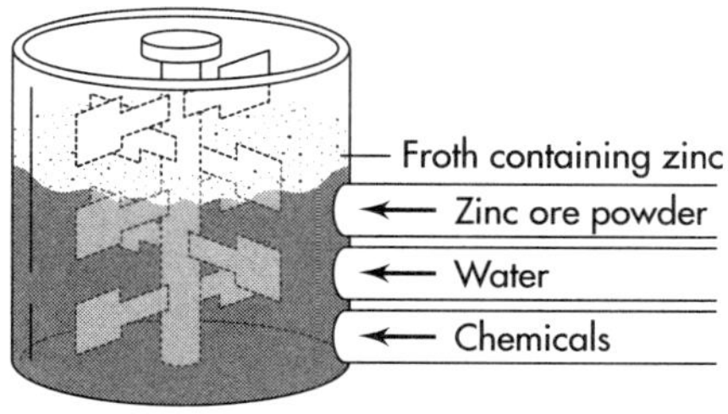

Zinc can be separated out of the ore by a process called froth flotation.

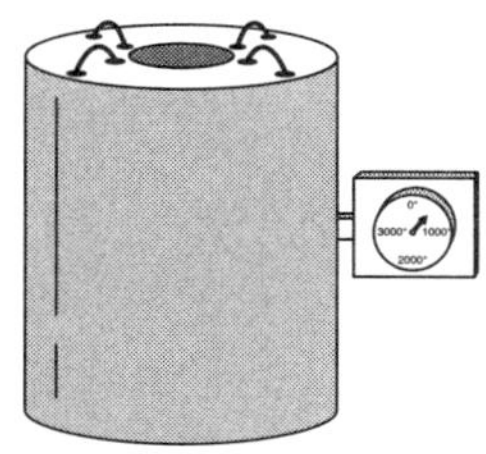

The filtered, paste-like zinc is mixed with lime and sent to a furnace. The heat fuses the minerals into solid chunks called sinter.

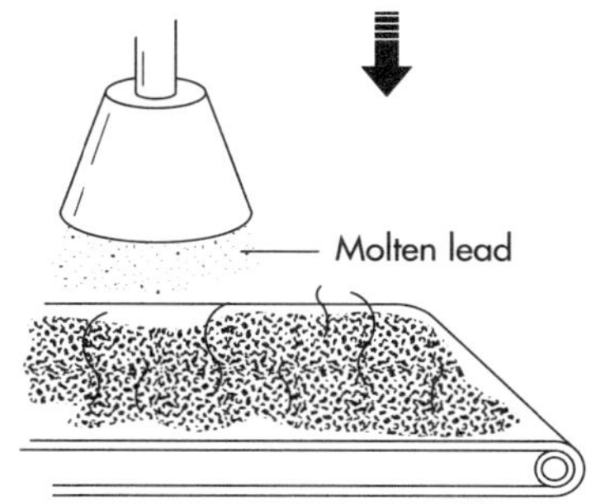

The sinter is sent to a blast furnace. To reduce the reformation of zinc oxide, the zinc is sprayed with molten lead. The lead dissolves the zinc, but in the cooling chamber, the zinc reappears and is separated out.

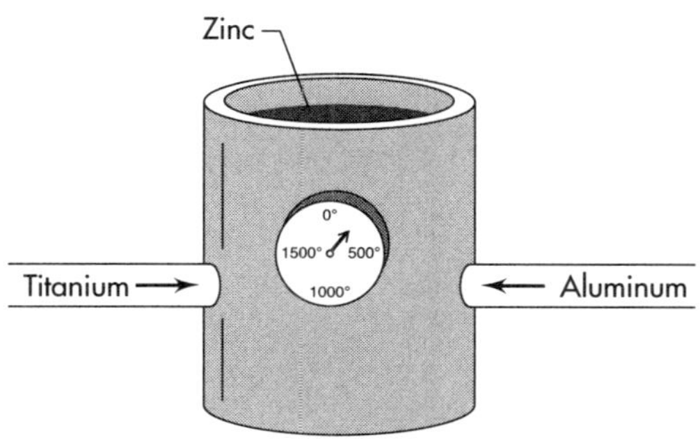

Most zinc is alloyed with other metals to improve its properties.

The molten metal is then poured into molds to form sows or ingots.

attach themselves to the bubbles. The operation is performed inside a vat and agitated with an impeller. The rotating impeller draws the air down the standpipe that surrounds the impeller shaft and dissipates it throughout the mixture or "pulp." The zinc rises to the top and the residue stays in the bottom of the pulp, since it cannot adhere to the bubbles. Automatic scrapers remove the mineral-laden froth containing the zinc.

Filtering

3 The froth is filtered to remove the water and liquid oils. The paste-like remainder is mixed with lime and sent to a furnace.

The furnace roasts the mixture at 2500°F (1371°C), which fuses the minerals into solid chunks called sinter. At this point, the material has been completely converted to zinc oxide.

Smelting

4 The next reduction process uses a blast furnace to melt the prepared ore into its elemental components. The blast furnace is fueled by electricity, coke, or natural gas, which generate temperatures of up to 2200°F (1204°C). This, however, also generates carbon dioxide, which recombines with the zinc as it cools to re-form zinc oxide. To reduce this reformation, the zinc is sprayed with molten lead while it is still hot. The lead, at 1022°F (550°C), dissolves the zinc and carries it to another chamber, where it is cooled to 824°F (440°C). At this temperature, the lighter zinc separates out of the lead and is drained off the top. The lead is reheated and returned to the blast furnace.

Refining

5 Further metal improvement can be made by keeping the zinc molten and undisturbed for several hours. In this state, iron and other contaminants settle to the bottom, allowing the almost pure zinc to be carefully drawn off the top and cast into ingots.

Alloying

6 Most zinc is alloyed with other metals before use to improve its properties. Alloying involves remelting and mixing the zinc with other metals in precise proportions. For example, approximately 4% aluminum is added to improve casting quality and die life in the die-casting process. Other added alloys are small amounts of titanium, copper, and magnesium. After alloying, the molten metal is poured into sow molds and ingot molds. Sows can weigh several thousand pounds, while ingots weigh about 45 pounds (20 kg).

Quality Control

Metal alloys are inspected by a process called spectrographic analysis. The metal is burned under a protective cover using an electrical arc. The light emitted by the burning metal is passed though an apparatus much like a prism, which breaks the light into all of its individual colors. Every element has a different set of colors, or spectrum, which is like a fingerprint. Any foreign material will alter the spectrum, and in doing so show its unique color spectrum, identifying it. The computer in the spectrograph uses sensors to pick up these colors. The computer program then produces a printout that identifies each element in the spectrum and the concentration within the metal. Elements can be reduced or increased to alter the composition.

The Future

Because of the strength to weight ratio of zinc, its use by the automotive industry as a die casting has been diminishing in the past few years. Magnesium, aluminum, and plastics have taken over many zinc applications. The use of zinc to galvanize automobile body parts has been increasing, however. Many vehicles today are protected by zinc galvanizing which allows the manufacturer to offer extended warranties for body rust problems with new cars.

Where To Learn More

Periodicals

Queneau, Paul B. and Jerome P. Downey. "Secondary Zinc Production Minimizes Waste." *Pollution Engineering*, November 1994, pp. 42-44.

Yates, Edward M. "Zinc: Major Mine Production Cuts in 1993." *Engineering and Mining Journal*, March 1994, pp. 19-21.

—Douglas E. Betts

Index

B

G

H

N

O

P

Q

R

S

T

U

V

W

Y

Z